Pollution Prevention

Fundamentals and Practice

Paul L. Bishop
University of Cincinnati

Long Grove, Illinois

For information about this book, contact:
Waveland Press, Inc.
4180 IL Route 83, Suite 101
Long Grove, IL 60047-9580
(847) 634-0081
info@waveland.com
www.waveland.com

10-digit ISBN 1-57766-348-9
13-digit ISBN 978-1-57766-348-5

Printed in the United States of America

8 7 6 5 4

I dedicate this book to my wife, Pam, for her patience, understanding, and encouragement, and for putting up with all of the time I spent on the preparation of this book.

ABOUT THE AUTHOR

PAUL L. BISHOP is Associate Dean for Graduate Studies and Research and the Herman Schneider Professor of Environmental Engineering at the University of Cincinnati. Dr. Bishop received a B.S. in civil engineering from Northeastern University and an M.S. and Ph.D. in environmental engineering from Purdue University. He spent 16 years in the Department of Civil Engineering at the University of New Hampshire as professor and department chair, and for the past 16 years he has been in the Department of Civil and Environmental Engineering at the University of Cincinnati, serving for 5 years as William Thorns Professor and head of the department and now as Associate Dean for Graduate Studies and Research in the College of Engineering. He spent one year as visiting professor at Heriot-Watt University in Edinburgh, Scotland, and another in the same capacity at the Technical University of Denmark, Lyngby, Denmark. Dr. Bishop's specialties are pollution prevention, biological waste treatment, and hazardous waste management. He is the author or co-author of 4 books and more than 300 technical papers. He is a diplomate in the American Academy of Environmental Engineers, for which he served two terms as a member of the Board of Trustees, was a member of the Board of Directors of ABET, completed a term as president of the Association of Environmental Engineering and Science Professors (AEESP), is a member of the Governing Board of the International Water Association (IWA), and chairs the IWA U.S.A. National Committee (USANC). Dr. Bishop has had a long history of involvement in pollution prevention activities. This includes significant research on pollution prevention opportunities, presentations on pollution prevention at national and international conferences, consulting with industry on pollution prevention topics, and serving for 10 years on the Science Advisory Board of the U.S. EPA National Center for Clean Industrial and Treatment Technologies.

PREFACE

It is estimated that our materials-dominated society consumes about 10 metric tons of raw materials per person per year in the production of consumer goods. Within six months of extraction or production of these materials, 94 percent of them become residual material that is disposed of as waste. More efficient practices for using materials in manufacturing are needed to lessen the demands for raw materials and to reduce the amounts and toxicity of waste materials. It is estimated that 70 percent of this waste material could be eliminated through better design decisions and reuse of materials.

As currently structured, engineering education has evolved into fairly segregated disciplines; each focuses on a narrowly defined design and manufacturing function without consideration of its environmental consequences. This is no longer the case in industry, however, where pollution prevention and waste minimization have become very important. This rapidly changing industrial emphasis was initially in response to regulatory pressure, but now it is driven primarily by economics. Industries are striving to minimize waste generation at the source, to reuse more of the waste materials that are generated, and to design products for easier disassembly and reuse after their useful life is completed. The overall objective is to minimize "end-of-pipe" treatment, although some waste treatment will always be needed. This new environmental ethic in manufacturing is labeled "pollution prevention," "green engineering," or "environmentally conscious engineering." It is a collection of attitudes, values, and principles that result in an attempt by the engineering profession to reduce the rate at which we adversely impact the environment.

Industry has accepted the concept of pollution prevention, because management has seen the economic benefits resulting from it. However, most of our engineering graduates are not prepared to step into a role where green engineering principles are espoused. It is essential that we quickly incorporate the green engineering principles into the engineering curriculum in all disciplines to ensure that all engineering graduates are aware of environmental issues and understand the environmental and economic consequences of engineering decisions. The goal of this educational change should be to reduce the necessity for end-of-pipe treatment by incorporating, at all stages of engineering, measures that minimize wastes and permit recycling and reuse. A knowledge of pollution prevention principles should allow the engineer to include environmental consequences in decision processes in the same way that economic and safety factors are considered. Eventually, we must extend this way of thinking to others in the decision-making process, including management, but this probably will not be successful until engineers embrace it.

The objective of this book is to introduce the principles of pollution prevention, environmentally benign products, processes and manufacturing systems. Students will learn the impacts of wastes from manufacturing and post-use product disposal, environmental cycles of materials, sustainability, and principles of environmental economics. Materials selection, process and product design, and packaging are addressed.

ORGANIZATION

This book is intended for use by novices to the field of pollution prevention as well as by students majoring in environmental engineering or chemical engineering. Sufficient background information is provided to those new to the field to understand the concepts discussed in later chapters.

The book is divided into 14 chapters. The first chapter introduces the concept of pollution prevention, gives a historical perspective, provides definitions that will be used throughout the book, and discusses the important, but often overlooked subject of environmental ethics and its role in pollution prevention. Chapters 2 and 3 are background chapters, providing information on properties and fates of environmental contaminants and the impacts of industry on the environment. Knowledge of environmental regulations is essential to proper implementation of pollution prevention programs; regulations are covered in Chapter 4. Chapter 5, "Improved Manufacturing Operations," is intended to describe general design and manufacturing processes that are used in industry and to show how changes in the manufacturing process can minimize pollution generation. The next three chapters deal with how we can assess the effectiveness of a proposed process change and how effective pollution prevention programs are constructed. Chapter 6 describes the life-cycle assessment process, while Chapter 7 is an overview of pollution prevention economics. Chapter 8 focuses on pollution prevention planning. Chapters 9 and 10 then investigate in more detail technologies that can be used to minimize pollution. Chapter 9 focuses on such topics as green chemistry, design for disassembly/demanufacturing, and improved packaging. Chapter 10 describes new procedures for minimizing the use of water, energy, and reagents in a manufacturing process through the application of a procedure called "pinch analysis." No matter how effective an industrial pollution prevention program is, there will always be some waste that can't be eliminated and must be disposed of. Chapter 11 discusses options for disposal of these residuals. Chapter 12 addresses another form of industrial pollution—fugitive emissions—that result from unintentional equipment leaks or releases. Chapter 13 discusses what can be done at the municipal level to regulate industrial pollution emissions. The book culminates in Chapter 14, which is a philosophical discussion of the subject of sustainability and the role of pollution prevention in maintaining a more sustainable society.

USE OF THIS BOOK

Pollution Prevention: Fundamentals and Practice contains enough material to allow flexibility in its use. This book is intended for engineering students from any engineering discipline, but it should also be useful to practicing engineers needing a comprehensive book on pollution prevention. This includes both environmental engineers who are entering the pollution prevention consulting arena and engineers in industry who need to bring their knowledge of available pollution prevention options up to date. With selective reading, it will also be of use to nonengineers in industrial management who must make intelligent choices on implementation of pollution prevention alternatives or learn how to sell these alternatives to upper management.

The book is specifically designed for senior- or graduate-level engineering students from all engineering disciplines, but it may be used by junior-level students as well. It assumes no prior knowledge of pollution prevention or related concepts, instead providing all necessary background for the reader. By careful selection of the topics covered, the book can be used in a general course on pollution prevention intended for all engineering students, or in a pollution prevention course designed specifically for environmental engineering students.

The material in this textbook can be used in a variety of ways, depending on the discipline and educational background of the students and the intent of the course. Suggested outlines are presented below for courses intended (1) as a general introduction to pollution prevention for students with little environmental engineering background, (2) as a more rigorous course for environmental engineering students, and (3) as a course for nonengineering students, such as business or management majors. In addition, portions of this book may be used in a basic freshman-level engineering course to introduce the need for the environmentally conscious engineering ethic to newly developing engineers. This might include Chapter 1, parts of Chapter 3, and Chapter 6.

A suggested outline for a cross-disciplinary course on pollution prevention for nonenvironmental engineering students is presented in outline A. This introduces the concepts of pollution prevention, describes the consequences of pollution emissions, and presents methods for setting up a pollution prevention program and assessing its effectiveness. Students from any engineering discipline should have the necessary background knowledge for this course.

Outline A

Topic	Chapter	Sections
Introduction	1	All
Properties and fates of environmental contaminants	2	All
Industrial activity and the environment	3	All
Environmental regulations	4	All
Improved manufacturing operations	5	5.1, 5.2.2, 5.4.2 through 5.4.4
Life-cycle assessment	6	All
Pollution prevention economics	7	Depends on students' backgrounds
Waste audits	8	All
Design for disassemble/demanufacturing	9	9.1, 9.3, and 9.4
Fugitive emissions	12	12.1 and 12.2
Sustainability	14	All

A course on pollution prevention specifically designed for environmental engineering students would differ from outline A. Depending on their educational level, these students may already be familiar with the materials in Chapters 2, 3, 4, and 11. These chapters could be assigned as background reading. The course could focus on the organizational and technical aspects of pollution prevention. Outline B suggests a plan for such a course.

Outline B

Topic	Chapter	Sections
Introduction	1	All
Improved manufacturing operations	5	All
Life-cycle assessment	6	All
Pollution prevention economics	7	Depends on students' backgrounds
Waste audits	8	All
Design for disassemble/demanufacturing	9	9.1, 9.3, and 9.4
Water, energy, and reagent conservation	10	All
Fugitive emissions	12	All
Toward a sustainable society	14	All

Ultimately, the decision on whether to implement a pollution prevention program is made by the industry management. It is essential that managers become familiar with the problems created by pollution and what the opportunities associated with pollution prevention are. It is unlikely that business schools can fit in a full course on pollution prevention, but the essential material could be covered in a short minicourse, as suggested in outline C.

Outline C

Topic	Chapter	Sections
Introduction	1	All
Industrial activity and the environment	3	All
Environmental regulations	4	All
Life-cycle assessment	6	All
Pollution prevention planning	8	All
Municipal pollution prevention programs	13	All
Toward a sustainable society	14	All

A teacher's manual that accompanies this textbook is available. It contains solutions to all problems in the text. I would appreciate any comments, suggestions, corrections, and contributions of problems for future revisions.

ACKNOWLEDGMENTS

This book could not have been written without the valuable assistance of a number of people. This textbook grew out of a grant I received from the Ohio Environmental Protection Agency to develop a cross-disciplinary pollution prevention program for engineering students. The agency's support is gratefully acknowledged. Amit Gupta, a former graduate student, provided invaluable assistance in developing the course, creating many of the overhead teaching slides presented in the accompanying teacher's manual, and reviewing the entire manuscript. I would especially like to thank him for contributing much of the writing for Chapter 14, "Toward a Sustainable Society." His insights and perspectives on this subject were very perceptive. I also extend my gratitude to Anthony Dunams, another former graduate student, for writing much of Chapter 13, "Municipal Pollution Prevention Programs," and for serving as a sounding board for much of the other material. Tatsuji Ebihara and Hassan Arafat, two current doctoral students, contributed several of the design examples. I have taught several courses using drafts of this textbook to students from a multitude of academic backgrounds. Their helpful comments and corrections of the text have vastly improved it, and their asssistance is gratefully acknowledged. I would especially like to thank the outside reviewers of this text for their very valuable contributions. These include Dr. C. P. L. Grady (Clemson University), Dr. Steven Safferman (University of Dayton), and Dr. Angela Lindner (University of Florida).

I would also like to thank Eric Munson, Carole Schwager, and Jean Lou Hess for their great support and assistance with this project.

Paul L. Bishop

BRIEF CONTENTS

1 **Introduction to Pollution Prevention** 1

2 **Properties and Fates of Environmental Contaminants** 22

3 **Industrial Activity and the Environment** 78

4 **Environmental Regulations** 147

5 **Improved Manufacturing Operations** 180

6 **Life-Cycle Assessment** 251

7 **Pollution Prevention Economics** 296

8 **Pollution Prevention Planning** 329

9 **Design for the Environment** 353

10 **Water, Energy, and Reagent Conservation** 421

11 **Residuals Management** 454

12 **Fugitive Emissions** 509

13 **Municipal Pollution Prevention Programs** 533

14 **Toward a Sustainable Society** 572

Appendixes 621

- **A** The Elements 621
- **B** Properties of Selected Compounds 624
- **C** Hazardous Waste Lists 635
- **D** Toxic Release Inventory Chemicals and Chemical Categories 662
- **E** Present Worth of a $1.00 Investment 668
- **F** Physical Constants 670
- **G** Physical Properties of Water at 1 Atmosphere 672
- **H** Properties of Air 674
- **I** Useful Conversion Factors 675

Index 680

CONTENTS

1 Introduction to Pollution Prevention 1

1.1 The 3M Experience 1

1.2 Pollution Prevention 3

1.3 The Historical Perspective 6

1.3.1 The Industrial Revolution / 1.3.2 Impacts of Industrialization

1.4 What Is Pollution Prevention? 9

1.4.1 Waste Definition / 1.4.2 Pollution Prevention Definition / 1.4.3 Other Terms / 1.4.4 Sustainability

1.5 The Pollution Prevention Hierarchy 13

1.6 Recycling vs. Pollution Prevention 14

1.7 Environmental Ethics 15

References 19

Problems 20

2 Properties and Fates of Environmental Contaminants 22

2.1 Organic Chemicals 23

2.1.1 Nomenclature of Organic Chemicals

2.2 Metals and Inorganic Nonmetals 39

2.2.1 Arsenic / 2.2.2 Cadmium / 2.2.3 Chromium / 2.2.4 Lead / 2.2.5 Mercury / 2.2.6 Cyanides

2.3 Contaminant Transport and Transformation in the Environment 44

2.3.1 Contaminant Concentrations / 2.3.2 Transport Processes / 2.3.3 Partitioning Processes / 2.3.4 Transformation Processes

References 72

Problems 73

3 Industrial Activity and the Environment 78

3.1 Introduction 78

3.2 Air Pollution 79

3.2.1 The Atmosphere / 3.2.2 Smog Formation / 3.2.3 Acid Rain / 3.2.4 Global Warming / 3.2.5 Ozone Depletion

3.3 Solid Wastes 102

3.3.1 Sources and Composition / 3.3.2 Solid Waste Management / 3.3.3 Conservation

3.4 Hazardous Wastes 114

3.4.1 Superfund Sites / 3.4.2 Resource Conservation and Recovery Act

3.5 Water Pollution 118

3.5.1 Minimata Disease / 3.5.2 The Kepone Incident / 3.5.3 Industrial Wastewater Treatment

3.6 Energy Usage 121

3.6.1 Historical Perspective / 3.6.2 Energy Consumption / 3.6.3 Energy Reserves / 3.6.4 Fossil Fuels / 3.6.5 Nuclear Energy / 3.6.6 Renewable Energy Sources / 3.6.7 Electricity / 3.6.8 Energy Conservation

3.7 Resource Depletion 138

3.7.1 Earth's Structure / 3.7.2 Recoverable Resources / 3.7.3 Mineral Resources

References 142
Problems 144

4 Environmental Regulations 147
4.1 Introduction 147
4.2 The Regulatory Process 149
4.3 Environmental Regulations 150
4.3.1 Laws Pertaining to Clean Air / 4.3.2 Laws Pertaining to Clean Water / 4.3.3 Laws Pertaining to Hazardous Materials and Wastes / 4.3.4 Laws Pertaining to Products / 4.3.5 Laws Pertaining to Pollution Prevention
References 178
Problems 178

5 Improved Manufacturing Operations 180
5.1 Introduction 180
5.2 The Manufacturing Process 183
5.2.1 Sequential Engineering / 5.2.2 Concurrent Engineering / 5.2.3 Manufacturing Processes
5.3 Process Development and Design 215
5.3.1 Computer Tools
5.4 Process Changes 220
5.4.1 Advanced Process Technologies / 5.4.2 Product Changes / 5.4.3 Storage / 5.4.4 Management
5.5 Pollution Prevention Examples 239
5.5.1 Acrylonitrile Manufacturing / 5.5.2 Maleic Anhydride Production
References 247
Problems 248

6 Life-Cycle Assessment 251
6.1 Overview of Life-Cycle Assessment 251
6.2 History of Life-Cycle Assessment Development 253
6.3 Life-Cycle Assessment and the Regulatory Process 254
6.4 Life-Cycle Assessment Methodology 255
6.4.1 Goal Definition and Scoping Stage / 6.4.2 Inventory Analysis / 6.4.3 Impact Analysis / 6.4.4 Improvement Analysis
6.5 Streamlining Life-Cycle Assessments 269
6.6 Pollution Prevention Factors 271
6.7 Application of Life-Cycle Assessment 277
6.7.1 Corporate Strategic Planning / 6.7.2 Product Development / 6.7.3 Process Selection and Modification / 6.7.4 Marketing Claims and Advertising / 6.7.5 Ecolabeling
6.8 Use of Computer Models in Life-Cycle Assessment 287
6.9 Life-Cycle Assessment in Waste Management Operations 288
References 292
Problems 294

7 Pollution Prevention Economics 296
7.1 Overview of Economics 296

7.2 Microeconomics 298
7.2.1 Market Mechanisms / 7.2.2 Supply and Demand / 7.2.3 Marginal Cost and Marginal Benefit / 7.2.4 Market Externalities / 7.2.5 Control Measures

7.3 Engineering Economics 311
7.3.1 Discount Rate / 7.3.2 Present Worth / 7.3.3 Comparing Investment Alternatives

7.4 Estimating Long-Term Cleanup Liability 317

7.5 Total Cost Assessment 320
7.5.1 Life-Cycle Costing / 7.5.2 Life-Cycle Cost Assessment Process / 7.5.3 Life-Cycle Cost Assessment Case Study / 7.5.4 Summary

References 325

Problems 326

8 Pollution Prevention Planning 329

8.1 Introduction 329

8.2 Structure of the Pollution Prevention Process 330
8.2.1 Organizing the Program / 8.2.2 Preliminary Assessment / 8.2.3 Pollution Prevention Program Plan Development / 8.2.4 Developing and Implementing Pollution Prevention Projects / 8.2.5 Implementing the Pollution Prevention Plan / 8.2.6 Measuring Pollution Prevention Progress

8.3 Environmental Management Systems 340

8.4 Environmental Audits 343
8.4.1 Emissions Inventory

8.5 Toxic Release Inventory 346
8.5.1 Toxic Release Inventory Reporting Requirements / 8.5.2 Toxic Release Inventory Chemicals / 8.5.3 Problems with Toxic Release Inventory Data

References 350

Problems 351

9 Design for the Environment 353

9.1 Introduction 353
9.1.1 Design for X / 9.1.2 Chapter Focus

9.2 Green Chemistry 357
9.2.1 Sources of Wastes / 9.2.2 Alternative Synthetic Pathways / 9.2.3 Alternative Reaction Conditions / 9.2.4 Design of Safer Chemicals / 9.2.5 Green Chemistry Research Needs

9.3 Design for Disassembly/Demanufacturing 386
9.3.1 Recycle versus Reuse / 9.3.2 Recycle/Reuse Hierarchy / 9.3.3 Recycle Legislation / 9.3.4 Requirements for Effective Reuse/Recycling / 9.3.5 Disassembly Strategy / 9.3.6 Computer-Aided Design / 9.3.7 Waste Exchanges / 9.3.8 Recovery through Composting or Energy Reuse / 9.3.9 Barriers to Reuse

9.4 Packaging 410
9.4.1 Minimizing Packaging / 9.4.2 Degradable Packaging

References 417

Problems 419

10 Water, Energy, and Reagent Conservation 421

10.1 Introduction 421

10.2 Reduction in Water Use for Cleaning 422
10.2.1 Case Study

10.3 Pinch Analysis 430
10.3.1 Thermal Pinch Analysis / 10.3.2 Pinch Analysis for Water Use / 10.3.3 Pinch Analysis for Process Emissions / 10.3.4 Summary

References 450

Problems 450

11 Residuals Management 454

11.1 Introduction 454

11.2 Wastewater Treatment 455
11.2.1 Treatment Options / 11.2.2 Physicochemical Processes / 11.2.3 Biological Waste Treatment / 11.2.4 Sludge Management

11.3 Air Pollution Control 499
11.3.1 Particulate Control / 11.3.2 Gas Removal

11.4 Solid Waste Disposal 504

References 504

Problems 505

12 Fugitive Emissions 509

12.1 Introduction 509

12.2 Sources and Amounts 511

12.3 Measuring Fugitive Emissions 512
12.3.1 Average Emission Factor Approach / 12.3.2 Screening Ranges Approach / 12.3.3 EPA Correlation Approach / 12.3.4 Unit-Specific Correlation Approach

12.4 Controlling Fugitive Emissions 519
12.4.1 Equipment Modification / 12.4.2 Leak Detection and Repair Programs

12.5 Fugitive Emissions from Storage Tanks 524
12.5.1 Emissions Estimation / 12.5.2 Emissions Control

12.6 Fugitive Emissions from Waste Treatment and Disposal 530

References 531

Problems 532

13 Municipal Pollution Prevention Programs 533

13.1 Introduction 533

13.2 Regulatory Basis for Pollution Prevention Programs 534
13.2.1 Resource Conservation and Recovery Act and Hazardous and Solid Waste Amendments / 13.2.2 Emergency Planning and Community Right-to-Know Act / 13.2.3 Clean Water Act / 13.2.4 Pollution Prevention Act / 13.2.5 Regional Pollution Prevention Initiatives: Great Lakes Water Quality Initiative

13.3 Source Control and Pretreatment Permit Programs 539
13.3.1 Mandatory Programs / 13.3.2 Local Discharge Standards, Limits, and Authority

13.4 Types of POTW-Administered Pollution Prevention Programs 542
13.4.1 Voluntary Programs / 13.4.2 Regulatory and Enforcement Programs / 13.4.3 Market-Based Programs and Pollution Prevention Incentives / 13.4.4 Measuring Pollution Prevention Progress

13.5 Desirable Qualities of a Municipal P2 Program 558
13.5.1 Industry Requests / 13.5.2 Publicly Owned Treatment Works Requests

13.6 Development of Publicly Administered Pollution Prevention Programs 561
13.6.1 Program Recommendation / 13.6.2 Program Implementation

13.7 Internal Pollution Prevention in Publicly Owned Treatment Works 567

13.8 Summary 568

References 568

Problems 570

14 Toward a Sustainable Society 572

14.1 Introduction 572

14.2 Defining the Problem 574
14.2.1 Biodiversity / 14.2.2 Impediments to Achieving Sustainability

14.3 The History of Sustainability 576
14.3.1 Origin of the Sustainability Concept / 14.3.2 United Nations Conference on Environment and Development

14.4 Sustainability and What It Means 581
14.4.1 Definitions of Sustainability / 14.4.2 What Is Sustainable Development? / 14.4.3 Conceptualization of Sustainability / 14.4.4 Hurdles to Sustainability

14.5 Achieving Sustainable Development 588
14.5.1 Sustainable Development Framework / 14.5.2 Application of Sustainability Strategies / 14.5.3 Indicators of Sustainability / 14.5.4 What Is Being Done to Achieve Sustainability?

14.6 Sustainability in the United States 597
14.6.1 President's Council on Sustainable Development / 14.6.2 Role of Local Governments

14.7 Sustainability in the Third World 601
14.7.1 Barriers to Sustainability in the Third World / 14.7.2 Models of Macroeconomic Management for Third-World Countries

14.8 A Framework for Sustainability 603
14.8.1 The Role of Individuals / 14.8.2 The Role of Industry / 14.8.3 The Four Elements of Clean Manufacturing

14.9 Industrial Ecology 608
14.9.1 Ecoindustrial Parks / 14.9.2 Industrial Ecology Principles

14.10 Measures of Economic Growth 613
14.10.1 Gross Domestic Product and Gross National Product / 14.10.2 Green Accounting
14.11 Steps for Adopting a Sustainability Approach 616
14.12 Summary 617
References 618
Problems 620

Appendixes 621
A The Elements 621
B Properties of Selected Compounds 624
C Hazardous Waste Lists 635
D Toxic Release Inventory Chemicals and Chemical Categories 662
E Present Worth of a $1.00 Investment 668
F Physical Constants 670
G Physical Properties of Water at 1 Atmosphere 672
H Properties of Air 674
I Useful Conversion Factors 675
Index 680

“Waste is a terrible thing to waste.”

The X-Files

CHAPTER 1

INTRODUCTION TO POLLUTION PREVENTION

> We have learned the inherent limitations of treating and burying wastes. A problem solved in one part of the environment may become a new problem in another part. We must curtail pollution closer to its point of origin so that it is not transferred from place to place.

With these words, William Reilly, then administrator of the U.S. Environmental Protection Agency (EPA), announced in 1990 the new EPA policy to decrease reliance on end-of-pipe treatment of industrial wastes and to promote elimination of waste production at the source. Although new from a regulatory standpoint, this philosophy was not necessarily new from a practice standpoint. Several industries had begun to put this paradigm into practice years before, but it was revolutionary as an agency policy. Before examining in detail what pollution prevention is and how it can be more effectively accomplished, let us take a look at what one company, 3M, has accomplished.

1.1 THE 3M EXPERIENCE

The 3M Company, a major multinational corporation with more than 130 manufacturing sites in the United States as well as others in 41 countries, produces everything from Magic Tape and Post-it Notes to heart-lung machines. In addition to being one of the largest producers of consumer products, 3M was also one of the largest producers of wastes, both toxic and nontoxic. Not only were wastes produced during manufacturing processes at 3M, but they were also produced during the processing and manufacture of the goods and chemicals that went into 3M's products, during the transportation of these raw materials to the manufacturing plant and of the finished products from manufacturing to the consumer, and after the consumer had finished with the product and discarded it.

3M, as well as many other companies, began examining their waste management practices as a result of public pressure. In the late 1960s and early 1970s, there was a major outcry by the public to clean up our environment and prevent further degradation. Congress quickly passed several pieces of important legislation designed to do just that. These included the Clean Air Act Amendments (CAAA) in 1967, the National Environmental Policy Act (NEPA) in 1969, the Federal Water Pollution Control Act (FWPCA) in 1972, the Safe Drinking Water Act (SDWA) in 1974, and the Toxic Substances Control Act (TSCA) and Resource Conservation and Recovery Act (RCRA) in 1976. These and other environmental statutes are described in more detail in Chapter 4. As a result of this regulatory pressure, industries across the nation began to examine ways to treat their wastes or, better yet, to minimize the amounts of waste they were generating.

When 3M started looking at the company's waste, management realized that they could never reach their goal of a clean environment through treatment of these wastes. Most treatment technologies do not destroy wastes, but rather move them from one medium to another, only delaying the eventual pollution problem. Consequently, 3M decided that preventing the wastes from being created in the first place was its only viable solution. In 1975 the company became the first to initiate a companywide pollution prevention program and adopted a new corporate policy preventing pollution at the source wherever and whenever possible. The policy asserts that 3M will:

- Solve its own environmental pollution and conservation problems.
- Prevent pollution at the source wherever and whenever possible.
- Develop products that will have a minimum effect on the environment.
- Conserve natural resources through the use of reclamation and other appropriate methods.
- Assure that its facilities and products meet and sustain the regulations of all federal, state, and local governments.
- Assist, wherever possible, governmental agencies and other official organizations engaged in environmental activities.

Pollution prevention thus became the foundation of 3M's approach to all environmental policies (Zosel, 1993). Rather than using waste treatment as the basis for waste management, 3M first looks at the feasibility of process modifications, recycling, reuse, reclamation, and augmenting efficiency. The goal of 3M's Pollution Prevention Pays (3P) program is to make pollution prevention a way of life throughout 3M, from the boardroom to the laboratory to the manufacturing plant. It is based on the premise that prevention does pay—in terms of environmental benefit, lower disposal and treatment costs, operating savings, improved product quality, and a more positive corporate image (Bringer and Benforado, 1989).

The 3P program has been a success. It has evolved into a fully integrated, high-quality environmental management system, creating an environmentally sensitive corporate culture at 3M. Environmental engineers are assigned to business unit facilities to assist in 3P implementation, employees are given awards for identifying ways to prevent waste generation or ways to recover and recycle materials, and meetings and conferences are held throughout the company by employee groups to exchange ideas on pollution prevention. Each year, 3M budgets approximately $150 million for research and development related to environmental issues, such as reducing the environmental impacts of products and processes. These activities have resulted in a 20 percent cut in energy consumption and a 35 percent cut in waste generation; by the year 2000, 3M plans to cut waste generation by 50 percent and release of pollutants to the environment by 90 percent. Not only has this resulted in a major reduction in environmental stress, but the company has realized savings of more than $150 million in lower costs for energy, process chemicals, and waste treatment (Bringer and Benforado, 1992).

1.2 POLLUTION PREVENTION

3M is only one example of how a reexamination of the way a company deals with its pollution problems can lead to significant environmental improvement, as well as cost savings to the company. Many other companies have also come to the realization that pollution prevention does indeed pay (see Table 1.1 for some other examples of early

TABLE 1.1
Selected industrial pollution prevention programs and goals

Company and program	Scope	Goal
Amoco: Waste Minimization Program	Primary focus on minimizing hazardous waste disposal as well as minimizing and tracking nonhazardous wastes	Eliminate the generation and disposal of hazardous wastes
BP America: Waste Minimization Program	Adopts EPA's environmental management hierarchy, with source reduction preferred	Annual waste minimization goals for all facilities
General Dynamics: Zero Discharge	Industrial source reduction, toxic chemical use substitution, recycling, treatment, and incineration	Eliminate all RCRA-manifested wastes leaving company facility
3M: Pollution Prevention Pays	Eliminate pollution sources through product reformulation, process modification, equipment redesign, recycling, and recovery of waste materials for resale	By 2000, cut all hazardous and nonhazardous releases to air, land, and water by 90% and reduce the generation of hazardous wastes by 50%, with 1987 as the base year
Monsanto: Priority One (TRI waste)	Source reduction, reengineering, process changes, reuse, and recycling to reduce hazardous air emissions and TRI solid, liquid, and hazardous wastes	A 90% reduction in hazardous air emissions from 1987 to 1992; a 70% reduction in TRI solid, liquid, and gaseous wastes from 1987 to 1995
Xerox	Toxic chemical use substitution, materials recovery and recycling	Reduce hazardous waste generation by 50% from 1990 to 1995

Source: U.S. EPA, 1991.

proponents of pollution prevention). This book examines the sources and impacts of industrial pollution and how industry can minimize the negative environmental effects of manufacturing processes in a cost-effective way. These effects are not limited to a company's internal activities, but also include those of their suppliers and those of the consumer through use and disposal of the product.

The production of waste throughout the United States and the rest of the world is accelerating at a rapid pace. As countries become more industrialized and the wealth of their citizens increases, there is an increased demand for goods and services. Corporations looking to increase sales and therefore profits often feed this craving with advertising for their products. The result is an ever-spiraling demand for goods, resulting in more and more wastes (see Figure 1.1). The environmental impacts of this have been all too evident: despoiled air and water, hazardous waste dumps leaking their toxins into our groundwater supplies, increased rates of cancer and other diseases from exposure to these chemicals, rapidly depleted resources, global warming, and damage to the protective ozone layer over Earth. Many people have come to the realization that this desecration of Earth cannot continue without causing insurmountable problems. Attempts to eliminate the problem through end-of-pipe treatment of wastes after they are produced have helped, but the problems are too large for this to be the solution. What is needed is a major overhaul of the way we manage wastes and the environment.

Pollution prevention requires a holistic approach to waste management. Rather than waiting until after a waste is produced and then attempting to make it innocuous, the pollution prevention approach considers the entire life of a product, from extraction of the raw materials from the earth through manufacturing to product use and finally to product disposal and possible recycling or reclamation, in order to find ways to minimize all environmental impacts. This may mean using materials that are less toxic or that are less scarce in the earth, finding more efficient manufacturing processes or processes that demand less energy, designing new products that make recycling after use easier, or creating new packaging materials that reduce the amount of packaging going to landfills or incinerators.

In the past, industry showed little concern for the types or amounts of wastes generated, and the public had little knowledge of the impacts of these wastes on the environment. These wastes were usually just discharged into the air or a nearby river, or they were dumped or buried on land (see Figure 1.2*a*). Disposed materials that were thought

FIGURE 1.1
"Mutts's" version of capitalism. (Source: P. McDonnell, "Mutts." King Features Syndicate, October 18, 1996. Reprinted with special permission of King Features Syndicate.)

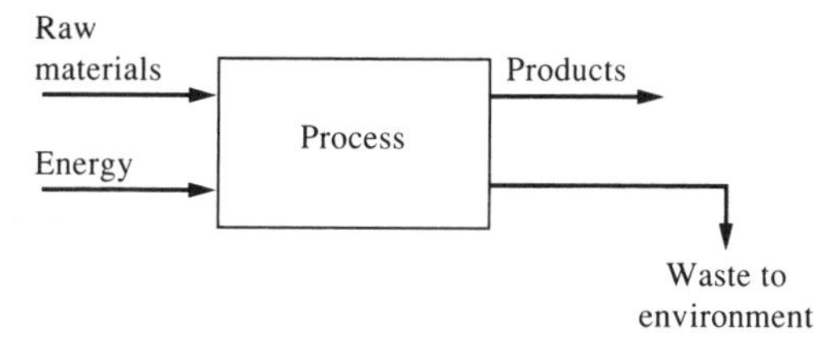

(*a*) **Past industrial practices**

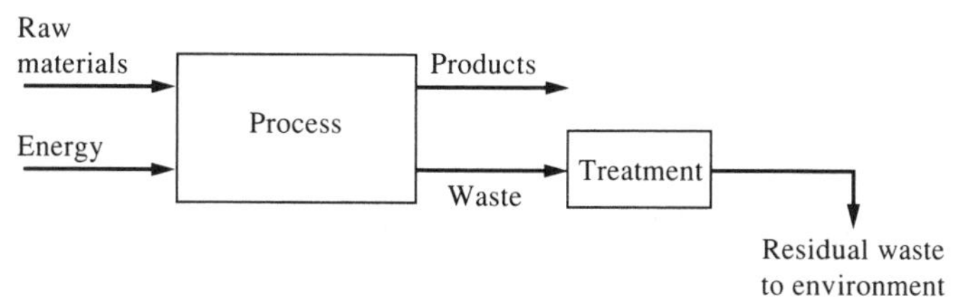

(*b*) **Recent industrial practices**

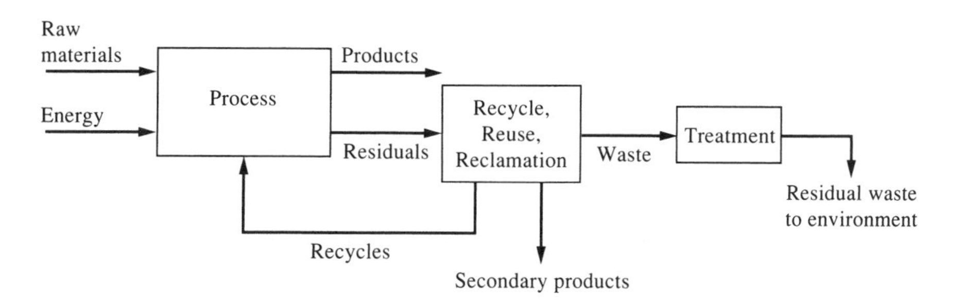

(*c*) **Current pollution prevention practices**

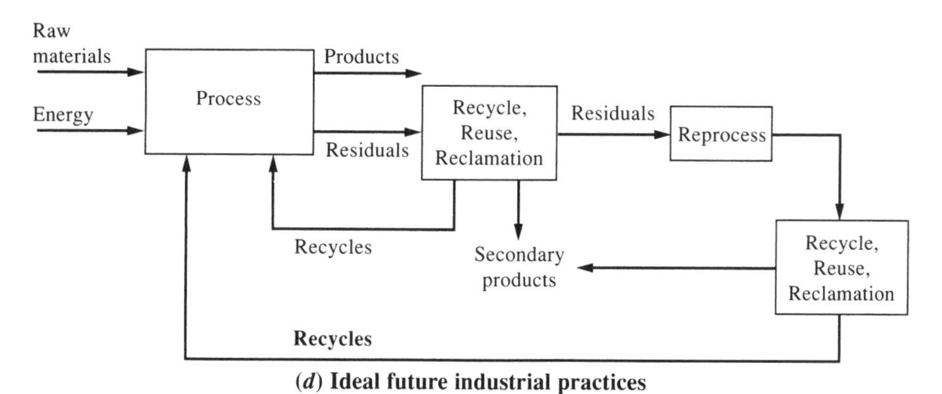

(*d*) **Ideal future industrial practices**

FIGURE 1.2
Recent and proposed industrial waste management practices.

FIGURE 1.3
Zero pollution is a goal, but it is usually not totally achievable.

to be gone forever through dilution in air or water or by burial in the ground came back to haunt us (see Figure 1.3). As these impacts became known, industries began to treat their wastes to remove the most egregious ones (Figure 1.2*b*). Eventually, some industries began recycling and reusing some of their waste materials (Figure 1.2*c*). Other industries, recognizing that product marketing and pollution prevention are intimately related, began marketing the environmental "greenness" of their products as a way to attract new customers. This was the beginning of the pollution prevention era. We can do more, though. What is needed is a goal of "zero pollution," in which most process waste production is eliminated through process changes, and as much remaining waste as possible is recycled, reused, or reclaimed, at either the facility of origin or another facility (Figure 1.2*d*). The little residue remaining after recycling and reuse can be treated and disposed of in an environmentally acceptable fashion.

The goal of pollution prevention is zero pollution, but this is a goal only; not all waste can be prevented or recycled and there will always be some waste to finally be disposed of. The objective should be to make the volume of this waste small enough that it can be managed effectively in an environmentally safe manner.

Before investigating the workings of a pollution prevention program, it is useful to look at how industry has grown over the last few hundred years, bringing us to the point where we are now.

1.3 THE HISTORICAL PERSPECTIVE

The process of industrialization describes the transition from a society based on agriculture to one based on industry. Modern industrialization is often dated as having its origins in the Industrial Revolution, but environmental pollution can be traced to manufacturing in ancient times. Goods have been produced to some extent since the dawn of civilization. Pottery works and factories for the manufacture of glassware and bronze ware have been discovered in Greece and Rome. In the Middle Ages, large silk factories were operating in the Syrian cities of Antakya and Tyre. During the late medieval period, textile factories were established in several European countries. These were all fairly small operations, though, and had little impact on the environment beyond their immediate area.

During the Renaissance (fourteenth to seventeenth centuries), industrialization increased in many areas, primarily following advances in science and the development of new trading partners in Asia and the New World. Factories were created to produce such goods as paper, firearms, gunpowder, cast iron, glass, clothing, beer, and soap (Kaufman and Farr, 1995). These factories differed from those found today, though; generally they were large workshops where each laborer functioned independently. Industrial processes were largely carried out by means of hand labor and simple tools; mechanization or machinery was rare. Organized factories could be found, but home production was still the norm. The guilds were very strong at that time and resisted any attempts to increase the expansion of factories. Consequently, environmental impacts due to goods production were still minor and spread out over a large area.

1.3.1 The Industrial Revolution

This all changes with the onset of the Industrial Revolution, which began with the application of power-driven machinery to manufacturing. The Industrial Revolution brought many changes to the way people lived and worked. It led to the movement of people from rural to urban areas and a shift from home to factory production. It also was the impetus for the creation of a new working class. The Industrial Revolution is considered to have begun in Britain in the early 1700s and then to have spread rapidly throughout much of Europe and North America in the early nineteenth century. A form of Industrial Revolution is still under way or is just beginning in many less developed countries.

By the early eighteenth century, Britain had burned up much of its forests to provide heat for its inhabitants and for its limited industry. However, large deposits of coal were available as a fuel, and there was an abundant labor supply to mine the coal and iron. What was needed was a way to transform the energy in coal into a form that could be used in manufacturing. Machines were being used in manufacturing in England at that time, but on a limited basis only. Matthew Boulton built a factory in 1762 which employed more than 600 workers to run a variety of lathes and polishing and grinding machines. Josiah Wedgwood and others used waterwheels and windmills in Staffordshire to turn machines which mixed and ground materials for making chinaware (Rempel, 1995).

The first major use of mechanization in industry, though, came in the British textile industry. The industry was fraught with severe inefficiencies: it took 4 spinners to keep up with the demand of 1 cotton loom and 10 persons to prepare yarn for 1 woolen weaver. Weavers were often idle because of the lack of needed yarn. A way to spin yarn more quickly was required. In 1764, James Hargreaves invented the spinning jenny.

This was quickly followed by the water frame for spinning yarn invented by Richard Arkwright in 1769, and the spinning mule by Samuel Crompton in 1779. The ready supply of yarn created by these inventions led to the invention of the power loom by Edward Cartwright in 1785, which enabled women to do the weaving in place of men. At about the same time, a machine was patented that printed patterns on the surface of cloth by means of rollers. Prior to these developments, most spinning, weaving, and textile manufacturing was done in the home. However, the new machines were generally too large to operate in a home and were often waterwheel driven, so textile manufacturing had to move to factories. The result of the use of all these machines was a 50 percent to 90 percent drop in the cost of manufacturing textiles, reflecting the reduced labor requirement, and an explosive growth in the sale of textiles, both in England and abroad. Another revolution in the textile and garment industries occurred in 1846 when the American Elias Howe invented the sewing machine. The American Civil War soon created another impetus for growth in the industry—the need for uniforms. Clothing manufacturers were forced to develop standardized sizes for uniforms, and eventually other clothing, to meet the needs of mass production practices.

The textile industry may have been the one that ushered in the Industrial Revolution, but the invention that usually is most associated with it is the steam engine. Because waterpower had been the only way to run machines, factories had to be located next to a river or stream. Steam engines released company owners from this restriction, allowing them to locate factories near the raw materials supplies, manpower, and markets, rather than basing factory location on the water supply. The first piston engine, developed in 1690 by the French physicist Denis Papin for pumping water, was never practical. The first modern steam engine was built by an English engineer, Thomas Newcomen, in 1705 to pump seepage water from coal, tin, and copper mines. It was not very efficient, but it was used to pump water from mines. The breakthrough in steam engine design came in 1763 when James Watt, a Scottish engineer, invented the reciprocating steam engine, changing it from one that operated on atmospheric pressure to a true "steam engine." He also added a crank and flywheel to provide rotary motion (Rempel, 1995). The value of these engines was quickly recognized, and Watt produced hundreds of them over the next several years, freeing industry from the need for water power. The development of the steamship by Robert Fulton in 1807 and of steam locomotives in the 1830s, both of which were driven by steam engines, vastly expanded the markets for industrially produced goods and helped to speed the spread of the Industrial Revolution.

Other major events that sped the Industrial Revolution along were the development of ways to inexpensively produce electricity, which could be used to power machines in factories that did not have ready access to coal for steam production; the internal combustion engine; the automobile along with the assembly line techniques developed by Henry Ford in 1913 to mass produce autos; aviation; and rapid worldwide communications systems. In recent years, there has been a new revolution in manufacturing practices with the advent of sophisticated factory equipment, often run by computers. Automation has significantly and rapidly changed the size and the skills of the factory work force. These advances in manufacturing have greatly reduced manufacturing costs and produced major increases in consumerism.

Before 1860, approximately 36,000 patents for new inventions had been issued in the United States. Between 1860 and 1890, an additional 440,000 patents were issued. In 1899, Charles H. Duell, U.S. commissioner of patents, stated that "everything that can be invented has been invented." However, an additional 900,000 patents were issued in the first quarter of the twentieth century. The total is currently in excess of 5,500,000. Obviously, Commissioner Duell was wrong and industrialization has continued to grow at an ever-increasing rate.

1.3.2 Impacts of Industrialization

The result of all this improvement in mechanization was a great increase in industrial productivity, lower costs for manufactured products, and usually an increase in the standard of living of the population. However, this did not come without cost. Factories dumped waste materials from manufacturing processes into the water, air, and land. Because industries are usually clustered together near or in cities, these discharges have a negative cumulative effect on a small area. The rapid growth in production also greatly increased the demand for energy, raw materials, and natural resources, often taxing our supply of them and causing environmental damage due to resource extraction processes. Typically, industries have been set up to use only virgin materials because of the expense and complexity of renovating recycled materials into usable quality for manufacturing new products. Therefore, there is often no market for the materials in a product after it has served its useful purpose and it is discarded into a landfill, creating more pollution. It has only been in about the last decade that this practice has begun to change.

The Industrial Revolution has also brought with it a change in the social structure of industrialized countries. Industrialization usually results in increased wealth for everyone, although it is not equally distributed, resulting in new class distinctions. Because of the increased wealth, populations tend to increase. The population rapidly shifts from a mainly rural one to an urbanized culture as people move to where the jobs are. The clustering of industries in a common area and the resulting urbanization creates severe stress on the environment. Areas that could previously assimilate the wastes from the rural community can no longer handle them and environmental degradation occurs. Some of these impacts are discussed in Chapter 3. In many cases, the environmental threats can be reduced by more efficient utilization of resources by industry and by better design of products so that they can be more easily recycled after use. The intent of this book is to show how pollution prevention can often be easily accomplished and how it will benefit industry as well as the environment.

1.4 WHAT IS POLLUTION PREVENTION?

Congress, the EPA, and environmental professionals came to the conclusion in the 1980s that a new industrial waste management philosophy was needed if the ever-expanding industrial pollution and resource depletion problems were to be solved. Indiscriminate use of virgin resources in manufacturing and subsequent end-of-pipe treatment of resulting wastes would not provide the resource sustainability and environmental quality demanded by the public. As a result, a new paradigm was developed which emphasized minimizing the use of harmful or overexploited resources and eliminating or minimizing waste production at the source in the industry's production area. This philosophy became known by many names, including waste minimization, source reduction, waste reduction, green engineering, and sustainable engineering, but the name that is most often associated with it is *pollution prevention.* Pollution prevention, or P2 (said as *P-two,* not P-squared, because we do not want to square the amount of pollution!), is the term adopted by EPA and the term usually used in federal legislation.

1.4.1 Waste Definition

Before we can discuss how waste reduction or pollution prevention programs should be structured, *waste* must be defined. Congress developed a legal definition, included in the Resource Conservation and Recovery Act, which is described in detail in Chapter 3, but there are many other definitions that can be used more readily.

We usually tend to think of waste as a solid product left over at the end of a process or action, but waste is a much broader issue than that. It encompasses wastage of energy or water in producing or using a product. We must focus on the total picture when we are describing waste. For example, home beverage can recycling programs can be very beneficial in conserving natural resources and decreasing the amount of landfill space required, but driving several miles to deposit a few newspapers, empty cans, and glass or plastic bottles in collection bins can be very wasteful of gasoline. The resources consumed in doing this, together with the further resources needed to take the materials from the collection point to a reprocessing center, could exceed the resources saved by not throwing them away. It is wasteful to allow food, which has consumed resources and energy in its production, to be damaged or spoiled. Extreme measures to reduce packaging may have the effect of reducing the use of paper, metals, glass, and plastics at the expense of the food they would protect, despite the value of the wasted food being many times greater than the value of the now-avoided packaging (World Resource Foundation, 1996). Thus we must be careful in how we define waste.

Industrial waste is usually described as materials coming from a manufacturing process that are not directly used within the corporation and that are marked for disposal or release to the environment (Graedel and Allenby, 1995). They may be a waste to that process or that company, but they may still have value to someone else. For example, the spent pickle liquor from a steel mill is considered a significant waste problem by the steel industry, but it has great potential as a neutralizing agent and coagulant in other applications. The problem is that the cost of marketing and transporting it to the potential user often makes its use uneconomical. Thus the same industrial process by-product could be categorized as a waste or a usable commodity, depending on its quality and the ready accessibility of a market for it. A waste may not necessarily have to be a waste. There is a better definition of a waste:

> A waste is a resource out of place.

It is the responsibility of pollution prevention personnel to find the right place to turn the waste into a resource.

1.4.2 Pollution Prevention Definition

Pollution prevention is a term used to describe production technologies and strategies that result in eliminating or reducing waste streams. The EPA defines pollution prevention as

> the use of materials, processes, or practices that reduce or eliminate the creation of pollutants or wastes at the source. It includes practices that reduce the use of hazardous materials, energy, water or other resources and practices that protect natural resources through conservation or more efficient use.

Thus pollution prevention includes both the modification of industrial processes to minimize the production of wastes and the implementation of sustainability concepts to conserve valuable resources.

Pollution prevention activities range from product changes to process changes to changes in methods of operation. This wide variety of activities is depicted in Figure 1.4.

The main premise underlying pollution prevention is that it makes far more sense for a generator not to produce waste than to develop extensive treatment schemes to ensure that the waste poses no threat to the quality of the environment (Freeman, 1995).

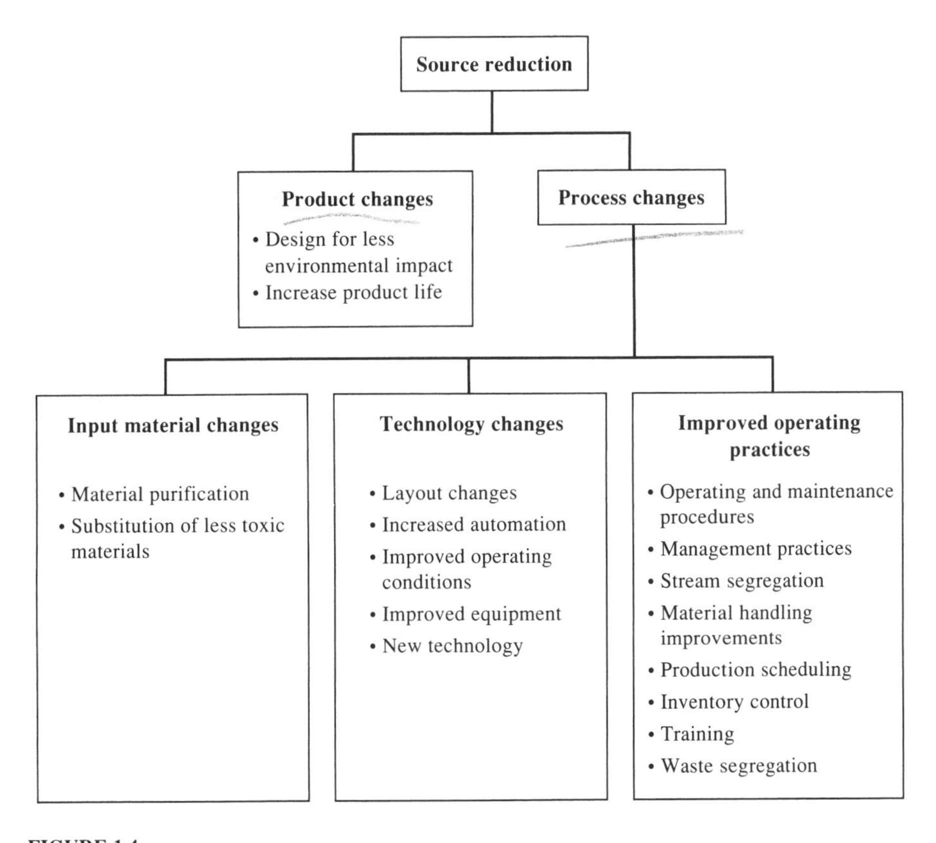

FIGURE 1.4
Typical source reduction methods. (Adapted from U.S. EPA, 1992)

1.4.3 Other Terms

We can avoid confusion by defining a few more terms that are commonly used to describe pollution prevention before we move on. *Source reduction* is an activity that reduces or eliminates the waste at the step where the pollution is created. *Waste minimization* and source reduction are often used interchangeably. *Emission reduction* is an activity that reduces or eliminates pollutants within the industry boundary limits so that they are not emitted into the environment. *Waste reduction* is any activity that reduces the amount of waste that is generated at any step of manufacture, use, or disposal. Thus, changes to an industrial process to increase efficiency of process chemicals utilization is deemed source reduction, while treatment of the residual chemicals leaving the process to either destroy them or recycle them back to the process is emission reduction. Waste reduction encompasses both of these. *Recycling* refers to the recovery and direct reuse of a material from a waste stream. For example, chromium can be recovered from the drag-out water from a plating bath rinse system by reverse osmosis or ion exchange and recycled back to the plating bath for reuse. *Reclamation* generally indicates that the recovered chemical is used in some other application. An example is the recovery of spent pickle liquor from a steel mill operation and its use as a neutralizing agent in another industry.

1.4.4 Sustainability

A final definition is that of *sustainability* or *sustainable waste management*. Another synonymous term that is enjoying current favor is *integrated waste management*. These are difficult terms to define and have different meanings to different people. Discussion of how sustainability should be defined was initiated by the Bruntland Commission, a group assigned to create a "global agenda for change" by the General Assembly of the United Nations in 1984. They defined sustainable very broadly:

> Humanity has the ability to make development sustainable—to ensure that it meets the needs of the present without compromising the ability of future generations to meet their own needs. (World Commission on Environment and Development, 1987)

Professor Robert K. Ham of the University of Wisconsin, a leading authority on solid waste management, states that "sustainable" means an action or process can continue indefinitely. No resources are used to extinction or faster than they are naturally replenished. "Sustainable waste management" implies that there would be no degradation of land, water, or air by wastes; however, this is not feasible and regulatory authorities generally allow some degradation to levels deemed to be acceptable. Ham finds the term "integrated waste management" to be more practical in that it requires use of multiple waste management techniques to minimize resource, environmental, and economic impacts. It includes waste reduction, recycling, treatment, and environmentally safe disposal (Thurgood, 1996).

N. C. Vasuki, chief executive officer of the Delaware Solid Waste Authority, says that a society has attained its sustainable development goal when the material and economic aspirations of its members are satisfied through optimum materials management and minimum use of natural resources. Essential prerequisites for attaining such a goal include the following (Thurgood, 1996):

- A free and democratic system to develop consensus on materials use policies.
- A relatively free market system which adjusts to supply and demand of materials.
- Timely use of appropriate technology to minimize environmental degradation.
- Recognition that a risk-free society is not feasible.
- Elimination of hidden subsidies and equitable allocation of costs.
- A regulatory framework for establishing a rule of law and an adjudication system for equitably enforcing laws.

Finally, Dr. Peter White, principal scientist at Procter & Gamble, states that an integrated waste management system should manage all of the waste in an environmentally and economically sustainable way. He defines *environmental sustainability* as reducing overall environmental burdens by optimizing both consumption of resources and generation of emissions. *Economic sustainability* means that the overall costs are acceptable to all sectors of the community that are served—householders, businesses, institutions, and government. Integrated waste management considers both of these to achieve the most acceptable result based on overall environmental burdens and costs. It involves the use of a range of different waste minimization schemes and treatment options (White, 1996).

The topic of sustainability is a vital one to any discussion on pollution prevention. It is discussed in more detail in Chapter 14.

1.5 THE POLLUTION PREVENTION HIERARCHY

Integrated waste management provides the flexibility to use an almost limitless variety of waste minimization, waste treatment, and waste disposal techniques. Until recently, only the last two were seriously considered. This changed with passage of the Pollution Prevention Act (PPA) of 1990 (see Chapter 4 for more details). The preamble to the PPA says:

> The Congress hereby declares it to be the national policy of the United States that pollution should be prevented or reduced at the source whenever feasible; pollution that cannot be prevented should be recycled in an environmentally safe manner, whenever feasible; pollution that cannot be prevented or recycled should be treated in an environmentally safe manner whenever feasible; and disposal or release into the environment should be employed only as a last resort and should be conducted in an environmentally safe manner.

Thus Congress made pollution prevention a national policy rather than just a desired goal, and it established a hierarchy for determining how pollution should be managed. It established source reduction as the preferred method to be used for waste management, if it is feasible and cost effective. Congress realized that all pollution could not be eliminated through source reduction alone, and it set recycling (and presumably other methods of waste reuse such as reclamation) as the preferred alternative

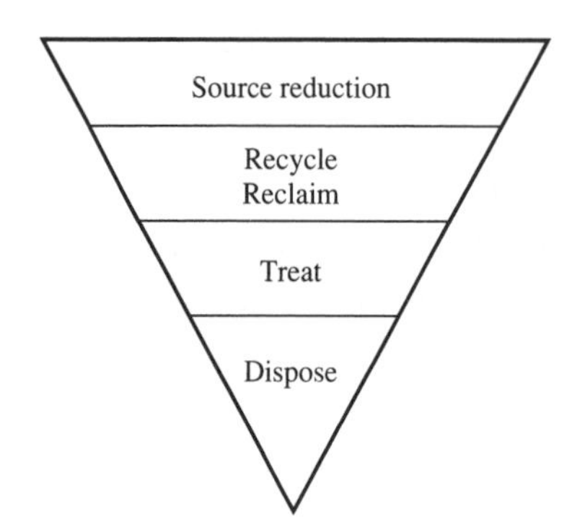

FIGURE 1.5
The pollution prevention hierarchy.

for management of residuals that remain after all viable source reduction measures are taken. Anything that remains after these steps should be treated to render it less hazardous and more compatible with the environment. Disposal into secure chemical landfills or direct release into the environment is allowed only as a last resort. These options were depicted in Figure 1.2.

In almost all cases, total pollution elimination through source reduction or recycling will not be possible. There will always be some residuals that cannot be prevented or reclaimed. Recovery systems are not 100 percent efficient, and residual streams may not be pure enough to recycle directly or to recover economically. The remaining pollution requiring treatment after source reduction and recycling should be greatly reduced in volume, however, thus making treatment easier and much less expensive. Again, it is unlikely that treatment will detoxify all pollution. Some will always need to be disposed of, either to a secure chemical landfill or directly to the environment at levels that the environment can safely assimilate. This amount should be much reduced from the amount of the original waste, though. Thus the pollution prevention hierarchy can be depicted as an inverted triangle, where the area of the band denoting the management option is indicative of the amount of pollution involved (see Figure 1.5). The objective of pollution prevention is to make the pointed base as small as possible.

1.6 RECYCLING VS. POLLUTION PREVENTION

A close examination of the EPA definition of pollution prevention indicates that it refers only to elimination of pollution at its source. Recycling or reprocessing of a recovered material into a form that can be used in another process is not deemed pollution prevention by the EPA. This is a much narrower definition than that used by many states, which include recycling and reuse in their definition of pollution prevention, because all three options lead to a reduction in the amount of material wasted to the environment or needing treatment before discharge.

In this book we use the broader definition of pollution prevention that includes recycling and reuse of residuals from industrial processes. Most of the book is devoted to source reduction, recycling, and reuse, but treatment and final waste disposal are discussed as necessary to complete the picture of integrated or sustainable waste management. The reader desiring more knowledge on industrial waste treatment can find it elsewhere (Conway, 1980; Eckenfelder and Dasgupta, 1989; Nemerow and Dasgupta, 1991; Sell, 1992).

1.7 ENVIRONMENTAL ETHICS

Most pollution prevention activities are initially begun because of regulatory pressure, an understanding that there could be cost savings associated with minimizing waste generation, or a desire to improve a corporate image. In recent years, however, some companies have moved beyond these impetuses to espouse pollution prevention on a more moral basis. They base their pollution management decisions on environmental ethics.

Environmental ethics is a systematic account of the moral relationships between human beings and their natural environment. It assumes that moral norms can and do govern human behavior toward the natural world (Des Jardins, 1997). Environmental ethics is concerned with humanity's relationship to the environment, its understanding of and responsibilities to nature, and its obligation to leave some of nature's resources to posterity. Pollution, population control, resource use, food production and distribution, energy production and consumption, the preservation of the wilderness and of species diversity all fall under the purview of environmental ethics (Pojman, 1994).

The inspiration for the recently developed study of environmental ethics was the first Earth Day in 1970, when environmentalists started urging philosophers who were involved with environmental groups to do something about environmental ethics. Discussions on environmental ethics go back much further than this, however, and include the writings of Henry David Thoreau, John Pinchot, and John Muir. Many books have been written recently discussing environmental ethics. Among the better ones for developing an insight into this rapidly developing field are those by Callicott and Da Rocha (1996), Des Jardins (1997), Gunn and Vesilind (1986), Hayward (1994), and Pojman (1994). The book *Management for a Small Planet,* by Stead and Stead (1996), should be required reading for anyone interested in the need for corporations to develop an environmental ethic for their corporate practices.

Rather than attempt to condense the philosophy into a few paragraphs, the reader is directed to the references just mentioned. We only briefly discuss the impact of these philosophies on corporate management.

There are several philosophies used to describe environmental ethics. One of the early ones, defined by Gifford Pinchot, is *conservationism.* Its basis is the view that wilderness is a resource that must be utilized and protected at the same time. The value system that is expressed in this philosophy puts people's needs above all others. Wilderness, as such, receives no moral standing beyond needing to be available for human consumption in a sustained manner. Conservationists work to protect natural and wilderness areas from nonessential forms of permanent destruction, ensuring the future existence of the areas for human utilization. The Sierra Club is an example of an organization that espouses this philosophy. Its mission statement includes the line "to practice and promote the responsible use of the earth's ecosystems and resources."

Another philosophy, which dates back to the writings of Henry David Thoreau and John Muir, is that of *preservationism.* This philosophy promotes the ethic that nature is meant to be enjoyed and experienced by humans and that it is our duty to protect the wilderness for our future enjoyment. This philosophy again places the value of nature on human utilization of it, but it is directed at benign uses of nature (enjoyment), rather than on sustainable extraction of resources.

A third environmental ethics philosophy, which has developed only over the past decade or so, is often referred to as *deep ecology,* a term coined by Arne Naess. Deep ecology extends the base of morality to include all life on Earth, including plants and animals. According to this philosophy, decisions should be made from a viewpoint that incorporates other positions in addition to human ones. In some respects this is not really a "new" ethic; as it is similar to the philosophy long espoused by many Native Americans and by many other cultures around the world. The most well known, and most notorious, organization espousing deep ecology is the group Earth First! Its slogan is "No compromise in defense of mother earth," and its belief, often expressed violently, is that Earth must be actively defended from development for its own sake and for the health of its ecosystems.

Recently, a new philosophy, termed *social ecology,* has begun to develop. Social ecology places a strong value on human existence while still recognizing the uniqueness of nature. It identifies human interactions as the main problem to be solved. The ultimate goal is to create an environment where humans interact with one another and the environment in a responsible manner (Knauer, 1997).

All of these philosophies have a common element—there is a responsibility for all people to minimize their impact on the environment as much as they can. This includes organizations and businesses. Since the Industrial Revolution, societies all over the world have based their hopes on the concept of unlimited economic wealth. The desire for economic growth has been raised to almost the same status as religion (Stead and Stead, 1996). As eloquently stated by Campbell and Moyers (1988), you can tell which institution a society considers most important by the relative height of its buildings. In medieval times, the churches were the tallest buildings. After the Renaissance, the tallest buildings were the seats of government. Today, the tallest buildings are the centers of economic activity. This philosophy of unlimited economic growth must now be tempered by a need to maintain sustainability. This necessitates a rethinking by industry of acceptable methods of doing business. In other words, industry must develop an underlying set of environmental ethics upon which corporate decisions are made.

Corporate ethics involves the moral issues and decisions confronting corporations and the individuals working in those corporations, including moral conduct, character, and ideals. Corporate environmental ethics concerns the way corporations conduct business in relation to their impact on the environment. Values are the key ingredients people use to judge right and wrong. Thus a person's ethical system is the sum total of the values he or she holds dear. Corporations have ethical systems that are primarily composed of the dominant values of the key strategic decision makers in the firm. This means that ethical considerations are an inherent part of corporate strategic decision-making processes. Effectively incorporating Earth into strategic decision-making processes means extending the firm's ethical reasoning to include the planet (Stead and Stead, 1996).

Sustainability should be the core value for a corporation's ethical system; it allows for positive economic success and environmental responsibility. It is based on an understanding that economic success and ecosystem survival are both worthy and necessary goals for individuals and organizations. Acceptance of this tenet would allow environmentalists and industrialists to cooperate to achieve common goals of adequate industrial growth within the constraints of resource preservation.

One way of achieving this cooperative style of industrialism is for businesses to use what is referred to as *stakeholder management*. Stakeholders are persons or groups that can affect or are affected by the achievements of the business's objectives. Stakeholders include customers, shareholders, suppliers, competitors, activists, and advocacy groups, all of whom have an interest in the practices of the corporation. Stakeholder management refers to serving the varied, often conflicting, needs of these multiple stakeholders (Stead and Stead, 1996). It requires giving weight to the ethical, social, and political dimensions of a situation along with the economic dimensions, bringing ethical considerations to the forefront of strategic decision making. It is based on the theory that organizations that serve the needs of the greater society are more likely to prosper than are self-serving organizations.

Consumers are becoming more demanding and are insisting on high-quality goods that save time and energy and preserve the environment. They are searching out manufacturers that they perceive as socially and environmentally responsible. About one in seven Americans is actively involved in "green organizations" (Bhat, 1996). These people are concerned about the impact that products have on the environment. They are more likely to buy products in recyclable or biodegradable packages, and they are willing to pay for this. We now have large numbers of "green consumers" buying "green products" and then recycling the wastes. Green products are those that are of high quality, durable, made with nontoxic materials, produced and delivered using energy-efficient processes, packaged in small amounts of recyclable material, not tested on animals, and/or not derived from threatened species. As environmental standards being developed by the International Organization for Standardization (under ISO 14000, to be described more fully in later chapters) become widely adopted, it will become easier for the public to decide what are and are not environmentally sound products. There are now several investment groups specializing in "ethical stock funds," investing only in companies that practice environmentally conscious manufacturing.

In 1990, the members of the Chemical Manufacturers' Association (CMA) agreed to a set of guidelines designed to improve the way chemical manufacturers manage the environmental aspects of their businesses. The following nine guidelines are contained within the Responsible Care program:

1. To safely develop, produce, transport, use, and dispose of chemicals.
2. To make health, safety, and the environment priority considerations in planning for both current and new products.

3. To promptly report any chemical or health hazards and to be prepared to deal with them if they occur.
4. To inform customers how to safely transport, store, and use chemicals.
5. To always operate plants in a safe manner.
6. To support research on the environmental impacts of products, processes, and wastes.
7. To contribute significant efforts to resolve problems caused by past practices.
8. To participate with the government to develop laws and regulations that promote a safer, more environmentally sound industry.
9. To share environmental management experiences and information with other firms in the industry.

Many members of the chemical industry have signed on to these guidelines and are making a serious effort to comply, as are organizations from other industrial sectors.

Based on their environmental performance, Bhat (1996) classifies companies as red, yellow, or green (see Table 1.2). *Red companies* are those that, because of recent negative experiences such as permit violations or accidents, have decided to support improved compliance with environmental laws and regulations. They usually have a short-term planning horizon of a week to a month, and they concentrate on waste treatment or incineration as their mode of pollution prevention. All company decisions are based solely on cost. *Yellow companies* are more proactive. Companies typically go from red to yellow as a result of regulations that significantly increase pollution control costs or when they are attempting to counteract bad publicity arising from some incident. Yellow companies have moved beyond a firefighting mode and have set a goal of zero violations. Management is oriented toward preventing environmental problems. The planning time frame is usually about a year. Yellow companies try to prevent environmental problems by engaging in preventive activities. *Green companies* represent the ultimate stage in environmental responsibility. Companies become green not only because they want to comply with laws, improve their image, and reduce costs, but also because they believe that it is the right thing to do. The company management has clearly articulated its environmental policies. Rules and regulations are not seen as constraints, but rather as cost-reduction and profit-improvement measures. Source reduction is the primary waste reduction technique. Life-cycle assessment, benchmarking (i.e., comparison of P2 results with established goals), and environmental audits are well-established practices in the company. Several corporations have developed environmental codes of conduct that are used in all corporate decision making, for example, the 3M Corporation, cited earlier.

The International Chamber of Commerce has adopted the Business Charter for Sustainable Development in cooperation with the United Nations Environmental Programme and other international business organizations. The charter contains 16 principles; the first is to "make environmental management a corporate priority," and the second is to "integrate environmental management into all corporation levels and functions." The goal of this book is to demonstrate how this can be accomplished.

TABLE 1.2
Classification of companies on the basis of environmental performance

Feature	Red Corporation	Yellow Corporation	Green Corporation
Management approach	Problem solving	Problem preventing	Opportunity seeking
Time horizon	Short term	Medium term	Long term
Top management involvement	Nonexistent	Not wholehearted	Long and continuous involvement
Organization	Unit operation	Plant	Companywide
Manager responsible for environment	Legal department	Plant or legal department	Chief executive officer
Company policy	Nonexistent	Not well written	Comprehensive
Driving force	Laws and regulations	Costs of compliance	Opportunities
Strategy	Cure	Prevention	Prevention
Pollution-reduction techniques	Treatment, landfilling, and incineration	Recycling and better housekeeping	Source reduction and housekeeping
Public relations	Responds to accidents	Responds to accidents	Systematic and well organized
Supplier selection	Price only	Price and selected green factors	Price and selected green factors
Training	Nonexistent	Not well organized	Lifelong and systematic
Management style	Paternal	Somewhat paternal	Egalitarian
Research and development	Nonexistent	Compliance-oriented	Focused on new opportunities
Use of resources	Ineffective	Somewhat effective	Efficient and effective
Communication	Top-down	Mostly top-down	Two way
Green performance	Violations, permit denials, discharges, spills	Amount of waste reduced, disposal costs	Cradle-to-grave, audits, life-cycle assessment, benchmarking
Reward system	Blame for violations and accidents	Blame for violations and small reward for compliance	Green performance major factor
Technology	Compliance-oriented	Reduced compliance costs in products and services	Continuous integration

Source: Bhat, 1996.

REFERENCES

Bringer, R. P.; and Benforado, D. M. "Pollution Prevention as Corporate Policy: A Look at the 3M Experience." *Environmental Professional* 11 (1989): 117–26.

Bringer, R. P.; and Benforado, D. M. 3P Plus: Total Quality Environmental Management. 85th Annual Air and Waste Management Association Meeting, Reprint 92-54.03, 1992.

Bhat, V. N. *The Green Corporation: The Next Competitive Advantage.* Westport, CT: Quorum Books, 1996.

Callicott, J. B.; and da Rocha, F., Jr. *Earth Summit Ethics*. Albany: State University of New York Press, 1996.
Campbell, J.; and Moyers, B. *The Power of Myth*. Garden City, NY: Doubleday, 1988.
Conway, R. A. *Handbook of Industrial Waste Disposal*. New York: Van Nostrand Reinhold, 1980.
Des Jardins, J. R. *Environmental Ethics: An Introduction to Environmental Philosophy*, 2nd ed. Belmont, CA: Wadsworth, 1997.
Eckenfelder, N. L.; and Dasgupta, A. *Industrial Water Pollution Control*, 2nd ed. New York: McGraw-Hill, 1989.
Freeman, H. M. "Pollution Prevention: The U.S. Experience." *Environmental Progress* 14 (1995): 214–23.
Graedel, T. E.; and Allenby, B. R. *Industrial Ecology*. Englewood Cliffs, NJ: Prentice Hall, 1995.
Gunn, A. S.; and Vesilind, P. A. *Environmental Ethics for Engineers*. Chelsea, MI: Lewis Publishers, 1986.
Hayward, T. *Ecological Thought*. Cambridge, England: Cambridge Press, 1995.
Kaufman, J. J.; and Farr, G. N. "Factory System." In *Microsoft Encarta 96 Encyclopedia*. Ramsey, NJ: Funk and Wagnalls, 1995.
Knauer, J. Environmental Ethical Theory Applied in the Modern Environmental Movement. http://www.envirolink.org/elib/enviroethics/essay.html (1997).
Nemerow, N. L.; and Dasgupta, A. *Industrial and Hazardous Waste Treatment*, 2nd ed. New York: Van Nostrand Reinhold, 1991.
Pojman, L. P. *Environmental Ethics: Readings in Theory and Application*. Boston: Jones and Bartlett, 1994.
Rempel, G., The Industrial Revolution. http://mercury.acnet.wnec.edu/~grempel/rempel/wc2/industrial.rev.html (1995).
Sell, N. J. *Industrial Pollution Control: Issues and Techniques*, 2nd ed., New York: Van Nostrand Reinhold, 1992.
Stead, W. E.; and Stead, J. G. *Management for a Small Planet*, 2nd ed., Thousand Oaks, CA: Sage Publications, 1996.
Thurgood, M. "Definitions." *Warmer Bulletin* 49 (1996): 2–6.
U.S. EPA. *Pollution Prevention 1991*. EPA 21P-3003. Washington, DC: U.S. EPA, 1991.
U.S. EPA. *Facility Pollution Prevention Guide*. EPA/600/R-92/088. Washington, DC: U.S. EPA, 1992.
White, P. "So What Is Integrated Waste Management?" *Warmer Bulletin* 49 (1996): 6.
World Commission on Environment and Development. *Our Common Future*. Oxford: Oxford University Press, 1987.
World Resource Foundation. Waste Minimisation. http://www.wrfound.org.uk/WasteMin-IS.html (1996).
Zosel, T. W., Pollution Prevention Research Directions and Opportunities. Paper presented at the AEEP Research Opportunities Conference, Ann Arbor, Michigan, 1993.

PROBLEMS

1.1. Waste can be defined as "a resource out of place." Examine the contents of your home trash and determine the potential uses for the materials being discarded. How difficult would it be to recycle/reuse these materials? What fraction of your waste is potentially recyclable/reusable?

1.2. The quantities of waste generated by almost all activities can generally be reduced. Consider a typical business office, such as your university engineering department office. What are the sources of waste in the office and how could they be reduced? Consider all waste sources.

1.3. Many photographic processes use chromium salts. Washing the photographic plates results in excess hexavalent chromium being discharged into wash waters and eventually into our waterways. Considering the pollution prevention hierarchy, what options are available to reduce these discharges and which would be the most preferable?

1.4. Describe what you feel the term "environmental ethics" means.

1.5. Using the Internet, find a corporation that has established a corporate environmental ethics policy, then describe and critique the main components of the policy. Into which Bhat classification would you place this company?

1.6. The concept of sustainability maintains that present practices should be designed to ensure that future generations will have the ability to meet their own needs. However, based on past experience, it can be argued that people are adaptable and find ways to cope with a changing world. For example, known petroleum reserves may soon (in a few decades) run out. Some argue that this is not a reason to cut back on gasoline use now, because by then other undiscovered petroleum reserves may be found or alternative fuels will have been developed. Petroleum-based fuels may be of little value in the future. Discuss the pros and cons of this argument and give your views.

1.7. Underground injection wells used to be one of the major forms of ultimate waste disposal for many industrial wastewaters in the United States, but their use has drastically declined in the 1990s. Research the use and operation of injection wells, describe their drawbacks, and explain why their use has declined.

1.8. You are an engineer for a large manufacturing facility. One of your responsibilities is to evaluate pollution emissions at your plant and to suggest ways to minimize them. The company previously produced and emitted large quantities of a toxic chemical from one of its processes. To reduce emissions, the company recently replaced the process with a new system that produces only negligible quantities of the chemical. You are told that since this is a new low-polluting process, you do not need to investigate it. However, during your inquiries into another process line, you discover that the new process is emitting another potentially harmful compound. This compound has not yet been proven harmful and is not on any EPA lists, but recent literature you have read indicates that it may be a potential endocrine disrupter. Should you report this to your supervisor? The process cost millions of dollars to install and its shutdown would cause irreparable damage to the company. If you do report it and your supervisor tells you to ignore it until firm evidence proves its harmful, what should you do?

1.9. For the scenario presented in Problem 1.8, the company is appealing a fine for violations due to its previous process, based on the fact that the company has already spent millions of dollars to install a new, nonpolluting process. You are called on to describe the merits of the new process before EPA, and to state that the process is environmentally benign. You are instructed by your upper management to say nothing about your concerns relative to the new chemical being emitted. What do you do? If you do testify as instructed, and are asked about the new chemical by an EPA official, what do you say?

1.10. Obtain an annual report from a chemical company (e.g., Du Pont, Eastman Kodak, Dow Chemical, Monsanto). These can be found in your library or on the Internet. Find the company's environmental policy statement and comment on its adequacy.

CHAPTER 2

PROPERTIES AND FATES OF ENVIRONMENTAL CONTAMINANTS

Essentially all industrial processes yield by-products that become waste materials. When a product reaches the end of its useful life, it is often considered a waste material and is disposed of to a landfill or incinerator. Even when wastes are recycled or reprocessed into other goods, there are usually residual materials that must still be disposed of, and these become waste. All of these wastes can contribute to the contamination of our environment.

Pollution prevention and waste minimization techniques may be successful in reducing the amount of contaminants that enter the environment. However, for these techniques to be useful and to be applied in an effective way, they must be sound from economical, environmental, and public health standpoints. No pollution prevention activity will be implemented by industry if it prices the product out of competition with similar products or if the activity makes only a minimal impact on environmental quality or public health safety. Thus one of the main objectives of this book is to assist the engineer or business manager in deciding whether a particular waste minimization strategy will be beneficial to the company involved.

To effectively evaluate the usefulness of a proposed product design or production line change, the environmental impacts of the changes must be known. This means that the decision maker must have some basic understanding of the materials involved and their environmental and public health impacts. He or she must also understand the fate of these materials in the environment, since these fates will often dictate the long-term risks and costs associated with disposal of a particular material. Disposal of a more toxic material will not always lead to greater long-term costs to the industry because that material may be easily degradable and short-lived in the environment. Moreover, the contaminant, although very toxic, may be essentially immobile at the disposal point in the environment, minimizing any potential risk. In addition to evaluating the economic advantages of making process or process chemical changes, anyone contemplating such changes should investigate the properties of the existing by-products and any new by-products anticipated from the changes and determine their environmental implications.

Many of these environmental impact assessments will need to be made by a qualified engineer or scientist, but the plant engineer or business manager should have enough understanding of the by-product properties and environmental fates to be able to make an informed judgment as to the best decision for the plant. This decision maker does not have to be an environmental engineer or chemist. The impacts and fates in the environment of the contaminants commonly encountered in industrial wastes and some of their pertinent properties are discussed in this chapter. With an understanding of these concepts, the person contemplating a waste minimization initiative can make a more informed decision.

This chapter discusses common classes of industrial contaminants, including solvents, plasticizers, hydrocarbons, dioxins and furans, and plating metals. It is not intended to be an exhaustive discussion of these contaminants, but rather a brief introduction to the nomenclature used in naming these compounds, their properties and toxic effects, and their environmental fates. The discussion assumes the reader has had only an introductory course in chemistry. Readers interested in a more detailed description of environmental contaminants are referred to the book by Watts (1998).

2.1 ORGANIC CHEMICALS

Organic compounds are those that contain carbon and usually hydrogen. In addition, they may contain oxygen, nitrogen, sulfur, phosphorus, halogens, metals, or other elements. They run the gamut from simple (e.g., methane, CH_4) to highly complex (e.g., fulvic acids). A general knowledge of the nomenclature of organic compounds is critical to an understanding of environmental regulations and waste minimization assessments.

2.1.1 Nomenclature of Organic Compounds

Because of the many hundreds of thousands of different organic compounds available or potentially available, it is necessary to carefully name them to avoid confusion as to what compound is being described. Some compounds have common names associated with them, but in general it is essential that a commonly accepted nomenclature system be used. The system most often used is referred to as the IUPAC (International Union of Pure and Applied Chemists) system. Except for commonly used terms such as "methylene chloride" (the IUPAC name is "dichloromethane"), the IUPAC nomenclature is used here. Organic compound nomenclature can become very involved. Only a simplified version is presented here. More detailed descriptions can be found elsewhere (Cahn, 1979; IUPAC, 1961; Watts, 1998).

The carbon atom normally has four electrons to share with other atoms when making a compound. The carbon atom can link together in a wide variety of ways, forming straight or branched chains or rings. Other types of atoms can also be included in these chains or rings.

```
    H   H   H   H   H
    |   |   |   |   |
H — C — C — C — C — C — H
    |   |   |   |   |
    H   H   H   H   H
```

Pentane

```
            H
            |
        H — C — H
            |
        H — C — H
            |
    H   H   |   H   H
    |   |   |   |   |
H — C — C — C — C — C — H
    |   |   |   |   |
    H   H   H   H   H
```

3-Ethylpentane

```
          H
          |
H         C         H
 \      / | \      /
  C       H       C
  | \  H     H  / |
  |  H        H   |
  | /    H     \  |
  C      |        C
 /   \   C    /    \
H        |          H
         H
```

Cyclohexane

```
         H
         |
H        C        H
 \     /   \\    /
  C             C
  ||            |
  C             C
 /   \       //  \
H        C        H
         |
         H
```

Benzene

The result of these many types of bonds is the possibility for a multitude of different compounds.

Organic compounds can be divided into two main groupings by their structure. *Aliphatic compounds* contain straight or branched chains of carbon atoms, or are formed into rings containing single bonds between the carbons. Pentane, 3-ethylhexane, and cyclohexane are examples. Aliphatic compounds can be further classified as alkanes, alkenes, alkynes, and their derivatives, depending on the degree of saturation of the carbon bonds. This will be explained later. *Aromatic compounds* are a special group of organic compounds which contain carbon-based rings or multirings with alternating single and double carbon-carbon bonds. Benzene (shown above), phenols, and dioxins are examples of aromatic compounds.

ALIPHATIC COMPOUNDS. *Alkanes* are aliphatic hydrocarbon compounds in which all bonds between carbon atoms are single bonds. A carbon atom with all four bonds connected to different atoms is considered to be *saturated.* A compound, such as an alkane, with all saturated bonds is considered to be a saturated compound. The carbon chains may be straight or branched. They have the general formula C_nH_{2n+2}. The structure of alkanes can be drawn with all bonds shown, as with 2-methylpentane:

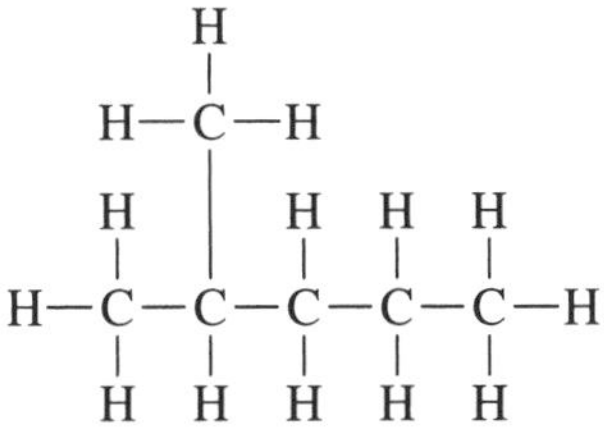

```
            H
            |
         H—C—H
            |
      H     |   H   H   H
      |     |   |   |   |
   H—C — C — C — C — C—H
      |     |   |   |   |
      H     H   H   H   H
```

2-Methylpentane

or, more commonly, without the carbon-hydrogen bonds shown:

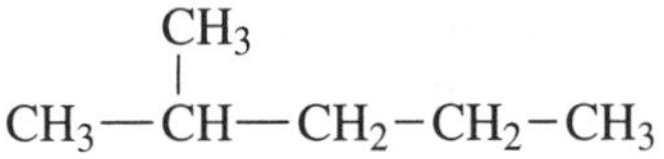

```
         CH3
          |
   CH3—CH—CH2–CH2–CH3
```

2-Methylpentane

or, in another style:

$$CH_3CH(CH_3)CH_2CH_2CH_3$$

2-Methylpentane

The alkane compounds form a class of compounds also known as *paraffins.* Methane (CH_4) is the simplest alkane and serves as the base of the alkane series. The second member of the series, containing two saturated carbon atoms, is ethane (CH_3—CH_3). The alkane series through 12 carbon atoms is shown in Table 2.1. The names of all alkanes end in -ane.

Alkanes with four or more carbon atoms can be branched, rather than straight-chained. Therefore, two or more compounds can have the same number of carbon atoms but have totally different structures and properties. These compounds are referred to as *isomers.* There are 4 carbon alkanes that can have 2 isomers, while there are 75 possible isomers of decane, a 10-carbon alkane. As the number of carbons increases, the number of possible isomers goes up rapidly. Proper nomenclature is essential if these compounds are to be described properly.

TABLE 2.1
Straight-chain alkanes

Name of compound	Number of carbons
Methane	1
Ethane	2
Propane	3
Butane	4
Pentane	5
Hexane	6
Heptane	7
Octane	8
Nonane	9
Decane	10
Undecane	11
Dodecane	12

Straight-chain alkanes are generally referred to as *normal* compounds. The symbol *n*- is usually put in front of the compound name to denote that it is a normal compound. Branched compounds are named in terms of the longest continuous chain of carbon atoms in the molecule. The names of the side chains are appended to this. The main chain is numbered to show where other groups are attached. When naming the compound, start numbering the chain at the end that allows use of the lowest possible numbers. For example, the compound

$$
\begin{array}{ccccccccccc}
 & & CH_2-CH_3 & & & & & & & & \\
 & & | & & & & & & & & \\
CH_3- & CH_2- & CH- & CH_2- & CH_2- & CH_2- & CH_2- & CH_2- & CH_3 & & \\
(1) & (2) & (3) & (4) & (5) & (6) & (7) & (8) & (9) & & \text{(Carbon number)}
\end{array}
$$

is named 3-ethylnonane. It is an isomer of undecane (11 carbons), but it is named as a nonane (9-carbon chain) with an ethyl group at the 3 position on the nonane skeleton. When more than one group is attached to the main chain, begin naming the compound with the groups listed in alphabetical order. Again, choose a numbering system that gives the lowest numbers to the side chains. Therefore, this compound is *not* 5-ethyl-7-methyloctane, but rather is named 4-ethyl-2-methyloctane:

$$
\begin{array}{cccccccc}
 & CH_3 & & CH_2-CH_3 & & & & \\
 & | & & | & & & & \\
CH_3- & CH- & CH_2- & CH- & CH_2- & CH_2- & CH_2- & CH_3
\end{array}
$$

4-Ethyl-2-methyloctane

More complicated compounds can be named in a similar fashion:

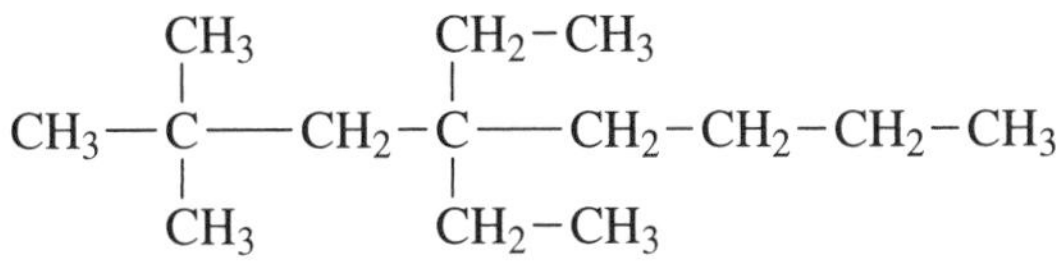

$$
\begin{array}{ccccc}
 & CH_3 & & CH_2-CH_3 & \\
 & | & & | & \\
CH_3- & C & -CH_2- & C & -CH_2-CH_2-CH_2-CH_3 \\
 & | & & | & \\
 & CH_3 & & CH_2-CH_3 &
\end{array}
$$

4,4-Diethyl-2,2-dimethyloctane

There are many alkanes that have other atoms or radicals substituted for one or more of the hydrogen atoms. These may include halides (Cl^-, F^-, Br^-, I^-), nitrogen groups (amines, $—NH_2$; amides, $—CO(NH_2)$; nitriles, —CN; or nitrosamines, —N—N=O), phosphorus, sulfur, and metals. Many solvents used in industry are halide-substituted (i.e., chlorine) alkanes. These will be described in more detail later. A group of aliphatic compounds that contain sulfur are called mercaptans. Mercaptans have a very disagreeable odor. They are commonly found in industrial wastes and are usually toxic. Butyl mercaptan is the compound that gives skunk emissions their distinctive odor. Compound naming for substituted alkanes is done by the same procedure as above. For example,

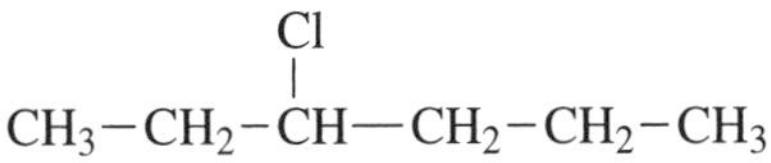

$$
\begin{array}{c}
\qquad\qquad Cl \\
\qquad\qquad | \\
CH_3-CH_2-CH-CH_2-CH_2-CH_3
\end{array}
$$

3-Chlorohexane

$$
\begin{array}{ccccc}
 & Cl & & NO_2 & \\
 & | & & | & \\
CH_3- & CH & -CH_2- & CH & -CH_2-CH_2-CH_3
\end{array}
$$

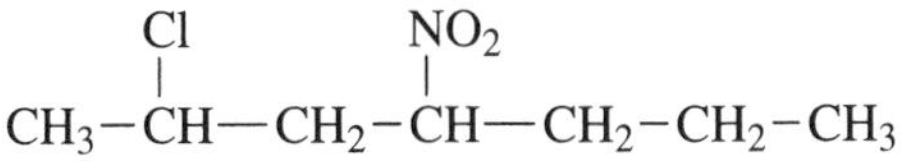

2-Chloro-4-nitroheptane

$$3HC—SH$$

Methyl mercaptan
(IUPAC name: Methanethiol)

One class of substituted alkanes that has been of particular environmental concern in recent years are the chlorofluorocarbons (CFCs), which have been implicated with ozone destruction in the upper atmosphere. These are described in more detail in Chapter 3. One CFC, known as Freon 11, is shown below:

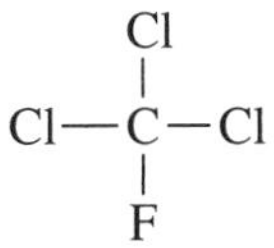

$$
\begin{array}{ccc}
 & Cl & \\
 & | & \\
Cl- & C & -Cl \\
 & | & \\
 & F &
\end{array}
$$

Freon 11
(IUPAC name: Trichlorofluoromethane)

Alkenes are aliphatic compounds in which double bonds exist between two adjacent carbon atoms. They have the general formula C_nH_{2n}. Because of the double bond, these compounds are often termed unsaturated compounds. The names of alkenes all end in -ene. In the past, the names ended in -ylene, and this terminology is still commonly used today. For example, the common solvent trichlorethene is usually called trichloroethylene (its shorthand abbreviation is 1,1,1-TCE). Another example of an alkene is 2-butene, commonly known as 2-butylene:

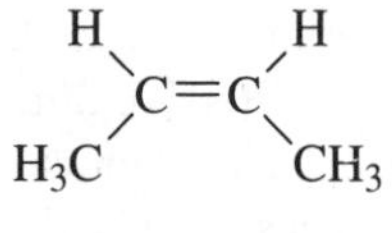

$$\begin{array}{ccc} H & & H \\ & C{=}C & \\ H_3C & & CH_3 \end{array}$$

2-Butylene
(IUPAC name: 2-Butene)

Again, the longest carbon chain serves as the basis for the compound name. The double-bonded carbon is designated numerically to have the lowest possible number and the double bond site is included in the name. Substituents are numbered as before.

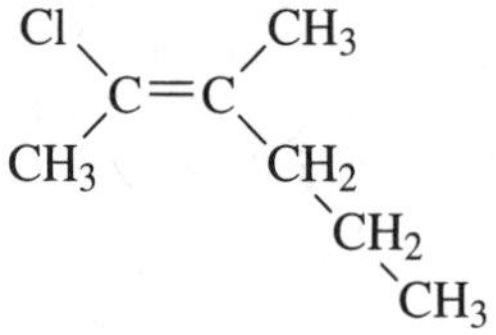

$$\begin{array}{ccc} Cl & & CH_3 \\ & C{=}C & \\ CH_3 & & CH_2{-}CH_2{-}CH_3 \end{array}$$

2-Chloro-3-methyl-2-hexene

Alkynes are denoted by a triple bond between two carbon atoms. These compounds are very unstable and, except for acetylene ($HC_3{\equiv}C_3H$; the IUPAC name is ethyne), are not commonly found as waste products. Names of alkynes end in -yne. The numbering system is as it was for alkenes.

$$CH_3{-}C{\equiv}C{-}CH_2{-}CH_3$$

2-Pentyne

Organic acids are commonly used in industry as process chemicals; they are also often found as by-products from chemical processes. Organic acids usually have a carboxylic acid group (—COOH) attached to one end of the molecule. The compound's name ends in -anoic acid. For example,

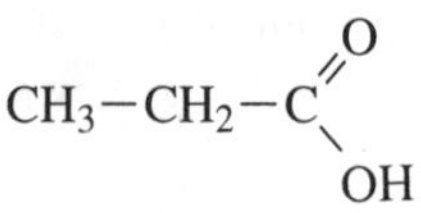

$$CH_3{-}CH_2{-}C({=}O){-}OH$$

Propanoic acid

Common names are generally used more often for organic acids than are their proper IUPAC names. Table 2.2 lists the common names and IUPAC names for the common organic normal carboxylic acids.

TABLE 2.2
Names of the more common normal saturated organic acids

Common name	IUPAC name	Formula
Formic	Methanoic	$HCOOH$
Acetic	Ethanoic	CH_3COOH
Propionic	Propanoic	C_2H_5COOH
Butyric	Butanoic	C_3H_7COOH
Valeric	Pentanoic	C_4H_9COOH
Caproic	Hexanoic	$C_5H_{11}COOH$
Enanthic	Heptanoic	$C_6H_{13}COOH$
Caprylic	Octanoic	$C_7H_{15}COOH$
Perlagoric	Nonanoic	$C_8H_{17}COOH$
Capric	Decanoic	$C_9H_{19}COOH$
Palmitic	Hexadecanoic	$C_{15}H_{31}COOH$
Stearic	Octadecanoic	$C_{17}H_{35}COOH$

Unsaturated organic acids are also commonly found in nature and are used as process chemicals. For example, they are used in the production of plastics such as Plexiglas and in a variety of oils such as linseed oil. The common unsaturated organic acids are listed in Table 2.3.

Many saturated and unsaturated monocarboxylic acids occur in nature as constituents of fats, oils, and waxes. Thus they are often referred to as *fatty acids*. Many have very objectionable odors, such as butyric acid, which gives rancid butter its noxious odor.

Esters are compounds formed by the reaction of alcohols and organic acids. The general formula of an ester is R—COO—R′, where R and R′ refer to organic groupings. They are named based on the alkyl nomenclature and the unprotonated salt of the carboxylic acid involved in their formation. An exception is when the alkyl group contains only two carbons; it is then named acetate rather than ethanate. The name of the alkyl group is placed first, followed by the name of the salt of the carboxylic acid, ending in -ate. For example, the compound

$$H_3C-\overset{\overset{\displaystyle O}{\|}}{C}-O-CH_2-CH_3$$

Ethyl acetate
(IUPAC name: Acetic acid ethyl ester)

is composed of the salt of acetic acid and an ethane group from ethanol. It is named ethyl acetate. Esters are used widely in industry. Many have pleasing aromas and are used in perfumes and flavoring extracts. Others are used as solvents.

TABLE 2.3
Common unsaturated organic acids

Name	Formula
Oleic acid	$CH_3(CH_2)_7CH{=}CH(CH_2)_7COOH$
Linoleic acid	$CH_3(CH_2)_4CH{=}CHCH_2CH{=}CH(CH_2)_7COOH$
Linolenic acid	$CH_3(CH_2CH{=}CH)_3CH_2(CH_2)_6COOH$

Ethers are formed by combining two alcohols; during the reaction a molecule of water is removed. Ethers have the general formula R—O—R′. They are generally named by combining the names of the two alcohols involved in the synthesis reaction.

$$CH_3-CH_2-O-CH_2-CH_3$$

Diethyl ether

Ethers are widely used in industry as solvents. Many are highly flammable, and some may become explosive if exposed to air for a prolonged period. Diethyl ether has been used widely as an anesthetic.

Aldehydes are the oxidation products of primary alcohols (R—OH), and *ketones* are the oxidation products of secondary alcohols (R—(C(OH)H)—R′); they have two alkyl groups attached to the carbonyl group (—CO—). The aldehydes contain a double-bonded oxygen and a hydrogen on the terminal carbon. They are named by adding -al to the basic alkane name. Ketone names end in -one.

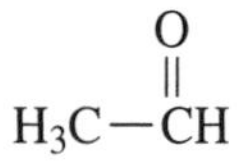

$$H_3C-\overset{\displaystyle O}{\overset{\|}{C}}H$$

Acetaldehyde
(IUPAC name: Ethanal)

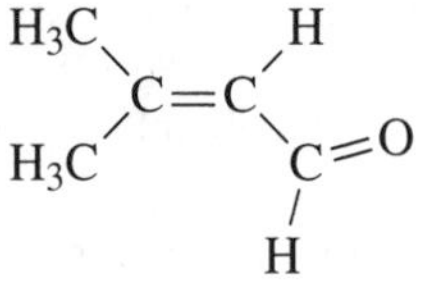

$$(H_3C)_2C{=}CH-CH{=}O$$

Senecialdehyde
(IUPAC name: 3-Methyl-2-butenal

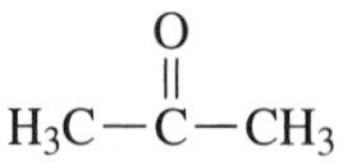

$$H_3C-\overset{\displaystyle O}{\overset{\|}{C}}-CH_3$$

Acetone
(IUPAC name: 2-Propanone)

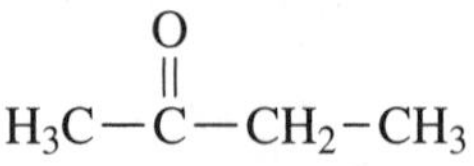

$$H_3C-\overset{\displaystyle O}{\overset{\|}{C}}-CH_2-CH_3$$

Methyl ethyl ketone
(IUPAC name: 2-Butanone)

Aldehydes and ketones are usually referred to by their common names. Table 2.4 lists the aldehydes and ketones common in industrial use.

Acetaldehyde and formaldehyde are two commonly used chemicals in organic synthesis reactions. Ketones are commonly used as solvents in industry and as a precursor in the synthesis of many organic chemicals.

TABLE 2.4
Common aldehydes and ketones

Common name	IUPAC name	Formula
Aldehydes		
Formaldehyde	Methanal	$HCHO$
Acetaldehyde	Ethanal	CH_3CHO
Propionaldehyde	Propanal	C_2H_5CHO
Butyraldehyde	Butanal	C_3H_7CHO
Acrolein	2-Propenal	$CH_2{=}CHCHO$
Ketones		
Acetone	Propanone	CH_3COCH_3
Methyl ethyl ketone (MEK)	Butanone	$CH_3COC_2H_5$
Methyl isopropyl ketone	3-Methyl-2-butanone	$CH_3COC_3H_7$
Methyl isobutyl ketone (MIBK)	4-Methyl-2-pentanone	$CH_3COCH_2CH(CH_3)_2$

A number of *cyclic aliphatic compounds* are found in petroleum. These are usually five- or six-ring saturated aliphatic compounds, although cyclic alkenes are also known. The most common cyclic aliphatic is cyclohexane.

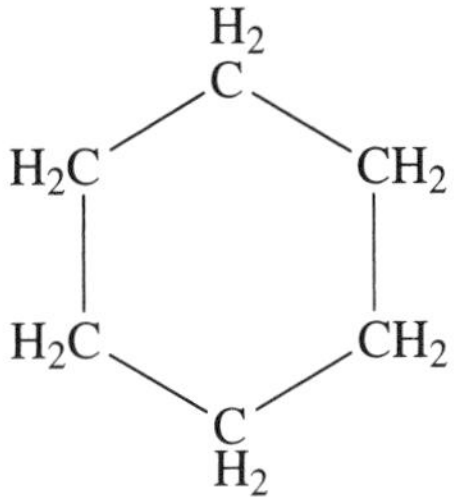

Cyclohexane

Cyclic alcohols and ketones are also known, including the ketone cyclohexanone.

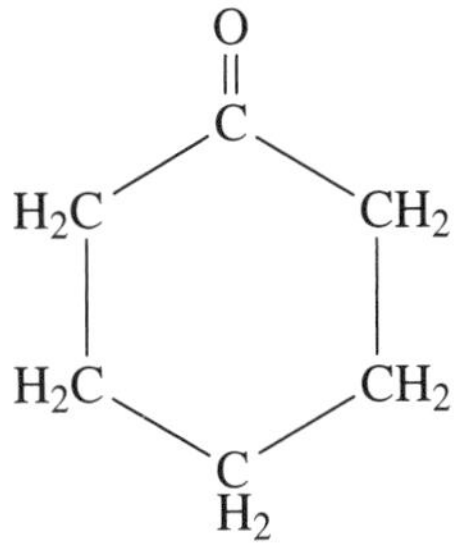

Cyclohexanone

Example 2.1. Draw the structures of the following compounds:
(*a*) 3-Methyl-5-propyloctane.
(*b*) 1,1,2,2-Tetrachloroethylene (a commonly used solvent also known as perchloroethylene or Perc).

Solution.
(*a*) The base chain is heptane, an 8-carbon chain. A methyl group is attached to the third carbon in the chain and a propyl group is attached to the fifth carbon. Therefore, the structure is:

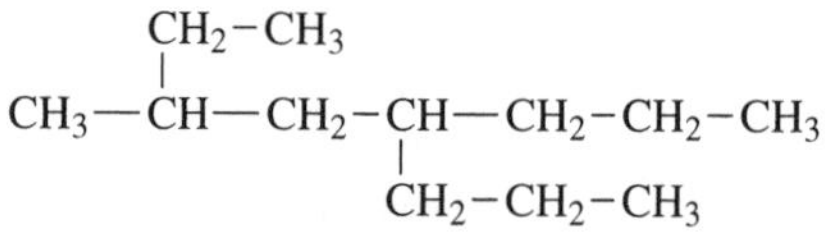

$$
\begin{array}{l}
CH_2-CH_3 \\
| \\
CH_3-CH-CH_2-CH-CH_2-CH_2-CH_3 \\
| \\
CH_2-CH_2-CH_3
\end{array}
$$

3-Methyl-5-propyloctane

(*b*) Ethylene contains two carbons joined by a double bond. Two chlorines are attached to each carbon, replacing all of the hydrogens:

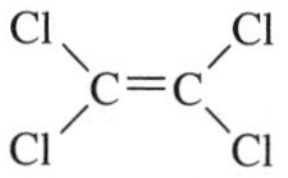

$$
\begin{array}{ccc}
Cl & & Cl \\
 & C=C & \\
Cl & & Cl
\end{array}
$$

1,1,2,2,-Tetrachloroethylene

Example 2.2. Name the following compound:

$$
\begin{array}{l}
Br \\
| \\
CH_3-CH-CH-CH=CH-CH_2-CH_3 \\
| \\
NH_2
\end{array}
$$

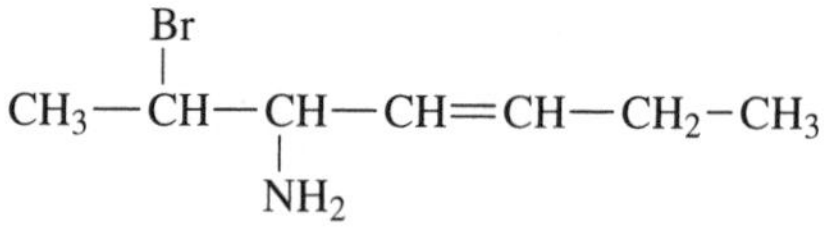

Solution. The base chain has seven carbons, so it is in the heptane series. There is a double bond between the numbers 3 and 4 carbons (starting at the end that makes the lowest number for the double bond), so this is an alkene, named 3-heptene. Following the same direction for numbering, there is an amino group attached to the 5 position and a bromine attached to the 6 carbon. Functional groups are listed alphabetically, so this compound is named 5-amino-6-bromo-3-heptene.

AROMATIC COMPOUNDS. All aromatic compounds are ring compounds with alternating single and double bonds between the ring carbons. The simplest aromatic compound is benzene (C_6H_6).

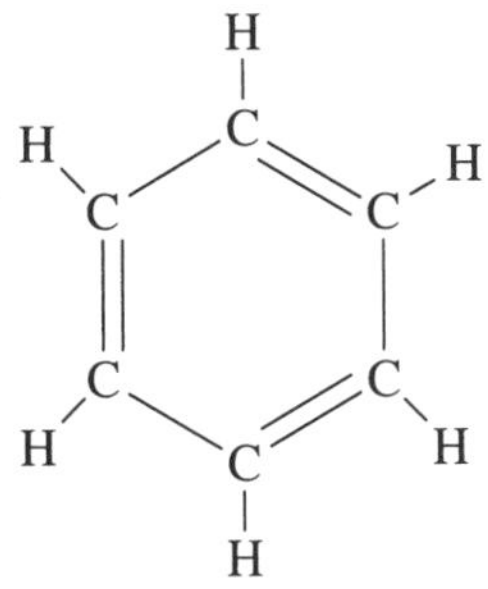

Benzene

The alternate bonds are shown as double bonds, but in actuality the bonds between any two carbon atoms resonate between being double and single bonds. There is actually a cloud of electrons encircling the 6 carbon ring. The aromatic ring bonds usually do not act like the typical covalent bonds in aliphatic compounds. It is difficult to add an atom across this bond, making these compounds quite stable.

For simplicity, aromatic compounds are usually shown in a shorthand form, as in **(a)** below without the carbon-hydrogen bonds or, more commonly, without the carbons and hydrogens associated with the aromatic ring structure, as in **(b)**:

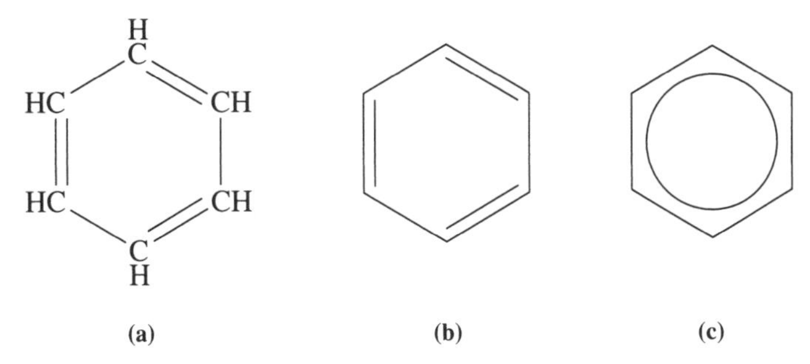

Simplified diagram **(c)** denotes the fact that the bonds resonate between single and double ones and are not located between any particular pair of carbon atoms. A fourth way of describing benzene is with the symbol phi ϕ.

Proper nomenclature is important when naming aromatic compounds. When substitution of one of the hydrogens on benzene occurs, the compound is named by placing the name of the substituent first followed by -benzene. For example,

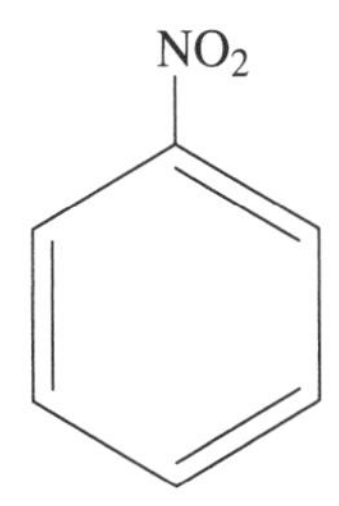

Nitrobenzene

When benzene is attached to an aliphatic chain, the term phenyl is used to denote the aromatic group. For example,

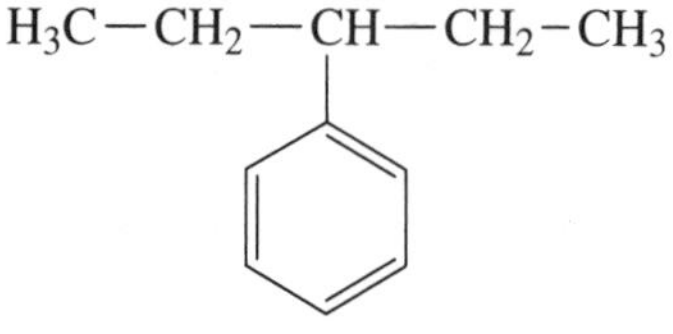

$$H_3C-CH_2-CH-CH_2-CH_3$$

3-Phenyl pentane
or 3-pentylbenzene

As can be seen, this compound could also be referred to as 3-pentylbenzene.

Some monosubstituted benzenes are traditionally called by common names, rather than by their proper IUPAC name:

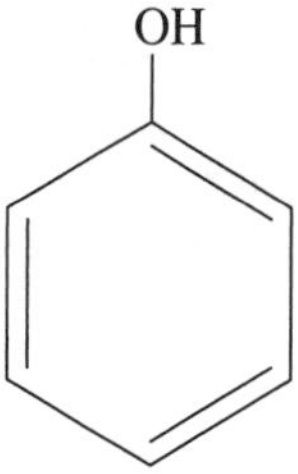

OH CH_3 NH_2

Phenol 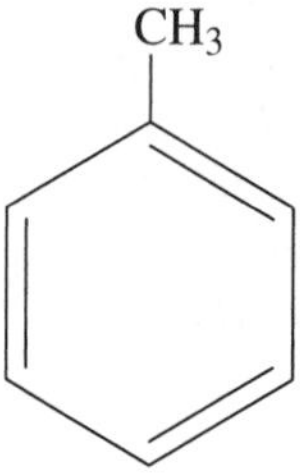 Toluene 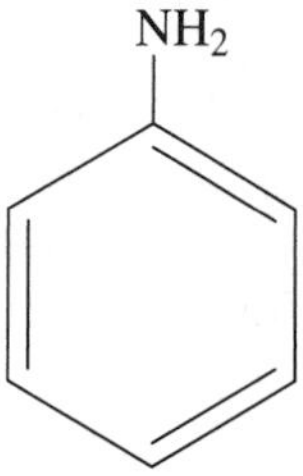 Aniline

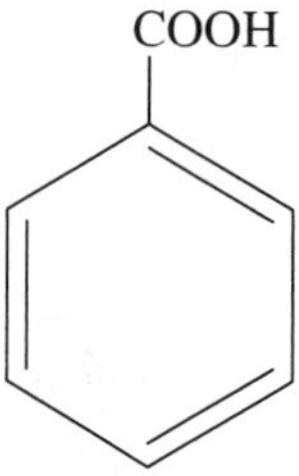

$COOH$ $HC=CH_2$

Benzoic acid 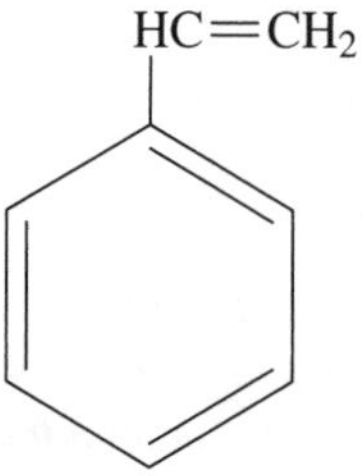 Styrene

When substitution of the benzene's hydrogens occurs at two or more locations, it becomes necessary to describe where these substituents are on the ring. There are two systems for doing this. The formal IUPAC procedure is to use a numbering system for positions on the ring:

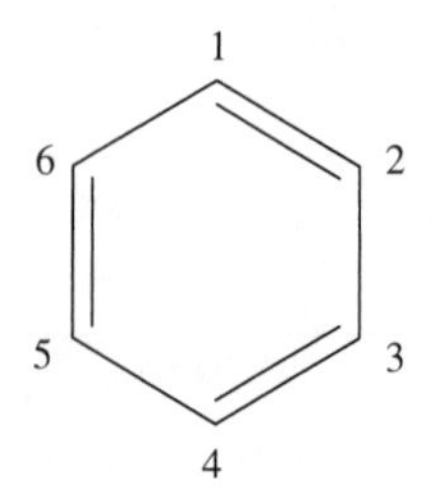

An example of this is 1,3-dichloro-5-nitrobenzene:

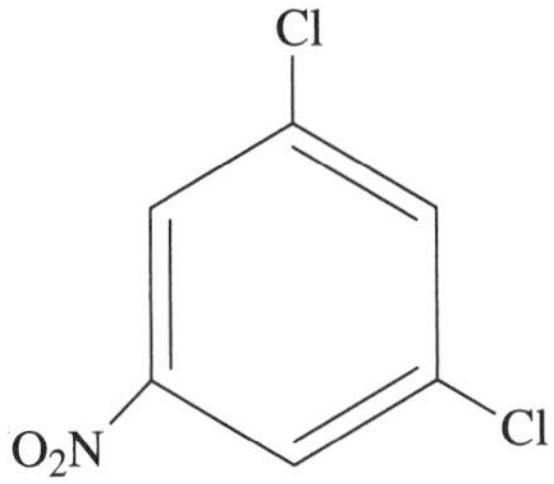

1,3-Dichloro-5-nitrobenzene

If there are only two substituents, their locations may be denoted by using the terms *ortho- (o-), meta (m-),* or *para- (p-)*:

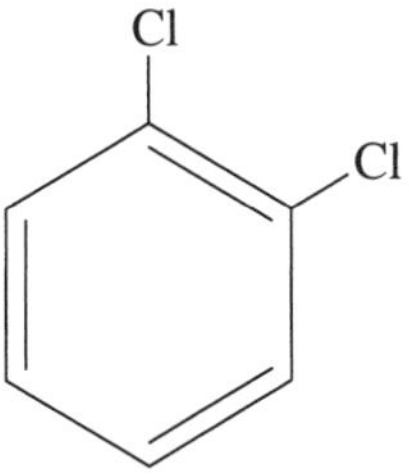

o-Dichlorobenzene

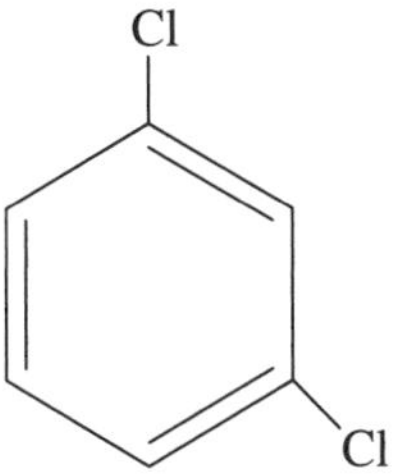

m-Dichlorobenzene

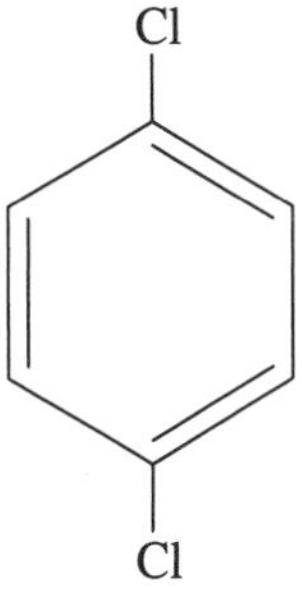

p-Dichlorobenzene

Ortho- indicates that the substituents are adjacent to each other (1,2 position), *meta-* denotes that they are displaced by one carbon on the ring (1,3 position), and *para-* indicates that they are opposite each other on the ring (1,4 position). The compound is named by putting the substituents in alphabetical order.

If the base ring compound is one that uses a common name (e.g., toluene, aniline), the substituent conferring the name is assumed to be in the 1 position. For example,

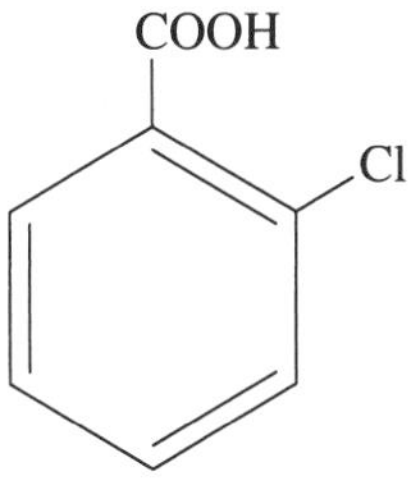

2-Chlorobenzoic acid
or *o*-chlorobenzoic acid

There are also some disubstituted benzene derivatives that more conventionally go by common names, rather than by their proper IUPAC names. The three possible isomers of each of these can be denoted using the *o-*, *m-*, and *p-* designation. Among these are xylenes, cresols, and phthalic acids:

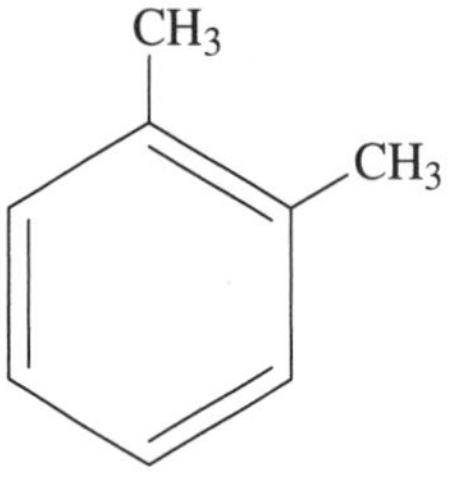

o-Xylene 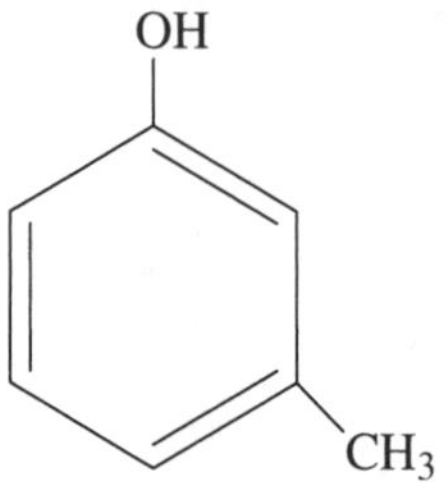 *m*-Cresol 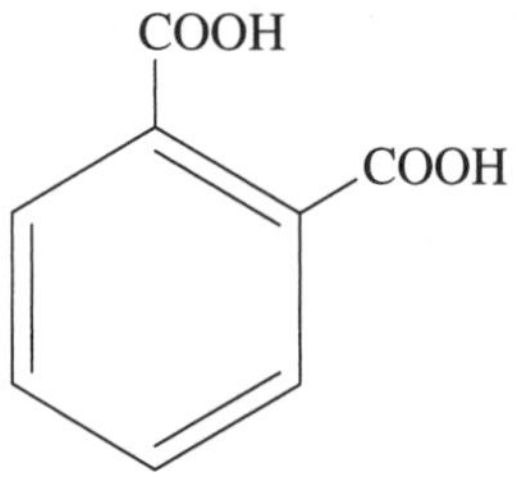 Phthalic acid

Two or more benzene rings can be fused together, sharing pairs of carbon atoms. These compounds are referred to as *polycyclic aromatic hydrocarbons* (PAHs) or *polynuclear aromatic compounds* (PNAs). The simplest two- and three-ring PAH compounds are naphthalene and anthracene:

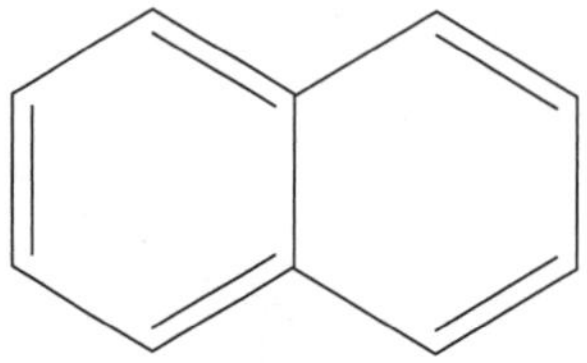

Naphthalene 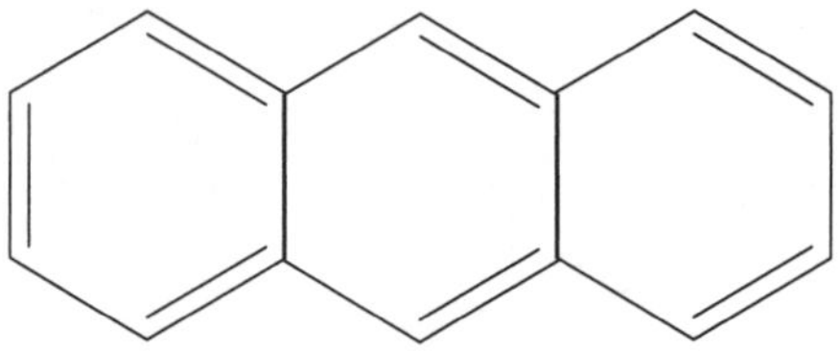 Anthracene

Another possible configuration for a three-ring aromatic is phenanthrene:

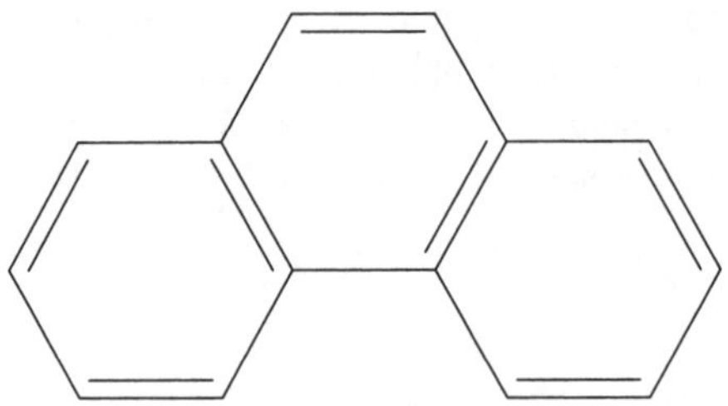

Phenanthrene

Many other PAH compounds are known. A major source of PAHs is incomplete combustion of organic compounds (automobile exhaust, cigarette smoke, etc.). They are also a component of the heavier fractions of petroleum products. Many are very stable in the environment, toxic, and potentially carcinogenic. Of particular concern recently is benzo[*a*]pyrene:

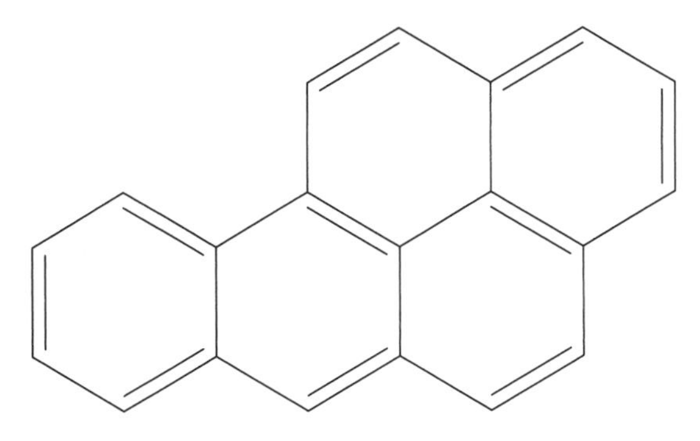

Benzo[*a*]pyrene

Detailed descriptions of PAHs and their environmental significance can be found elsewhere (Chakrabarty, 1982; Chaudry, 1994; Rochkind-Dubinsky, Sayler, and Blackburn, 1987; Watts, 1998). The procedure for naming substituted PAHs is complicated and won't be described here. The reader is referred to Cahn (1979), IUPAC (1961), and Watts (1998).

Many chlorinated aromatic compounds have important industrial uses. Chlorinated benzenes (e.g., chlorobenzene, hexachlorobenzene) are widely used as industrial solvents and as insecticides. *Polychlorinated biphenyls* (PCBs) have the structure

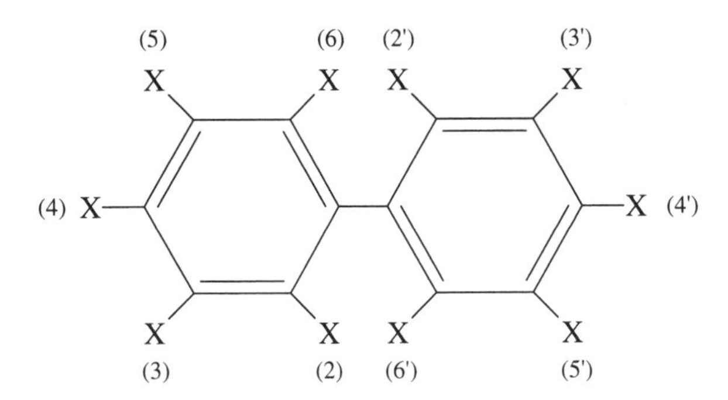

where parentheses denote the carbon number and X can be either a chlorine or a hydrogen. PCBs are heat-stable oils used extensively as transformer cooling oils and hydraulic fluids and as solvents and plasticizers.

There are 10 sites on the biphenyl molecule where chlorine can be substituted for hydrogen atoms. This results in 209 possible PCBs (called congeners), depending on the number and arrangement of chlorines on the molecule. PCBs are formally named using the same nomenclature as previously cited. For example,

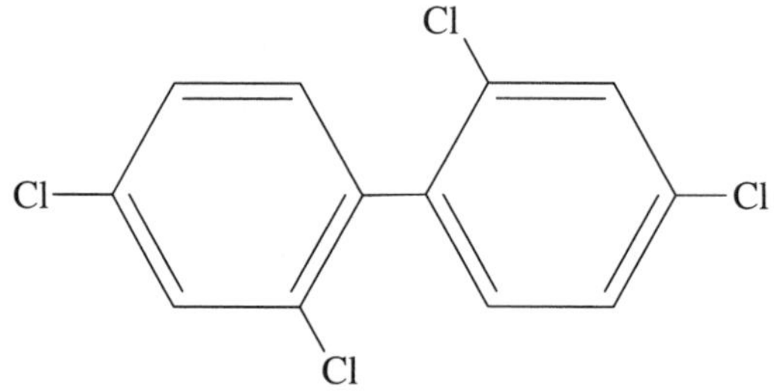

2,2',4,4'-Tetrachlorobiphenyl

In the United States, PCBs were marketed by Monsanto under the trademark Aroclor, followed by a four-digit number, for example, Aroclor 1221. The number denotes the number of carbons and the degree of chlorination. The first two digits refer to the 12 carbons contained in the biphenyl. The last two digits represent the average weight percentage of chlorine in the PCB mixture. Aroclor 1221 is a 12-carbon biphenyl mixture containing 21 percent chlorine, by weight. Production of PCBs has been banned in the United States since 1977, but much of it is still in use and it continues to be a significant environmental and public health problem.

The last groups of organic compounds we will discuss are the *polychlorinated dibenzodioxins* (PCDDs) and the *dibenzofurans* (PCDFs). These are not compounds that are produced for any useful purpose, but rather are unwanted by-products formed during the manufacture of other organics or during combustion of chlorinated organic materials such as plastics. Chlorinated dioxins are derivatives of dibenzo-*p*-dioxin. Between one and eight chlorine atoms may be substituted on the dioxin molecule, resulting in 75 possible congeners. An example of a PCDD is a compound commonly called TCDD:

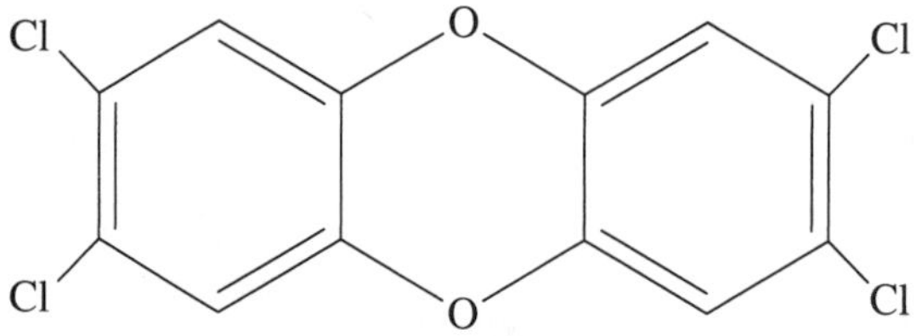

2,3,7,8-Tetrachlorodibenzo-*p*-dioxin (TCDD)

TCDD is a major health risk because it is highly persistent in the environment, bioaccumulates in animals, and is extremely toxic. Polychlorinated dibenzofurans are similar in structure to dioxins, but with only one oxygen between the benzene rings. A typical furan compound is

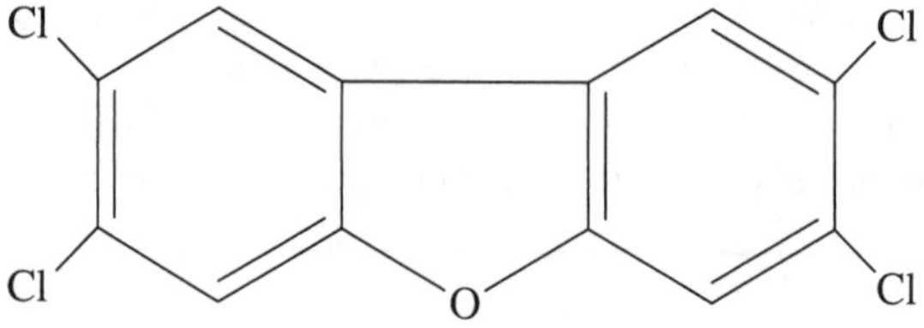

2,3,7,8-Tetrachlorodibenzofuran

PCDFs form in a similar fashion to PCDDs and pose similar environmental and public health problems.

Example 2.3. Draw the structure of the following compounds:
(*a*) 2,4-Dichloroaniline.
(*b*) 2-Bromo-3-chlorobenzoic acid

Solution.

(*a*) The base compound, aniline, is a benzene with an amino group (—NH_2) at the 1 position. Chlorines are substituted for hydrogens at the 2 and 4 positions. Therefore, the structure is

NH_2
Cl
Cl

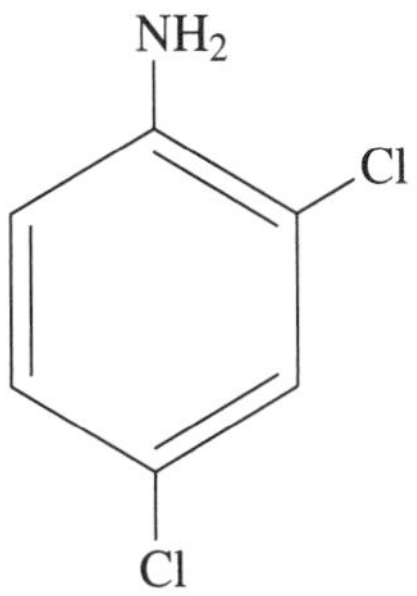

(*b*) The base compound is benzoic acid, which is a benzene with a carboxyl group (— COOH) at the 1 position. A chlorine is substituted at the 3 position and a bromine is substituted at the 2 position. Therefore, the structure is

COOH
Br
Cl

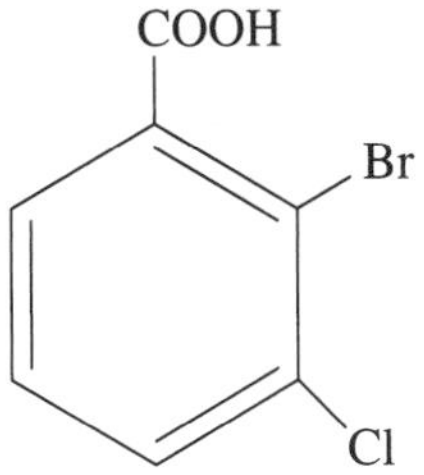

2.2 METALS AND INORGANIC NONMETALS

The division between metals and inorganic nonmetals is not well defined. In general, those elements that easily lose electrons to form positive ions are considered to be metals. Metals usually conduct electricity readily. Elements that hold electrons firmly and tend to gain electrons to form negative ions are called nonmetals (Sawyer, McCarty, and Parkin, 1994). The periodic table of the elements seen in Appendix A is divided into seven horizontal rows (periods) and 16 vertical columns (families). The heavy line divides metals from nonmetals.

Heavy metals are defined as those with atomic numbers greater than that of iron and with densities greater than 5.0 g/cm^3. Many heavy metals (e.g., lead, cadmium, chromium, mercury) are of great environmental concern because of their toxicity.

The most significant source of anthropogenic metals in the environment is metal finishing operations, either from rinsing metals after plating or from disposal of spent metal plating baths. Often these wastewaters are discharged into receiving streams near the industry or to municipal wastewater treatment plants, where some passes through to the receiving stream without being removed. The plating bath sludges or the sludges from the municipal treatment plant are often disposed of in landfills, where the metals can leach into the underlying groundwater. Waste metals in discarded products can also eventually solubilize and enter groundwater or surface waters. Metals in wastes undergoing incineration, such as tin cans and other metallic refuse, may volatilize under the high temperatures present and become an air pollutant. The nonvolatilized metals will accumulate in the fly ash or bottom ash and can threaten groundwaters after landfilling.

Small quantities of many metals are essential nutrients for humans and other animals, but too much of them can be toxic. Metal toxicity is often associated with the species of metal present. Forms that are essentially insoluble generally pass through the human body and are excreted without doing any damage. More soluble forms, though, can be retained in tissue or the blood stream and may cause severe toxicity. Because of chemical or biochemical transformations that can occur in the environment or in the body, the metal species present may not be the one that was discharged into the environment as a waste material. For example, under anaerobic conditions in sediments or in the human gut, metallic mercury can be combined with methyl groups to form methylmercury or dimethylmercury. Both are much more soluble than the parent mercury and can cause severe biological damage. Thus knowledge of all known species of a particular metal contaminant is important because the metal may be transformed into much more hazardous compounds than what an industry is discharging.

A major reason for concern about inorganic contaminants in general, and heavy metals in particular, is that they often bioaccumulate in nature. Concentrations in organisms increase as one goes up the food chain. These compounds are more soluble in tissues than they are in water or in the tissue of the lower order organism that has been consumed. Thus concentrations found in tissue can be orders of magnitude greater than found in the water at the initial industrial discharge point. What may appear to be innocuous amounts in an industrial waste effluent entering a receiving water may be concentrated to toxic levels in fish or later in humans that consume that fish. For example, oysters and mussels can contain mercury or cadmium at concentrations hundreds of thousands times greater than that of the water in which they are growing.

In this section, we examine only a few inorganic contaminants of significant industrial usage. Those chosen for discussion are discussed in other parts of this book. More detailed descriptions can be found elsewhere (Meyers, 1977; Sax, 1987, 1992) for these and other inorganics.

2.2.1 Arsenic

Arsenic compounds are extremely toxic. Anyone who remembers the old movie *Arsenic and Old Lace* is aware of its lethal nature. Less than 0.1 g of arsenic is usually lethal. It is also a known carcinogen. Copper chromium arsenate (CCA) was formerly widely used as a wood preservative for such things as wood decks and outdoor furniture; its use has now been largely curtailed because of its toxicity. Although the industrial use of arsenic has decreased over time, arsenic is still widely used in agriculture as a herbicide and as an animal disinfectant. Arsenic and its compounds are commonly used in industry as process chemicals.

Arsenic is not a true metal, but rather is classified as a semimetal or metalloid; its properties lie between those of a metal and those of nonmetals. Arsenic has five electrons in its outer shell. It can share three electrons with another atom, making it a 3+ ion (arsenite), or it can lose all five electrons, making it a 5+ ion (arsenate); it can also exist in an elemental form in the 0 state, but only under very low oxidation-reduction conditions not generally found in nature. Under reducing conditions (anaerobic), arsenic may exist in the 3− state (arsine), accepting three electrons to fill its outer electron shell. Under aerobic conditions, arsenate (AsO_4^{3-}) (As(V)) is the most common form; under low oxidation-reduction potential conditions ($E_h < 100$ mV) the most common arsenic species is arsenite (AsO_3^{3-}) (As(III)). (Note: the Roman numeral refers to the oxidation or valence state of the element.) Under very strong reducing conditions, highly toxic arsine gas (AsH_3) may be formed. As(III) is considered to be more toxic than As(V); unfortunately, As(V) is often converted to As(III) in the human body.

The average inorganic arsenic content of drinking water is about 2.5 μg/L. The maximum allowable concentration for arsenic in drinking water is 0.05 mg/L.

2.2.2 Cadmium

Cadmium is a metal that usually occurs in nature together with zinc, and most of the cadmium in use in industry is produced as a by-product of zinc smelting. Cadmium is widely used in metal plating because it is very resistant to corrosion. It is also commonly used as a stabilizer in polyvinyl chloride (PVC) plastics. Cadmium compounds have been used for hundreds of years as pigments in paints, and more recently in plastics. They have also been used in television screens. Cadmium has also found widespread use as an electrode in rechargeable "ni-cad" (nickel-cadmium) batteries, but these are being phased out because of concern over their significant environmental impacts. When municipal solid wastes are incinerated, the cadmium in the batteries is volatilized and is emitted from the exhaust stack as an air pollutant. Because of the great toxicity of cadmium, its use has declined over recent years.

Cadmium is almost always found in the 2+ valence state. Cadmium compounds with simple anions such as chloride are salts rather than molecules. Cadmium is highly toxic. Only about 1 g is a lethal dose. Cadmium is not quickly eliminated from the human system and can easily bioaccumulate to toxic levels. Its half-life in humans is estimated to be 20–30 years.

Due to its similarity to zinc, an essential nutrient for plants, plants absorb cadmium from irrigation waters and from soils where cadmium may have accumulated from atmospheric deposition or from land spreading of wastes containing cadmium. Eating crops contaminated with cadmium can lead to serious health problems. Strict regulations govern the amount of cadmium that can contact crops used for human consumption. Most serious episodes of cadmium contamination, though, have been associated with smelting operations.

Because of its great toxicity, the drinking water standard for cadmium has been set at 5 μg/L.

2.2.3 Chromium

Chromium is a transition metal exhibiting valence states from 2− to 6+, although the 3+ and 6+ species are generally the only ones of interest. It is commonly present as Cr(VI) in the chromate (CrO_4^{2-}) and dichromate ($Cr_2O_7^{2-}$) forms, and as Cr(III) in the chromic (Cr^{3+}) state. Cr(VI) compounds are generally soluble in water, whereas Cr(III) compounds are insoluble.

Chromium is used primarily in metal plating because of its resistance to acid attack and as pigments, but it is also used in leather tanning, as a catalyst in chemical processes, and in the manufacture of electronic equipment. All chromium in the United States comes from foreign sources, primarily from South Africa.

Chromic and chromate salts are irritating to exposed tissues. Some chromium compounds are known carcinogens. Because of their toxicity, the total chromium concentration is limited to 0.1 mg/L in drinking water.

2.2.4 Lead

Lead is widely used in industry because of its unique properties. It has a low melting point (327°C), high density, malleability so that it can easily be shaped into pipes, and high acid resistance. Its primary use is in automobile batteries (about 65 percent of all lead used), but it is also used in electroplating, pigments, plastics, glass, and electronics equipment. Because of its toxicity and ability to bioaccumulate, the use of lead in paints for residential purposes has been banned, as has its addition to automobile gasoline. Lead has been commonly used for water distribution pipes since Roman times, but new lead pipes are no longer installed; many are still in use, however, and probably will be for a long time because of their durability. It is still used in construction for roofing and flashing, and in many types of electrical solder.

In pure form, lead does not usually cause environmental or health problems. It is only when it dissolves yielding ionic forms that it becomes toxic. Lead ions usually exist in the 2+ state. It can also exist in the tetravalent form, particularly when complexed with organic radicals. Leaded gasolines used in the past contained lead in the form of tetraethlylead ($Pb(C_2H_5)_4$), added to prevent premature ignition, or "knocking." Automobile emissions of this lead resulted in serious environmental degradation and many cases of lead poisoning, particularly in children. Although it has a low boiling point, its vapor point is quite high (1740°C) and its vapor pressure is low, so volatilization of elemental lead is not usually a major problem.

At high levels, lead is a general metabolic poison. At lower concentrations, it can interfere with the production of hemoglobin, leading to anemia. It can also cause kidney dysfunction, high blood pressure, and permanent brain damage. Lead poisoning is of particular concern with young children who are prone to eat paint chips or soil around buildings, which may contain residues of lead paint or lead from automobile exhaust. This practice is called *pica*. Chronic exposure to lead in this way in young children can lead to serious and potentially permanent neurological damage.

Fortunately, a major recycling infrastructure has been developed for lead-containing materials, particularly lead batteries. Consequently, most lead-bearing materials are now recycled, and the amount entering the environment is much reduced from what it was in the past. The amount of lead recycled in 1988 was approximately 79 percent of the amount of lead refined that year.

2.2.5 Mercury

Mercury is unusual among metals in that it is a liquid at room temperature. It can exist in valence states of 0, 1+, and 2+. It can exist in both organic and inorganic forms. Its ability to expand uniformly with changes in temperature makes it ideal for use in thermometers. In the elemental form, its ability to conduct electricity is used in fluorescent light bulbs and mercury lamps used for exterior lighting. It is also widely used in miniature batteries, as an industrial catalyst in the production of chlorine and chlorinated compounds, in the pharmaceutical industry, and in fungicides and insecticides.

Mercury is highly volatile and its vapor is very toxic. Because of this, there has recently been a movement to replace mercury vapor lamps with somewhat less damaging sodium vapor lamps. These newer lamps pose less of a toxicity hazard and are more efficient.

The common ion of mercury is the 2+ species, Hg^{2+}, known as the mercuric ion. Mercuric sulfide (HgS) is highly insoluble in water. Thus the presence of sulfide in water often makes the mercury immobile. Mercuric nitrate ($Hg(NO_3)_2$), on the other hand, is water soluble. At one time it was used to treat the fur used in the manufacture of felt for hats. The workers exposed to the felt often developed nervous disorders, depression, and insanity. This is the derivation of the term "Mad Hatter" in *Alice in Wonderland.*

Mercury salts, such as mercuric chloride ($HgCl_2$), can become methylated in anaerobic waters or sediments by anaerobic bacteria. The result is the formation of the highly toxic volatile liquids methylmercury (CH_3HgX), where X is an anion, usually a halide, and dimethylmercury ($Hg(CH_3)_2$). These compounds are much more toxic than the inorganic mercury salts. They can quickly evaporate into the atmosphere above the liquid, becoming an air pollutant; they can also be rapidly taken up by fish and other organisms in water and be bioaccumulated in fatty tissue, where they become harmful to that organism or to others that consume it. During the 1960s it was discovered that large fish such as tuna and swordfish could have mercury concentrations in their tissues that were one million times higher than the concentrations in the waters through which they swam. Residents of the Japanese fishing village of Minimata became exposed to serious methylmercury poisoning through the fish they consumed from Minimata Bay. A chemical plant using mercury as a catalyst in polyvinyl chloride production discharged mercury-laden wastewaters into the bay. The mercury bioaccumulated in the fish to levels as high as 100 ppm, far in excess of the recommended limit of 0.5 ppm mercury in fish for human consumption. The result was mercury poisoning of thousands of people and the deaths of hundreds of them.

Mercury has also been widely used in the production of chlorine and chlorinated compounds and sodium hydroxide. In this process, metallic sodium is produced by reduction of NaCl in solution in an electrolysis cell using mercury as the cathode:

$$Na^+(aq) + Hg + e^- \rightarrow Na \qquad \text{(as a Na} - \text{Hg amalgam)}$$

The freed chlorine from the sodium chloride is recovered. The metallic sodium that is produced is in the form of an amalgam with mercury. The amalgam is then separated from the salt solution and reacted with water in a separate reactor to produce high-purity sodium hydroxide that is free of any salts. The overall reaction is

$$2NaCl(aq) + 2H_2O \rightarrow 2NaOH(aq) + Cl_2(g) + H_2(g)$$

The mercury is recovered and reused, but some escapes into the air and into the plants' cooling water. The amounts of mercury lost to the environment in the past were enormous. In recent years, this process is being phased out and replaced by a membrane process that does not require the use of mercury. The NaCl solution and the chloride-free solution are separated by a membrane through which the Na^+, but not the Cl^-, can pass.

Because of its severe toxicity, the maximum allowable concentration of mercury in drinking water is 2 μg/L.

2.2.6 Cyanides

Cyanide is not an element but rather an inorganic nonmetallic anion with the structure

$$C{\equiv}N^-$$

It is the conjugate base of the weak acid hydrogen cyanide (HCN):

$$HCN \rightleftharpoons CN^- + H^+ \qquad pK_a = 9.1$$

Under neutral or acidic conditions, the highly toxic HCN gas is formed. In high concentrations, this gas can be lethal. Cyanide binds with metal-containing enzymes (cytochromes) that participate in respiration. Asphyxia can result from exposure to hydrogen cyanide. Cyanide solutions should never be mixed with acids because of this potential for a life-threatening situation, but it has happened accidentally on numerous occasions.

Cyanide salts are commonly used in metal plating baths. The cyanide in the metal-cyanide salt helps the metal ions to be plated to reach hard to get at places in pieces being plated. Considerable effort has been expended in recent years to find ways to eliminate the need for cyanide in plating baths, but much more research is needed. Cyanides are also used as industrial intermediates and in the recovery of gold and silver during ore refining.

The maximum allowable concentration of cyanides in drinking water is 0.2 mg/L.

2.3 CONTAMINANT TRANSPORT AND TRANSFORMATION IN THE ENVIRONMENT

Because of the potential hazard that exposure to hazardous compounds poses to humans and the environment, the levels of toxic and carcinogenic substances in the environment have become important criteria for evaluating environmental quality. The amount of a material which enters the environment, though, is not always indicative of the amount that will be found there. The concentration of a contaminant at any point in the environment depends on the quantity added and the processes that influence its fate. This can include both *transport* and *transformation* mechanisms.

Transport processes tend to move materials from one point to another and may involve intermedia exchanges between atmospheric, aquatic, and soil environments, as well as movement within each of these media. Transport may be due to advective, dispersive, or diffusional processes.

Associated with transport processes are partitioning processes, which dictate how much of the material will be in the gas, liquid, and solid phases. Partitioning is dependent on the compound's solubility, density, polarity, ionic state, and vapor pressure. Another form of partitioning is bioaccumulation, in which compounds are taken up by living organisms and concentrated in their tissues.

Transformation processes within each media chemically alter the contaminants to new compounds that may have lower, equal, or greater toxicity. Transformations may be due to chemical, photochemical, or biological processes. The rates at which chemicals are transformed are critical to an understanding of the seriousness of a particular pollution incident. For example, a small spill of sodium acetate may have minimal environmental impact because it is rapidly biodegraded by bacteria in soil and water, whereas a small spill of creosote may have much greater implications because of its greater persistence in the environment.

Many books have been dedicated to the topic of contaminant fate and transport in the environment (Clark, 1997; Hemond, 1994; Knox, 1993; LaGrega, Buckingham, and Evans, 1994; Schnoor, 1997; Thibodeaux, 1996). In the following sections, we will only superficially describe these processes so that the reader will better understand why avoiding introduction of these materials into the environment through pollution prevention techniques may be desirable. In addition, the reader will develop a better understanding of the process of establishing priorities for waste reduction.

Before beginning this discussion, though, it is essential to describe how concentrations of materials in the environment are expressed.

2.3.1 Contaminant Concentrations

Except when a major spill occurs, toxic materials in the environment are usually found in very low, although possibly still harmful, concentrations. Concentrations of materials can be expressed in several ways, depending on whether they are present in air, water, or soil. In most cases, chemical concentrations in aqueous solutions are expressed in terms of mass per unit volume, usually as mg/L (one-thousandth of a gram per liter). When concentrations are very dilute, as in groundwaters, they may be expressed as μg/L (one-millionth of a gram per liter). In some cases, aqueous concentrations are expressed as parts per million (ppm) or parts per billion (ppb), a mass/mass designation, and considered equivalent to mg/L or μg/L. Technically, this is incorrect because the unit basis is not equivalent. However, if the solvent is water, which has a density of 1.00 g/cm^3 at 4°C, a 1.0 mg/L solution does essentially equal 1.0 ppm.

$$1.0 \text{ ppm} = \frac{1 \text{ mg contaminant}}{10^6 \text{ mg media}}$$

$$1.0 \text{ mg/L} = \frac{1.0 \text{ mg contaminant}}{10^3 \text{ mL solvent}}$$

Assuming the solvent is water, with a density of 1.00 g/cm^3 (or 1×10^3 mg/mL),

$$1.0 \text{ mg/L} = \frac{1.0 \text{ mg contaminant}}{(10^3 \text{ mL solvent})(10^3 \text{ mg/mL})} = \frac{1.0 \text{ mg contaminant}}{10^6 \text{ mg solvent}} = 1.0 \text{ ppm}$$

Therefore, for practical purposes,

$$\frac{\text{mg}}{\text{L}} = \frac{\text{mg}}{\text{kg}} = \text{ppm}$$

At other temperatures, the density of water varies from 1.00 g/cm^3 and this relationship does not hold. Therefore, it is usually safer to use mg/L units.

Contaminant concentrations in soils or sludges are usually expressed on a mass/mass basis. The most common expression is mg contaminant/kg soil, which is equivalent to ppm.

Air contaminant concentrations are expressed either on a mass/volume (μg contaminant/m^3 air) or a volume/volume (ppm) basis, where

$$\text{ppm} = \frac{1 \text{ part contaminant by volume}}{10^6 \text{ part air by volume}}$$

2.3.2 Transport Processes

Prediction of the fate of toxic pollutants in the environment requires knowledge of which processes act on the pollutants. Figure 2.1 illustrates the transport and transformation processes which are potentially important. Transport processes will be discussed first. They can be subdivided into loading processes, advective processes, dispersive processes, and diffusive processes.

LOADING PROCESSES. Loading processes are mechanisms by which contaminants are introduced into the environment. Contaminants can reach the environment through emissions into air or water or application to land. Once the contaminant reaches the environment, it is acted on by many other processes that tend to cause it to move within that medium or to move to another medium. For example, particulates emitted into the air from a smoke stack will move from the stack due to wind, thermal density gradients, and so on. Gravity will act on them, possibly carrying them to the ground or to water, converting them from an air contaminant to a water pollutant or soil contaminant. The same can be said for contaminants added to soil. They may volatilize into the atmosphere, becoming an air pollutant; they may be washed by overland stormwater flow into a water body, creating a water pollution problem; or they may seep into the groundwater, creating a pollution problem there.

Industrial contaminants can enter the environment from a wide variety of sources. They may come directly from industry in the form of air emissions from smoke stacks or process reactors exposed to the atmosphere, or from process wastewaters discharged into receiving streams with or without prior treatment. In other cases, wastewaters may be directed to a municipal treatment plant (often termed a pub-

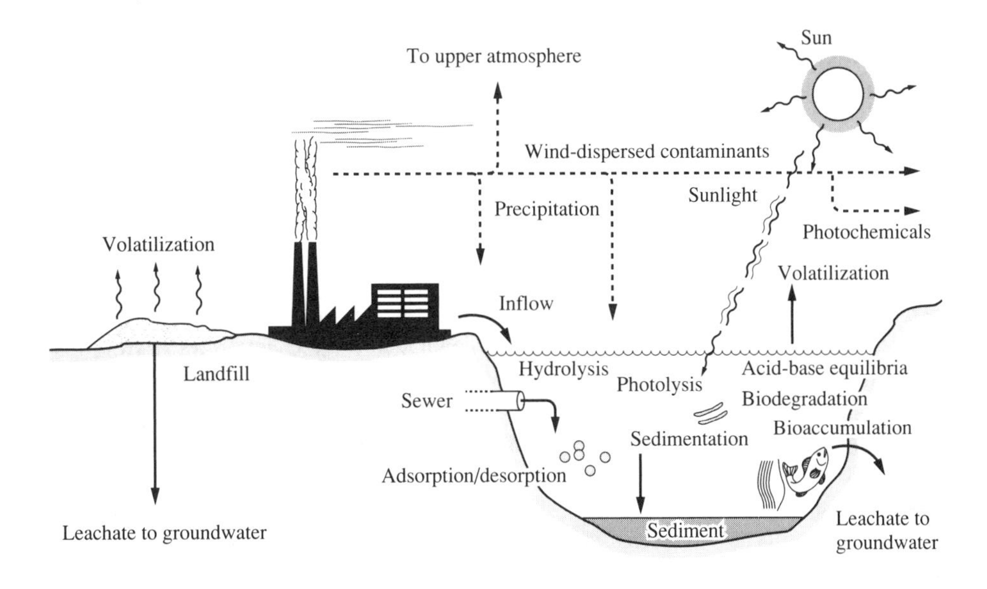

FIGURE 2.1
Fate of industrial contaminants in the environment.

licly owned treatment works, or POTW), but the POTW effluent may still contain some of the contaminants. Solid wastes from the industry may be disposed of in a landfill, creating possible land or groundwater pollution, or may be incinerated, creating a potential air pollution problem; the resulting fly ash and bottom ash often contain heavy metals and must still be properly landfilled. These can all be referred to as *point sources*. Another significant source of contaminants released to the environment from many industrial sources, particularly chemical and petrochemical industries, is what is termed *fugitive emissions*. They can be described as unintentional emissions from leaking process equipment such as valves and pumps. These can be in the form of gases or liquids. (Fugitive emissions are described in more detail later.)

Other sources of pollution due to a specific industry may not be obvious. Transportation of the process ingredients to the factory and of the finished products from the factory require the use of fuels and energy, the production and use of which also create pollution. The same can be said for the process chemicals and raw materials themselves. Use of the product by the consumer may also involve creation of polluting materials, as may disposal of the product after it fulfills its useful life. The design and composition of these products may dictate their eventual environmental impacts.

Thus there are many sources of waste loads on the environment associated with a particular industrial process. It is the responsibility of industry and its pollution prevention experts to evaluate all potential environmental inputs caused by its products, from the raw materials processing stage through production and use (and possible reuse) of the product to final disposal, in order to minimize the total life-cycle environmental impacts.

The path that the contaminant will take can sometimes be predicted by analyzing wind patterns or groundwater gradients. In a surface stream, the contaminant will be transported with the stream.

ADVECTIVE PROCESSES. Advection is the movement of a contaminant away from the source due to the physical movement of the medium in which it is contained. Contaminants in the air are moved from the source by winds; those in water by currents, and those in groundwater by pore water movement. These factors tend to move the contaminant away from the source and prevent the buildup of high concentrations at that point. If the emission is a pulse release, the concentration at the source will reach a peak quickly, and then the mass of contaminant will be transported away from the source, leaving no contaminant behind. Assuming no dispersion occurs, the mass of contaminant will move with the carrying medium (wind, groundwater, etc.) and the contaminant concentration will remain constant. However, if the contaminant emission is continuous, new contaminant will continually replace the material moving away from the source, and the concentration all along the transport path will be constant. The main consequence of advective processes is the transport of contaminants from the contamination source to other areas.

DISPERSIVE PROCESSES. Pure advection never occurs. There is always some mixing of contaminants in a unit of air or water with surrounding uncontaminated air or water, resulting in a reduction of contaminant concentration in the original unit of medium. This spreading of a contaminant as it moves downwind, downstream, or downgradient is termed dispersion. In the atmosphere, this dispersion is usually due primarily to atmospheric turbulence, which causes units of air to mix with surrounding units of air. It may also be caused by thermal or density differences between the air in the contaminant plume and the surrounding air. Similar effects can be seen in surface or groundwaters. The most significant consequences of dispersive processes are the spreading of the contaminant over a greater area and the dilution (reduction) of the contaminant concentration in the plume (see Figure 2.2).

Computer models are available for predicting the impact of dispersive processes on a contaminant plume, but discussion of these is beyond the scope of this book. The reader is referred to Clark (1997), Schnoor (1997), and Thibodeaux (1996) for more information.

DIFFUSIONAL PROCESSES. Dispersion processes are concerned with the movement and mixing of the air or water carrier of the contaminants. However, the individual contaminant molecules or ions will also move in response to concentration gradients in the air or water. They will tend to move from a point of higher concentration to one of lower concentration. This movement is termed *diffusion*. Molecular diffusion can be described using Fick's law:

$$f = D\frac{dC}{dx}$$

where f = mass flux of contaminant ($g/cm^2 \cdot sec$)
D = contaminant diffusion coefficient in the particular medium (cm^2/sec)
dC/dx = contaminant concentration gradient ($g/cm^3 \cdot cm$)

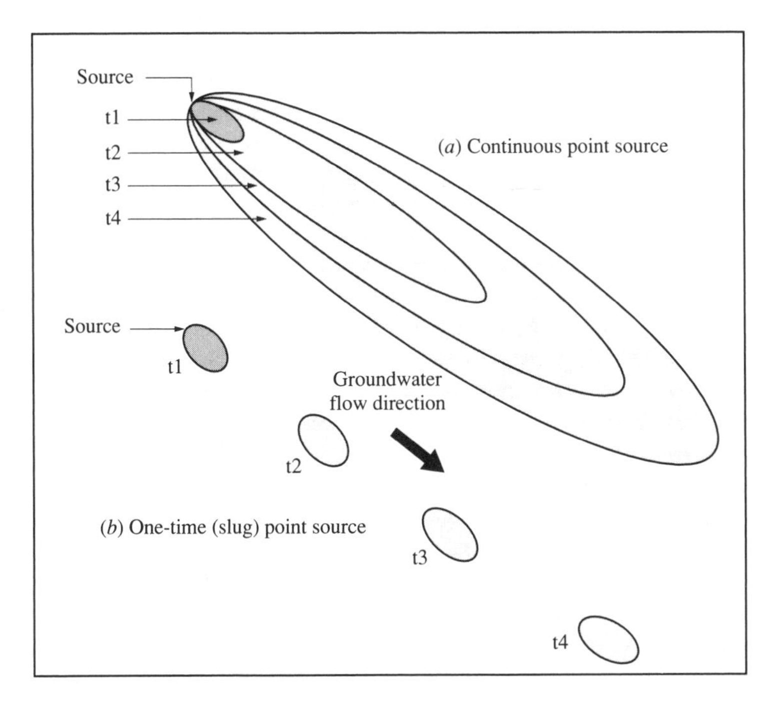

FIGURE 2.2
Plume migration affected by dispersion and source type. (Adapted from LaGrega, Buckingham, and Evans, 1994)

Diffusive effects can also be modeled, but again discussion of this is beyond the scope of this book. The effects of molecular diffusion are again to dilute the contaminant over a greater area and to reduce its concentration within the plume.

To assess the impact of a contaminant release into the environment and the potential effects on the general public, it is essential that valid predictions of actual exposure concentrations be made. This requires that the combined effects of advection, dispersion, and diffusion of the contaminants in the plume be determined. The analysis is made more complicated by the fact that the form and properties of the contaminant may change during transport. For example, a particulate air contaminant may settle to earth and be removed from the airstream. Gaseous contaminants may become dissolved in water droplets in the air and be removed by precipitation as rain. In both cases, they may undergo chemical reaction in the atmosphere, changing to a new compound with new diffusive properties. Similar actions may occur with contaminants in surface or groundwaters. The effects of partitioning and transformation processes on contaminants are described in the following sections.

2.3.3 Partitioning Processes

Compounds in the environment are rarely found in their pure form. Rather, they dissolve and diffuse through media, trying to achieve a minimum concentration difference with the surrounding material. Many factors govern the rate of dispersion of the compound, as described above. Other factors that govern dispersion are based on the tendency of a material to want to be associated with one phase or another. This division between two phases is termed *partitioning.* Partitioning is dependent on the properties of the compound and media it is in contact with. There are many properties that can affect partitioning, but we limit this discussion to solubility, acid-base effects, adsorptive effects, and volatilization effects. The impacts of these on the transport processes discussed earlier are described.

ACID-BASE IONIZATION. Many acids and bases are used in industry, as are the salts of these acids and bases. These run the gamut from very strong acids and bases, such as sulfuric acid and sodium hydroxide, to very weak ones, such as carbonic acid and sodium sulfide. They may be in the form of gases, liquids, or solids. These materials may be very corrosive and may cause severe personal injuries and environmental damage. As will be seen, the degree of ionization of these materials may play a significant role in their transport and eventual fates.

The classical definition of an *acid* is a compound that yields a hydrogen ion (H^+) upon addition to water. A *base* yields a hydroxyl ion (OH^-) upon addition to water. These definitions are fairly simplistic and not totally accurate, but they suffice for our purposes. In the case of a strong acid, the bond between the hydrogen atom and the anionic group is weak so that essentially all of the hydrogen atoms leave the molecules when the acid is placed in water. Ionization is essentially 100 percent complete. The same holds true for strong bases, where all of the base ionizes, liberating hydroxyl ions. Weak acids and bases hold on to their H^+ and OH^- groups better, and they only partially ionize.

For a monoprotic acid (one that contains only one ionizable hydrogen), such as acetic acid, Arrhenius's theory of ionization states that the dissociation ratio can be described as

$$HAc \rightleftharpoons H^+ + Ac^-$$

$$\frac{[H^+][Ac^-]}{[HAc]} = K_a = 1.75 \times 10^{-5} \qquad \text{at } 25°\text{C}$$

where Ac^- is used to denote the acetate ion (CH_3COO^-), $[H^+]$ is the concentration of dissociated hydrogen ions in moles, $[Ac^-]$ is the moles of dissociated acetate ions, and [HAc] is the amount of undissociated acetic acid at equilibrium. The larger the value of K_a, the *dissociation constant,* the greater is the ionization and the stronger is the acid. Diprotic acids act in a similar fashion but have two dissociation constants, one for each ionizing hydrogen atom.

Water acts as a weak monoprotic acid. Its dissociation can be depicted as

$$H_2O \rightleftharpoons H^+ + OH^-$$

$$\frac{[H^+][OH^-]}{[H_2O]} = K_a$$

The dissociation of water is very small, so the term $[H_2O]$ changes almost infinitesmally with repect to the ions. Consequently, it is usually considered a constant and the expression is rewritten as

$$[H^+][OH^-] = K_a \cdot [H_2O] = K_W = 1 \times 10^{-14} \qquad \text{at } 25°C$$

where K_W is the ionization constant of water.

The hydrogen ion concentration present can be described in terms of its negative logarithm; this designation is termed the pH of the solution:

$$pH = -\log[H^+] \qquad pH = \log \frac{1}{[H^+]}$$

In the absence of any other materials besides H_2O, and assuming that K_W is 1×10^{-14}, $[H^+] = [OH^-] = 1 \times 10^{-7}$. Therefore, the pH of pure water is

$$[H^+][OH^-] = 1 \times 10^{-14}$$

$$[H^+] = 1 \times 10^{-7}$$

$$pH = -\log[1 \times 10^{-7}] = 7.0$$

If the hydrogen ion concentration increases because of the addition of acidic materials, the pH will go down; if bases are added, the pH will go up.

The pH of a solution, or its hydrogen ion concentration, has a direct bearing on the speciation of many contaminants added to water. For example, assume that a small quantity of pentachlorophenol (PCP) enters a body of water. PCP is an alcohol and as such has the ability to ionize by giving up the hydrogen from its alcohol group:

OH, Cl, Cl, Cl, Cl, Cl ⇌ O⁻, Cl, Cl, Cl, Cl, Cl + H^+ $pK_a = 4.74$

The dissociation reaction can be written as

$$C_6Cl_5OH \rightleftharpoons C_6Cl_5O^- + H^+ \qquad pK_a = 4.74$$

$$\frac{[C_6Cl_5O^-][H^+]}{[C_6Cl_5OH]} = 10^{-4.74}$$

where pK_a is the negative logarithm of the ionization constant. Typical ionization constants for weak acids are listed in Table 2.5. When the pH of the solution and the pK_a are equal, 50 percent of the acid will have donated its ions to the solution and will exist as charged anionic species. Thus at pH 4.74 half of the PCP will be in the un-ionized form and half will be ionized. As the solution pH increases, the ratio will shift to a greater fraction being ionized. In general, a pH one unit above the pK_a will result in 90 percent of the material being ionized, while a pH two units higher will cause 99

TABLE 2.5
Ionization constants for selected organic acids and bases

Compound	Equilibrium equation	K_a	pK_a
Acids			
Acetic	$CH_3COOH \rightleftharpoons H^+ + CH_3COO^-$	1.8×10^{-5}	4.74
Ammonium	$NH_4^+ \rightleftharpoons H^+ + NH_3$	5.56×10^{-10}	9.26
Carbonic	$H_2CO_3 \rightleftharpoons H^+ + HCO_3^-$	4.3×10^{-7} (K_{a1})	6.37
	$HCO_3^- \rightleftharpoons H^+ + CO_3^{2-}$	4.7×10^{-11} (K_{a2})	10.33
Hydrocyanic	$HCN \rightleftharpoons H^+ + CN^-$	4.8×10^{-10}	9.32
Hydrogen sulfide	$H_2S \rightleftharpoons H^+ + HS^-$	9.1×10^{-8} (K_{a1})	7.04
	$HS^- \rightleftharpoons H^+ + S^{2-}$	1.3×10^{-13} (K_{a2})	12.89
Phenol	$C_6H_5OH \rightleftharpoons H^+ + C_6H_5O^-$	1.2×10^{-10}	9.92
2,4-Dichlorophenol	$C_6H_3ClOH \rightleftharpoons H^+ + C_6H_3ClO^-$	1.4×10^{-8}	7.85
Pentachlorophenol	$C_6Cl_5OH \rightleftharpoons H^+ + C_6Cl_5O^-$	1.8×10^{-5}	4.74
2-Nitrophenol	$C_6H_4(NO_3)OH \rightleftharpoons H^+ + C_6H_4(NO_3)O^-$	6.2×10^{-8}	7.21
Bases			
Ammonia	$NH_3 + H_2O \rightleftharpoons NH_4^+ + OH^-$	1.8×10^{-5}	4.74
Carbonate	$CO_3^{2-} + H_2O \rightleftharpoons HCO_3^- + OH^-$	2.13×10^{-4} (K_{b2})	3.67
	$HCO_3^- + H_2O \rightleftharpoons H_2CO_3 + OH^-$	2.33×10^{-8} (K_{b1})	7.63
Calcium hydroxide	$CaOH^+ \rightleftharpoons Ca^{2+} + OH^-$	3.5×10^{-2} (K_{b2})	1.46

percent of the weak acid to be ionized. Thus for pentachlorophenol at a pH of 6.74 (slightly acidic), 99 percent of the PCP will be ionized. This can be seen clearly in Figure 2.3, where the fraction of un-ionized PCP is plotted against pH. As will be seen later, this has a significant impact on its fate.

Example 2.4. An industry plans to discharge wastewater containing 25 mg/L acetic acid into a holding pond. Because of the requirements of subsequent reclamation processes, it is desired to keep at least 95 percent of the acetic acid in the un-ionized form. What pH should the wastewater be buffered to?

Solution. The acetic acid will dissociate to the acetate and hydrogen ions, the extent of dissociation dependent on the pH present:

$$CH_3COOH \leftrightarrow CH_3COO^- + H^+ \qquad pK_a = 4.74$$

$$\frac{[CH_3COO^-]\,[H^+]}{[CH_3COOH]} = 10^{-4.7}$$

Next determine the total concentration of acetic acid in the wastewater at 25°C:

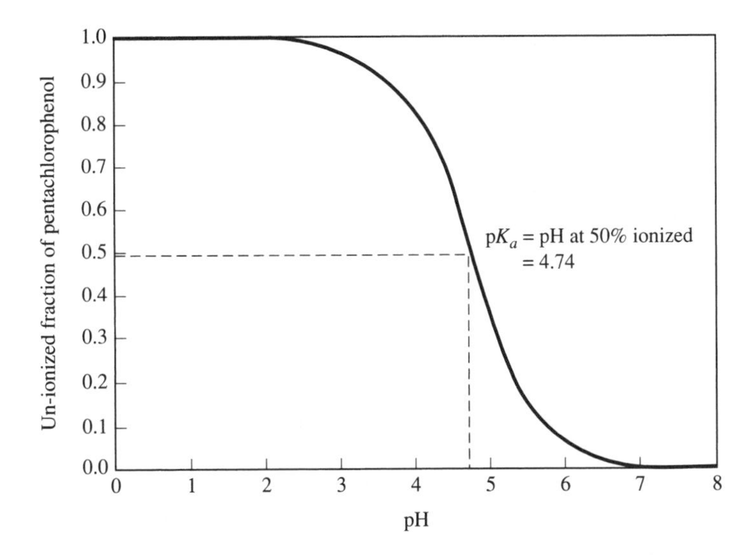

FIGURE 2.3
Typical distribution of pentachlorophenol.

$$[CH_3COOH] = \frac{25 \text{ mg/L}}{(1000 \text{ mg/g})(46 \text{ g/mol})} = 5.44 \times 10^{-4} \text{ mol/L}$$

where mol wt of HAc = 46.

Finally, at 95 percent ionized,

$$[CH_3COO^-] = (0.05)\,(5.44 \times 10^{-4} \text{ mol/L}) = 2.72 \times 10^{-5} \text{ mol/L}$$

$$\frac{(2.72 \times 10^{-5} \text{ mol/L})\,[H^+]}{5.44 \times 10^{-4} \text{ mol/L}} = 10^{-4.7}$$

$$[H^+] = 4.0 \times 10^{-4}$$

$$\text{pH} = 3.4$$

The pH must be maintained at 3.4 to keep 95 percent of the acetic acid in the un-ionized form.

SOLUBILITY. The water solubility of a hazardous material often dictates its fate in the environment. Water solubility is defined as the maximum (or saturation) concentration of a substance that will dissolve in water at a given temperature. Solubility is very important because dissolved and undissolved fractions of a material act quite differently. For example, assume that naphthalene and anthracene, both polyaromatics, are added to water in an amount capable of producing 5 mg/L of each compound. Naphthalene has a water solubility of 32 mg/L, while anthracene has a solubility of only 0.031 mg/L. Essentially all of the naphthalene will dissolve into the water and its fate will largely be that of a dissolved compound, but only a very small fraction of the anthracene should dissolve in the water. The remainder will stay in the free form, sinking to the bottom of the container because of its greater density than water. Its fate will largely be determined by different factors than those for the naphthalene.

The undissolved fraction of a material in water will float or sink, depending on its density. Undissolved fractions are called *Nonaqueous Phase Liquids* (NAPLs). Dense NAPLs (those with a density greater than water) are called DNAPLs. They will sink to the bottom of a container filled with water or to the bottom of a groundwater aquifer (see Figure 2.4). Light NAPLs (LNAPLs) will float on the surface of water, whether in a container or in groundwater. In either case, both are very difficult to remove from the water. NAPLs may also volatilize or sorb onto solids at a greater rate than the dissolved fraction. In some cases, the undissolved material may form an emulsion in the water, particularly if any turbulence is imposed in the water.

The water solubility of a compound is controlled by a number of factors, especially its size and structure. Table 2.6 summarizes some generalized relationships between a molecule's properties and its water solubility. These are only generalizations, though, and there are many exceptions to these rules. Solubilities of selected compounds cited in the text are listed in Appendix B.

Materials that are highly polar are more soluble in water. Polar compounds, in general, are compounds that can have a charge upon dissociation. A polar material such as phenol has a water solubility of 82,000 mg/L at 20°C, whereas benzene, which is structurally similar to phenol but without the polar hydroxyl group, has a solubility of 1780 mg/L. The solubilities of materials of similar structure but with varying sizes tend to decrease as the molecular size increases, as shown in Table 2.7.

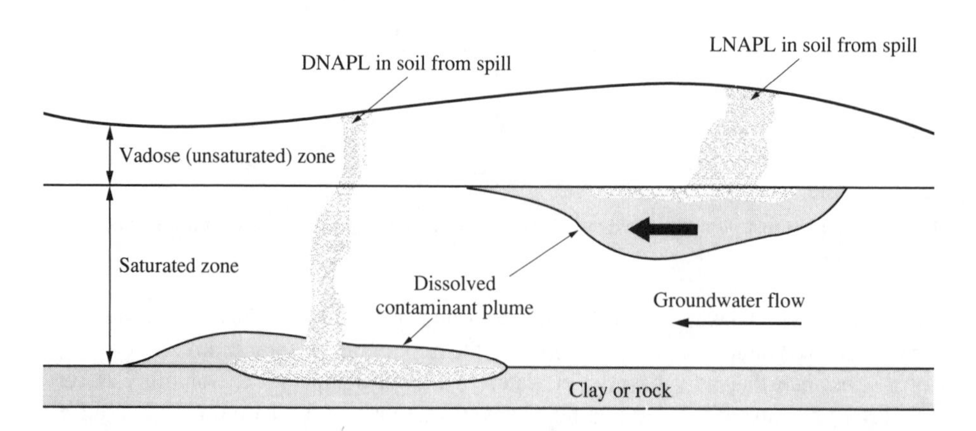

FIGURE 2.4
Migration and fate of LNAPLs and DNAPLs in the subsurface. (Adapted from La Grega, Buckingham, and Evans, 1994)

TABLE 2.6
Factors affecting solubility of organic compounds

As molecular size increases, solubility decreases
As molecule polarity increases, solubility increases
As the number of double or triple bonds increases, solubility decreases
The order of decreasing solubility is Aliphatic > Aromatic > Cycloaliphatic
As branching increases, solubility decreases
As halogenation increases, solubility decreases
As temperature increases, solubility increases
Polar substituents such as carboxyl, amine, alcohol, and nitrite increase the solubility of the base compound
Acid/base salts are more soluble than the undissociated species

TABLE 2.7
Water solubility of selected normal aliphatics of the same class

Compound	Molecular weight	Water solubility, mg/L
Butane	58.12	61.0
Pentane	72.15	39.6
Hexane	86.18	10.9
Heptane	100.20	2.0
Octane	114.23	1.46
Nonane	128.26	0.122
Decane	142.29	0.021
Dodecane	170.34	0.005

ADSORPTIVE EFFECTS. Sorptive properties of a material are very important in describing the fate of the material in the environment. *Sorption* can be defined as the transfer of a material from one phase to another. Sorption can be divided into two categories: absorption and adsorption. *Absorption* involves the movement of one material into another, for example the dissolution of oxygen into water. *Adsorption* involves the condensation and attachment of one material onto the surface of another material, for example, the accumulation of a toxicant on the surface of an activated carbon particle. Here we concentrate on adsorptive properties.

Sorption is an equilibrium process in which compounds partition between the liquid (or in some cases gaseous) phase and a solid surface, based on their affinity for the two phases. It is an equilibrium process, so some of the material will be found in each phase. The stronger the affinity for the solid, the less will remain in solution. Sorption occurs when the net sorbent-solute attraction overcomes the solute-solvent attraction. (Note that the solute is the contaminant being sorbed, the sorbent is the solid surface doing the sorbing, and the solvent is the water.)

There are three general types of adsorption: physical, chemical, and electrostatic. Physical adsorption is due to the weak forces of attraction between molecules, or van der Waals' forces. This type of adsorption is fairly weak and is easily reversible if the contaminant concentration in the liquid phase decreases. It is usually the dominant mechanism for sorption of organics to solid surfaces. Chemical adsorption involves much stronger forces, resulting from bonding interactions between the organic compound and constituents on the solid surface. These bonds are relatively irreversible. Electrostatic adsorption is caused by electrical attraction between the adsorbate and the surface. It is of particular importance for ionic materials such as metals and their salts. Materials of opposite charge are attracted to one another. The greater the charge on the ion, the greater the attraction. For example, Al^{3+} will be attracted to a negatively charged surface much more strongly than will be Na^+. For ions of equal charge, smaller ions will be bound more tightly than larger ones because of their higher charge-to-mass ratio.

Partitioning of a material between two phases contributes, to a large degree, to the ultimate fate of the material. A contaminant in aqueous solution that is discharged into a river will be partitioned between the water phase and any solid material present (suspended solids, bottom sediments, fish, etc.), and its future movement will be dictated by movement of the solid rather than that of the water. The same holds true for contaminants discharged onto land, where the contaminant will partition between the groundwater and the soil. The movement of a material that has a strong tendency to sorb to solids may be significantly impeded by its attachment to the solid surface. In addition, adsorptive forces may seriously alter its rate of volatilization or transformation in the environment. Generally, sorbed materials are less available for volatilization, biodegradation, or photochemical attack than are dissolved species. Sorption effects can also be used to remove a toxic material from water. Many organics have a stronger affinity for the surfaces of activated carbon than they do for the water phase that they are in. By bringing the water into contact with activated carbon, the toxic organic will transfer from the water phase onto the surface of the activated carbon, leaving a relatively contaminant-free water.

The tendency for a material to sorb to a surface is dependent on the characteristics of the solid surface, the hydrophobicity of the solid and the sorbent, and the charge on the solid and the sorbent. Since adsorption is a surface phenomenon, the rate and extent of adsorption are related to the surface area of the solid involved. Smaller particles, such as clays, will have a greater tendency to sorb organics than sand particles because of clay's greater specific surface area (particle surface area/particle volume). Activated carbon is particularly suited for adsorption because of its enormous specific surface area, ranging from about 600 to 1200 m^2/g (see Table 2.8). The more hydrophobic (nonpolar) the sorbate is, the less likely it will be to remain in the polar aqueous phase. Therefore, unsubstituted alkanes will have a greater tendency to be attracted to a hydrophobic solid surface than will their alcohol analogues. The hydrophobicity of the sorbent is also important. The higher the organic content of a soil, for example, the greater is the tendency for adsorption of aqueous phase organics to occur. Finally, as described previously, charged ions, such as metals, may be removed by sorptive processes if the solid surface contains oppositely charged locations.

TABLE 2.8
Properties of selected adsorbents

Material	Particle diameter, mm	Particle density, g/cm^3	Specific surface area, m^2/g
Watco 517 (12 × 30)	1.2	0.42	1050
Darco	1.05	0.67	600–650
Calgon Filtrasorb 300 (8 × 30)	1.5–1.7	1.3–1.4	950–1050
Westvaco Nuchar W-L (8 × 30)	1.5–1.7	1.4	1000
Calgon RB (powdered activated carbon)	0.008	1.4	1100–1300

A number of empirical mathematical expressions have been developed to describe sortion equilibrium concentrations. The resulting plots, called *isotherms,* show the relationship between the amount of a compound that will sorb onto a surface relative to the aqueous concentration (or gaseous concentration for gas-solid sorption) at equilibrium. Several sorption models are in common use, including the *Langmuir isotherm* (assumes the surface is homogeneous and the adsorbed layer is only one molecule thick) and the *Freundlich isotherm* (assumes a heterogeneous surface with different types of adsorption sites). Each has its own governing assumptions and each produces a differently shaped plot. The appropriate one to use is usually decided on through analysis of experimental data using each model, selecting the one that provides the best fit. The equations for the two models are:

Langmuir:
$$C_s = \frac{x}{m} = \frac{abC_e}{1 + bC_e}$$

Freundlich:
$$C_s = \frac{x}{m} = K_F \cdot C_e^{1/n}$$

where C_s = amount of contaminant sorbed on the solid per unit of solid (g sorbate/g sorbent)
C_e = concentration of contaminant remaining in solution at equilibrium (g/cm^3)
x = mass of contaminant adsorbed (g)
m = mass of solid sorbent (g)
a, n = empirical coefficients
b = saturation coefficient (m^3/g)
K_F = Freundlich isotherm constant (mg/g)

It should be stressed that these models are all equilibrium models, and the values of C_s and C_e are the concentrations achieved only after equilibrium is reached. This may be a slow process, and equilibrium may never be reached in a natural system.

Figure 2.5 shows a typical isotherm for a hydrophobic pollutant on a river sediment using each of the isotherm models. As can be seen, at low aqueous equilibrium concentrations (which is typical of most environmental situations) all give essentially the same solid loading. It can also be seen that all are essentially linear at these low concentrations. Therefore, for many cases, where low equilibrium concentrations are desired, a *linear isotherm* can be used to approximate the sorption process.

Linear isotherm: $$C_s = \frac{x}{m} = k_p \cdot C_e$$

where k_p is the partition coefficient:

$$k_p = \frac{C_s}{C_e}$$

A high k_p indicates a high tendency to adsorb. The value of k_p for any combination of aqueous contaminant and sorbent (i.e., soil, activated carbon) can be measured experimentally.

The Langmuir and Freundlich isotherm equations can be linearized to make their solution easier. For the Langmuir isotherm, taking the reciprocal of both sides yields

$$\frac{1}{x/m} = \frac{1 + bC_e}{abC_e} = \frac{1}{a} + \frac{1}{abC_e}$$

A plot of $1/(x/m)$ versus $(1/C_e)$ should yield a straight line (if Langmuir-type adsorption is present), with the slope of the line equal to $1/ab$ and the intercept equal to $1/a$ (see Figure 2.6*a*). From this, the constants a and b can be determined.

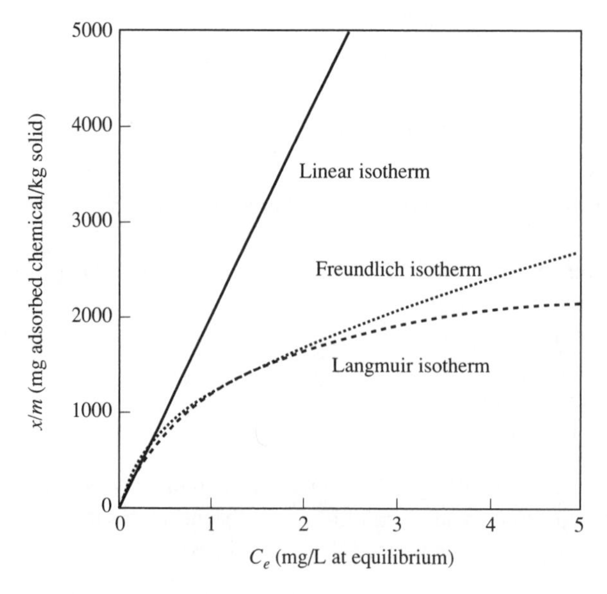

FIGURE 2.5
Typical isotherms for adsorption of an organic contaminant to a solid surface.

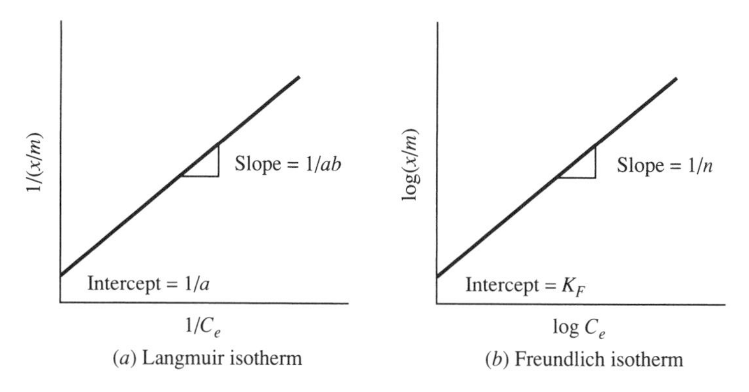

FIGURE 2.6
Linearization of the Langmuir and Freundlich isotherms.

The Freundlich isotherm can be linearized by taking the log of both sides of the equation:

$$\log\left(\frac{x}{m}\right) = \log K_F + \frac{1}{n}\log C_e$$

A plot of log (x/m) versus log C_e will yield a straight line (if Freundlich-type adsorption is present), with the slope equal to $1/n$ and the intercept equal to K_F (see Figure 2.6*b*).

Example 2.5. An isotherm experiment was performed to determine the adsorption of phenol by activated carbon. From a 1000-mg/L solution of phenol, 100 mL was added to each of six beakers containing different amounts of activated carbon. The containers were shaken for five days to establish equilibrium between the aqueous phenol and that sorbed to the activated carbon. The samples were then filtered and analyzed for phenol. Following are the results of these analyses:

Container	Activated carbon added (mg)	Phenol (mg/L)
1	50	1000
2	100	1000
3	200	1000
4	300	1000
5	400	1000
6	500	1000

Determine the isotherm constants for the (a) Langmuir, (b) Freundlich, and (c) linear isotherm models and plot their isotherms.

Solution. First develop a spread sheet to solve for the required constants for each isotherm:

m	C_0	C_e	x	x/m	$1/(x/m)$	$1/C_e$	$\log x/m$	$\log C_e$
0.05	600	450	15	300	0.0033	0.0022	2.477	2.653
0.1	600	300	30	300	0.0033	0.0033	2.477	2.477
0.2	600	100	50	250	0.0040	0.0100	2.398	2.000
0.3	600	33	56.7	189	0.0053	0.0303	2.276	1.519
0.4	600	20	58	145	0.0069	0.0500	2.161	1.301
0.5	600	15	58.5	117	0.0085	0.0667	2.068	1.177

Note: $x = (C_0 - C_e)V$, where $V = 100$ mL $= 0.1$ L.

Next plot the results for each of the three possible isotherms, using the linear transform equations (see Figure 2.7). Finally, calculate the relevant isotherm constants and write the isotherm expression, then plot the resulting isotherms.

(*a*) Langmuir isotherm:
Intercept $= 1/a = 0.0031$
$a = 322.6$
Slope $= 1/ab = 0.079$
Therefore, $b = 1/(322.6)(0.079) = 0.039$
Langmuir isotherm equation:

$$C_s = \frac{x}{m} = \frac{(322.6)\,(0.039)\,C_e}{1 + 0.039\,C_e} = \frac{12.6\,C_e}{1 + 0.039\,C_e}$$

The resulting line is very linear ($r^2 = 0.995$), indicating that the Langmuir isotherm is a good predictor of adsorption.

(*b*) Freundlich isotherm:
Slope $= 1/n = 0.264$
Intercept $= K_F = 1.821$
Freundlich isotherm equation:

$$C_s = \frac{x}{m} = 1.821\,C_e^{0.264}$$

The resulting line is not very linear ($r^2 = 0.92$), and therefore the Freundlich isotherm is not as good a predictor of adsorption in this case.

(*c*) Linear isotherm: The resulting plot is obviously not linear. However, if only the low C_e concentration data are considered, a linear isotherm can be used.
Slope $= 4.0$

$$C_s = \frac{x}{m} = 4.0\,C_e$$

As can be seen in Figure 2.8, the Langmuir plot gives the best results, but any of the three may be acceptable at low C_e.

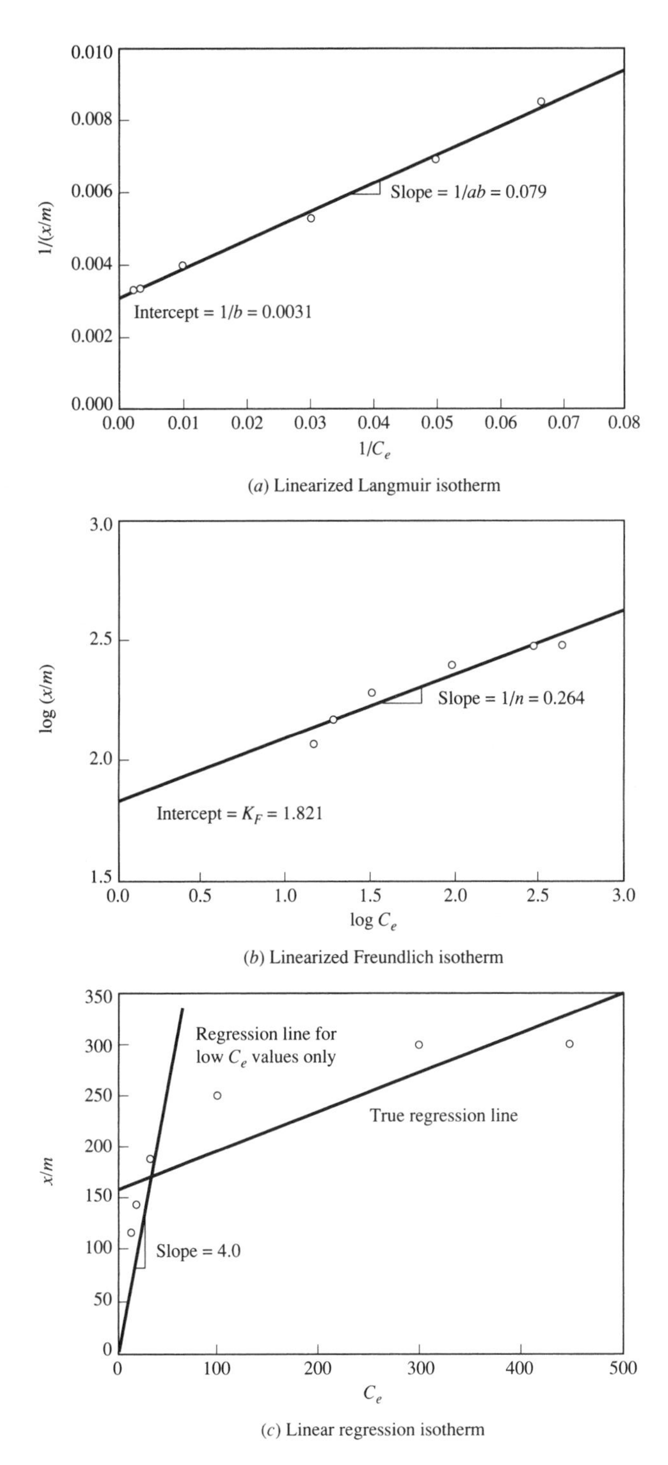

(*a*) Linearized Langmuir isotherm

(*b*) Linearized Freundlich isotherm

(*c*) Linear regression isotherm

FIGURE 2.7
Plots of the three possible isotherms.

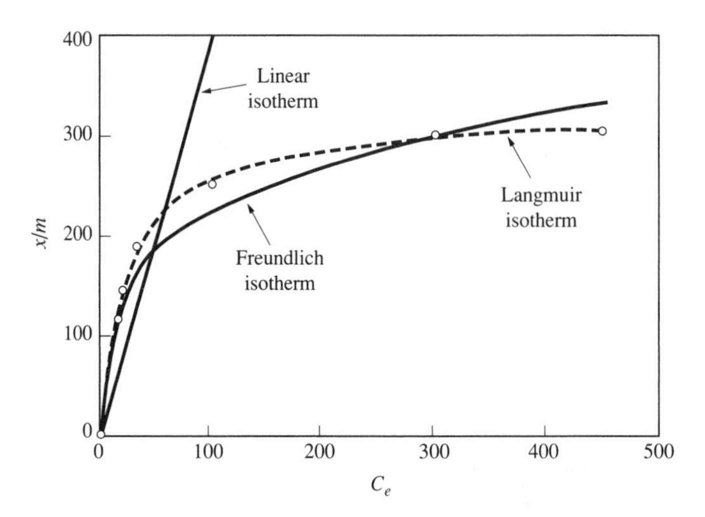

FIGURE 2.8
Langmuir, Freundlich, and linear regression isotherms.

The relative amount of pollutant sorbed and dissolved on suspended solids in a receiving stream depends on both the suspended solids concentration and the partition coefficient. At equilibrium, this relationship is empirically depicted as

$$\frac{C_e}{C_T} = \frac{1}{1 + K_p \cdot S \cdot 10^{-6}}$$

where C_T = total aqueous phase concentration = $C_e + C_s$ (mg/L)
S = suspended solids concentration (mg/L)

Table 2.9 shows the relationship of the dissolved and sorbed phase concentrations of a contaminant to the partition coefficient and the sediment concentration. In the table, C_W refers to the total dissolved phase contaminant concentration if C_T = 1 mg/L, X = the mass of sorbed pollutant per mass of suspended solids, and C_S = the total suspended phase concentration if C_T = 1 mg/L. As can be seen, sorption to suspended sediment is minimal if the partition coefficient is small or if there is little suspended sediment in the water. However, for highly hydrophobic compounds with a high K_p (possibly greater than 10^5), partitioning to the solids can become significant.

Example 2.6. Coal Tar Conversion, Inc. (CTC), treats its wastewater using the activated sludge process, in which the wastewater is mixed with a concentrated suspension of bacteria (activated sludge) in an aerated basin. The objective is for the bacteria to rapidly biodegrade the organics present, following which the bacteria will be separated from the wastewater by gravity settling and recycled to the aeration tank to treat more waste. The CTC wastewater, among other things, contains 10 mg/L of phenanthrene (solubility = 1.0 mg/L; $k_p = 1 \times 10^4$). Assume that the aeration tank is operated at a suspended bacterial concentration of 2500 mg/L and that the biodegradation rate of the phenanthrene is low. Where would you expect to find the phenanthrene after wastewater treatment?

TABLE 2.9
Relationship of dissolved and sorbed phase contaminant concentrations to the partition coefficient and suspended sediment concentration in a water body for C_T = 1 mg/L

K_p	S, mg/L	C_e/C_T	C_W, mg/L	X, mg/kg	C_S, mg/L
1	1	1.0	1.0	1.0	0.0
	10	1.0	1.0	1.0	0.0
	100	1.0	1.0	1.0	0.0
	1,000	1.0	1.0	1.0	0.0
	10,000	1.0	0.99	0.99	0.01
10	1	1.0	1.0	10.0	0.0
	10	1.0	1.0	10.0	0.0
	100	1.0	0.999	9.99	0.001
	1,000	1.0	0.99	9.9	0.01
	10,000	0.9	0.91	9.09	0.09
100	1	1.0	1.0	100.0	0.0
	10	1.0	0.999	100.0	0.001
	100	1.0	0.99	99.0	0.01
	1,000	0.9	0.91	91.0	0.09
	10,000	0.5	0.50	50.0	0.50
1,000	1	1.0	0.999	1,000	0.001
	10	1.0	0.99	990	0.01
	100	0.9	0.91	910	0.09
	1,000	0.5	0.50	500	0.50
	10,000	0.1	0.09	90	0.91
10,000	1	1.0	0.99	9,900	0.01
	10	0.9	0.91	9,100	0.09
	100	0.5	0.50	5,000	0.50
	1,000	0.1	0.09	910	0.91
	10,000	0.0	0.01	100	0.99

Note: K_p = partition coefficient, S = suspended solids concentration, C_e/C_t = fraction of the pollutant in the dissolved phase at equilibrium, C_W = total dissolved phase concentration when C_T = 1 mg/L, X = mass of sorbed pollutant per mass of suspended solids, $C_S = X \cdot S$ = total suspended phase concentration when C_T = 1 mg/L.

Solution. Use the expression

$$\frac{C_e}{C_T} = \frac{1}{1 + K_p \cdot S \cdot 10^{-6}}$$

In this case, C_T = 10 mg/L, $k_p = 1 \times 10^4$, and S = 2,500 mg/L. Therefore,

$$\frac{C_e}{C_T} = \frac{1}{1 + (10^4)(2500)(10^{-6})} = 0.04$$

Only 4 percent, or 0.4 mg/L, of the phenanthrene will be found in the effluent wastewater. The remaining 9.6 mg/L will be sorbed onto the suspended solids that were settled out.

VOLATILIZATION. Volatilization is defined as the transfer of matter from the dissolved phase to the gaseous phase. Many toxic materials volatilize in the natural environment, with the rate dependent on the properties of the contaminant and the characteristics of the water body. Volatilization may decrease the concentration of the pollutant in the water body, but it results in an increased concentration in the atmosphere. Volatilization is also a major cause of fugitive emissions at industrial plants.

The volatility of a compound is a function of its vapor pressure. The vapor pressure of a liquid is the pressure exerted by the vapor over the liquid on the liquid at equilibrium. Compounds with high vapor pressures have a greater tendency to transfer into the gas phase. Vapor pressure for a given compound generally increases with temperature; therefore, the tendency to volatilize increases as the temperature of the aqueous solution increases.

In a closed container, a contaminant will partition itself between the aqueous and gaseous headspace phases until equilibrium is reached. In many ways, this is analogous to the partitioning occurring during adsorption or the partitioning between two liquid solvents, described earlier. The partitioning can be described using *Raoult's Law,* assuming the liquid and contaminant form an ideal, binary mixture:

$$P_a = P_{vp} \cdot x_a$$

where P_a = partial pressure of contaminant a (atm)
P_{vp} = vapor pressure of the pure compound (atm)
x_a = mole fraction of contaminant a in water = (mol a)/(mol a + mol water)

The partitioning between the gas and liquid phases of a dilute solution in a closed container can be more conveniently described using *Henry's law.* This approach cannot be used for open containers because the volatilized compound disperses away from the liquid, and equilibrium may never be reached until all of the organic has evaporated. Henry's law also cannot be used for highly concentrated contaminants; instead, Raoult's law must be used. For most cases that concern us, though, contaminant solutions can be considered to be dilute.

Henry's law is an expression that relates the concentration of a chemical dissolved in the aqueous phase to the concentration (or pressure) of the chemical in the gaseous phase when the two phases are at equilibrium with each other. This can be expressed as

$$P = K_H C_W$$

where P = contaminant equilibrium partial pressure in the atmosphere above the liquid (atm)
K_H = Henry's law constant for that contaminant (atm·m^3/mol)
C_W = equilibrium concentration of the contaminant in water (mol/m^3)

As stated earlier, Henry's law is valid only for dilute solutions. This expression is good up to an aqueous phase concentration of 0.02 mol fraction. For compounds with molecular weights greater than 50 g/mol, which encompasses most of the compounds of interest to us, a mole fraction of 0.02 represents a concentration of at least 55,000 mg/L, which is much higher than normally found in the environment, so Henry's law can normally be safely used.

Appendix B lists Henry's law constants for selected compounds of interest to us. More complete lists can be found elsewhere (LaGrega, Buckingham, and Evans, 1994; Metcalf & Eddy Inc., 1991; Watts, 1998). Care must be taken when looking up values for Henry's law constants because several different units are used in different references. The units used here are atm · m^3/mol. In other references, the value may be given in dimensionless units. These are used with the following relationship:

$$K'_H = \frac{C_a}{C_w}$$

where K'_H = Henry's law constant (dimensionless)
C_a = contaminant molar concentration in air (mol/m^3)
C_w = contaminant concentration in water (mol/m^3)

The two values are related to one another and can be equated using the following:

$$K'_H = \frac{K_H}{R \cdot T} = 41.6 K_H \qquad \text{at } 20°\text{C}$$

where R = universal gas constant (8.2×10^{-5} m^3 · atm/mol · K)
T = temperature (K)

Henry's law constant can also be estimated if it is not available using the equation

$$K_H = \frac{P_s \cdot \text{MW}}{760 \cdot S_W}$$

where P_s = saturation vapor pressure of the pure compound (atm)
MW = compound molecular weight
S_W = water solubility of the compound (mol/m^3)

This expression reflects the importance of molecule size and water solubility on volatilization potential.

Example 2.7

(*a*) Estimate the Henry's constant for trichloroethylene in water at 20°C.
(*b*) Compare this value with the actual value of 0.0073 atm · m^3/mol.
(*c*) Determine the Henry's constant in dimensionless terms.

Solution.

(*a*) From Appendix B:

$$P_s = 58 \text{ mm Hg} = 58 \text{ mm Hg}/(760 \text{ mm Hg/atm}) = 0.076 \text{ atm}$$

$$\text{MW} = 131.78$$

$$S_w = 1312 \text{ mg/L} = 1312 \text{ g/m}^3 = \frac{1312 \text{ g/m}^3}{131.38 \text{ g/mol}} = 9.986 \text{ mol/m}^3$$

(*b*) Use the Henry's constant approximation equation:

$$K_H = \frac{P_s}{S_w} = \frac{0.0763 \text{ atm}}{9.986 \text{ mol/m}^3} = 0.0076 \text{ atm·m}^3\text{/mol}$$

The estimated Henry's constant is 0.0076 atm · m^3/mol, which is nearly identical to the actual value of 0.0073 atm · m^3/mol.

(*c*) In dimensionless terms,

$$K'_H = \frac{K_H}{RT} = \frac{0.0763 \text{ atm·m}^3\text{/mol}}{(8.32 \times 10^{-5}\text{atm/mol·K})\ (293 \text{ K})} = 3.13$$

Some generalizations can be made concerning the tendency to volatilize, based on the value of K_H. When the Henry's constant is greater than 10^{-3} atm · m^3/mol, volatilization is rapid. When K_H is less than 3×10^{-7} atm · m^3/mol, the contaminant is essentially nonvolatile and will remain in the water. In fact, its volatilization rate may be less than that of water, and the concentration may become greater as the water evaporates leaving the contaminant behind. Between these two limits, volatilization rate increases with increase in K_H.

It must be reiterated that the foregoing analysis holds only for closed containers where the air and liquid concentrations have come to equilibrium. In many cases, these boundary conditions will also hold for volatilization in soils because of the limited exchange of air with the surface. For open containers or evaporation from surface impoundments, equilibrium will probably never be reached because of escape of the contaminant into the larger gas phase. In this case, the rate of evaporation and rate of transport of the contaminant away from the site are much more important. Detailed analysis of volatilization kinetics is beyond the scope of this book. For those who are interested, there are a number of good references (Clark, 1997; Hemond, 1994; Schnoor, 1997; Thibodeaux, 1996).

The solubility of a contaminant is also important in determining the importance of volatilization. For example, anthracene has a volatilization rate constant of 18 cm/h compared with a rate constant of 20 cm/h for carbon tetrachloride. Even though the two rate constants are very close, the solubility of anthracene in water is only 0.06 mg/L versus 785 mg/L for carbon tetrachloride. Therefore, if each of these were to volatilize from saturated solutions, the flux of carbon tetrachloride volatilization would be 15,000 times as great as for anthracene.

Similar impacts can be described for the effect of sorption on volatilization. Only dissolved species can volatilize from aqueous systems. If a large part of a contaminant in an aqueous solution is bound to the surfaces of solids by adsorption, the amount of the contaminant in solution will be greatly reduced and volatilization fluxes will be much less. As dissolved material volatilizes, the aqueous concentration will go down and more of the compound will desorb off the solids to maintain equilibrium, but the flux will always be governed by the soluble concentration.

The solution pH also plays a significant role in the volatilization fluxes of some compounds. Only electrically neutral species are directly volatile. Therefore, volatilization rate expressions for organic acids and bases must use only the non-ionic concentration of the contaminant. The fraction of an acid or base which is in the non-ionic form can be determined by

$$\alpha_{A_0} = \frac{A_0}{A} = \frac{1}{1 + 10^{(pH - pK_a)}}$$

$$\alpha_{B_0} = \frac{B_0}{B} = \frac{1}{1 + 10^{(pK_w - pK_b - pH)}}$$

where α_{A_0} = decimal fraction of the organic acid which is in the non-ionic form
α_{B_0} = decimal fraction of the organic base which is in the non-ionic form
A_0 = concentration of the organic acid which is in the non-ionic form
A = total *dissolved* concentration of the organic acid
B_0 = concentration of the organic base which is in the non-ionic form
B = total *dissolved* concentration of the organic base

Thus compounds that readily dissociate at the pH encountered in the environment will have a low volatilization flux, even though they may have a high volatilization rate constant.

Example 2.8. While analyzing a water sample from a holding pond at the Quirky Chemical Co., which discharges into the Ohio River, it is found that the water contains 20 μg/L of 2-nitrophenol (total dissolved form). The water has a pH of 8.0. Further analysis shows that the volatilization rate constant for 2-nitrophenol is 2 cm/h. Determine the rate of volatilization (volatilization flux) of the 2-nitrophenol from the holding pond surface on a per unit area basis.

Solution. From Appendix B, the pK_a of 2-nitrophenol is 7.21. The fraction present in the non-ionic form is

$$\alpha_{A_0} = \frac{1}{1 + 10^{(pH - pK_a)}} = \frac{1}{1 + 10^{(8.0 - 7.2)}} = 0.14$$

Therefore, the 2-nitrophenol present in the non-ionic form is

$$C_{A_0} = \alpha_{A_0} \cdot C_T = (0.14)\,(20\ \mu g/L) = 2.8\ \mu g/L$$

The volatilization flux is then

$$R_v = h_v \cdot C_{A_0} = (2\ cm/h)\,(2.8\ \mu g/L)\,(1000\ L/m^3)\,(1\ m/100\ cm) = 56\ \mu g/h \cdot m^2$$

2.3.4 Transformation Processes

Many organic compounds undergo chemical or biochemical transformations upon entering the environment. Transformation reactions involve the breaking of chemical bonds in the original compound, creating new compounds. Metal compounds can also undergo transformation reactions into new compounds, but the metal itself cannot be transformed into another element. Depending on the extent of degradation, the reaction products may be more, less, or as toxic or harmful to the environment or public health as the original material.

A number of transformation reactions can occur. They may be biotic (biological) or abiotic (physical or chemical). Among the more important transformation processes are hydrolysis, elimination, oxidation-reduction, photolysis, and biodegradation reactions. Following is a brief overview of each of these.

HYDROLYSIS. Hydrolysis reactions involve the attack of an electron-rich water molecule on an electron-poor organic bond in the compound. During attack, the water molecule is added across the bond and the constituent originally at the bond leaves the molecule, often with either the H^+ or OH^- from the water molecule. Compounds that are susceptible to hydrolysis include those that act as acids or bases as well as those that contain S^{2-}, Cl^-, or NO_3^- or are halogenated aliphatics, esters, amides, carbamates, or phosphoric esters. Examples of hydrolysis reactions are:

$$Cl{\cdot}CH_2-CH_2-CH_2-CH_2-CH_2-CH_3 + H_2O \longrightarrow HO-CH_2-CH_2-CH_2-CH_2-CH_2-CH_3 + HCl$$

1-Chlorohexane → 1-Hexanol

$$CH_3-C(=O)-O-CH_2-CH_3 + H_2O \longrightarrow CH_3-COOH + CH_3-CH_2-OH$$

Ethyl acetate → Acetic acid, Ethanol

$$C_{10}H_7-O-C(=O)-N(H)-CH_3 + H_2O \longrightarrow C_{10}H_7-OH + NH_2-CH_3 + CO_2$$

Carbaryl → Naphthanol, Methylamine

Hydrolysis reactions result in the alteration of the original compound, but the resulting products may not necessarily be less toxic than the original compound. For example, the very toxic pesticide 2,4-D (2,4-dichlorophenoxyacetic acid) is often produced in nature from the hydrolysis of 2,4-D esters.

Hydrolysis reactions are often catalyzed by hydrogen or hydroxyl ions, so the reactions are strongly pH dependent. For example, the hydrolysis rate of the insecticide carbaryl increases logarithmically with pH (the rate at pH 8 is 10 times greater than that at pH 7 and 100 times greater than at pH 6).

ELIMINATION REACTIONS. Elimination reactions involve the removal of adjacent constituents from an organic molecule chain, leaving a double bond between the affected carbon atoms. The most important elimination reaction from an environmental viewpoint is *dehydrohalogenation,* in which adjacent halogen and hydrogen atoms are removed, leaving an alkene. An example is the dehydrochlorination of DDT to DDE (You, 1995):

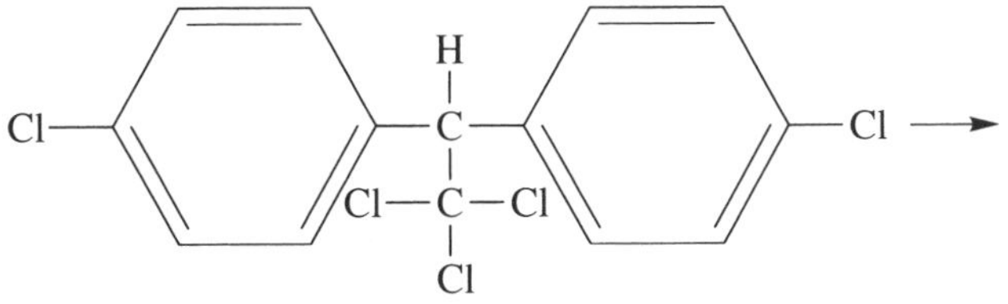

DDT

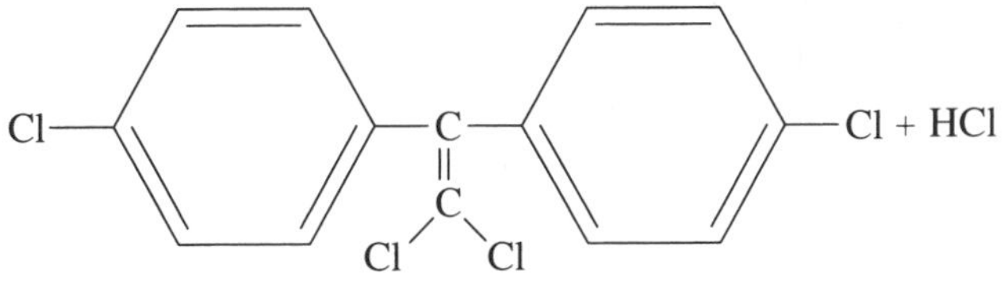

DDE

These reactions can be carried out both abiotically (usually at a very slow rate) and biotically by a limited number of microorganisms. They are a principal mechanism for removal of chlorinated organics from the environment.

OXIDATION-REDUCTION REACTIONS. The most common organic compound transformation reactions are oxidation or reduction. These reactions involve the transfer of electrons. An example of an oxidation reaction is the biologically mediated conversion of glucose to carbon dioxide and water. Both biotic and abiotic oxidation-reduction pathways are known, but natural abiotic reactions in aquatic systems are generally very slow and of little significance. Biological reactions are much more important from an environmental standpoint. In air, though, abiotic oxidation-reduction reactions can be significant.

Oxidation occurs when an organic compound (electron donor) loses one or more electrons to an oxidizing agent (electron acceptor). The most common natural oxidizing agents in aquatic systems are molecular oxygen, ferric(III) iron, and manganese(III/IV). In the atmosphere, rapid oxidations can occur due to the very strong photochemically produced oxidants ozone (O_3) and hydrogen peroxide (H_2O_2) and various free radicals such the hydroxyl radical ($\cdot OH$) and the peroxy radical ($\cdot OOR$). Atmospheric reactions involved in the destruction of the ozone layer are discussed in Chapter 3. Biological oxidations are discussed later.

Reduction occurs when an organic compound gains one or more electrons from a reducing agent (electron donor). Again, these reactions may be abiotic (usually occurring under anoxic conditions) or biotic. Oxidized compounds such as chlorinated aliphatics or nitroaromatics can be reduced in anaerobic soils or groundwaters. Common environmental reducing agents include pyrite (FeS), ferrous carbonates, sulfides, and natural organic matter. An example of a reduction reaction is the reductive dehalogenation of perchlorethylene to ethylene by sequential removal of chlorines:

$$\underset{\text{PCE}}{Cl_2C{=}CCl_2} \xrightarrow{H^+ \; Cl^-} \underset{\text{TCE}}{HClC{=}CCl_2} \xrightarrow{H^+ \; Cl^-} \underset{\text{DCE}}{HClC{=}CHCl} \xrightarrow{H^+ \; Cl^-} \underset{\text{Vinyl chloride}}{HClC{=}CH_2} \xrightarrow{H^+ \; Cl^-} \underset{\text{Ethylene}}{H_2C{=}CH_2}$$

In abiotic systems, the electrons necessary for carrying out these reactions must come from highly reducing environmental chemicals in contact with the contaminant. In biotic systems, the electrons can be provided by microbial oxidation of a wide variety of materials, both organic and inorganic.

PHOTOCHEMICAL REACTIONS. Photochemical transformations are sometimes of importance. Since interaction of sunlight with the contaminant is required for photochemical reactions to occur, they will generally be significant only in the atmosphere or the top layer of a water body. The interaction of sunlight and contaminant may either be direct, in which the contaminant directly receives the light energy, or indirect, in which another compound such as humic acids in the water receives the energy and then transfers it to the contaminant in question. In the direct case, the contaminant absorbs the light energy and is converted into an excited state, which then releases this energy in conjunction with a conversion into a different compound.

The best known example of photochemical reactions is the formation of smog. Photochemical smog is produced in the lower atmosphere by reaction between hydrocarbons, emitted from combustion sources such as automobiles and industrial processes, and oxides of nitrogen (NO and NO_2, collectively termed NO_x) in the presence of sunlight. Ozone, nitroderivatives of the hydrocarbons, and other organic compounds such as formaldehyde are formed. The nitro compounds react with moisture in the air to form the colloidal dispersion known as smog:

$$NO_2 + \text{uv radiation} + \text{VOC} + O_2 \rightarrow NO + O_3 + \text{PAN} + \text{aldehydes}$$

Smog is known to cause respiratory problems, and the ozone produced is a strong oxidizing agent and can attack many materials. The formation of smog and its environmental consequences are described in greater detail in Chapter 3.

Another instance where photochemical reactions are important from a pollution prevention perspective is in the development of photodegradable plastics for use in packaging. This is described further in Chapter 9.

BIOLOGICAL TRANSFORMATIONS. Most organics are, to some extent, biodegradable by microorganisms in the environment. Since microorganisms are ubiquitous in nature, biodegradation, particularly by bacteria, is often the most important process for the destruction of organic matter in aquatic environments and in soils. Over the eons of life on earth, bacteria have developed pathways for oxidizing essentially all *naturally occurring* organics as a way to obtain the building blocks and energy they need to build new cells. Biodegradation is carried out in small steps, using a variety of enzymes as reaction catalysts for each step. During many of these steps, energy released from the reaction is recovered and used by the microorganism to create new needed molecules. Many microorganisms are also capable of at least partially biodegrading *synthetic* organics, either directly or in conjunction with the degradation of more common organics. These reactions may not be intentional; rather, they may derive from the ability of an enzyme used on a naturally occurring organic to catalyze a reaction with the xenobiotic (not naturally occurring) compound. The microorganism may get no benefit in the way of energy recovery from this reaction. This type of reaction is termed *cometabolism.* While most organics can be at least partially biodegraded, the rate of biodegradation may be very slow for some xenobiotic compounds. These compounds are often classified as nonbiodegradable or recalcitrant.

Many biodegradation pathways lead to the conversion of complex organic compounds to carbon dioxide and water:

$$C_aH_bO_cN_dP_eS_fCl_g \rightarrow iCO_2 + jH_2O + kNH_4^+ + lPO_4^{3-} + mSO_4^{2-} + nCL^- + \text{Energy}$$

This is called *mineralization.* Biodegradation of recalcitrant compounds does not always lead to mineralization, though; often organic by-products result which may not degrade further at an appreciable rate under the existing environmental conditions. An example of this is the oxidation of trichloroethylene (TCE) by a group of bacteria that utilize methane (called *methanotrophs*) as their source of carbon and energy. The methanotrophs produce a methane monooxygenase enzyme that by chance also has the ability to convert TCE to a TCE epoxide. This latter reaction provides no benefit to the methanotroph and is a form of cometabolism. These cometabolic by-products may be more or less toxic than the original compound.

Biodegradation transformations are often similar to the abiotic reactions described earlier. Of particular importance are oxidation reactions. These generally form polar compounds from nonpolar ones. An example is the dioxygenase enzyme–mediated hydroxylation of benzene to form catechol:

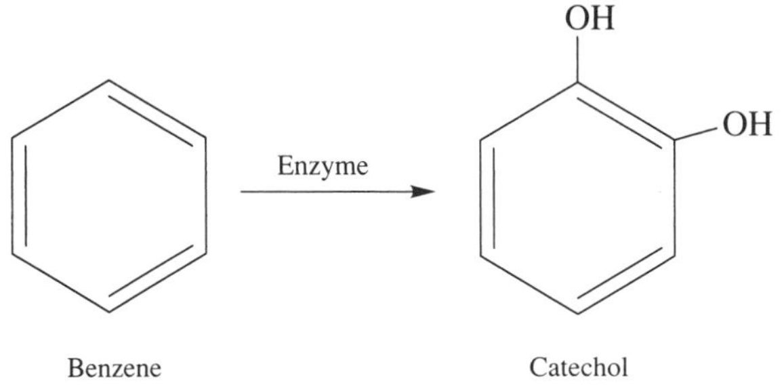

OH

OH

Enzyme

Benzene Catechol

The ring structure of the polar catechol molecule can now be cleaved, producing an aliphatic compound that is readily biodegradable:

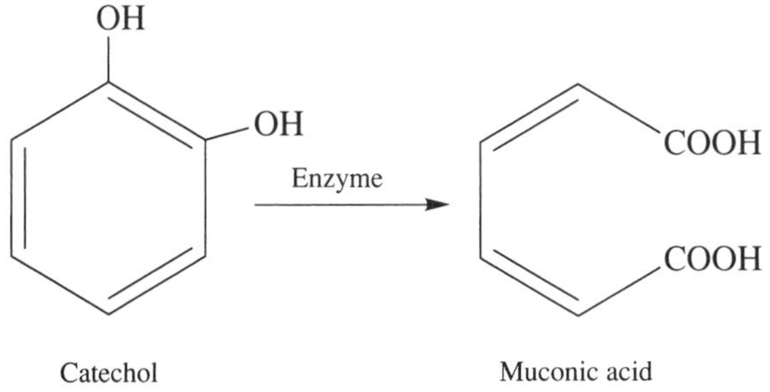

OH

OH

COOH

Enzyme

COOH

Catechol Muconic acid

TABLE 2.10
Biodegradation preferences

Nonaromatic compounds are preferred over aromatic ones
Materials with unsaturated bonds (alkenes, alkynes, tertiary amines, etc.) are preferred over saturated compounds (e.g., alkanes)
The *n*-isomers (straight chains) of lighter weight molecules are preferred over branched isomers and complex, polymeric substances
Soluble organics are usually more readily degradable than insoluble ones
Straight-chain hydrocarbons of more than nine carbons are more easily biodegraded than those with less than nine carbons
Heavily chlorinated compounds such as PCBs and TCE are more readily biodegraded under anaerobic conditions; chlorinated compounds with only a few chlorines such as vinyl chloride are more readily biodegraded aerobically
The presence of key functional groups at certain locations in the molecule can make a compound more or less amenable to biodegradation
Alcohols, aldehydes, and acids are more biodegradable than their alkane or alkene homologues
Halogenation or addition of a nitro-group makes a hydrocarbon more resistant to biodegradation
Meta substitution to a benzene ring generally makes it more resistant to biodegradation than to the ortho or para position

Other biodegradation mechanisms include hydrolysis, dehalogenation, and reduction processes.

The literature on biodegradation of organics is vast. We discuss only some general rules that can be applied to the analysis of biodegradation potential. For more detailed information, the reader is referred elsewhere (Alexander, 1994; Chakrabarty, 1982; Chaudhry, 1994; Levin and Gealt, 1993; Pitter and Chuboda, 1990; Rochkind-Dubinsky, Sayler, and Blackburn, 1987; Stoner, 1994; Young and Cerniglia, 1995).

Most research on biodegradation has involved pure cultures of microorganisms exposed to a single organic substrate. However, in most cases, this does not occur in nature. Microbial populations are mixed and there usually is a mixture of organics available to the microorganisms. The presence of more easily biodegradable substrates may delay the use of other, more recalcitrant, organic compounds. Table 2.10 lists some generalizations that can be made concerning preferences of organics as growth substrates.

REFERENCES

Alexander, M. *Biodegradation and Bioremediation.* San Diego: Academic Press, 1994.

Cahn, R. S. *Introduction to Chemical Nomenclature,* 5th ed., London: Butterworth, 1979.

Chakrabarty, A. M. *Biodegradation and Detoxification of Environmental Pollutants.* Boca Raton, FL: CRC Press, 1982.

Chaudry, G. R. *Biological Degradation and Bioremediation of Toxic Chemicals.* Portland, OR: Dioscorides Press, 1994.

Clark, M. M. *Modeling for Environmental Engineers and Scientists.* New York: Wiley, 1997.

Hemond, H. F. *Chemical Fate and Transport in the Environment.* San Diego: Academic Press, 1994.

IUPAC. *Rules for I.U.P.A.C. Notation for Organic Chemicals,* 5th ed. London: Longmans, 1961.

Knox, R. C. *Subsurface Transport and Fate Processes.* Boca Raton, FL: Lewis Publishers, 1993.

LaGrega, M. D.; Buckingham, P. L.; and Evans, J. C. *Hazardous Waste Management.* New York: McGraw-Hill, 1994.

Levin, M. A.; and Gealt, M. A. *Biotreatment of Industrial and Hazardous Waste.* New York: McGraw-Hill, 1993.

Metcalf & Eddy Inc. *Wastewater Engineering.* New York: McGraw-Hill, 1991.

Meyers, E. *Chemistry of Hazardous Materials.* Englewood Cliffs, NJ: Prentice-Hall, 1977.

Pitter, P.; and Chuboda, J. *Biodegradability of Organic Substances in the Aquatic Environment.* Boca Raton, FL: CRC Press, 1990.

Rochkind-Dubinsky, M. L.; Sayler, G. S.; and Blackburn, J. W. *Microbiological Decomposition of Chlorinated Aromatic Compounds.* New York: Marcel Dekker, 1987.

Sawyer, C. N.; McCarty, P. L.; and Parkin, G. F. *Chemistry for Environmental Engineering,* 3rd ed. New York: McGraw-Hill, 1994.

Sax, N. I. *Hazardous Chemicals Desk Reference.* New York: Van Nostrand Reinhold, 1987.

Sax, N. I. *Sax's Dangerous Properties of Industrial Materials.* New York: Van Nostrand Reinhold, 1992.

Schnoor, J. L. *Environmental Modeling.* New York: Wiley, 1997.

Stoner, D. L. *Biotechnology for the Treatment of Hazardous Wastes.* Boca Raton, FL: Lewis Publishers, 1994.

Thibodeaux, L. J. *Environmental Chemodynamics: Movement of Chemicals in Air, Water and Soil.* New York: Wiley, 1996.

Watts, R. *Hazardous Wastes: Sources, Pathways, Receptors.* New York: Wiley, 1998.

You, G. *Anaerobic DDT Biodegradation and Enhancement by Nonionic Surfactant Addition.* Ph.D. diss., University of Cincinnati, 1995.

Young, L.Y.; and Cerniglia, C. E., *Microbial Transformation and Degradation of Toxic Organic Chemicals.* New York: Wiley-Liss, 1995.

PROBLEMS

2.1. Name the following compounds using IUPAC rules:

(*a*) $CH_3—CH_2—Cl$

(*b*)

$$\begin{array}{ccccccc} & & H & & H & & H & \\ & & | & & | & & | & \\ H & — & C & — & C & — & C & — H \\ & & | & & | & & | & \\ & & Cl & & Cl & & H & \end{array}$$

(*c*)

$$\begin{array}{ccccccc} & & COOH & & & & \\ & & | & & & & \\ CH_3 & — & CH & — & CH_2 & — & CH_3 \end{array}$$

(*d*)

$$\begin{array}{ccccccc} & & Cl & & Cl & & \\ & & | & & | & & \\ H & — & C & — & C & — & H \\ & & | & & | & & \\ & & Cl & & Cl & & \end{array}$$

(*e*)

$$\begin{array}{ccccccccccccc} & & CH_3 & & & & & & & & & & \\ & & | & & & & & & & & & & \\ CH_3 & — & CH & — & CH & — & CH_2 & — & CH_2 & — & CH_2 & — & CH_3 \\ & & & & | & & & & & & & & \\ & & & & CH_2-CH_3 & & & & & & & & \end{array}$$

(*f*)

$$\begin{array}{ccccccccccccccc} & & & & & & & & NH_2 & & & & & & \\ & & & & & & & & | & & & & & & \\ CH_3 & — & CH_2 & — & CH_2 & — & CH & — & CH & — & CH & = & CH & — & CH_3 \\ & & & & & & | & & & & & & & & \\ & & & & & & Cl & & & & & & & & \end{array}$$

2.2. Draw the structures of the following compounds:

(*a*) 3-Chloro-2-nitrohexane

(*b*) 2-Phenyl-1-propanol

(*c*) Acetylene tetrabromide

(*d*) 2-Amino-3-hydroxybutanoic acid

(*e*) 3-Hydroxybutyric acid

(*f*) 1,3-Dichloropropene

2.3. Name the following compounds using IUPAC rules:

(*a*)

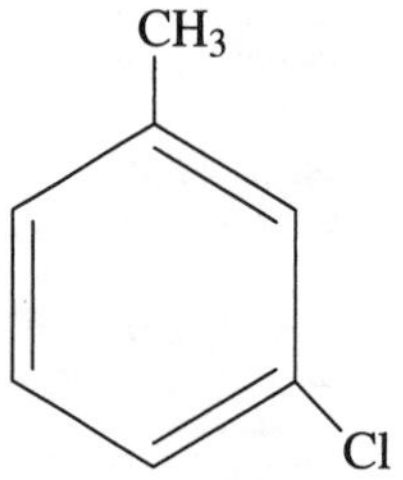

CH_3

Cl

(*b*)

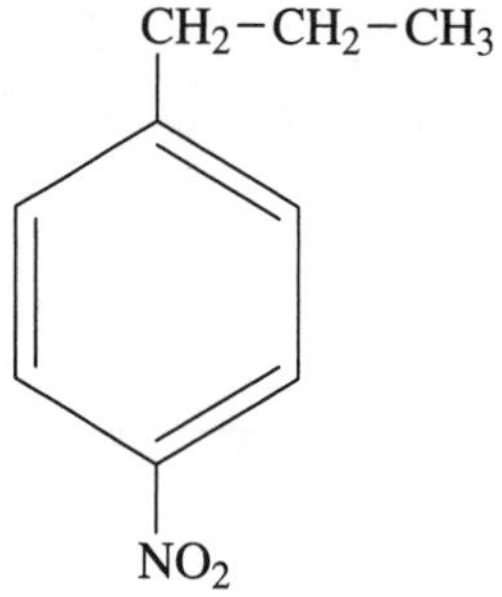

$CH_2-CH_2-CH_3$

NO_2

(*c*)

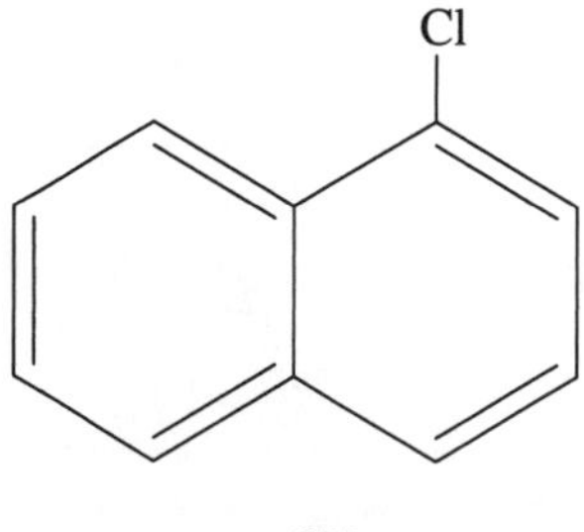

Cl

(*d*)

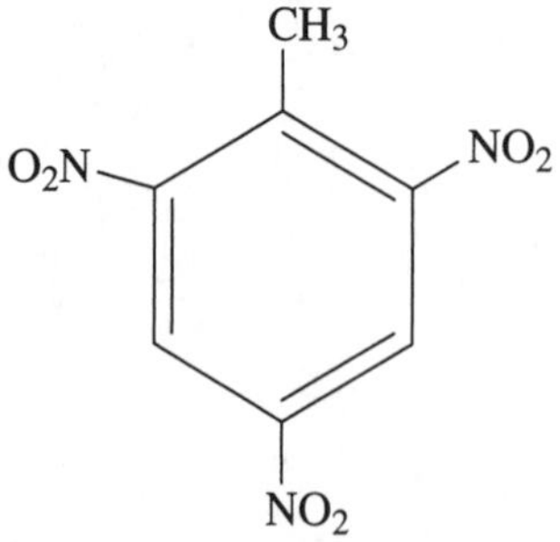

CH_3

O_2N NO_2

NO_2

(*e*) Cl Cl

Cl Cl

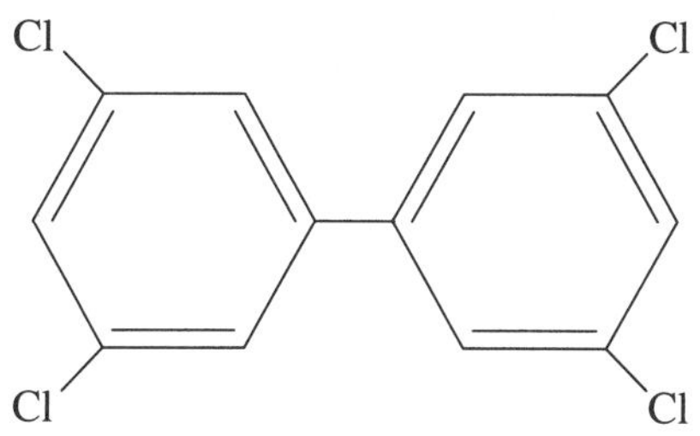

2.4. Draw the structures of the following aromatic compounds:
(*a*) Isobutylbenzene
(*b*) *p*-Ethyltoluene
(*c*) 2,4,6-Trichlorophenol
(*d*) 2,4′-Dichlorobiphenyl
(*e*) 1-Aminonaphthalene

2.5. Using the *Merck Index* or other sources, find the structures, properties, and uses of the following compounds (many can also be found using the Internet):
(*a*) Acetylaminobenzene
(*b*) Chrysene
(*c*) 1,4-Benzenediol

2.6. Benzene has commonly been used as an industrial intermediate, but its use has declined in recent years. Why? What is it primarily used for? What has it been replaced by in many synthesis reactions?

2.7. A large dry cleaning operation has closed. A potential buyer would like to convert the facility to a day care center. You have been called in by the buyer to determine whether the facility is safe for this use. How would you proceed with your investigation? What chemicals should be analyzed for and where? Why are they of concern?

2.8. Cyanide is commonly used in metal plating baths, but it is currently being replaced in plating solutions because of its extreme toxicity. Using various sources, determine what the role of cyanide is in these plating baths, how the bath formulations are being changed to minimize cyanide use, and what potential problems arise from using these substitutes.

2.9. An industry discharges 500 m^3/day of wastewater containing 12 mg/L of cadmium into a river that serves as a drinking water source for a downstream community. The stream flow is 3500 m^3/min. Assuming that the cadmium acts in a conservative fashion (i.e., it disperses in the water and is not removed by precipitation, adsorption, volatilization, biouptake by organisms in the river, or by water treatment processes) and that the average daily per capita consumption of water is 2 L/day, what is the average daily intake of cadmium due to water consumption?

2.10. A 5-g sample of pentachlorophenol is added to 1 L of water at pH 7.5. What will be the resulting concentration of un-ionized pentachlorophenol?

2.11. If 50 mg of hydrogen cyanide is added to 1 L of buffered water and the resulting pH is measured as 7.0, what percentage of the HCN will be in the volatile (and lethal) undissociated form? What if the pH is 9.0?

2.12. Equal volumes of water and the solvent methyl isobutyl ketone (MIBK) are mixed in a drum. Where would you expect to find these materials in the drum?

2.13. Repeat Problem 2.12, but assume the organic contaminant is naphthalene.

2.14. In general, water solubility of organic compounds increases as the polarity of the compounds increases. Why?

2.15. Describe the mechanistic bases of the Langmuir, Freundlich, and linear isotherms. How do you decide which is more appropriate to use for a given situation?

2.16. Granular activated carbon will be used to remove nonrecovered products after crystallization in an industrial process. The crystallizer effluent contains a total organic carbon content of 550 mg/L. An isotherm study was run for a particular type of activated carbon using 100-mL samples, resulting in the following results. Determine the Freundlich isotherm constants for this application.

m, g	C_e, mg/L
0.0	550
0.1	270
0.2	90
0.3	30
0.4	12
0.5	2

2.17. Repeat Problem 2.16, solving for the Langmuir constants.

2.18. Compare the results from Problems 2.16 and 2.17 with the linear isotherm for these data. Assuming that an activated carbon column effluent concentration of 5 mg/L or less is desired, which isotherm would be most appropriate? How much activated carbon would be needed per day to treat 1000 m^3/day of crystallizer effluent?

2.19. Why is it important to know the solubility and adsorptability of a compound when estimating its potential to volatilize?

2.20. What is the difference between Henry's law and Raoult's law? Under what circumstances would each be used? Which law would you use to determine volatilization from a drum of pure TCE? From process water containing TCE contaminants?

2.21. For 2,4,6-trichlorophenol, estimate Henry's law constant from its water solubility and vapor pressure, using values from the appendixes. Compare the estimated value to the tabulated value.

2.22. A closed 208-L (55-gal) drum is half-filled with pure perchlorethylene (PERC). The top half of the drum is empty. The drum has not been opened for several months. When it is opened, what will be the PERC partial pressure and concentration in the head space? The temperature is 25°C and the atmospheric pressure is 1 atm.

2.23 Repeat Problem 2.22, assuming the drum is half-filled with an aqueous solution containing 50 mg/L PERC.

2.24. Select one of the following compounds: nitric acid, urea, styrene, terephthalic acid, or isopropyl alcohol. Write a report describing the raw materials used to produce this chemical, the production methods (including flow sheets), the annual production rate, market price for the chemical, and types and amounts of by-products and wastes.

2.25. For the chemical selected in Problem 2.24, qualitatively evaluate the environmental impact of doubling the annual production of this chemical.

CHAPTER 3

INDUSTRIAL ACTIVITY AND THE ENVIRONMENT

3.1 INTRODUCTION

Every industrial operation produces some amount of waste. No operation is 100 percent efficient in conversion of base materials into finished products or in its use of energy to produce these materials. One objective of this book is to show how industry can use materials more efficiently and reduce the amount of material going to waste.

The effect of industrial wastes on the environment and ultimately on human health depends on where the wastes go. Some may enter the atmosphere through exhaust stacks or from fugitive emissions, some may enter water bodies through wastewater effluent pipes or overland runoff, and some may enter the ground through burial or indiscriminate dumping and eventual leaching of the contaminants into the soil. In addition to these direct environmental entry methods, a myriad of intermedia processes are at work. Contaminants that enter the atmosphere may precipitate to Earth due to gravity or with rain, then becoming soil or water contaminants. They may undergo chemical reactions in the atmosphere, changing to new compounds. Water pollutants may settle to the bottom of the water body, leaving the aqueous phase and becoming soil contaminants. They may also volatilize into the atmosphere, becoming air pollutants. Some may be biodegraded in the water, often detoxifying, but others may be taken up by aquatic organisms such as fish and may be bioconcentrated, possibly creating a human health hazard if the fish is eaten. Wastes buried or deposited on land can go through many of the same processes. They may volatilize and become air pollutants or leach into groundwater becoming water pollutants. Thus wastes cannot be classified solely by the media into which they are placed. A multimedia approach must be taken to describe the fates of particular contaminants. However, a discussion of what occurs within a specific medium is useful toward understanding the processes that act on contaminants there.

This chapter examines the fates of contaminants that enter the air, water, and land environments. It is divided into sections on (1) air pollution and the effects of air pollution on smog formation, acid rain, global warming, and ozone depletion; (2) solid waste disposal and the potential for resource recovery; (3) hazardous waste management; (4) water pollution; (5) energy consumption; and finally (6) natural resources depletion. These all have strong ties to industrial processes. Examples of how industry has affected the environment in each of these categories are presented, along with descriptions of how environmental impacts can be minimized. These are expanded on throughout the book.

3.2 AIR POLLUTION

On a global scale, one of the more important impacts of industrial systems on the environment is the emission of gaseous, liquid, and particulate materials to the atmosphere. The resulting air pollution is a complex problem, because of the complicated transport processes acting on it, as well as the multitude of physical and chemical processes that may change its character or the media in which it is found. Air pollution can affect the environment or the general population at the local, regional, national, or global level. For example, an industry that burns coal and emits from its boiler exhaust gases that contain sulfur dioxide and hydrogen sulfide may cause serious odor problems in the immediate area as well as materials damage through acids that form upon reaction between the contaminants and water vapor in the air. The acids may also react with other materials in the atmosphere and contribute to smog formation. These impacts may not be limited to the local area. Winds may carry the contaminants great distances, causing an acid rain problem on a regional or even a national or international basis. Some of the pollutants may rise to the upper reaches of the atmosphere, where they may participate in a wide variety of reactions, eventually causing a depletion of the ozone layer above Earth, which protects us from the harmful effects of ultraviolet radiation. They may also contribute to a general increase in Earth's temperature, a process known as *global warming*. The impact of these can be felt on a global scale.

Air pollutants have many damaging effects. For example, hazardous airborne contaminants were accidentally released at Bhopal, India, in 1984, when methyl isocyanate leaked into the atmosphere from a Union Carbide plant, killing 2000 and injuring thousands more. In many cases industrial air emissions become trapped in valleys due to unusual weather conditions, leading to serious medical problems; this occurred in the Meuse Valley in 1930 when over a three-day period several thousand people became ill and 60 people died. Another case occurred in Donora, Pennsylvania, in 1948, when half the population was stricken with severe eye, nose, and throat irritations and chest pains, and 20 people died. This chapter focuses on four major effects—smog formation, acid rain, global warming, and ozone depletion—because of their significance from an industrial input viewpoint and because of their environmental impacts. Following is a brief discussion of what these environmental problems are and how they are created. What industry can do to minimize these impacts is discussed later.

Before considering environmental concerns related to air pollutants, it is necessary to describe the atmosphere in which they are acting.

3.2.1 The Atmosphere

The atmosphere is a mixture of gases that forms a layer about 400 km (250 mi) thick around Earth. Most of the atmosphere is held close to Earth by the force of gravity, so the atmosphere gets thinner with increasing distance from Earth. The atmosphere can be divided into four distinct layers of contrasting temperature due to differential absorption of solar energy, as seen in Figure 3.1.

The layer of air immediately adjacent to Earth's surface is called the *troposphere*. The troposphere, ranging in depth from 16 km (10 mi) over the equator to about 8 km (5 mi) over the poles, contains about 75 percent of the total mass of the atmosphere. This is where most weather events occur. The air is well mixed because of winds. Air temperature drops rapidly with increase in elevation, reaching about −60°C (−76°F) at the top of the troposphere.

The layer of air immediately above the troposphere is the *stratosphere*. It extends upward for about 50 km (31 mi). Air temperature in the stratosphere is relatively con-

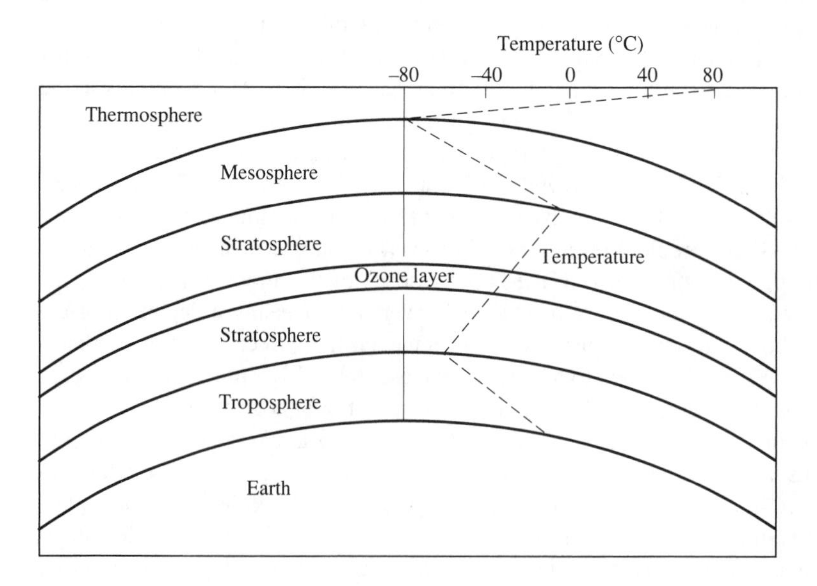

FIGURE 3.1
Layers of Earth's atmosphere showing temperature profile.

stant, or may even increase slightly with increase in elevation. This zone is of greatest importance to our environmental quality because it contains the ozone layer, which protects life on Earth by absorbing ultraviolet radiation from the sun. The ozone concentration in the stratosphere is about 1000 times greater than in the troposphere.

Above the stratosphere are the *mesosphere* and the *thermosphere*. There is little interchange of materials between the troposphere, where human activity has an impact, and these two zones, so they won't be discussed further here.

As Earth absorbs heat from the sun, it transfers some of this heat to the air above it. The heated air expands and rises. When its heat content is radiated into space, the air cools, becomes more dense, and settles back toward Earth. These vertical convection currents mix the gases in the lower atmosphere and transport heat from one area to another. Because of the rotation of Earth, the air also moves horizontally over Earth's surface. This combination of air movements creates the specific wind patterns characteristic of different parts of the world (see Figure 3.2). More detailed descriptions of weather can be found elsewhere (Baird, 1995; Cunningham and Saigo, 1995).

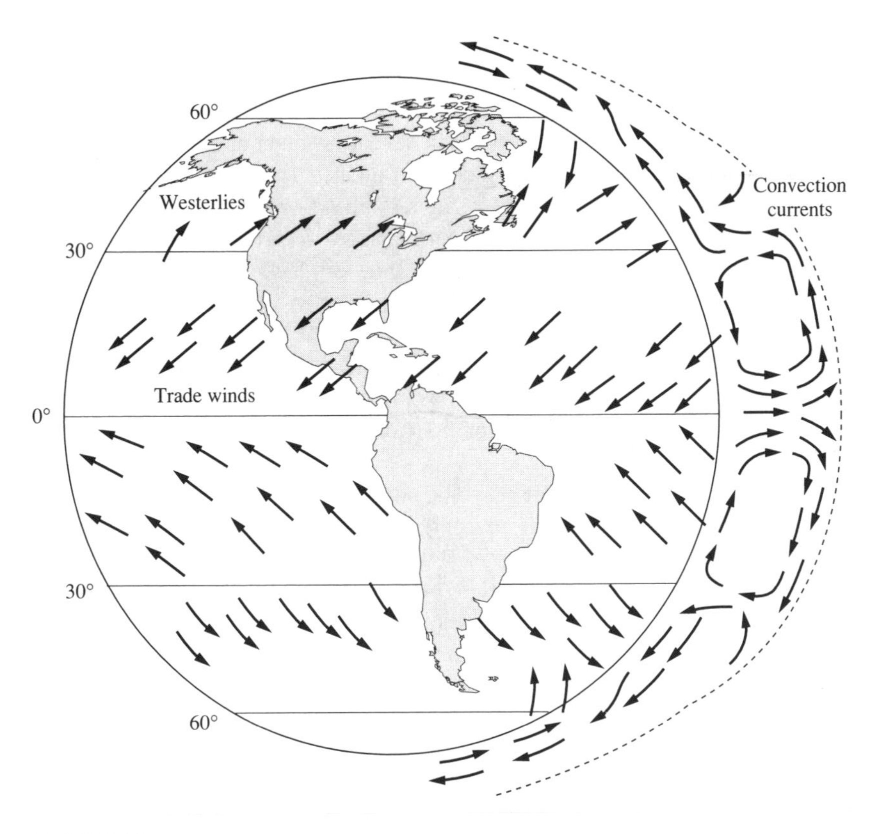

FIGURE 3.2
Global wind patterns. (Source: Enger and Smith, 1995)

TABLE 3.1
Composition of the troposphere

Gas	Formula	Percent by volume*
Nitrogen	N_2	78.08
Oxygen	O_2	20.94
Argon	Ar	0.934
Carbon dioxide	CO_2	0.033
Neon	Ne	0.00182
Helium	He	0.00052
Methane	CH_4	0.00015
Krypton	Kr	0.00011
Hydrogen	H_2	0.00005
Nitrous oxide	N_2O	0.00005
Xenon	Xe	0.000009

* Average for dry, clean air.

An important concept related to this predictable movement of air is that Earth has *airsheds*, which are analogous to watersheds. An airshed is the land area that contributes to the air—and the things found in the air—that flows over a particular geographical area. Watersheds are defined by geographical features that control where water will flow. Airsheds are far less distinct, but they still have a major impact on air quality in a particular region, and they may be very large in size. For example, much of the air quality in New England is dictated by the upwind industrial activities in the midwestern United States. To improve air quality in one region may require changes in air emissions in another region. Many problems may arise because airsheds do not conform to political boundaries, unless a regional approach to the problems is taken.

COMPOSITION OF THE ATMOSPHERE. The atmosphere is composed primarily of nitrogen (78 percent) and oxygen (21 percent), but many other trace gases are also present. Table 3.1 shows the composition of clean, dry air. Water vapor concentrations vary from near zero to 4 percent. In addition, other materials may be found in air as a result of human activities. These may be gases or particulates (liquid or solid particles suspended in the air, known collectively as *aerosols*). These anthropogenic materials are the subject of this book.

AIR POLLUTANTS. Hundreds of materials are discharged into the atmosphere as a result of human activities. Those that produce adverse health or environmental effects are termed *air pollutants*. Air pollutants may be classified as *primary pollutants,* which are released directly into air in a harmful form, or *secondary pollutants,* which are modified into a hazardous form after they enter the atmosphere due to chemical reactions or which are formed in the atmosphere. Typical primary pollutants are carbon monoxide, hydrocarbons, particulates, sulfur dioxide, and nitrogen compounds. Examples of secondary pollutants are photochemical oxidants and atmospheric acids formed in the atmosphere due to solar energy–activated reactions of less harmful materials.

The Clean Air Act of 1970 designated seven major pollutants as *conventional* or *criteria pollutants* because of their serious threat to public health and environmental quality. These include carbon monoxide, hydrocarbons, sulfur dioxide, particulates, nitrogen oxides, photochemical oxidants, and lead. *Maximum ambient air standards* have been set for these materials.

Carbon monoxide (CO) is a colorless, odorless, nonirritating poison that can quickly cause death at quite low concentrations. Inhaled carbon monoxide will tightly attach to hemoglobin in blood, reducing the capacity of the blood to carry oxygen to bodily tissues. The result is headaches, drowsiness, and eventually asphyxiation. Carbon monoxide is produced when organic materials such as gasoline, coal, or wood are incompletely burned. The automobile is responsible for most of the carbon monoxide produced in cities (67 percent), but stationary fuel combustion (20 percent) and industrial processes (6 percent) also contribute.

Hydrocarbons constitute a large group of volatile organic compounds. The major anthropogenic sources of hydrocarbons emitted to the atmosphere are evaporation of petroleum-based fuels and remnants of the fuels that did not burn completely. The automobile is the major source of hydrocarbons, followed by refineries and other industries. About 28 million tons of these materials are emitted annually in the United States. In addition to naturally occurring hydrocarbons, many synthetic organic compounds are also emitted, including benzene, toluene, formaldehyde, vinyl chloride, phenols, and trichloroethylene. These total over 2.5 million metric tons per year. Some of these are carcinogenic and others contribute to other health-related problems. The most significant impact of airborne hydrocarbons, though, is the part they play in the formation of ozone in the lower atmosphere, as described later in this chapter.

Sulfur dioxide (SO_2) is a colorless corrosive gas that is a respiratory irritant and a poison. It is also of major concern because it can react in the atmosphere with ozone, water vapor, and other materials to form sulfuric acid (H_2SO_4). Sulfuric acid, one of the strongest oxidizing agents known, can cause substantial damage to construction materials, metals, and other materials. It can also form very small aerosols or attach to minute aerosols, which can travel deep into the respiratory system when inhaled, doing serious tissue damage. The major anthropogenic sources of sulfur dioxide emissions to the atmosphere are combustion of sulfur-containing fuels (coal and oil) and industrial processes. More than 70 percent of the sulfur dioxide in our air in urban areas comes from coal-burning electrical power plants.

Particulates, small pieces of solid or liquid materials dispersed in the atmosphere, constitute the third largest category of air pollutant. These include dust, ash, soot, smoke, and droplets of pollutants either emitted to the air or formed in the atmosphere. Particulates are characterized as 0.005 to 100 μm in size. The major human sources of particulates are from unburned fuels from stationary fuel combustion and transportation as well as industrial processes. The impacts of particulates range from reduced visibility and soiling of exposed surfaces to respiratory problems and carcinogenicity. Respirable particles (smaller than 2.5 μm) are especially problematic because they can be drawn deep into the lungs, where they can damage respiratory tissues. Because of their small size (80 percent are less than 2.5 μm), particulates from vehicles are particularly insidious. Hydrocarbons are also of major concern because they can participate in photochemical reactions producing smog, as will be described later.

Nitrogen oxides (NO and NO_2) are highly reactive gases formed from oxidation of the nitrogen in air during combustion. When combustion takes place in air at a temperature above about 2000°F, the nitrogen and oxygen molecules in air may react with each other forming oxides of nitrogen:

$$N_2 + O_2 \rightarrow 2NO \qquad \text{(nitrogen oxide)}$$

$$2NO + O_2 \rightarrow 2NO_2 \qquad \text{(nitrogen dioxide)}$$

The designation NO_x is often used to signify mixtures of NO and NO_2 in air. The nitrogen dioxide is a critical component in smog-forming reactions. Nitrogen oxides can also react with water vapor, forming nitric acid (HNO_3), and contribute to acid precipitation problems. The primary source of NO_x is the automobile engine, although combustion of coal, oil, or natural gas at high enough temperatures can also contribute to the problem. Ironically, an excellent way to minimize emissions of hydrocarbons and carbon monoxide is to burn fuels at high temperatures with an abundance of air, but this then increases the amount of NO_x emitted.

Photochemical oxidants are products of secondary atmospheric reactions driven by solar energy. One of the most important of these is ozone (O_3). Ozone is a strong oxidant and can destroy lung tissue and chlorophyll in plants. It will be described in detail in the next section. Other photochemical oxidants include peroxyacetal nitrate (PAN) and acrolein, both of which are strong oxidants and can damage materials. PAN is also a severe eye irritant.

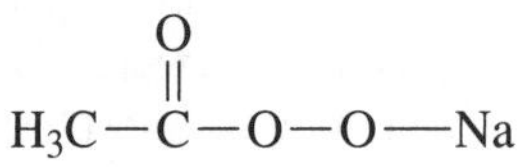

$$H_3C-\overset{\overset{\large O}{\|}}{C}-O-O-Na$$

Peroxyacetyl nitrate (PAN)

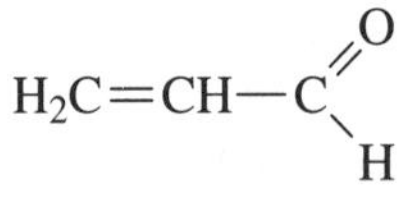

$$H_2C{=}CH-C\overset{\large O}{\underset{\large H}{\lessgtr}}$$

Acrolein

Lead and other volatile metals such as mercury and cadmium are mined and used in manufacturing processes; some also occur as trace elements in fuels. They are released to the air as metal fumes or suspended particulates from fuel combustion, ore smelting, and incineration of wastes. Worldwide, a major source of lead in the atmosphere is the burning of gasolines containing tetraethyllead, which is added to gasoline to reduce knock and engine wear. Approximately 2 million metric tons of lead are emitted to the atmosphere worldwide per year. Air in urban areas typically contains 5–10 times as much lead as does air in rural areas in the many countries where leaded gasoline is used, and several times as much as in air found over midocean areas. Since leaded gasoline was phased out in the United States in the late 1970s, atmospheric lead concentrations in the United States have declined substantially.

Hazardous air pollutants (HAPs) are another class of air contaminants. A HAP is defined in the Clean Air Act as "an air pollutant to which no ambient air quality standard is applicable and which . . . causes or contributes to . . . an increase in mortality or an increase in serious irreversible, or incapacitating reversible illness." Currently, there are nearly 200 chemicals regulated as HAPs. The HAP emission control program is technology-based and requires the installation of state-of-the-art pollution control equipment.

Carbon dioxide (CO_2) is another compound of concern in the atmosphere. It is usually considered nontoxic and innocuous, and it is not generally listed as an air pollutant. However, in the upper atmosphere, increasing concentrations of carbon dioxide have been implicated in global warming. This will be discussed in more detail later.

3.2.2 Smog Formation

The term "smog" comes from a combination of the words "smoke" and "fog." It was coined to describe air conditions in London. Smog can be produced by the action of fog with high concentrations of smoke particles or fly ash, but what we are discussing here is another form of air condition better termed *photochemical smog*. Photochemical smog is the result of the interaction between nitrogen oxides and hydrocarbons under the influence of sunlight. These reactions produce a mixture of photochemical oxidants (ozone, PAN, acrolein, etc.) which can readily react with and harm other materials. Automobile exhaust is the major source of photochemical smog in most urban areas.

As described previously, automobile exhaust contains large amounts of nitrogen oxide (NO) because of the high temperatures used to efficiently combust gasoline in the automobile's engine:

$$N_2 + O_2 \rightarrow 2NO$$

The NO can react with oxygen in air to form nitrogen dioxide:

$$2NO + O_2 \rightarrow 2NO_2$$

The nitrogen dioxide created gives the air a reddish-brown color and reduces visibility. This would be the extent of the problem if further photochemical reactions did not take place. However, they often do.

In the upper atmosphere, where there is abundant ultraviolet radiation, oxygen molecules readily absorb uv radiation from the sun and split into two oxygen radicals:

$$O_2 \rightarrow 2O\cdot$$

These have very high energy and readily react with oxygen molecules to form ozone:

$$O_2 + O\cdot \rightarrow O_3$$

This reaction does not occur to any appreciable extent near Earth's surface because most of the shortwave uv radiation is absorbed before it reaches the surface. Some uv radiation does penetrate to the lower atmosphere, though. Unfortunately, nitrogen dioxide is a very efficient absorber of this uv radiation. Upon absorbing the radiation, it becomes energized and breaks down into nitrogen oxide and an oxygen radical:

$$NO_2 + \text{uv radiation} \rightarrow NO + O\cdot$$

The oxygen radical can then react with oxygen to form the highly reactive ozone molecule. The ozone produced is so highly reactive that it is very short-lived and would quickly revert back to an oxygen molecule if no other materials were present in the air. However, if hydrocarbons (volatile organic carbon compounds, or VOCs) are present, the ozone will react with them, forming PAN and other strong photochemical oxidants. The net reaction is

$$NO_2 + \text{uv radiation} + VOC + O_2 \rightarrow NO + O_3 + PAN + \text{aldehydes}$$

These oxidizing compounds are much longer lived and can do substantial damage. The aldehydes produced (formaldehyde, acetaldehyde, benzaldehyde) are poisons and sometimes carcinogens.

Thus a combination of NO_x, hydrocarbons, and sunlight is needed for photochemical smog to form. In addition, climatological or geographical conditions must be such that the reactive compounds and products accumulate near Earth's surface, rather than dissipating into higher reaches of the atmosphere where they would become diluted and do little harm. These conditions often prevail in valleys or other areas where little air mixing occurs, particularly under thermal inversion conditions. Cities like Los Angeles, Denver, and Phoenix are often plagued by smog because they each have mountains that block prevailing winds from mixing surface air with upper-level air.

Control or elimination of smog would require that one of the essential ingredients be removed. The amount of sunlight reaching Earth's surface in these locations cannot be controlled, and control of NO_x is very difficult and expensive. The NO_x is produced in the automobile's internal combustion engine as a result of the need for high temperatures to ensure proper combustion of the gasoline. Therefore, attempts to control smog have usually centered on reducing hydrocarbon emissions from the automobile and from automobile fueling operations where much gasoline vapor may escape to the atmosphere. Positive crankcase ventilation (PCV) valves and leakproof gasoline filler caps are now required on all automobiles to reduce the loss of hydrocarbons, and engines are being tuned to more efficiently burn the fuel. Catalytic converters, which use a platinum catalyst to complete the combustion of organic compounds to carbon dioxide and water at lower temperatures than would otherwise be required, are now mandatory to remove unburned hydrocarbons from the engine exhaust and to reduce NO_x emissions. Other atmospheric hydrocarbon sources are also being regulated (lawn mowers, wood stoves, refineries, etc.). California has gone so far as to prohibit the use of lighter fluid or charcoal with outdoor grills in some parts of the state because of the harmful hydrocarbons released from the lighter fluid and charcoal before high enough temperatures are reached to consume the volatile hydrocarbons.

Example 3.1. Gasoline vapor recovery systems are now required in all filling stations to collect fumes escaping from automobile fuel tanks during filling operations. As the gasoline is pumped into the tank, it displaces fumes in the empty portion of the tank. Previously, these were emitted into the atmosphere. Determine the amount of gasoline that would enter the atmosphere in the United States annually if refilling operations were uncontrolled, assuming that there are 100 million vehicles filled weekly with 10 gallons of gasoline. Also assume that the vapor temperature averages 20°C, that the displaced gas contains gasoline at a mole fraction of 0.4, and that the gasoline has an average molecular weight of 70 and an average liquid specific gravity of 0.65.

Solution. Calculate the vapor volume per tank:

$$V_v = 10 \text{ gal } (0.0038 \text{ m}^3/\text{gal}) = 0.038 \text{ m}^3 \text{ vapor space/tank/refill}$$

Use the universal gas law to determine the amount of vapor in the tank:

$$PV = nRT$$

$$n = \frac{PV}{RT}$$

where $P = 1$ atm
$R = 8.34 \times 10^{-5}$ m$^3 \cdot$atm/mol$\cdot$K
$T = 20°\text{C} = 293$ K

$$n_v = \frac{(1 \text{ atm})(0.038 \text{ m}^3)}{(8.34 \times 10^{-5} \text{ m}^3\cdot\text{atm/mol}\cdot\text{K})(293 \text{ K})} = 1.56 \text{ mol}$$

Calculate the gasoline mole fraction in the vapor:

$$n_g = (1.56 \text{ mol})(0.4) = 0.62 \text{ mol}$$

Calculate the liquid volume of this vapor:

$$V_1 = \frac{(0.62 \text{ mol})(70 \text{ g/mol})(264.2 \text{ gal/m}^3)}{(0.65 \times 10^3 \text{ kg/m}^3)(1000 \text{ g/kg})}$$

$$= 0.018 \text{ gal/vehicle/refill}$$

Calculate the annual losses of gasoline to the atmosphere in the United States:

$$V_{\text{lost}} = (0.018 \text{ gal/vehicle/refill})(100 \times 10^6 \text{ vehicles})(52 \text{ refills/yr}) = 94.6 \times 10^6 \text{ gal/yr}$$

The amount of gasoline lost per refill is small (only 0.018 gal, or 0.3 oz), but this adds up to a large amount over the entire nation.

The 1990 Clean Air Act required that urban smog be reduced 15 percent by 1996 and 3 percent per year beyond then until federal air quality standards for hydrocarbons, nitrogen dioxide, and ozone are met. In addition, passenger cars must be designed to emit 40 percent less hydrocarbons and 60 percent less nitrogen oxides by the year 2003. Cleaner-burning fuels are required in many polluted regions of the country, and special low-emission vehicles are being mandated for some fleets. These requirements are having a major impact on many industries, including the automobile and oil refining industries, but also on all industries requiring the use of transportation (i.e., essentially all industry).

3.2.3 Acid Rain

Smog problems usually affect a relatively limited region. An industrially based problem that often affects a larger region is that of acid deposition, more commonly referred to as acid rain. *Acid rain* is rain that is more acidic than normal because it contains sulfuric acid (H_2SO_4) or nitric acid (HNO_3) originating from sulfur dioxide and nitrogen oxides in the atmosphere. The acidity can also come from acids associated with particulate matter in the air. These acids may reach materials on Earth by direct deposition (settling, impact, etc.) as well as by the particulates being incorporated in rain droplets. The generic term "acid rain" is usually used to describe all these forms of acid precipitation, even if rain is not involved.

Atmospheric acids can occur naturally (vegetation, lightning, volcanoes), but most derive from coal burning and use of the internal combustion engine, as described earlier. Electric utility plants account for about 70 percent of annual SO_2 emissions and 30 percent of NO_x emissions in the United States. Overall, more than 20 million metric tons of SO_2 and NO_x are emitted into the atmosphere in the United States each year. The SO_2 and NO_x produced can react with water vapor in the air in the presence of oxidizing agents (ozone, hydrogen peroxide, hydroxyl ions) to form sulfuric and nitric acids.

Burning petroleum-based fuels can produce acid rain because most contain small amounts of sulfur, but most acid rain problems derive from the combustion of coal. Some coals, particularly those from West Virginia, Ohio, Kentucky, and parts of Pennsylvania, contain large concentrations of sulfur, often as high as 5 percent. When this coal is burned, the sulfur is emitted from the combustion chamber as sulfur dioxide. By contrast, natural gas usually contains almost no sulfur, and many coals are considered to be low-sulfur coals (less than 1 percent sulfur). The use of coal, particularly for generation of electricity, has greatly increased over the years (see Figure 3.3). This increase has been a significant contributor to acid rain problems. The nation's coal reserves are vast, and coal is a good source of energy. However, the sulfur contained in high-sulfur coal must either be removed from the coal before burning or from the exhaust gases after combustion. The latter is the approach usually taken.

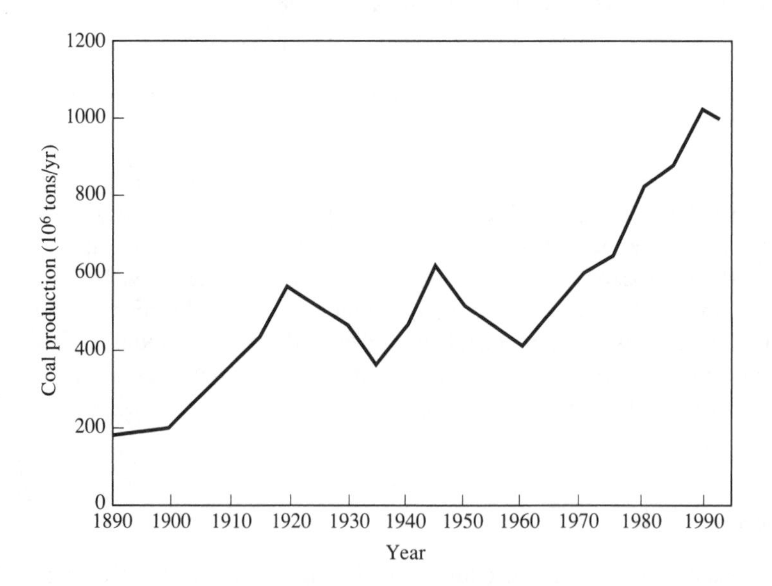

FIGURE 3.3
Coal production in the United States, 1890–1993. (Source: Department of Energy, 1995)

The effect of acid rain has been magnified in recent years by a policy of increasing the heights of stacks at power plants and industries to more effectively dissipate air pollutants at a greater height and thus lessen the impact at the ground level in the immediate area. Consequently, the pollutants are not immediately removed but instead can travel great distances in the upper troposphere winds, impacting previously unaffected areas.

Acid rain problems are not limited to the United States. Even more serious problems can be found in other parts of the world where high-sulfur coal is burned, including Scotland, Finland, and Germany. The vast forests in Germany have been severely affected by acid rain.

Unpolluted rain usually has a pH of about 5.6 due to the dissolution of carbon dioxide from the atmosphere, which reacts to form carbonic acid (H_2CO_3) in the water droplet:

$$CO_2 + H_2O \rightarrow H_2CO_3$$

The very small amount of acidity produced is enough to lower the pH because of the minimal acid buffering capacity present. This acidity can quickly be neutralized on contact with materials, and the carbonate acidity usually has negligible impact on the environment. However, if other acids are added to the air, the impacts can be serious. The average pH in rainfall in the northeastern part of the United States and in parts of Ontario, areas impacted by the burning of high-sulfur coal, is between 4.0 and 4.5. In extreme cases the pH can be much lower. A rainfall pH of 2.1 was recorded in 1969 in New Hampshire, an area that burns very little coal but is affected by airflow patterns from the industrial midwest, where high-sulfur coal is burned. Since pH is expressed on a logarithmic scale, this pH reading indicates that the rain contained about 5000 times more acid than normal clean air.

Acid rain has several negative effects. The acids are very corrosive and can attack many materials, including limestone, marble, and metals. Many of the world's most cherished buildings and monuments (the Taj Mahal near Agra, India, the Colosseum in Rome, the Lincoln Memorial in Washington, D.C.) are slowly dissolving away because of the action of deposited acids. The damage to structures in the United States alone due to acid deposition is estimated at over $4.8 billion per year. Sulfur dioxide emissions can also lead to the formation of sulfate particulates in the atmosphere, which can cause serious visibility reduction. This is a major problem in the Shenandoah Valley, the Great Smoky Mountains, and the Grand Canyon.

Ecosystem effects are potentially even more serious. Forests around the world are dying at an alarming rate, due in part to acid rain. The acids are toxic to many trees and plants, particularly to the tender shoots and roots. Some areas have seen 50 percent or greater mortality of trees, and even higher decreases in survival of new saplings. Forest ecology is very complex and the exact causes of these effects are not completely known, but most researchers implicate acid rain as a major cause. Considerable research is still under way into this topic.

Effects of acid rain on aquatic systems are more established. Acid precipitation and overland runoff into many lakes has caused substantial decreases in lakewater pH. The effects of acid precipitation are compounded in areas without alkaline soils or rock. For example, the northeastern part of the United States has acidic soils and granitic bedrock, which cannot neutralize the acid precipitation. The acidity in over 75 percent of acidic lakes and over 50 percent of acidic streams is caused by acid rain, according to the National Surface Water Survey. These areas do not have the acid-neutralizing capabilities of carbonate or limestone-based soils found in other places, such as the western United States. In parts of eastern Canada and in Sweden, thousands of lakes can no longer support the growth of fish, and other aquatic life (frogs, plants, etc.) is also affected. In addition to the direct effect of the acids on aquatic organisms, the lowered pH can also change the speciation of toxic metals in the water and sediment, making them more available to the organisms. The economic loss due to acidified forests and lakes is enormous.

To control acid rain problems, we obviously need to reduce the amounts of sulfur dioxide and nitrogen oxides entering the atmosphere. This means abandoning the use of high-sulfur coal, which may not make economic sense, or improving our sulfur recovery processes at coal combustion facilities. Also, we must look at new ways to reduce the amount of coal needed by industry by optimizing the efficiency of energy-demanding industrial processes and increasing the use of recycled materials while reducing reliance on energy-intensive processes such as smelters, aluminum mills, and pulp and paper mills. Enhanced industrial pollution prevention activities could go a long way toward minimizing acid rain problems.

Progress in reducing SO_2 and NO_x is already being seen. The Clean Air Act of 1990 established incentives to industry for reductions in emissions of these two gases. As can be seen in Figure 3.4, SO_2 emissions were down markedly by 1994, while NO_x emissions had dropped slightly (U.S. EPA, 1996a).

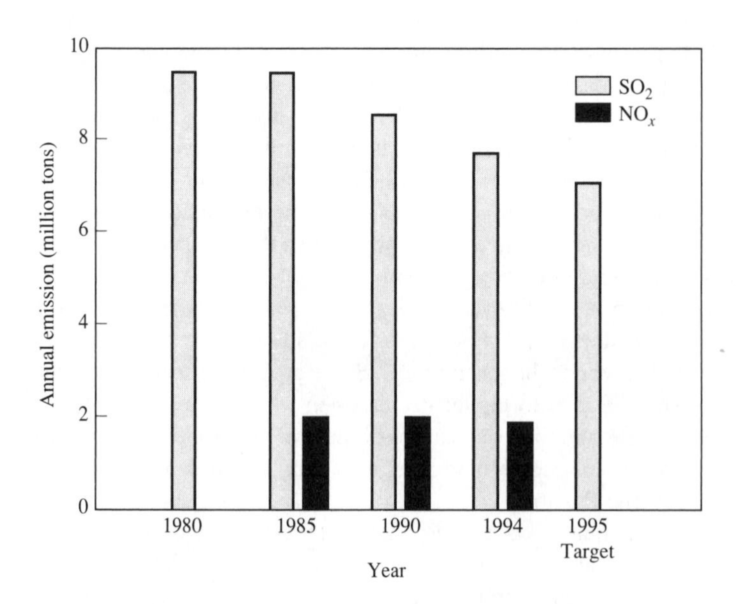

FIGURE 3.4
Sulfur dioxide and nitrogen oxide emissions in the United States, 1980–1994. (Source: U.S. EPA, 1996a)

3.2.4 Global Warming

As we saw in the previous section, air pollution from one region can impact another, otherwise pristine, region. This can also occur on a global scale. Pollutants emitted from a relatively small portion of Earth's surface can accumulate and spread in the upper atmosphere to the extent that weather can be affected globally.

The world's climate is controlled by many factors, as shown schematically in Figure 3.5. One of the major factors is Earth's temperature. During the late 1970s and into the 1980s, scientists became concerned that increases in atmospheric carbon dioxide levels could cause Earth's temperature to increase, which could lead to vast changes in weather patterns. This could result in melting of the ice caps, rising sea levels, coastal flooding, shifting of crop-producing regions, and adverse impacts on populations of humans and other life. Their predictions were that these outcomes could result from only a few degrees increase in global temperatures. Other scientists have discounted these predictions, saying that climatic changes have occurred throughout the history of the planet and that the planet has ways to cope with increasing carbon dioxide production. Not all of the carbon dioxide added to the atmosphere remains there. Some is dissolved in the oceans and some is incorporated in plants and animals as biomass. But most scientists believe the predictions to be true. If correct, the effects on life on Earth would be devastating. Planning needs to be done to combat this increasing temperature trend if it is proven to be correct.

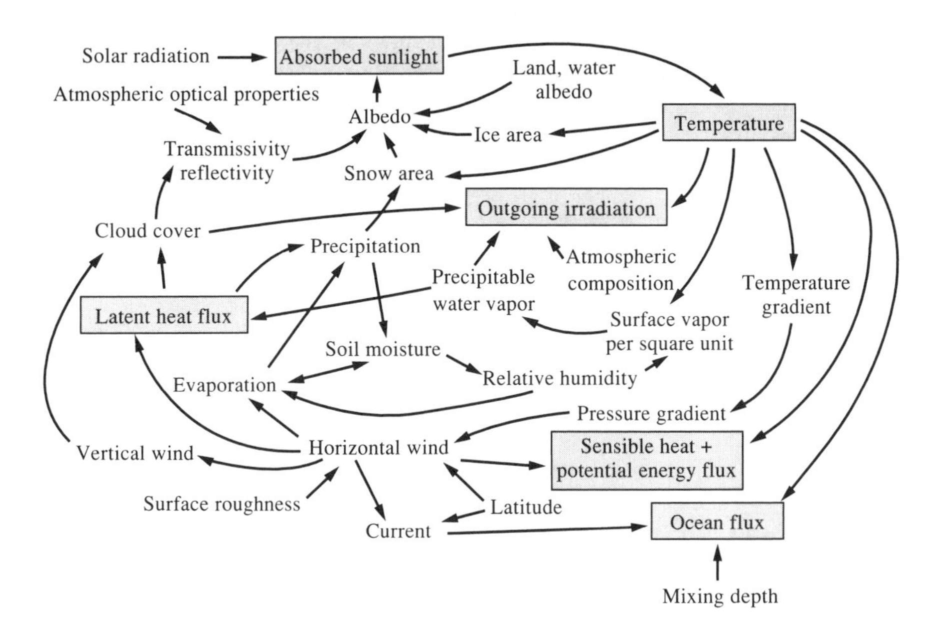

FIGURE 3.5
Schematic illustration of some of the many interactions that control the global climate. (Source: Schneider, 1974)

Why has this concern been raised? After all, the predicted increases in atmospheric carbon dioxide concentrations are quite small. The answer is that even small changes can have a significant impact because of the way in which carbon dioxide functions in the atmosphere.

For the thousands of years of civilization before the Industrial Revolution, human activities affected only local atmospheric conditions. Burning of wood for heat and cooking added carbon dioxide to the atmosphere, but the amounts were minor and of little consequence. Earth's physical, chemical, and biological systems were capable of keeping CO_2 levels in balance, and atmospheric CO_2 concentrations remained stable over tens of thousands of years. This, in turn, kept Earth's climate and the average global temperature in balance. Since the last ice age more than 10,000 years ago, global average variations from the long-term mean temperature have generally been less than 1°C. This long period of stable temperatures and climate has led to the current distributions of populations on Earth and the infrastructure necessary to support them.

With the Industrial Revolution came the rapid increase in the use of fossil fuels for energy and the conversion of forest lands to agriculture and then to urbanized areas to support the rapidly growing populations. The CO_2 generated by burning carbon-based fuels has led to a 25 percent increase in atmospheric CO_2 over the last 200 years. Concentrations have gone from about 280 ppm at the start of the Industrial Revolution to about 350 ppm now. In only 30 years (from 1959 to 1989), the atmospheric CO_2 concentration at Mauna Loa Observatory, Hawaii, increased steadily from 316 ppm to 351 ppm, an increase of 11 percent (see Figure 3.6). This has been compounded by the deforestation, which has reduced the amount of vegetation available to remove CO_2 from the atmosphere and keep it in balance. It is now estimated that atmospheric CO_2 concentrations are increasing by more than 0.5 percent per year. The burning of fossil fuels for heat and energy and the burning of forests to make them ready for agriculture are now generating about 8.5 billion metric tons of CO_2 annually. Projections are that

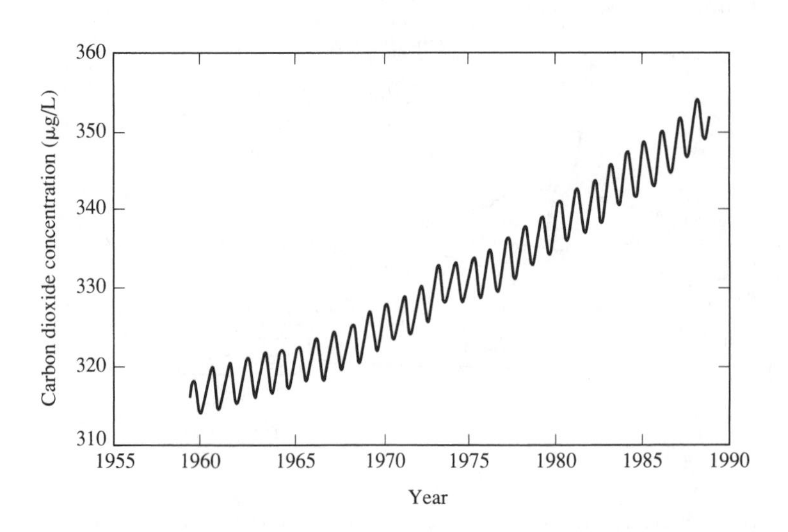

FIGURE 3.6
Change in atmospheric carbon dioxide at Mauna Loa Observatory, Hawaii. (Source: White, 1990)

increased use of coal, oil, and natural gas to support our growing populations and continued deforestation will cause this rate of atmospheric CO_2 increase to accelerate. This could lead to a global temperature increase of 1.5–4.5°C (3–7°F) by the year 2075, resulting in severe climatological changes.

Carbon dioxide is the gas generally associated with global warming because it is being added to the atmosphere in such great quantities, but other gases also play a role. Methane, nitrous oxides, chlorofluorocarbons (CFCs), and other trace gases are also involved. These gases trap heat much better than CO_2 (methane traps 20 to 30 times as much as CO_2, and CFCs trap about 20,000 time as much), but they are generally present in much smaller concentrations. Together, though, their impact is still significant. They may increase the rate of predicted temperature increase due to CO_2 alone by a factor of up to 2.

As would be expected, most of the increase in gases associated with global warming is coming from the highly industrialized nations (United States, 19 percent; the former Soviet Union, 14 percent; Europe, 12 percent; China, 8 percent; and Japan, 4 percent), amounting to over one-half of the CO_2, followed by countries where extensive deforestation by burning is occurring (e.g., Brazil, Indonesia).

GREENHOUSE EFFECT. Why does atmospheric carbon dioxide concentration play such a major role in determining our weather? The answer is that it increases the ability of Earth to retain heat, much as the glass does in a greenhouse. Consequently, the process has been termed the *greenhouse effect.* Greenhouses are much warmer inside than the air is outside because the glass is transparent to light and allows short-wavelength light to pass through and heat the contents of the greenhouse; however, the glass also reflects the longer wavelength heat radiating from inside the greenhouse, preventing it from passing back out (see Figure 3.7). Thus heat becomes trapped in the greenhouse and the temperature increases.

The same happens in the upper atmosphere due to the presence of carbon dioxide and other heat-trapping gases. About half of the sun's energy reaching the upper atmosphere is reflected away or absorbed by the atmosphere and by particulates in the atmosphere, but the other half passes through, warming Earth (see Figure 3.8). Visible light from the sun generally passes through almost undiminished, while the ultraviolet portion of the sun's energy is almost entirely absorbed by ozone in the upper atmosphere. The infrared portion is absorbed mostly by carbon dioxide and water in the troposphere. Carbon dioxide is transparent to visible light from the sun and allows the light to pass through to Earth. Eventually, all of the energy absorbed at Earth's surface is reemitted back into space as longer wave infrared heat. If this was not the case, Earth would act as an energy sink and it would continually heat up.

Carbon dioxide, while transparent to visible light, is very efficient at absorbing the infrared heat radiation emitted by Earth's warm surface. It traps the heat in air close to Earth's surface and can reemit it back toward Earth. The result is a steady increase in the amount of energy reaching Earth's surface and an increase in temperature. The greenhouse effect is essential to life on Earth. Without it, temperatures on Earth would be comparable to those on the moon, which has no atmosphere and therefore no greenhouse effect. It is estimated that without the greenhouse effect, the average surface temperature on Earth would be about 35°C (63°F) colder, precluding most living organisms.

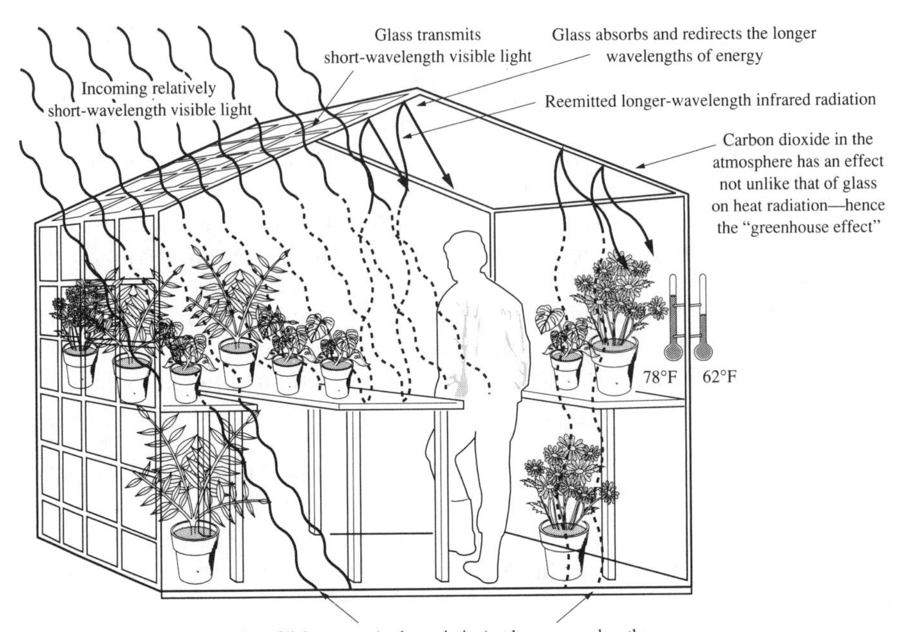

FIGURE 3.7
The greenhouse effect in a greenhouse. (Source: Kupchella and Hyland, 1986)

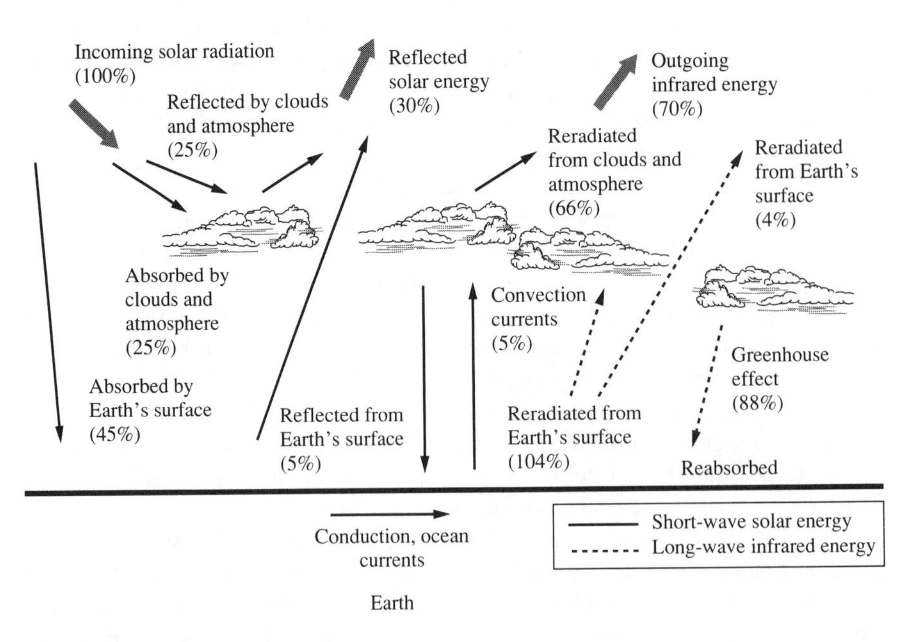

FIGURE 3.8
The greenhouse effect. The numbers show the percentage of incoming solar energy involved in a particular process. Note that Earth reemits more energy than received from the sun because of the greenhouse effect. (Adapted from Cunningham and Saigo, 1995)

TABLE 3.2
Effect of major greenhouse gases on global warming

Gas	Contribution to global warming, %
Carbon dioxide	57%
Chlorofluorocarbons	25
Methane	12
Nitrous oxide	6

OTHER GREENHOUSE GASES. As mentioned, carbon dioxide is often considered to be the main greenhouse gas affecting global climate, but it is not the only greenhouse gas. Methane, CFCs, and nitrous oxide (N_2O) are also greenhouse gases. The emissions of methane to the atmosphere on a global scale have doubled since the Industrial Revolution brought increased cultivation of wetland rice, increased numbers of cattle and sheep, and use of natural gas systems. Chlorofluorocarbons, developed in 1928, rapidly gained favor for use as refrigerants, foaming agents, degreasers, aerosol propellants, and fire retardants. They were considered to be inert and nontoxic, and therefore environmentally safe. CFCs are also used as the propellants in inhalers used to relieve symptoms of asthma; 440 million of these inhalers are used per year. These CFC compounds, commonly known as Freon, can be cheaply produced. CFCs are not naturally occurring, and therefore there were no CFCs in the atmosphere until recently. They are now found in the atmosphere at concentrations of about 1 ppb. Their high infrared energy-trapping capacity is equivalent to that of carbon dioxide, but at a level about 20,000 times greater (20 ppm CO_2 or 0.002 percent). CFCs in the atmosphere are also deleterious because of their effect on ozone, as will be described later. Most CFC use has been banned in the United States and in many other Western countries, but they are still widely used in other parts of the world. Atmospheric nitrous oxide concentrations are also increasing. This derives primarily from combustion and from biodegradation of nitrogen fertilizers in soils. Still, the amount of carbon dioxide in the atmosphere generally swamps the effects of these other gases. Table 3.2 shows the estimated impacts of these gases on global warming (Enger and Smith, 1995).

TECHNOLOGICAL OPTIONS. Can anything be done to stop this seemingly inevitable change in the world's climatic conditions? Short of reducing the world's population, it appears that the only solution is to reduce, or at least maintain, the current levels of carbon dioxide in the atmosphere. There are two ways of doing this—reducing CO_2 emissions into the atmosphere or increasing the removal of atmospheric CO_2 by plants—neither of which will be easy. Any option undertaken will need a lot of time before its impact is seen, because of the long residence times of gases already emitted into the atmosphere. As can be seen in Table 3.3, the turnover rate for CO_2 already in the atmosphere is about 250 years; that for chlorofluorocarbons is on the order of 100 years (U.S. DOE, 1990).

TABLE 3.3
Concentrations and lifetimes of greenhouse gases

Gas	Nonurban tropospheric concentration, ppm	Annual increase in atmospheric concentration, %	Atmospheric lifetime, yr	Primary removal processes
CO_2	351.0	0.4	250	Uptake by the oceans
CH_4	1.7	1	10	Chemical transformation to CO_2 and H_2O
N_2O	0.31	0.3	150	
$CFCl_3$	2.6×10^{-4}	5	70	Photodissociation to chlorine and hydrochloric acid that is subsequently removed by precipitation
CF_2Cl_3	4.4×10^{-5}	5	120	Photodissociation to chlorine and hydrochloric acid that is subsequently removed by precipitation
$C_2Cl_3F_3$	3.2×10^{-5}	10	90	Photodissociation to chlorine and hydrochloric acid that is subsequently removed by precipitation

Source: Adapted from U.S. DOE, 1990.

Global warming is driven primarily by CO_2 produced by fossil fuel energy usage and by industry (see Figure 3.9). Other contributors include agricultural practices (use of nitrogen fertilizers) and deforestation (reduction in CO_2 removal from the atmosphere). Table 3.4 shows the contributions of each toward global warming.

Carbon dioxide emissions can be reduced by reducing total energy consumption, but this will mean a reduction in our standard of living or a major effort to make our energy-consuming processes much more efficient. Another option is to switch from fossil fuels to other energy sources such as solar energy, nuclear energy, or fusion processes. With the exception of nuclear energy, which has its own potential problems, all of these require much more research and development before they can become technologically viable.

Industries seeking to minimize pollution from their processes should seriously examine how they are using energy in their industrial processes and how that use can be reduced. More energy-efficient products, such as automobiles, electrical appliances, motors, lighting, and home insulation, should also be a goal. For example, new automobiles average about 28 miles per gallon of gasoline, but prototype cars have been developed that achieve up to 80 miles per gallon. Use of improved technologies would reduce CO_2 emissions substantially. Improving home insulation would greatly reduce the amount of energy required for home heating, up to 90 percent by some estimates. Commercially available lighting technology can reduce energy requirements by up to 70 percent. Some devices consume energy even when they are not in use. Redesigning

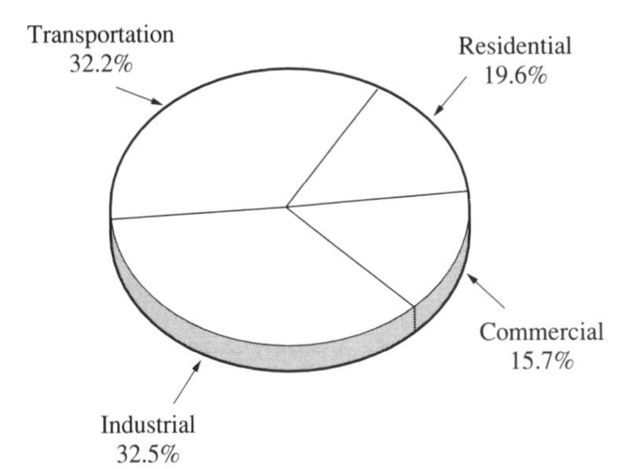

FIGURE 3.9
Sources of carbon dioxide emissions in the United States, 1996.

TABLE 3.4
Contributions toward global warming

Activity	Contribution, %
Energy use	49%
Industrial processes	24
Deforestation	14
Agriculture	13

televisions, VCRs, and computers so that they do not draw power while in the standby mode would save a lot of energy. Switching from coal to natural gas as a fuel would reduce CO_2 emissions because of the lower carbon content of natural gas per unit of energy. Currently available technology could cut energy consumption, and thus CO_2 emissions, substantially and at little increase in cost, if they were more widely adopted.

To stabilize atmospheric CO_2 concentrations at their current levels, we will need to reduce emissions by about 8 to 10 percent. This certainly can be done, and it should be a goal of all industry. Stabilizing CO_2 levels will not prevent future increases, but it will help until we find other sources of inexpensive, nonpolluting fuels.

The second option is to increase the removal of carbon dioxide from the atmosphere through uptake by vegetation. Increasing the world's vegetative biomass could possibly offset the increased CO_2 production through use during photosynthesis. This could be done by planting millions of new trees around the world. There have even been proposals to fertilize the oceans in order to stimulate increased plankton growth and the resultant photosynthesis. At the present time, though, the trend is the opposite to this, with huge portions of Earth's surface being deforested daily. For this option to be useful, these deforestation practices would have to be reversed and new forests would have to be planted. With the rapid increases in population in less developed countries, which demand new agricultural land, this is not likely unless significantly improved sharing of resources between developed and less developed countries occurs. Moreover, even if forest resources can be increased, the beneficial effect may be of only short duration, because eventually this new biomass will die, decay, and release additional CO_2 into the atmosphere. This problem could be overcome by harvesting mature trees and using them as fuel before they die and decay. In this way, the use of other fossil fuels could also be reduced.

That said, reforestation should still be a major part of any global warming policy. It appears, though, that the primary goal should be to reduce CO_2 global emissions by improving technology to decrease energy requirements and by promoting use of nonfossil fuels. This needs to be done at the local level by industry's design engineers and pollution prevention experts, as well as at the national and global levels.

THE KYOTO PROTOCOL. In December 1997, world leaders met in Kyoto, Japan, to develop a strategy for controlling carbon dioxide and other greenhouse gas emissions (methane, nitrous oxide, hydrofluorocarbons, perfluorocarbons, and sulfur hexafluoride) on a global scale. This is not an easy task. Most of the current energy consumption is by the industrialized nations, and developing nations argue that countries currently producing the most CO_2 should be required to produce the largest part of the emissions reductions needed. The developed countries argue that although they are producing the most CO_2, they are doing this using highly energy-efficient manufacturing processes. They produce the most CO_2 because they are producing more goods. As the developing countries expand, they argue, these countries must also participate in the necessary CO_2 emission reductions and be required to move from energy-wasting to energy-conserving processes. Current estimates indicate that greenhouse gas emissions from developing countries will surpass those from industrialized countries before the year 2000.

After much discussion and bartering, an agreement—known as the Kyoto Protocol—was reached. As part of the agreement, an overall goal of reducing greenhouse gases by at least 5 percent below 1990 levels by 2010–2012 was set. Unfortunately, reductions do not need to begin until 2008. The reduction requirement per country varied. The United States agreed to a 7 percent reduction from 1990 emissions, while Japan agreed to a 6 percent reduction; most European countries will be required to achieve 8 percent reductions. Figure 3.10 indicates the impact of these proposed requirements on CO_2 emissions from the United States. Emission levels would need to be decreased to 7 percent below 1990 levels, but this represents nearly about a 30 percent reduction below what is predicted would be the emission levels without the restrictions. Obviously, industry and public utilities are objecting vigorously.

Developing countries objected to any restrictions and were exempted. There is a fear now that companies in industrialized nations will move their operations to these developing countries to eliminate the need to reduce CO_2 emissions. If this occurs, the objective of the Kyoto Protocol will have been defeated.

Because developing countries are not required to reduce their greenhouse gas emissions, and because their industrial development is continuing to expand, it is likely that atmospheric CO_2 levels will continue to rise, even with reductions by industrialized nations (see Figure 3.11).

The provisions in the Kyoto Protocol must be ratified in Congress before they are accepted as law in the United States. Only time will tell if this will occur.

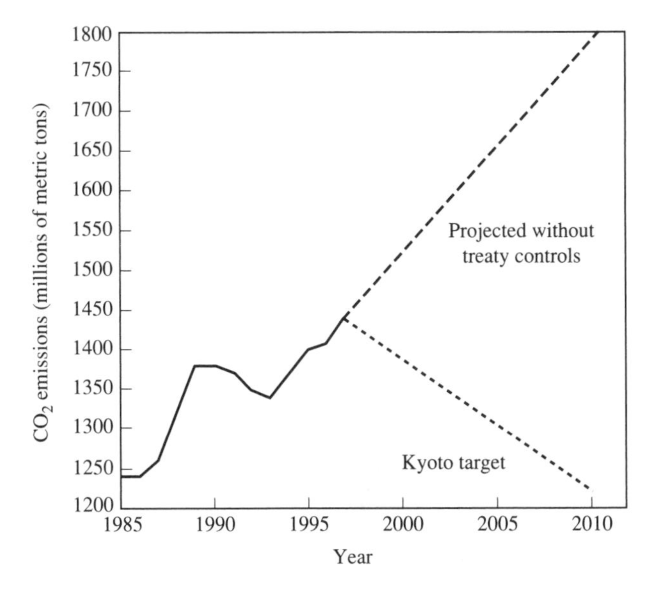

FIGURE 3.10
Projected carbon dioxide emissions in the United States with and without adherence to the Kyoto Protocol.

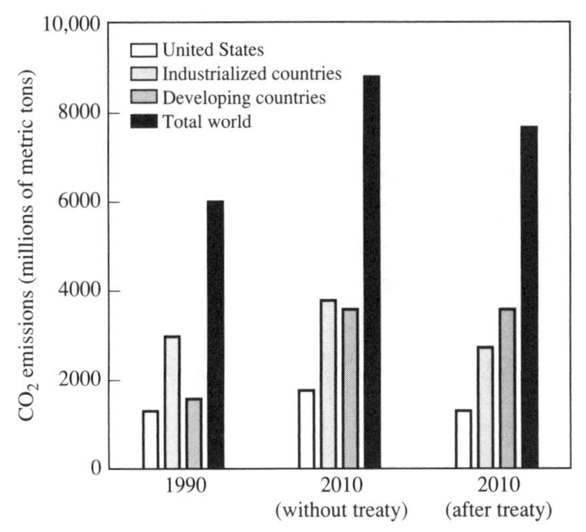

FIGURE 3.11
Projected carbon dioxide emissions by the United States, industrialized nations, developing nations, and the world, with and without the Kyoto Protocol treaty.

3.2.5 Ozone Depletion

In the mid-1970s, scientists became concerned about the possibility that chemicals emitted into the atmosphere could react with ozone in the upper atmosphere and reduce its concentration there. Ozone plays an important role in the stratosphere, absorbing all but a small fraction of the harmful ultraviolet radiation from the sun. As the ozone absorbs ultraviolet light, it is split into an oxygen molecule and an oxygen radical:

$$O_3 + \text{uv radiation} \rightarrow O_2 + O\cdot$$

Recombination of oxygen radicals and oxygen molecules allows ozone to be reformed, available to absorb more uv radiation:

$$O_2 + O\cdot \rightarrow O_3$$

The ozone prevents the uv radiation (radiation with a wavelength less than 340 nm) from reaching Earth's surface, where it could cause skin cancer, mutations, or cataracts in humans or other living organisms. Reduction in the amount of ozone present in the upper atmosphere would result in more uv radiation reaching Earth's surface. It is estimated that a 1 percent loss of ozone can result in a 2 percent increase in the amount of uv radiation reaching Earth, which could cause about 1 million extra human cancers per year worldwide if no protective measures are taken (Cunningham and Saigo, 1995). In fact, increased cancers and cataracts in animals living in ozone-depleted areas have been documented.

In 1985, researchers at the South Pole discovered that ozone levels in the stratosphere over the pole drops precipitously for a few months each year when air over the pole becomes stagnant. The ozone level appears to return to normal months later when air in the stratosphere over the pole mixes with other global air. Ozone becomes nearly nonexistent in the atmosphere between 9 and 23 km (5.4–14 mi) above Earth during these periods. The size of this "ozone hole" has grown larger each year, now extending over the southern tip of South America. Currently, the hole covers about 20×10^6 km^2 (7.7×10^6 mi^2), nearly the size of North America. Stratospheric ozone depletion is now being found in other locations around the world as well, with up to 25 percent depletion over parts of northern Canada and Siberia. In the stratosphere over the United States and England, average ozone levels have decreased 12 percent to 14 percent over the last two decades.

Research has strongly implicated chlorofluorocarbons, described in the previous section, as the primary cause of ozone depletion in the stratosphere. At Earth's surface, CFCs are nearly inert, but when impacted by uv radiation in the upper atmosphere, the CFCs release chlorine atoms:

$$CF_2Cl_2 + \text{uv radiation} \rightarrow CF_2Cl + Cl\cdot$$

The chlorine atom then quickly reacts with an ozone molecule, breaking it down to oxygen:

$$Cl\cdot + O_3 \rightarrow ClO + O_2$$

As described above, ozone protects us from uv radiation reaching us by absorbing the radiation and breaking into an oxygen atom and an oxygen radical. The oxygen radical and oxygen molecule can then recombine to form a new ozone molecule, available to absorb more uv radiation. However, if CFCs are present, the ClO produced upon the reaction of the chlorine atom and ozone can react with the oxygen radical, forming more chlorine atoms that can react with more ozone:

$$ClO + O\cdot \rightarrow Cl\cdot + O_2$$

In addition to producing another Cl atom, this reaction removes the oxygen radical and prevents it from recombining with an oxygen atom to form an ozone molecule. As can be seen, the chlorine radical is not consumed through this process and acts essentially as a catalyst. A little chlorine radical can be responsible for the destruction of a lot of ozone. It is estimated that 1 chlorine atom can destroy more than 100,000 ozone molecules before finally being removed from the stratosphere. The net result of the presence of CFCs in the upper atmosphere is a substantial decrease in ozone concentrations.

International concern about this problem has led to legislation designed to eliminate, or at least limit, the production and use of CFCs. The United States banned all nonessential use of CFCs as aerosol propellants in 1978 and planned to phase out all use of CFCs by the year 2000. This date was later moved up to 1996 because of the seriousness of the problem. Many other countries have since followed suit. In 1989, in the *Montreal Protocol on Substances that Deplete the Ozone Layer,* 81 nations agreed to phase out all use of CFCs by the year 2000. Currently, 140 countries are parties in the Montreal Protocol. Use of CFC propellants could be quickly banned because other suitable propellants were available. This was not the case, though, for replacement of its use as a refrigerant. Research was needed to develop new refrigerants for use in the millions of refrigerators, home and automobile air conditioning units, industrial cooling units, and so on. The goal was to replace Freon-type refrigerants in new products with less harmful materials by 1996. At the present time, most refrigeration systems are being manufactured using an alternative refrigerant called hydrochlorofluorocarbons (HCFCs). These compounds release much less chlorine per molecule. Many older air conditioning units are designed to use CFCs, though, and require expensive retrofitting to be able to use the new coolants. Consequently, an extensive black market in CFCs has developed in the United States. There is a critical need for industry to come up with inexpensive nonhalogenated coolants so that the ozone depletion problem can be eliminated.

An effort is also under way to reduce the use of chlorine in other products and industrial processes as well, because this chlorine could also get into the atmosphere. Included is use of chlorine for water disinfection. Elimination of use of these compounds poses a significant challenge to engineers to come up with safe alternatives that can do as good a job.

Indications are that the CFC ban is already having a positive effect, much more quickly than was previously believed possible. Concentrations of CFCs in the upper atmosphere are already coming down, and the size of the Antarctic ozone hole in 1996 was slightly smaller than in previous years. If all countries abide by their agreements, the ozone loss should peak between 2001 and 2005, and then ozone levels should gradually return to normal (see Figure 3.12).

The ozone depletion problem is a good example of how nations can work together to quickly overcome a serious environmental concern. It is also a good example of why engineers must thoroughly study the materials they are producing or using to ensure that they will create no adverse environmental effects. Chlorofluorocarbons were used because they were perceived to be innocuous in the immediate environment, but no effort was made to examine their impact in other parts of our environment. Much more research is needed on how to predict potential environmental threats from our industrial and consumer activities.

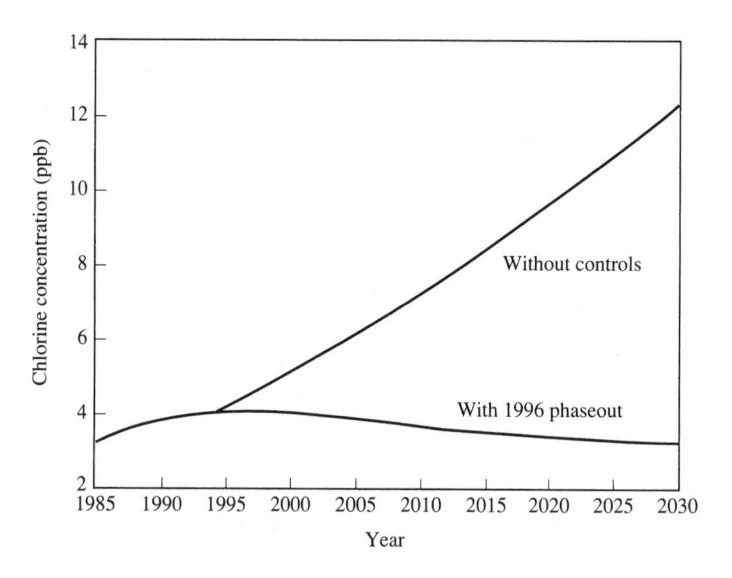

FIGURE 3.12
Impact of the Montreal Protocol on chlorine content of the stratosphere. (Source: U.S. EPA, 1996c)

3.3 SOLID WASTES

Every year, Americans generate about 210 million tons of municipal solid waste. To this figure must be added another 400 million tons per year from industry. These amounts are huge and pose a monumental management problem, but they pale in comparison with the 3 billion tons per year of mining wastes and the 500 million tons of agricultural wastes (see Figure 3.13). Even with current recycling activities, the per capita production of solid wastes continues to climb each year (Figure 3.14). Many believe that the solid waste management problem facing our cities is one of the most critical environmental threats of our time, but little is being done to overcome it. Solutions to the problem are often more political than technological (no one wants a landfill or incinerator in his backyard, even if a safe one could be built), but even the technological hurdles have not been overcome. Unfortunately, the expense of research to solve these problems is too much for individual cities to bear, and the federal government has largely taken a hands-off policy, saying that solid waste management is a local rather than a national problem.

Finding safe and effective ways to manage our solid wastes will always be a challenge to us, but a significant component of any program must be reductions in the amount of solid waste generated. By examining what we throw out in our trash daily, it is easy to see that a substantial reduction in solid waste could be made by scaling down packaging materials used with consumer goods and by using recyclable packaging. Recycling activities have increased in many areas of the country over the past several years, and this has had a significant impact on our waste generation activities. Without these, the rate of increase shown in Figure 3.14 would have been even higher.

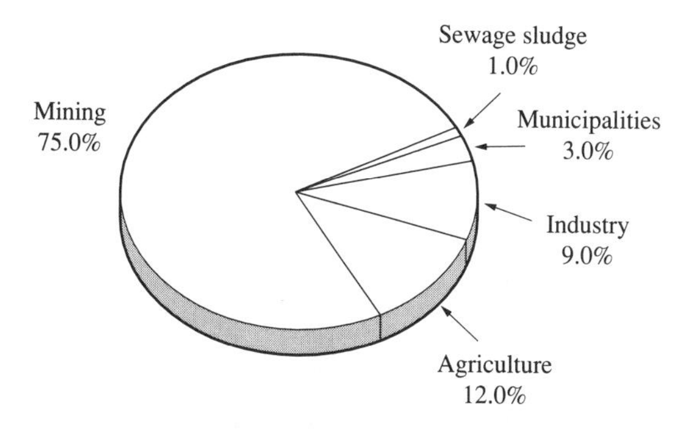

FIGURE 3.13
The major sources of solid wastes in the United States.

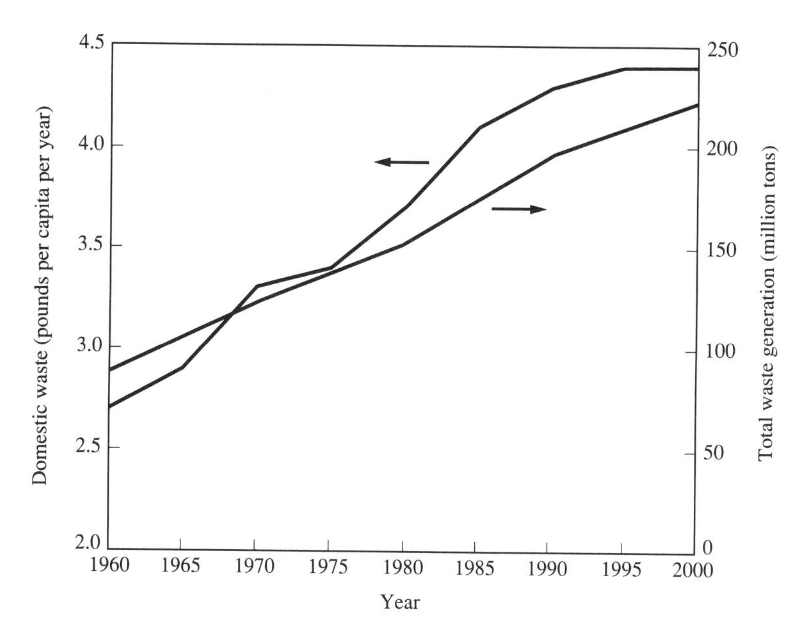

FIGURE 3.14
Total and per capita solid waste production in the United States, 1960–2000. (Adapted from Levy, 1996)

We can still do much more. Solid waste generation rates in Europe and other industrialized areas with a standard of living comparable to ours are lower than in the United States [approximately 4.0 pounds per capita per day (lb/cap · day)], ranging from 2.7 lb/cap · day in Germany to 3.3 lb/cap · day in the Netherlands. However, the recent recycling wave in the United States makes it the current leader in recycling activities (see Figure 3.15).

In this section, we examine the sources and composition of solid wastes and methods for managing them. Solid wastes present a significant opportunity for industrial pollution prevention. Throughout the remainder of the book, we evaluate options available to industry for solid waste reductions. In addition to examining the industrial

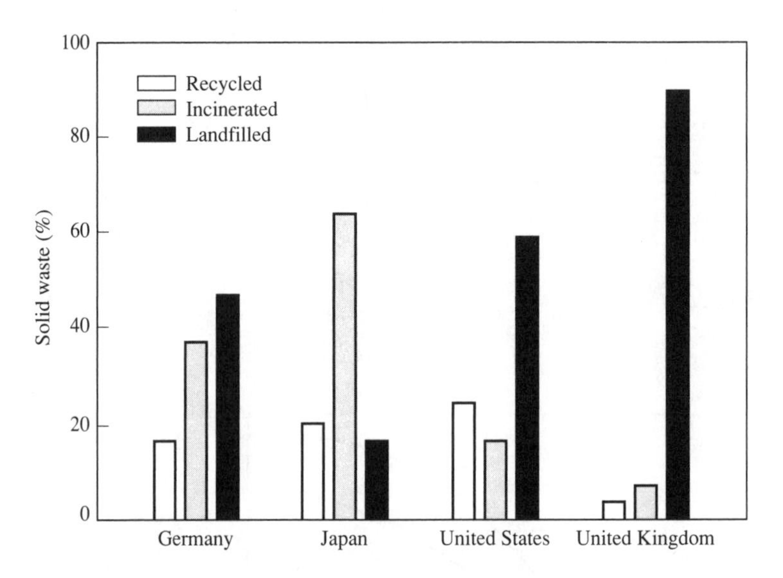

FIGURE 3.15
Percentages of solid wastes incinerated, recycled, and landfilled in Germany, Japan, the United States, and the United Kingdom. (Adapted from Levy, 1996)

workplace setting, we consider how industry can impact solid waste reductions in the home and in the raw materials processing sectors. A subset of solid wastes is hazardous wastes generated by industry. These are discussed in Section 3.4.

3.3.1 Sources and Composition

In general terms, solid waste is any solid material that is disposed of because it has no further use to society in its present form. In more specific terms, the U.S. EPA has defined a solid waste as "any discarded material, including solid, liquid, semi-solid, or contained gaseous material resulting from industrial, commercial, mining and agricultural operations, and from community activities." This is a very general definition that includes basically anything that is not directly disposed of into a waterway or into air. Even gases contained within an aerosol can or an acetylene tank are considered solid wastes if they are deposited in a landfill.

For ease of discussion, solid wastes can be divided by source into five categories: municipal solid wastes, industrial solid wastes, sewage sludges, agricultural wastes, and mining wastes.

MUNICIPAL SOLID WASTES. Municipal solid wastes consist of residential wastes (garbage, trash, yard wastes, ashes from fireplaces or heating units, and bulky wastes such as furniture or appliances), commercial and institutional wastes (construction and demolition wastes and special wastes such as waste from hospitals), street refuse, dead animals, abandoned vehicles, and so on.

Projections of the actual amounts of municipal solid wastes generated in the United States vary widely, depending on the source of the data and the definitions used. Estimates range from 1.4 kg/cap·day (3 lb/cap·day) to 4.5 kg/cap·day (10 lb/cap·day). Whatever number is used, it is much higher than in most other countries. The estimated solid waste production rate in the United Kingdom and other European countries is about 0.7 kg/cap·day (1.5 lb/cap·day), about half that of the United States. While per capita production rates cannot be accurately measured, indications are that they have increased over the years (see Figure 3.14).

The composition of municipal solid wastes has changed dramatically over the last century. In 1900, municipal refuse consisted primarily of combustion residues from burning coal (see Table 3.5). The remaining material had essentially no value; people recycled what little useful material was present, such as packaging. This changed after the Depression when the general public became more affluent, switched to noncoal fuels for heating, purchased more packaged goods, used more disposable items, and decreased their recycling practices. The result was a significant change in both the types and quantities of materials discarded.

The current composition of urban municipal solid waste is shown in Table 3.6. As can be seen, the largest component by far is paper, much of which could be recycled. Glass, metals, and plastics, again materials that could be recycled, make up about another 25 percent of the waste. Because of the negative impact of yard wastes on landfill capacity and because most of it can be mulched and reused, many states and municipalities have recently banned the disposal of yard wastes in landfills. This should help lower its fractional contribution to the solid waste problem in the future.

The composition of municipal refuse varies from location to location within the United States, due to socioeconomic status, climate, degree of urbanization, and recycling activities. This disparity in solid waste composition is particularly evident when comparing different countries. Table 3.7 compares typical solid waste generated in the United States and in four other countries. As can be seen, paper, plastics, glass, and metal components are considerably higher in industrialized countries than in less industrialized ones, while the solid waste from the latter is composed largely of food

TABLE 3.5
Refuse composition circa 1900

Constituent	Composition, %
Cinders	50
Ash	12
Dirt and dust	20
Paper, straw, vegetable refuse	13
Miscellaneous (tins, 0.7%; metal, 0.2%; bottles, 1.5%)	5

Source: Pfeffer, 1992.

TABLE 3.6
Current composition of typical municipal solid waste

Constituent	Composition, wt %
Paper and paper products	38.9
Yard and garden wastes	14.6
Metal	7.6
Plastic	9.5
Glass	6.3
Wood	7.0
Food wastes (garbage)	6.7
Miscellaneous (rubber, textiles, wood, etc.)	9.4

Source: U.S. EPA, 1996b.

TABLE 3.7
Typical composition of municipal solid waste by country

	Composition, wt %					
Constituent	**U.S.**	**Denmark**	**U.K.**	**Poland**	**China**	**Egypt**
Paper	40	35	36	11	2	13
Yard wastes	16	—	—	—	—	—
Metal	9	4	8	2	1	3
Plastic	9	6	5	2	2	2
Glass	8	8	11	6	1	2
Food wastes	7	35	27	24	36	60
Ashes, dust	—	4	5	45	57	—
Miscellaneous	11	8	8	10	1	20

wastes. This does not necessarily mean that the less affluent countries are throwing away more food; it means that the overall per capita production of solid wastes is much lower than in richer countries and that the food wastes make up a greater percentage of this smaller waste amount.

It is evident that greater emphasis on solid waste recycling could go a long way toward reducing the amount of solid wastes that must be disposed of. A significant public education effort will be required to get the population to participate, and our recycling infrastructure must be greatly improved. We currently lack the facilities to effectively reprocess all of the potentially recyclable materials. This includes both recyclable collection and reprocessing facilities. In addition, the technologies are not currently available to economically produce quality materials from all of these recyclable materials. Discarded plastics can and are being recycled into new products such as patio chairs, but there is a limit to how many plastic patio chairs can be sold. Processes are needed to efficiently separate mixed plastics into like components so that better quality recycled plastics can be made; alternatively, systems are needed to economically break the plastic polymers down to their base monomers for reformulation into new plastics. In addition, industry needs to develop better packaging to reduce the amount of packaging surrounding products. Much of the paper goods, glass, and plastics in our solid waste stream, amounting to almost three-fourths of our solid waste, is used for packaging and is disposed of immediately after delivery of the product to the user. Even small reductions in the amount of packaging used, or conversion to more readily reusable packaging, would go a long way toward reducing our solid waste crisis. It would also help to reduce our global problems associated with resource depletion. Product packaging is discussed in detail in Chapter 9.

Use of more readily recyclable materials in the products would also make recycling more viable when the product has outlived its useful life and is disposed of. This includes adapting methods of construction of the products to make them more easily disassembled into reusable components or materials. This is also discussed in Chapter 9.

If industry does not adapt these philosophies voluntarily, they will probably be mandated by the government in the future. This is now the case in Germany, where industries must take their products back from the consumer after the product's useful life is over.

INDUSTRIAL WASTES. Industrial solid wastes generally arise from two sources: process wastes remaining after manufacturing a product; and commercial/institutional wastes from office activities, cafeterias, laboratories, and the like. In addition, there are wastewater sludges created during wastewater treatment. To a large extent, industrial solid wastes are handled in a fashion similar to that of municipal solid wastes. Minimization of industrial solid waste generation is discussed throughout this book.

SEWAGE SLUDGES. The sludges left over after treating water or wastewater must be handled properly to ensure public safety and to minimize environmental damage. The amount of sludge currently requiring disposal in the United States is enormous. This number will only increase in the future as more of our drinking water and wastewaters are treated. Proper disposal of these wastes is a serious problem, but it is not discussed in this book. Disposal of industrial waste sludges is briefly discussed in Section 3.4.

AGRICULTURAL WASTES. Agricultural wastes, which comprise a significant amount of the total solid waste generated in the country, often pose a significant problem in rural areas. Both crop residues that cannot be returned to the soil (e.g., orchard trimmings, vineyard wastes, discarded crop material, and soil at canneries) and manure from animal feeding facilities have created serious problems in many areas.

MINING WASTES. Mining wastes comprise the largest component of our solid waste stream. Whether surface mining (strip mining) or underground mining (shafts and tunnels) techniques are practiced, a tremendous amount of overburden materials and mine tailings (material left over after extraction of the desired ore) is produced. These waste piles are unsightly and often lead to water pollution problems. Current law in the United States requires that the mined area be reclaimed after mining operations cease, an essential and beneficial—but very costly—practice.

3.3.2 Solid Waste Management

Historically, most solid waste was just dumped in depressions on vacant land or was burned in open barrels. When population densities and per capita generation were low, this was marginally acceptable. However, as societies developed, land-disposed solid waste attracted vermin; posed public health, air pollution, and odor problems; and led to the pollution of our surface and groundwaters. Safer and more efficient disposal practices were needed. This led to the use of sanitary landfills and later to such technologies as mass-burn incineration, composting, and eventually resource recovery. At one time, ocean disposal was also used by many coastal communities, but this practice was banned in the United States in the late 1970s because of the severe pollution problems it created; many other countries have also banned ocean disposal of refuse, but it is still practiced in a large number of countries because it is so inexpensive.

Currently, approximately 61 percent of the 210 million tons of solid wastes generated per year in the United States is disposed of in sanitary landfills, 15 percent is burned in incinerators or refuse-to-energy facilities, and 24 percent is recovered through resource recovery operations.

In addition to the cost and complexity of operations associated with disposal of solid wastes is the collection of those solid wastes for disposal. Particularly in rural areas, where there may be long distances between pickup points, the cost of collection can be very high. Increased recycling and decreased waste generation could greatly reduce these costs.

SANITARY LANDFILLS. Approximately 61 percent of the solid wastes generated in the United States go to sanitary landfills. Sanitary landfills are not the open dumps often used in the past for solid waste disposal. In a sanitary landfill, the waste is deposited in compacted layers and covered with Earth at the end of each day. Modern landfills are usually designed with an impervious liner (clay or synthetic membrane) beneath the fill material and with leachate collection systems to prevent any leachate from entering the subsoil or groundwater. Gas collection systems are also often incorporated into the design to trap odorous and sometimes toxic or explosive gases generated in the decaying refuse. These gases may be treated to remove pollutants or may be burned to produce energy. When the site is finally filled, it is covered with another impervious barrier to prevent rainwater from seeping into the refuse.

A properly designed and sited landfill is safe. However, because of the stigma attached to them in the past, the number of operating landfills in the United States is rapidly declining (see Figure 3.16). It has now become extremely difficult, if not almost impossible, to site a new landfill to replace existing ones as they are closed because they have reached capacity or because they are old and do not meet environmental and public safety requirements. Without new landfill capacity, other alternatives for disposing of our wastes are needed.

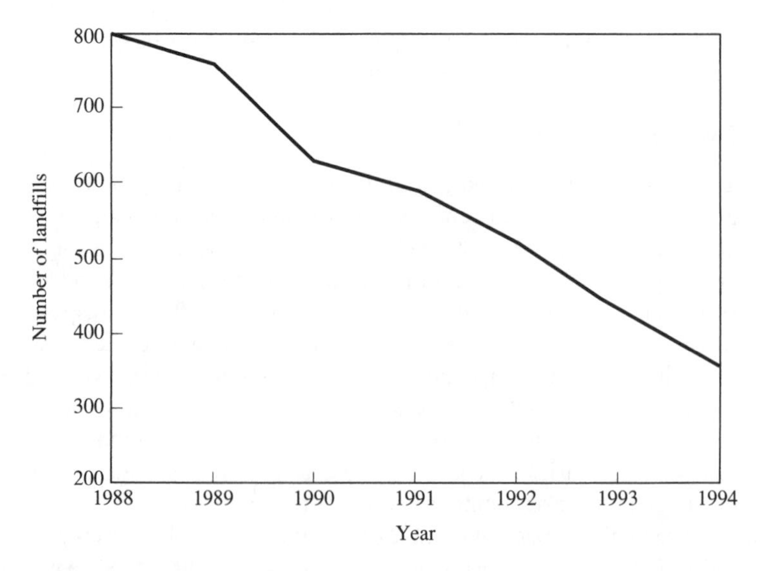

FIGURE 3.16
Number of operating landfills in the United States. (Adapted from Levy, 1996)

INCINERATION. Wastes have been incinerated in one form or another since before recorded time. Wastes were frequently burned in the open or in barrels or other open containers. The first engineered incinerator was built in the United Kingdom in the 1870s. By the early twentieth century, most solid wastes in urban areas were incinerated because incineration required little land area and it was felt that burning the trash would eliminate public health problems. With the introduction of air pollution control laws in the United States in 1970s, many incinerators were phased out because they became uneconomical to operate within acceptable environmental and public health guidelines. There are currently about 110 incinerators in use in the United States, burning about 45,000 tons of refuse per day. This is only about 9 percent of the solid wastes generated in the United States. Incinerators are still widely used, though, in other parts of the world (see Figure 3.17).

By the 1970s, the general public was becoming concerned about the odors, soot, and other air pollutants coming from many incinerators. The Clean Air Act forced the utilization of very expensive pollution control equipment, both on new incinerators and existing ones. In many cases, retrofitting existing incinerators was cost prohibitive and they were shut down. When it was discovered that many incinerators were emitting minute quantities of dioxins and furans, the pressure to place increasingly stringent controls on them increased rapidly. Although it is possible to design a clean burning incinerator with few or no air emissions, siting a new incinerator in an area acceptable to the public is nearly impossible today.

Another problem with incinerators is that they do not totally consume the wastes. Residual ash and unburnable residues constitute 10 percent to 20 percent of the original waste volume. This remaining material must still be landfilled, so incineration does not eliminate the need for a landfill. While incineration does greatly reduce the landfill capac-

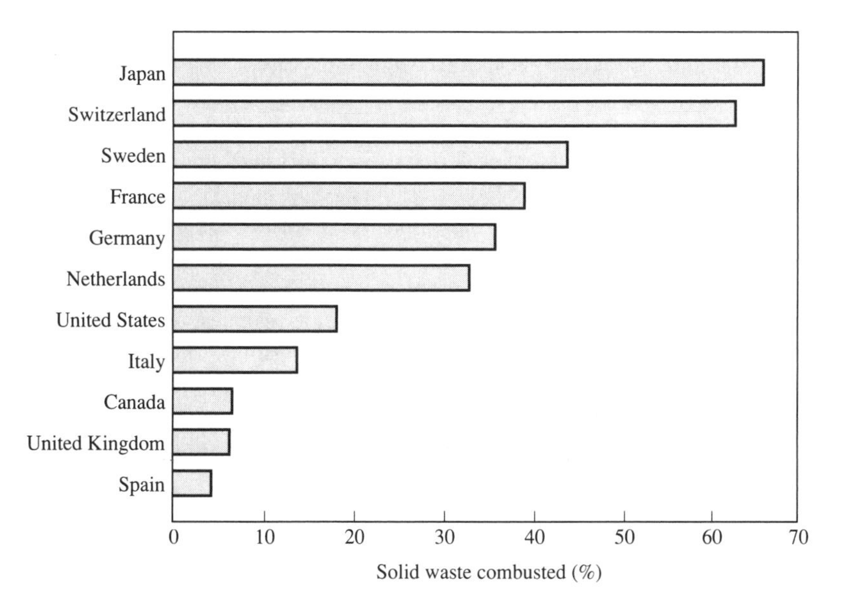

FIGURE 3.17
Combustion of municipal waste in various countries. (Adapted from Levy, 1996)

ity required, the residue contains many contaminants (e.g., metals) in much more concentrated form than the original refuse and may create more difficulties for final disposal.

Because of the stringent requirements placed on incinerator design and operation, the cost of incineration is generally very high. A typical landfill cost in the United States is $30 per metric ton, whereas incineration costs in waste-to-energy facilities are typically $50 per metric ton or more.

One advantage of incinerators is that they can be designed for "cogeneration" of power (also called waste-to-energy) while burning the refuse. Most new incinerators are designed with boilers to heat water and produce steam from the waste heat in the incinerator's exhaust gases. The steam in turn drives a turbine to produce electricity, which is sold to a utility. This greatly helps to offset the cost of incineration.

COMPOSTING. Faced with a lack of landfill capacity, many communities have banned the disposal of yard wastes in existing landfills. Most of these have instituted composting programs for their community's yard waste materials. Some have also investigated composting of garbage or shredded mixed refuse along with the yard waste. Procter & Gamble has funded extensive research that shows that disposed paper diapers are ideal for composting.

Composting is an aerobic process where microorganisms decompose the organic waste materials, usually at elevated temperatures, converting them to carbon dioxide, water, and a humuslike residual material. The finished product often can be used as a soil conditioner that aids the soil in retaining water and improving crop yields. Composting usually works best at a temperature of 50–60°C. It is not economically feasible to heat the large mass of material involved, so self-heating in an insulated system is typically used. The biological action usually provides more than enough heat to maintain the elevated temperatures long enough for complete composting, provided the heat is contained within the composting mass. Insulation can be provided by an insulated reactor or by layers of finished compost placed over the composting material. Mixing of the waste material during composting is also needed in most cases. This can be done by composting in a mixed, aerated reactor; by placing the material in static piles and force aerating it; or by forming the waste into long windrows (long rows of approximately triangular cross section) and mixing the piles periodically to mix and aerate it.

Composting of yard waste has become established in many places across the nation. Composting of garbage or other municipal solid wastes has received mixed reviews, however. It can be done, and it has been done successfully in some places, but the cost is often high, there are often odor problems, and the resulting material may be contaminated with unwanted materials (metals, pesticide residues, or pathogens if high enough temperatures were not maintained during composting). This may make it unacceptable for use as a soil conditioner, thereby negating any useful benefits of composting. These problems can be minimized if proper refuse screening is done initially. More work is needed in this area before municipal waste composting will become viable on a widespread basis.

RESOURCE RECOVERY. We have already discussed energy recovery during the incineration of solid wastes. This is a form of resource recovery. However, there are many materials in the waste that can be recovered directly. Indeed, recovery of materials is becoming mandatory in many places. In California, 50 percent of the waste must be diverted from landfills by the year 2000. In Europe, work is under way to require that 90 percent by weight of packaging materials be recovered, and at least 60 percent must be recycled (Kiely, 1997). Resource recovery may involve fairly simple manual sorting of aluminum cans, paper, and glass, or it may involve very complex automated materials sorting and recovery facilities (see Figure 3.18).

Materials that are commonly recovered include aluminum cans, paper and cardboard, glass, plastics, and miscellaneous materials such as waste oil, car batteries, iron from heavy appliances, and electrical components. Because of separation difficulties due to contamination with other materials or to size restrictions, not all of the potentially recoverable material in refuse is actually recoverable. Table 3.8 presents typical

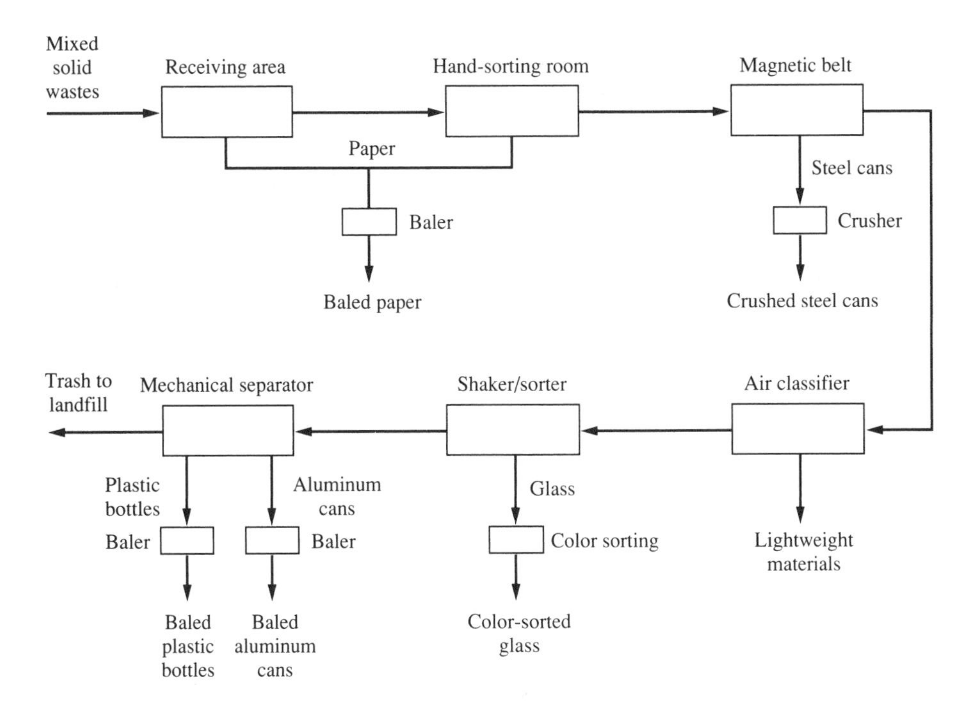

FIGURE 3.18
Schematic of a typical resource recovery facility.

TABLE 3.8
Potential recovery of municipal solid waste (MSW) components

Component	Typical MSW composition, %	Potential recoverability of this component, %	Amount of MSW recoverable, %
Aluminum cans	2%	90%	1.8%
Metal cans	4	80	3.2
Plastic (mixed)	9	50	4.5
Glass (mixed)	9	65	5.9
Paper (mixed)	40	50	20.0
Percentage of total MSW	64		35.4

potential materials recovery rates. As can be seen, efficient recovery of useful solid waste materials could reduce the amount needing disposal by over one-third.

The key to a successful resource recovery program is having an adequate infrastructure for collecting, sorting, transporting, and reusing the recovered materials. In particular, markets for the recovered materials must be there or the materials will just accumulate. This has occurred many times in the past. In the late 1970s there was a demand for wastepaper and paperboard materials. As the price for these materials increased, many communities established paper recycling programs. The large amount of recovered paper that became available then sent the prices for these materials tumbling to the point that transport costs exceeded possible profits from selling the paper. Unsalable recovered paper stocks built up all over the country, forcing many recycling programs to shut down (see Figure 3.19). These price versus supply fluctuations occur regularly now, not only for paper but for all other commodities as well. A greatly expanded and more stable market for recycled materials is needed if resource recovery is to be relied on as a major part of our nation's solid waste management program.

There has been a significant increase in the amount of materials recycled over the last several decades. This can be vividly seen by the data in Table 3.9. Paper recycling has essentially doubled over that period, while glass and metals recycling has gone from essentially nothing to 23.4 percent and 35.9 percent, respectively. Over 17 percent of the total municipal waste amount is now recycled.

Figure 3.20 shows past and predicted recovery rates through the year 2010. Examination of the potential for recycling of materials constitutes a large part of the remainder of this book.

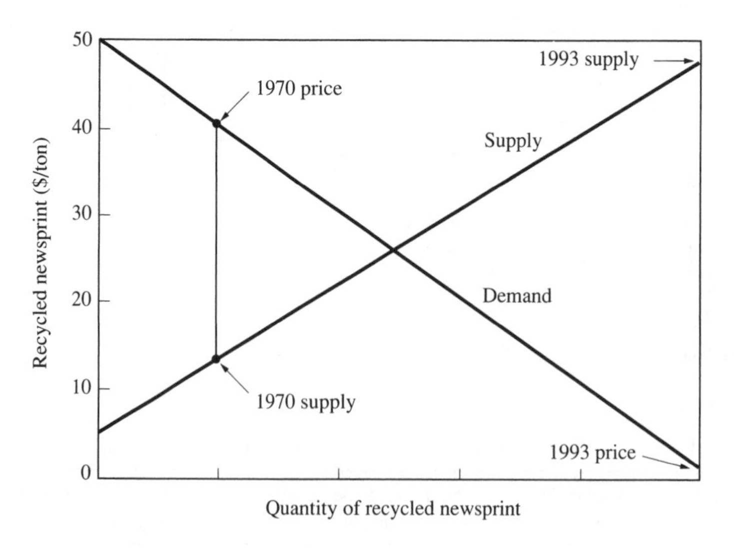

FIGURE 3.19
Impact of newsprint supply on price. (Adapted from Enger and Smith, 1995)

TABLE 3.9
Recovery of municipal solid waste in the United States, 1960–1990

	Amount of material generated, %							
Materials	**1960**	**1965**	**1970**	**1975**	**1980**	**1985**	**1990**	**1994**
Paper and paperboard	18.1	15.0	16.7	19.1	21.8	21.3	28.6	35.3
Glass	1.5	1.1	1.6	3.0	5.3	7.6	19.9	23.4
Metals								
Ferrous	1.0	1.0	0.8	1.6	3.4	3.7	15.4	32.3
Aluminum	Neg.	Neg.	Neg.	9.1	16.7	26.1	38.1	37.6
Other nonferrous	Neg.	60.0	42.9	44.4	45.5	50.0	67.7	66.1
Total metals	1.0	3.6	2.8	4.9	8.3	10.6	23.0	35.9
Plastics	Neg.	Neg.	Neg.	Neg.	Neg.	0.9	2.2	4.7
Rubber and leather	15.0	11.5	9.4	5.1	2.3	5.3	4.4	7.1
Textiles	Neg.	Neg.	Neg.	Neg.	Neg.	Neg.	4.3	11.7
Wood	Neg.	Neg.	Neg.	Neg.	Neg.	Neg.	3.2	9.8
Other materials	Neg.	100	37.5	23.5	17.2	14.7	23.8	20.9
Total materials in products	10.9	10.1	10.2	11.3	13.4	13.8	20.2	25.9
Other wastes								
Food wastes	Neg.	Neg.	Neg.	Neg.	Neg.	Neg.	Neg.	3.4
Yard trimmings	Neg.	Neg.	Neg.	Neg.	Neg.	Neg.	12.0	22.9
Miscellaneous inorganic wastes	Neg.	Neg.	Neg.	Neg.	Neg.	Neg.	Neg.	Neg.
Total other wastes	Neg.	Neg.	Neg.	Neg.	Neg.	Neg.	8.2	15.7
Total MSW recovered	6.7	6.6	7.1	7.7	9.6	10.0	17.1	23.6

Source: Adapted from U.S. EPA, 1996b.

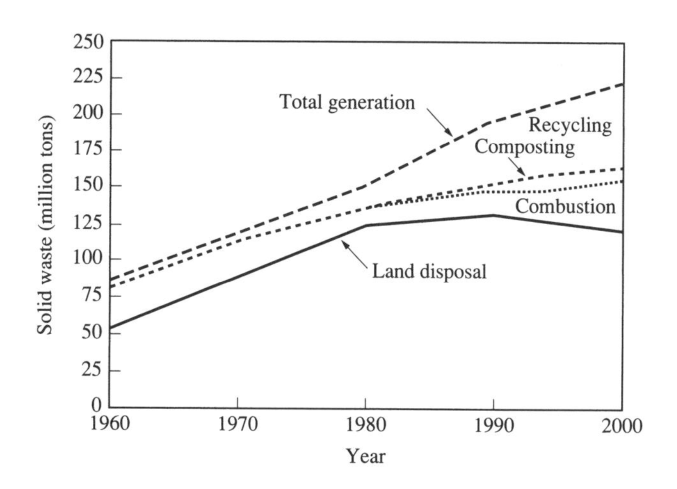

FIGURE 3.20
Municipal solid waste management, 1960–2000. (Adapted from U.S. EPA, 1996b)

3.3.3 Conservation

In the end, the only way to reduce the amount of materials being disposed of is to reduce the amount of solid wastes generated by society. This means that emphasis must be placed on reducing packaging, substituting new materials for old, and making products more easily recyclable. This is not current practice with most industries, but all industry must adopt this general philosophy if we are to preserve our precious resources and reduce the solid waste disposal crisis to manageable proportions.

3.4 HAZARDOUS WASTES

Hazardous wastes are legally a subset of solid wastes, even though they may be gaseous or liquid. A hazardous waste is one that is harmful to the environment or public health. According to the Resource Conservation and Recovery Act of 1976 (RCRA), a hazardous waste is a waste that can "cause or significantly contribute to an increase in mortality or an increase in serious irreversible, or incapacitating reversible, illness; or pose a substantial present or potential future hazard to human health or the environment when improperly treated, stored, transported or disposed of, or otherwise mismanaged." In abbreviated form, it is defined as a solid waste that exhibits any one of the four following characteristics:

Ignitability
Reactivity
Corrosivity
Toxicity

The bases for these hazardous waste characteristics were described in Chapter 2; the regulatory framework for them is described in Chapter 4.

A total of about 265 million metric tons of RCRA-defined hazardous wastes are generated annually by more than 20,000 large-quantity industrial generators in the United States. This amount includes only RCRA-defined hazardous wastes and does not include toxic or hazardous wastes regulated under other acts, such as the Clean Water Act or the Clean Air Act; wastes generated by small-quantity generators; household hazardous wastes; or hazardous wastes that are exempt under RCRA. A more likely figure for hazardous waste generation in the United States is about 700 million metric tons per year. This works out to about 1000–2700 kg (2200–5940 lb) of hazardous waste generated per person per year in the United States alone. It must be kept in mind that the amount of hazardous waste generated by industry is only a small fraction of the total industrially generated solid waste. About 11 billion metric tons of nonhazardous industrial waste is also generated each year.

The staggering amount of hazardous waste has created severe environmental stress. The primary industrial producers of hazardous wastes are the chemical industry, the electronics industry, the petroleum refining industry, and the primary metals industry. As can be seen in Table 3.10, the chemical industry generates over half of all RCRA-hazardous wastes in the United States.

TABLE 3.10
Largest industrial producers of hazardous wastes, 1992

Industry group	Waste produced, metric tons/yr	Percentage of total waste produced
Chemical and allied products	345	51%
Electronic and other electrical equipment	62	9
Petroleum refining and other related industries	61	9
Primary metal industries	52	8
Transportation equipment	46	7
All other industries	109	16

Hazardous wastes are not evenly generated across the country. States with the largest numbers of hazardous waste generators are, in order, New Jersey, New York, California, Ohio, and Texas. States producing the largest quantities are Texas, Tennessee, Louisiana, Michigan, and New Jersey (U.S. EPA, 1995).

3.4.1 Superfund Sites

For many years, hazardous waste management by many industries consisted merely of finding an empty site and dumping the wastes there, or of dumping them into a waterway. This sounds extremely negligent, but in many cases the harmful effects of these materials on the environment and human health were not known. It was often assumed that these wastes would diffuse into the environment, become dilute, and cause no problems. Waste dumping was a generally accepted industrial practice. It was not until the 1960s that concerns were expressed about the potential damage these materials were doing. Rachel Carson's *Silent Spring* (1962), while not the first book to indict industrial chemicals for causing environmental harm, became a best-seller and created a public clamor about the uncontrolled release of toxic materials into the environment. This eventually led to the first Earth Day on April 22, 1970, and the birth of the environmental movement. Protecting our natural resources quickly became a high priority of the majority of the public. Congress began debate on a number of new laws to deal with pollution. Unfortunately, the "energy crisis" soon came along and environmental protection took a back seat to new energy production. This was soon to end, though, with the revelations of the effects of hazardous waste dumping at Love Canal, Times Beach, and other locations in the United States and around the world.

LOVE CANAL. Love Canal in Niagara Falls, New York, was originally built in the late nineteenth century by William T. Love for a proposed hydroelectric power project. The project was never completed and the canal stood abandoned for many years. In 1942, the city began dumping municipal refuse in it. Hooker Chemical Co., a local chemical manufacturer, soon got approval to dump chemical wastes (pesticides, herbicides, and other chemicals, some contaminated with dioxins) there. At the time, no one lived nearby and it was believed that these practices would not be detrimental. Hooker Chemical Co. eventually dumped more than 20,000 metric tons of various chemical wastes, including dioxins, into the canal.

In the 1950s, the city of Niagara Falls allowed development of the filled canal and adjacent land for housing, a school, and a playground. Hooker was concerned about the site and would allow the city to acquire it only if the company was absolved of any future liabilities. The canal was sold to the city for one dollar for development. Soon residents began complaining of odors from the site and reported foul liquids oozing to the surface. They also became alarmed about skin and other types of diseases local children were coming down with after playing at the playground. In 1977, the groundwater and pooled surface waters at the site were discovered to be highly contaminated with a wide variety of toxic chemicals. Epidemiological studies on the nearby residents soon found that there was a higher than normal incidence of birth defects, miscarriages, and chronic medical problems. Approximately 10,000 people lived within a mile of the canal. Fear approaching panic set in among the residents. Approximately 950 families were evacuated from a 10-square-block area surrounding the canal. Many of the homes were later demolished. Eventually, in 1988, Occidental Petroleum Co., the parent company of Hooker Chemical, agreed to pay over $250 million in damages to the former residents and to clean up the site. The fact that Hooker's waste disposal practices in the past were in accordance with accepted norms at the time, and that the land had been developed by the city, knowing the wastes were there, was of little import considering the magnitude of the problem.

The landfill contaminants were contained by constructing a barrier drain and a leachate collection system, capping the site with a synthetic membrane, and demolishing the nearby houses and school. Contaminated soil and other materials were incinerated. Currently, much of the site has been cleaned adequately for surviving residences to be reoccupied or for industrial development. Even with the stigma attached to the site, people have begun to move back because of the low cost of the housing.

The Love Canal incident focused the nation's attention on the problems that can accrue from indiscriminate dumping of hazardous wastes. The result was the passage of the Comprehensive Environmental Response, Compensation, and Liability Act (CERCLA)—also known as the Superfund Act—and the control and cleanup of hazardous wastes sites around the country.

TIMES BEACH. In 1983, Times Beach, Missouri, a community of 2000 located on the Meramec River southwest of St. Louis, was found to be heavily contaminated with dioxins, to the extent that it was no longer habitable. Authorities at first could not determine the source of the dioxins, but extensive research revealed that they were contained within waste oils sprayed on the local dirt roads as a dust suppressant. Mixed in with the waste oil were liquid wastes from a factory that had produced hexachlorophene. Dioxins were produced in minute concentrations along with the hexachlorophene and were removed with the process wastes. The U.S. EPA declared the site uninhabitable and relocated all residents at a cost of $30 million.

A cleanup effort costing $118 million ensued. A dioxin incinerator was built near Eureka, Missouri, to burn the 100,000 cubic yards of dioxin-contaminated soil, buildings, and other materials from Times Beach, as well as from 26 other sites in eastern Missouri.

TABLE 3.11
Common contaminants at Superfund sites

Chemical	Average occurrence, %	Average concentration, ppm
Lead	51.4%	309
Cadmium	44.7	2.19
Toluene	44.1	1120
Mercury	29.6	1.38
Benzene	28.5	16.58
Trichlorethylene	27.9	103
Ethylbenzene	26.9	540
Benzo[*a*]anthracene	12.3	148
Bromodichloromethane	7.0	0.02
Polychlorinated biphenyls (PCBs)	3.9	128
Toxaphene	0.6	12.36

OTHER SITES. The U.S. EPA currently estimates that there are about 36,000 abandoned hazardous waste disposal sites in the United States. The General Accounting Office estimates that the true total, including undiscovered sites and those from exempted sources, may be as high as 425,000. The National Priority List (NPL) of hazardous waste sites which have been certified for funding for cleanup from Superfund moneys totals over 1300, with up to 3000 more probably eventually eligible. Estimated costs for these cleanups totals $100 billion to $500 billion.

Cleanup problems are often compounded by the wide variety of materials found at these sites, particularly at sites where different types of waste were disposed. Table 3.11 depicts the most commonly found contaminants at Superfund sites.

3.4.2 Resource Conservation and Recovery Act Sites

Not all hazardous waste problems are associated with abandoned disposal sites. In fact, the problems caused by currently produced industrial process wastes often far exceed those of Superfund wastes, even though they may not receive as much attention. Essentially all industrial processes produce hazardous waste materials. Management of these wastes is dictated by the Resource Conservation and Recovery Act (RCRA). Quantities by industry were presented in Table 3.10.

Many of these hazardous materials pose very difficult treatment and disposal problems. A variety of treatment alternatives are available, but many are quite expensive and some are only marginal in effectiveness for a particular waste. Among those treatment technologies most frequently used are biological treatment, chemical oxidation, activated carbon adsorption, ion exchange, air stripping, and combinations of these. Among the disposal alternatives are landfilling in a chemically secure landfill; disposal into subsurface injection wells; waste immobilization by solidification/stabi-

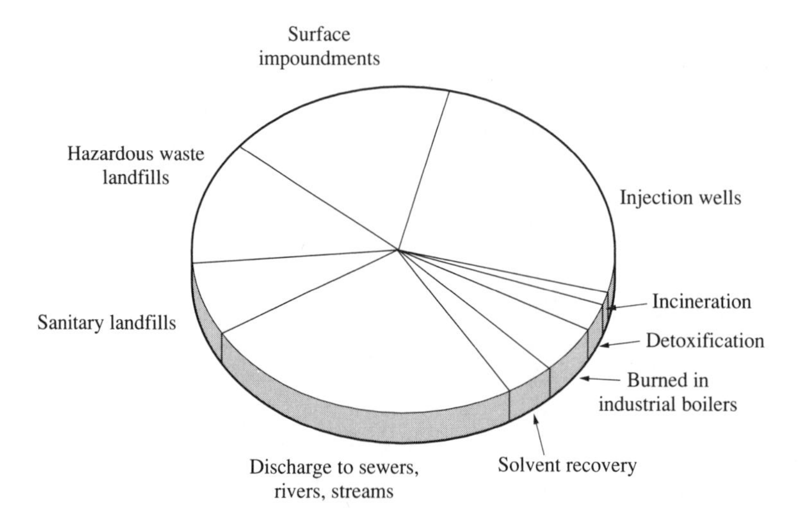

FIGURE 3.21
Hazardous waste management methods in the United States.

lization, followed by burial; and incineration. Incineration will still require ash disposal by some safe means. Figure 3.21 shows a breakdown of currently used hazardous waste disposal practices.

Proper treatment and disposal of these wastes cost industries billions of dollars per year, costs which are usually passed on to the consumer. Those industries that can reduce the amount of hazardous wastes generated, though, do not have to pay these high fees and have a cost savings that means lower costs to the consumer. This results in a greater market share for the industry, and greater profits to the company. Reduction of waste generation at the industrial source should be the aim of all industry.

Much of the remainder of this book deals with how to minimize the generation of hazardous wastes by industry.

3.5 WATER POLLUTION

Industries have always been a major source of water pollution. Many industries dispose of their process wastewaters, cooling waters, spent process chemicals, and other liquid wastes into surface waters, either directly, by piping them to a nearby river, stream, lake, or other water body, or indirectly, by adding them to a municipal sewer which eventually leads to a water body. Disposal of these wastes is largely controlled now, whether it is through a surface water discharge permit [a National Pollution Discharge Elimination System (NPDES) permit] or a municipality's industrial users discharge permit for use of its wastewater treatment plant.

The cost of properly treating these wastes so that they no longer are potentially harmful to the environment or public health is enormous. Consequently, most industries are endeavoring to minimize their use of water and their discharge of wastes to aquatic systems. Industries consume about half of all water used in the United States. Most of this, nearly 90 percent, is used by industry as cooling water. Care must be taken to prevent this water from becoming contaminated. If it does become contaminated from leaks, spills, or commingling of waste streams, the wastewater management problem becomes much greater and the cost for waste treatment rapidly escalates.

The wide range of industrial waste inputs into our nation's waters makes general categorization of quantities or components almost impossible. Many rivers and streams in the country have been severely impacted by industrial discharges. Massive fish kills have occurred, sediments are heavily laden with metals and PCBs, and waters are tainted with toxic organics to the extent that they sometimes cannot be used for drinking, fishing, or even swimming. Significant progress has definitely been made on combating industrial pollution over the past two decades, but much more remains to be done.

It must be stated that nonpoint sources of pollution (e.g., stormwater runoff and agricultural runoff) are also significant sources of stream pollution, requiring much more attention than they are currently receiving. Even with more control of industrial inputs, major additional stream quality improvement will not be realized until nonpoint sources are controlled.

Instead of examining industrially derived water pollution problems in general, we examine how industrial discharges can have a significant effect on local water bodies through two case studies.

3.5.1 Minimata Disease

In the 1950s, about 100 people living in the small village of Minimata, Japan, located on Minimata Bay, were killed and thousands more were seriously disabled by a neurological disease. For several years prior to this, pet cats had succumbed to a similar illness, but for some time no one made a connection. After prolonged study, it was determined that both the residents and their pets were victims of mercury poisoning.

The residents, mainly fishermen and their families, ate fish as a major part of their diet, and also fed fish to their cats. Analysis of the fish tissue showed inordinately high concentrations of mercury. The source of the mercury was traced to the Chisso Chemical Plant, which for years had been discharging mercury-laden wastes into Minimata Bay. The plant had been using elemental mercury as a catalyst in the production of formaldehyde from acetylene. It was assumed that the mercury, being in elemental form, would become trapped in the sediment and be innocuous to humans. The elemental mercury, which is insoluble in water, did precipitate into the sediment, but it was later discovered that some anaerobic benthic microorganisms can convert the metallic mercury into soluble methyl- or dimethylmercury:

$$Hg + 2CH_4 + \text{anaerobic bacteria} \rightarrow (CH_4)_2Hg$$

The methylated mercury is soluble and is easily transportable from the sediment to the water column, where it can be absorbed into fish and shellfish tissue. In fatty tissues it can be concentrated to concentrations up to thousands of times higher than in the surrounding water. When people ate the fish, they were exposed to these high concentrations of mercury. To repair this environmental damage, it was necessary to stop all inputs of mercury into Minimata Bay and to dredge up all of the mercury-contaminated sediment.

This episode showed that industrial contaminants in their discharged form could be converted from relatively innocent chemicals into very harmful ones through processes occurring in the environment. One cannot be concerned only with what is in the industrial effluent; one must also be able to predict what the fate of that compound will be in the environment.

3.5.2 The Kepone Incident

In 1975, environmental officials in Virginia discovered that Kepone, a potent insecticide, was being produced illegally in Hopewell, Virginia, and that many workers were being poisoned. The workers reported tremors, chest pains, and other problems. Allied Chemical Corp. created Kepone in 1949 and manufactured it in Hopewell from 1966 to 1974. This was the only Kepone manufacturing site in the United States. The estimated annual production was 882,000 lb. More than 99 percent of the production was exported. In 1974 Life Science Products Co., run by two former Allied employees, set up a makeshift Kepone production facility and ran it for 16 months before being shut down.

Kepone (chlordecone or decachlorotetracyclodecanone) is a very toxic synthetic chlorinated insecticide, similar to DDT, linked to cancer in laboratory animals and possibly to humans. It can cause tremors, erratic eye movements, and other problems if ingested in sufficient amounts. Most of it is used in Europe and Central America, but it was also used in the United States in ant and roach traps and as a larvicide for flies. It is very long-lived, lasting hundreds of years in the environment. The use of Kepone in the United States was banned by the U.S. EPA in 1978.

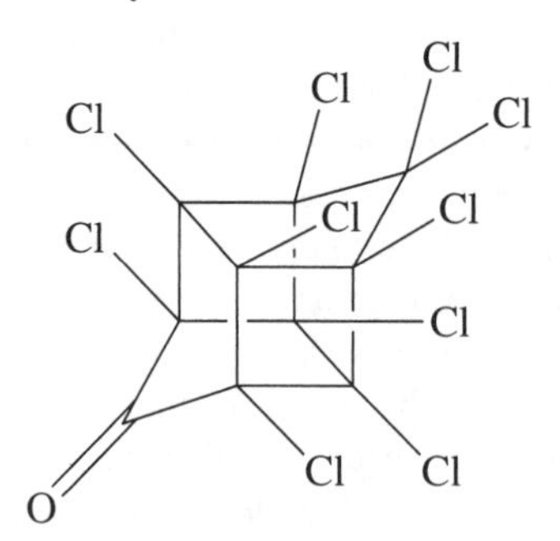

Kepone (chlordecone)

The process wastes were being discharged into the James River. Investigations showed that the river sediments, as well as air, soil, and wellwater near the plant, were laced with Kepone. As much as 100,000 lb of Kepone was estimated to have been discharged into the James River. After finding high levels of Kepone in fish, state officials banned all fishing in the James River and its tributaries from Richmond to Hampton Roads. This ban lasted for 13 years. The cost to the public, including lost revenue to fishermen, is estimated at tens of millions of dollars. Allied was fined $13.2 million for violating pollution laws.

Numerous cleanup studies were conducted, including dredging of the river to remove the Kepone-contaminated sediments, but estimates placed the cost of effective cleanup in excess of $1 billion. Remediation never was performed and the Kepone still resides in the sediments. Kepone is still found in the tissues of fish and in nearby groundwater wells, but levels are declining. It will probably remain buried in the sediment for hundreds of years, unless some catastrophic event such as a hurricane stirs the sediment up and resuspends Kepone-contaminated particulates (Springston, 1996).

3.5.3 Industrial Wastewater Treatment

Many other examples of gross contamination of our waterways by industrial wastes are available. To overcome these problems, many industries successfully treat their wastes before discharge to remove contaminants before they reach the receiving stream. This is often very costly, however, so other industries have opted to discharge their wastes, with or without prior treatment, into municipal wastewater collection facilities. Providing that the Publicly Owned Treatment Works (POTW) has been designed to handle these industrial wastes, this is often a safe, economic solution for industry. It is also often of benefit to the municipality as well because the number of discharge points into the river is reduced and it is easier to manage the waste load. Often the POTW employees are much better trained to treat wastes than are the industrial personnel. Chapter 13 discusses available options for combining industrial pollution prevention with treatment of industrial wastes in POTWs.

Discharge to a POTW isn't always the best solution, however. First, the POTW treatment plant must be designed to have the capacity and the ability to remove the contaminants. Some materials may pass through the POTW unchanged if, for example, they are not biodegradable under the conditions present in the wastewater treatment plant. Other materials may be removed from the wastewater stream but may not be destroyed; they may just change phases. For example, metals may sorb onto suspended solids particles and be removed with the particles as they settle out in a clarifier, forming sludge. This contaminates the sludge and may make it unusable for land application. Volatile organic compounds may be stripped from the wastewater in aeration tanks, thus cleansing the wastewater, but in the air they may be an air pollutant. Thus it is essential to examine the total fate of industrial contaminants as they pass through a POTW before deciding if it is an acceptable treatment solution. It may be necessary for industry to pretreat wastes to remove offending contaminants before discharge of the industrial wastewater to the municipal sewer. This will be expensive to the industry, but it is often less expensive to treat these at the source in a small flow than to try to remove them from the much larger stream after they have mixed with the municipal wastewater and become diluted.

As we shall see later, eliminating the waste altogether by efficient pollution prevention practices and thus not having to treat it at all is usually much less expensive than treating the wastes at the source or at a POTW.

3.6 ENERGY USAGE

3.6.1 Historical Perspective

Any civilized society needs some form of energy supply. Even the cavedwellers burned wood to provide warmth and heat for cooking. Over most of recorded history, the options for obtaining energy were limited. Muscle power (human and, later, domestic animal) has always been available. Wood and later coal and charcoal could be burned, or water and wind power could be used. As humans developed, cultures changed from a hunter-gatherer existence to dependence on agriculture and domesticated animals. This life-style required increased amounts of energy. Still, the amount of energy required per capita was low, and since populations were small, total energy demand was minor. Wood was by far the major source of energy. Until the Industrial Revolution, most energy sources were used for heating and cooking, with only modest amounts used in industry. The rapid increase in industrial production beginning with the Industrial Revolution forced the increased use of these conventional fuels and required the development of new ones. For example, large quantities of iron and steel were required to supply the burgeoning industries, and their production necessitated the switch from burning wood, a resource that was quickly dwindling as forests were cut to supply industry, to burning huge quantities of coal. This meant that many new coal mines had to be developed to supply the steel mills. Figure 3.22 shows the change in wood and coal consumption between 1850, when wood supplied 90 percent of U.S. energy, and 1900, when wood supplied only 20 percent of the energy and coal supplied 70 percent (Enger and Smith, 1995).

The first oil well was drilled in 1859 in Pennsylvania. The high-energy properties of oil led eventually to the replacement of much of the coal demand by oil. When the automobile was invented, the demand for oil skyrocketed. Oil now provides about 40 percent of the energy in the United States. Natural gas, a by-product of oil wells, was available almost since oil was first drilled, and was usually recovered along with the oil, but it was usually considered a nuisance and 90 percent of it was flared off to get rid of it. It was not until the early 1940s, and the enormous demand for energy created by World War II, that gas was really seen as a useful material. Natural gas now provides almost 30 percent of our energy needs (see Figure 3.23). Energy consumption per capita has increased over time from about 2000 calories per capita per day for primitive humans, to about 26,000 cal/cap·day for preindustrial people, to about 70,000 cal/cap·day after the Industrial Revolution. Energy consumption in the United States is now about 230,000 cal/cap·day. In recent years, the repertoire of energy sources has had to increase to meet these new demands and now includes nuclear and solar energy. On the horizon are new technologies such as fusion and hydrogen.

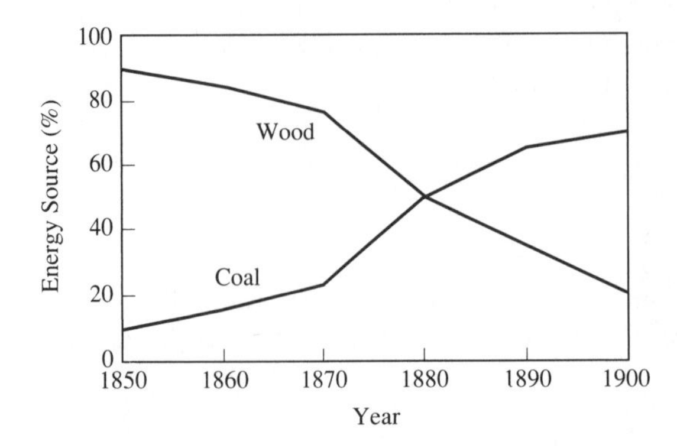

FIGURE 3.22
Change in usage of wood and coal between 1950 and 1900 in the United States.

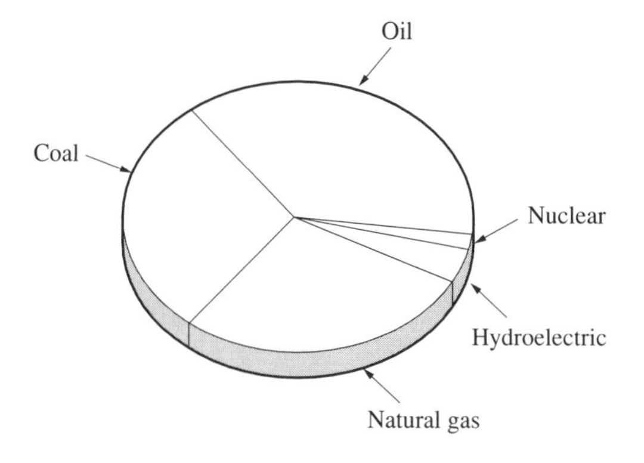

FIGURE 3.23
Worldwide commercial energy consumption, 1989. (Source: World Resources Institute, 1992)

3.6.2 Energy Consumption

Fossil fuels now provide about 95 percent of all the commercial energy in the world (see Figure 3.23). Renewable or sustainable energy resources—solar, biomass, hydroelectric—currently provide only about 2.5 percent. The remaining 2.5 percent comes from nuclear energy. As fossil fuel reserves are used up, a shift toward either nuclear energy or renewable resources will be necessary.

Energy consumption is far from equal around the world. The developed nations, accounting for only 20 percent of the world's population, consume 78 percent of the natural gas, 65 percent of the oil, and 50 percent of the coal produced each year. The United States and Canada alone make up only about 5 percent of the world's population but consume about 25 percent of the available energy. Figure 3.24 indicates that there is a direct relationship between per capita energy consumption and gross national product (GNP) of a country (World Resources Institute, 1990). On average, each person in the United States uses about 300 gigajoules (GJ) of energy per year, equivalent to about 60 barrels of oil (1 GJ $= 947 \times 10^9$ Btu). In poorer countries, such as India, the per capita consumption is less than 1 barrel per year. The relationship is not absolute, however. Some developed countries, such as Denmark and Switzerland, have very effective energy conservation programs and have a low per capita consumption rate relative to their high GNP and high standard of living. They prove that the standard of living does not have to decrease if energy conservation is practiced.

These per capita consumption figures may be misleading, though, because many of the developed countries use this energy to produce much of the food and consumer goods used in the less developed countries. A better indicator of per capita energy consumption would be based on total energy consumed, but these data are not readily available. Interestingly, over the last decade, per capita energy consumption in the member countries of the Organization for Economic Cooperation and Development (OECD), which consists of the most developed nations, has remained constant or has risen only slightly. These countries have seen a shift toward more service-based economies, with energy-intensive industries moving to less developed countries where energy consumption is now rising more rapidly. This can be seen in Figure 3.25, which shows that use of energy in the United States has actually decreased over the last two decades. Total energy use has increased, but only slightly, over this period, due to population growth.

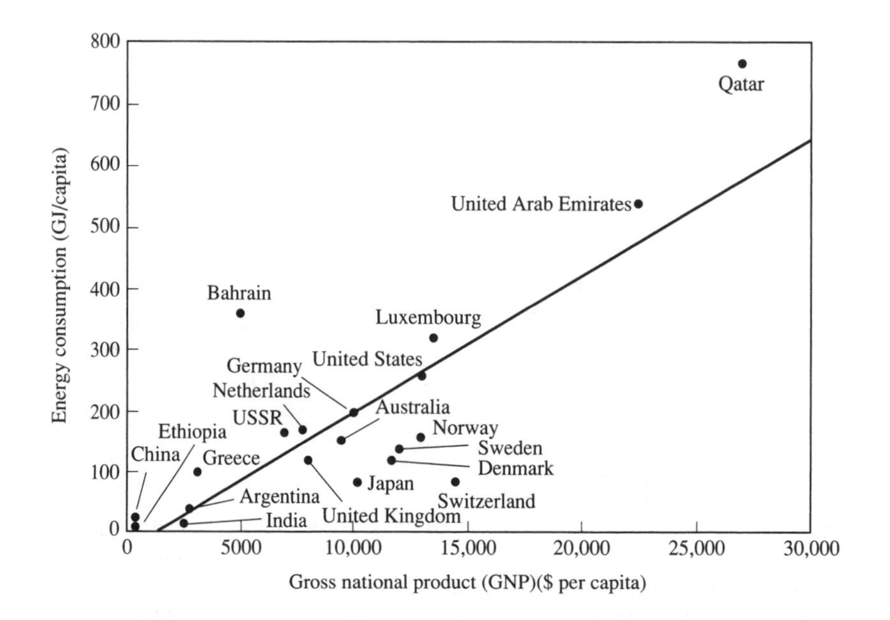

FIGURE 3.24
Per capita energy use and GNP in various countries, 1989. (Source: World Resources Institute, 1990)

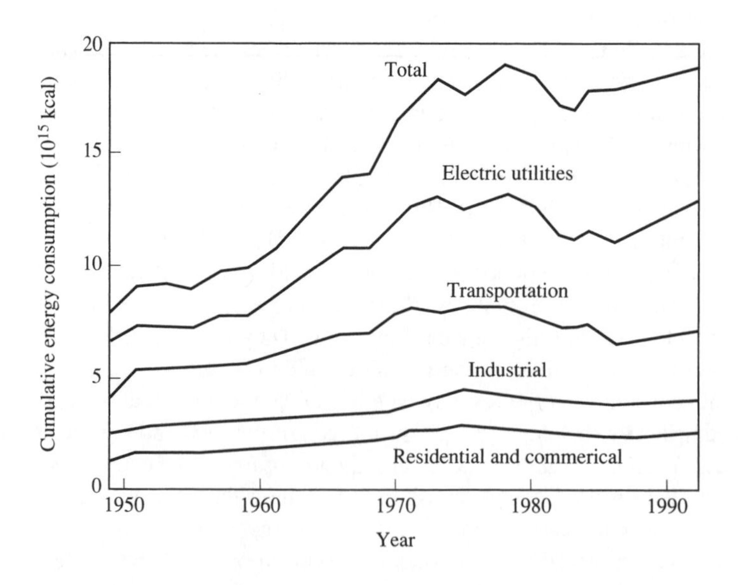

FIGURE 3.25
Changes in U.S. energy consumption. (U.S. Department of Energy, 1995)

In the United States, the largest user of energy (about 37 percent) is industry. This is followed by energy for buildings (heating, ventilating, air conditioning, lighting, hot water heating) at 35 percent, and transportation (mainly automobiles) at 28 percent (see Figures 3.26 and 3.27). The primary metals industry consumes over one-fourth of all energy used in the United States. The chemical industry is the second largest user, with nearly 20 percent. Many initiatives have been undertaken to reduce energy consumption. This includes improving gasoline consumption per mile in automobiles, reducing heating losses in homes and other buildings by using better insulation and more energy-efficient windows, and better lighting. Developing more energy-efficient construction materials, home appliances, automobiles, and so on, has become a major

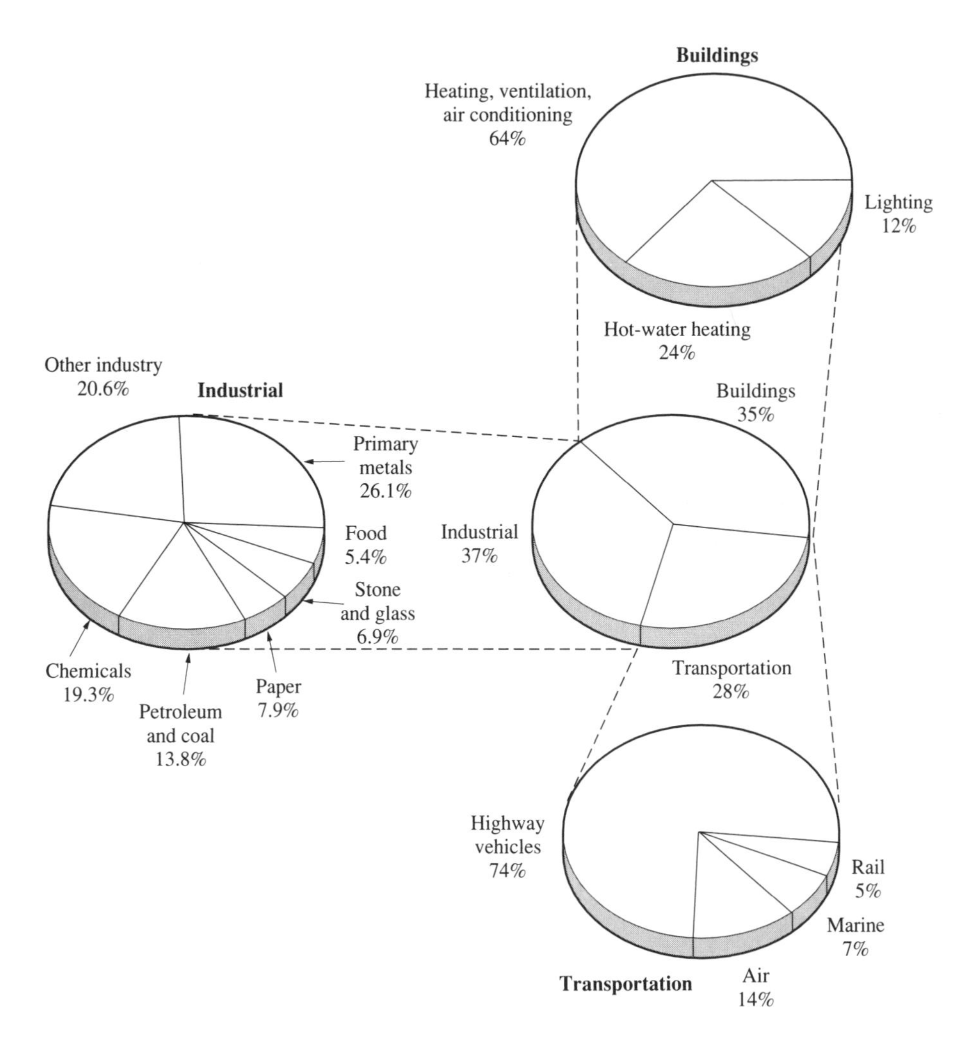

FIGURE 3.26
Energy usage in the United States. (Updated from Cunningham and Saigo, 1995)

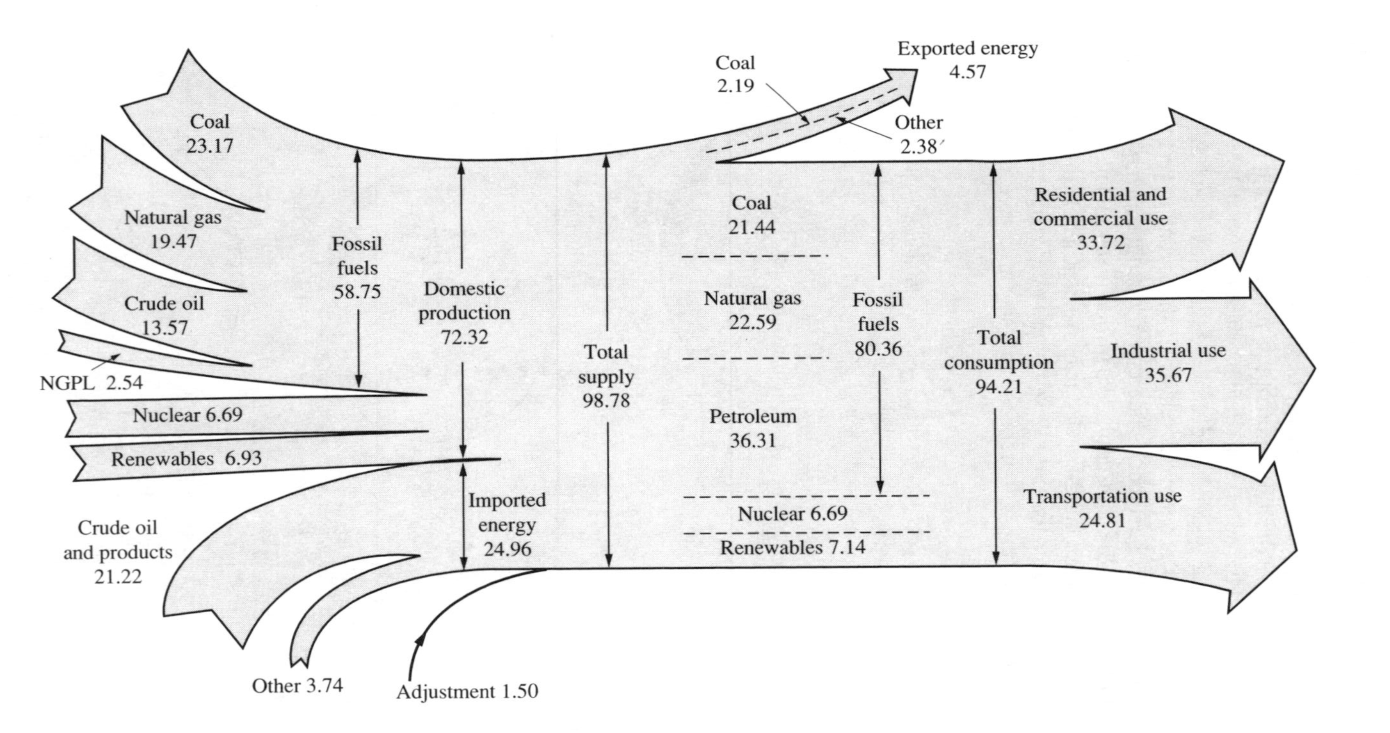

FIGURE 3.27
Energy flow in the United States, 1997 (10^{15} Btu). (Source: U.S. Department of Energy, 1998)

growth industry. Many industries have also improved their energy efficiency, mainly for economic reasons, but much more can be done. The biggest hindrance is the initial capital cost of changing to less energy-demanding processes; once in place, they usually pay for themselves quickly.

3.6.3 Energy Reserves

The amount of fossil fuels (oil, gas, coal, peat, etc.) in the world is enormous, but these fuels are being consumed at an enormous rate. Unfortunately, fossil fuel resources are not equitably distributed around the world, and not all resources are easily available to acquire. Energy *reserves,* those resources that can profitably be extracted using current technology, make up a small fraction of the total resources. As extraction technology improves, more of these resources may become available, but this is a difficult area to predict. It must also be remembered that a considerable amount of energy must be expended to obtain energy materials. The excavation of coal, for example, requires consumption of a considerable amount of energy. Processing the coal to make it suitable for use (cleaning, crushing, possibly removing sulfur) takes more energy. The coal must then be transported to the user, another energy-intensive activity. Thus one should not assume that all of the material available as a reserve is available for use in industry or in the home.

Depletion of our available energy resources is described in Section 3.7.

3.6.4 Fossil Fuels

COAL. Fossil fuels were produced over hundreds of millions of years. In prehistoric times massive swamps covered Earth's surface. Plant material continually died, collected underwater, and slowly decayed, forming *peat.* Over time, many of these peat deposits became very deep. Deposits are sometimes up to 100 m thick and can cover areas of tens of thousands of square kilometers. Layers of sediment often accumulated over them. The weight of the overburden material compressed the spongy peat into a more compact, denser material called *lignite.* Lignite and peat are both used as a low-grade fuel in some places. Occasionally, geological conditions were such that the pressure from the overlying material and the heat from Earth were great enough that further compression took place and the lignite was converted to *bituminous coal.* This took millions of year to occur. Eventually, if the heat and pressure were applied for long enough, the bituminous coal could be converted to *anthracite coal,* a very hard, high-energy material. Nearly all of our coal was laid down during the Carboniferous period (286 million to 360 million years ago) when Earth's climate was warmer and wetter than it is now, and more conducive to massive plant growth. For all practical purposes, we can assume that no new coal deposits will be created on Earth because of the long time required to produce it. Coal must be considered a nonrenewable resource.

Coal provides a major portion of the energy used in the United States and is used to produce over half the electricity. Figure 3.28 shows the sources and uses of coal in the United States in 1995.

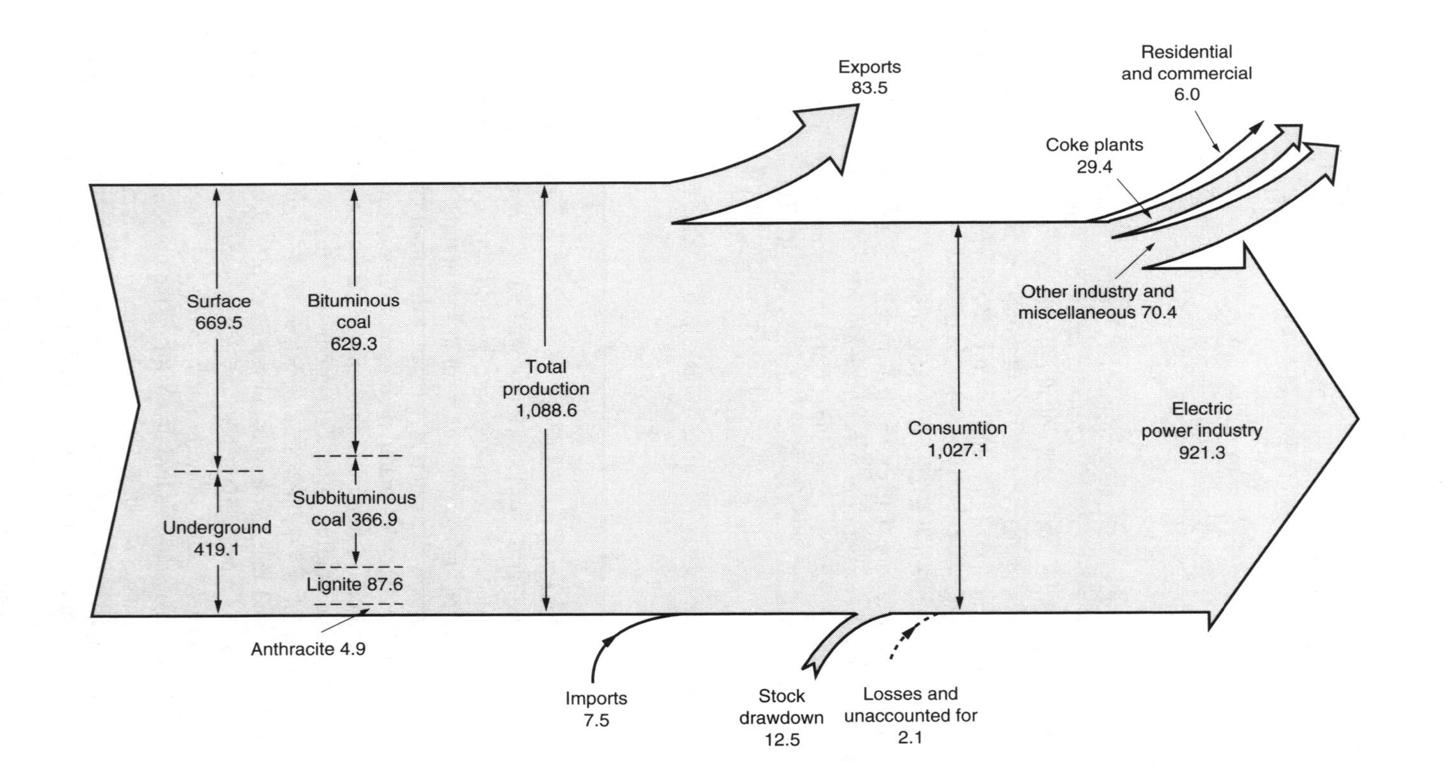

FIGURE 3.28
Coal flow in the United States, 1997 (million tons). (Source: U.S. Department of Energy, 1998)

The total coal resources in the world are estimated to be 10 trillion metric tons. Most is located in North America (one-fourth of the known coal supply in the world), China, and the former Soviet Union countries. Assuming current coal use rates, this amount of coal could last for several thousand years. Unfortunately, much of it is of low quality, contains large amounts of sulfur or other contaminants that can create severe air pollution problems when burned, or is not accessible using current technology. Actual coal reserves will last only about 200 years at present consumption rates. Many of these reserves are not economically recoverable using current mining techniques, so the actual projected life of available coal is much less than this. Thus we are facing a major shortage of coal in the world, possibly within the next century. We cannot count on coal as a major supplier of energy in the future.

Ultimately, coal use in the United States may be reduced because of environmental concerns, rather than reductions in coal supplies. Air pollution (sulfur dioxide, unburned hydrocarbons, particulates, metals, etc.), coupled with the impact of coal burning on global warming, the visual degradation of the environment caused by mining, the impacts on natural waters due to acid mine drainage, and the health effects on the miners who contract black lung disease, are all creating a political environment that is forcing stricter controls on the mining and use of coal.

OIL. Oil, like coal, is derived from organisms living millions of years ago. In most cases, it originated from marine microorganisms that grew in vast seas that covered much of what is now land. As these microorganisms died, they accumulated on the ocean floor, were covered by sediments, were compressed, and with time were converted to oil. The oil is not usually present as a pool as is often depicted; rather, it is contained as droplets within the pores of sand, sandstone, or shale, where it accumulates. Usually, the deposit is a mixture of liquid oil, natural gas, and tars.

The total amount of oil in the world is about 600 million metric tons (about 4 trillion barrels). Only about half of this is easily recoverable. Some is located in inaccessible areas or under deep oceans, and some is not economically recoverable using existing techniques. As oil is pumped from a reserve, the concentration of oil diminishes and recovery of additional oil becomes increasingly difficult and expensive. Typically, only about 30 percent to 40 percent of the oil can be pumped before the cost of extraction becomes prohibitive. Thus the proven oil reserves in the world are only about 150 million metric tons, enough to last only about 50 years at current consumption rates. This could increase as newer technology makes extraction of more oil available or as the depletion of oil reserves makes oil increasingly more expensive; this may decrease the rate of oil consumption and may make recovery possible for some oil that is too expensive to recover now.

Middle Eastern countries (Saudi Arabia, Iraq, Iran, United Arab Emirates, Kuwait) have most of the world's supply of oil (over 61 percent combined). The United States has only 3.5 percent of the world's oil reserves (see Figure 3.29). It has already used about 40 percent of its original recoverable oil reserve. The amount remaining would last only about 10 years at current consumption rates. Consequently, the United States

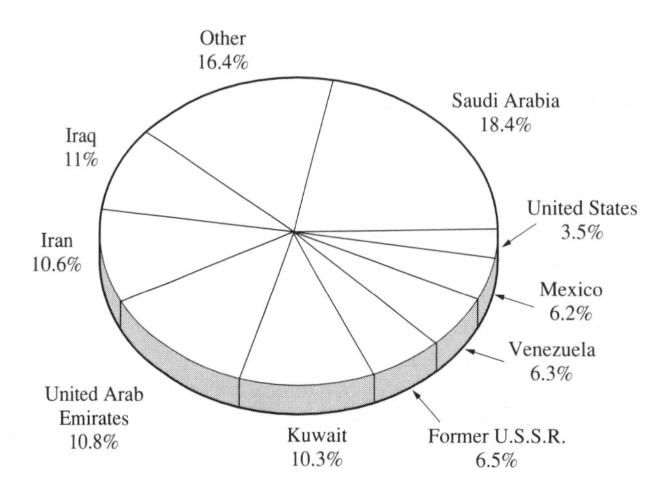

FIGURE 3.29
World proven oil reserves by country, 1991. (Source: World Resources Institute, 1991)

has had to rely on imported oil for many years to meet its great demand. This has led to several political crises in the recent past (e.g., the OPEC crisis in the mid-1970s, the Gulf War in 1991), brought about by our need to maintain the constant flow of oil.

It must be remembered that oil is used for more than just as an energy source. During the refining process to separate the components of oil that go into fuels (gasoline, kerosene, heating oils, diesel fuel, jet fuel, etc.) other hydrocarbons are also recovered. These often become the feedstocks for many industrial processes and synthetic materials, including plastics. As demand for these materials inevitably increases, the demand for oil will also increase. Figure 3.30 shows the sources and uses of petroleum in the United States in 1995.

Other sources of oil may be discovered in the future, and economical ways may be found to recover oil from vast oil shale and tar sand resources that are not currently exploitable. These would increase the relatively short estimated life of our oil reserves given above. However, none of these can be achieved without potentially devastating environmental impacts. In the end, the possibility of further environmental degradation due to oil recovery may actually shorten the lifetime of oil use. Other forms of renewable energy are needed, and this need will encourage much research in the future.

NATURAL GAS. Natural gas deposits were formed in a similar fashion to oil, and they are often found together. During the accumulated organic material transformation processes, some of the organics were converted to lightweight, volatile materials, primarily methane, but also other volatile hydrocarbons such as propane and butane. Natural gas is often recovered along with oil as the oil is pumped. Where a natural gas recovery, distribution, and use infrastucture system has not been developed, as is the case in many developing countries, and even some of the major oil-producing Middle Eastern countries, the natural gas is often considered a nuisance and is flared off, wasting a very valuable resource.

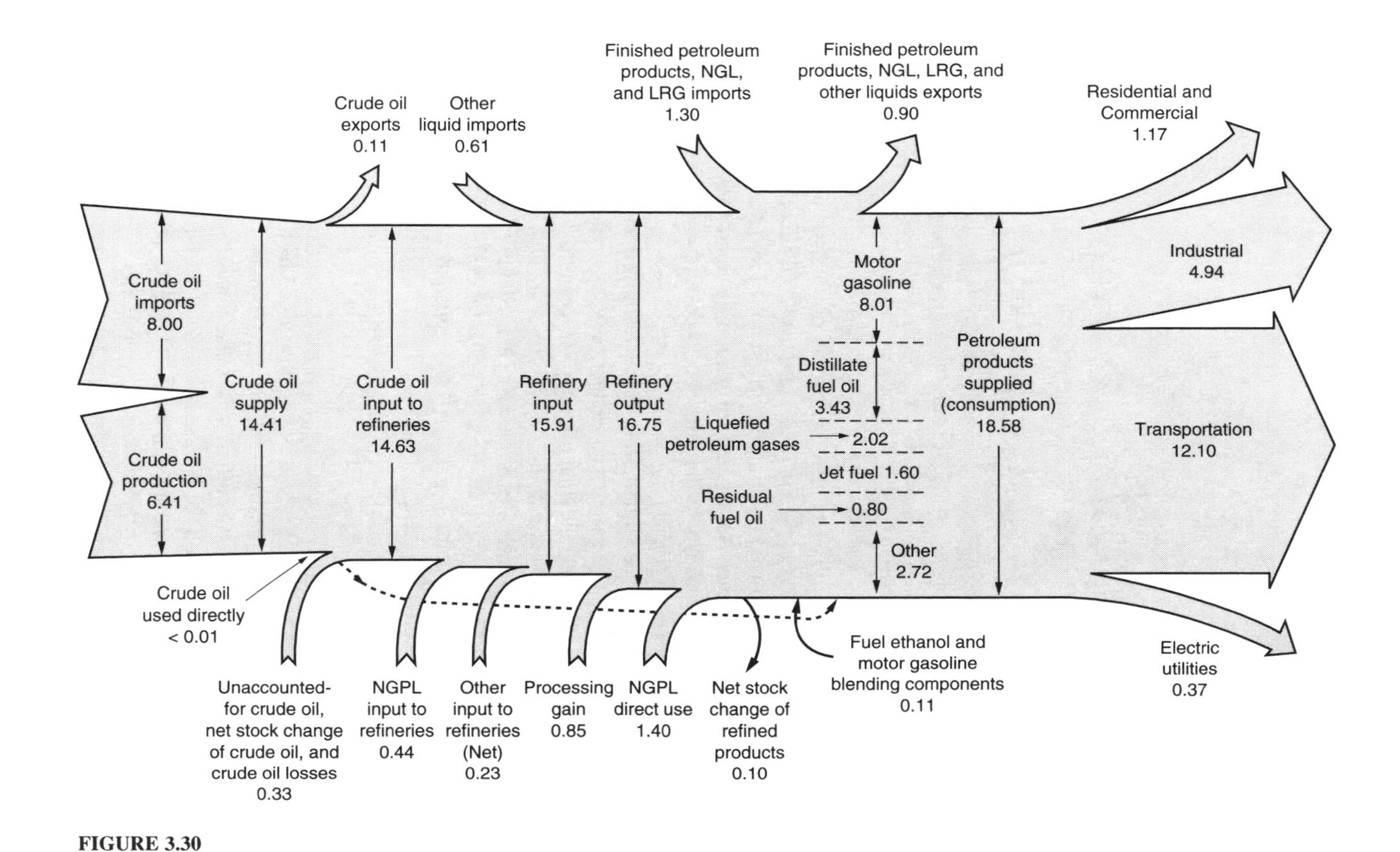

FIGURE 3.30
Petroleum flow in the United States, 1997 (million barrels per day). (Source: U.S. Department of Energy, 1998)

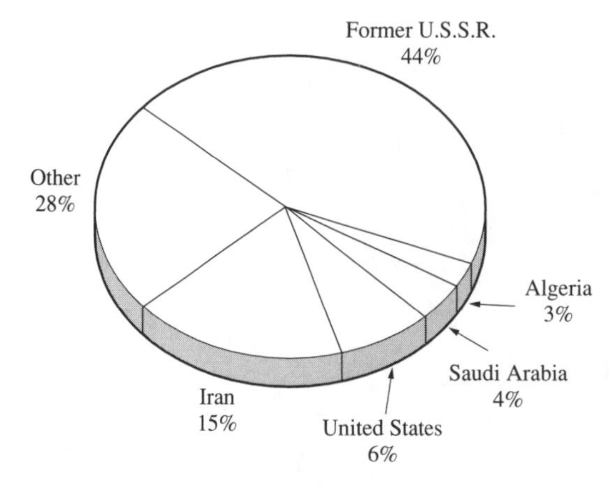

FIGURE 3.31
Percentage of natural gas reserves by country, 1990. (Source: World Resources Institute, 1990)

Most of the known natural gas reserves are in the former Soviet Union countries (about 44 percent) and in Iran. The United States has only about 6 percent of the world's proven natural gas reserves (see Figure 3.31). This is only about a 10-year supply at current consumption rates. The currently proven world reserves of natural gas will last the world about 60 years. As demand for natural gas increases, there will be an increased effort to find undiscovered deposits and to develop the infrastructure to recover more of the known reserves.

Natural gas companies in the United States have developed an effective strategy to increase the use of natural gas, based on its cleaner burning characteristics. Natural gas provides more energy per molecule than oil and produces only about half as much carbon dioxide as does an equivalent amount of coal, so switching from coal to natural gas would help reduce global warming problems. This will only increase the rate of natural gas depletion, though, and will provide an impetus for development of new natural gas fields. Other sources of natural gas, such as municipal solid waste landfills, will also be exploited to a greater extent.

3.6.5 Nuclear Energy

Several decades ago, it was believed that nuclear energy would be the salvation for our increasing energy plight. Indeed, in the mid-1960s it was commonly stated that the cost of producing energy by nuclear power would be so low that it would not be worth charging for it. Estimates were that by the year 2000, essentially all of the world's electricity would be produced using nuclear energy. Unfortunately, these predictions have not come to pass. Many nuclear reactors were built in the 1960s and 1970s and many more were planned, but since then construction of new reactors has essentially ceased (see Figure 3.32). Construction costs ballooned due to increased safety and fail-safe requirements, and restrictions placed on the reactors for environmental and safety reasons made their implementation economically and politically impossible. Of the 167 nuclear reactors in the planning stages in 1974, more than 130 were subsequently canceled. The Three Mile Island incident in Pennsylvania in 1979 and the disaster at Chernobyl in Ukraine in 1986 only served to exacerbate the problems for the nuclear power industry.

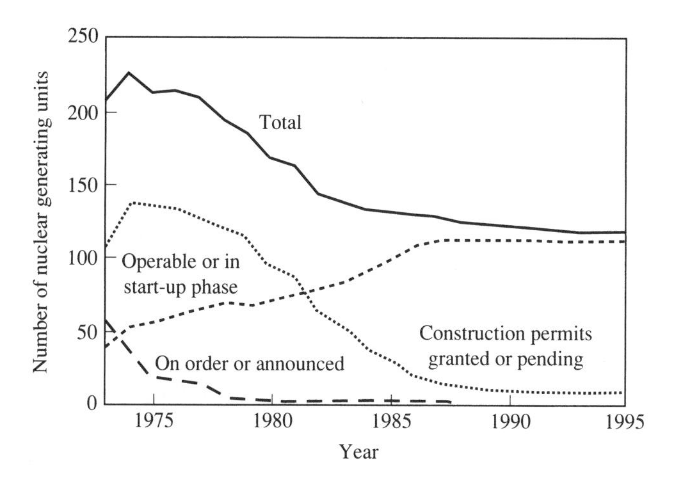

FIGURE 3.32
Nuclear generating units in the United States, 1973–1975. (Source: U.S. Department of Energy, 1995)

There are indications that some nuclear generating facilities may again enter the design stages in the near future, but the golden age of nuclear power is probably over until technology makes them truly safe in the eyes of the public and the nuclear industry finds a way to safely dispose of the resulting radioactive waste materials.

Research has progressed on how to produce safer reactors. Currently, most nuclear reactors are *boiling water reactors* (BWRs). In the BWR, water is used both to moderate the nuclear reaction by absorbing energy and slowing neutrons and to serve as a coolant for the reactor. Steam formed in the reactor is used directly to turn an electricity-generating turbine. The resulting steam contains radioactive materials and must be treated. A second type of nuclear reactor is the *pressurized water reactor* (PWR), in which water is pumped past the nuclear core rods, heated by them to about 317°C, and pressurized to 2235 psi. This water is then pumped to a steam generator where it heats a separate water line, producing steam which turns a turbine generator. This steam is separated from the heated water line and is not radioactive. The pressurized water is recycled to the reactor after giving up its heat to the steam line. Because of the very high pressures involved, the reactor containment building must be very strong to contain any radioactive materials in case of an accident. This has driven their construction costs up considerably. Other nuclear reactor designs have used low-pressure cooling water flowing through the core in combination with graphite to moderate the reaction. These low-pressure reactors were thought to be relatively safe, but the Chernobyl reactor used this design, showing that it is not always safe. Newer, potentially safer designs are now being developed. Even if successful, though, only a major effort will overcome the stigma already attached to nuclear energy facilities.

An even more difficult problem to overcome is management of the radioactive wastes produced from nuclear reactors. Until 1970, most countries dumped their low-level radioactive wastes into the oceans. Although this has now essentially ceased, Russia still disposes of large amounts of radioactive wastes this way. Most low-level nuclear wastes are now stored in tanks or buried underground. Of more concern to the nuclear power industry is what to do with the spent fuel assemblies from the reactors. These high-level wastes are currently being stored underwater at the power plants, because there are as yet no acceptable disposal sites available. A repository for these wastes has been planned for construction deep underground in Yucca Mountain, Nevada, for many years, but its start-up appears to be far in the future, if ever. The need for a safe disposal site is becoming critical. In addition to the spent fuel cores, we will soon need to determine how to safely decommission, dismantle, and dispose of the highly radioactive buildings and reactors from phased-out power plants. It is unlikely that the public will accept the construction of new nuclear power facilities until a solution is found for the disposal of these wastes.

This is unfortunate, because nuclear energy is much less polluting than other forms of energy. It is estimated that the 110 nuclear reactors in current service prevent the burning of 268 million tons of coal, 62 million barrels of oil, and 983 billion cubic feet of natural gas. This has kept 147 million tons of carbon dioxide out of the atmosphere, not to mention tons of sulfur dioxide and oxides of nitrogen, which cause photochemical smog.

3.6.6 Renewable Energy Sources

Coal, oil, and gas provide about 88 percent of the world's commercial energy; nuclear power provides another 5 percent. The remaining 7 percent is provided by renewable energy sources—water power, wind power, solar energy, and biomass burning.

WATER POWER. Water power has been used for thousands of years in the form of water wheels. These have been used to run sawmills and gristmills, provide power for industrial applications, and more recently to produce electricity at hydroelectric facilities. Hydropower facilities are usually found on rivers, but in some cases they can be found on tidal waters, where they rely on the rise and fall of the tides to supply the energy needed to turn a turbine. In some cases, hydroelectric or other water-powered operations can be considered nonpolluting, but in most cases a reservoir is needed in order to provide the head needed to efficiently drive the electrical generator or industrial equipment. These reservoirs can be very environmentally damaging. In addition to destroying the ecosystem in the area affected and disrupting the people living in the area, sedimentation problems behind the dam can seriously affect water quality (e.g., the Aswan Dam project in Egypt) and the changes in flow patterns in the river can affect people both upstream and downstream of the dam.

Construction of new hydroelectric facilities is on the increase around the world. In the United States, however, most of the available water power sites have already been developed. Most of the remaining undeveloped sites are on rivers designated by the Department of the Interior as "Wild and Scenic Rivers," where dams cannot be constructed. Care must be taken to ensure that the added benefits of any newly generated water power outweigh the damages done by construction and operation of the facility.

GEOTHERMAL POWER. Geothermal power has the potential to supply essentially nonpolluting energy to regions of the world where underground temperatures are hot.

Volcanic areas and areas with hot springs and geysers often have hot subterranean masses that can be tapped to provide heat for running steam generators or for heating buildings. Unfortunately, these areas are very limited around the world. The United States has about half of all developed geothermal electrical generating capacity. Others using this technology to produce significant quantities of electricity include the Philippines, Italy, Mexico, Japan, New Zealand, and Iceland. Future significant development of geothermal energy is limited, unless economical methods are developed to tap much deeper pockets of heated Earth.

WIND POWER. Wind power has been used for thousands of years to run mills and pump water. In recent years, it has also been used to run electrical generators. There are now thousands of wind generators in use in the United States, and many more around the world. In some countries, such as the Netherlands and Denmark, wind power is used to supply a significant part of the nation's electricity. Wind power cannot be used everywhere, however. A steady and dependable source of wind is required. Many people feel that row after row of modern wind generators is aesthetically objectionable. Wind power will probably continue to increase in use, but it is unlikely that it will ever produce more than about 5 percent of our energy needs.

SOLAR ENERGY. Some day, the sun may become the primary source of power on Earth. It is available in essentially unlimited supply and is free for the taking. What isn't free at this time is a method for converting solar energy into easily used, storable, and transportable electrical energy. Because of this, less than 1 percent of the world's energy currently comes from the sun. Solar energy already provides heat for many of our homes and swimming pools. Much more work is needed, though, to make it universally usable.

One problem with solar energy is that it is available only part of the time. Methods are needed to use stored energy at night and on heavily overcast days, when solar energy cannot be used directly. For home heating, it is possible to store solar-derived heat in rocks or water for later use, but this is not practical on a larger basis.

Solar energy can be used to generate electricity. In one concept, solar energy can be used to heat oil to very high temperatures; the oil can then be used to produce steam to be used in a conventional electrical generator. Many researchers are examining the use of photovoltaic cells for the production of electricity. Here, solar energy is captured and converted directly to electrical current by separating electrons from their parent atoms and accelerating them across an electrostatic barrier. Photovoltaic cells have been used successfully to power devices as small as pocket calculators, and they have been used to power houses. For large applications, cost of power generation is still prohibitive, but these costs are coming down. At some time in the future, they may become economical to power automobiles or remotely located homes. Solar energy capture efficiencies have already increased from less than 1 percent 25 years ago to about 10 percent today; in the laboratory, efficiencies of 30 percent have been achieved. A greater problem may be developing compact, high-capacity storage batteries for use when solar energy is not available. Until the cost and size of photovoltaic cells and storage batteries come down significantly, solar energy will probably play only a minor role in supplying our energy needs.

BIOMASS. Wood burning was probably humans' first method for obtaining and using energy. In less developed countries, it is often still the principal source of energy. There was renewed interest in the United States in the use of wood for fuel during the energy crisis in the 1970s. Several wood-fired power plants were built and many people installed wood-burning stoves in their homes to provide heat. This interest soon waned, however, as costs for other fuel sources eventually decreased, and the problems associated with wood burning sank in. Maintaining a wood stove is a lot of work, the wood is heavy, and the process is dirty with a lot of smoke, soot and creosote produced. Air pollution is a serious problem, with many carcinogenic compounds produced. The air pollution is not confined to the outside air; indoor air pollution is significant in many homes using wood stoves. Finally, because of the high cost of transporting wood from forests to users, wood burning can be economical only in rural areas with a ready source of timber. Many less developed countries are now suffering energy crises because their land has been stripped bare and all usable wood has already been burned.

Biomass conversion is also being used on a limited basis to provide energy. Organic waste products, such as animal manure and crop residues, can be fermented to produce methane and other organic fuels. This may be successful on a local basis, such as a farm, but its potential for large-scale use is far off.

3.6.7 Electricity

Electricity is a major form of energy used in the United States, as well as in other parts of the world. It is different from all other forms of energy discussed previously, because it must be produced; it is not a primary form of energy. About 35 percent of the energy used in the United States is electricity. Its popularity is largely due to its ease of delivery directly to electrical appliances and machines. Electricity allows the consumption of fuel (coal, natural gas, oil, etc.) at a remote location and the delivery of the resulting energy to another location for use. An enormous infrastructure has been developed for delivery of electrical energy to just about any point in the country.

Industrial use accounts for about 40 percent of the electricity demand in the United States. Figure 3.33 depicts the sources and uses of electricity in the United States.

When looking at such things as the efficiency of an electrical appliance or the efficiency of space heating using electricity (often as much as 90 percent efficient), it is necessary to remember that the energy conversion efficiency to produce that electricity is only about 35 percent. Thus the overall energy efficiency will be much lower than the listed efficiency.

3.6.8 Energy Conservation

By far the best option for reducing our energy requirements is to practice energy conservation. Much of the energy we use is wasted. Most energy conversion processes we use are very inefficient, with much of the energy in the fuel being lost as unusable heat. For example, burning oil to produce electricity yields only about 38 percent as much electrical energy as was the original energy in the oil. Heating a home with a wood stove is only about 40 percent efficient, whereas an energy-saving gas furnace is up to

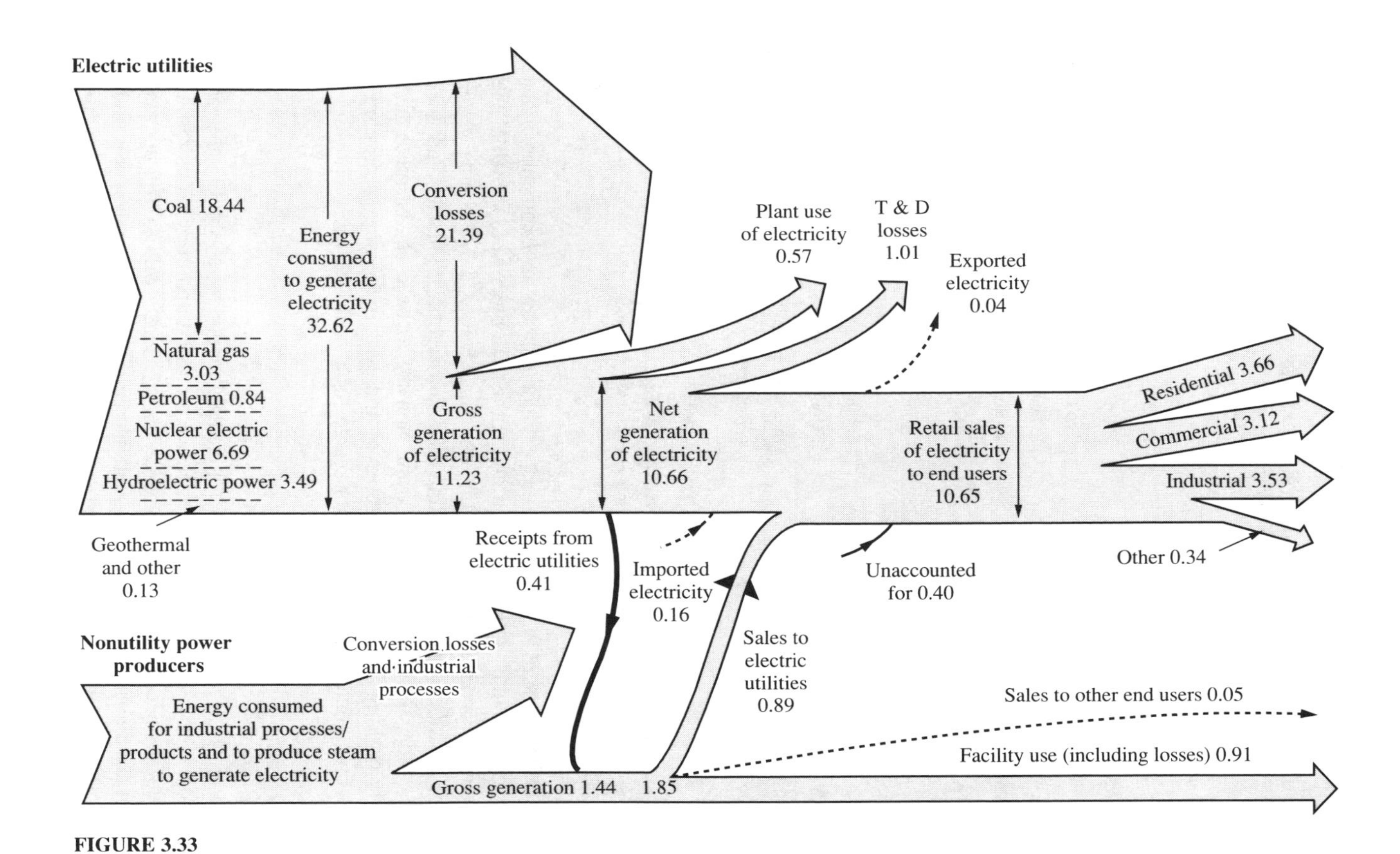

FIGURE 3.33
Electricity flow in the United States, 1997. (Source: U.S. Department of Energy, 1998)

TABLE 3.12
Typical net efficiencies of energy-conversion processes

Process	Yield, %	Process	Yield, %
Electric power plants		**Space heating**	
Hydroelectric	90	Electric resistance	99*
Fuel cell (hydrogen)	80	High-efficiency gas furnace	90
Coal-fired generator	38	Typical gas furnace	70
Oil-burning generator	38	Efficient wood stove	65
Nuclear generator	30	Typical wood stove	40
Photovoltaic cell	10	Open fireplace	−10
Transportation		**Lighting**	
Pipeline (gas)	90	Sodium vapor lamp	60*
Pipeline (liquid)	70	Fluorescent bulb	25*
Diesel electric train	40	Incandescent bulb	5*
Diesel engine automobile	35	Gas flame	1
Gasoline engine automobile	30		
Jet engine plane	10		

* Note that 60–70% of the energy in the original fuel is lost in electric power generation.

90 percent efficient (see Table 3.12). It is evident that by applying more energy-efficient systems to our daily activities and industrial processes, valuable energy resources could be conserved, lengthening the time before they become depleted. Requiring the use of energy-saving building materials, lighting, household appliances, automobiles, and capturing waste heat from industry to be used for space heating would greatly prolong the life of these reserves and give us the time to develop new technologies that would not depend on our scarce fossil fuel reserves.

3.7 RESOURCE DEPLETION

Possibly the most critical long-term environmental, economic, and political problem facing our societies is the rapid rate at which we are using up our precious resources. As these materials become more and more scarce, their costs will go way up, making products that use them much more costly; nations will fight to maintain access to them, as was recently the case in the Gulf War, which was fought primarily to maintain the flow of oil from Kuwait; and alternative, potentially more polluting techniques will be used to try to extract resources that are currently economically unattractive. We may have to switch to new materials that may be more polluting.

The world's resources can be divided into *nonrenewable resources* (or exhaustible resources) and *renewable resources*. Renewable resources include sunlight, wind, and biomass. These have been discussed previously as a potential source of energy for society. Nonrenewable resources include the minerals, fossil fuels, and other materials present in essentially fixed amounts in the environment. There is a limit to how much of these materials we can use before the supply becomes exhausted. We are already approaching the exhaustion point for some of these materials. Only enhanced recycling and more efficient use of these materials will allow them to be available for use in the future.

Scores of books have been written on the subject of resource depletion. Here we concentrate on the highlights of concern for a few resources. We have already discussed the plight of fossil fuels and its impact on energy and synthetic materials production in Section 3.6. This section is devoted to a brief discussion of what is happening to our mineral resources.

3.7.1 Earth's Structure

Earth is composed of a number of layers (see Figure 3.34). It consists of a *core* of dense, intensely hot, molten metal—mostly iron and nickel—about 4000 km in diameter, surrounded by a layer of hot, pliable rock called the *mantle*. The mantle extends up to about 2900 km below Earth's surface. It is composed primarily of oxygen, silicon, and aluminum. The outermost layer of Earth, called the *crust,* floats on the mantle. Parts of the mantle rise above the oceans, forming continents and islands, while most of it remains below water. Table 3.13 lists all the components of Earth (dominated by the dense core) and Earth's crust.

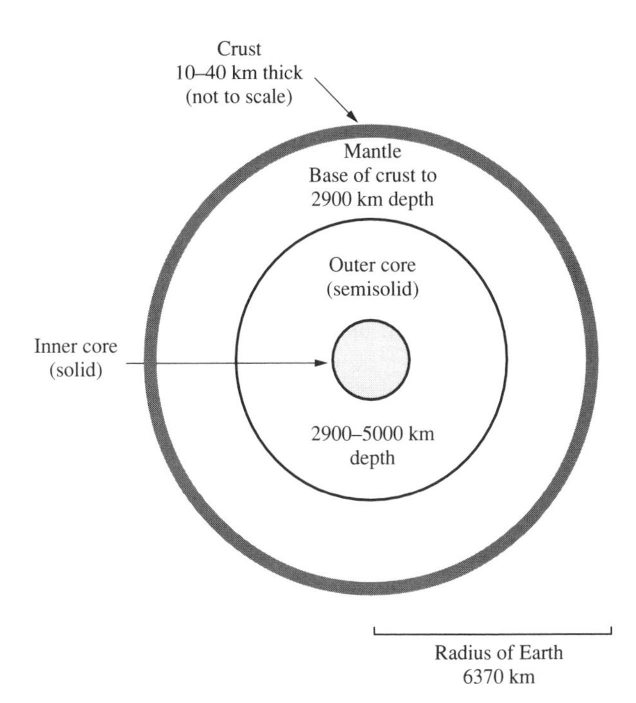

FIGURE 3.34
Layers of Earth.

TABLE 3.13
The most common elements in Earth and Earth's crust

Whole Earth		Crust	
Element	**Distribution, %**	**Element**	**Distribution, %**
Iron	33.3%	Oxygen	45.2%
Oxygen	29.8	Silicon	27.2
Silicon	15.6	Aluminum	8.2
Magnesium	13.9	Iron	5.8
Nickel	2.0	Calcium	5.1
Calcium	1.8	Magnesium	2.8
Aluminum	1.5	Sodium	2.3
Sodium	0.2	Potassium	1.7
Total percent	98.1	Total percent	98.3

Source: Adapted from Cunningham and Saigo, 1995.

The rocks making up Earth's crust are classified according to their internal structure, chemical composition, physical properties, and mode of formation. *Minerals* are the basic materials of all rocks. A mineral is an inorganic, crystalline solid that has a definite chemical composition and physical characteristics. There are thousands of minerals, but relatively few have economic importance.

3.7.2 Recoverable Resources

Earth is comprised of a vast number of minerals, but not all are easily recoverable with existing or potential technologies, or at acceptable environmental and economic costs. Some may be buried deep within Earth, where they are not extractable; some may be present at concentrations too low to make extraction economically viable; and some may just not have been discovered yet. Figure 3.35 depicts the categories of natural resources with respect to their degree of geological assurance that they are present and to the degree of economic feasibility that they are, if present, recoverable.

Proven reserves are those resources that have been thoroughly mapped and are economical to recover at current prices with available technology. *Known resources* are those that have been located but are not completely mapped. Their recovery now may or may not be economical, but they should be recoverable in the future. *Undiscovered resources* are only speculative or inferred. *Recoverable resources* are accessible with current technology but are not likely to be economically feasible in the foreseeable future, whereas *nonrecoverable resources* are so diffuse or remote that they are not ever likely to be technologically accessible. Unfortunately, almost all of Earth's resources fall into the latter categories, leaving less than 0.01 percent of the minerals in the upper 1 km of Earth's crust economically recoverable.

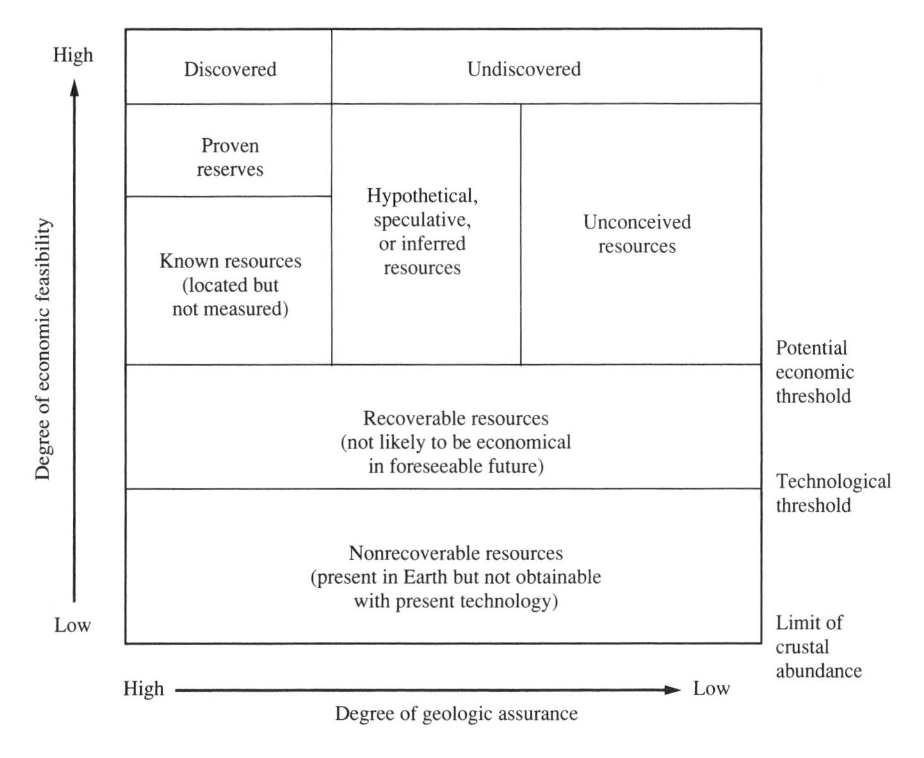

FIGURE 3.35
Categories of natural resources. (Source: Cunningham and Saigo, 1995)

3.7.3 Mineral Resources

Metal ores are the most widely used minerals in the world. Among these are iron, manganese, copper, chromium, aluminum, and nickel (see Table 3.14). These are used primarily in the United States, Japan, and Europe. Unfortunately for these countries, their supplies of many of these metals are limited. Much of their demand for these metals must come from countries where they are more abundant—South Africa, countries in South America, and the former Soviet Union countries. There are about 80 minerals that are considered to be of significant economic importance. Of these, at least 18 are considered to be in very short supply. Within this group are tin, platinum, gold, silver, and lead. Table 3.15 shows the major sources of several minerals imported into the United States.

The rate at which we are consuming mineral resources is increasing at an alarming rate. The entire metal production in the world over all of history before World War II was less than what has been consumed in the past five decades. Figure 3.36 illustrates the substantial increase in material consumption in the United States and other countries over the last few decades.

TABLE 3.14
World consumption of major metals

Metal	Amount consumed annually, million metric tons
Iron	740.0
Manganese	22.4
Copper	8.0
Chromium	8.0
Aluminum	4.8
Nickel	0.7

TABLE 3.15
Major world sources of mineral resources

Mineral	Amount consumed in U.S. that was imported, %	Major supplier countries
Columbium (niobium)	100%	Brazil, Canada, Thailand
Mica (sheet)	100	India, Belgium, France
Strontium	100	Mexico, Spain
Manganese	100	South Africa, Gabon, France
Bauxite	100	Jamaica, Guinea, Australia
Platinum	88	South Africa, former Soviet Union, United Kingdom
Tantalum	85	Thailand, Brazil, Germany
Cobalt	82	Zaire, Zambia, Canada
Chromium	80	South Africa, Zimbabwe, Turkey
Nickel	74	Canada, Australia, Norway
Tin	73	Malaysia, Thailand, Brazil
Potassium	67	Canada, Israel, former Soviet Union
Cadmium	54	Canada, Australia, Mexico
Zinc	30	Canada, Mexico, Peru

Source: U.S. Department of Commerce, 1995.

REFERENCES

Baird, C. *Environmental Chemistry.* New York: W. H. Freeman, 1995.

Cunningham, W. P., and Saigo, B. W. *Environmental Science: A Global Concern.* Dubuque, IA: Wm. C. Brown, 1995.

Enger, E. D., and Smith, B. F. *Environmental Science: A Study of Interrelationships.* Dubuque, IA: Wm. C. Brown, 1995.

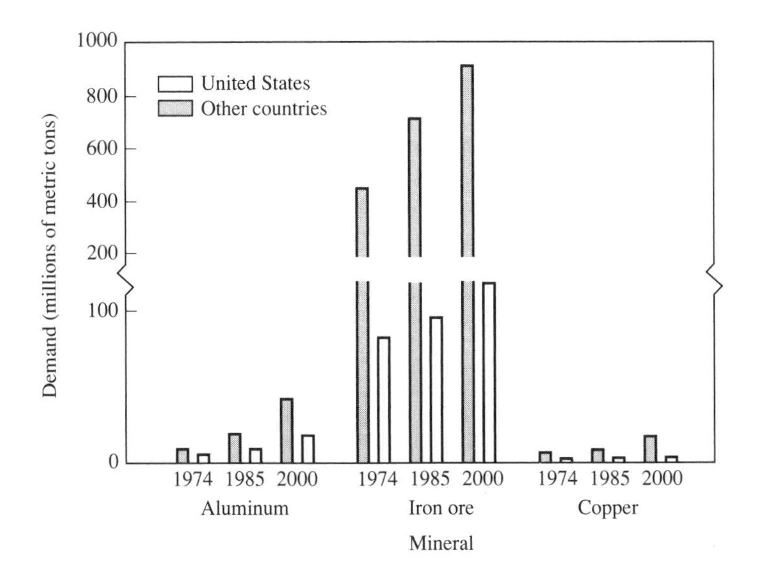

FIGURE 3.36
Mineral consumption in the United States and the rest of the world. (Source: Kupchella and Hyland, 1986)

Kiely, G. *Environmental Engineering*. New York: McGraw-Hill, 1997.

Kupchella, C. E., and Hyland, M. C. *Environmental Science: Living within the System of Nature*. Boston: Allyn and Bacon, Inc., 1986.

Levy, S. *The Municipal Solid Waste Factbook*. Washington, DC: U.S. EPA, 1996.

Pfeffer, J. T. *Solid Waste Management Engineering*. Englewood Cliffs, NJ: Prentice Hall, 1992.

Schneider, S. "The Population Explosion: Can It Shake the Climate?" *Ambio* 3 (1974): 150–55.

Springston, R. Signs of Life after Disaster. http://www.vcu.edu/cesweb/news/kepone.html (1996).

U.S. Department of Commerce. *Statistical Abstract of the United States, 1992*. Washington, DC: Bureau of the Census, 1995.

U.S. Department of Energy. *Annual Energy Review 1995*. Washington, DC: U.S. Department of Energy, 1995.

U.S. Department of Energy. *Energy and Climate Change*. Chelsea, MI: Lewis Publishers, 1990.

U.S. Department of Energy. Energy Overview. http://www.eia.doe.gov/fueloverview.html (1998).

U.S. EPA. Acid Rain Program: Emissions Scorecard. http://www.epa.gov/acidrain/scorcard/es1994.html (1996a).

U.S. EPA. *Characterization of Municipal Solid Waste in the United States: 1995 Update*. EPA530-R-96-001. Washington, DC: U.S. EPA, 1996b.

U.S. EPA. Protection of the Ozone Layer. http://www.epa.gov/ozone/science/indicat/indicat.html (1996c).

U.S. EPA. *The National Biennial RCRA Hazardous Waste Report*. Washington, DC: U.S. EPA, 1995.

White, R. M. "The Great Climate Debate." *Scientific American* 263, no. 1 (1990): 36–45.

World Resources Institute. *1991 Information Please Environmental Almanac*. Boston: Houghton Mifflin, 1990.

World Resources Institute. *1992 Information Please Environmental Almanac*. Boston: Houghton Mifflin, 1991.

World Resources Institute. *1993 Information Please Environmental Almanac*. Boston: Houghton Mifflin, 1992.

PROBLEMS

3.1. What are the boundaries of the airshed in which your university is located?

3.2. Carbon dioxide is probably the largest primary pollutant emitted into our atmosphere, but it is not a regulated pollutant, no limits have been set for its emission, and there is no ambient air standard. What reasons could be given to explain this apparent anomaly?

3.3. Until recently, atmospheric particulates of concern were defined as those greater than 10 μm in size, but this has now been lowered to 2.5 μm. Why? Investigate the documentation for the change found in the *Federal Register.*

3.4. Highway runoff used to contain high concentrations of lead due to the lead in gasoline used to power automobiles. Since lead in gasoline was substantially reduced in the late 1970s, these concentrations have substantially decreased. But lead is still found in highway runoff because of lead pollution associated with automotive uses. What are some of these?

3.5. Trichlorethylene (TCE) is a commonly used industrial solvent. From an air pollution perspective, it is of concern because of its potential toxicity, but it may be of more concern when emitted to the atmosphere for a broader scale secondary impact. What is this and what role does TCE play in it?

3.6. An industry burns large quantities of fossil fuels for its manufacturing processes. Its boilers are old and not very efficient. Consequently, they emit more unburned hydrocarbons and carbon monoxide than they should. The company is contemplating installing more efficient burners that will operate at much higher temperatures and with excess oxygen, which should greatly reduce these emissions. However, this will probably increase the emissions of another regulated pollutant. Which one is it? Would you recommend modifying the equipment as proposed? Why?

3.7. The average pH in acid rain in the northeastern United States is between 4.0 and 4.5, but it can be as low as 2.5–3.0. Compare these pH values with those of vinegar (acetic acid), Coca Cola (phosphoric acid), and soda water (carbonic acid). Note that you may want to check the pH of these with a pH meter.

3.8. A power plant burns 4000 kg of coal containing 5.0 percent sulfur per hour. Assuming that all of the S is converted to SO_3 during combustion and that there is enough moisture in the air to convert all of the SO_3 to H_2SO_4, how much acid is produced per hour in the atmosphere?

3.9. Research and report on the Kyoto Protocol, which emanated from the international meeting on global warming in Kyoto, Japan, in December 1997.

3.10. There is considerable debate as to the role of CO_2 emissions on global warming. One side of the debate says that the CO_2 contributes to the greenhouse effect and will soon lead to increased global temperatures. The other side says that the oceans serve as a sink for atmospheric CO_2 and will moderate any increases in atmospheric CO_2 in the long term. Research and analyze these two sides of the debate. What is your opinion on the subject?

3.11. There is a paradox in that ozone is essential in the upper atmosphere, but it is harmful in the lower atmosphere. Explain why this is so. Why don't the two interact with each other and balance?

3.12. McDonald's has sold about 100 billion hamburgers over the past years.
(*a*) If the paper packaging for these hamburgers had been reduced by only about 10 g per hamburger, how much solid waste would have been prevented?
(*b*) If it takes 17 trees to produce a metric ton of paper, how many trees would have been saved if the reduced packaging had been used?
(*c*) If it takes 200,000 L of water to produce 1 metric ton of paper, how much water would have been saved and how much less wastewater would have needed to be treated?

3.13. Many businesses receive goods and materials on wooden pallets. These are usually discarded after use. List ways that this significant source of industrial solid wastes could be minimized.

3.14. Evaluate, quantify, and list all of the solid waste that you generate over a seven-day period. How much of this material is potentially recyclable? What and how much of it could have been reduced by improved packaging?

3.15. Contact your local solid waste management authority and report on recycling activities in your community, both for residential and commercial/industrial solid wastes. Are the residential programs voluntary or mandatory? How much material is recycled? How is the material collected, sorted, and processed? What are the markets?

3.16. Using the Internet, determine the current prices for recycled (*a*) aluminum cans, (*b*) ferrous metal, (*c*) copper, (*d*) brass, (*e*) newsprint, (*f*) paper board, and (*g*) plastic bottles.

3.17. What are the regulatory limits for a waste to be declared hazardous due to (*a*) ignitability, (*b*) reactivity, (*c*) corrosivity, and (*d*) toxicity?

3.18. Research and prepare a report on the Pristine, Inc., Superfund site in Reading, Ohio. What did the site process, why was it placed on the Superfund list, how much waste is located there, what remediation processes were specified, and what is the current status of the cleanup effort?

3.19. We generally think of sulfur emissions and acid rain when we think of coal-burning power plants. However, several other environmental problems are associated with coal burning. Describe three of them.

3.20. Heating process water in an electric heater consumes much more total energy than in a coal or oil-fired boiler. Why?

3.21. The automobile industry often argues that it has done about all it can to improve gasoline efficiency in automobiles, but critics claim that the industry can do more. Using Figure 3.26 as the basis, calculate the amount of energy that could be saved in the United States if the miles per gallon of fuel for highway vehicles could be increased by 25 percent.

3.22. Some argue that we shouldn't develop nuclear energy power plants because of the potential danger from leaks at the plants and pollution from the resulting nuclear wastes. Others argue that fossil fuel power plants are much more polluting and dangerous to human health. Research the basis for these two positions and give your views on the argument.

3.23. Some environmentalists argue that we could vastly improve the environment by replacing our coal- and petroleum-fired power plants with nonfossil fuel plants (hydroelectric, geothermal, wind, etc.). If we were to triple the output of these nontraditional power sources, what would be the impact on fossil fuel use? What would be the impact on the environment?

3.24. Explain how we can be "running out" of some minerals, even though they are conservative and not destroyed upon use or disposal; that is, the total amount in the ecosystem remains constant.

3.25. Resources are materials that have value. These values change from culture to culture. For example, sea shells are of little value to most people today, except for their aesthetic appeal, but they were of great value to some cultures where they were used as a form of money and for jewelry. Iron ore would have been of little value to early humans but is of great value today. Name some shifts in mineral values that have occurred just since 1990.

CHAPTER 4

ENVIRONMENTAL REGULATIONS

4.1 INTRODUCTION

The multitude of potential sources of environmental contamination by industry or by products of industry has, over recent years, led to an ever-increasing level of regulation aimed at minimizing environmental damage and threats to public health. A series of recently enacted laws—such as the Clean Water Act, the Clean Air Act, the Safe Drinking Water Act, the Resource Conservation and Recovery Act, and the Superfund Act—form the cornerstone of the nation's attempt to preserve the environment for current and future generations. The phenomenal rate at which environmental laws have been enacted over the past few years can readily be seen in Figure 4.1. Laws are generally written to be very broad in scope and provide only the framework for enforcement; the actual implementation of what is in the law is carried out through regulations. The number of regulations and their rate of increase of the past several years have also been staggering (see Figure 4.2).

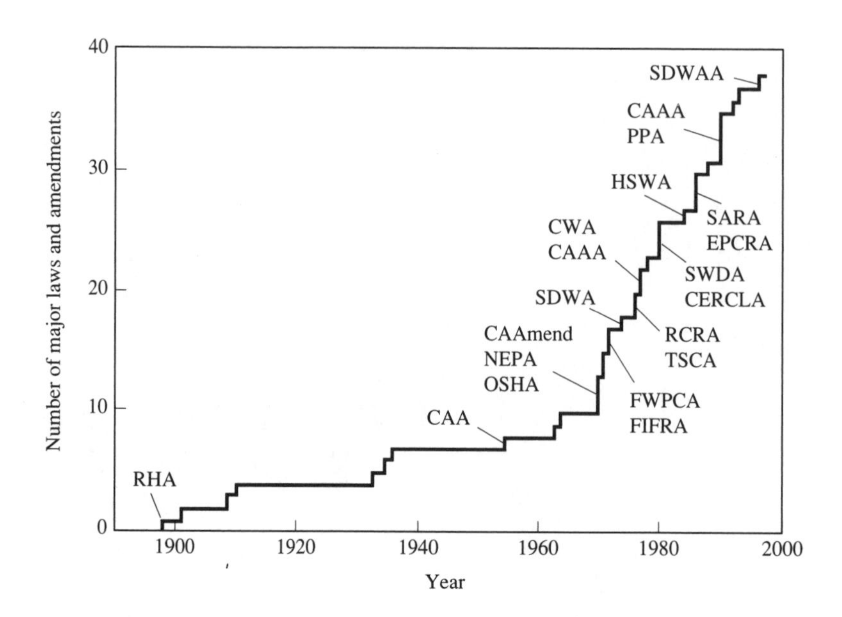

FIGURE 4.1
Growth of major environmental laws and amendments in the United States.

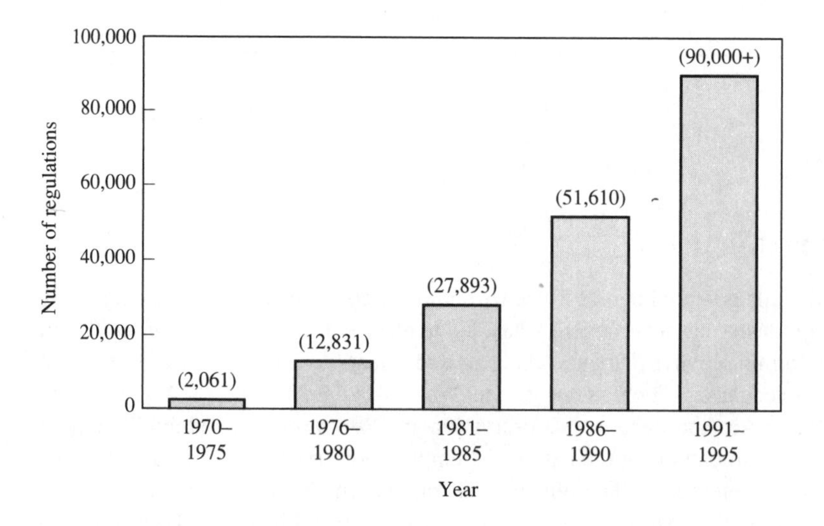

FIGURE 4.2
Changes in U.S. environmental regulations.

4.2 THE REGULATORY PROCESS

The process for enacting laws, developing regulations that articulate those laws, and enforcing the regulations varies from country to country. In this chapter, we concentrate on the regulatory process in the United States.

The United States is a democracy where bills are developed by Congress and signed into law by the president. A law is a body of official rules that is used to govern a society and to control the behavior of its members. There are several types of laws. There are natural laws, which are often based on religious beliefs, such as the concept that all life is sacred and shouldn't be destroyed; common laws, which derive from legal rulings and interpretations by previous court actions; and statutes, the laws that are written by Congress or state legislatures. All federal statutory law is compiled in the United States Code. Statutes are often general in content, describing the goals and policies that are intended by the governing body. Because of the technical complexity of the law, the language of the law is often deliberately broad. Congress provides only the goals; a regulatory agency provides the details on how the law is to be implemented. These detailed procedures are referred to as regulations. They are published daily in the *Federal Register* and are compiled in the *Code of Federal Regulations* (*CFR*). In addition to these formalized procedures, the president has the authority under certain circumstances to issue executive orders, essentially laws that do not need congressional approval.

Congressional action to create a law is thus only the first stage in the regulatory process. Development of regulations to implement the law is a very time-consuming process. Once a law is enacted, it may be years before approval of the regulations that are used to enforce that law. For example, when Congress passed the Resource Conservation and Recovery Act (RCRA) in December 1976, they gave the U.S. EPA 18 months to develop and begin implementing regulations to enforce the statute. The EPA immediately began drafting regulations and held hearings on them. However, because of the complexity of the problem and the large number of comments both for and against the regulations, the deadline passed without any regulations in place. Love Canal surfaced a few months later, and increased pressure was placed on EPA to respond. Following several federal court mandates, the EPA finally published its proposed regulations on December 14, 1978. Many objections to these regulations were voiced, again from both sides of the issue. Finally, on May 19, 1980, three and a half years after passage of RCRA, the final regulations were published. These were still incomplete, and many new regulations and revisions of existing regulations have been added since then. The long delay was not the fault of the EPA; the need to develop fair, equitable, and achievable regulations that would accomplish the goals of the law without severely harming industry was overwhelming. In many cases, the science to support the regulations did not exist at that time and had to be produced. The EPA had to develop regulations that were strict enough to safeguard the public, while at the same time ensuring that they were legally and scientifically correct enough to stand up in court. The process was much more time consuming than Congress had anticipated.

The regulatory process does not end with regulations. The EPA also issues guidance documents providing directions for carrying out a regulation. These theoretically are intended to provide guidance, as the term implies, and not be legally binding, but over time many of them become de facto binding.

In addition to federal statutes pertaining to pollution, many state and local environmental laws are intended to control industrial wastes. Normally, these must be at least as stringent as the federal laws, but they may be stricter.

In most cases, environmental laws are under the administrative jurisdiction of the Environmental Protection Agency. Other federal agencies do play a role, though. For example, the Department of Agriculture administers many forest and pesticide programs. The Department of the Interior administers most of the laws dealing with public lands, such as mining. By and large, though, the EPA generally is responsible for environmental protection. The EPA has the authority to delegate some of these responsibilities to state agencies that meet certain criteria, and in many cases this has been done. In some areas, such as solid waste management, the EPA has generally taken a hands-off attitude and left enforcement to state and local governments.

International environmental laws are becoming increasingly important to industry. Many international laws and treaties now govern the discharge of wastes into international waters, the shipment of waste materials from one country to another, and the use of environmentally harmful materials. The ban on use of chlorofluorocarbon compounds is a good example of this (see Chapter 3). Many pesticides have also been banned, and there is a general ban on shipment of wastes to underdeveloped countries for disposal. The North American Free Trade Agreement (NAFTA) has established environmental standards for products traded among the United States, Canada, and Mexico. A new international regulation governing the environmental soundness of a product (called ISO 14000) will have wide-ranging implications for industry in the near future. This is discussed further later in this chapter and in Chapter 7.

4.3 ENVIRONMENTAL REGULATIONS

As seen in Figure 4.1, a multitude of laws have been designed specifically to control environmental degradation. Most are pertinent to industries and other businesses (see Figure 4.3). The laws and regulations take up tens of thousands of pages, and interpretations of these take many more. More regulations are being added daily. Even lawyers who are experts in environmental law cannot truly understand all of them. Obviously, the pollution prevention engineer for a company, or the process or production people within the company, cannot know all the nuances of these laws and regulations. However, they should know their basics in order to comply with them and to be able to intelligently understand their impacts on business and industry.

4.3.1 Laws Pertaining to Clean Air

CLEAN AIR ACT. The first laws established to protect and enhance the quality of the nation's air resources were passed in 1955 (the Clean Air Act, or CAA), but these laws were completely replaced in 1967. The CAA was amended in 1970 and 1977. In 1990, the air quality laws were again totally rewritten to provide for attainment and maintenance of national ambient air quality. Following is a brief synopsis of the law as currently enforced.

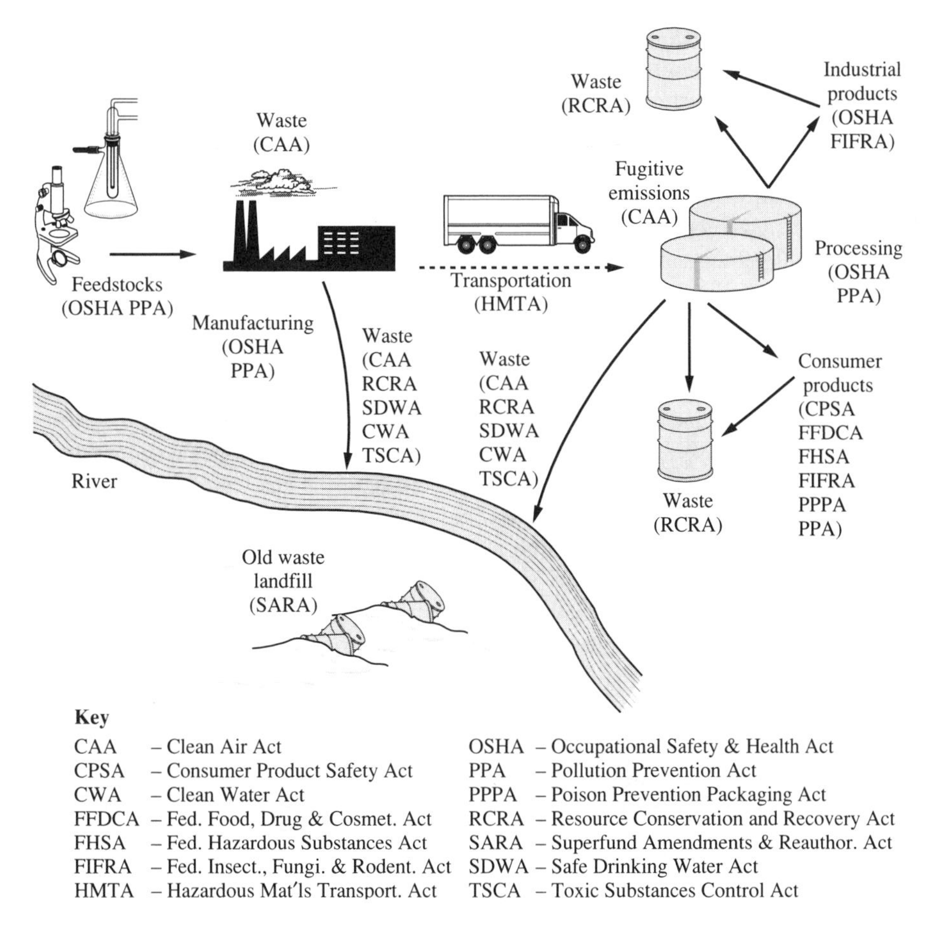

FIGURE 4.3
Laws affecting the production and use of a product.

The objective of the Clean Air Act is to protect and enhance our air quality to promote the public health and welfare and the productive capacity of the general population. Regulations generally are enforced through emission standards on stationary and mobile sources of air pollution. Initially, most efforts were placed on developing end-of-pipe treatment systems to remove contaminants from airstreams before discharge into the atmosphere, but recently much of the effort has been in reducing the amount of contaminants produced.

Criteria pollutants. When the CAA was first passed, the U.S. EPA was directed to establish *National Ambient Air Quality Standards* (NAAQSs) for certain air pollutants. An NAAQS is the concentration of the contaminant that should not be exceeded anywhere in the United States. These are ambient air quality standards and should not be confused with emission standards, which regulate source emissions to achieve or maintain the ambient air standards set by the NAAQS. Pollutants for which NAAQS are set are classified as *criteria pollutants*. *Primary standards* were established for protection of public health, including the health of "sensitive" populations such as asthmatics, children, and the elderly; *secondary standards* protect public welfare, including protection against decreased visibility and damage to animals, crops, vegetation, and buildings. Some pollutants have standards for both long-term and short-term averaging times. The short-term standards are designed to protect against acute, or short-term, health effects, whereas the long-term standards were established to protect against chronic health effects. They are listed in Table 4.1.

To enforce these standards, the states were divided into Air Quality Control Regions. Each region is assessed to determine if the criteria pollutant standards are being met. If a region persistently exceeds one or more of the NAAQSs, it is designated a *nonattainment region*. Nonattainment classifications for ozone are marginal, moderate, serious, severe, and extreme; those for carbon monoxide or particulates are listed as moderate or serious. These regions come under much more stringent requirements for achieving reductions in air emissions, with the requirements based on the classification. For example, marginal areas are required to conduct an inventory of their ozone-causing emissions and institute a permit program. Nonattainment areas with more serious air quality problems must implement various control measures. The worse the air quality, the more controls will be needed. Designating a region as being in nonattainment is a formal rule-making process and normally is done only after air

TABLE 4.1
National Ambient Air Quality Standards

	Standard	
Pollutant	**Value**	**Type**
Ozone	0.12 ppm (1-h ave.)*	Primary and secondary
	0.08 ppm (8-h ave.)	Primary and secondary
Carbon monoxide	9 ppm (8-h ave.)	Primary
	35 ppm (1-h ave.)	Primary
Particulate matter < 10 μm (PM-10)	50 $\mu g/m^3$ (annual mean)	Primary and secondary
	150 $\mu g/m^3$ (24-h ave.)	Primary and secondary
Particulate matter < 2.5 μm (PM-2.5)	15 $\mu g/m^3$ (annual mean)	Primary and secondary
	65 $\mu g/m^3$ (24-h ave.)	Primary and secondary
Sulfur dioxide	0.14 ppm (24-h ave.)	Primary
	0.03 ppm (annual mean)	Primary
	0.50 ppm (3-h ave.)	Secondary
Nitrogen dioxide	0.053 ppm (annual mean)	Primary and secondary
Lead	1.5 $\mu g/m^3$ (quarterly ave.)	Primary and secondary

* The ozone 1-h standard applies only to areas that were designated nonattainment when the ozone 8-h standard was adopted in July 1997.

TABLE 4.2
Decreases in emissions and air concentrations of criteria pollutants, 1986–1995

Pollutant	Decrease in emissions, %	Decreases in ambient air concentration, %
Carbon monoxide	16	37
Lead	32	78
Nitrogen oxides	3	14
Ozone	—	6
Particulates	17	22
Sulfur dioxide	18	37

Source: U.S. EPA, 1995.

quality standards have been exceeded for several consecutive years. Many regions are in attainment, but as of December 1998, there were 130 nonattainment regions in the United States, with a total population of more than 100 million (nearly half the total population). Most violations were due to ozone, followed by carbon monoxide, particularly in urban areas with a high density of automobiles.

Enforcement of the Clean Air Act has resulted in a significant improvement in the nation's air quality. Since its passage, air pollution has decreased 30 percent nationally, even as the gross domestic product has doubled (Hanson, 1997). Table 4.2 shows reductions in air sampling network concentrations and emissions for the criteria pollutants. As can be seen, notable improvements have been made in lead, carbon monoxide, and sulfur dioxide concentrations in the air.

Hazardous air pollutants. The Clean Air Act also established a list of hazardous air pollutants (HAPs) that must also be controlled. A hazardous air pollutant is defined as ". . . an air pollutant to which no ambient air quality standard is applicable and which . . . causes or contributes to . . . an increase in serious irreversible, or incapacitating reversible illness." They are commonly defined as those pollutants that are known or suspected of causing cancer or other serious health effects such as birth defects or developmental effects.

The majority of HAPs come from industrial sources, with most of the remainder originating from automobiles (see Figure 4.4). In 1990, approximately 8.8 billion pounds of hazardous air pollutants were emitted into the nation's air. Exposure to such quantities of air toxics may result in 1000–3000 cancer deaths per year. Estimates are that emissions of HAPs are now decreasing and this is leading to a decrease in ambient air concentrations. For example, the average concentration of benzene in air in nonattainment areas decreased by 38 percent between 1994 and 1995, possibly as a result of the use of reformulated gasoline in those areas.

Originally, the EPA listed 189 chemicals as HAPs that needed emission standards (one compound, captrolactam, was recently removed from the list of toxic air

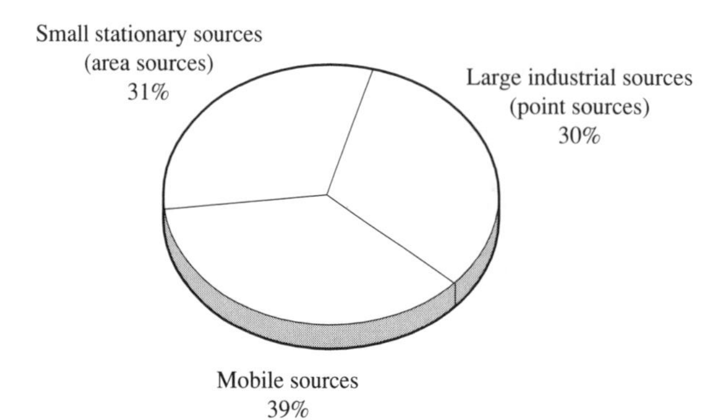

FIGURE 4.4
Total toxic air pollution emissions in the United States by source, 1990. (Source: U.S. EPA, 1995)

pollutants, so there are now 188 toxic air pollutants). The list includes such chemicals as benzene, chromium, cadmium, and vinyl chloride. By 1990, however, the EPA had established emission standards for only seven of them (arsenic, asbestos, benzene, beryllium, mercury, radionuclides, and vinyl chloride). Consequently, the Clean Air Act Amendments of 1990 greatly strengthened the requirements on toxic air pollutants and mandated the EPA to regulate new and existing sources of these 188 pollutants through development of *Maximum Allowable Control Technology* (MACT) standards. The MACT standards are being established for each of 174 industrial source categories. The industries must achieve "the maximum degree of reduction in emission of hazardous air pollutants" determined by the EPA to be achievable through application of specified processes, systems, or techniques. These may include installation of air pollution control equipment, industrial processes changes, materials substitutions, and worker training. The EPA was directed to have requirements in place for all 188 chemicals by the year 2000. The first MACT standards were put in place in 1994. As of October 1996, the EPA had issued air toxics standards for 47 source categories, such as chemical plants, oil refineries, and steel mills, as well as area sources such as dry cleaners and chromium electroplating. By the year 2005, EPA projects that the toxic air program will reduce toxic emissions by 75 percent.

As an example of a hazardous air pollutant rule, we will examine the MACT standard for chromium electroplating and anodizing operations. These operations coat metal parts and tools with a thin layer of chromium to protect them from corrosion and wear. Chromium, a toxic air pollutant that causes lung cancer in humans, is released during this process. The MACT rule is designed to reduce the chromium emissions from this process by 99 percent. The rule affects 2800 decorative chromium electroplating facilities, 1500 hard chromium electroplating facilities, and 700 chromium anodizing facilities nationwide. The MACT values vary depending on the type of process used and whether it is an existing or a new facility. For example, existing facilities using hard chromium electroplating tanks must meet an emission limit of 0.015 mg of chromium per dry standard cubic meter (mg/dscm) of ventilation air. To achieve this, the operators will need to use a packed-bed scrubber. All new and existing chromium anodizing operations must meet an emission limit of 0.01 mg/dscm or use a wetting agent–type fume suppressant in the plating bath and maintain a surface tension of no greater than 45 dynes/cm. The latter allows a foam blanket to form on the surface of the tank contents, minimizing volatilization of the chromium. Thus MACT standards vary from process to process and may require either pollution control equipment or changes to operating procedures.

Permits. A significant change to the Clean Air Act brought about by the Clean Air Act Amendments of 1990 was the institution of a permit program for larger sources of air emissions. Under the program, permits are issued by states or, when the state fails to carry out the responsibility satisfactorily, by the EPA. The permit includes information on which pollutants are being released, how much may be released, and what kind of steps the source's operator is taking to reduce pollution, including plans to monitor the pollution. Because of the Bhopal tragedy in India (discussed in Chapter 3), businesses must also develop plans to prevent accidental releases of highly toxic chemicals. The permit system unifies many of the legal responsibilities of the business. For example, an electric power plant may be covered by the acid rain, hazardous air pollutant, and nonattainment (smog) parts of the Clean Air Act; the permit gathers all applicable data into one place.

The 1990 CAA Amendments give businesses a lot of latitude in how they reach their pollution cleanup goals. Several market-based approaches are available, including pollution allowances that can be traded, bought, and sold. There are also economic incentives, such as credits to gasoline refiners if they produce cleaner gasoline than required; refiners can use those credits when their gasoline doesn't quite meet cleanup requirements. A company that plans to increase production, which would lead to the emission of a greater amount of pollution, may use an *offset*. This means that the company can reduce the criteria air pollutant by an amount somewhat greater than the planned increase somewhere else in the plant or even in another plant within the nonattainment area, so that the permit requirements are still met. These approaches are described in more detail in Chapters 8 and 13.

Mobile sources. While motor vehicles built today emit much less pollution than those built in the 1960s (60 percent to 80 percent less, depending on the pollutant), cars and trucks still account for almost half the smog-forming volatile organic compounds and nitrogen oxide compounds as well as up to 90 percent of the carbon monoxide in urban areas. These improvements in engine performance have been offset by the increased number of vehicles in use. The number of miles driven has increased by about a factor of 4 over the past three decades.

New provisions under the CAA are focused on reducing emissions from cars and other vehicles and on the development of cleaner fuels. Ironically, the ban on use of leaded gasoline led to the production of more highly polluting fuels. As lead was being phased out, gasoline refiners changed gasoline formulas to make up for octane loss. These changes made gasoline more likely to release smog-forming VOC vapors into the air. The CAA requires that fuels be less volatile. Gasoline refiners will have to reformulate gasoline in smoggy areas so that they contain less VOCs such as benzene. In areas where car ignitions cause a carbon monoxide problem in cold months, refiners will have to sell oxyfuel, gasoline with alcohol-based oxygenated compounds added to make the fuel burn more efficiently, thereby reducing carbon monoxide release. All gasolines must contain detergents which keep engines running more smoothly and burn fuel more cleanly. Refiners must reduce the amount of sulfur in diesel fuel. In addition, all gasoline filling stations must use vapor recovery nozzles on their gas pumps to reduce vapor release during filling. A major step forward was the institution of regular automobile emission inspection and maintenance programs.

The CAA also mandates production of cars that can burn cleaner fuels, such as alcohol or natural gas, particularly for vehicle fleets. In 1996, Ford Motor Company introduced the first electric car for use in parts of California.

Acid rain. Approximately 20 million tons of sulfur dioxide are emitted annually in the United States, mostly from the burning of high-sulfur coal by electric utilities. The resulting acid rain has caused enormous damage to the environment and possibly to public health (see Chapter 3). The CAA contains provisions that will reduce sulfur dioxide emissions by about 40 percent.

Reductions will be carried out in two phases, with the first beginning in 1995. In this phase, 110 coal-burning boilers were mandated to reduce their emissions to a level equivalent to 2.5 lb $SO_2/10^6$ Btu times the average of their 1985–1987 fuel use. In 2000, Phase II goes into effect, requiring approximately 2000 utilities to reduce their emissions to a level equivalent to the product of an emissions rate of 1.2 lb $SO_2/10^6$ Btu times the average of their 1985–1987 fuel use. The law permits the use of *emissions allowances,* pollution allowances established by the EPA for each utility that proscribe how much SO_2 may be emitted by that plant. Allowances may be traded within a company or bought and sold between utilities. A company that is doing well and is emitting less than allowed by the EPA may sell its excess allowances to a utility that cannot meet its limits. Thus, in the end, the goal of acid rain reduction can be met without severe hardships to those utilities that must rely on high-sulfur coal.

The law also includes specific requirements for reducing emissions of NO_x.

Global climate protection. The CAA uses a market-based approach to phase out the production of substances that deplete the ozone layer. The 1990 CAA Amendments set a schedule for ending production of these chemicals, in accordance with the Montreal Protocol (see Chapter 3), with the most damaging chemicals being phased out first. Table 4.3 shows the current phaseout schedule.

The CAA also requires recycling of CFCs and labeling of products containing ozone-destroying chemicals. This includes CFCs from automobile air conditioning systems, the largest single source of ozone-destroying chemicals. The development of "ozone-friendly" substitutes for ozone-destroying chemicals is encouraged. As CFCs are phased out and new chemicals replace them, many new designs for appliances and industrial processes will be needed. Although this changeover will be expensive, it opens up many opportunities for industry to develop new environmentally safe products and processes.

TABLE 4.3
Phaseout schedule for ozone-destroying chemicals

Compound	Use	Production ends
CFCs	Solvents, aerosol sprays (most spray can uses banned in 1970s), foaming agents in plastic manufacture	January 1, 1996
Halons	Fire extinguishers	January 1, 1994
Carbon tetrachloride	Solvents, chemical manufacture	January 1, 1996
1,1,1-Trichloroethylene (methyl chloroform)	Very widely used solvent; in many workplace and consumer solvents	January 1, 1996
HCFCs (hydro CFCs)	CFC substitutes; chemicals slightly different from CFCs	January 1, 2003

Source: U.S. EPA, 1993b.

4.3.2 Laws Pertaining to Clean Water

CLEAN WATER ACT. The Federal Water Pollution Control Act is the primary federal legislation that protects the nation's surface waters, such as lakes, rivers, and coastal areas. The law was originally passed in 1948, but it was totally rewritten and expanded in 1972 because of the growing public outcry about the state of our nation's waters. This law became known as the Clean Water Act (CWA). It has since been revised several times, with major revisions in 1977, 1981, and 1987. It is currently under review again.

The objective of the Clean Water Act is to improve and maintain water quality by restoring the physical, chemical, and biological integrity of the nation's waters. The CWA has five main elements: (1) a system of minimum national effluent standards for each industry, (2) water quality standards, (3) a discharge permit program that translates these standards into enforceable limits, (4) provisions for special wastes such as toxic chemicals and oil spills, and (5) a revolving construction program for publicly owned treatment works (POTWs). The CWA requires the EPA to establish effluent limitations for the amounts of specific pollutants that may be discharged by industrial facilities and municipal wastewater treatment plants. A system of nationwide, base-level standards has been set, but in some cases specific discharge limits can be stricter than national norms; they are then based on the intended use of the receiving water (swimming, boating, fishing, etc.).

The primary method by which the goals of the act are met is through the implementation of a nationwide permit program called the *National Pollutant Discharge Elimination System* (NPDES). Anyone who wishes to discharge any pollutant into any waters in the United States from any point source must first apply for and obtain a permit. The permit dictates the amounts of pollutants allowed to enter the receiving stream and how these levels are to be reached. Treatment of industrial wastes to reach discharge standards dictated in an NPDES permit is usually very costly. Therefore, many industries prefer to discharge to a municipal sewer and have their wastes treated at the

POTW. Industries that discharge their wastes to a POTW do not need to receive an NPDES permit, but they must follow the pretreatment regulations spelled out in the act and administered by the POTW. These pretreatment regulations require that industrial dischargers remove or treat all pollutants that could pass through a POTW untreated or could adversely affect the performance of the municipal system. There is usually a significant cost involved with pretreatment as well as a user fee the discharger must pay for treatment of its wastes at the POTW. In many cases, this combined cost is still less than the cost of fully treating the waste for direct discharge into a surface water, making municipal treatment of the waste the most economical way to go. In either case, using pollution prevention techniques to minimize the amount of waste that must be treated can substantially reduce waste management costs.

The foregoing requirements are specifically directed at protection of surface waters. The act also contains permitting provisions for stormwater and other non–point source pollution. The intent is to prevent contaminated stormwater runoff from polluting surface waters. Stormwaters that flow across contaminated land, particularly paved areas, have been shown to be a major source of pollution in our rivers and streams. Industries are required to develop stormwater pollution prevention plans and to construct and operate stormwater treatment facilities when needed.

Other provisions of the Clean Water Act control the discharge of dredged or fill materials into the nation's waters in order to protect wetlands and other valuable aquatic habitat. The EPA has established guidelines for dredge and fill operations, but issuance of dredge and fill permits has been delegated to the U.S. Army Corps of Engineers.

OIL POLLUTION ACT. Oil spills cause great environmental and economic damage. The Oil Pollution Act (OPA) of 1990 was passed largely as a result of the public outcry following the *Exxon Valdez* incident. This very wide-reaching law is designed to prevent any oil spills from reaching any navigable waters of the United States or adjoining shorelines. It is a comprehensive statute designed to expand the oil prevention, preparedness, and response capabilities of the federal government and industry. The OPA amends Section 311 of the Clean Water Act to clarify the authority of the EPA in preventing and containing oil spills. This prevention program combines planning and enforcement measures.

To prevent oil spills, the EPA requires owners and operators of certain facilities that drill, produce, gather, store, process, refine, transfer, distribute, or consume oil to prepare and implement spill prevention, control, and countermeasure (SPCC) programs that detail the facility's spill prevention and control measures. Any facility with a total above-ground oil storage capacity of greater than 1320 gallons (or greater than 660 gallons in a single above-ground tank) or total underground oil storage capacity greater than 42,000 gallons must comply with the law.

The EPA also enforces the oil spill liability and penalty provisions under the act, which provides incentives to owners/operators to prevent oil spills from occurring. Under the OPA, the owner or operator of a facility from which oil is discharged is liable for the costs associated with the containment and cleanup of the spill and any damages resulting from the spill. An Oil Spill Liability Trust Fund has been set up to cover removal costs and damages resulting from oil discharges from unknown sources. Originally, a five-cent per barrel tax on imported and domestic oil was used to establish the fund, but this tax expired at the end of 1994. The fund now is supported by interest on the existing fund and by any fines and civil penalties collected.

NATIONAL ENVIRONMENTAL POLICY ACT. The National Environmental Policy Act (NEPA) of 1969 established national policies and goals for the protection of the environment. It also provided the framework for the government to assess the environmental effects of its actions and the progress made toward improving environmental quality. The two main provisions of the act were (1) to create the president's Council on Environmental Quality (CEQ) to advise the president on environmental policy and to prepare an annual environmental quality report to Congress and (2) to require federal agencies to prepare environmental impact statements (EISs) to assess the environmental effects of all proposed projects and legislation. The contents of the EIS must be considered when making a decision on a proposed action. The agency must prepare a Record of Decision (ROD), which is a statement of its decision discussing its choice among alternatives and the means that will be employed to mitigate or minimize environmental harm. Although NEPA requires federal agencies to take a detailed look at the environmental consequences of their actions, it does not force them to take the most environmentally sound alternative.

SAFE DRINKING WATER ACT. The Clean Water Act was designed primarily to protect the nation's surface waters. Following passage of the CWA, there was still an absence of regulatory control over the quality of our groundwaters. Therefore, in 1974, Congress enacted the Safe Drinking Water Act (SDWA) to manage potential contamination threats to groundwater. In 1986 and again in 1996, the act was substantially rewritten through the Safe Drinking Water Act Amendments (SDWAA). The original act instructed the EPA to establish a national program to prevent underground injections of contaminated fluids that would endanger drinking water sources. The resulting standards apply to drinking water at the tap as supplied by a public water distribution system. The groundwater standards are also used to determine groundwater protection regulations under a number of other statutes (e.g., RCRA), as will be described later.

The SDWA covers water quality at all public water systems. The law defines a *public water system* as a water supply system that provides piped water for human consumption which has at least 15 connections or regularly serves at least 25 people. This includes essentially all community water systems, workplaces, hospitals, campgrounds, and gas stations. Except for some very small systems serving very small trailer parks or subdivisions, all public water supplies are covered.

National primary drinking water standards. The Safe Drinking Water Act requires the EPA to establish National Primary Drinking Water Regulations for contaminants that may cause adverse public health effects. The 1996 amendments emphasize the use of sound science and risk-based standard setting. The regulations include both mandatory levels (Maximum Contaminant Levels, or MCLs) and nonenforceable health goals, or Maximum Contaminant Level Goals (MCLGs), for each included contaminant. Although they are only recommended limits for drinking water, many of the MCLGs have become the standards used under Superfund regulations for hazardous waste site cleanups. Table 4.4 lists the current MCLs and MCLGs.

Groundwater protection. A second part of the Safe Drinking Water Act, greatly expanded under the 1996 amendments, is directed toward source water protection. States are now required to maintain a program for delineating underground source water areas of public water systems and to assess the susceptibility of each source water to contamination. They must also prepare a plan to protect existing and potential new groundwater sources and implement the plan.

Industries can be impacted by the SDWA in several ways. They may be involved in wellhead protection programs if they are located in the proximity of source waters of a public water supply. This could affect the way in which they manage their wastes. If they operate a hazardous waste disposal facility, they may have to meet the national drinking water standards for groundwaters in the vicinity of their disposal facility. There are also provisions governing the production and sale of bottled water.

4.3.3 Laws Pertaining to Hazardous Materials and Wastes

TOXIC SUBSTANCES CONTROL ACT. The Toxic Substances Control Act (TSCA) was passed in 1976 to give the EPA power to obtain information on all new and existing chemical substances and to control any of these substances determined to cause an unreasonable risk to public health or the environment. Prior to this, the EPA could act to control a toxic substance only after damage had occurred. There were no provisions for prescreening chemicals before they entered the marketplace. Now, all new chemicals must be evaluated before they are manufactured for commercial purposes. Any company planning to manufacture or import a new chemical must first submit to the EPA a *Premanufacture Notification* containing information on the identity, use, anticipated production or import volume, workplace hazards, and disposal characteristics of the substance. The manufacturer may also be required by the EPA to conduct toxicological tests and to report unpublished health and safety studies on listed chemicals. Following review of the application, the EPA may allow the sale of the substance, or it may prohibit or limit its manufacture and sale or require special labeling. Several chemical substances are exempted from these regulations, usually because the exempted substance is regulated under other laws. Among these exempted substances are pesticides, tobacco and tobacco products, nuclear materials and by-products, food, food additives, drugs, and cosmetics.

A separate part of the act prohibits the manufacture, processing, and distribution of polychlorinated biphenyls (PCBs). The EPA has established regulations on the handling and disposal of PCBs using this act for enforcement. Any PCBs found at hazardous waste sites are managed under the guidance of this act, rather than under SARA or RCRA.

In 1986, TSCA was amended to include the Asbestos Hazard Emergency Response Act. This law orders school systems throughout the country to inspect their buildings for asbestos, identify areas where asbestos-containing materials pose hazards

TABLE 4.4
National Primary Drinking Water Standards

Contaminant	MCLG, mg/L	MCL, mg/L	Contaminant	MCLG, mg/L	MCL, mg/L
Fluoride	4.0	4.0	Organics		
Microorganisms			Acrylamide	0	TT
Giardia lamblia	0	TT	Adipate [di(2-ethylhexyl)]	0.4	0.4
Legionella	0	TT	Alachlor	0	0.002
Standard plate count	N/A	TT	Aldicarb	0.001	0.003
Total coliform	0	<5%+	Aldicarb sulfone	0.001	0.002
Turbidity	N/A	TT	Aldicarb sulfoxide	0.001	0.004
Viruses	0	TT	Atrazine	0.003	0.003
Inorganics			Carbofuran	0.04	0.04
Antimony	0.006	0.006	Chlordane	0	0.002
Arsenic (interim)	0.05	0.05	Chlorobenzene	0.1	0.1
Asbestos (>10 μm)	7 MFL	7 MFL	Dalapon	0.2	0.2
Barium	2	2	Dichloromethane	0	0.005
Beryllium	0.004	0.004	2,4-D	0.07	0.07
Cadmium	0.005	0.005	*o*-Dichlorobenzene	0.6	0.6
Chromium (total)	0.1	0.1	1,2-Dichloroethylene	0.07	0.07
Copper	1.3	TT**	Dibromochloropropane	0	0.0002
Cyanide	0.2	0.2	1,2-Dichloropropane	0	0.005
Lead	0	TT*	Dinoseb	0.007	0.007
Mercury (inorganic)	0.002	0.002	Diquat	0.02	0.02
Nickel	0.1	0.1	Dioxin	0	0.00000003
Nitrate	10	10	Endothal	0.1	0.1
Selenium	0.05	0.05	Endrin	0.002	0.002
Sulfate (proposed)	400/500	400/500	Epichlorohydrin	0	TT
Volatile organics			Ethylbenzene	0.7	0.7
Trihalomethanes	0	0.10	Ethylene dibromide	0	0.00005
Benzene	0	0.005	Glyphosate	0.7	0.7
Carbon tetrachloride	0	0.005	Heptachlor	0	0.0004
p-Dichlorobenzene	0.075	0.075	Heptachlor epoxide	0	0.0002
1,2-Dichloroethane	0	0.005	Hexachlorobenzene	0	0.001
1,1-Dichloroethylene	0.007	0.007	Hexachlorocyclopentadiene	0.05	0.05
Trichloroethylene	0	0.005	Lindane	0.0002	0.0002
1,1,1-Trichloroethane	0.2	0.2	Methoxychlor	0.04	0.04
Vinyl chloride	0	0.002	Oxamyl (Vydate)	0.2	0.2
Radioactive			PAHs (benzo[a]pyrene)	0	0.0002
Beta/photon emitters	0	4 mrem/yr	PCBs	0	0.0005
Alpha emitters	0	15 pCi/L	Pentachlorophenol	0	0.001
Combined radium 226/228	0	5 pCi/L	Phthalate, [di(2-ethylhexyl)]	0	0.006
Radium 226 (proposed)	0	20 pCi/L	Picloram	0.5	0.5
Radium 228 (proposed)	0	300 pCi/L	Simazine	0.004	0.004
Uranium (proposed)	0	0.02	Styrene	0.1	0.1
			Tetrachloroethylene	0	0.005
			1,2,4-Trichlorobenzene	0.07	0.07
			1,1,2-Trichloroethane	0.003	0.005
			Toluene	1	1
			Toxaphene	0	0.003
			2,4,5-TP	0.05	0.05
			Xylenes (total)	10	10

Key: TT = treatment technique required; MFL = million fibers per liter; pCi = picocurie; mrem = millirems.
* Action level = 0.015 mg/L ** Action level = 1.3 mg/L
Source: U.S. EPA, 1994.

to humans, and abate those hazards. In 1990, amendments were enacted that extended this requirement to all public and commercial buildings.

RESOURCE CONSERVATION AND RECOVERY ACT. Until 1976, hazardous wastes were regulated under the Solid Waste Disposal Act of 1965, which dealt primarily with the disposal of nonhazardous wastes. That year, Congress passed the Resource Conservation and Recovery Act (RCRA), which totally rewrote the previous legislation and provided the framework for national programs to achieve environmentally sound management of both hazardous and nonhazardous wastes. The act established a permit program to manage potentially hazardous materials from their point of origin to their point of ultimate disposal and beyond. This came to be known as the *cradle-to-grave* concept. For the first time, legislation was also enacted that promoted resource recovery to reduce the generation of waste materials. The Hazardous and Solid Waste Amendments of 1984 (HSWA) expanded the scope of RCRA and became an integral part of it. A major component of the HSWA was legislation dealing with leaking underground storage tanks (USTs), which are responsible for a large fraction of our groundwater pollution problems.

The RCRA, as amended, contains 10 subtitles. Of these, the most important from the standpoint of pollution prevention are Subtitle C, "Hazardous Waste Management"; Subtitle D, "State and Regional Solid Waste Plans"; and Subtitle I, "Regulation of Underground Storage Tanks." Subtitle C provides regulatory authority for hazardous wastes; Subtitle D specifies the regulatory requirements for nonhazardous solid wastes.

Hazardous waste definition. Much of the RCRA is devoted to defining what a hazardous waste is and how to regulate it. Hazardous wastes are classified as a subset of solid wastes. Subtitle A of RCRA defines a *solid waste* as

> any garbage, refuse, sludge, from a waste treatment plant, water supply treatment plant or air pollution control facility and other discarded material including solid, liquid, semi-solid, or contained gaseous materials resulting from industrial, commercial, mining and agricultural activities and from community activities but does not include solid or dissolved material in domestic sewage, or solid or dissolved materials in irrigation return flows or industrial discharges which are point sources subject to permits under Section 402 of the Federal Water Pollution Control Act, as amended, or source, special nuclear, or byproduct material as defined by the Atomic Energy Act of 1954, as amended.

In general, a solid waste is any solid, liquid, or contained gaseous material that is discarded by being disposed of, incinerated, or recycled. Even materials that are recyclable or can be reused in some way (such as burning used oil for fuel) may be considered waste.

A solid waste may be classified as a hazardous waste if it exhibits certain hazardous waste characteristics or is categorically listed as a hazardous waste by the EPA. The RCRA defines a *hazardous waste* as

> a solid waste, or combination of solid wastes, which because of its quantity, concentration, or physical, chemical or infectious characteristics may . . . cause, or significantly contribute to an increase in mortality or an increase in serious irreversible, or incapacitat-

ing reversible, illness; or . . . pose a substantial present or potential hazard to human health or the environment when improperly treated, stored, transported or disposed of, or otherwise managed.

The regulations developed to enforce this definition consist of a series of hazardous waste lists, hazardous waste characteristics, and lists of exempted wastes. To be classified as a hazardous waste, the waste must first be a "solid waste." A number of solid wastes have been exempted from further consideration as hazardous wastes by decree, even if they meet the definition of a hazardous waste by other provisions. These include:

Household solid wastes
POTW sludges
Agricultural wastes returned to the ground
Mining overburden
Oil and gas exploration drilling wastes
Utility wastes from coal combustion
Wastes from the extraction and processing of ores and minerals
Cement kiln wastes
Arsenic-treated wood wastes generated by end users of such wood
Certain chromium-bearing wastes
Radioactive wastes

These exemptions were provided in some cases because of the magnitude of the waste problem and the enormous cost associated with meeting the RCRA requirements (e.g., coal ash, wastewater treatment plant sludges), because they are covered by other statutes (e.g., radioactive wastes), and in other cases for political reasons.

The EPA has designated four characteristics of a hazardous waste: (1) ignitability, (2) corrosivity, (3) reactivity, and (4) toxicity. *Ignitable wastes* are those that can catch fire under certain circumstances, such as some paints and many degreasers and solvents. *Corrosive wastes* corrode metals or have a very high or low pH. These include such materials as rust removers, alkaline or acid cleaning fluids, and battery acids. *Reactive wastes* are those that are unstable and explode or produce toxic fumes, gases, and vapors when mixed with water or under other conditions such as high heat or pressure. Examples are certain cyanides and sulfide-bearing wastes. *Toxic wastes* are those that are harmful when ingested or absorbed or that leach toxic chemicals into soil or groundwater when disposed of on land. Many heavy metals and some organics fit into this category.

The first three waste types have well-defined testing procedures, but analysis for the toxicity characteristic has always been controversial. The test procedure currently in use, the Toxicity Characteristic Leaching Procedure (TCLP), is designed to simulate leaching under landfill conditions. The resulting leachate concentrations are compared to a list of 31 organic chemicals and 8 inorganics (see Table 4.5). If the waste leachate concentration is greater than the TCLP list concentration, the waste is characterized as toxic. This is a pass–fail test; there is no degree of toxicity allowed.

TABLE 4.5
Toxicity characteristic constituents and their regulatory levels

Constituent	Regulatory level, mg/L	Constituent	Regulatory level, mg/L
Arsenic	5.0	Hexachlorobenzene	0.13
Barium	100.0	Hexachloro-1,3-butadiene	0.5
Benzene	0.5	Hexachloroethane	3.0
Cadmium	1.0	Lead	5.0
Carbon tetrachloride	0.5	Lindane	0.4
Chlordane	0.03	Mercury	0.2
Chlorobenzene	100.0	Methoxychlor	10.0
Chloroform	6.0	Methyl ethyl ketone	200.0
Chromium	5.0	Nitrobenzene	2.0
o-Cresol	200.0	Pentachlorophenol	100.0
m-Cresol	200.0	Pyridine	5.0
p-Cresol	200.0	Selenium	1.0
Total cresol	200.0	Silver	5.0
2,4-D	10.0	Tetrachloroethylene	0.7
1,4-Dichlorobenzene	7.5	Toxaphene	0.5
1,2-Dichloroethane	0.5	Trichloroethylene	0.5
1,1-Dichloroethylene	0.7	2,4,5-Trichlorophenol	400.0
2,4-Dinitrotoluene	0.13	2,4,6-Trichlorophenol	2.0
Endrin	0.02	2,4,5-TP (silvex)	1.0
Heptachlor (and its epoxide)	0.008	Vinyl chloride	0.2

A solid waste that is not excluded from the hazardous waste category may be classified as hazardous if it is a *listed waste*. The EPA has generated four lists of hazardous wastes, which are published in the *Code of Federal Regulations* (40 *CFR* pt. 261). These are wastes that have been tested from many sources and generally found to exhibit at least one characteristic of a hazardous waste. Currently, more than 400 wastes are listed. The F List—hazardous wastes from nonspecific sources—describes wastes that are found coming from many industrial processes. They describe generic wastes, such as spent halogenated solvents used in degreasing and spent cyanide-plating solutions from electroplating. The K List—wastes from specific sources—includes wastes generated from specific sources unique to specific industrial groups, such as distillation bottoms from aniline production and spent pickle liquor from steel finishing operations. The F List and K List are used for wastes generated by industrial processes. The P List and U List are used for products discarded by industry because they are off-specification or out-of-date. The P List is for chemical products considered to be acutely hazardous; the U List is for those classified as toxic.

One important provision of the hazardous waste classification criteria is the *mixture rule,* which states that if a listed hazardous waste is mixed with a nonhazardous waste, the entire mixture must be classified as a hazardous waste. The intent of this rule is to prohibit industries from diluting their hazardous wastes with other materials to make them pass the toxicity characteristic criteria. The implications of this rule are significant. For example, if an industry allows hazardous waste materials to contaminate stormwater crossing its site, then all of the stormwater must be considered to be hazardous and must be managed as such. Thus it is to the great benefit of industry to minimize the amount of hazardous waste generated and to isolate the hazardous wastes that are created from all other materials as much as possible.

Another important provision of the statute is referred to as the *derived-from rule.* This rule states that a waste generated from a hazardous waste treatment, storage, or disposal facility is categorically a hazardous waste. Thus incineration ash from a hazardous waste incinerator or landfill leachate from a hazardous waste landfill is a hazardous waste and must be managed as such, even if it no longer contains hazardous materials. Therefore, an industry that incinerates its hazardous wastes does not see its responsibility and liability for that waste end there; it must still ensure that the ash is managed properly in perpetuity.

The law does contain a mechanism for "delisting" wastes for a particular generator if the generator can prove that it is truly nonhazardous. The procedure is long, laborious, and expensive, but the expense may be worthwhile if the industry can get out from under the restrictions imposed by the RCRA. Several wastes have been delisted in the last few years.

Treatment, storage, and disposal. Hazardous waste management facilities receive hazardous wastes for treatment, storage, or disposal. They are often referred to as treatment, storage, and disposal facilities (TSDFs). These facilities are closely regulated by the EPA to ensure that they operate properly and protect human health and the environment. They may be owned and operated by waste generators to manage their own wastes, or they may be run by an independent firm receiving wastes from a number of companies.

Treatment, storage, and disposal facilities include landfills, incinerators, lined surface impoundments, holding tanks, biological and chemical treatment units, and a myriad of other processes designed to safely and efficiently manage waste. The EPA has developed very detailed regulations on design, construction, and operation of these facilities.To operate a TSDF, the operator must first obtain a permit from the EPA (or from the state if regulatory powers have been entrusted to that state), which dictates how it is to be operated, personnel training, fire and spill control plans, and record-keeping procedures. The operator must also provide insurance and adequate financial backing.

It should be noted that even small businesses that plan to store their wastes, such as spent solvents at a dry cleaning operation, must obtain a permit if the generator produces more than 100 kg of waste per month or if the wastes are to be stored over 90 days (regulations are less stringent for small-quantity generators). The expense of shipping wastes to a TSDF or recycler on a more frequent basis may be high, but it is probably less expensive than obtaining a storage permit and then meeting all the requirements contained in it.

Permit system. The RCRA provided the framework for the cradle-to-grave management of hazardous wastes. The backbone for this is the permitting system, which includes everyone involved in the generation, transport, storage, treatment, and final disposal of the hazardous waste. The RCRA delineates the responsibilities of each party and the records that must be maintained to fully account for all hazardous wastes generated, now and into the future.

Every owner or operator of a TSDF is required to obtain a permit to operate, and every shipment of waste to a TSDF must be manifested. The manifest system is designed to allow the waste to be clearly tracked from the time of creation to placement in its final resting place. The generator is responsible for testing the waste when it is created to determine if it is hazardous (unless it is a listed waste, which is automatically a hazardous waste). If it is hazardous, the generator must obtain an EPA identification number by filing the Notification of Hazardous Waste Activity form with the EPA. When it is time to ship the wastes to a TSDF, the generator must properly package, label, and mark all hazardous shipments; must use a licensed hazardous waste transporter; and must send the waste to a permitted TSDF. A Uniform Hazardous Waste Manifest form must be prepared detailing the classifications and quantities of wastes. This manifest must accompany the waste until it is properly disposed of, and all parties must keep a copy of the records.

Even though the waste generator has complied with all regulations and has sent its waste to a permitted TSDF using the manifest system, the generator is not clear of any liability or damages that occur during transport, storage, treatment, disposal, or even years after final disposal, for example, due to a leak in the liner at a hazardous waste landfill. The generator is responsible for its waste for eternity. Thus it is important that the generator minimize its waste production as much as possible and do all it can to ensure that the wastes that are produced are managed properly. The costs of remediating environmental damage from improperly disposed hazardous wastes are enormous.

Land disposal restrictions. An important provision of the 1984 HSWA was directed at minimizing contamination of groundwaters by hazardous wastes. Congress placed a restriction on the land disposal of hazardous wastes unless it could be demonstrated that there would be no migration of hazardous constituents from the disposal unit for as long as the wastes remain hazardous. As a result, the EPA banned land disposal of dioxins and solvents, and the "California List" of chemicals (a list of chemicals previously banned in California). The EPA also restricted the disposal of uncontained liquid hazardous wastes in landfills. Contained liquid wastes can be landfilled only if there is no acceptable alternative. All other liquid wastes must be solidified before landfilling or must be treated by some other process. These land disposal restrictions were further strengthened by the Land Disposal Program Flexibility Act of 1996.

Underground storage tank regulations. Underground storage tanks containing hazardous wastes were regulated by the Resource Conservation and Recovery Act of 1976, but those storing petroleum or hazardous materials were not. The Clean Water Act of 1972 required owners of large underground tanks (> 42,000 gal) to take certain measures to prevent corrosion and to test tanks periodically. These requirements, however, applied to those tanks that could potentially pollute only navigable surface waters, not groundwaters, as is the case with most leaking storage tanks. The Superfund Act authorized the EPA to respond whenever a hazardous substance is released into the environment, but Superfund cannot be used to respond to releases from leaking underground storage tanks (USTs) because petroleum is specifically exempt from the list of hazardous substances defined under CERCLA (U.S. EPA, 1996). New legislation was required to address the hundreds of thousands of underground storage tanks that may potentially leak, contaminating groundwaters all across the nation.

In 1984, Congress passed the Hazardous and Solid Waste Amendments to RCRA. This act added Subtitle I to RCRA to provide for regulation of new and existing UST systems. This subtitle was amended in 1986 as part of the Superfund Amendments Reauthorization Act to provide federal funds for corrective actions on petroleum releases from UST systems and to require compensation by the polluter to anyone harmed by the release.

The magnitude of the leaking UST problem is huge. As of August 1996, more than 314,700 releases from leaking underground storage systems had been reported. Many of these were at gasoline service stations. On average, about 30,000 new releases are reported each year. Cleanups have been initiated at over 242,000 of these sites and have been completed at over 141,000 of them.

The regulations define a UST system as any one or a combination of tanks that have 10 percent or more of their volume below the surface of the ground in which they are installed. This definition includes the tank, connected underground piping, underground ancillary equipment, and the containment system. Tanks that contain less than 110 gallons, that are used to store heating oil for consumptive use on the premises, and some that are used on farms, at wastewater treatment plants, for storm water control, or as septic tanks, are exempted. The regulations pertain to UST systems that contain regulated substances such as solvents, methanol, and ethylene glycol, as well as petroleum materials. Tanks containing hazardous wastes are exempt because they are covered under other RCRA regulations.

Underground storage tank owners and operators must register their tanks with the state or local agency that implements the UST program in their state. The agency must also be notified at least 30 days before a new tank is put into service. All tanks were required to meet leak detection requirements and upgrade requirements by December 22, 1998. In addition, owners and operators were required to meet financial responsibility requirements; perform a site check and provide corrective action in response to any leaks, spills, or overfills; replace or close any USTs that did not meet the upgrade requirements by December 22, 1998; follow regulatory rules during installation of new tanks and closure of existing tanks; and maintain records on all activities related to the UST system.

All underground tanks and associated piping are required to have *leak detection* installed. Piping must have devices that automatically shut off or restrict flow or have an alarm that indicates a leak, and the piping system must be checked regularly for tightness. Tanks must be monitored monthly using such methods as automatic tank gauging, monitoring for vapors in the soil adjacent to the tank, monitoring for liquids in the groundwater, or secondary containment with monitoring of the interstitial space between the tank and the secondary container.

Upgrading of UST systems consists of adding corrosion protection and installing devices to prevent against spills and overfills. *Corrosion protection* is required on all existing steel tanks and piping to prevent leakage. Three options are available: interior lining, cathodic protection, and a combination of the two. New tanks must be constructed of coated and cathodically protected steel, fiberglass, or steel with a fiberglass clad. *Spill protection* consists of construction of a catchment basin (a bucket sealed around the fill pipe) to contain spills and adherence to industry standards for correct filling practices. Overfill protection devices that are required include automatic shutoff devices, overflow alarms, and ball float valves.

Costs associated with these UST upgrades or replacements are not small. The EPA estimates that the cost to upgrade an existing three-tank facility at a filling station with spill, overfill, and corrosion protection is approximately $12,700. If the tank does not have high enough integrity to warrant ungrading, it must be replaced. Replacement of a three-tank facility would cost roughly $80,000–$100,000 (including closing the existing USTs and putting in the new ones), assuming no cleanup is needed (U.S. EPA, 1996).

COMPREHENSIVE ENVIRONMENTAL RESPONSE, COMPENSATION, AND LIABILITY ACT. In response to public concern about Love Canal and other recently discovered major abandoned hazardous waste sites, Congress passed the Comprehensive Environmental Response, Compensation, and Liability Act (CERCLA) in 1980. Also known as *Superfund,* CERCLA has become a significant weapon against hazardous waste problems across the country. Under CERCLA, Congress gave the EPA broad authority to regulate hazardous substances, to respond to hazardous substance emergencies, and to develop long-term solutions to serious hazardous waste problems. While RCRA deals with the cradle-to-grave management of hazardous wastes as they are produced, CERCLA focuses on uncontrolled releases of historic wastes, mainly those that were dumped prior to institution of RCRA, and on spills of hazardous materials. The definition of a hazardous material is much broader under CERCLA than under RCRA; it includes uncontrolled spills or other discharges of hazardous process chemicals as well as hazardous wastes.

Hundreds of abandoned hazardous waste sites were discovered, often with little evidence as to who the parties responsible for the dumping were. In other cases, the responsible parties were no longer in business or did not have the financial resources to remediate the sites. As a consequence, a major part of the CERCLA legislation was the creation of a $1.6 billion trust fund, supported by an excise tax on feedstock chemicals and domestic or imported petroleum, to pay for cleanup activities at abandoned waste sites. The responsible parties are still required to conduct or pay for the cleanup, but if this is not possible, the federal government can clean up the site using Superfund trust funds.

In 1986, CERCLA was reauthorized and amended by the Superfund Amendments and Reauthorization Act (SARA), which expanded the authority of the EPA to respond to needed hazardous waste site remediation and increased the cleanup trust fund to $8.5 billion. Because of the slow pace of cleanups to that point, SARA also mandated schedules for initiation of cleanup work and studies. Another significant provision of SARA is the Emergency Planning and Community Right-to-Know Act, which will be described later.

National Contingency Plan. One requirement placed on the EPA by CERCLA was the development of a National Contingency Plan (NCP). The NCP describes the steps that responsible parties must follow in reporting and responding to hazardous substance releases into the environment. Industries are responsible for reporting all releases to the EPA's National Response Center. The EPA then investigates the severity of the release and decides on appropriate corrective action. The goal of the NCP is to select remedies that protect human health and the environment, that maintain protection over time, and that minimize untreated waste left at the site.

As part of the NCP, the EPA was required to develop a method for assessing and ranking hazardous waste sites based on hazard potential. The resulting list of sites, which is constantly changing as new sites are discovered and old sites are remediated, is known as the Superfund Site Inventory (CERCLIS). Currently, more than 40,000 sites are listed on CERCLIS. High-ranking sites are listed on the National Priorities List (NPL) and are eligible for Superfund funds (there are currently approximately 1300 sites on the NPL). The EPA has developed a detailed, thorough grading system for numerically ranking sites on the NPL.

Each site is subjected to a Remedial Investigation/Feasibility Study (RI/FS), which defines the contamination at the site, the degree of contamination, environmental pathways, potential adverse effects to human health and the environment, and feasible remedial designs. The EPA selects one of the proposed remediation plans and presents it as a Record of Decision (ROD). This plan must then be carried out by the potentially responsible parties (PRPs, those companies deemed by EPA to be responsible for the contamination), or it will be done under the direction of the EPA and the costs will be billed to the PRPs.

The time required to go from discovery of a hazardous waste site to initiation of cleanup can be many years (often 10 years or more) because of the complexity of the process. The actual cleanup effort may take one to two decades or more. The costs to industry associated with these cleanups is obviously enormous. In the future, as the more critical sites are cleaned up, greater emphasis will be placed on uncontrolled releases of hazardous materials from municipal landfills, where many industries have disposed of their wastes and where most consumer products go after they have served their useful purpose by the public. It is certainly in the best interest of industry to take all appropriate actions to minimize the creation of waste materials, either during processing or from post-consumer use of a product, in order to minimize these costs.

Brownfields Initiative. One result of the public concern over hazardous waste sites is the difficulty businesses have selling or redeveloping property that may have been contaminated by hazardous materials. According to CERCLA, property owners are partially responsible for the cost of cleanup of a contaminated site, even if they were not the owners or involved in any way at the time of the contamination. Most people are reluctant to purchase a piece of industrial property, with the potential to have to assume this liability. Also, industries may be reluctant to sell property, either because immediate cleanup may be needed or because they don't want to assume any future liability from improper practices by the new owners which may contaminate the site; it may be impossible to determine who was responsible for the pollution and both parties may be deemed responsible. The result is a rapidly increasing inventory of industrial property across the nation that is considered unusable for sale or redevelopment. To overcome this problem, the EPA, using its powers under CERCLA, recently launched the Brownfields Initiative.

Brownfields are abandoned, idle, or underused industrial and commercial facilities where expansion or redevelopment is complicated by real or perceived environmental contamination. The Brownfields Initiative was developed to empower states, communities, and others in economic redevelopment to work together in a timely manner to assess, safely clean up, and sustainably reuse brownfields. The goal of the program is to sufficiently clean up brownfields sites so as to remove 27,000 of the 40,000 sites on the Superfund site inventory and allow for their redevelopment. It is hoped that the result will be more use of brownfields for redevelopment, rather than the development and potential contamination of pristine "greenfields." Cleanup activities have to be sufficient to protect the health and well-being of surrounding communities. However, they need not be as rigorous as is usually required in Superfund RODs, where the objective is to bring the site back to essentially pristine conditions so that there will be no restrictions on use of the land in the future.

The EPA is currently running a pilot program to show how brownfields can be returned to productive use and to develop streamlined cleanup procedures. Federal funds, up to $200,000, are provided for environmental activities preliminary to cleanup, such as site identification, site assessment, site characterization, and site remediation planning and design.

Several incentives are being provided or are proposed for brownfields developers, including indemnification from groundwater contamination caused by discharges occurring before cleanup and redevelopment, possible tax incentives, and the possible designation of the community as an Empowerment Zone and Enterprise Community (EZ/EC).

EMERGENCY PLANNING AND COMMUNITY RIGHT-TO-KNOW ACT. Partly in response to the 1984 chemical tragedy in Bhopal, India, where a release of toxic gas killed or injured thousands of people (see Chapter 3), Congress enacted the Emergency Planning and Community Right-to-Know Act (EPCRA) in 1986. It was passed as a stand-alone provision, Title III, of SARA. To reduce the likelihood of such a disaster occurring in the United States, Congress mandated the development of emergency plans for responding to chemical accidents and required businesses and industries to provide local governments and the public with information about possible chemical hazards in their communities. All facilities must also immediately notify the local emergency planning districts regarding all hazardous materials releases at their sites. The law is based on the assumption that the more citizens know about chemical hazards in their communities, the better equipped they and their local governments will be to make decisions and take actions that will better protect them from unacceptable risks.

The EPCRA contains three subtitles. Subtitle A, Emergency Planning and Notification, establishes mechanisms to enable states and communities to prepare to respond to unplanned releases of hazardous materials. Subtitle B, Reporting Requirements, requires submission of inventory-related data on hazardous chemicals at the site and annual reporting to the EPA and the state on environmental releases of listed toxic chemicals manufactured, processed, or otherwise used at the facility. Subtitle C, General Provisions, lists a variety of general provisions, including penalties for violations of the reporting requirements.

Each community must have a local emergency plan, developed by a committee representing a broad cross section of the community, including elected officials, law enforcement officials and firefighters, health and environmental workers, community groups, news reporters, owners and operators of industrial plants, and other users of chemicals in the community. The emergency plan must include (U.S. EPA, 1993a) the following:

- Information on facilities and transportation routes where hazardous substances are present.
- Emergency response procedures, including evacuation routes, for dealing with accidental chemical releases.
- Notification procedures for those who will respond to the emergency.
- Methods for determining the occurrence and severity of a release and the areas and populations likely to be affected.
- Ways to notify the public of a release.
- A list of emergency equipment available.
- A program and schedules for training local emergency response and medical workers to respond to chemical emergencies.
- Methods and schedules for conducting "exercises" (simulations to test elements of the plan).
- A designated community coordinator and facility coordinators at each affected business to carry out the plan.

If there is a chemical accident at a facility or on a transportation route, and if the accident results in the release of a hazardous substance, the facility or transporter must immediately notify the community and the state of the release. Chemicals covered by this regulation include not only the 366 chemicals listed as "extremely hazardous substances," but also more than 700 other hazardous substances subject to the Superfund hazardous waste cleanup law. For some of the more hazardous materials, releases of more than 1 pound must be reported. For others, the reporting quantities range from 10 to 10,000 pounds.

The law also requires that facilities must report to the public the amounts, locations, and potential effects of hazardous chemicals being used or stored in designated quantities in the community. All companies, large or small, manufacturing or nonmanufacturing—totaling more than 4.5 million facilities—are subject to this requirement. This information can be used to help businesses reduce inventories; it is also invaluable to fire departments and public health officials in case of an emergency. To meet this requirement each facility must submit copies of all Material Safety Data Sheets (MSDSs) for chemicals at the facility, which provide information on a chemical's physical properties and health effects; and each must submit an annual inventory of all chemicals.

A major component of EPCRA is the requirement for an annual report of all routine releases of any of some 320 toxic chemicals into the air, water, or soil. Each facility must estimate the total amount of these chemicals that it releases, either accidentally or as a result of routine plant operations, or as waste transported to another location. These data are compiled by the community and the state and published annually as the Toxics Release Inventory (TRI). The TRI can be used as an indicator of public exposure to toxic materials and as a gauge of industrial efforts to minimize waste. This is discussed in more detail in Chapter 7.

HAZARDOUS MATERIALS TRANSPORTATION ACT. The Hazardous Materials Transportation Act of 1975 (HMTA), as amended, is designed to improve the federal regulatory and enforcement authority to protect the public adequately against risks to life and property which are inherent in the transportation of hazardous materials in commerce. The statute is managed and enforced by the Department of Transportation. Any material that may pose a risk to health and safety or property may be designated as a *hazardous material.* Note that hazardous materials are not necessarily hazardous wastes; indeed, the vast majority of chemicals classed as hazardous materials are process chemicals in transit to industry rather than wastes leaving industry for disposal.

The HTMA regulations apply to

> any person who transports, or causes to be transported or shipped, a hazardous material; or who manufactures, fabricates, marks, contains, reconditions, repairs, or tests a package or container which is represented, marked, certified, or sold by such person for use in the transportation in commerce of certain hazardous materials.

Thus not only is the transport firm liable to these regulations, but so also are the firm shipping the material and the manufacturer, reconditioner, and tester of the containers in which the material is shipped.

In addition to meeting DOT and EPA requirements, transporters also have to meet a myriad of state and local regulations. To bring sense to this maze of often conflicting regulations, Congress enacted the Hazardous Materials Transportation Uniform Safety Act (HMTUSA) in 1990. The statute includes provisions to encourage uniformity among different state and local highway routing regulations and criteria for issuing permits.

Figure 4.5 is a flow chart summarizing the steps a generator must take before the actual transport of hazardous materials, including a brief description of each requirement and the associated CFR citation (Kaster and Zdelar-Bush, 1996). These steps apply to the transport of *all* hazardous materials.

Placards and labels are used to identify potential dangers from hazardous materials in case of an accident. The DOT classifies materials based on ten hazard classes:

1. Explosives.
2. Gases.
3. Flammable liquids.
4. Flammable solids, spontaneously combustible materials, and materials dangerous when wet.

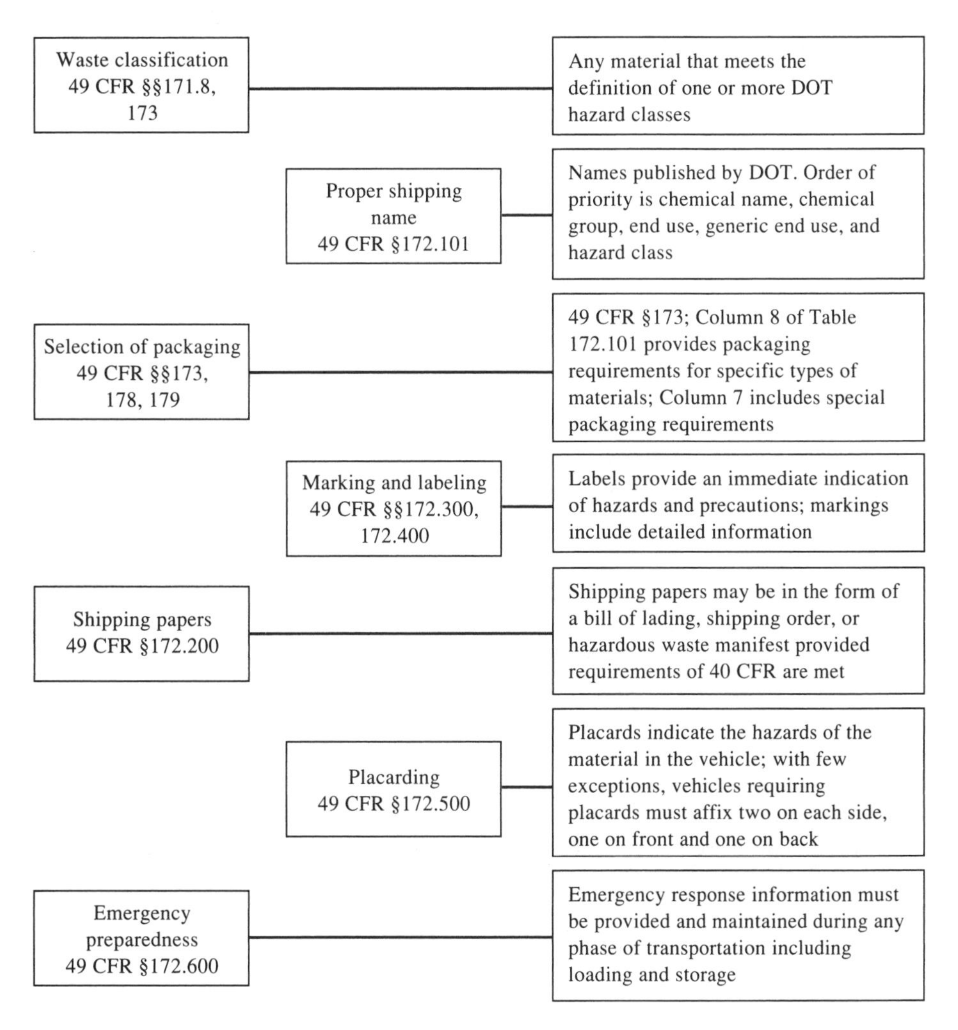

FIGURE 4.5
Department of Transportation requirements governing the shipment of hazardous materials. (Adapted from Kaster and Zdelar-Bush, 1996)

5. Oxidizers and organic peroxides.
6. Poisonous materials.
7. Biohazards.
8. Radioactive hazards.
9. Corrosives.
10. Other regulated materials.

Each hazardous material is placed in one class based on its most significant hazard, even though it may exhibit more than one hazard property.

FEDERAL INSECTICIDE, FUNGICIDE, AND RODENTICIDE ACT. The first pesticide control law was enacted in 1910 and was replaced in 1947 with a law requiring that pesticide products be registered with the Agriculture Department. In 1970, Congress moved authority for enforcing the statute to the EPA. The current statute, known as FIFRA, was enacted in 1972 and amended in 1975, 1978, 1980, 1990, and 1994. The objective of FIFRA is to regulate the distribution, sale, and use of pesticides in the United States. No manufacturer or importer may produce or sell a product for pest control unless the compound is registered with the EPA. A pesticide product may be registered or remain registered only if it performs its intended functions without causing unreasonable adverse effects on the environment.

Pesticides are widely used in industry to control vermin and other pests in the workplace. They are also commonly used in industrial processes; for example, in paper pulping operations they are used to control the growth of fungi in the pulp slurry. Industries must be cognizant of the pesticide use restrictions imposed by FIFRA and ensure that any pesticides used are used properly.

4.3.4 Laws Pertaining to Products

OCCUPATIONAL SAFETY AND HEALTH ACT. The Occupational Safety and Health Act (OSHA) was enacted by Congress in 1970 and amended in 1990. This was not the first time the workplace was regulated. Inspection of factories by regulatory agencies began in England in the early nineteenth century in response to public protest against the working conditions for women and child laborers. Later governments adopted regulations against unhealthful and dangerous working conditions. Factory codes soon became standard in every industrialized country. Before 1970, each U.S. state regulated the inspection of factories within its own borders, but acceptable conditions varied from state to state. In the late 1960s and early 1970s, though, there was an awakening not only to the need to uniformly protect public health and the general environment, but also to the need to monitor workplace conditions in order to improve worker health and well-being. In a sense, there was a desire to improve the indoor environment in the workplace. There was also an economic incentive because the personal injuries and illnesses arising out of workplace situations impose a substantial burden: lost production, wage loss, medical expenses, and disability compensation payments. The intent of the OSHA legislation is to ensure safe and healthful working conditions for working men and women by enforcing standards developed under the act and by providing for research, information, education, and training in the field of occupational safety and health.

The Occupational Safety and Health Administration (also called OSHA) and its 25 state partners have jurisdiction over more than 100 million working men and women and their 6.5 million employers. They set standards for workplace health and safety, conduct inspections to ensure that standards are being carried out, provide guidance on better operating practices, and penalize those who violate the standards.

When OSHA was enacted, the National Institute for Occupational Safety and Health (NIOSH) was established to develop recommended occupational safety and health standards. These standards, based on extensive scientific testing, are then enforced by OSHA.

Many people feel that OSHA has been driven too often by numbers and rules, rather than by results. Businesses often complain about overzealous enforcement and burdensome rules, in some cases rules that make no sense. However, since 1970, the overall workplace death rate has been cut in half, reducing the death toll by 100,000; brown lung disease among textile workers due to cotton dust has been virtually eliminated, and deaths from trench cave-ins have declined by 35 percent. Inspections by OSHA are generally successful: in the three years following an OSHA inspection that results in penalties, injuries and illnesses at that workplace drop on average by 22 percent. Overall injury and illness rates have declined in the industries where OSHA has concentrated its attention, whereas those rates have remained unchanged or have actually increased in the industries where OSHA has had less presence (OSHA, 1996).

Even with these successes, there is still a long way to go to ensure worker safety. More than 6000 Americans still die each year from workplace injuries, an estimated 50,000 people die from illnesses caused by workplace chemical exposures, and 6 million people suffer nonfatal workplace injuries. The injuries alone cost the economy more than $110 billion a year.

Industrial pollution prevention programs should be instituted in conjunction with OSHA standards in order to minimize threats to worker health and safety.

4.3.5 Laws Pertaining to Pollution Prevention

POLLUTION PREVENTION ACT. Essentially all of the regulations on waste management described to this point have dealt with end-of-pipe requirements. These all assume that creation of the pollution is inevitable and that the waste must be treated after it is produced. Commonly, this meant that the waste was just shifted from one receiving medium to another to satisfy the most pressing environmental regulation at that time. Aqueous contaminants were removed from water and applied to land or to the air; air pollutants were scrubbed out of the air, creating a water pollution problem; contaminated soils were washed, with the contaminants moving into the water phase; and so on. Since many contaminants cannot be destroyed, at least economically, waste management was essentially a shell game, in which the waste was constantly being moved from one medium to another, with no real solution to the problem.

For many years, though, progressive environmental engineers have taken a different position. They have stated that many wastes can be prevented from being produced in the first place, and that the costs associated with process changes to minimize waste production are often lower than the cost of treating the waste after it is produced. This eventually became common practice in some industries, such as metal finishing. Recovery and reuse of plating bath solutions and rinse waters spread throughout the industry during the 1970s and 1980s, resulting in significant decreases in the amount of wastes these industries had to treat, while at the same time reducing the cost of process chemical purchases because of the reuse of reclaimed chemicals. This philosophy was slow to spread across all industries, though, and end-of-pipe treatment remained the generally accepted practice until passage of the Pollution Prevention Act of 1990.

The Pollution Prevention Act (PPA) of 1990 finally established a pollution management hierarchy system. Pollution prevention was declared to be the nation's primary pollution control strategy. The goal of the act is to shift the nation's waste strategy from the control of waste after its generation to the reduction of waste at its source. With this as the goal, Congress decreed that pollution should be prevented or reduced at the source whenever feasible. Recognizing that not all waste can be prevented, Congress declared that pollution that cannot be prevented should be recycled if possible. Treatment and disposal in an environmentally safe manner should be the last resort.

To support the pollution prevention hierarchy, the EPA was directed to establish a Pollution Prevention Office, whose responsibility is to develop and implement a strategy to promote source reduction. The Congress also authorized a Source Reduction Clearinghouse to compile information on source reduction and make it available to the public as well as a grant program, which enables states to obtain EPA matching grants for providing source reduction technical assistance to businesses.

The Pollution Prevention Act is a significant departure from previous environmental legislation in that it relies almost entirely on *voluntary compliance* by industry, rather than on mandatory regulatory controls. The only major requirement of industry is to annually file a toxic chemical source reduction and recycling report with EPA.

Other features of the Pollution Prevention Strategy are directed at providing incentives to industry to implement pollution prevention programs. Among these are the Industrial Toxics Program (the 33/50 Program) and the Green Lights Program. Under the 33/50 Program initiative, the EPA targeted 17 toxic pollutants and asked industry to voluntarily commit to reducing the amount of these pollutants they release by 33 percent from 1988 to 1992 and by at least 50 percent by the end of 1995. The goal of this initiative was to determine if voluntary pollution prevention programs could work. Many industries voluntarily complied, and the initiative was deemed a major success. The Green Lights Program is also voluntary. By joining Green Lights, more than 2000 businesses have agreed to install energy-efficient lighting. Since lighting accounts for 20 percent of all electricity sold in the United States, the program not only leads to a monetary savings to business, but also reduces the production of carbon dioxide, sulfur dioxide, and nitrogen oxides from electricity generation. The EPA provides technical assistance to the participants to help them assess what needs to be done and how to most cost-effectively accomplish it. These and other initiatives are described in detail in Chapter 13. Much of the strategy is directed toward educating industry concerning the economic benefits to industry of pollution prevention.

Many individual states have also instituted pollution prevention laws to supplement the federal law. Some of these are also voluntary, but many contain strict mandatory provisions for reducing waste generation. For example, the states of Massachusetts and New Jersey established goals of 50 percent reduction in toxic waste generation by 1997. Coverage of all these laws is beyond the scope of this book; the reader should consult the state pollution prevention office for regulations pertaining to the state in question.

EXECUTIVE ORDERS. In addition to federal statutory law, a number of executive orders have been issued that pertain to pollution prevention. Executive orders are executed by the president without passage by Congress, but they are binding on federal agencies and affiliated entities. Among these are requirements for energy and conservation programs at federal facilities; mandatory compliance with EPCRA at federal facilities, which previously had been exempted; revisions of procurement procedures to ensure purchase of energy-saving equipment and use of alternatives to ozone-depleting materials; encouragement of the use of recycled materials; and the installation of recycling and reuse programs at federal facilities. The overall goal is to have the federal government set an example for business as to the efficacy of pollution prevention programs. A more detailed description of these executive orders can be found in Thurber and Sherman (1995).

ISO 14000. The widespread globalization of business and industry in recent years has led to a need to be intimately aware of international laws and regulations, as well as federal, state, and local ones. International standards for many products have been in place for many years, but it has only been since the creation of the European Union (EU) and its demand for conformity with internationally agreed-upon standards that U.S. companies that export products or services have had to seriously heed these regulations. Most of these international standards are set by the International Organization for Standardization (ISO), a worldwide federation of national standards bodies from more than 110 countries. Created in 1947, the ISO is located in Geneva, Switzerland. Conveniently, the acronym ISO is reminiscent of the Greek word *isos,* meaning "equal." The American National Standards Institute (ANSI) is the U.S. representative to ISO.

The mission of the ISO is to promote the development of standardization and related activities in the world with a view to facilitating the international exchange of goods and services and to developing cooperation in the areas of intellectual, scientific, technological, and economic activity. The objective is to facilitate trade through enhanced product quality and reliability; simplification of product design for improved usability; improved health, safety, and environmental protection; and reduction of waste. The ISO's work results in international agreements which are published as International Standards (ISO, 1996). Some of the more commonly known ISO standards are the SI system of units of measurements, the ISO film speed code on photographic film, the universal symbols used on automobile controls no matter the country of manufacture, and the ISO metric screw thread sizes.

A very important activity in recent years was the publication of ISO 9001, which sets quality standards for manufacturing—from concept to implementation—whatever the product or service. This includes requirements on everything from manufacturing processes to test equipment calibration procedures. The standard has led to the development of a concept termed *total quality management,* now used by many businesses to ensure production of quality products, whatever they may be. Complying with these standards is very expensive and time-consuming, but it is worthwhile in that it ensures a quality product. In addition, it is now a requirement for export of goods into many countries.

In 1995, ISO 14001, the proposed Environmental Management Standard, was elevated to a Draft International Standard. It was accepted by ANSI in 1996 as a national standard, and it is expected to be accepted internationally soon. ISO 14000 is a series of standards that are intended to provide businesses and other organizations with a system for managing the impact that they have on the environment. Similar to ISO 9001, ISO 14001 contains the specifications needed for a product to be certified as being in compliance with environmental safety standards. Other ISO 14000 series standards are guidance documents to assist businesses develop and implement the required Environmental Management System (EMS). These include guidelines on principles, environmental auditing, environmental performance evaluation, ecolabeling, life-cycle assessment, and environmental aspects in product standards.

The Environmental Management System is designed to address all facets of an organization's operations, products, and services. It covers environmental policy, resources, training, operations, emergency response, audits, measurements, and management reviews. The EMS requires an organization to focus its efforts on establishing reliable, affordable, and consistent approaches to environmental protection that engage all employees. The environmental protection system becomes part of the total management system, receiving the same attention as quality, personnel, cost control, maintenance, and production (Woodside and Cascio, 1996).

More details on the preparation and implementation of an EMS are provided in Chapter 7.

REFERENCES

Hanson, D. "EPA Reports Continued Air Quality Improvement." *Chemical & Engineering News* 75, no. 1 (1997): 22.

ISO. Welcome to ISO Online. http://www.iso.ch/welcome.html (1996).

Kaster, S. K., and Zdelar-Bush, D. L. "Putting Controls in Pre-Transportation Planning." *Environmental Technology* 6, no. 6 (1996): 38–40.

LaGrega, M. D., Buckingham, P. L., and Evans, J. C. *Hazardous Waste Management*. New York: McGraw-Hill, 1994.

OSHA. The New OSHA: Reinventing Worker Safety and Health. www://osha.gov/oshinfo/reinvent/reinvent.html (1996).

Thurber, J., and Sherman, P. "Pollution Prevention Requirements in United States Environmental Laws." *Industrial Pollution Prevention Handbook*, ed. H. M. Freeman. New York: McGraw-Hill, 1995, pp. 27–49.

U.S. EPA. *EPA Regulatory Agenda*. Washington, DC: U.S. EPA, 1989.

U.S. EPA. *Chemicals in Your Community: A Guide to the Emergency Planning and Community Right-to-Know Act*. EPA 550-K-93-003. Washington, DC: U.S. EPA,1993a.

U.S. EPA. *The Plain English Guide to the Clean Air Act*. EPA-400-K-93-001. Washington, DC: U.S. EPA, 1993b.

U.S. EPA. *National Primary Drinking Water Standards*. EPA810-F-94-001. Washington, DC: U.S. EPA, 1994.

U.S. EPA. *1995 National Air Quality: Status and Trends*. Washington, DC: U.S. EPA, 1995.

U.S. EPA. *UST: Options and Costs for 1998*. Washington, DC: Office of UST Publications, 1996.

Woodside, G., and Cascio, J. "Total Environmental Management." *Industrial Wastewater* 4, no. 5 (1996): 53–56.

PROBLEMS

4.1. Distinguish between a *law* and a *regulation*.

4.2. Review recent copies of the *Federal Register* in the library or on the Internet and select a proposed hazardous waste regulation. Prepare a brief synopsis of the regulation and indicate why you think the regulation should or should not be approved. What changes would you make to it to make it a better regulation?

4.3. How many regions in the United States are currently in nonattainment with respect to air quality? Is the region in which you reside in attainment? If not, for what reason?

4.4. Find and describe the Maximum Allowable Control Technology (MACT) for dry cleaners.

4.5. MTBE (methyl *tert*-butyl ether) is an oxygenated additive now commonly added to gasoline to make the fuel burn cleaner. However, recent studies have indicated potential environmental and public health problems from the use of MTBE. Research the subject and write a two-page paper describing your findings.

4.6. Contact your local wastewater treatment authority and obtain a copy of its industrial pretreatment requirements. Examine the requirements for oil and grease discharges. What is the limit and on what is the limit based? Do the limits appear reasonable to you?

4.7. Research the mechanisms for *delisting* an industrial waste under RCRA and write a paper summarizing the procedure.

4.8. You work for a company that makes wooden furniture (e.g., desks, chairs, tables) and are responsible for ensuring that the company is in compliance with all environmental regulations. What federal regulations apply to this industry?

4.9. The following wastes are being generated at industrial facilities. Determine whether they should be classified as hazardous wastes, and if so, what the hazard code should be:
(*a*) Spent cyanide plating bath solutions from electroplating operations.
(*b*) Discarded unused wood preservatives containing pentachlorophenol.
(*c*) Wastewater treatment sludges from production of creosote.
(*d*) Discarded off-specification butyric acid.
(*e*) Wastewater treatment sludges from a POTW containing 15 mg/L cadmium.

4.10. You are the environmental compliance officer at a company that is setting up a new paint coating line. Describe the steps you would take to determine if any hazardous wastes will be generated in the process that will need disposal. If there are any, what steps would you take to properly dispose of them?

4.11. Investigate what your university has done to meet underground storage tank regulations.

4.12. Determine which Superfund site is located closest to you and write a summary of the site characteristics, the cleanup status, and the schedule for remediation. This information is available on the Internet.

4.13. Obtain a Material Safety Data Sheet (MSDS) from one of your laboratories and describe the information contained in it.

4.14. A pesticide manufacturer's wastewater analysis, using the TCLP test, shows the following concentrations: arsenic, 3 mg/L; chlordane, 0.05 mg/L; lindane, 0.3 mg/L; and toxaphene, 0.4 mg/L.
(*a*) What responsibility does the company have for disposal of this waste to a receiving stream?
(*b*) What would be the responsibility if it was to be discharged into a municipal sewer?
(*c*) If it was sent to a commercial waste disposal facility, would it be considered a hazardous waste and require a manifest?

CHAPTER 5

IMPROVED MANUFACTURING OPERATIONS

5.1 INTRODUCTION

Businesses are increasingly finding themselves in a dilemma created by stockholders' demanding greater profits and regulations demanding more stringent control of pollution from their operations. Increased competition for consumers, both national and international, demands for inexpensive high-quality products, and rapid technological changes force businesses to do everything possible to reduce costs of production. In the past, this often meant minimizing pollution control activities to the least necessary to meet environmental regulations. The philosophy that any more pollution control than was essential was a waste of money is now coming to an end as more and more companies realize the overall costs of not minimizing their wastes. As environmental regulations become stricter and more contaminants must be removed from the waste stream, treatment costs are skyrocketing (see Figure 5.1). Continually retrofitting an industrial waste treatment plant to keep up with these constantly changing requirements pushes up the cost of the industry's products, making them less competitive. In addition, the costs of raw materials that go into the product are going up as suppliers try to meet environmental regulations. More efficient use of raw materials during processing and recycling of materials lost from the process are becoming essential in a well-operated industry. Finally, many companies have found that "going green" is not just a public boon but makes economic sense. The public is increasingly demanding that products, and processes used to make them, be more environmentally "friendly." Businesses are finding that it is necessary to become more environmentally benign, even if initial costs are high. In most cases, though, companies have found that they can do much to improve their operations at little or no long-term cost; indeed, in most cases they find that significant cost savings can be achieved.

The production of some waste materials during manufacturing is generally inescapable (see Figure 5.2). The objective for progressive firms is to minimize the amount of waste produced while maximizing output productivity at minimum cost. This may seem like an impossible challenge, but in reality it is not.

Many large industries are already taking strides to minimize pollution from their manufacturing operations. A recent survey of 212 large firms in the United States,

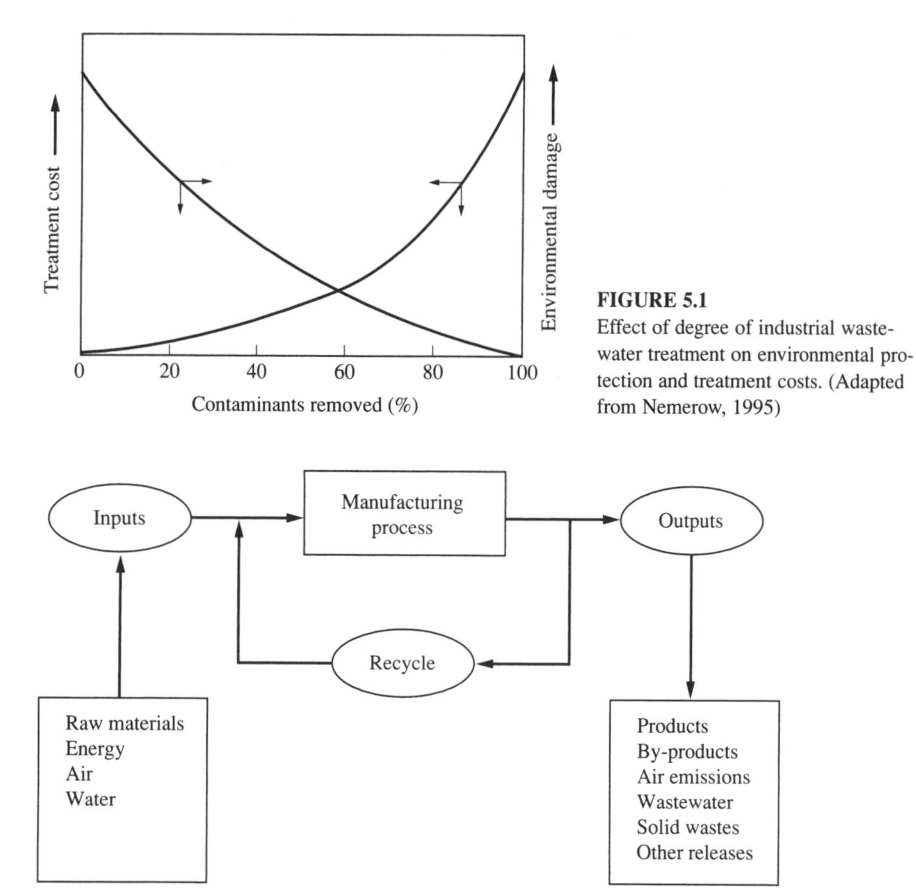

FIGURE 5.1
Effect of degree of industrial wastewater treatment on environmental protection and treatment costs. (Adapted from Nemerow, 1995)

FIGURE 5.2
A typical industrial process. (Adapted from Shen, 1995)

representing a variety of industrial sectors, showed that pollution prevention is a central element of the shift from a traditional regime of end-of-pipe treatment or remediation technology to a new regime which emphasizes technological and organizational improvements, while simultaneously improving industrial and environmental performance (Florida, 1996). More than three-quarters of the survey respondents indicated that pollution prevention was important to overall corporate performance, with 34.4 percent of those indicating that it was very important. They are backing that conviction with substantial capital expenditures. Of those firms that provided a figure, 84 percent are spending 1 percent to 10 percent of their total capital expenditures on pollution prevention activities, while 16 percent are spending in excess of this. The survey respondents strongly favor source reduction, recycling, and production process improvements over treatment, as shown in Table 5.1. More than three-quarters used recycling to improve their production processes, while about two-thirds upgraded their existing process technologies or introduced wholly new technologies. These efforts to reduce waste and prevent pollution have been successful. More than 40 percent of the respondents reported emission reductions of greater than 10 percent over the previous year, with 30 percent achieving reductions in the 11 percent to 25 percent range, 9 percent reporting 26 percent to 50 percent reductions, and 5 percent reporting in excess of 51 percent. Clearly, these industries feel that pollution prevention does pay.

Before examining ways of achieving source reduction and recycling/reuse, we will discuss the typical manufacturing processes and what can be done to realize waste reduction through proper equipment selection and use. We will then discuss how source reduction can be achieved through process changes and materials substitutions. Finally, we will examine how fairly low-tech options such as proper housekeeping, record keeping, and training can play a major role. These techniques may be considered rather mundane, but they often result in substantial reductions in wastes needing treatment, as well as in cost savings to the industry. In Chapters 9 and 10, we examine some new techniques, either in current use or being developed, for achieving additional pollution prevention.

TABLE 5.1
Elements of pollution prevention efforts by 212 industries

Type of pollution prevention effort	Percentage of respondents	Number of respondents
Source reduction	89.6%	190
Recycling	85.9	182
Production process improvements	77.8	165
Treatment	41.0	87
End-of-pipe control technology	24.5	52
Facility downsizing	7.1	15
Other	8.5	18

Source: Florida, 1996.

5.2 THE MANUFACTURING PROCESS

The objective of the manufacturing process is to transform an idea into a salable product (Halevi and Weill, 1995). The typical methodology used to accomplish this is to divide the manufacturing process into several activities, arranged serially. Figure 5.3 outlines the overall structure of an industrial enterprise. As can be seen, production is only one component of the manufacturing process. Other important segments are design, product development, quality assurance, and general management.

The steps involved in the design of a new product can be divided into three phases (Weck, Eversheim, König, and Pfeifer, 1991):

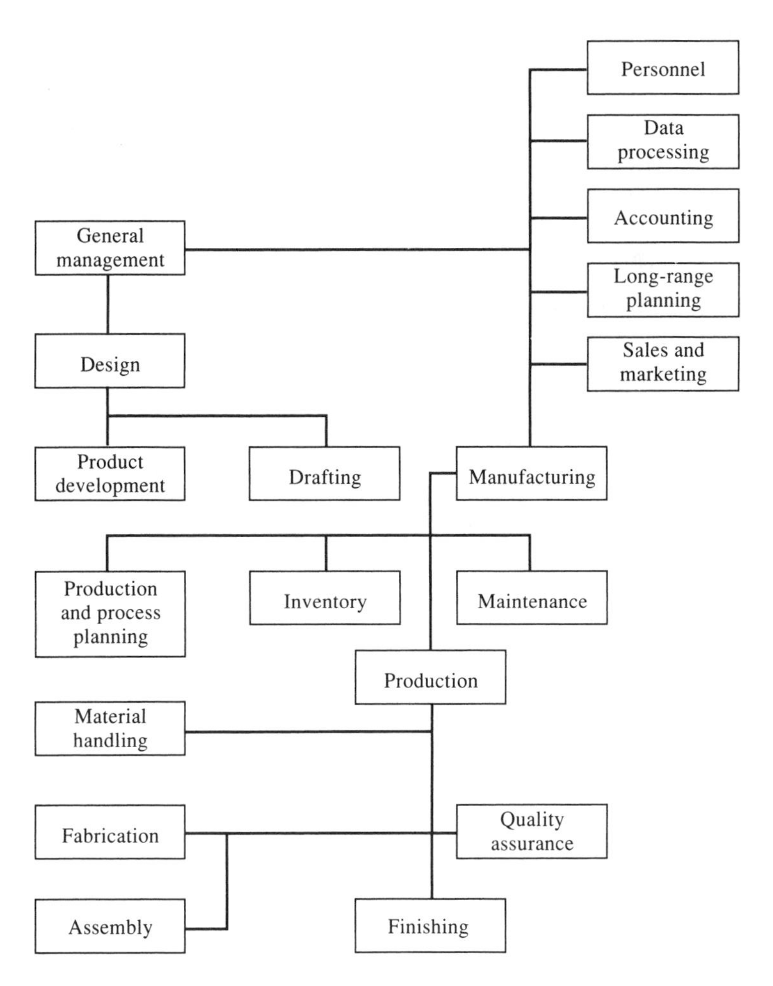

FIGURE 5.3
Overall structure of an industrial enterprise. (Adapted from Halevi and Weill, 1995)

- *Product planning* is the long-term process of identifying the product areas of interest to a company.
- *Product development* is the further examination of these product areas in relation to possibilities for the future of the company.
- *Product design* is the creation of a design for the product on the basis of market-related, functional, technical, manufacturing, and aesthetics requirements.

The objective of this process is to achieve the highest overall quality for the product at the lowest cost.

There are many areas where the design of the product can greatly influence the amount of waste produced, both during production and during and after use of the product. The designer has many considerations to ponder in designing the product: the materials used during manufacture (Are they toxic?), the conformation of the product (Will the shape of a part drag out copious quantities of plating bath chemicals, causing them to be wastes?), the method of assembly (Will welded parts make disassembly/reuse difficult?), the final appearance of the product (Is chrome plating necessary?). These topics are discussed in detail in Chapters 9 and 10, as we describe the rapidly evolving field of "design for the environment."

5.2.1 Sequential Engineering

Typically in a manufacturing organization, marketing identifies the need for new products, price ranges, and the performance customers or potential consumers expect. Design and engineering departments commonly work alone developing the technical requirements (e.g., materials and size) and final design details. The manufacturing, testing, quality control, and service groups generally see the design only in its almost complete state. As the process is sequential in progression, it is commonly known as *sequential engineering* (see Figure 5.4). This type of engineering, although the most commonly used practice in industry, is inefficient because any changes required in a later stage will cause delay and additional cost in upstream stages, and subsequent stages will be delayed until the current stage has been completed (Syan and Menon, 1994). Because there is little communication between the design, engineering, and manufacturing sectors, there is little opportunity for design or specification changes to a product, which could result in source reduction of wastes.

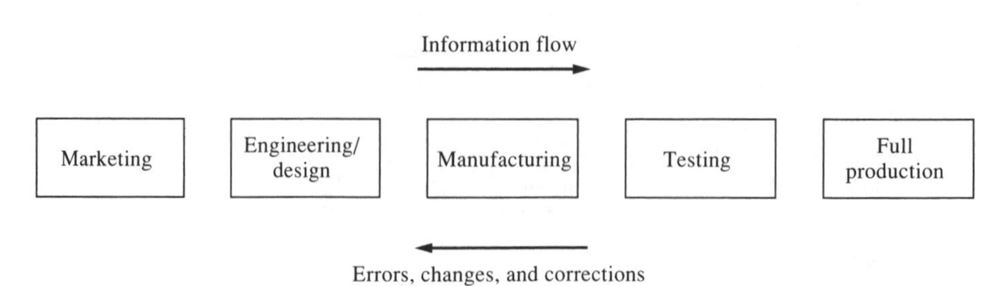

FIGURE 5.4
The sequential engineering process. (Adapted from Syan and Menon, 1994)

5.2.2 Concurrent Engineering

To overcome these deficiencies, a new mode of manufacturing, which is a dramatic departure from the past, is now being implemented by some firms. Known as concurrent engineering (also referred to as simultaneous engineering, life-cycle engineering, integrated product development, or team design), it arose from a 1982 Defense Advanced Research Projects Agency (DARPA) study into improving concurrency in the design process. Concurrent engineering (CE) provides a systematic and integrated approach to introduction and design of products (see Figure 5.5). The purpose is to ensure that the decisions made in the design stage result in a minimum overall cost during a product's life-cycle. The main objectives of concurrent engineering may be summarized as follows (Syan and Menon, 1994):

- Decreased product development lead time.
- Improved profitability.
- Greater competitiveness.
- Greater control of design and manufacturing costs.
- Close integration of departments.
- Enhanced reputation of the company and its products.
- Improved product quality.
- Promotion of team spirit.

The subsets of CE include *design for manufacture,* the consideration of how well a design can be integrated into factory processes such as fabrication and assembly; *design for assembly,* the consideration of ease of assembly, error-free assembly, common part assembly, and the like; *design for serviceability,* to facilitate initial installation, as well as repair and modification of products in the field; and *design for reliability,* the consideration of such topics as electrostatic discharge, corrosion resistance, and operation under variable ambient conditions (Graedel and Allenby, 1996). The term "*design for X*" has been coined to describe these various segments, where *X*

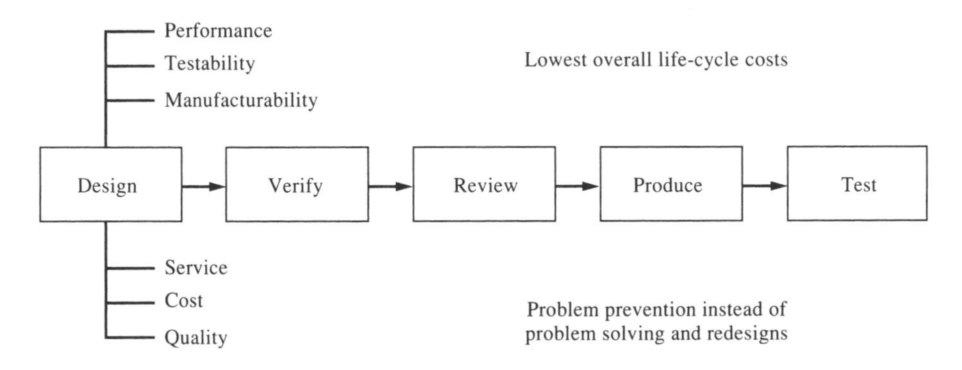

FIGURE 5.5
The concurrent engineering process. (Adapted from Syan and Menon, 1994)

stands for the particular segment. These are all intended to improve the efficiency and reduce the cost for manufacturing the product. To these have been added design subsets intended to minimize waste, both during manufacture and after the product has served its useful life: *design for the environment* and *design for disassembly,* respectively. These topics are discussed in more detail in Chapter 9.

5.2.3 Manufacturing Processes

The objective of most industrial processes is to maximize the amount of output while minimizing production costs. Until recently, the production of sidestreams (which typically have been sent to waste) has received little attention. As an example, a dog food production plant in Illinois in the 1970s purchased meat, grease, grains, and other materials and blended them together to produce the dog food. The process was not very efficient and large amounts of grease, which had been purchased as a dog food ingredient, were wasted to the sewer. The company was not interested in modifying the process to minimize the grease loss because the expense involved was not worth the savings in purchase of the inexpensive grease. Only after the grease totally clogged the wastewater treatment plant, requiring expensive remediation, did the company decide to improve the blending process. This may seem like a sloppy piece of management, but in actuality it is the way decisions are commonly made.

In many cases, simple changes to a process can result in substantial savings in process chemicals and waste treatment. Putting continuously flowing rinse tanks on an automated switch so that water flows only when needed is one example. In other cases, much more elaborate and high-tech changes are needed. Although often expensive, they still commonly pay off in a reasonably short period of time.

Following is a very brief description of some of the more common industrial processes used. A more complete description of industrial processes can be found elsewhere (Theodore and McGuinn, 1992; Higgins, 1989). The processes discussed here are those cited in later parts of this book. As the pollution prevention specialist for a particular company examines his/her processes, he or she should be cognizant of the advantages and disadvantages, pollution potential, and ease of upgrading that each type of process presents. The engineer should also consider whether batch processes can be replaced by continuous processes, which are usually less polluting, as will be discussed later. The selection of materials used in industrial processes, and whether substitution of more benign chemicals can be beneficial, is presented in Chapter 9.

CHEMICAL REACTORS. The reactor is the heart of the manufacturing process. It is the place where raw materials are converted into products. This is usually the key unit process in a production facility requiring reaction of chemicals or synthesis of new ones. It is also often the primary source of waste materials. Reactors may be operated in batch or continuous mode; may be a single unit or may be multiple units operated in series or in parallel; may involve various phase combinations (liquid-liquid, liquid-solid, liquid-solid-gas, or gas-solid); and may provide contact between phases by one of several methods (stirring, packed bed, fluidized bed).

In a *batch process,* all reagents are added to a stirred tank at the beginning of the reaction; no material is fed into or removed from the reactor during the rest of the reaction period. Stirring is commonly used to initially mix the ingredients, to maintain complete mixing during reaction, and to increase heat transfer. After some time period, the reaction is stopped and the product is withdrawn. The reaction may not be carried out to its stoichiometric endpoint, because this may take too long for small gains in product recovery near the end of the reaction. For example, if the reaction follows first-order kinetics (where the reaction rate is proportional to the concentration of one of the reactants), conversion will be rapid at first because of the high concentration of the reactants, but the rate of reaction will continually slow down over time as the reactants are consumed and their concentrations decrease (see Figure 5.6). It may not be practical or economical to continue the reaction for a long period in order to achieve a marginally better conversion efficiency. Thus the contents extracted from the batch reactor will still contain some unreacted reagents and possibly also small amounts of other chemicals created during side reactions in the reactor. A separation step may be needed following the reactor to purify the desired product. The remaining residual material usually must be treated as a waste if it cannot be recycled or reprocessed.

Batch processes are typically used when an extended reaction period is needed, when only small quantities of a material need to be produced, or when the same reactor is needed to produce different products or different grades of a product. In addition, batch processes are more efficient than continuous flow processes when reactants might precipitate on reactor walls, because the batch reactor will be cleaned after each use; this is not the case with a continuous-flow reactor. A drawback to batch reactors is that there are usually sizable downtimes at the beginning and end of each batch reaction, while reactants are added, products are removed, and the reactor is cleaned. Pharmaceuticals are commonly made using batch reactors; dyeing of leather is also normally done in batches because the color of the dyes used may change frequently.

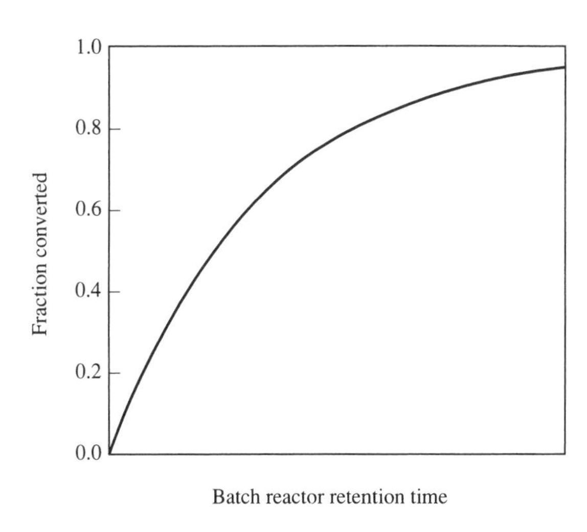

FIGURE 5.6
Rate of first-order conversions of reagents into product in a batch reactor.

The amount of time necessary for a batch reaction depends on the desired conversion efficiency and on the reaction kinetics involved. The reaction rate is usually expressed as the rate of change of one component in the reaction:

$$C + D \rightarrow E$$

$$\frac{-d[C]}{dt} = k[C]^n$$

where C = concentration of a reactant (mg/L)
k = proportionality constant (units depend on the reaction order)
n = reaction order

The reaction order is based on the value of the exponent n that provides the best fit for the empirical data. A *zero-order reaction* is independent of the concentration, and the rate constant is constant with time:

$$\frac{-dC}{dt} = k(C)^0 = kC$$

This equation can be integrated to determine the amount of reactant, C_t, remaining at time t:

$$C_t = C_0 - k \cdot t$$

where C_0 = initial concentration of the reactant
C_t = concentration of the reactant at time t

This indicates that the reaction rate is independent of reactant concentration, and that the reactant concentration does indeed decrease linearly with time.

A *first-order reaction* is dependent on the concentration remaining of one of the reactants:

$$\frac{-dC}{dt} = k(C)^1 = k \cdot C$$

The integrated form of this expression is

$$C_t = C_0 \cdot e^{-kt}$$

Many chemical reactions are first order. The rate of the reaction begins high but decreases over time as the concentration C_t decreases.

In some cases, the reaction rate is dependent on the concentrations of both reactants; these are referred to as *second-order reactions:*

$$\frac{-dC}{dt} = k \cdot C \cdot D$$

The preceding equations can be used to determine the reaction time required to convert a desired percentage of reactants into products. In most cases, the longer the reaction time, the greater will be the percent conversion, resulting in smaller amounts of contaminating unreacted reactants in the effluent process stream. Thus pollution could be minimized by increasing reaction times. However, if undesired by-products are also produced, the amounts of these may also increase with increased reaction time. Moreover, longer reaction times mean that reactors may need to be larger and production rates (units of product/time) will be lower, which will result in higher cost. All these factors must be weighed against each other to arrive at the optimum reaction conditions.

Example 5.1. It is desired to convert reactant A into product B. A laboratory reaction rate study was performed by carrying out the reaction and measuring the disappearance of reactant A over time. The following data resulted:

t, min	*A*, mg/L
0	1000
1	225
2	50
3	10
4	2
5	1

(*a*) Determine whether the reaction is zero order or first order and the appropriate reaction rate constant k.
(*b*) Assuming that 98 percent conversion is desired, how long should the reactants be held in the reactor?

Solution.
(*a*) Determine k for each reaction time, assuming first that the reaction is zero order and then that it is first order.
For zero order:

$$C_t = C_0 - k \cdot t$$

$$k = \frac{C_0 - C_t}{t}$$

For first order:

$$C_t = C_0 e^{-kt}$$

$$k = \frac{\ln C_0 - \ln C_t}{t}$$

Solve the above equations and put the results in a spread sheet.

t, min	*A*, mg/L	*k* (zero order), mg/L · min	*k* (first order), min^{-1}
0	1000	—	—
1	225	777	1.49
2	50	475	1.50
3	10	330	1.54
4	2	250	1.55
5	1	200	1.38

It is apparent from these results that the reaction is first order (all k values are approximately equal), and that the reaction rate constant equals 1.5 min^{-1}.

(*b*) The time required to achieve 98 percent conversion can be determined as follows:

$$C_t = C_0 e^{-kt}$$

$$C_t = 0.02\,(1000\text{ mg/L}) = 20\text{ mg/L}$$

$$20\text{ mg/L} = 1000\text{ mg/L} \cdot e^{-1.5t}$$

$$t = 2.6\text{ min}$$

The reaction time required is 2.6 minutes.

Continuous-flow reactors normally operate under selected steady-state conditions. Reagents are continually fed into the reactor and the resulting product is continually removed (see Figure 5.7). The reactors may be designed to be plug flow or completely mixed reactors.

The contents of a plug flow reactor change along the length of the tubular reactor as the reactants are transported through it. For a first-order reaction, the change in reactant concentrations with time can be expressed as

$$\frac{C_t}{C_0} = e^{-kt}$$

The contents of a completely mixed reactor (commonly referred to as a CSTR, for continuously stirred tank reactor) are constant throughout the tank and are the same as the discharge material. The concentration of the product is dependent on the theo-

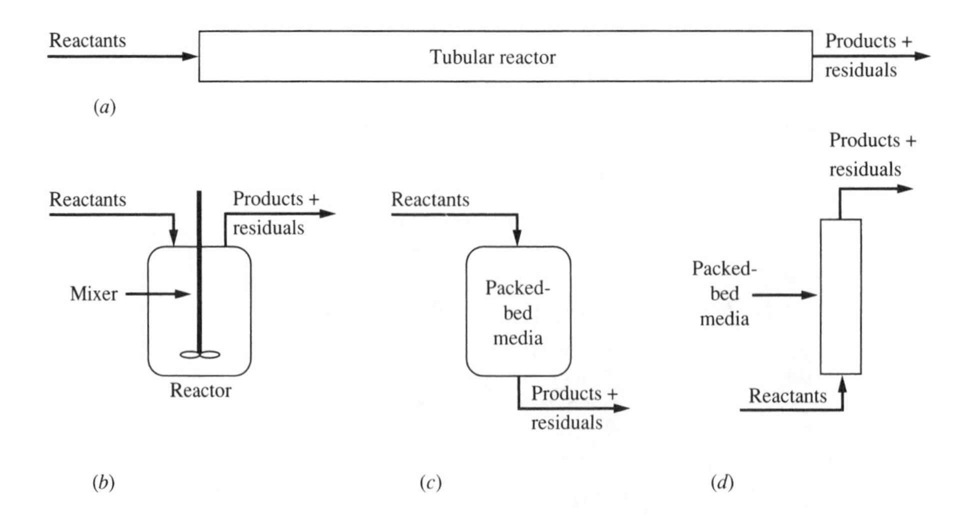

FIGURE 5.7
Continuous-flow reactors. (*a*) Plug flow reactor; (*b*) completely mixed reactor; (*c*) packed-bed reactor; (*d*) fluidized-bed reactor.

retical hydraulic retention time of the reactor (reactor volume divided by the flow rate). The longer the retention time, the higher the rate of conversion. The effluent concentration (or reactor content concentration) from a CSTR can be determined using a system mass balance:

$$\frac{C_t}{C_0} = \frac{1}{1 + k \cdot \theta}$$

where θ = hydraulic retention time.

Example 5.2. Using the data and the calculated k from Example 5.1, determine the reaction time necessary for 98 percent conversion of reactant A using: (*a*) a plug flow reactor, and (*b*) a completely mixed continuous-flow reactor.

Solution.
(*a*) For a plug flow reactor,

$$\frac{C_t}{C_0} = e^{-kt}$$

$$\frac{20 \text{ mg/L}}{1000 \text{ mg/L}} = e^{-1.5\theta}$$

$$\theta = 2.6 \text{ min}$$

The required reaction time for a plug flow reactor is the same as was found in Example 5.1 for a batch reactor, because the plug flow reactor contents flow through the reactor without mixing with any of the influent reactants.
(*b*) For a CSTR,

$$\frac{C_t}{C_0} = \frac{1}{1 + k \cdot \theta}$$

$$\frac{20 \text{ mg/L}}{1000 \text{ mg/L}} = \frac{1}{1 + 1.5\theta}$$

$$\theta = 32.7 \text{ min}$$

The required reaction time in a completely mixed reactor is considerably longer, because the reaction rate throughout the reactor is governed by the low (20 mg/L) concentration of reactant A. The reaction rate in the plug flow reactor starts high (based on a concentration of A equal to 1000 mg/L) and then decreases to a minimum rate based on 20 mg/L of A only at the reactor exit.

Plug flow reactors and continuously mixed reactors are designed to produce the desired product, but the resulting process stream also contains unreacted residues in their effluents; that from the CSTRs can usually be controlled more easily. Generally, continuous processes are less expensive to operate, particularly when large and continuous quantities of a product are needed, but they are less flexible and often require significant downtimes when desired reactions are changed.

Although CSTRs must be larger than plug flow reactors to achieve the same conversion efficiency, they have the advantages of being more controllable, having constant reactor contents throughout, having the ability to produce large quantities of one or two products at high rate, and requiring much less downtime. Both types of reactor have advantages for particular applications, and both are widely used in industry.

Stirred-tank reactors are commonly used for liquid-liquid and liquid-solid reactions. Mixing is provided to keep the reactor contents homogeneous throughout, to increase contact between reacting molecules, and to keep solid particles in suspension. Tubular reactors are generally used for gaseous reactions, but they may also be used for some liquid phase reactions. In a packed-bed reactor, the reactor is filled with a solid granular material through which the flow passes, usually in a downflow fashion. The solid phase in a packed-bed reactor can be either a catalyst or one of the reactants. It is normally used for gas or gas-liquid reactions. In a fluidized-bed reactor, the solids are buoyantly held in suspension by the upward flow of the reacting fluid. This provides better contact with all of the solid surfaces, high mass and heat transfer rates, and good mixing.

Process reactor design is an extensive subject which requires mastery of all of the fundamentals of chemical engineering: materials balances, energy balances, heat transfer, mass transfer, and momentum transfer (Hopper, 1995). The efficiency of a reactor can often be improved, thus reducing the amount of unreacted reagents going to waste, by improving the physical mixing in the reactor. Modifications to the reactor such as adding or improving baffles, installing a high-rpm motor on the agitators, or using a different mixer blade design (or multiple impellers) can improve mixing. Better distribution of the feed into the reactor, particularly in packed beds, can often greatly improve reactor efficiency. Finally, better control of the process operating conditions can also help. This control may be complex, particularly in batch reactors where conditions may be constantly changing due to the reactions, but it can often yield major improvements (Nelson, 1995).

Continuous-flow reactors need to be cleaned periodically, resulting in waste materials. Batch reactors generally need to be cleaned whenever a process changes and commonly are cleaned between each batch. This can lead to significant amounts of waste material. In either case, leaking reactors or reactor piping and appurtenances can also lead to waste generation.

HEAT EXCHANGERS. Heat exchangers are used in many chemical processes to deliver heat to or from the reactants. They may be either a heater or a cooler, depending on whether the transfer of heat is from or to the process line. Shell and tube exchangers, in which a bundle of tubes carrying one of the fluids is enclosed in a cylindrical shell containing the other fluid, are the most commonly used (see Figure 5.8). The tubes may be smooth walled or finned to increase heat transfer. Plate and frame heat exchangers, which are similar to the hot water radiators used to heat homes, are also used (Walker, 1990). Another type of exchanger is direct-contact heat exchange, in which there is no barrier separating the hot and cold streams. An example is a water cooling tower, where water is cooled by direct contact with air.

In a typical heat exchanger, heat passes from one fluid to another through one or more materials separating the two fluids (such as tubes or baffles). The amount of heat to be transferred to bring the temperature of the process fluid from its initial temperature to a new one can be described as

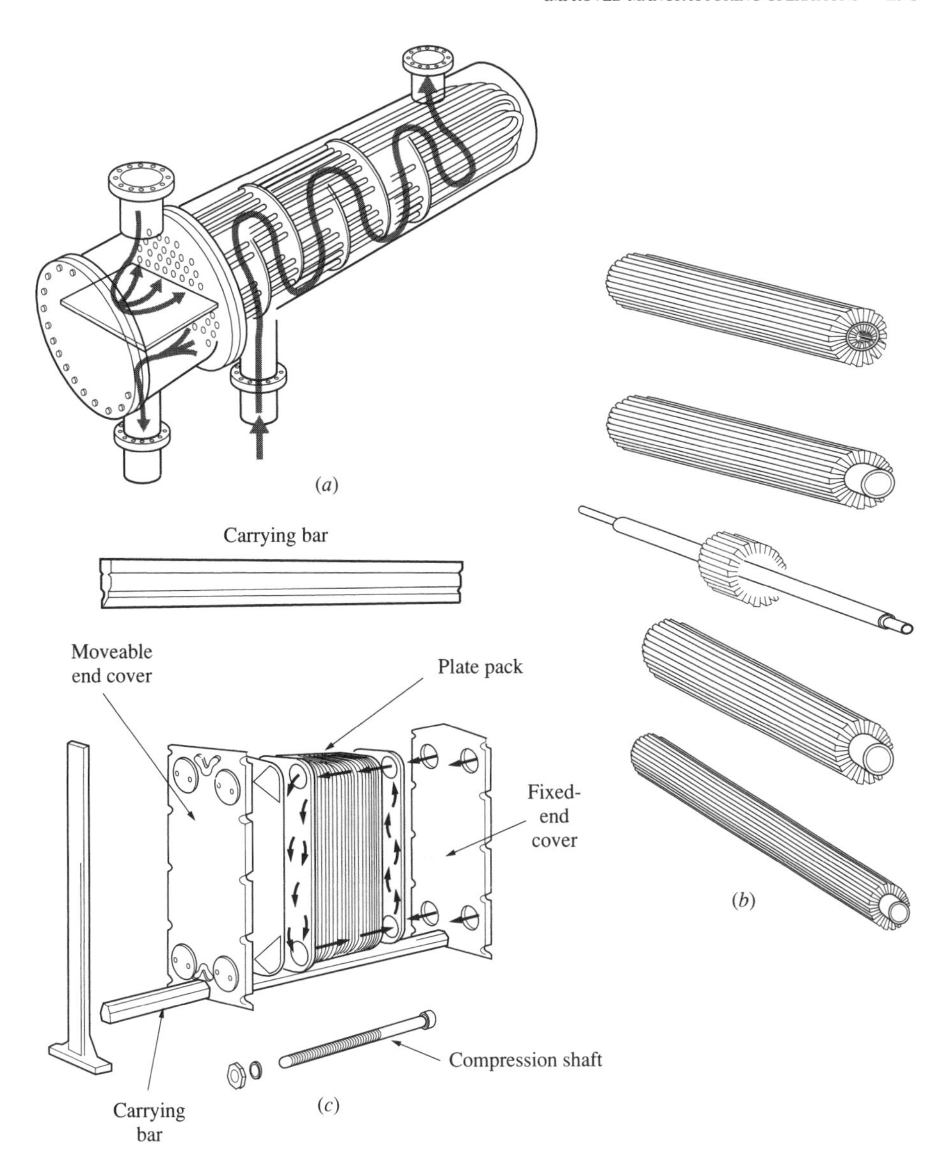

FIGURE 5.8
Typical heat exchangers. (*a*) Tube and shell heat exchanger; (*b*) longitudinal finned tubes for tube and shell heat exchanger; (*c*) plate-and-frame heat exchanger.

$$H = W \cdot C \cdot \Delta T$$

where H = rate of heat addition required to achieve the desired temperature increase, J/h
W = mass rate of material to be heated, kg/h
C = mean specific heat of material being heated, J/kg · °C
ΔT = difference between the temperatures of the process stream entering and leaving the heat exchanger, °C

The simplest form of an equation for describing the heat transfer operation may be written as

$$Q = U \cdot A \cdot \Delta T_m$$

where Q = heat transferred per unit time, W
A = area available for the transfer of heat, m^2
U = heat transfer coefficient for the material separating the two flows, $W/m^2 \cdot °C$
ΔT_m = mean temperature difference of the two fluids, °C

Typical heat transfer coefficients are listed in Table 5.2.

Heat transfer can be accomplished by passing the fluids by each other in either a cocurrent or countercurrent fashion. However, the temperature profiles, as seen in Figure 5.9, indicate that the heat exchange driving force, ΔT, decreases along the length of the heat exchanger. Therefore, heat transfer normally occurs by passing the fluids by each other in a countercurrent fashion. Thus the temperature difference between the two fluids varies along the length of the heat exchanger.

TABLE 5.2
Typical heat transfer coefficients

Hot fluid	Cold fluid	U, $W/m^2 \cdot °C$
	Shell-and-tube exchangers	
Heat exchangers		
Water	Water	800–1500
Organic solvent	Organic solvent	100–300
Gases	Gases	10–50
Coolers		
Organic solvent	Water	250–750
Gases	Water	20–300
Water	Brine	600–1200
Heaters		
Steam	Water	1500–4000
Steam	Organic solvent	500–1000
	Air-cooled exchangers	
Water		300–450
Light organics		300–700
Heavy organics		50–150
	Jacketed vessels	
Steam	Dilute aqueous solution	500–700
Water	Dilute aqueous solution	200–500

Source: Adapted from Sinnott, 1993.

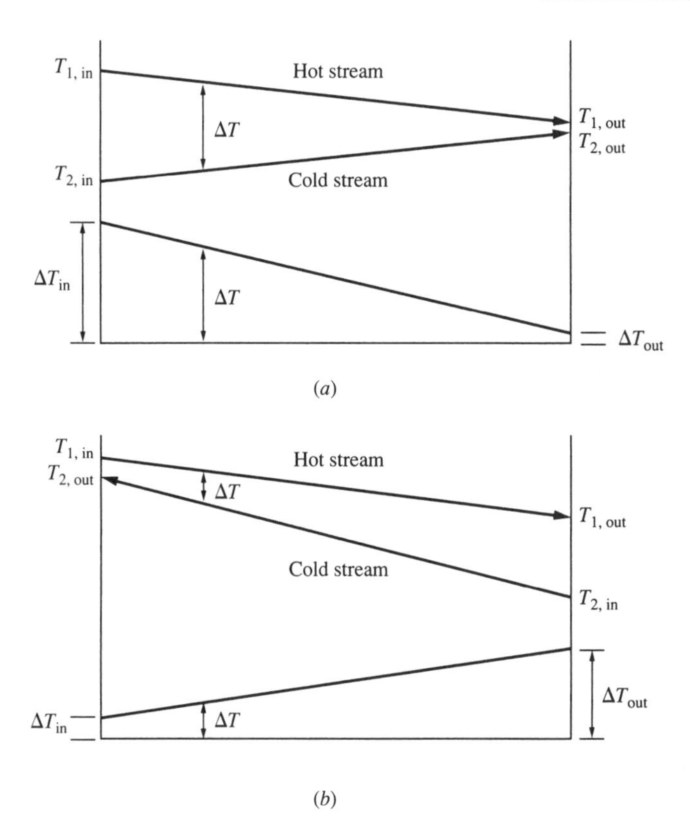

FIGURE 5.9
Temperature profiles for (*a*) cocurrent and (*b*) countercurrent heat exchangers.

The *logarithmic mean temperature difference* can be determined using (Sinnott, 1993)

$$\theta_m = \frac{\theta_1 - \theta_2}{\ln(\theta_1/\theta_2)}$$

where θ_m = logarithmic mean temperature difference, °C
θ_1 = temperature difference at the inlet to the heat exchanger, °C
θ_2 = temperature difference at the outlet to the heat exchanger, °C

Heat exchangers can contribute to waste generation in several ways. Scale builds up on surfaces and must be removed frequently and sent to waste to avoid negatively affecting heat transfer. This scale results in a decrease in the heat transfer coefficient, thus requiring an increase in the heat exchanger area or a greater temperature difference between the two fluids to effect a desired temperature change. When heated, many chemical systems form precipitates or sludges, which must be removed and properly disposed of. If there are any small leaks in the system, the heat transfer medium can become contaminated, possibly necessitating replacement. Direct-contact heat exchangers can be a significant source of air pollutants if the stream being cooled contains any volatile organic compounds.

The methods for reducing heat exchanger waste include reducing sludge formation, using scale inhibitors (antifoulants) to prevent scale buildup, and improving cleaning procedures (Allen and Rosselot, 1997). Sludge formation can be reduced by lowering the temperature of the steam or other heated stream used in the heat exchanger. In some cases, this can be done by selective use of other heated streams in the facility that need to be cooled, and that have a lower temperature than the available steam. Not only does this allow for a better match between temperatures of the streams to be heated and cooled, but it also reduces the factory's steam requirement, thus reducing energy costs and the pollution resulting from steam generation. The use of heat exchanger networks is discussed in more detail in Chapter 10.

Example 5.3. A company needs to raise the temperature of a process stream from 25°C to 50°C before placing it in a reactor for the next process step. The process stream flow rate is 10,000 kg/h and the stream's heat capacity is 4187 J/kg·°C.

(*a*) How much heat must be supplied to the process stream to achieve the desired temperature?

(*b*) Assuming that there is a 15,000 kg/h waste heat source (at 85°C) that could be used to heat the process stream, and that the heat exchanger has a heat transfer coefficient of 2 kW/m^3·°C, calculate the heat exchanger surface area required.

(*c*) If the heat exchanger is used, what will be the annual energy cost savings from not having to directly heat the process stream? Assume a heating energy cost of \$0.06/kW/h.

Solution.

(*a*) Calculate the required heat exchange to bring the process stream from 25°C to 50°C:

$$H = W \cdot C \cdot \Delta T$$

$$H = (10{,}000\ \text{kg/h})(4187\ \text{J/kg} \cdot {}^\circ\text{C})(50 - 25^\circ\text{C})$$

$$= 1.047 \times 10^9\ \text{J/h} = 290.8\ \text{kW}$$

(*b*) The process stream inlet temperature is 25°C and the outlet temperature is 50°C. The cooling water inlet temperature is 85°C. The cooling water outlet temperature can be calculated as follows, because the heat transferred to raise the temperature of the process water came from the cooling water:

$$H = W \cdot C \cdot \Delta T$$

$$1.047 \times 10^9\ \text{J/h} = (15{,}000\ \text{kg/h})(4187\ \text{J/kg} \cdot \text{h})\Delta T$$

$$\Delta T = 16.7^\circ\text{C}$$

Therefore, the cooling water outlet temperature = 85 − 16.7 = 68.3°C. The logarithm mean temperature can now be determined:

$$\theta_m = \frac{43.3 - 30}{\ln(43.3/30)} = 36.2^\circ\text{C}$$

The required heat transfer area is

$$Q = U \cdot A \cdot \Delta T = U \cdot A \cdot \theta_m$$

$$290.8\ \text{kW} = (2\ \text{kW/m}^2 \cdot {}^\circ\text{C})(36.2^\circ\text{C})(A)$$

$$A = 4.0\ \text{m}^2$$

(*c*) At a cost of \$0.06/kWh, the annual energy savings achieved by using the heat exchanger is

$$(290.8\ \text{kW})(8760\ \text{h/yr})(\$0.06/\text{kWh}) = \$152{,}844$$

EVAPORATION/DRYING. Evaporation and drying are both industrial processes in which heat is used to remove water from a material. *Evaporation* is used to concentrate a solution consisting of a nonvolatile solute and a volatile solvent. It is normally used to produce a concentrated liquid, often prior to crystallization of the solute. With respect to pollution prevention, evaporation can also be used as a means of volume reduction, as in the recovery of metals from spent plating baths for reuse. The residue of evaporation is usually still a liquid, although sometimes a highly viscous one. In most cases, the solvent being evaporated is water. The resulting thick liquor is usually the valuable product, while the vapor is condensed and discarded. *Drying,* on the other hand, is used to remove relatively small amounts of water or other liquids from solids to produce a dry product. It is usually necessary to remove as much water as possible by mechanical means (filtration, pressing, etc.) before using thermal drying. Drying is often the final step in a series of operations producing a solid product, leaving the product ready for packaging or use. Evaporation and drying are both very energy-intensive operations, creating all of the environmental stresses associated with the use of energy.

Several types of evaporators are available. Among these are *long-tube falling-film evaporators,* in which the liquid flows as a thin film on the walls of a long, vertical, steam-heated tube (Figure 5.10*a*); *long-tube evaporators with upward flow,* in which the feed flows upward through tubes because of forced pumping or because of the decrease in liquid density with increase in temperature as it rises up in the steam-heated tubes (Figure 5.10*b*); and *direct-heated evaporators,* such as solar pans and submerged combustion units.

Some evaporator units are designed with a single evaporator, but in many cases a series of evaporators is used; these are called *multiple-effect evaporators.* Connections are made so that the vapor from one effect serves as the heating medium for the next. This significantly reduces the amount of steam required.

Drying equipment can operate in a continuous or batchwise fashion, can be agitated or unagitated, can be operated under atmospheric pressure or vacuum to reduce the drying temperature, and can provide heat to the solids directly or indirectly by conduction, convection, or radiation. The solids to be dried may be in many forms—flakes, granules, crystals, powders, slabs, or continuous sheets (McCabe, Smith, and Harriott, 1993). Heat must be applied to the dryer to accomplish the following:

1. Heat the feed (solids and liquid) to the vaporization temperature.
2. Vaporize the liquid.

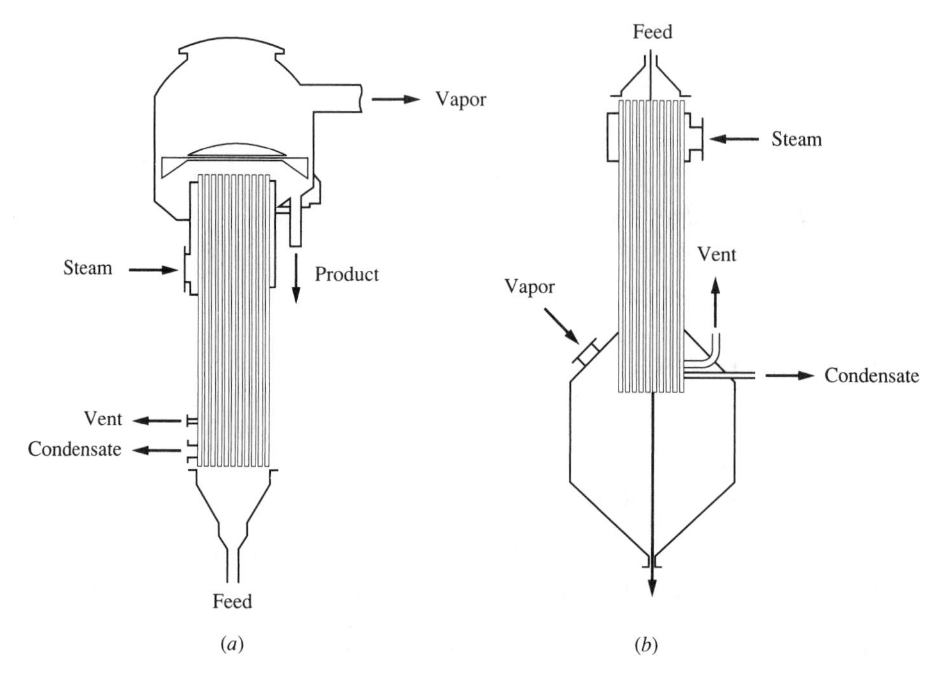

FIGURE 5.10
Long-tube evaporators. (*a*) Rising film; (*b*) falling film. (Adapted from Sinnott, 1993)

3. Heat the solids to their final temperature.
4. Heat the vapor to its final temperature.

In almost all cases, hot air is used as the heating and mass transfer medium in industrial dryers. Several types of dryers are available, including tunnel dryers, rotary dryers, drum dryers, spray dryers, and flash evaporators. Detailed descriptions of drying processes can be found elsewhere (Long, 1995).

As stated earlier, the drying process is a very energy-intensive one. Other potential environmental problems can be associated with drying. The evaporated gases must be condensed and reprocessed or must be treated as a waste. Improper operation of the dryer could lead to attrition of the solids and loss as waste in the airstream.

CRYSTALLIZATION. Crystallization is the formation of solid particles within a homogeneous phase, usually aqueous but sometimes gaseous. It is used for the production, purification, and recovery of crystalline solids. Crystallization is an important industrial process used in a wide variety of industries, including the production of specialized chemicals, pharmaceuticals, fertilizers, sugar, and salt.

When a hot, saturated solution of a material that can crystallize is cooled, crystals will form because the solution becomes supersaturated at the lower temperature. Usually, crystallization occurs at a surface that can serve as a nucleus for crystal growth. This can be the reactor wall, which makes recovery difficult. Therefore, nuclei, in the form of seeds of a previously purified batch, are added to the solution as it is cooled. The rate of cooling has to be controlled to obtain the crystal size desired. The rate of agitation can also be used to control crystal size. From a marketing standpoint, it is critical that the crystals produced be clear, of high purity, and of uniform size and that they be produced at a high yield. This usually requires that crystals be grown at a slow rate. When the rate of crystal growth is slow, considerable time may be required to reach equilibrium, though. Therefore, in an industrial setting, shorter reaction times are used, resulting in lower yield and a final process liquor that is still supersaturated with the crystalline material. The effects of this on pollution prevention are obvious. In addition, removing crystalline material from the reactor walls adds to the industry's waste load.

Several types of crystallizers are available. The simplest type consists of a *crystallization tank* in which the mother liquor is cooled to induce crystallization. These are usually batch-operated systems used to produce small quantities of crystals, but they may be continuous. In some cases, the walls (cooling surfaces) are continuously scraped to prevent fouling by deposited crystals and to promote heat transfer. *Circulating liquor crystallizers* are the most widely used industrial crystallizers. A stream of supersaturated solution is passed through a fluidized bed of growing crystals, within which supersaturation is released by nucleation and crystal growth. *Circulating magma crystallizers* are similar, except that both the mother liquor and the growing crystals are circulated through the heating and cooling equipment (see Figure 5.11). In most cases, a vacuum unit is used to provide the evaporative cooling necessary to create supersaturation. The circulating crystallizers are used when high output is needed.

Whatever the method of crystal formation, the product removed from the crystallizer is in the form of a slurry of crystals in the saturated mother liquor. The crystals must be separated from this slurry by filtration or centrifugation. The rejected mother liquor is usually returned to the crystallizer for reprocessing.

DISTILLATION. Distillation is widely used by industry and is one of the most important processes for the separation of chemical mixtures. Major users are the petroleum and petrochemical industries, which fractionate crude oil into many compounds, wine and liquor producers, and the pharmaceutical industry. It is also used in waste minimization to recover used oils and solvents for recycling and to remove toxic or hazardous materials from waste streams.

The basis for distillation is that compounds have different volatilities (vapor pressures) and can be separated from a mixture by gradually varying the temperature of the mixture. As the temperature is increased, compounds will volatilize from the mixture at their boiling points. The vapors can be collected and condensed to recover relatively pure compound. The boiling points of the liquids can be adjusted by changing the pressure at which the distillation is run. By raising the pressure, the boiling point of condensed gases can be raised to permit convenient condenser temperatures. The boiling points of high-boiling liquids can be reduced to prevent thermal decomposition by running under vacuum (Long, 1995) or by injecting steam into the liquid to form an azeotropic mixture that has a lower boiling point (Higgins, 1989).

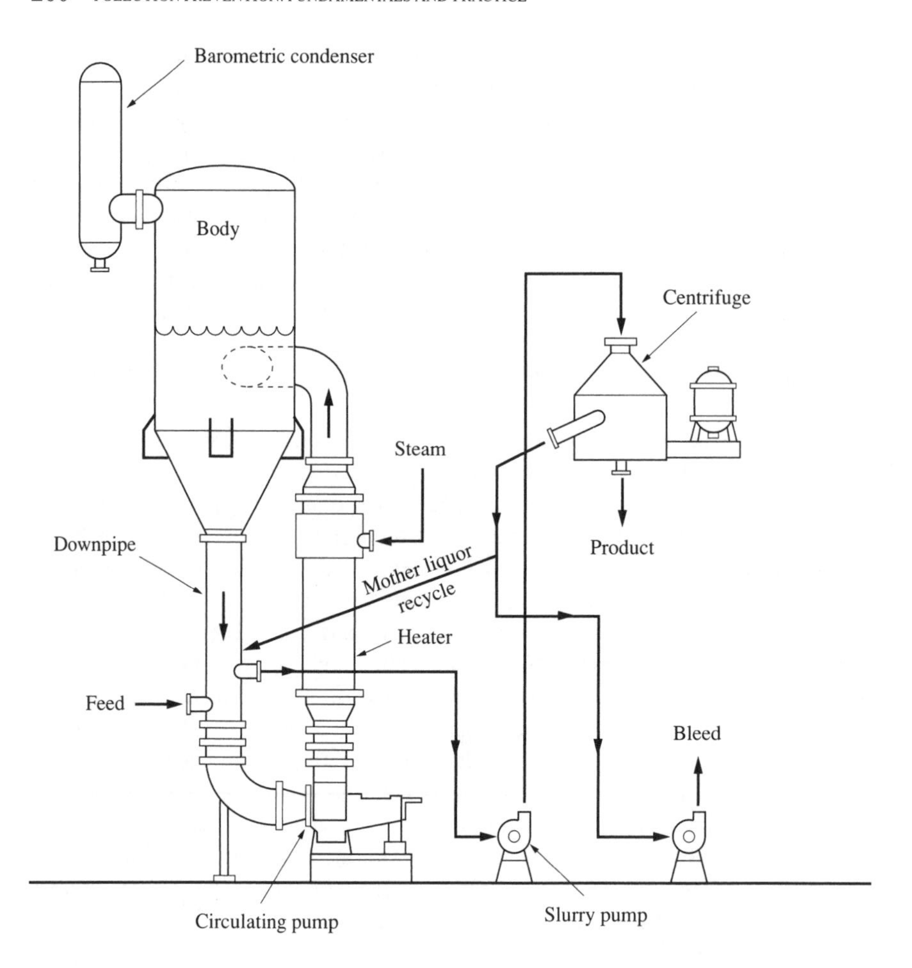

FIGURE 5.11
Circulating magma crystallizer.

Distillation can be done under either batch or continuous modes. A simple *batch distillation* involves placing a quantity of feedstock in a boiling vessel, heating it to generate vapors, and then condensing the vapors in a separate vessel by cooling them to below their condensing temperature. A simplified batch distillation column is shown in Figure 5.12*a*. Batch distillation is primarily used for recovery of solvents or used oils. In *continuous distillation,* the feed stream is continuously fed to the distillation column. The continuous distillation unit can vary in complexity from a simple column providing one-stage equilibrium separation to a compound column having multiple enriching and stripping stages in a single column. A simple continuous distillation column is depicted in Figure 5.12*b*. Continuous distillation is usually used when feed rates exceed about 200 L/h (50 gal/h).

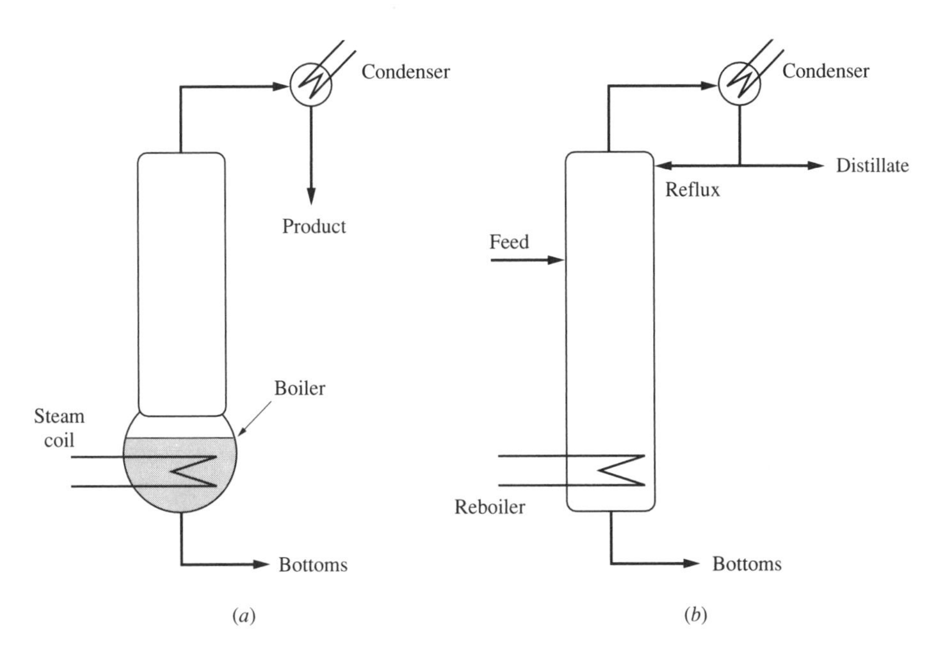

FIGURE 5.12
Distillation columns. (*a*) Batch distillation; (*b*) continuous distillation.

More detailed discussions of distillation operation are provided by Rogers and Brant (1989), McCabe, Smith, and Harriott (1993), Heaton (1996), and Sinnott (1996). The detailed design of distillation columns is beyond the scope of this text, particularly when the feed contains more than two components. However, a simple binary liquid-liquid system with two components will be briefly described here to indicate the usefulness of distillation and the most significant design and operation factors.

Figure 5.13 is a schematic diagram of a continuous distillation column. The column contains a series of perforated metal trays (also called plates or stages) or packing material. As shown in the schematic, vapor flows upward through the column and liquid flows down the column, in a countercurrent fashion. The vapor and liquid are brought into contact with each other on the surfaces of the plates or packing material. As vapor passes through the liquid, there is mass transfer between phases and their components approach equilibrium. Vapor moving up the column becomes enriched in the more volatile component and the liquid becomes enriched in the less volatile component. At the top of the column, the vapor is condensed. If the vapors are simply condensed and removed from the system, the process is called *flash distillation.* However, a portion of the condensate is commonly returned to the top of the column to provide the liquid reflux, which flows down the column. This process is termed *continuous distillation with reflux.* Also, part of the liquid at the bottom of the column is heated in a heat exchanger (reboiler) to provide the vapor that moves up the column. The liquid mixture to be separated is fed into the column at midheight. In the column below the

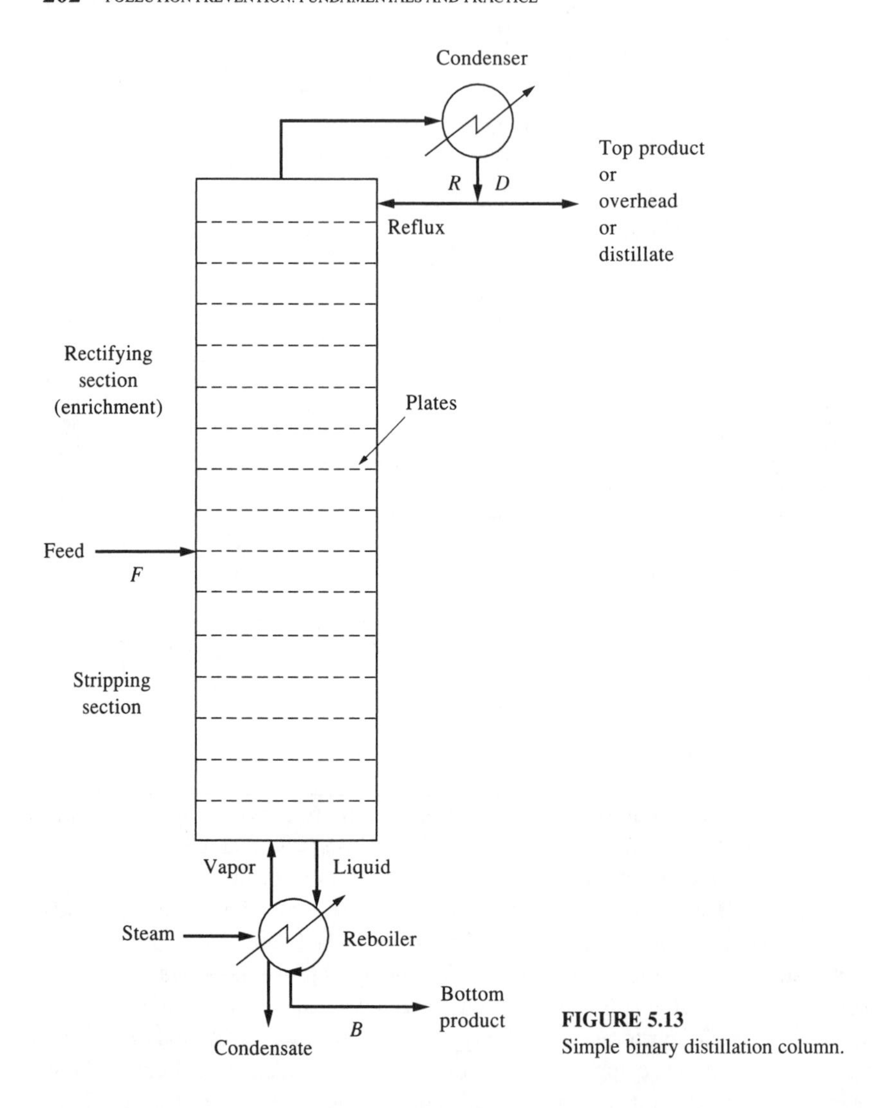

FIGURE 5.13
Simple binary distillation column.

feed, the more volatile components are stripped from the liquid; this section of the column is known as the *stripping section.* Above the feed, concentration of the more volatile component is increased; this part of the column is called the *enrichment* or *rectifying section.*

The column seen in Figure 5.13 is designed to recover only two products: the tops, or distillate, and the bottoms. If more components are to be separated, sidestreams can be withdrawn at different heights up the column.

The countercurrent flow of liquid down and vapor up the column produces the series of equilibrium stages which separate the components. The amount of separation, or fractionation, depends on the number of trays, the vapor and liquid flow rates, and the relative volatilities of the components. One measure of the vapor and liquid flow rates is the *reflux ratio*, which is defined as the ratio of the reflux flow rate R to the distillate flow rate:

$$\text{Reflux ratio} = \frac{R}{D}$$

The number of plates or stages required for a given separation will be dependent on the reflux ratio used. Total reflux is the condition when all of the condensate is returned to the column as reflux, that is, no product is recovered. This is obviously not useful or practical, but it does serve to establish the theoretically minimum number of plates required to effect the desired separation. As the reflux ratio is reduced, a pinch point will occur at which the separation can be achieved only with an infinite number of stages. This is also not practical, but it sets the minimum possible reflux ratio for the desired separation. The optimum reflux ratio will be somewhere between these two limits. This will be the point where the specified separation occurs at minimum cost. Increasing the reflux reduces the number of stages required, and therefore the capital cost, but it increases steam and water requirements and the operating costs. Typically, the optimum reflux ratio will lie between 1.2 to 1.5 times the minimum reflux ratio (Sinnott, 1993).

The design of a distillation column is a multistep process (see Figure 5.14). Calculation of the operation of the distillation column can be accomplished by performing mass balances on the entire system and on each plate. A mass balance on flows for the entire system can be written as

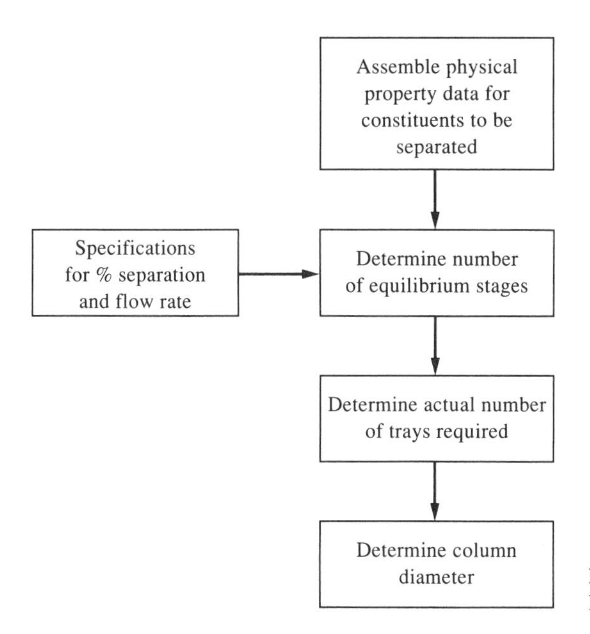

FIGURE 5.14
Distillation column design procedure.

$$F = D + B$$

where F = flow rate of the feed to the column
D = flow rate of the distillate product
B = flow rate of the bottoms product

Similarly, a mass balance on the recoverable component in the feed can be written as

$$F \cdot x_F = D \cdot x_D + B \cdot x_B$$

where x_F = concentration of the recoverable component in the feed
x_D = concentration of the recoverable component in the distillate
x_B = concentration of the recoverable component in the bottoms

Normally, we know the feed flow rate F, the feed composition x_Z, and the desired product compositions x_D and x_B. Solving the preceding equations simultaneously yields

$$D = \frac{F(x_F - x_B)}{(x_D - x_B)}$$

$$B = F - D$$

Material balances can also be written for any stage in a multistage column. We will use the nomenclature shown in Figure 5.15, where

n = any stage in the rectifying section, numbered from the top
m = any stage in the stripping section, numbered from the feed plate
V_n or V_m = vapor flow from stage n or m
V_{n+1} or V_{m+1} = vapor flow into stage n or m from below
L_n or L_m = liquid flow from the stage above
L_{n-1} or L_{m-1} = liquid flow into stage n or m from above
x = mole fraction of component in the liquid stream
y = mole fraction of component in the vapor stream

For this analysis, we will assume that we have equilibrium stages or theoretical stages, meaning that the liquid and vapor streams leaving a stage are in equilibrium ($y_n = K_n x_n$).

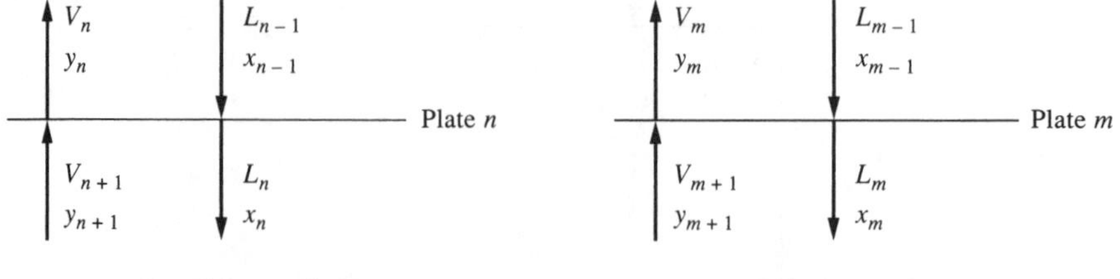

FIGURE 5.15
Nomenclature for distillation column plate mass balances.

In actuality, this does not occur, but the relationship can easily be corrected by applying a theoretical plate efficiency factor.

Starting with the top and bottom plates, we can relate the liquid and vapor stream compositions for each successive plate (Heaton, 1996). Consider a mass balance for the component in the upper rectifying section of the column. For the section of column down to plate n, the mass balance is

$$y_{n+1}V_{n+1} = x_n L_n + x_d D$$

Therefore,

$$y_{n+1} = \left(\frac{L_n}{V_{n+1}}\right)x_n + \left(\frac{D}{V_{n+1}}\right)x_D$$

Since $R = L_n/D$ and $D = V_{n+1} - L_n$, this can also be written as

$$y_{n+1} = \left(\frac{R}{R+1}\right)x_n + \left(\frac{1}{R+1}\right)x_D$$

This equation represents the *top operating line,* indicating the relationship between liquid and vapor compositions between stages. Assuming that the vapor and liquid flow rates do not change with depth in the column (which is usually the case), then

$$L_n = L$$

$$V_n = V_{n+1} = V$$

Since $V = L + D$, the equation for the top operating line can be written as

$$y_{n+1} = \left(\frac{L}{V}\right)x_n + \left(\frac{D}{V}\right)x_D$$

This is the equation of a straight line, with slope L/V. A similar analysis can be done for the bottom stripper section below the feed plate. The equation for the bottom operating line, relating compositions on successive plates in the stripping section, is

$$y_m = \left(\frac{L}{V'}\right)x_{m-1} - \left(\frac{B}{V'}\right)x_B$$

where V' = vapor flow rate in the bottom of the column.

These equations can be solved numerically for each plate, but they are more commonly solved by the graphical method developed by McCabe and Thiele in 1925. The procedure is depicted in Figure 5.16. First, draw the equilibrium line and the diagonal ($y = x$) line. For a binary system, the *equilibrium line* can be determined based on the relative volatilities (based on Henry's constants) of the lighter component with respect to the heavier component, denoted as α ($\alpha = K_{\text{more volatile}}/K_{\text{less volatile}}$), as follows:

$$y = \frac{\alpha x}{1 + (\alpha - 1)x}$$

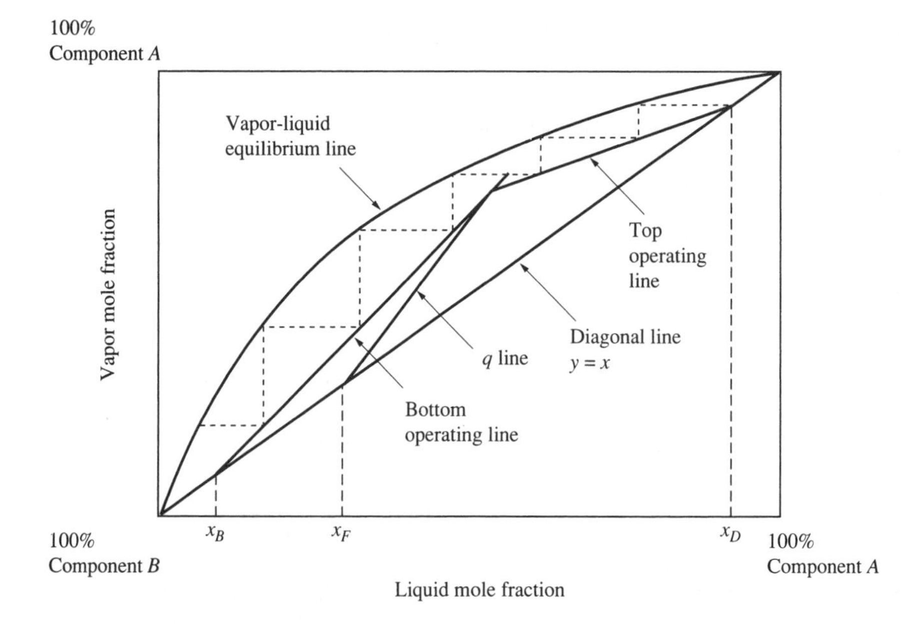

FIGURE 5.16
McCabe-Thiele diagram.

Next draw the top operating line with a slope of L/V and intersecting the diagonal at $y = x_D$. The bottom operating line will intersect the diagonal at $y = x_B$. The point of intersection of the two operating lines is dependent on the feed composition, the feed thermal condition q, and the reflux ratio. A mass balance around the feed plate (plate at which the feed enters the column) gives the locus of intersection points as follows:

$$y = \left(\frac{q}{q-1}\right)x - \frac{x_F}{q-1}$$

where q = the fraction of the feed that is liquid
x_F = concentration of the component in the feed

Determine q and plot the q-line from the point of intersection of $y = x_F$ and the diagonal with a slope of $q/(q-1)$ to intersect with the top operating line. In many cases, the feed mixture will be a liquid at the boiling point. If this is the case, the equation for q becomes $q = 1$, and the q line is simply a vertical line. If one of the components in the feed is a gas at the feed temperature, then thermodynamic relationships must be applied to determine the liquid and gas phase compositions. Now draw the bottom operating line from the point on the diagonal at $y = x_B$ to the intersection of the top operating line and the q line. Finally, starting at x_B or x_D, step off the theoretical plates between operating lines and the equilibrium line, shifting from the top to the bottom operating line after the intersection. The number of steps indicates the number of theoretical plates required.

There are two limiting cases to this procedure. The first is when the minimum number of theoretical plates corresponds to infinite reflux conditions. At infinite reflux, the two operating lines are coincident with the $y = x$ diagonal. The second is the minimum reflux ratio—the opposite extreme—which occurs when the two operating lines intersect on the equilibrium line, rather than below it. At this point, the reflux ratio is the minimum possible, but an infinite number of plates are required to effect separation because the driving gradient for interphase mass transfer goes to zero. As indicated earlier, the optimum design often occurs when the reflux ratio lies between 1.2 to 1.5 times the minimum reflux ratio.

Example 5.4. A continuous distillation column is to be used to separate a mixture of benzene and toluene. The feed contains 35 percent benzene and 65 percent toluene. The desired result is enrichment of the benzene to 95 percent in the distillate and the toluene to 95 percent in the bottoms. Assume that the equilibrium relationship between the mole fraction of benzene in the vapor (y) and in the liquid (x) is

x	0.1	0.2	0.3	0.4	0.5	0.6	0.7	0.8	0.9
y	0.19	0.34	0.48	0.59	0.69	0.77	0.84	0.90	0.96

Also assume that the feed is a liquid at the boiling point.

(*a*) Plot the equilibrium line and the operating lines and determine the number of theoretical plates needed to achieve the desired outputs, assuming a reflux ratio of 4.

(*b*) Determine the minimum reflux ratio and estimate the number of plates required if the reflux ratio is 1.3 times the minimum reflux ratio.

Solution.

(*a*) Plot the equilibrium line using the data above and draw the diagonal ($y = x$), as shown in Figure 5.17. Locate the points $y = x_B$, $y = x_F$, and $y = x_D$ on the diagonal. For this feed, $q = 1.0$. Draw the vertical q line, beginning at the point where $x = x_F$ on the diagonal.

Next plot the top operating line:

$$y_{n+1} = \left(\frac{R}{R + 1}\right)x_n + \frac{x_D}{R + 1}$$

$$y_{n+1} = \left(\frac{4}{4 + 1}\right)x_n + \frac{0.95}{4 + 1} = 0.8x_n + 0.19$$

The slope of the top operating line is 0.8, and the line intersects the diagonal at $x = 0.95$. Draw the bottom operating line between the intersection point of the diagonal and $y = x_B$, and draw the intersection point between the q line and the top operating line. Finally, step off the stages between the operating lines and the equilibrium line. The total number of theoretical stages required is 13.7.

(*b*) The minimum number of stages corresponds to the top operating line and the q line intersecting on the equilibrium line. From the McCabe-Thiele diagram, the intersection point in this example will be at $x = 0.35$, $y = 0.52$. Thus the slope of the top operating line will be:

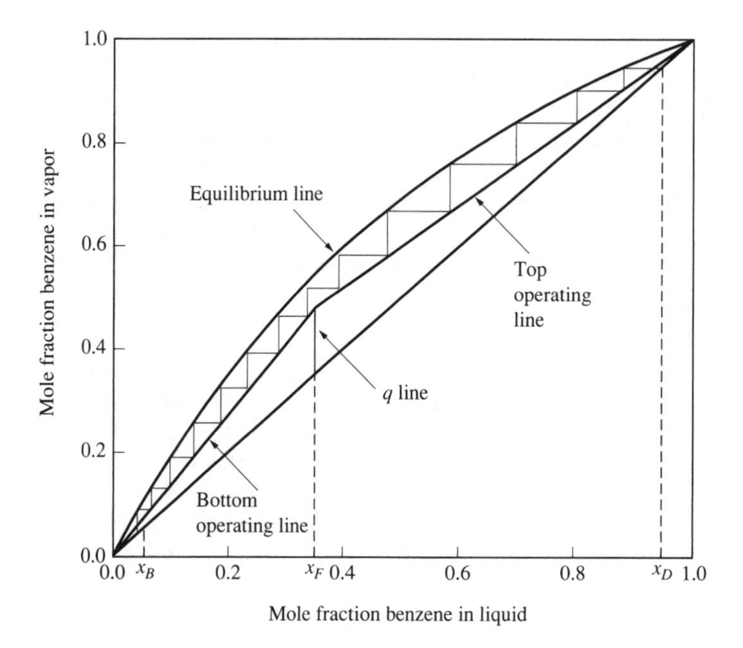

FIGURE 5.17
Equilibrium and operating lines.

$$\frac{R_{\min}}{R_{\min} + 1} = \frac{0.95 - 0.52}{0.95 - 0.35} = 0.72$$

$$R_{\min} = 2.6$$

For $R = 1.3R_{\min}$, the top operating line has the equation

$$y_{n+1} = \left(\frac{R}{R + 1}\right)x_n + \frac{x_D}{R + 1}$$

$$= \left(\frac{3.4}{3.4 + 1}\right)x_x + \frac{0.95}{3.4 + 1}$$

$$= 0.77x_n + 0.22$$

A McCabe-Thiele plot can be constructed for these conditions, as in Figure 5.17. From the plot, a requirement for 15.7 theoretical plates is found. Either design will give the same outputs, but the latter design should be approximately the optimum design for this system, from an economic standpoint, even though more plates are required.

Distillation is a very energy-intensive process. Efficient distillation systems should be used to minimize the amount of energy needed and, concomitantly, the amount of pollution generated in producing that energy. Distillation efficiency can often be improved by (Nelson, 1995):

1. Increasing the reflux ratio (the ratio of condensate from the top of the tower that is returned to the still for redistillation to the distillation column feed flow rate).
2. Increasing the height of the column.
3. Improving the feed distribution to the column.
4. Changing the type of packing media.
5. Insulating or preheating the column feed.
6. Reducing the pressure drop in the column.

Distillation columns are often used to recover materials, and thus prevent pollution, to improve the quality of the raw materials used during processing, or to remove unwanted materials from the product, but they can also contribute to pollution in several ways. Incomplete separation leads to impurities in the product, which must be removed later to waste. In some cases, waste materials may be formed in the column itself, usually because of high reboiler temperatures, which cause polymerization. Inadequate condensing can result in loss of product, which must be vented or flared (Nelson, 1995). In some cases, the bottom product from the distillation column is a useful product, but often it is a varying mixture of unwanted materials and is disposed of as waste, called *"bottom stills."* Some of the recovered sidestreams or the top product may also be of no use to the industry. Unless markets or other uses are found for them, they may also become waste materials. In some cases, these materials are flared off, creating a potential air pollution problem. It behooves industry to find uses for as much of the material as possible, or their waste generation rate will be much higher than necessary. All distillation columns have a propensity to leak, allowing VOC emissions to the air from condenser vents, accumulator tank vents, and storage tank vents. The still bottoms can consist of solids, tars, and sludges. All distillation columns require reboilers or heat exchangers to heat the liquid to the boiling point. Thus they are prone to all the problems described earlier for heat exchangers.

ABSORPTION. Absorption, or stripping, involves the selective transfer of one or more soluble gases in a gas process stream into a liquid. If the objective is to recover the gas, the dissolved gas is subsequently removed from the liquid by distillation, while the absorbing liquid is discarded or reused. If the objective is to remove a contaminant from the process stream, then all of the absorbent stream will be waste.

The absorption can be chemical or physical. Chemical absorption involves a reaction between the compound and the solvent, whereas physical absorption merely involves dissolution of the compound (Palepu, Chauhan, and Ananth, 1995).

A number of vapor-liquid contacting schemes can be used. Packed towers are most commonly used, but tray towers or spray towers are also available. Two principal types of packing are used: randomly dumped packing material of various shapes (6-75-mm dimension) made from clay, ceramic, porcelain, or plastic; and structured packing consisting of units of corrugated sheeting that are stacked in the column. Typically, a countercurrent flow tower is used, in which the gas flows upward through the tower while the liquid phase moves downward as a thin film around the packing material. Highly efficient transfer usually occurs in these systems.

Even though transfer rates may be high, transfer is never complete. Transfer is dependent on good contact between the gas and liquid and the relative solubilities of the compound of interest in the two fluids. An equilibrium is established between the two fluids, meaning that the stream that the compound is being transferred from will always have some of the compound remaining.

It is essential that the absorption column be well designed and properly operated to avoid excessive pollution. If transfer is not efficient, the volume of waste absorbent could become excessive and the efficiency of gas recovery may be reduced, leading to loss of the desired product. An inefficient absorber will also increase energy and water consumption.

A good discussion on the design of gas absorbers can be found in McCabe and colleagues (1993).

EXTRACTION. Extraction, or leaching, involves the transfer of solutes from a liquid or solid phase into a liquid solvent. Liquid-liquid extractions are used to transfer from one liquid to another compounds that are immiscible with each other; solid-liquid extraction (leaching) requires that the solid containing the desired compound not be soluble in the liquid.

Leaching differs little from washing of solids. It is usually accomplished by percolation of the extraction fluid through a stationary or moving bed of the solids being leached. In many cases, the system is operated in a countercurrent fashion, which heightens extraction efficiency. Examples of industrial leaching include the extraction of gold from its ore by use of cyanide solutions and the leaching of cottonseed oil from cottonseeds by hexane.

Liquid-liquid extractions are commonly used to separate materials with similar boiling points, which negates the use of distillation. However, because solvent extraction necessitates that the solvent be recovered for reuse, an operation that is often energy intensive, it is common practice to select distillation over extraction if distillation is viable. An example of extraction is the separation of an aromatic, such as benzene with a boiling point of 80.1°C, from a paraffin, such as cyclohexane with a boiling point of 80.7°C. Because their boiling points are nearly the same, separation by distillation is nearly impossible. However, solvent extraction with the solvent tetraethylene glycol can be used because the benzene will dissolve more readily in the solvent than will the cyclohexane; the solvent will become enriched with the benzene, leaving the cyclohexane behind. After extraction, the benzene-laden solvent can be sent to a distillation column for recovery of the benzene and regeneration of the solvent for reuse. Another example is the recovery of penicillin from its fermentation broth by extraction with butyl acetate, after lowering the pH to get a favorable partition coefficient. The solvent is then treated with buffered phosphate solution to extract the penicillin from the solvent and give a purified aqueous solution, from which the penicillin is eventually produced by drying (McCabe, Smith, and Harriott, 1993).

Extraction requires that the two phases be efficiently brought into contact with one another. This can be accomplished in one or more stirred reactor-settler units (called decanters) in series to separate the phases, in packed bed columns, or in centrifugal extractors.

For solvent extraction to be effective, the species to be removed should have high affinity for the solvent, the solvent should be highly selective for the species to be removed relative to other materials present, and the solvent should have low solubility in the feed material or solution (Palepu, Chauhan, and Ananth, 1995). Thus selection of the appropriate solvent is critical.

In recent years, solvents other than water have been used for extraction. Various organic solvents, chosen based on their selectivity for the compound desired, are now commonly used. For example, the acetone in an acetone-water mixture can be separated from the water by adding carbon tetrachloride to the mixture and mixing them. After mixing, the water and carbon tetrachloride will separate into two phases, because they are not miscible within each other. Acetone has a stronger affinity for carbon tetrachloride and will mostly move into that phase, leaving only a little in the water. A second extraction of the water phase will remove more of the acetone.

Extraction efficiency can be calculated based on the partitioning coefficients, K_D, of the desired species to be recovered with the original solvent (S_1) and the extracting solvent (S_2):

$$K_D = \frac{S_1}{S_2}$$

If the two fluids are totally immiscible with each other, the design of an extraction system is relatively straightforward, as can be seen in Example 5.5.

Example 5.5. A lab-scale extraction study is being performed. A 100-mL aqueous solution contains two compounds, A and B. To extract the sample, 100 mL of the solvent amyl acetate is used. The distribution coefficients for A and B between water and amyl acetate are $K_D(A) = 0.2$ and $K_D(B) = 10$. What will be the composition of the water and the solvent phases after extraction?

Solution. Determine the amount of each compound, X, transferred to the amyl acetate:

$$K_D(A) = 0.2 = \frac{X_A/100 \text{ mL amyl acetate}}{(1 - X_A)/100 \text{ mL water}} \qquad K_D(B) = 10 = \frac{X_B/100}{(1 - X_B)/100}$$

$$X_A = 0.167 \text{ g } A \text{ in amyl acetate} \qquad X_B = 0.909 \text{ g } B \text{ in amyl acetate}$$

Therefore, after extraction, the amyl acetate solvent contains 0.167 g A and 0.909 g B, while the aqueous solution contains 0.833 g A and 0.091 g B. Separation was successful. A second extraction of each phase would improve the results even more.

Unfortunately, the two liquids are often slightly miscible and the solution becomes very complicated; after extraction, the extracted feed stream and the extract will both contain all three compounds. For example, assume that you have a mixture of acetone and water. You would like to extract the acetone from the water using methyl isobutyl ketone (MIBK). After extraction, most of the acetone will have transferred to the MIBK phase, but there will still be some remaining in the water phase because of partitioning. However, some of the MIBK will also be found in the water phase and some water will be in the MIBK phase because they are also slightly miscible with each other. Analysis of these systems is complex and will not be described here. The procedure can be found in most standard chemical engineering texts (Luyben and Wenzel, 1988; McCabe, Smith, and Harriott, 1993). The design approach for a solvent extraction system is similar to that for a distillation system. The McCabe-Thiele method can be used here with information on the partition coefficients and miscibility of the fluids to determine the number of ideal stages required for a countercurrent extraction system. This procedure can also be found elsewhere (McCabe, Smith, and Harriott, 1993; Sinnott, 1996).

More recently, supercritical extraction has come into play. *Supercritical extraction* is essentially a liquid extraction process employing compressed fluids under supercritical conditions instead of normal solvents. Fluids are usually either liquids or gases. However, at elevated temperatures and pressures, a point is reached where the differences between liquid and gas are no longer discernible. This is termed the *critical point* (see Figure 5.18). A supercritical fluid is a single phase at temperatures and pressures above the critical point and has properties between those of a liquid and a gas. When water becomes a supercritical fluid, the hydrogen-bonded structure largely breaks down, making it much less polar; thus it can dissolve relatively large amounts of non-polar materials (Clifford, 1995).

Supercritical fluids generally have solubilities and densities approaching those of the liquid phase and viscosities approaching those of the gas phase. The result is that organic compounds will have the high solubilities of the liquid phase and the high mass transfer characteristics of the gas phase. Solubilities and diffusivities of organic constituents in supercritical fluids are typically several orders of magnitude greater than under normal conditions (Jain, 1993). The supercritical fluids easily penetrate porous solids and diffuse into liquids. Supercritical fluids are thus excellent extractants, as organic compounds will readily dissolve in them at elevated temperature and pressure conditions and be released from them at lower temperatures and pressures. In addition, under supercritical conditions, solvent characteristics can be varied over a wide range by means of pressure and temperature changes (Shen, 1995). When total extraction of all components is necessary, the extractions are usually carried out at higher pressures because this increases solubilities, while extractions of specific compounds are often carried out at lower pressures (but still above the supercritical point) because the supercritical solvent is more selective at lower pressures.

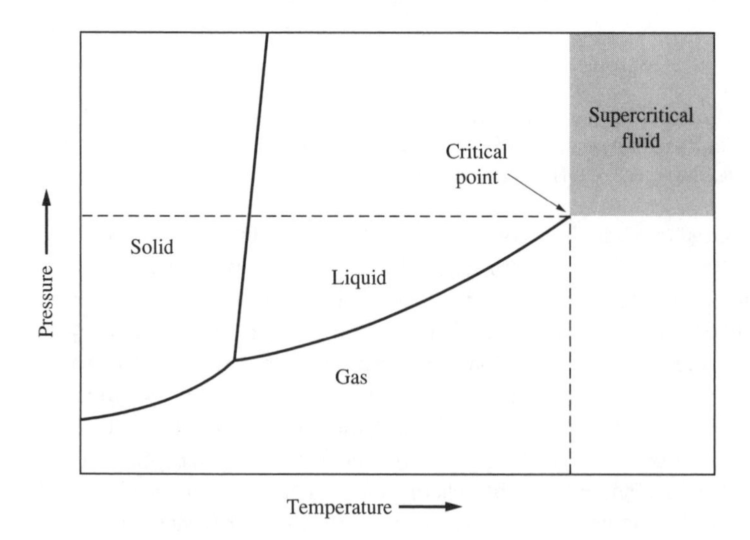

FIGURE 5.18
Typical phase diagram of a solvent showing its supercritical point.

TABLE 5.3
Supercritical fluid constants

Solvent	Temperature, °C	Pressure, atm	Density, g/cm^3
Carbon dioxide	31.1	73.0	0.460
Water	374.15	218.4	0.323
Ammonia	132.4	111.5	0.235
Benzene	288.5	47.7	0.304
Toluene	320.6	41.6	0.292
Cyclohexane	281.0	40.4	0.270

Carbon dioxide is most commonly used as the supercritical extraction fluid, especially in the food processing industry. Its critical point is at a much lower temperature (31.1°C) than other supercritical solvents, and its pressure (73.0 atm) is easily achievable in many reactors (see Table 5.3), making it much easier to use. It is also much less toxic, less corrosive, and less flammable than other choices.

A good example of supercritical extraction is the decaffeination of coffee beans. This used to be done with toxic chlorinated hydrocarbon solvents, but now most companies use supercritical carbon dioxide to extract the caffeine.

Extraction processes are commonly used in the pharmaceutical industry to recover desired materials. A process called *natural product extraction* is used to recover useful drugs, such as Taxol (paclitaxel), from natural materials such as roots and leaves. A wide variety of solvents are used, including chlorinated solvents, ketones, and alcohols. Solvent extraction is also commonly used to remove oil and grease from parts (as in cleaning of metal parts in preparation for metal plating) and in decontamination of solutions, soils, and sludges. Examples are the removal of phenols from petroleum refining effluents using methyl isobutyl ketone (MIBK) as the solvent and recovery of acetic acid from industrial wastewaters using ethyl acetate as the solvent (Palepu, Chauhan, and Ananth, 1995).

In extraction processes, as in adsorption or absorption, a partition is set up between the two phases. The original stream will still contain some product and process residuals after extraction that may need to be removed, wasted, or reprocessed. The process stream will also contain small quantities of the solvent used for extraction.

ADSORPTION. Adsorption occurs when a molecule is brought into contact with a surface and held there by physical and/or chemical forces. The quantity of a compound adsorbed depends on the balance between the forces that keep the compound in the fluid phase and those that attract it to the adsorbing surface (Palepu, Chauhan, and Ananth, 1995).

In industry, adsorption is usually carried out in a column. The intent is to separate one or more gaseous or liquid components from a fluid mixture by transferring them to a solid adsorbent. Usually a packed-bed column is used, but fluidized bed columns are also available. Activated carbon is a commonly used adsorbent, but others, such as activated silica or resins, can also be used. Adsorption can be used to separate and concentrate a desired product, in which case another process step will be needed to recover it from the adsorbent; this is often difficult, reducing the usefulness of this process for product recovery. Adsorption is widely used in industry, though, to remove contaminants from industrial waste streams (air or water) before they are discharged. In either case, the adsorbent will eventually need to be replaced or regenerated, creating another waste material.

OTHER UNIT OPERATIONS. Other unit operations commonly used by industry include size reduction (grinding, shearing, etc.) and various solid-liquid separation processes such as filtration, settling, membrane filters (reverse osmosis, ultrafilters, emulsion liquid membranes), electrodialysis, diffusion dialysis, ion exchange, and centrifuges (see Figure 5.19 for a description of the size ranges each of these separation processes is good for). Figure 5.20 lists additional separation technologies (Palepu, Chauhan, and Ananth, 1995). Each of these is a potential source of wastes, and each should be evaluated to determine how wastes from the operation can be minimized.

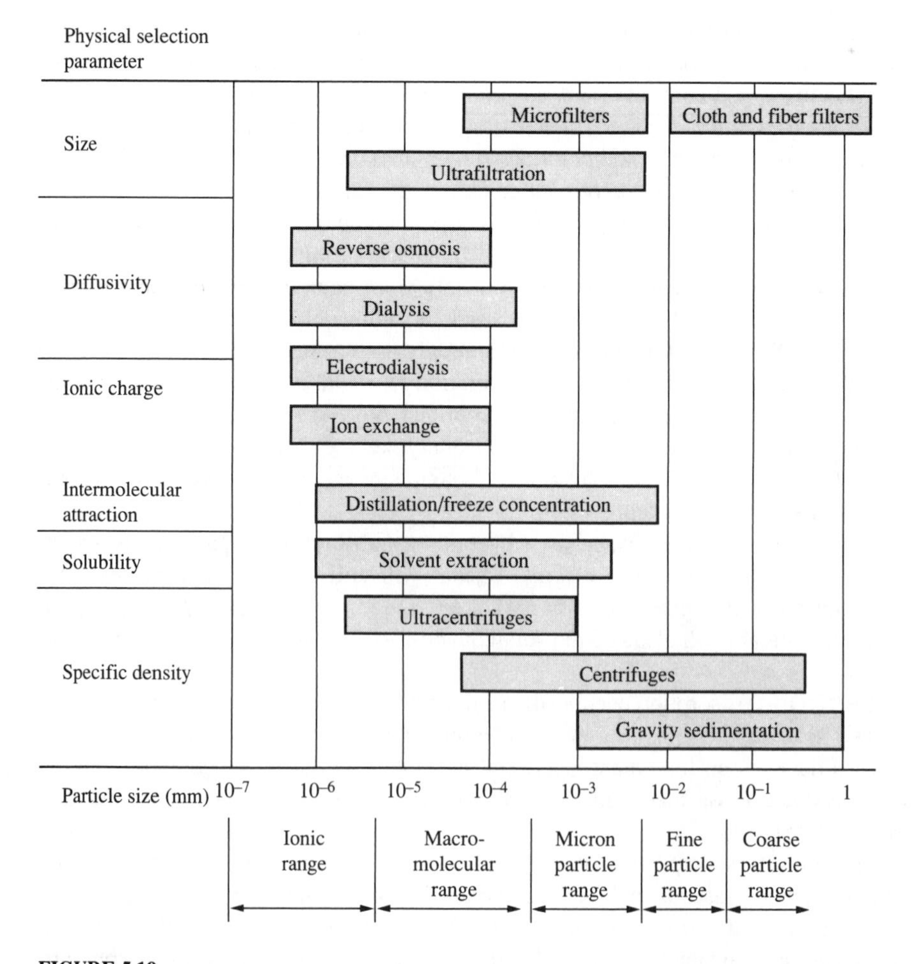

FIGURE 5.19
The relationship of various separation technologies to particle size. (Adapted from Shen, 1995)

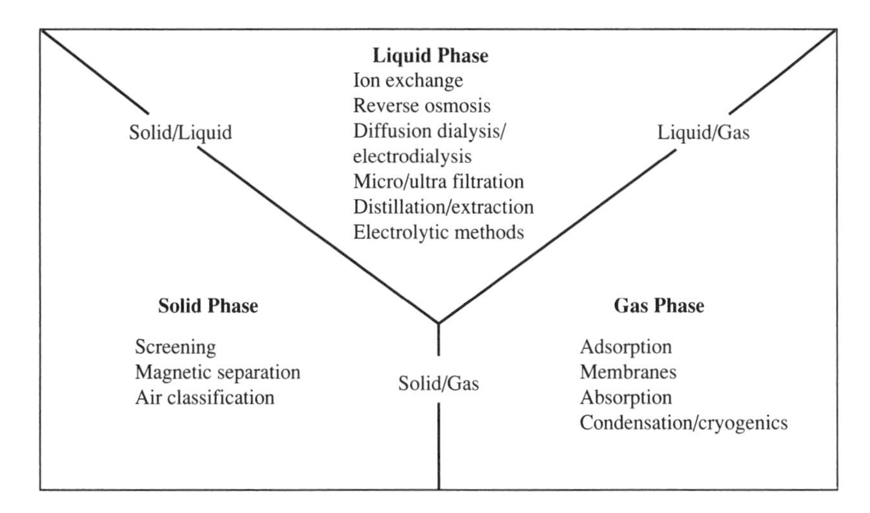

FIGURE 5.20
Key separation technologies in pollution prevention. (Adapted from Palepu, Chauhan, and Ananth, 1995)

ANCILLARY EQUIPMENT. Other parts of the manufacturing plant that require study are the devices for transporting gases and liquids to, from, or between unit process equipment (pipes, ducts, fittings, stacks); pumps, fans, and compressors; and storage facilities. These are often subject to leaks and overflows, and they can unnecessarily contribute a significant amount of material to the plant's waste load. Leaking pump seals, valves, and pipe flanges are particularly notorious for producing copious quantities of waste materials. Termed "fugitive emissions," these losses are often the first point of attack in a pollution prevention program. The topic of fugitive emissions and their control is covered in more detail later in this chapter and again in Chapter 11.

5.3 PROCESS DEVELOPMENT AND DESIGN

Process development describes the refinement of a process concept from early conceptual stages (articulation of process objectives, selection of process steps, determination of constraints) through preliminary engineering (development of preliminary economic analysis, piping and instrumentation diagrams, and process flow diagrams) (Butner, 1995). As we have seen, the processes used in the manufacture of a product play a key role in determining the amount of waste that will result during production. Waste generation can often be greatly minimized through proper design and operation of the process system. Developing and implementing pollution prevention process improvements, though, are far more demanding than traditional process design efforts. They require significant knowledge of the entire plant design, plant operations, maintenance procedures, and applicable and potentially applicable environmental regulations (Hertz, 1995).

A critical first step in pollution prevention process development is to clearly and explicitly identify the environmentally related process development objectives and constraints (see Table 5.4). Commonly, design objectives may conflict with one another.

TABLE 5.4
Typical environmental design constraints and objectives

Constraints	Objectives
Compliance with all applicable environmental regulations	Minimal use of toxics in process
Compliance with existing permit requirements for discharge and emissions	Minimize life-cycle impact within acceptable financial parameters
Process loadings not to exceed existing treatment capacity	Implement all pollution prevention options meeting investment hurdles
Zero discharge of regulated wastes	Maximize use of recycled raw materials

Source: Butner, 1995.

For example, heating a reactor may increase the rate and extent of a reaction, but the heated process stream may increase loss of volatile contaminants as air pollutants and will certainly increase energy costs. These trade-offs must be balanced, usually on an economic basis.

Most of the pollution prevention benefits derive from the earliest stages of the process design (see Figure 5.21). Hertz (1995) distinguishes between long-term and short-term pollution prevention options, defining long-term options as those process changes that either satisfy environmental requirements expected in the next 5 to 10 years or that will take at least 5 years to implement, and *short-term options* as those that can take effect within 5 years.

Much of the data needed to support pollution prevention decisions later in the development cycle can be generated during bench-scale testing. This includes such things as reaction stoichiometry and equilibrium yield; reaction by-products; effects of recycle; chemical data on ingredients, products, and intermediates; and corrosion rates and products. In most cases, though, process development involves revamping or upgrading existing processes, where bench-scale testing may not be possible. Here, the most significant pollution prevention impact will come at the conceptual design stage, the point where the designer is asked to come up with a process that is both technically and economically viable. This is where decisions will be made concerning the type of technology to be used for separation, purification, utilities, heat exchange, and end-of-pipe treatment. Beyond this point, the design gets into details such as equipment selection, and little additional pollution prevention can be achieved. Once the process is designed and constructed, the possibilities for additional source reduction are small. Thus it is important that pollution prevention design begin at the earliest stage of the project.

While significant emissions reductions can be achieved through the direct application of well-known engineering principles to existing process plants, major emission reductions from industry will be the result of business decisions that involve the closing of heavily polluting obsolescent manufacturing sites and their replacement with modern, less-polluting ones. To a large extent, such business decisions can be made only if cleaner and more efficient technology has been developed to replace the offending antiquated processes (Cortright, Crittenden, and Rudd, 1996). Because of the large

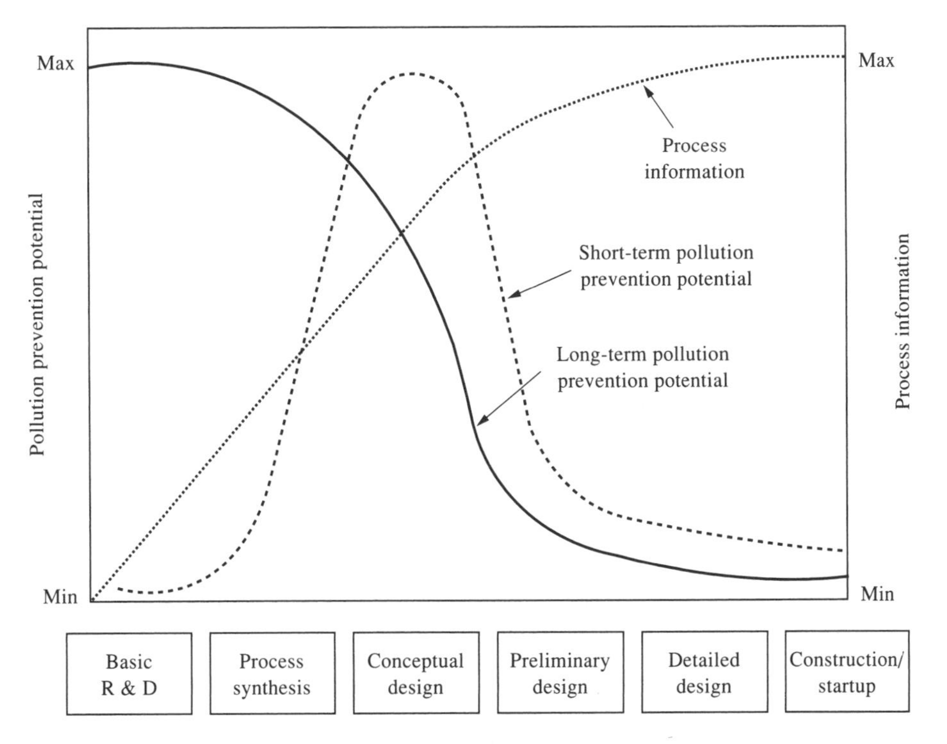

FIGURE 5.21
Pollution prevention potential in the design process. (Adapted from Hertz, 1995)

and expensive infrastructures in place in most industries, these decisions are not made lightly and are not made often. It is estimated that construction of entirely new plants accounts for only about 5 percent of capital expenditures in any given year; industrial process design and revamping in-plant production accounts for the remaining 95 percent (Radecki, 1994). Thus most pollution prevention activities are limited to small, in-house changes to modify an existing system.

5.3.1 Computer Tools

Currently, most process redesign is done using traditional engineering procedures. However, computer tools, such as molecular description, synthesis pathway identification, and synthesis reactor scale-up, are being developed to assist the engineer in the preliminary stages of design (see Figure 5.22). Computer tools can also be used in the design phases. Generic models are being developed that will allow engineers to experiment with reaction conditions, the inclusion of additives or inhibitors, and other variables. With such tools, engineers could relatively easily evaluate a multitude of different reactor types, reactor conditions, and chains of processes to determine which might work best with the least environmental impact for a given synthesis pathway. Figure 5.23 shows some

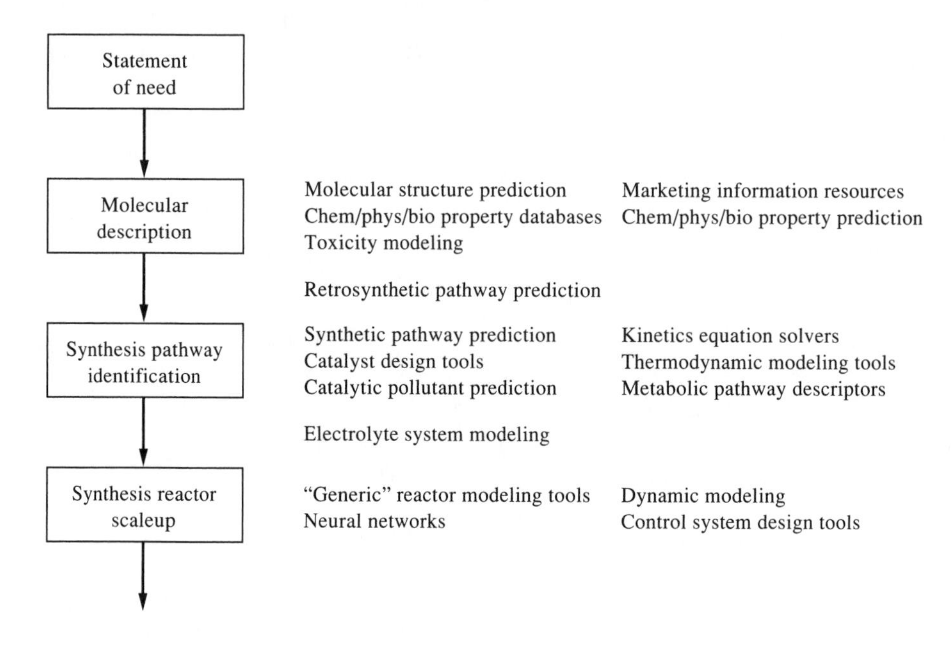

FIGURE 5.22
Computer-based tools for use in early planning stages of process design. (Adapted from Radecki, 1994)

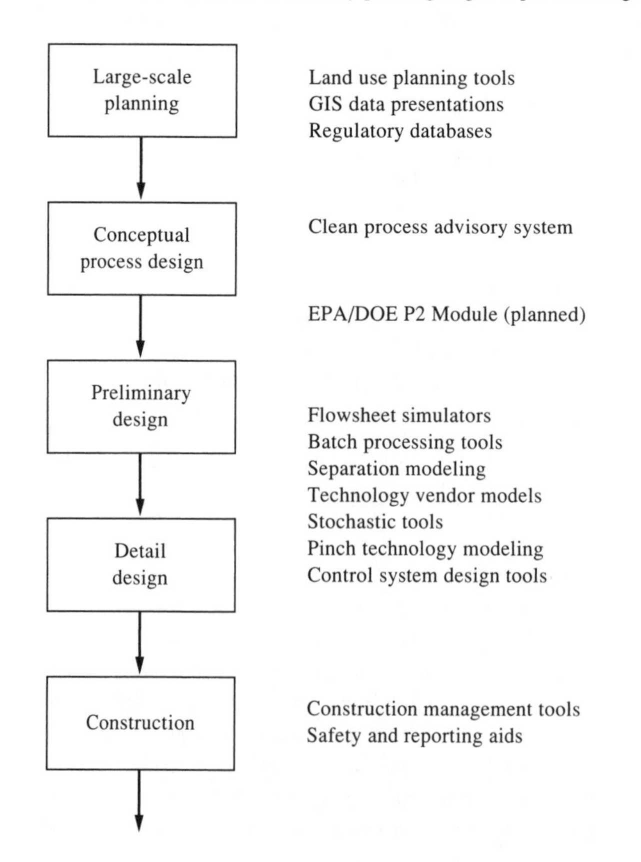

FIGURE 5.23
Computer-based tools for use in industrial-scale process design. (Adapted from Radecki, 1994)

computer tools for design that either exist or are in the development stage; Figure 5.24 shows some computer tools that can be used during production to minimize waste.

The Clean Process Advisory System (CPAS) is one such program being developed to integrate pollution prevention concepts into process design (Radecki, 1994; Radecki, Hertz, and Vinton, 1994). This is a collaborative effort of the Center for Clean Industrial and Treatment Technologies (CenCITT, a U.S. EPA–sponsored consortium of Michigan Technological University, the University of Wisconsin, and the University of Minnesota) and two industrial consortia, the Center for Waste Reduction Technologies (CWRT) and the National Center for Manufacturing Sciences (NCMS), which together comprise more than 230 companies. The purpose of the CPAS is to furnish engineers with a means to routinely find, simulate, and compare various design approaches on the basis of pollution prevention and to do so in a way which integrates with more conventional cost and safety comparisons. The target is the conceptual stage of design, where the greatest pollution prevention dividend can be identified. The CPAS is made up of a number of integrated software packages containing (1) technology descriptions and expert guidance to find design options, (2) numerical simulations and property data resources to simulate those options, and (3) data resources and algorithms to quantify and compare their pollution prevention dividends in conjunction with cost and safety implications (Crittenden, 1996).

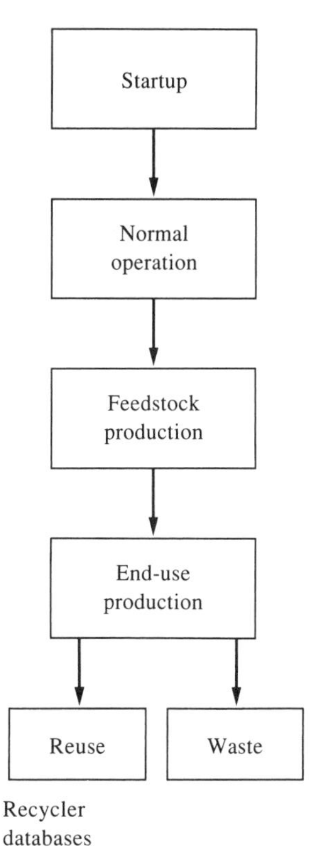

FIGURE 5.24
Computer-based tools for use during the industrial production age. (Adapted from Radecki, 1994)

Other models are also available. The U.S. EPA in 1992 issued a software package called SWAMI (Strategic WAste Minimization Initiative), which was designed to identify waste minimization opportunities within an industrial setting (U.S. EPA, 1992). Based on user-provided information for process definition, as well as material inputs and products for each unit operation and outputs associated with waste streams, it provides a scheme for identifying and prioritizing (on a cost or volume basis) waste reduction opportunities in process units and treatment operations, performs mass balance calculations, draws process flow diagrams, and directs the selection of candidate waste minimization strategies. This software is far less sophisticated than CPAS, but it does provide a simplified approach that may be sufficient for many small projects.

A software program that can be useful for pollution prevention analyses is SuperPro Designer, by Intelligen, Inc. It allows process simulation, economic evaluation, and environmental impact assessment for pollution control, wastewater treatment, and unit operations used in the manufacturing of biochemicals, pharmaceuticals, specialty chemicals, and food products. It has the added advantage of having all costs associated with each unit operation included in the analysis. Therefore, in addition to being able to analyze the technical aspects of a particular process configuration, the designer can see the impact of process changes on capital and operations and maintenance costs, thus facilitating the best design choice.

The Department of Energy is testing another computer-based pollution prevention tool that will also allow designers and engineers the opportunity to see the potential impact of environmental strategies, manufacturing processes, or products before investing significant amounts of time or money (Kratche, 1997; Sandia National Laboratories, 1996). The software, known as EcoSys, is a decision support tool developed for a manufacturing setting that uses life-cycle assessment principles to help designers and management make product/process/material decisions based on environmental factors. EcoSys4 can be used either as a design for environment (DfE) or life-cycle assessment (LCA). It includes a database with the environmental characteristics of more than 500 substances and four models of environmental thinking from which users can choose: (1) environmental protection (environmental compliance approach), (2) resource management (a sustainability approach), (3) ecodevelopment (a global approach), and (4) user-defined (user selects weighting factors). These four views differently weight the importance of environmental risk, resource consumption, and waste production in the decision-making process.

Many of these tools are still in the development stage and are not yet ready for routine use. In the near future, though, computer simulations of industrial processes will be routinely used to predict the impact of design decisions and to assist in the selection of the most environmentally sound design and manufacturing processes.

5.4 PROCESS CHANGES

Improving the efficiency of a production process often can greatly minimize the amounts of pollutants generated. These are commonly the most effective and cost-efficient modes of attack in the pollution prevention arena. Such modifications may include adopting more advanced process technologies, switching to less polluting reagents, changing cleaning processes and chemicals, using catalysts to increase reaction efficiency, segregating waste and process streams, and improving operating and maintenance procedures. In many cases, more than one of these approaches will be used in an integrated fashion to achieve optimum production while minimizing waste generation. These approaches are discussed further in the following sections.

5.4.1 Advanced Process Technologies

The first point of attack should usually be to investigate the operation and control of the reaction processes. Can anything be done to increase both the efficiency of the reaction so as to reduce the amount of process chemicals required and the conversion efficiency to product? Any improvements here will lead to reductions in the amounts of unreacted chemicals or by-products requiring further processing. In many cases, fairly simple changes to operating conditions such as reaction temperature or pressure can greatly increase operating efficiency. It has often been found that reagents are used in great excess over that necessary because "it has always been done that way." Production personnel routinely state that a particular reaction mix has been used for dozens of years and they are not about to "mess with success." However, that particular mix may have been developed long before there was a concern for the wastes produced, or when the economics was such that industry could afford to waste chemicals to ensure a good product. With the advent of easy-to-use process control equipment and a better understanding of the chemistry of many of these processes, it is often now possible to modify the process to pollute less without impairing the quality of the product.

An example of how process changes can be of benefit is the rapidly developing use of powder coating in place of traditional paint. Solvent-based paints have commonly been applied to parts' surfaces, generally by liquid spray, for corrosion protection, surface protection, identification, and aesthetic appeal (Higgins, 1989). An organic solvent is usually used as a carrier for the paint. Painting produces two significant sources of hazardous wastes: paint sludges and waste solvents. Paint sludges result from paint overspray; only about 50 percent of paint sprayed actually adheres to the piece being painted, while the remaining 50 percent is removed from the air in a water scrubber, resulting in a hazardous sludge deposit. The solvents in the paint evaporate into the air, causing another hazardous waste source. Solvents are also used to clean painting equipment after painting, again creating pollution. Water-based paints are becoming more prevalent, but their properties are often not as good as solvent-based paints.

Powder coating, or "dry powder painting," is now changing this situation. This technique is based on depositing specially formulated thermoplastic or thermosetting heat-fusible powders on the metal parts. In most cases, the dry paint powder is applied electrostatically by ionizing the air used to spray the paint, which in turn charges the dry powder particles. The surface to be coated carries the opposite charge, and the powder is electrostatically attracted to the surface. The coating is then fused to the surface and cured in conventional ovens. No solvents are used, so problems associated with solvent-based paints are eliminated and no solvents are needed for cleanup. The final product is often superior to that of conventional paints. The only drawback is that powder coating can be applied only to parts that can withstand the curing temperatures of about 350°F required. This means that it cannot be used on most plastics or aluminum.

By better controlling the temperature and pressure in a reactor through enhanced feedback control mechanisms, improved metering of chemicals to the reactors, use of sensors to monitor the reaction and the creation of product, and improved automation of the process, the efficiency of reagent conversion can be greatly enhanced and waste production can be more nearly minimized.

During the conceptual design stage for a process, several strategies should be considered to minimize pollution (Butner, 1995):

- Avoid adsorptive separations where adsorbent beds cannot be readily regenerated.
- Provide separate reactors for recycle streams, to permit optimization of reactions.
- Consider low-temperature distillation columns when dealing with thermally labile process streams.
- Consider high-efficiency packing rather than conventional tray-type columns, thus reducing pressure drop and decreasing reboiler temperatures.
- Consider continuous processing when batch cleaning wastes are likely to be significant (e.g., with highly viscous, water-insoluble, or adherent materials).
- Consider scraped-wall exchangers and evaporators with viscous materials to avoid thermal degradation of product.

IMPROVED REACTOR DESIGN. As we saw in Section 5.1.3, the reactor and the conditions under which it is operated are critical to source reduction. Stirred-tank reactors can often be modified to improve the degree of agitation achieved, and thus the mass transfer efficiency between reactants, through use of more efficient mixers and baffling systems. These can also help to improve heat transfer, making the reactor contents more homogeneous and increasing conversion efficiency. Insulating the reactors will also improve temperature control within the reactor. Adoption of high-efficiency heat exchangers, if they are not currently being used, can often greatly reduce energy costs.

It has already been mentioned that operation of reactors in a continuous-flow mode is often less polluting than using the batch mode. Because most reactions are not complete, some unreacted process chemicals usually remain at the end of the reaction; these must be separated and disposed of, and cleaning a batch reactor between each run can create a significant waste load. In addition, process control is much easier in a continuous-flow reactor, where the reactor contents and environmental conditions are the same throughout, than in a batch reactor, where the reaction may be causing environmental conditions to continually change. Reduction of the amount of pollution resulting from a chemical reaction may also be achieved by using something besides the common stirred-tank continuous-flow reactor. For example, plug flow reactors allow for staging, with different conditions, such as temperature, closely controlled in each stage to optimize output. Switching to a fluidized bed reactor from a stirred-tank reactor may also achieve significant waste reductions in some cases.

Other reactor design considerations include ensuring easy access in order to simplify cleaning of the interior of the reactor and implementing a design that allows complete draining of the vessel. The piping system should be designed such that piping run lengths and the number of valves and flanges are minimized. Drains, vents, and relief lines should be routed to recovery or treatment equipment.

In some cases, radical changes to a production process may be beneficial, but the improvements in efficiency must be balanced against the costs involved. For example, the steel industry has found that new electric arc furnaces are much less polluting than the more common open hearth and basic oxygen processes, but the cost of converting these plants to electric arc furnaces is enormous. These changes will come slowly as older units need to be replaced, but the dramatic improvements that would be realized through immediate use of this new technology cannot be achieved overnight.

Other process changes can have a significant impact on water use and can also have an impact on source reduction (Smith, 1995):

- Increase the number of stages in extraction processes that use water (to be described in more detail later).
- Use spray balls as a scouring agent to remove caked-on solids for more effective internal vessel cleaning.
- Change from wet cooling towers to air coolers.
- Improve control of cooling tower blowdown.
- Attach triggers to hoses to prevent unattended running.
- Improve energy efficiency to reduce steam demand and hence reduce the wastewater generated by the steam system through boiler blowdown, aqueous waste from boiler feedwater equipment, and condensate loss.
- Increase condensate return from steam lines to reduce boiler blowdown and aqueous waste from boiler feedwater equipment.
- Improve control of boiler blowdown.

IMPROVED REACTOR CONTROL. In most reaction systems, improving reactor environment controls can have a substantial impact. Often, when reactors are controlled by humans, who may not be properly attentive, accidents may occur, close reaction control may be difficult to regulate, and off-specification products may result. This may be alleviated by installing automated monitoring and control devices for such variables as temperature, pressure, feed rates, and reaction times. Robots used for welding and numerically controlled cutting tools are examples of successful changes. However, to be successful, these devices must be accurate and reliable and have a low failure rate. For example, humans can sometimes do a better job by eye of detecting defective parts than an optical sensor that does not perceive slight changes in a part. The use of automated controls can be a great benefit to quality control and waste minimization, but they must be used judiciously. Control system efficiency can be attributed to a combination of the following characteristics (Down, 1995):

- Measurement accuracy, stability, and repeatability.
- Sensor locations.
- Controller response action (proportional, integral, derivative, cascade, feedforward, stepped).

- Process dynamics.
- Final control element (valves, dampers, relays, etc.) characteristics and location.
- Overall system reliability.

Application of process controls to prevent pollution can be divided into several categories, including reduction of waste production through process efficiency improvements, pretreatment of pollutant-containing effluents through chemical reactions, capture and recycling of pollutant-containing waste by-products, and collection and storage of pollutant-containing waste by-products (Down, 1995). This may involve use of process controllers that continuously compare measured variables to desired values and regulate such things as reaction temperature and pressure and reaction time, or that regulate pumps and valves to recycle flow streams or to send off-specification materials to waste or to a reprocessing step.

IMPROVED SEPARATION PROCESSES. As we have previously seen, many industrial processes rely on separation units to recover products from a mixture of unreacted chemicals and unwanted by-products. These may involve separation of gases, liquids, or solids from gas, liquid, or solid streams. Many separation processes are available, some of which are more appropriate for a particular application than another. Table 5.5 lists some potentially useful separation processes for different applications.

Long (1995) presents several general rules of thumb for decreasing energy consumption in separation processes:

- Do mechanical separations first if more than one phase exists in the feed.
- Avoid overdesign, and use designs that operate efficiently over a range of conditions; favor simple processes.
- Favor processes transferring the minor rather than the major component between phases.
- Favor high-separation factors.
- Recognize value differences of energy in different forms and of heat and cold at different temperature levels.
- Investigate use of heat pump, vapor compression, or multiple effects for separation with small temperature ranges.
- Use staging or countercurrent flow where appropriate.
- For similar separation factors, favor energy-driven processes over mass separating-agent processes.
- For energy-driven processes, favor those with lower heat of phase change.

Although this list is aimed at energy reduction, these strategies often also reduce the amount of waste generated. For example, extracting the minor by-products from the process stream to another phase, rather than extracting the product itself, usually results in a much smaller volume of material to be reprocessed. Selecting processes that operate over a wide range of conditions allows for more steady operation and allows for operation with less operator interference and thus less chance of accidents or improper control settings.

TABLE 5.5
Potentially useful separation processes

	Recover from gas			Recover from liquid			Recover from solid		
Separation process	**Gas**	**Liquid**	**Solid**	**Gas**	**Liquid**	**Solid**	**Gas**	**Liquid**	**Solid**
Absorption	X	X							
Adsorption	X				X				
Centrifuge					X	X		X	X
Chemical precipitation						X			
Coalesce and settle		X		X	X	X			
Condensation		X							
Crystallization						X			
Cyclones		X	X		X	X		X	
Distillation					X				
Drying/evaporation						X		X	
Electromagnetic									X
Electrostatic precipitation		X	X						
Elutriation									X
Filtration			X			X		X	
Flotation					X	X			X
Gravity sedimentation		X	X		X	X		X	
Ion exchange					X				
Membrane	X				X	X			
Screening									X
Scrubbing	X	X	X						
Solvent extraction					X			X	
Sorting									X
Stripping				X	X		X	X	

IMPROVED CLEANING/DEGREASING. Cleaning and degreasing operations are commonly used by industry to remove dirt, oil, and grease from both process input materials and finished products. In the metal finishing industry, cleaning usually follows machining and comes before other surface finishing steps such as anodizing or metal plating. Circuitboards and other electronic equipment must be scrupulously clean to be effective. In paper and textile industries, cleaning of equipment to remove inks, dyes, oils, and other materials is frequently done. Solvent cleaning is the major process in the dry cleaning industry. Essentially all industrial production processes require cleaning or degreasing to some extent.

In some cases water is suitable for parts cleaning, but in many cases an organic solvent must be used. The most commonly used solvents are trichloroethylene (TCE) and perchloroethylene (PCE) because of their ability to dissolve a wide range of organic contaminants, their low flammability (flash point), and their high vapor pressure, which allows them to readily evaporate from coated surfaces (see Table 5.6). A density greater than water allows the solvent to be gravity separated from water after use, and a boiling point different from that of water allows the solvent to be separated by distillation. All of the solvents listed in Table 5.6 have these properties; however, all are halogenated solvents that are major environmental pollutants, posing significant potential public health threats. Thus all efforts must be expended to reduce the loss of these materials to the environment or to replace them with less polluting solvents.

Pollutants generated during cleaning and degreasing may include (1) liquid waste solvent and degreasing compounds containing unwanted film material; (2) air emissions containing volatile solvents; (3) solvent-contaminated wastewater from vapor degreaser-water separators or subsequent rinsing operations; or (4) solid wastes from distillation systems, consisting of oil, grease, soil particles, and other film material removed from manufactured parts (Thom and Higgins, 1995).

In some cases, modifying the manufacturing process may reduce the cleaning requirements for a particular process. For example, changing a piece of equipment to minimize carryover of material from one production step to another may minimize or eliminate the need for intermediate cleaning. These possible modifications should be investigated before looking at modifications to the cleaning process itself.

In a commonly used cleaning system, which usually uses water as the solvent, the part to be cleaned is placed in a *rinse tank* or consecutively in several rinse tanks containing either stagnant or flowing water. The objective is to transfer the contaminants on the surface of the manufactured part to the water by dissolution and dilution. This is the typical process used to remove excess plating solution dragged out of a plating bath. It is only useful, though, for removing materials that are water soluble. A single stagnant rinse tank can serve as a cleaning device, but it is not very effective and cleaning ability decreases dramatically with time as the level of contaminants in the rinse tank builds up. Large quantities of water are needed to effect cleaning. A series of stagnant rinse tanks improves the situation since subsequent rinse tanks in the

TABLE 5.6
Properties and characteristics of typical solvents

Property	Water	Trichloro-ethane (TCA)	Trichloro-ethylene (TCE)	Perchloro-ethylene (PCE)	Methylene chloride
Molecular weight	18.0	133.5	131.4	165.9	84.9
Boiling point (°C)	100	72.9	86–88	120–122	40
Density (g/cm^3)	1.00	1.34	1.46	1.62	1.33
Vapor pressure (mm Hg)	17.5	100	58	14	350
Flash point (°C)	None	None	None	None	None

progression contain fewer contaminants than the previous one, but they are still not very efficient. By using a flowing tank or series of tanks, cleaning efficiency can be significantly increased and water use decreased. Running the rinse water through the rinse tanks countercurrent to the flow of the pieces being cleaned provides the best and most efficient cleaning possible. Each additional cleaning step in the series contains cleaner water and improves the cleaning operation, but there is a limit to this improvement. Generally, little benefit is achieved by operating more than three countercurrent tanks in series. The use of rinse tanks is examined in detail in Chapter 10.

Some industries use *aqueous cleaning* with the addition of various detergents. The parts to be cleaned are either tumbled in an open, tilted vessel that rotates and that contains the cleaning solution, or they are dipped into a tank containing the cleaning solution. In either case, the parts must be rinsed thoroughly in a series of rinse tanks or under a continuously flowing stream of clean water to remove the cleaning solution and contaminants. In recent years, automated washers with low-volume, high-pressure water jets or water knives have been introduced to reduce the amount of water needed for cleaning and rinsing. Aqueous cleaning processes generally require copious quantities of water, and thus produce large volumes of dilute wastewater.

Most cleaning operations for removal of dirt or organic compounds such as oils and greases rely on organic solvents. Common solvent cleaning operations include cold cleaning and vapor degreasing. *Cold cleaning* is the simplest, least costly, and most widely used solvent cleaning process. Typically, this is done in open-topped dip tanks. The piece to be cleaned is dipped into a tank containing a liquid mineral spirit solvent. After transferring contaminants from the piece to the solvent, the piece is removed and solvent evaporates from it, leaving a clean surface. Unless collected and condensed, these vapors can be a major air pollution problem. Some industries use hoods over the cleaning bath and allow the cleaned pieces to dry within the hood, so that any volatilized solvent can be collected, condensed, and placed back into the cleaning tank. Eventually the solvent bath, which accumulates all of the dirt and grease that was on the part, must be cleaned.

Vapor degreasing uses chlorinated hydrocarbons, such as TCE, PCE, TCA, or methylene chloride, in the vapor phase to clean surfaces. Vapor degreasing occurs at an elevated temperature, necessary to produce the solvent vapor. A tank, filled to about one-tenth of its depth with liquid solvent, contains steam coils or some other heating mechanism to heat the solvent and produce the necessary vapors (see Figure 5.25). The vapors are heavier than air and remain in the tank. Pieces to be cleaned are placed in the vapor region of the tank. Because of the lower temperature of the pieces being cleaned, solvent vapors immediately condense on the surface of the piece, dissolve the contaminants, and then drip back to the bottom of the tank where they can be revaporized. Cleaning continues until the part is heated to the same temperature as the vapor, at which time no more vapor condensation occurs and the part is removed. To prevent loss of solvent vapors from the top of the cleaner, cooling coils are located in the top of the tank to condense any vapors that do not condense on the parts, returning them to the tank bottom. Solvent loss can also be decreased by increasing the freeboard in the tank. This allows more of the solvent vapors to be retained within the confines of the tank. Many solvent cleaning baths are now covered to minimize solvent loss. Other units are totally enclosed, preventing loss of nearly all of the solvent.

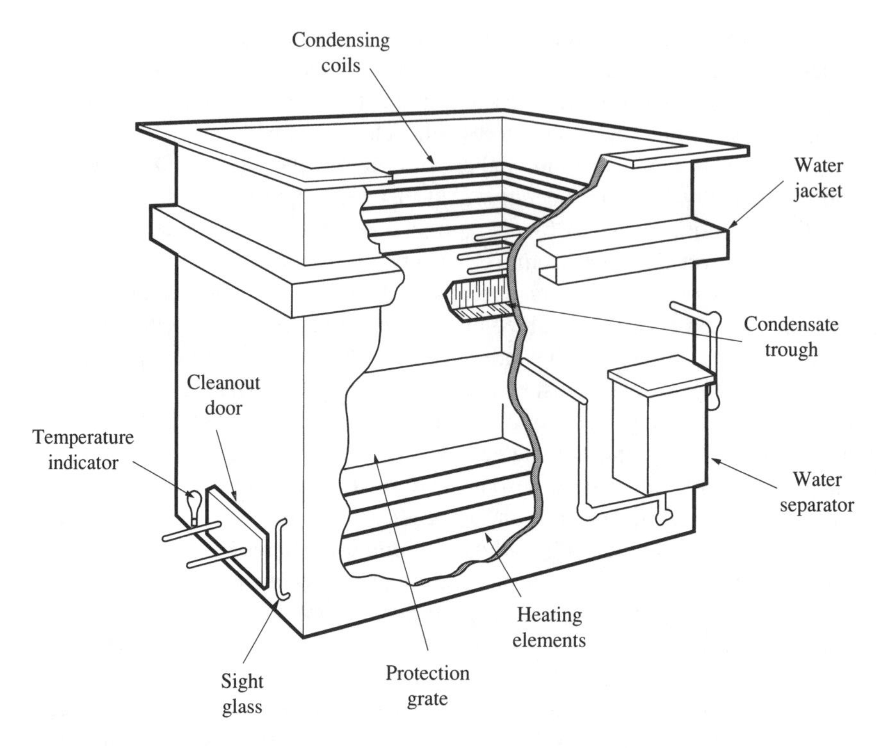

FIGURE 5.25
Solvent vapor degreaser. (Source: Thom and Higgins, 1995)

The solvent's vapors are generally pure solvent, but periodically the liquid solvent must be reprocessed to remove contaminants, and dirt that accumulates on the tank bottom must be removed. Some units have built-in recovery stills to clean the solvent. The volume of waste produced by a solvent cleaning system is only a small fraction of that produced by an aqueous cleaning system. However, solvent cleaning is generally suited only to cleaning of organic contaminants; this method is, for example, ineffective for removal of water-based metal plating dragout.

Ultrasonic cleaning devices are now becoming more widely used. These use ultrasonic waves in the 20–40 kHz range to induce vibrations that cause the rapid formation and collapse of microscopic gas bubbles in the solvent. The process, called *cavitation,* creates pressures as high as 10,000 psi and temperatures as high as 20,000°F on a microscopic scale that loosens contaminants and scrubs the workpiece. The vibrations are created by a transducer placed in the solvent bath. Ultrasonics can be used with a variety of cleaning solutions and will remove a wide range of organic and inorganic contaminants. The speed of cleaning is increased, and often a lower concentration of cleaning solution can be used. The workpiece must still be rinsed clean with clean water, creating an aqueous waste, but the amount produced is generally relatively low. Descriptions of the equipment and its operation can be found elsewhere (Gavaskar, 1995).

Other cleaning devices are also becoming available. These include a vacuum furnace to volatilize oils from parts and a laser ablation system, which uses a laser to rapidly heat and vaporize a thin layer of the part's surface and any contaminants on it.

When selecting a cleaning system, one should consider the entire spectrum of associated costs and environmental impacts. For example, vapor degreasing minimizes solvent use and solvent loss to the environment, thus minimizing environmental impacts, but it requires heating of the solvent and has all of the cost and environmental consequences of increased energy use. This decision-making approach, termed "life-cycle assessment," is described in detail in Chapter 6.

EQUIPMENT CLEANING. A primary method for reducing waste from equipment (chemical reactors, piping, heat exchangers, etc.) cleaning operations is to avoid unnecessary cleaning and to reduce cleaning frequency. This can be substantially achieved by switching from batch reactors to continuous-flow reactors, where possible. Improvements can also be realized by selecting reactors that have easy access for cleaning and that are designed for ease of cleaning (i.e., those with no internal parts that will accumulate contaminants).

Studies should be performed to determine how frequently cleaning is required. In the past, solvents commonly were reused until they were too contaminated to clean effectively then discarded. Unfortunately, this extended use often made them so contaminated that recovery of the solvent was infeasible. A better approach is to install a batch solvent recovery system and process the solvents more frequently on site. This allows almost limitless use of the solvent, minimizing the cost of solvent purchase, and greatly reduces the amount of hazardous waste that must be disposed of. Cost recovery for this system generally occurs within months (Thomas, 1995).

RECYCLING. Recycling or reuse of materials in a process effluent is often an attractive way of reducing waste streams needing treatment and their associated costs, while at the same time reducing the demand for virgin process chemicals and their associated costs. Many industries are finding that recovery and reuse of materials that were previously considered wastes can be cost effective. As described in Chapter 4, the EPA does not consider recycling or reuse to be pollution prevention, but both fall under this book's broader definition of pollution prevention because of their significant benefits to industry.

Following is a discussion of recycling within a process. In Chapter 9, we discuss in more detail recycling, and what can be done during product design to make recycling easier.

Materials can be recycled in several ways (Figure 5.26). Recovery may be on site or off site. It may be performed on site as a part of the process system or it may be accomplished in a separate step. The materials may be recycled to the process step where they were generated or to some other process in the facility. They may even be converted to a new material before reuse.

The effective use of recycle or material recovery options requires the efficient segregation of waste and process streams. Segregating wastes allows for minimization of contamination of process streams with wastes from other streams, simplifying recycling or reuse of the chemicals and reducing the volume of wastes requiring treatment. Segregation should take place as close to the source as possible. In the past, it was common practice to dump all waste streams into a common sewer, where wastes mixed

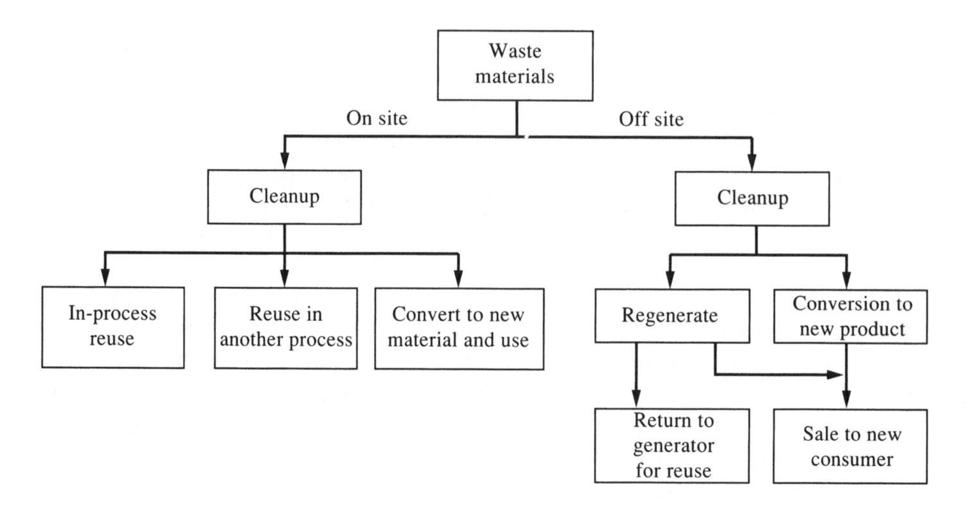

FIGURE 5.26
Recycling options.

together and flowed from the plant. Following combination with other waste streams, it is very difficult, if not economically impossible, to recover materials for reuse. The problem is compounded if a potentially recoverable waste stream is mixed with a high-flow dilute waste stream. The recoverable material will be diluted further, making recovery difficult, while the large-volume dilute waste stream, which may have needed only minor treatment before discharge, may now require extensive waste treatment. In the past, hoses used to wash equipment or fill reactors were run continuously because water was inexpensive and this practice was more convenient than constantly opening and closing valves. The unused water entered the waste sewer along with all other materials. This meant that all of this clean water became contaminated and had to be treated. If the contaminants were toxic, then the whole volume of waste would be deemed toxic, even though much of it was clean water that had never been used. Fortunately, most facilities have now stopped this practice.

For a recycling or reuse option to be viable, the chemical composition of the recycle stream must be compatible with the process to which it is directed; the cost of regenerating or cleaning up the material, or of modifying the process to accommodate recycle, must be economically justifiable; and the material must be available in a consistent quality and volume. Any of the techniques described earlier for process stream separation can be used to separate, clean up, and concentrate a chemical for recycle or reuse. In some cases, the material may be converted to a new chemical by chemical reaction before reuse in another process.

Recycling or reuse often occurs on site, with unused process chemicals being separated from a waste stream and returned to the same process where they originated. This is common in the metal plating industry, where the excess metal solution dragged out from a plating bath and removed from the plated workpiece during the rinse steps is concentrated, possibly using ion exchange or reverse osmosis, and returned to the plating bath. This not only reduces the amount of process chemicals being wasted to the sewer, but it also reduces the amount of chemicals needed to be purchased for plating.

The recovered materials may also be used on site in a different process or process step than where they originated. Usually, the chemical stream cannot be regenerated to as high a quality as virgin chemicals. For many uses, this is not a critical factor, and the recovered materials may be mixed with the new process chemicals going into the reactor. However, in some cases, impurities in the recycle stream may be detrimental to the process, negating the stream's direct recycle to that reactor. The recovered materials may be suitable for use in another process step, though, where the purity specifications may not be as exacting. For example, copper recovered from a printed circuitboard electroless plating process may contain impurities that would prohibit its return to the electroless process bath where tight quality control is maintained on the bath contents, but the recovered materials may be suitable for use in decorative electroless plating over plastic components where the specifications are not as critical. Another example is the use of recovered solvents used in solvent cleaning of parts. The recovered solvent may not be as clean as virgin solvent and may not be suitable for the final cleaning step on a component that must be scrupulously clean, but it may be fine for cleaning steps earlier in the manufacturing process. An excellent discussion of solvent recovery can be found elsewhere (Thom and Higgins, 1995).

A significant commodity for reuse that is often overlooked is water. In the past, water has been assumed to be a limitless low-cost resource. However, this attitude has changed as companies have had to deal with not only the increasing cost of water, but also the cost of treating contaminated water. Many industries use enormous quantities of water during processing. For example, paper mills use from 1000 to 100,000 gallons of process water per ton of product (averaging about 50,000 gal/ton of product). A typical pulp and paper facility manufactures about 1000 tons of paper per day. Thus about 50 million gallons of water are used per day in the typical plant. By reusing reclaimed water, many plants have reduced their water consumption by as much as 80 percent. This not only lowers their costs of process water purchase and treatment, but it also greatly reduces the volume of wastewaters requiring treatment and concentrates the constituents in the wastewater to the point where many of them are economically recoverable.

The preceding example concerns one of the largest industrial water users, but essentially all industries can reap major benefits by availing themselves of water minimization steps. Figure 5.27 shows typical water uses in an industrial chemical process facility. Water use can be minimized by (1) process changes, in which less water-demanding equipment is used; (2) reusing water directly in other operations, providing the level of previous contamination does not interfere with the process; or (3) treating the wastewater to remove harmful constituents that would prohibit the water's reuse and reusing the water in the original process or in another process (Smith, 1995).

Some industries are now exploring ways to integrate industrial processes in order to minimize water consumption. One way to do this is to use a technique called *"pinch analysis."* This procedure is discussed in detail in Chapter 10.

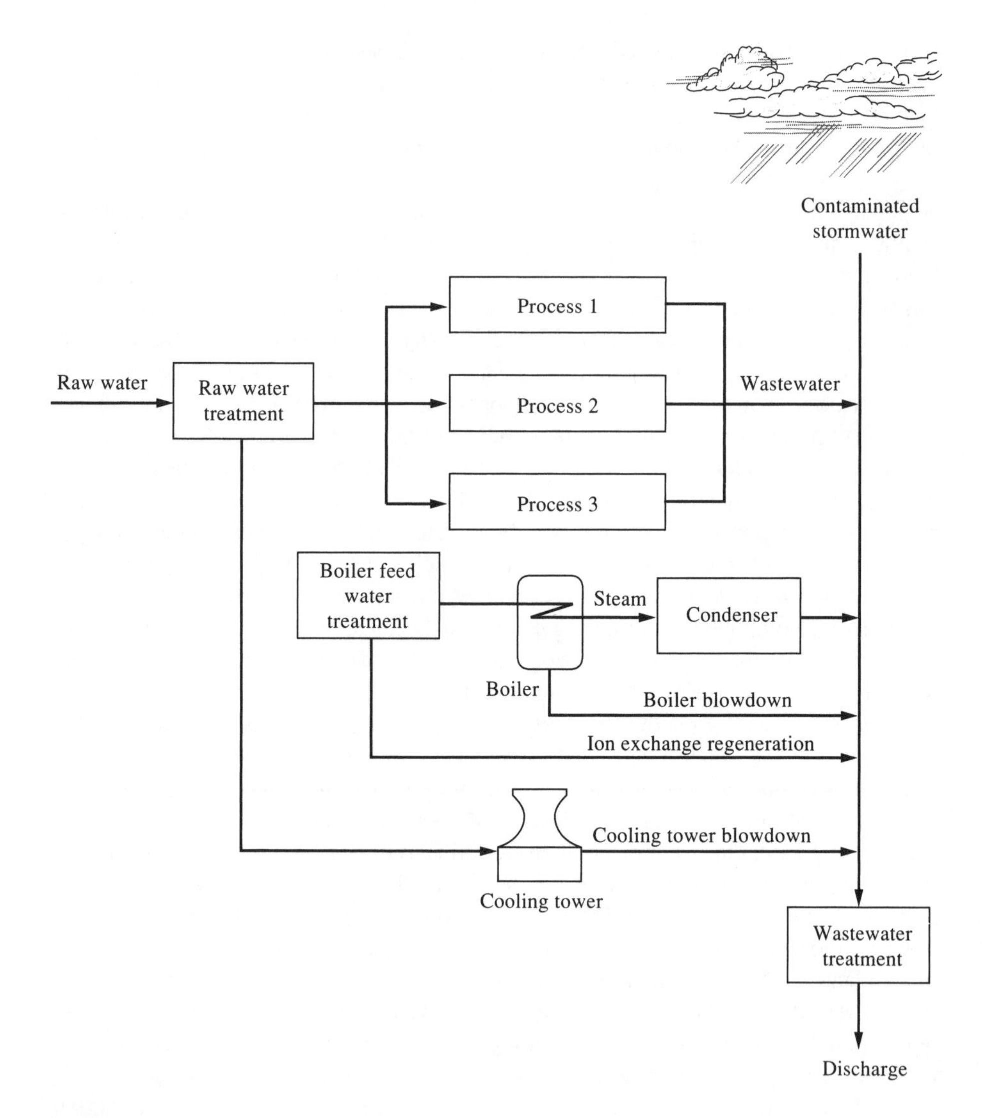

FIGURE 5.27
Typical water use on a chemical process site. (Adapted from Smith, 1995)

RECOVERED MATERIALS. Recycling or reuse of materials is generally desirable from the standpoint of resource conservation, but it may not be in terms of environmental life-cycle assessment. In many cases, more energy and resources are needed to recover the material or more wastes are generated during recovery than is the case when generating virgin materials. It may be preferable to dispose of the waste materials and use virgin materials for new products. A life-cycle analysis is needed to properly make this decision.

Recovered materials that are useless to the industry generating them may be of value to others; indeed, numerous companies have developed a market for recovered chemicals. They may either sell the chemicals to another company or trade them for chemicals that they can use that are recovered by the second company. Throughout the country, waste exchange services have been set up to assist companies in finding markets for their recovered materials for sale or for trade or to locate sources of recovered materials that a company needs. This may be a very economical transaction, because the recovered material can often be acquired for much lower cost than virgin materials. The company disposing of the material needs to charge only for the net cost of recovery after the anticipated cost of treatment and disposal of that material if it was not reused is factored in. Thus if wastewater containing a process chemical that cost $1.00 per pound would require a cost of $0.30 per pound for proper treatment and disposal although the chemical could be recovered for reuse at $1.10 per pound, it would be economical for the generator to sell the recovered chemical to another user at $0.80 per pound, the break-even point. Anything above this price would be a net profit to the generator, in spite of paying more than this to recover the material. Anything less than $1.00 per pound would be a savings to the purchaser, as long as virgin-quality chemicals were not required and the quality of the recovered materials was good enough.

A good example of how waste exchange can be successful is the partnership between Triad Energy Resources, Inc., and C&H Sugar Company. Triad ERI is an agricultural service company that tests and analyzes soils, crops, and water for farmers and then recommends nontraditional fertilizers such as surplus food and food by-products. C&H Sugar Company produces large quantities of sugar filtration by-products during the refining process. This material had historically been landfilled at high cost. C&H is now converting up to 200,000 tons of this filtration residue into an agricultural soil amendment and fertilizer to be marketed by Triad. The former landfill is being closed and converted into a 1300-acre park. This exchange was a major economic boost for both companies (CALMAX, 1996).

In many cases, the amount of waste generated by an industry is too small to economically justify on-site recovery, or the company may not have the expertise or the desire to operate a recovery system. A company may still eliminate the need to treat or dispose of these wastes, with all of the costs and headaches associated with this, by sending them to an off-site recovery center. Many such centralized facilities have been established to recover oils, solvents, scrap metals, process baths, plastic scrap, cardboard, and electroplating sludges. By combining wastes from many small industries, the recovery processes can be made economical. The recovered materials may be returned to the generator for reuse or may be sold to others.

5.4.2 Product Changes

One of the more effective ways to reduce pollution at the source is to make changes in the product itself or in the chemicals used to make the product. For example, new metal plating technology allows for the elimination of much of the cyanide that was required in cadmium plating baths. Water-based paints have replaced many solvent-based paints. Workpieces with fewer turnings and cavities will reduce the amount of dragout from process tanks. Replacement of steel automobile bumpers with aluminum or polymer composite bumpers makes them lighter and sometimes stronger and improves the automobile's fuel efficiency. However, when evaluating a change in a product to minimize pollution associated with its production, one must keep in mind that the paramount constraint is that the product, as constructed, serve its intended purpose in the best way possible. It makes no sense to switch to a new material or product configuration in order to reduce pollution if the resulting product cannot be economically produced or if it is inferior to the original product. A life-cycle assessment of the existing and proposed product can assist in deciding which will be the most cost-effective and environmentally benign.

The questions of product design, material and process selection, use of catalysts, and solvent substitution will be discussed in detail in Chapter 9 when we examine design for the environment.

5.4.3 Storage

Materials storage operations can be a major source of pollution in industrial processes, but these areas can be designed and operated to minimize possible emissions. As was described in Chapter 4, many industrial facilities that are used to store hazardous materials or hazardous wastes are regulated and must meet stringent requirements to mitigate against harmful releases.

Proper materials storage and delivery systems should be designed and managed with care to ensure that materials are not lost from storage tanks or piping systems through spills or leaks and to minimize losses during transfer operations. In many cases, commonsense precautions will suffice. Some commonly suggested approaches include the following (Hunt, 1995; Shen, 1995):

- Establish a spill prevention, control, and countermeasures (SPCC) plan.
- Use properly designed storage tanks and vessels for their intended purpose only.
- Install overflow alarms on all storage tanks.
- Install secondary containment areas to collect any spilled materials.
- Document all spillage.
- Space containers to facilitate their inspection for damage or potential leakage.
- Stack containers properly to minimize tipping, breaking, or puncturing.
- Raise containers off the floor to minimize corrosion from "sweating" concrete.
- Separate different hazardous substances to minimize cross-contamination and to separate incompatible materials.
- Use a just-in-time order system for process chemicals to minimize the amount of stored materials.
- Order reagent chemicals in exact amounts to minimize the amount of stored materials.
- Establish an inventory control program to trace chemicals from cradle to grave.
- Rotate chemical stock to prevent chemicals from becoming outdated, necessitating disposal.

Validate shelf life of chemical expiration dates, eliminate shelf-life requirements for stable materials, and test effectiveness of outdated materials.

Properly label all materials and containers with material identification, health hazards, and first aid recommendations.

Switch to less hazardous raw materials.

Switch to materials packaging and storage containers that are less susceptible to corrosion or leakage.

Use large containers where possible to minimize the tank surface-to-volume ratio, thereby reducing the area that has to be cleaned.

Use rinsable/reusable containers.

Empty drums and containers thoroughly before cleaning or disposing.

Eliminate storm drains in liquid storage and transfer areas.

Table 5.7 presents the results of many audits of industrial chemical storage facilities (La Grega, Buckingham, and Evans, 1994). As can be seen, many facilities had deficiencies. The most common problems were the existence of storm drains in liquid storage and transfer areas that could contaminate the relatively clean storm water, drains in storage areas that connected to wastewater sewers and could contaminate the wastewater with process chemicals, improper handling of empty drums, use of containers that were in poor condition, and inadequate secondary containment facilities.

Many of these problems could be eliminated if the industry instituted and enforced a spill prevention, control, and countermeasures (SPCC) plan. As described in Chapter 4, this is a plan developed by each industry to ensure that spills or accidental releases from storage areas will not occur and to outline reactive measures that will

TABLE 5.7
Most common deficiencies in storing hazardous wastes and hazardous materials

Deficiency	Audited facilities having this deficiency, %
Inadequate containment/drum/tank storage areas	68%
Lack of tank integrity testing	65
Lack of or inadequate inspection program for storage	48
Improper labeling	44
Inadequate training program	40
Lack of or inadequate spill prevention, control, and contingency plan	34
Incomplete wastewater analysis	24
Lack of or inadequate hazardous waste contingency plan	18
Inadequate handling of materials containing PCBs	18
Storm drains in liquid storage and transfer areas	18
Inadequate handling of empty drums	17

Source: LaGrega, Buckingham, and Evans, 1994.

be taken if they do. An SPCC plan is required for industrial oil storage facilities under the Oil Pollution Act of 1990 and should also be developed for other chemicals or toxic wastes stored on site. The SPCC plan should address at least the following three areas:

- Operating procedures that prevent spills and leaks.
- Control measures installed to prevent a spill or leak from reaching soil or water.
- Countermeasures to contain, clean up, and mitigate the effects of the spill or leak.

The plan should also address all of the items listed earlier for precautions to take to prevent spills and leaks.

The industry should also work closely with the local community committee responsible for the local emergency plan, as required by the Emergency Planning and Community Right-to-Know Act (EPCRA), to ensure that the community is prepared for any leaks or spills that do occur. This is described more fully in Chapter 4.

Many industries generate hazardous wastes during production processes. These wastes are usually stored for at least some time before treatment or shipment off site for disposal. The Resource Conservation and Recovery Act (see Chapter 4) mandates the regulation of virtually all facilities that store hazardous wastes. Any facility that stores more than 1000 kg of wastes for more than 90 days must have a storage permit and meet very strict requirements (facilities that store only 100 to 1000 kg can store for up to 120 days before they need a permit). Storage facilities are typically areas where drums or tanks of hazardous waste are held until they are processed or shipped, but they also include pits, ponds, lagoons, tanks, piles, and vaults.

In general, storage of hazardous wastes should be done in a way that minimizes accidental releases. All containers must be labeled properly, appropriate containers must be used, a hazard communication plan should be in place and the workers trained to implement it, and a detailed inventory of stored wastes should be maintained. Facilities must be designed to minimize the possibility of fire, explosion, or any unplanned sudden or nonsudden release of hazardous waste to the air, soil, or water. Materials should be stored at least 50 feet from property lines and away from any traffic, including forklift and foot traffic. An impermeable base should underlie all storage containers to hold any leaked or spilled material or accumulated rainfall. In some states, epoxy-coated secondary containment may be required to contain leaks. Adequate ventilation should be provided to prevent buildup of gases. Drums should not be stacked more than two high, and flammable liquids should not be stacked at all. Incompatible wastes should not be stored together and should be separated by barriers such as walls or berms. A drainage system should be installed so that spilled wastes or precipitation does not remain in contact with the containers. Storm water runoff into the containment area must be prevented. A more detailed description of storage requirements can be found elsewhere (Kuhre, 1995).

None of these actions to minimize leaks and spills during materials storage or preparations for emergency response to spills that do occur will be of use, though, unless they are meticulously carried out, their implementation is enforced, and all employees are well trained and informed of the need for close attention to the contents of the plan. Proper education and training are of paramount importance.

5.4.4 Management

No pollution prevention program will be successful unless the management is fully committed to its concepts. Employees will comply with a plan only if they feel that it is important to the company and that management provides proper support for it to be properly carried out. Among the factors that are critical to success are a well-conceived and functional preventive maintenance program, proper employee training, and good record keeping.

HOUSEKEEPING. Perhaps the pollution prevention activity with the greatest potential impact, and the one that should be addressed first, is improvement of housekeeping practices. This low-cost, and sometimes no-cost, action usually results in significant reductions in waste of process chemicals, water, and energy, and it usually improves worker health and safety. Many industries needlessly generate waste without even realizing it until they closely examine their housekeeping activities. A metal plater that allows rinse waters to continue to flow through a rinse tank when no workpieces are in the tank may be sending hundreds or thousands of gallons of clean water to a sewer where it mixes with wastes, becomes contaminated, and must be treated. A chemical processor that hoses down floors to remove spilled solid chemicals, rather than sweeping them up dry, is also increasing its wastewater pollution load and the resulting treatment costs. Inadequate equipment maintenance can lead to pumps and valves that leak process chemicals or oil, thereby wasting them and necessitating waste treatment; improper metering by chemical feed equipment, leading to waste of materials or possible production of off-specification products that must be disposed of; and cross-contamination of chemicals. Improper storage of materials, such as incorrect stacking or crowding of the storage area, can lead to leaks and spills. Lax labeling practices often lead to materials being used improperly or being discarded unnecessarily because the correct contents of a container are unknown.

In many cases, potentially polluting or harmful practices are performed solely because the employee is not aware of the consequences. For example, flushing out the solid residue in the bottom of a storage container with water before cleaning the container probably makes sense to the operator; moreover, this longstanding practice had been taught to the operator, who therefore is concerned only with ensuring that the tank is clean and that any residues won't contaminate the next batch. The employee is probably unaware of the impact of flushing the dry chemical into the sewer. The remaining material in the container could probably be vacuumed up and either reclaimed or disposed of in a dry form, but if the operator does not understand why this is preferable, the expedient practice of water flushing will likely continue to be used. All employees should be educated about the impacts of their practices on the overall operation of the facility and on the environment and should be shown ways to minimize harmful impacts. Many improper practices could be avoided solely through education.

The following simple housekeeping practices are often overlooked:

Close or cover solvent containers when not in use.

Isolate liquid wastes from solid wastes.

Turn off equipment and lights when not in use.
Eliminate leaks, drips, and other fugitive emissions.
Control and clean up all spills and leaks as they occur.
Develop a preventive maintenance schedule and enforce its use.
Schedule production runs to minimize cleaning frequency.
Improve lubrication of equipment.
Keep machinery running at optimum efficiency.
Dry sweep floors whenever possible.
Do not allow materials to mix in common floor drains.
Insist on proper labeling of all containers.
Educate all employees as to the need for proper housekeeping practices.
Include housekeeping reviews in all process inspections.
Adopt a total quality management philosophy.

Preventive maintenance is often the key to good housekeeping practices. Regular inspection and maintenance of all equipment, including replacement of worn or broken parts before the equipment actually fails, will lead to increased efficiency and longevity of equipment, fewer disruptions due to equipment failure, and less waste from leaking equipment or off-specification products. In particular, seals and gaskets in hydraulic equipment should be examined and replaced frequently to prevent leaks. A properly designed and implemented preventive maintenance program can often eliminate or minimize production shutdowns and prevent many potential sources of waste from becoming a problem. Although costly to perform, these programs usually pay for themselves many times over by avoiding production delays and unnecessary waste treatment.

Good housekeeping practices are generally based on common sense. It is essential, though, that all employees be trained in proper procedures and taught why they are essential. Their attention to good housekeeping practices must be monitored regularly; it is very easy to lapse into a mode of doing what is most convenient rather than what is right.

TRAINING. Once operating and maintenance procedures are established, they must be fully documented and made part of the employee training program. Proper and continual employee training is key to the success of any pollution prevention program. Without it, any plan is doomed to failure. The goal of the training program should be to make every employee, from the operators to the executive officers, aware of sources of waste generation in the facility, the impact of waste on the company and the environment, and ways to reduce the pollution. Training should be ongoing with frequent review updates.

Training should address spill prevention, response, and reporting procedures; good housekeeping practices; material management practices; and proper fuel and storage procedures (Shen, 1995).

The pollution prevention program should set goals for waste reduction. The training program should provide the information necessary for the personnel to achieve tnose goals. A variety of training techniques should be used, including seminars, written materials, and in-plant training. All plant personnel should be active participants in the training in order to ensure that everyone feels they are attacking the pollution problem as a team. Operators should play a major role in training preparation, as they are the most knowledgeable about the processes they work on.

A pollution prevention training program should be designed to accomplish the following (Osantowski *et al.*, 1995):

- Promote awareness of waste reduction initiatives.
- Establish quantifiable wastewater reduction goals.
- Train personnel in waste reduction techniques and water use minimization practices.
- Implement plantwide waste reduction techniques and water use minimization practices.
- Monitor progress toward attaining waste reduction goals and readjust objectives if they prove to be unattainable.
- Attain waste reduction goals.

Pollution prevention and worker safety go hand in hand. An additional benefit of a well-conceived pollution prevention training program is that it should result in improved worker safety.

RECORD KEEPING. Good record keeping can significantly help in both the development and the implementation of a sound pollution prevention program. Documentation of process procedures, control parameters, chemical specifications, chemical use, energy use, waste generation, and spill frequencies and causes will help to focus on areas where source reduction activities will be most useful. Inventory control, for both process chemicals and waste materials, will reduce the volumes of these materials that are kept on hand and thus reduce costs and potential losses to the environment from leaks or spills. It will also diminish the need to dispose of contaminated, off-specification, or out-of-date reagents. Good record keeping can also be used to ensure compliance with environmental regulations while showing the public that the company has a true concern for environmental quality.

5.5 POLLUTION PREVENTION EXAMPLES

Following are two examples of how modifying manufacturing operations can improve a process from a pollution prevention standpoint. The first case study involves the recovery process for acrylonitrile manufacture; the second involves reactor design for the production of maleic anhydride. These two case studies were developed by Tat Ebihara (1998) and were analyzed using the software SuperPro Designer.

5.5.1 Acrylonitrile Manufacturing

Acrylonitrile ($CH_2 = CHCHCN$) is a chemical feedstock used in the manufacture of plastics. Acrylonitrile is manufactured in a gas-phase fluidized bed reactor where propylene, ammonia, and oxygen are catalytically reacted at 450°C and 2 atm:

$$CH_3CH{=}CH_2 + NH_3 + 1.5O_2 \xrightarrow{400\text{–}460^\circ C} CH_2{=}CHCN + 3H_2O$$

The process also uses a catalyst containing bismuth and molybdenum oxides. The reaction products are acrylonitrile and water plus residual oxygen and ammonia. Small, but economically important, quantities of hydrogen cyanide (HCN) by-products are also produced. Approximately one-quarter of the hydrogen cyanide used in industry comes from recovery of HCN produced during the manufacture of acrylonitrile (Heaton, 1996).

Acrylonitrile is purified and recovered in a multiple-step process using water-vapor separators (flash drums), heat exchangers, a water scrubber (absorber), and a distillation column. Figure 5.28 is a schematic of the purification and recovery process.

As operated, the fluidized bed reactor produced an effluent process stream containing 5000 kg/h acrylonitrile, 5100 kg/h water vapor, 500 kg/h oxygen, and 50 kg/hr ammonia. A purification and recovery system, as depicted in Figure 5.28, was used to recover the acrylonitrile. The recovery plant was initially set to operate with the heat exchanger/cooler (unit E-101) having an exit temperature of 94°C, the heat exchanger/heater (unit E-102) with an exit temperature of 98°C. The absorber, with a water stream flow rate of 600 kg/h, was designed to recover 99.5 percent acrylonitrile, 99.0 percent ammonia, and 99 percent water. The continuous distillation column was operated at 1.03 atm, with a condenser temperature of 80°C and a reboiler temperature of 110°C. An optimization study of the process, using SuperPro Designer, showed that the percentage recovery in the products stream (stream S-112) was only 28.7 percent of that in the input stream (stream S-101); most of the acrylonitrile produced was actually lost to the wastewater system in the bottoms from the two flash drums. In addition, final product quality was low; the acrylonitrile product still contained 3.4 percent ammonia.

The company's engineers investigated ways to improve the recovery of the acrylonitrile produced in the fluidized bed reactor. Since the major losses were due to inefficient separation in the flash drums, they evaluated modification of them. The goal was to achieve at least 80 percent recovery of acrylonitrile, with less than 1.25 percent ammonia and less than 0.025 percent water in the final acrylonitrile product. Using the simulation software package, the engineers found that they could achieve the final product goal simply by increasing the outlet temperature of the cooler from 94 to 95.5°C and the outlet temperature of the heater from 98 to 99°C. This would increase the acrylonitrile recovery to 80.6 percent and would reduce the ammonia in the final product to only 1.2 percent. Thus a very small change in operating conditions resulted in a significant improvement in product recovery and in final product quality.

The engineers were also asked to determine whether it would be possible to reduce the amount of wastewater generated. In the process as originally operated, approximately 11,000 kg/h (70,000 gal/day) of wastewater was generated. This wastewater contained valuable acrylonitrile. They investigated methods to minimize wastewater production by reusing wastewater streams in the process, where appropriate. The only feedwater stream in the process is S-110, the adsorber feed water, at 5000 kg/h. Since the absorber is used to separate acrylonitrile from other residual gases (e.g., oxygen), recycle of the wastewater streams containing acrylonitrile should be feasible.

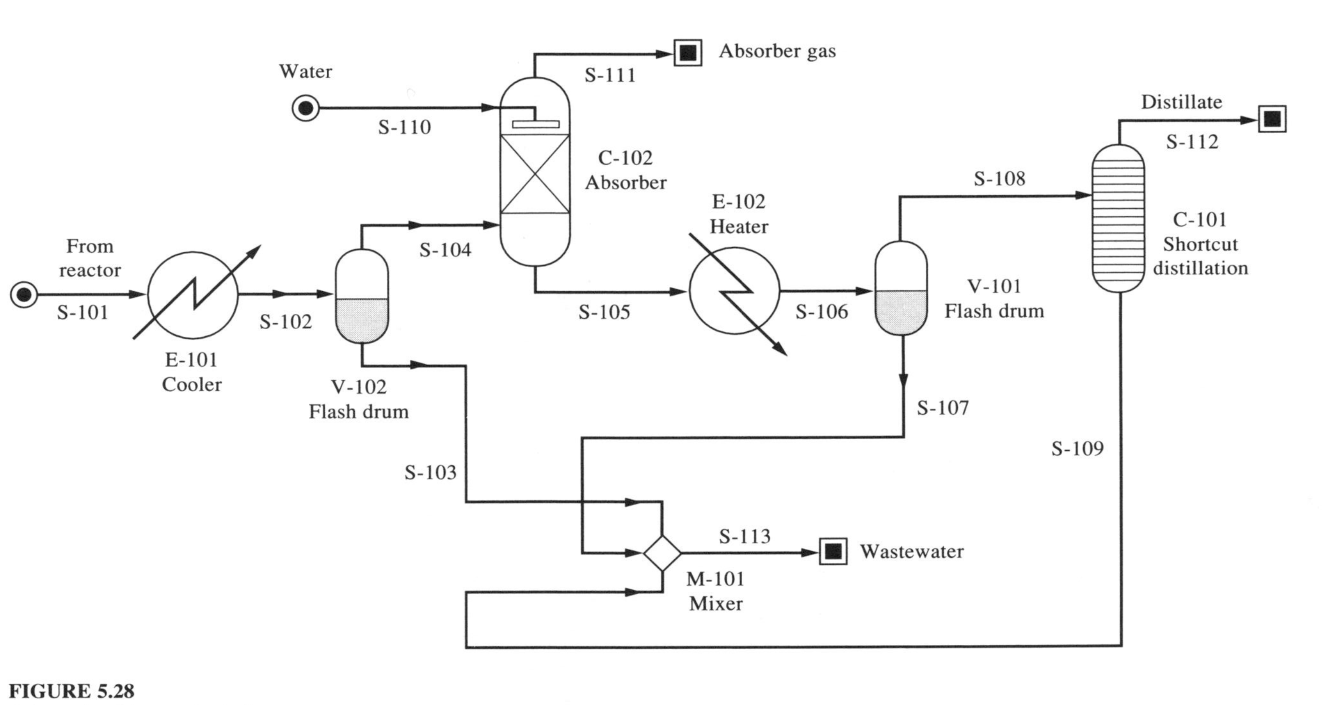

FIGURE 5.28
Schematic of the acrylonitrile recovery process.

Because a significant amount of water is generated in the reactor (5100 kg/h), this amount will need to be continuously removed from the process, so a zero wastewater discharge goal would not be feasible. However, we can significantly reduce the discharge from the present 11,000 kg/h.

The recommended water reuse concept is to install a two-way flow splitter after stream S-113 and return a fraction of the flow to the absorber tower (C-102) to replace the feedwater stream (S-110). Figure 5.29 is the new process flowsheet. This reduces the total wastewater effluent to 5363 kg/h, a reduction in flow of 51 percent. The product quality requirements (80 percent recovery, < 1.25 percent ammonia, < 0.025 percent water) are met if the percentage of flow at the flow splitter returned to the absorber tower is set at 50 percent [note that no liquid flow results in the second flash drum (V-102)]. If the percentage of flow returned to the absorber is set high at 80 percent to 90 percent, the product quality requirements would still be met. However, the water flow rate out of V-102 becomes unusually high (approximately 30,000 kg/h), which would likely overload the existing absorber tower design and result in very high flow velocities through existing piping.

5.5.2 Maleic Anhydride Production

Maleic anhydride (MAN) is a chemical intermediate used primarily in the manufacture of unsaturated polyester resin, fumaric and maleic acids, and lubricating oil additives. Production changes for MAN made over the last 15 years are examples of raw material substitution for pollution prevention. Before 1980 MAN was commercially produced by the vapor-phase catalytic oxidation of benzene. By the mid-1980s, commercial production of MAN had been converted entirely from benzene to butane feedstock for economic and environmental reasons.

The process design evaluation presented in this section is based on production of MAN from the catalytic conversion of *n*-butane and air (see Figure 5.30). The desired reaction for MAN production using a fixed bed of a vanadium-phosphor oxide catalyst is as follows:

$$C_4H_{10} + 3.5O_2 \rightarrow C_4H_2O_3 + 4H_2O \qquad \Delta H = -294.5\ \text{kcal/mol} \qquad (5.1)$$

Additional reactions that occur in the catalytic oxidation of butane which lower the yield of MAN are a total oxidation reaction (reaction 5.2) and a decomposition reaction (reaction 5.3):

$$C_4H_{10} + 5.5O_2 \rightarrow 2CO + 2CO_2 + 5H_2O \qquad \Delta H = -499.2\ \text{kcal/mol} \qquad (5.2)$$

$$C_4H_2O_3 + O_2 \rightarrow 4CO + H_2O \qquad \Delta H = +100\ \text{kcal/mol} \qquad (5.3)$$

Simulation software, such as SuperPro Designer, can be used to identify the most desirable operating conditions for the safe and efficient (maximum yield) manufacture of MAN. Assume that the following design constraints apply:

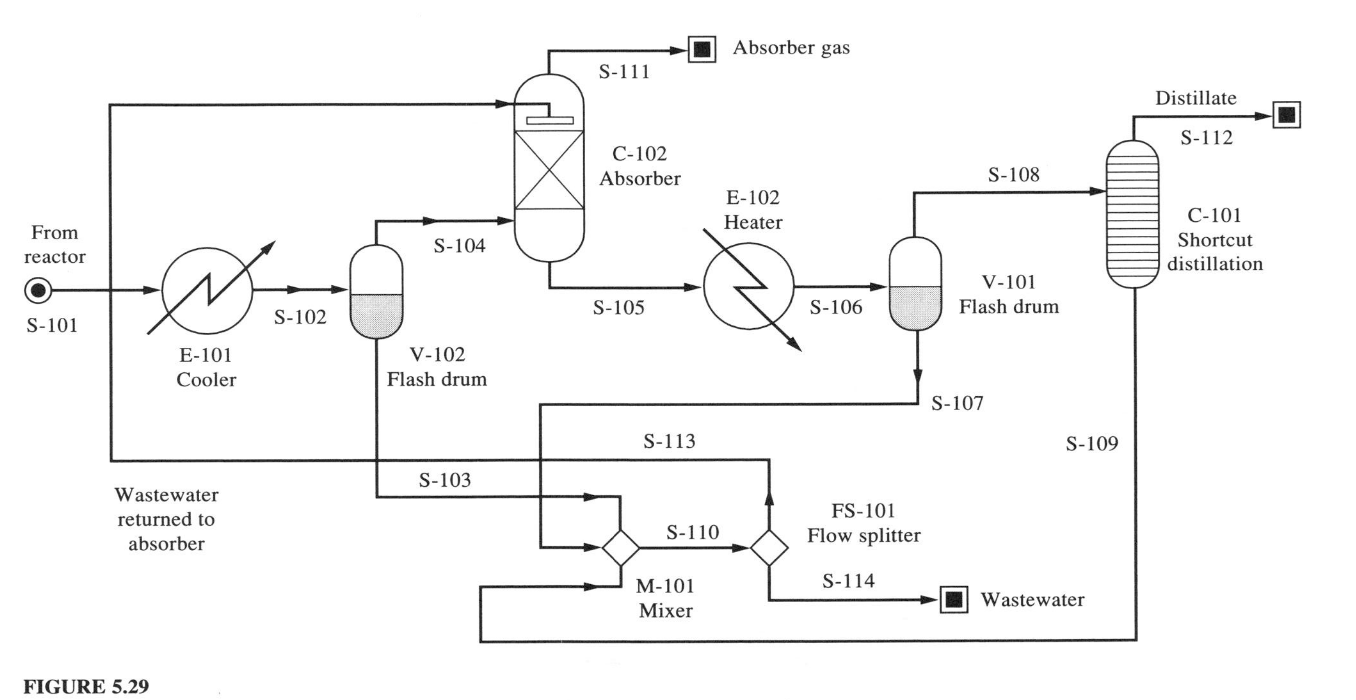

FIGURE 5.29
Revised schematic of the acrylonitrile recovery system for minimum wastewater production.

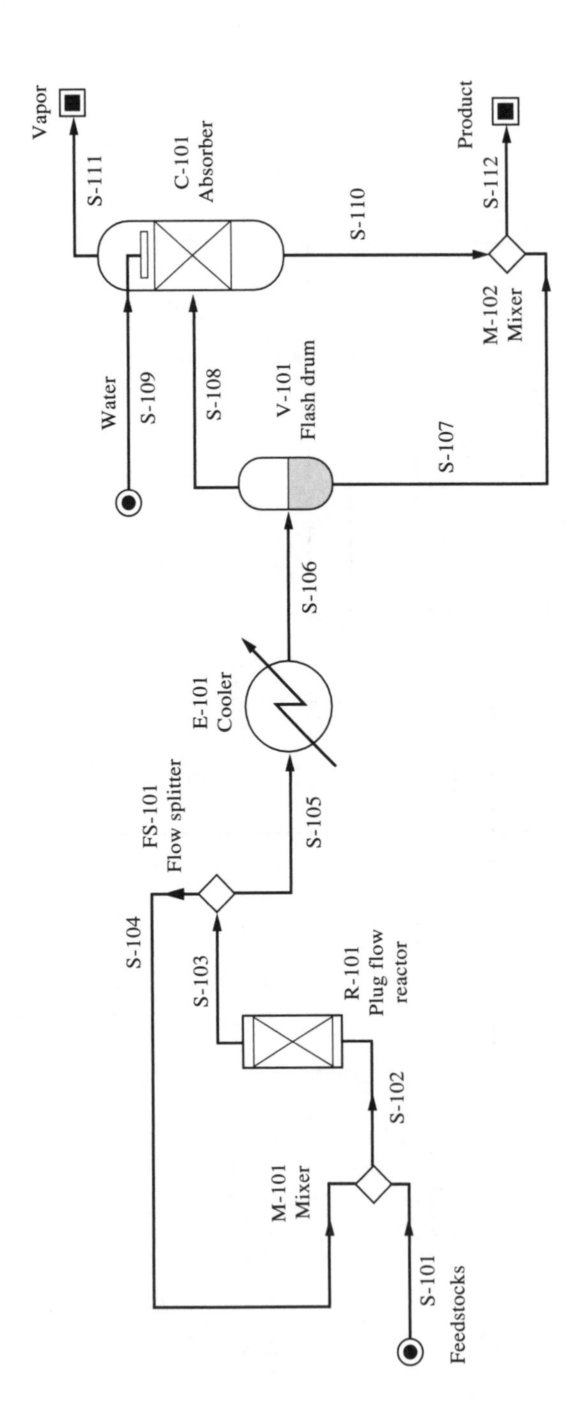

FIGURE 5.30
Schematic of the maleic anhydride production and recovery process.

1. Safety: Maximum butane concentration of 3.0 mol% in air to maintain concentrations below the ignition temperature. Assume a maximum temperature of 450°C for stable operation.
2. Design considerations: Assume a maximum space-time of 0.05 h for economical reactor design and maximum size of 2000 m^3 per reactor. Assume a reactor operating pressure of 1.00 bar.
3. Environmental considerations: Minimization of CO and CO_2 emissions.

The following kinetic parameters are assumed to apply for the specific vanadium oxide catalyst under consideration:

Desired reaction rate: $R_1 = k_1 p_1^{\alpha 1}$
Total oxidation reaction: $R_2 = k_2 p_1^{\alpha 2}$
Decomposition reaction: $R_3 = k_3 p_2^{\alpha 3}$

Note: k_1, k_2, k_3 are reaction rate constants
p_1 = concentration of C_4H_{10} p_2 = concentration of $C_4H_2O_3$
α_1, α_2, α_3 are the reaction order constants
It is assumed that the temperature dependence of the rate constant follows the Arrhenius equation as follows:

$$k_{i,T} = A_i \exp(E/RT)$$

The values of the reaction rates constants are provided in Table 5.8.

Modeling is begun by selecting a plug flow reactor to model the fixed-bed catalytic reaction, with a total feed flow rate of the butane-air mixture of 1000 kmol/h. Applying the design constraints provided in the problem statement, the reactor model can be run at various operating conditions to determine the "optimal" conditions to maximize yield and minimize CO/CO_2 emissions.

By inspection of the reaction information provided [based on Sharma and Cresswell (1991)] three facts are observed:

- The rate of the desired reaction 5.1 is greater than reaction 5.2: The rate expressions are equivalent except for the reaction rate constants; therefore, higher rates of MAN conversion will result per mole of butane than conversion to CO + CO_2.

TABLE 5.8
Kinetic constants for oxidation and decomposition reactions in maleic anhydride production

Reaction	k_I at T = 673 K	E, kJ/mol (Arrhenius eq.)	Reference rate compound and exponent in rate expression
Desired reaction	9.6×10^{-5}	93,100	Butane ($\alpha_1 = 0.54$)
Total oxidation reaction	1.5×10^{-5}	93,100	Butane ($\alpha_2 = 0.54$)
Decomposition reaction	2.9×10^{-6}	155,000	Maleic anhydride ($\alpha_3 = 1.00$)

- The rate of reaction 5.1 will increase with increasing temperature and mole percent butane: The highest yields of MAN will (discounting decomposition) be at the highest temperature and highest allowable mole fraction butane in the feed within safe operating parameters (i.e., below the explosive limit and ignition temperature). Since the reactor is isothermal and continuous, the reaction will not shift to favor the reactants according to LeChâtelier's principle.
- The rate of MAN decomposition to CO will increase with reaction temperature and increasing MAN concentration: The highest rates of decomposition will occur at the highest temperatures and highest yields; therefore, it is not certain that the highest overall yield will occur at the 3.0 mol % butane and $T = 450°C$. A reactor recycle stream, if used, may actually decrease yields of the desired product.

Since it is not intuitively clear which reactor conditions will provide the highest maleic anhydride yields, "response surface" methodology will be used to find the optimal conditions which maximize percentage conversion of butane to maleic anhydride and which yield the lowest amount of CO/CO_2 production per mole of butane.

The reactor model can be run at a number of reaction temperatures (375–450°C) and mole percents of butane in the feed (1–3 mol %) to determine their impacts on the molar flow ratios of MAN, CO, and CO_2 in the output. For this analysis, the reactor was set to run with a 0.05-h space-time, maximum reactor volume of 2000 m^3, and fixed molar flow rate of 1000 kmol/h. Results of this analysis are shown in Table 5.9.

Inspection of Table 5.9 indicates that the highest maleic anhydride yields occur at one of two points:

375°C and 3.0 mol % butane

450°C and 1.0 mol % butane

The lowest combined CO and CO_2 yields were also at these two points, except for 1.0% butane at temperatures of 375–425°C (at these points, residual butane remained in the product stream due to the lower rates of conversion resulting from lower concentration and temperature). It makes sense that maximum product yields would occur at the same points as the minimum CO and CO_2, since these are competing reactions.

TABLE 5.9
Maleic anhydride reactor outlet composition

	Mol% butane in feed					
	0.01		0.02		0.03	
Reactor temperature, °C	% conversion to MA	Moles CO or CO_2/mol butane	% conversion to MA	Moles CO or CO_2/mol butane	% conversion to MA	Moles CO or CO_2/mol butane
375	40.2	0.252	86.2	0.552	86.3	0.546
400	63.4	0.397	85.7	0.574	86.0	0.559
425	83.6	0.526	84.4	0.627	82.5	0.589
450	86.3	0.548	81.4	0.744	83.5	0.659

To determine whether increased yields would result at 3 mol % butane at temperatures lower than 375°C, the model was run at 3 mol % butane and 350°C, but incomplete butane conversion resulted in a yield of only 73.0 percent. (Note that temperatures greater than 450°C were not considered at 1.0 mol % butane, since temperatures exceeding 450°C were not considered within safe reactor operation.)

Therefore, the reactor condition of 3.0 mol % butane and reactor temperature of 375°C were selected as the desired operating point for the reactor. The higher butane concentration would provide a lower total reactor volume (hence lower capital and operating costs) than the 1.0 mol % butane and 450°C. This point also coincides with the region of stable operation provided in the kinetic model by Sharma and Cresswell (1991).

REFERENCES

Allen, D. T., and Rosselot, K. S. *Pollution Prevention for Chemical Processes.* New York: Wiley, 1997.

Butner, R. S. "Pollution Prevention in Process Development and Design." In *Industrial Pollution Prevention Handbook,* ed. H. M. Freeman. New York: McGraw-Hill, 1995, pp. 329–41.

CALMAX. The 1995 CALMAX Match of the Year. http://www.ciwmb.ca.gov/mrt/calmax/moty95.htm (1996).

Clifford, A. A. "Chemical Destruction Using Supercritical Water." In *Chemistry of Waste Minimization,* ed. J. H. Clark. London: Blackie Academic & Professional, 504–21 (1995).

Cortright, R., Crittenden, J., and Rudd, D. Research in Clean Reaction Technologies by the National Center for Clean Industrial and Treatment Technologies (CenCITT), unpublished manuscript, Michigan Technological University, Houghton, MI (1996).

Crittenden, J. *National Center for Clean Industrial and Treatment Technologies Activities Report: October 1994–September 1995.* Houghton: CenCITT, Michigan Technological University, 1996.

Down, R. D. "Pollution Prevention through Process Design." In *Industrial Pollution Prevention Handbook,* ed. H. M. Freeman. New York: McGraw-Hill, 1995, pp. 395–408.

Ebihara, T. Instructional Exercises on Pollution Prevention for Chemical Processes Using SuperPro Designer. University of Cincinnati, unpublished paper, 1998.

Florida, R. The Environment and the New Industrial Revolution: Toward a New Production Paradigm of Zero Defects, Zero Inventory, and Zero Emissions. Presented at the Annual Meetings of the Association of American Geographers, 1996.

Gavaskar, A. R. "Process Equipment for Cleaning and Degreasing." In *Industrial Pollution Prevention Handbook,* ed. H. M. Freeman. New York: McGraw-Hill, 1995, pp. 467–81.

Graedel, T. E., and Allenby, B. R. *Design for Environment.* Englewood Cliffs, NJ: Prentice Hall, 1996.

Halevi, G., and Weill, R. D. *Principles of Process Planning.* London: Chapman & Hall, 1995.

Heaton, A. *An Introduction to Industrial Chemistry.* London: Blackie Academic & Professional (1996).

Hertz, D. W. "Managing the Design Process." In *Waste Minimization through Process Design,* ed. A. P. Rossiter. New York: McGraw-Hill, 1995, pp. 265–87.

Higgins, T. *Hazardous Waste Minimization Handbook.* Chelsea, MI: Lewis Publishers, 1989.

Hopper, J. R. "Pollution Prevention Through Reactor Design." In *Industrial Pollution Prevention Handbook,* ed. H. M. Freeman. New York: McGraw-Hill, 1995, 343–60.

Hunt, G. E. "Overview of Waste Reduction Techniques Leading to Pollution Prevention." In *Industrial Pollution Prevention Handbook,* ed. H. M. Freeman. New York: McGraw-Hill, 1995, pp. 9–26.

Jain, V. K. "Supercritical Fluids Tackle Hazardous Wastes." In *Environmental Science & Technology* 27 (1993): 806–8.

Kratche, K. "Environmental Software Could Make Virtual Prototyping a Reality." In *Pollution Prevention Technical Bulletin* 2 (1997): 1–2.

Kuhre, W. L. *Practical Management of Chemical and Hazardous Wastes.* Englewood Cliffs, NJ: Prentice Hall, 1995.

LaGrega, M. D., Buckingham, P. L., and Evans, J. C. *Hazardous Waste Management.* New York: McGraw-Hill, 1994.

Long, R. B. *Separation Processes in Waste Minimization.* New York: Marcel Dekker, 1995.

Luben, W. L., and Wenzel, L. A. *Chemical Process Analysis: Mass and Energy Balances.* Englewood Cliffs, NJ: Prentice Hall (1995).

McCabe, W. L., Smith, J. C., and Harriott, P. *Unit Operations of Chemical Engineering*. New York: McGraw-Hill, 1993.
Nelson, K. E. "Process Modifications That Save Energy, Improve Yields, and Reduce Waste." In *Waste Minimization through Process Design,* ed. A. P. Rossiter. New York: McGraw-Hill, 1995, pp. 119–32.
Nemerow, N. N. *Zero Pollution for Industry: Waste Minimization through Industrial Complexes.* New York: Wiley, 1995.
Osantowski, R. A., Liello, J. C., and Applegate, C. S. "Generic Pollution Prevention." In *Industrial Pollution Prevention Handbook,* ed. H. M. Freeman. New York: McGraw-Hill, 1995, pp. 585–614.
Palepu, P. T., Chauhan, S. P., and Ananth, K. P. "Separation Technologies." In *Industrial Pollution Prevention Handbook,* ed. H. M. Freeman. New York: McGraw-Hill, 1995, 361–94.
Radecki, P. P. "Computer-Based Methods for Finding Green Synthesis Pathways and Industrial Processes for Manufacturing Chemicals: A View from 10,000 Feet." In *Proceedings: Workshop on Green Syntheses and Processing in Chemical Manufacturing.* EPA/600/R-94/125. Cincinnati: U.S. EPA, 1994, pp. 66–78.
Radecki, P. P., Hertz, D. W., and Vinton, C. "Build Pollution Prevention into System Design." In *Hydrocarbon Processing* 73 (8): 55–60 (1994).
Rogers, T. N., and Brant, G. "Distillation." In *Standard Handbook of Hazardous Waste Treatment and Disposal,* ed. H. M. Freeman. New York: McGraw-Hill, 1989, pp. 6.23–6.38.
Sandia National Laboratories. EcoSys: Life Cycle Information and Expert System. http://www.sandia.gov/EcoSys/eco.htm (1996).
Sharma, R. K., and Cresswell, D. L. "Kinetics and Fixed-Bed Reactor Modeling of Butane Oxidation to Maleic Anhydride." *American Institute of Chemical Engineers Journal* 37 (1991): 39–47.
Shen, T. T. *Industrial Pollution Prevention.* Berlin: Springer-Verlag, 1995.
Sinnott, R. K. *Coulson and Richardson's Chemical Engineering,* vol. 6. Oxford: Pergamon Press, 1993.
Sinnott, R. K. *Coulson and Richardson's Chemical Engineering,* vol. 1. Oxford: Butterworth-Heinemann, 1995.
Smith, R. "Wastewater Minimization." In *Waste Minimization through Process Design,* ed. A. P. Rossiter. New York: McGraw-Hill, 1995, pp. 93–108.
Syan, C. S., and Menon, U. *Concurrent Engineering.* London: Chapman & Hall, 1994.
Theodore, L., and McGuinn, Y. C. *Pollution Prevention.* New York: Van Nostrand Reinhold, 1992.
Thom, J., and Higgins, T. "Solvents Used for Cleaning, Refrigeration, Firefighting, and Other Uses." In *Pollution Prevention Handbook,* ed. T. Higgins. Boca Raton, FL: CRC Press, 1995, pp. 199–243.
Thomas, S. T. *Facility Manager's Guide to Pollution Prevention and Waste Minimization.* Washington, DC: Bureau of National Affairs, 1995.
U.S. EPA. *User's Guide: Strategic Waste Minimization Initiative (SWAMI) Version 2.0.* EPA/625/11-91/004. Washington, DC: U.S. EPA, 1992.
Walker, G. *Industrial Heat Exchangers.* New York: Hemisphere, 1990.
Weck, M., Eversheim, W., König, W., and Pfeifer, T. *Production Engineering: The Competitive Edge.* Oxford: Butterworth-Heinemann, 1991.

PROBLEMS

5.1. Select a company you have worked for (summer job, internship, co-op assignment) and determine whether it used a sequential engineering process or a concurrent engineering approach. Describe how the decision-making hierarchy could have been improved.

5.2. Under what circumstances is a batch reactor better than a continuous-flow reactor?

5.3. Consider the zero-order reaction of compound A to compound B, with a rate constant equal to 2.0 min^{-1}. How long will it take for 90 percent of a 100 mg/L solution of A to be converted to B in a batch reactor?

5.4. Repeat Problem 5.3, assuming it is a first-order reaction and the reaction rate constant is 0.06 mg/L · min.

5.5. Repeat Problem 5.3, assuming the reaction occurs in a plug flow reactor.

5.6. Repeat Problem 5.4, assuming the reaction occurs in a continuously stirred tank reactor.

5.7. List the advantages and disadvantages of plug flow and completely mixed reactors.

5.8. A company is using a heat exchanger to raise the temperature of a 5000-kg/h process stream (stream A) from 20°C to 50°C using another aqueous process stream (stream B) whose initial temperature is 80°C. How much of the hotter process stream is needed if its final temperature should be 40°C? Assume the heat capacity of stream A is 4500 J/kg · °C and stream B is 2700 J/kg · °C.

5.9. Determine the heat exchanger surface area required for a system with a log mean temperature difference of 80°C, an overall heat transfer coefficient of 2.6 kW/m^2 · °C, and a heat transfer requirement of 3×10^8 J/h.

5.10. A heat exchanger is to be used to cool 20 kg/s of nitric acid from 60°C to 30°C. Cooling water is available at 8°C and a flow rate of 30 kg/s. The overall heat transfer coefficient is 500 W/m^2 · K. Determine the required surface area for countercurrent flow through the heat exchanger. Assume the heat capacity of water is 4200 J/kg · K, and that of the nitric acid is 6000 J/kg · K.

5.11. Hydrogen in a process stream is to be reduced from 3.0 mol % to 0.2 mol % using a stripping column. The process stream enters the top of the column at a rate of 50 kg-mol/h and falls to the bottom. Air is passed upward through the column at a rate of 2 kg-mol/h to strip the hydrogen. The equilibrium relationship between the hydrogen in the vapor phase (y) and that in the liquid phase (x) is:

x	0.0	0.1	0.2	0.3	0.4	0.5	0.6	0.7	0.8	0.9	1.0
y	0.0	0.42	0.54	0.62	0.68	0.75	0.81	0.86	0.91	0.96	1.00

How many theoretical plates are required to achieve the desired purity in the liquid bottom product if a reflux of 3.0 is used? What is the hydrogen concentration in the overhead vapor product?

5.12. An aqueous process stream contains 45 percent ethylbenzene. It is desired to recover the ethylbenzene in a distillation column. The overhead product should contain 90 percent ethylbenzene and the bottom wastewater only 0.2 percent. The equilibrium relationship between the ethylbenzene in the vapor (y) and in the water (x) is:

x	0.0	0.1	0.2	0.3	0.4	0.5	0.6	0.7	0.8	0.9	1.0
y	0.0	0.42	0.58	0.67	0.73	0.78	0.83	0.87	0.92	0.96	1.00

Assuming a reflux ratio equal to 1.4 times the minimum reflux ratio, calculate the number of plates required.

5.13. Investigate the current status of using supercritical extraction for the production of decaffeinated coffee and write a two-page paper on its effectiveness and degree of use.

5.14. Visit a small plating operation. Evaluate the parts cleaning and rinsing operations and write a brief description of how the processes could be improved to minimize pollution.

5.15. Inspect one of your department's chemical storage areas and determine if the chemicals are being stored in a safe and efficient manner. Document your findings in a short paper.

5.16. Visit the hazardous waste storage facility on your campus. Write a paper describing the storage procedures, housekeeping practices, record keeping, and worker training.

5.17. Obtain a copy of a recent manifest from your university's hazardous waste storage facility and determine the types and quantities of wastes being shipped, the types of shipping containers used, any pretreatment performed, the name of the hazardous waste hauler used, and the location of the disposal facility.

CHAPTER 6

LIFE-CYCLE ASSESSMENT

6.1 OVERVIEW OF LIFE-CYCLE ASSESSMENT

The environmental impacts associated with the manufacture of a product are not limited to the emissions from the manufacturing line. Let us take a simple example—the manufacturing of a common construction nail. The nail factory probably purchases iron wire of the appropriate size, extrudes it, cuts it to length, sharpens or hammers the point, hammers on the end, and packages the resulting nail for shipping. A coating may or may not be applied before packaging. A cursory examination would indicate that the only wastes from the process are the metal filings resulting from cutting and forming nails. A more detailed analysis, however, finds that lubricants are used during the machining stages and water is used for equipment cooling. Electricity is needed to run the equipment and to provide heat and lighting for the factory. Significant environmental impacts result from production of this electricity. There are also wastes associated with the coating and packaging of the nails. A more broadly based impact study may focus on the iron used in the manufacturing process. Resources were destroyed, energy used, and air pollution created in mining and producing the iron for use in the nails. Additional energy was consumed and pollution produced in transporting the iron to the nail factory and again in transporting the finished nails to retailers and eventually to consumers. Finally, when the useful life of the building in which the nail was used is over, the nail must be disposed of in a landfill as solid waste. Thus the pollutive impact of a product depends on the scale of the process evaluated. The only way to fully understand the magnitude of these impacts is to use "life-cycle assessment."

A life-cycle assessment (LCA) is an evaluation of the environmental effects associated with any given activity from the initial gathering of raw material from the earth until the point at which all residuals are returned to the earth (Vigon et al., 1993). This evaluation includes all sidestream releases to the air, water, and soil from the production of the raw materials (including energy), the use of the product, and its final disposal, as well as from the processing of the product itself. Life-cycle assessments are used to identify and measure both "direct" (e.g., emissions and energy use during manufacturing processes) and "indirect" (e.g., energy use and impacts caused by raw material extraction, product distribution, consumer use, and disposal) (Nash and Stoughton, 1994). A simple depiction of this can be seen in Figure 6.1. The systematic approach of LCA provides a true measure of the impact of a particular product or process. Unlike an environmental audit of an industrial process, which focuses on one particular facility and usually only on the activities that occur on the site, LCA looks at the linked interactions of the firm with the actions of its suppliers and customers. The result is a total cradle-to-grave analysis of the environmental impact of a product.

Life-cycle assessment has been defined as an attitude through which manufacturers accept responsibility for the pollution caused by their products from design to disposal. This is a major change from the traditional philosophy that the responsibility begins with the raw material acquisition and ends with the sale of the finished products (Bhat, 1996).

All of the impacts of a product or process at all of its life stages are examined in LCAs. Thus they can be used to evaluate trade-offs between two possible options. For example, if only energy use impacts are considered when balancing the use of fluorescent versus incandescent light bulbs, the fluorescent bulbs will win hands-down because they use much less energy. However, fluorescent bulbs contain toxic mercury and would lose the comparison if toxic waste generation was the controlling factor. The LCA process can be used to ensure that all environmental impacts are accounted for and to help in the decision-making process to determine which product to use.

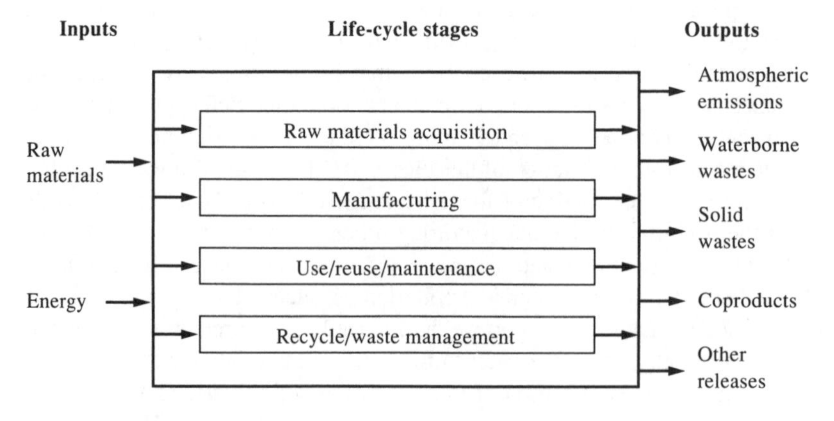

FIGURE 6.1
Life-cycle assessment stages and boundaries. (Source: Vigon et al., 1993)

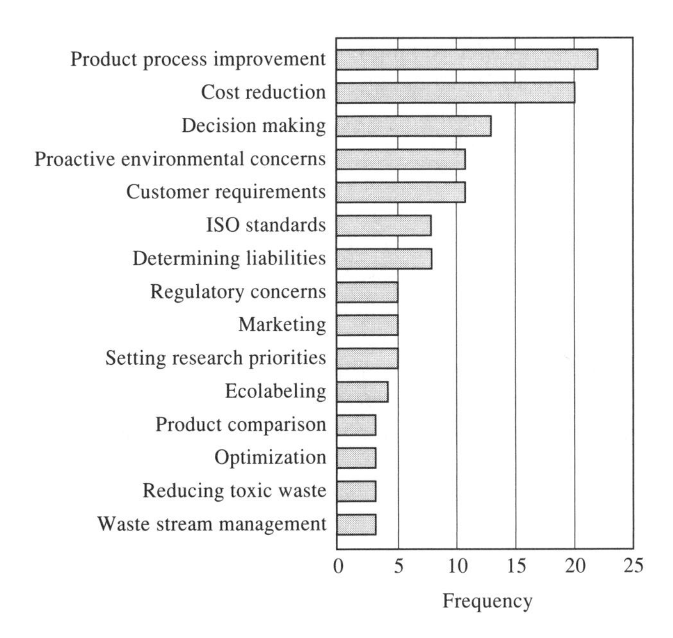

FIGURE 6.2
Motivations for implementing LCA. (Adapted from Foust and Gish, 1996)

Life-cycle assessments can be used for a number of purposes. A recent survey (Breville et al., 1994) showed the motivations for conducting LCAs as presented in Figure 6.2. Life-cycle assessments performed for product and/or process improvement and/or cost reduction will remain the primary drivers; LCAs performed for cost reduction reasons will likely increase in the future, as waste disposal costs continue to increase. The second tier of motivators—decisionmaking, proactive environmental positioning, and customer requirements—will continue as important drivers, with the last probably increasing. The other lower-tier drivers will vary slightly in significance but will probably remain as lower-tier drivers (Foust and Gish, 1996).

6.2 HISTORY OF LIFE-CYCLE ASSESSMENT DEVELOPMENT

The first work that can be considered to be a life-cycle assessment probably was done in the 1960s to 1970s in the field of energy systems. The U.S. Department of Energy commissioned several studies on "net energy analysis," focusing on calculating energy requirements, with some analysis of resulting environmental impacts of the use of energy. The oil shortages of the 1970s and the resulting energy crisis led to more intensive industrial energy analyses. Later in that decade, studies that focused on environmental issues were conducted by the research groups Arthur D. Little (Vigon et al., 1993) and Midwest Research Institutes (Curran, 1996a). Global modeling studies published in *The Limits to Growth* (Meadows, Randers, and Behrens, 1972) resulted in predictions of the effects of the world's growing population on the demand for finite materials and energy resources. In 1969, the Coca-Cola Company developed the current methods of life-cycle analysis to compare the environmental consequences of materials use and environmental releases for several different beverage containers.

Progress in the development of LCA techniques was slow, though, until the "green movement" in Europe in the mid-1980s brought renewed attention to the merits of recycling. Suddenly, pollution releases from industrial operations and both their environmental and economic impacts became important. This led to a rapidly increasing use of the LCA process by industry. These studies are now being used by industry to make management and production decisions and by environmental watchdog groups as a way of comparing the environmental performance of various competing products. Most LCAs have focused on product packaging (beverage containers, fast-food containers, and shipping containers) and have supported efforts to reduce the amount of packaging in the waste stream or to reduce the environmental emissions of producing the packaging. A few studies have looked at actual consumer products, such as diapers and detergents, while others have compared alternative industrial processes for the manufacture of the same product. Many LCAs from a wide spectrum of industry are currently available to the public. Extensive lists of sources for these can be found in Curran (1996b) and Bhat (1996). Several will be discussed more fully later.

One outcome of these studies is the finding that the indirect impacts, particularly those occurring after consumer use, often dwarf direct impacts. The American Fiber Manufacturers Association, for example, conducted an LCA of a polyester blouse and found that far more resources were used by consumers in washing and drying the blouse than were used in its manufacture. More environmental improvement would be realized by changing the fiber properties so that it could be washed in cold water, thus saving energy costs for the consumer, than by altering manufacturing processes (Nash and Stoughton, 1994).

Another finding from these studies is that recycling or switching to less toxic materials is not always advantageous. In many cases, more energy and resources are expended in recycling a product than is consumed to make the new product with virgin materials.

6.3 LIFE-CYCLE ASSESSMENT AND THE REGULATORY PROCESS

In Europe, use of life-cycle assessments is expanding rapidly, particularly as the basis for packaging recovery and recycling targets. In the United States, no current regulations or planned regulations mandate the use of LCAs by industry; however, many of the environmental laws and regulations described in Chapter 4 can be better met through use of LCA techniques. A variety of government actions address the use of LCAs in a nonregulatory sense (Foust and Gish, 1996). Executive Order 12873, signed on October 20, 1993, addresses federal acquisition, recycling, and waste prevention. The order calls for the use of LCA in federal purchases in order to safeguard natural resources through use of recycled products and through waste prevention. It defines as *environmentally preferable* products or services that have a lesser or reduced effect on human health and the environment when compared with competing products or services that serve the same purpose. Executive Order 12902, signed in 1994, is a directive dealing with energy efficiency and water conservation at federal facilities. This order directs federal facilities to reduce energy consumption by 30 percent within nine years, with 1985 as the baseline; to increase energy efficiency by 20 percent in nine years, with 1990 as the baseline; and to institute cost-effective water conservation whenever possible. All new federal facilities are to be designed and erected as show-case facilities, highlighting energy and water efficiency. Life-cycle approaches are to be used in all of these endeavors.

Congress has debated legislation that would encourage use of life-cycle methodologies, but to date no legislation of this type has been enacted. In Europe, though, LCA has already been mandated for use in areas such as ecolabeling. United States industries are beginning to use LCA on a voluntary basis in a limited way because of the usefulness of the results. Most large organizations have an LCA program in place or plan to implement one in the near future. Life-cycle assessment is a labor-intensive, data-intensive, complex process; wider use of its techniques probably awaits development of simplified procedures. It is the consensus of most experts that LCA will eventually become an integral part of environmental legislation (Foust and Gish, 1996).

Use of life-cycle assessments will undoubtedly expand rapidly in the near future because of ISO 14000, the Environmental Management Standards, now being implemented throughout the world. In 1993, ISO established TC-207 to develop environmental management tools and systems that would be applicable worldwide. Among the tools under development are environmental management systems, auditing, environmental performance evaluation, life-cycle assessment, and environmentally friendly labeling (Cascio, Woodside, and Mitchell, 1996; Fava and Consoli, 1996). Soon, ISO 14000 certification will be a requirement for doing business in the international marketplace, and LCAs will be an essential part of this certification.

6.4 LIFE-CYCLE ASSESSMENT METHODOLOGY

Life-cycle assessment uses a systems approach to identify the environmental consequences of various industrial alternatives. Ideally, it should take a holistic approach, considering choices of materials (premanufacturing), manufacturing processes (including use/reuse and maintenance), recycling, waste management, and product use; it should also evaluate energy use, resource consumption, and environmental releases in a cradle-to-grave scope.

This is usually beyond the ability of most companies, except for very limited projects. The data required may be voluminous and may not even be available, the analytical techniques required are still in the development stages, and the personnel costs to conduct the assessment may be excessive. Pollution prevention options selected may have far-reaching impacts on other industries that may not be easy to assess. For example, in a major study of the petrochemical industry, Rudd et al. (1981) found that simply changing the process route (and consequently the raw materials) used in the production of a particular type of plastic can have a significant impact on the entire petrochemical industry because of the overlapping uses for primary feedstocks. Figure 6.3 shows that many materials can be created using a variety of precursor materials. For example, ethylbenzene can be made from benzene, ethylene, or ethanol. Increasing the demand for one material, and thus lowering the demand for another, can affect the prices for those feedstocks across many other industries, the amounts of them produced, and the environmental impacts of their production. For example, an increase in

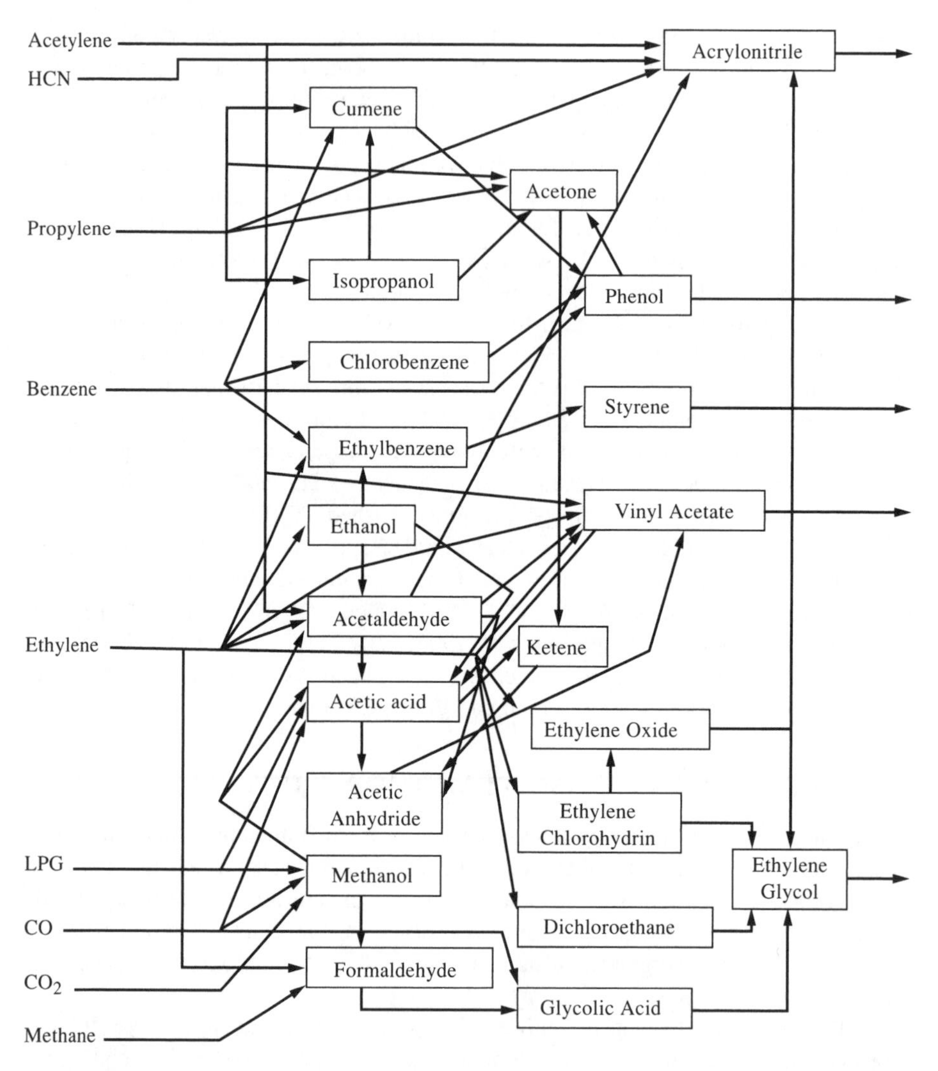

FIGURE 6.3
Part of the industrial system that converts primary feedstocks into end chemicals. (Source: Rudd et al., 1981)

the production of styrene would mean that the production of ethylbenzene would have to increase. Ethylbenzene production requires ethylene, ethanol, or benzene. Increasing the demands on these feedstocks will mean either that more of these will need to be produced or that the amounts of these used to produce cumene, isopropanol, acetaldehyde, chlorobenzene, and so on, would be reduced. This would be a market decision, with the cost of these feedstocks probably increasing, thus increasing the costs of all other associated chemicals. Ultimately, a wide range of chemicals unrelated to styrene will be affected. However, this does not mean that LCA is an unworkable methodology; rather, it means that realistic limits must be set on the objectives to be achieved.

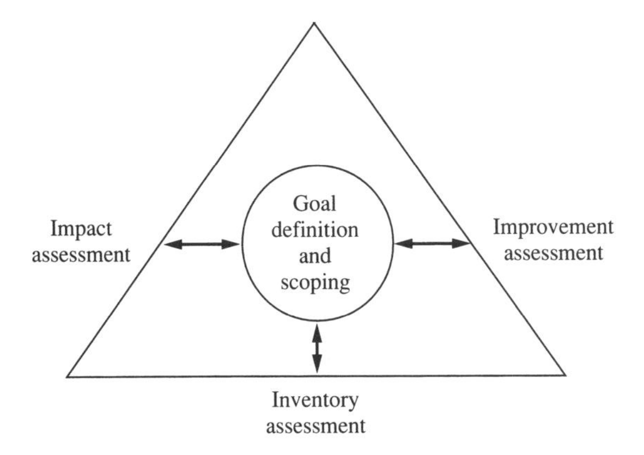

FIGURE 6.4
Components of a life-cycle assessment.

The principal components of life-cycle assessment have been established, and the procedures refined, by the Society of Environmental Toxicology and Chemistry (SETAC). In the early 1990s, SETAC established an LCA Advisory Group whose mission was to advance the science, practice, and application of LCAs to reduce resource consumption and environmental burdens associated with products, packaging, processes, and activities. The SETAC effort has resulted in a "Code of Practice" for life-cycle assessments (Consoli et al., 1993) and a framework for conducting life-cycle assessments (Fava et al., 1994). SETAC continues to lead the way in espousing the virtues of this method of environmental systems analysis.

To express the cradle-to-grave inputs and outputs (Fava et al., 1993), life-cycle assessments usually use three interdependent stages, or subsystems: life-cycle inventory assessment, impact assessment, and improvement assessment (see Figure 6.4). A fourth component, usually considered to be encompassed within the other three, is goal definition and scoping. These can be used independently or combined to achieve proper pollution prevention decisions. A complete LCA would include all of these, but because of the cost and complexity involved in some phases, not all are always done. We will discuss each of these phases individually, and then show how they can be combined for best decision making. However, of the three phases, only the inventory phase has been fully developed; the other phases will be primarily described in generalities. More complete descriptions of the process can be found in Boguski, Hunt, Chlokis, and Franklin (1996) and van den Berg, Huppes, and Dutilh (1996).

6.4.1 Goal Definition and Scoping Stage

This stage, as described by SETAC, defines the purpose of the study, the expected product of the study, the boundary conditions, and the assumptions. This is an essential first step in any LCA. In most cases, an LCA is undertaken to answer specific questions management has concerning a particular process or product. It may be used to compare products or to relate a product to a standard, to improve the environmental soundness of a product, or to simply provide more information on the product. The nature of those questions serves as the basis for the goals and scope of the study (Boguski et al., 1996).

This phase establishes what will or will not be included. Because of cost considerations, not all aspects of the LCA may be done. The extent of the LCA should be established at this stage, rather than later, although periodic reevaluation and modification of the goals and scope are often worthwhile.

The purpose of an LCA is usually to compare several options for making a product, modifying a process, changing packaging, and the like. Before an LCA can be properly undertaken, the company or organization conducting the LCA must have an understanding of the available options that are to be compared. It is incumbent upon the group undertaking the LCA to fully think out what options are available that should be compared, or it may overlook some that are worthwhile. Great thought should go into this phase, as it will set the framework for the LCA analysis.

Proper forethought can reduce the amount of work involved and make the LCA more valuable. If the main intent of the study is to compare the environmental consequences of changing the machinery used to make a product, and the product itself or its packaging is not going to change, then the scope of the project can be limited to an inventory analysis and impact assessment of the production stage alone; a full LCA for the product is not needed because the product and the packaging will be the same with either manufacturing option. However, if the question concerns whether to change the product itself (e.g., changing from steel to plastic for the product case), then a more complete assessment may be needed.

The scope of the LCA selected determines the life-cycle stages that need to be considered when setting the boundaries of the study. Among these stages are the following (Vigon et al., 1993):

- Raw materials and energy acquisition.
- Manufacturing, including intermediate materials manufacturing, materials transportation to the fabrication site, product fabrication, filling/packaging/distribution of the product, and transportation to the retail site.
- Use/reuse/maintenance by the consumer during the product's useful life.
- Recycle/waste management after use.

In all cases, all energy requirements to transport, produce, and use the product, and all environmental wastes from transporting, producing, and using the product must be considered, as must the post-consumer disposal of the product. Each step in the process should be evaluated, including production steps for ancillary inputs or outputs such as process chemicals, energy production and use, and packaging.

Life-cycle assessment systems are always complex, involving many steps and requiring a large amount of data. Even simple products may require hundreds of steps, when all input materials' processing steps are included, as well as all possible uses, reuse, and disposal options for the product. This level of detail may not always be needed, as when two alternative manufacturing processes are being compared. The product will be the same, so evaluation of postsales use of the product will be unnecessary. Thus many LCAs examine only subsets of the overall LCA. The more focused the scope of the analysis can be made, the more likely that meaningful results can be obtained. This must be done with caution, though, to ensure that important steps are not overlooked.

Because of the complexity of full LCAs, recent attempts have been made to reduce the workload burden by "streamlining" the process in a prescribed fashion, eliminating some of the LCA phases (such as raw materials acquisition or postconsumer product disposal). Depending on the purpose of the LCA, this may be perfectly acceptable. Streamlining is described more fully later in this chapter.

When the boundary conditions have been determined, a system flow diagram can be developed to better define the system. The process flow chart provides a qualitative graphical representation of all relevant processes involved in the system studied. It is composed of a sequence of processes (represented by boxes) connected by materials flows (represented by arrows).

6.4.2 Inventory Analysis

Inventory assessment is a systematic, objective, stepwise procedure for quantifying energy and raw materials requirements, atmospheric emissions, waterborne emissions, solid wastes, and other releases for the entire life cycle of a product, package, process, material, or activity (Curran, 1996b; Vigon, 1995; Vigon et al., 1993). Each applicable stage in the life cycle of a product or process is evaluated for each individual contaminant, energy requirement, resource requirement, and so on, and the results summed to determine the overall impact.

There are five basic step in an inventory analysis: (1) define the scope and boundaries; (2) gather data; (3) create a computer model; (4) analyze and report the study results; and (5) interpret the results and draw conclusions (Boguski et al., 1996).

Defining the scope and boundaries is an extension of the initial scoping process described earlier. Here the objective is to focus on those areas defined as being important for the inventory analysis. Resource constraints may be a factor here. Obtaining necessary data may be difficult, the cost may be prohibitive, or information, particularly concerning proprietary data from other companies (e.g., composition of ingredients purchased or their life-cycle assessment), may not be available. However, the accuracy and validity of the analysis should not be compromised by performing a faulty assessment. It may be better to forgo the analysis than to produce one that is in error but is still used for decision making.

Life-cycle inventories are data-intensive exercises and require a systematic approach to be meaningful and doable at reasonable cost. The scoping phase of the inventory analysis should set the framework for the data gathering phase, defining which process steps are of importance and to be analyzed and determining what data will be needed. Normally, it is best to organize gathering of the necessary information in an inventory checklist format (Vigon et al., 1993). Eight general decision areas should be addressed in the checklist:

- Purpose of the inventory.
- System boundaries.
- Geographic scope.
- Types of data used.
- Data collection and synthesis procedures.

- Data quality measures.
- Computational model construction.
- Presentation of the results.

An example of an inventory checklist can be seen in Figures 6.5 and 6.6. Figure 6.5 is the checklist for the project's scope and procedures definition; Figure 6.6 is used to quantitatively describe the accumulated data.

LIFE-CYCLE INVENTORY CHECKLIST PART I—SCOPE AND PROCEDURES
INVENTORY OF: ____________________

Purpose of Inventory: *(check all that apply)*

Private Sector Use

Internal Evaluation and Decision Making
- ☐ Comparison of Materials, Products or Activities
- ☐ Resource Use and Release Comparison with Other Manufacturer's Data
- ☐ Personnel Training for Product and Process Design
- ☐ Baseline Information for Full LCA

External Evaluation and Decision Making
- ☐ Provide Information on Resource Use and Releases
- ☐ Substantiate Statements of Reductions in Resource Use and Releases

Public Sector Use

Evaluation and Policy-making
- ☐ Support Information for Policy and Regulatory Evaluation
- ☐ Information Gap Identification
- ☐ Help Evaluate Statements of Reductions in Resource Uses and Releases

Public Education
- ☐ Develop Support Materials for Public Education
- ☐ Assist in Curriculum Design

Systems Analyzed

List the product/process systems analyzed in this inventory: ____________________

Key Assumptions: (list and describe)

____

Define the Boundaries

For each system analyzed, define the boundaries by life-cycle stage, geographic scope, primary processes, and ancillary inputs included in the system boundaries

Postconsumer Solid Waste Management Options: Mark and describe the options analyzed for each system.
- ☐ Landfill ____________
- ☐ Combustion ____________
- ☐ Composting ____________
- ☐ Open-loop Recycling ____________
- ☐ Closed-loop Recycling ____________
- ☐ Other ____________

Basis for Comparison
- ☐ This is not a comparative study.
- ☐ This is a comparative study.

State basis for comparison between systems; (Example: 1000 units, 1000 uses) ____________

If products or processes are not normally used on a one-to-one basis, state how equivalent function was established.

____

Computational Model Construction
- ☐ System calculations are made using computer spreadsheets that relate each system component to the total system.
- ☐ System calculations are made using another technique. Describe: ____________

Describe how inputs to and outputs from postconsumer solid waste management are handled. ____________

Quality Assurance: (state specific activities and initials of reviewer)

Review performance on:
- ☐ Data Gathering Techniques ______
- ☐ Coproduction Allocation ______
- ☐ Input Data ______
- ☐ Model Calculations and Formulas ______
- ☐ Results and Reporting ______

Peer Review: (state specific activities and initials of reviewer)

Review performed on:
- ☐ Scope and Boundary ______
- ☐ Data Gathering Techniques ______
- ☐ Coproduct Allocation ______
- ☐ Input Data ______
- ☐ Model Calculations and Formulas ______
- ☐ Results and Reporting ______

Results Presentation
- ☐ Methodology is fully described.
- ☐ Individual pollutants are reported.
- ☐ Emissions are reported as aggregated totals only. Explain why: ____________
- ☐ Report is sufficiently detailed for its defined purpose.
- ☐ Report may need more detail for additional use beyond defined purpose.
- ☐ Sensitivity analyses are included in the report. List: ____________
- ☐ Sensitivity analyses have been performed but are not included in the report. List: ____________

FIGURE 6.5

Life-Cycle Inventory Checklist—Scope and Procedures. (Source: Vigon et al., 1993)

LIFE-CYCLE INVENTORY CHECKLIST PART II—MODULE WORKSHEET

Inventory of: ______ Preparer: ______
Life-Cycle Stage Description: ______
Date: ______ Quality Assurance Approval: ______
MODULE DESCRIPTION: ______

	Data Value[(a)]	Type[(b)]	Data[(c)] Age/Scope	Quality Measures[(d)]
		MODULE INPUTS		
Materials				
Process				
Other[(e)]				
Energy				
Process				
Precombustion				
Water Usage				
Process				
Fuel-related				
		MODULE OUTPUTS		
Product				
Coproducts[(f)]				
Air Emissions				
Process				
Fuel-related				
Water Effluents				
Process				
Fuel-related				
Solid Waste				
Process				
Fuel-related				
Capital Repl.				
Transportation				
Personnel				

(a) Include units.

(b) Indicate whether data are actual measurements, engineering estimates, or theoretical or published values and whether the numbers are from a specific manufacturer or facility, or whether they represent industry-average values. List a specific source if pertinent, e.g., "obtained from Atlanta facility wastewater permit monitoring data."

(c) Indicate whether emissions are all available, regulated only, or selected. Designate data as to geographic specificity, e.g., North America, and indicate the period covered, e.g., average of monthly for 1991.

(d) List measures of data quality available for the data item, e.g., accuracy, precision, representativeness, consistency-checked, other, or none.

(e) Include nontraditional inputs, e.g., land use, when appropriate and necessary.

(f) If coproduct allocation method was applied, indicate basis in quality measures column, e.g., weight.

FIGURE 6.6
Life-Cycle Inventory Checklist—Module Worksheet. (Source: Vigon et al., 1993)

The data gathering step should be done using the system flow diagram for the process being studied and the checklist or worksheets. For all but the simplest projects, it is usually desirable to divide the system into a series of subsystems (an individual step or process that is part of the defined production system). For each subsystem, the analyst should determine all inflows (materials and energy) and all outputs (products, coproducts, and environmental emissions). Emissions should be quantified by type of pollutant. Transportation of materials from one process location to another should also be included.

Obtaining useful and accurate data is often difficult. Data on generic processes may be available from the literature and may be suitable for the study under way. It is essential, though, that the data used be derived from a comparable process. Manufacturing efficiencies often vary widely from business to business, and from one production operation to another within the same company. Numerous sources of data are available, including internal company reports; accounting or engineering reports; machine specifications; publicly available government documents and databases, such as census studies by the U.S. Department of Commerce and the U.S. Department of Energy or the Toxic Release Inventory (TRI) database published annually by the U.S. Environmental Protection Agency; technical books, such as the *Kirk-Othmer Encyclopedia of Chemical Technology* (Kroschwitz and Howe-Grant, 1991); and conference proceedings. A particularly vexing problem is getting data on proprietary products or processes supplied by others. For confidentiality reasons, most companies are reluctant to supply necessary information on these subsystems. Unless some type of confidentiality agreement can be reached, it may be necessary to estimate the required data; this could compromise the credibility of the final overall assessment. Obtaining data on the end-user phase of the life cycle can also be difficult. Market surveys and government reports may be helpful.

The assessment will only be as good as the data used. Care must be taken to ensure that all quantities used are accurate. If possible, materials balances should be performed on key materials to ensure that all material is accounted for. Inputs to a process should equal what is in the output plus what is waste, unless some of the material is created or destroyed during processing.

The data obtained will probably not be consistent in terms of units used to express them. Energy consumption may be in terms of kilowatts per month or year, water use in terms of thousands of cubic feet per year, raw materials in tons per year, air or water emissions in parts per million concentration and an average discharge volume per year, and so on. Before the LCA can be performed, the data must be reduced to a consistent format. This is often done based on the product's output level. All of the data are normalized to amount consumed, produced, or discharged per pound or ton of product produced. Because available plant data are often a composite of all production at the facility, the data need to be adjusted so that only that portion of the material associated with the process, material, or product being studied is considered. The objective should be to model, as closely as possible, what is actually occurring in the life cycle of the product, process, or activity (Boguski et al., 1996).

Once the data are put into a consistent form, the actual assessment can begin. Often extensive, complex computations are required to properly combine the normalized data with present and projected materials flows. If multiple subsystems are involved, they must all be integrated to provide an overall picture of the environmental impacts of the system being studied. Some of the input data are factual and accurate, whereas other data may be assumptions or based on estimates. Sensitivity analyses may be required to determine the impact of these assumptions on the results. Most LCAs are now done using computer modeling, either with a spreadsheet or with more sophisticated computer software. The resulting analysis should give the total energy and resource use and the environmental emissions resulting from the system being studied. Computer models are described in more detail later in this chapter.

Care should go into the preparation of the LCA report because the overwhelming amount of material that is usually gathered may easily confuse the reader unless essential information is separated from the chaff. The presentation should thoroughly describe the methodology employed in the analysis and explicitly define the system analyzed and the boundaries that were set (Vigon, 1995). Both tabular and graphical presentation of results are used, with enough text to make the results meaningful. Interpretation and drawing of conclusions should be done in such a way as to answer the questions set out in the original scoping of the project. Normally, the conclusions should address ways to reduce resource and energy use and minimize environmental emissions. The value of trade-offs (e.g., fossil fuels versus nonfossil fuels, or greater airborne emissions versus greater waterborne emissions) should be left to the impact assessment stage of the LCA (Boguski et al., 1996).

6.4.3 Impact Analysis

Procedures for the inventory analysis described previously have become relatively well established over the past few years. It is essentially a data gathering, accounting, and interpretation process that can be based on generally accepted quantitative procedures. The same cannot be said for the impact analysis stage. It is still in its infancy, and measures of actual impacts on human health, environmental quality, and resource depletion are still being developed.

Life-cycle impact analysis is a quantitative and/or qualitative examination of potential environmental and human health effects associated with the use of resources and environmental releases (Fava et al., 1993). The inventory assessment stage conducted previous to this analysis produced a large quantity of often complex data. A system is needed to convert these data to a form that can be used to assess the impacts of various possible production scenarios. This is the role of the impact analysis.

The conceptual framework for impact analysis consists of three phases: classification, characterization, and valuation (see Figure 6.7).

CLASSIFICATION. Classification is the assignment of items from the inventory assessment to a small number of impact categories, such as human health, ecological quality, and natural resource depletion. Each item from the life-cycle inventory is assigned to one or more of the categories so that impacts can be aggregated in a meaningful way. Potential production changes may also have an effect on the social environment. For example, changing a process so that it no longer requires a cooling tower may eliminate the resulting water vapor plume from the tower and possibly even the tall, sometimes unsightly, cooling tower. This may have some positive impacts on the environment, but a significant effect may be an improvement in the site's aesthetics, resulting in an increase in values of neighboring property and a general improvement in the quality of life in the area. This would be included in the social welfare category.

For each impact category, a list of stressors is developed. These may include such things as specific contaminants emitted (acids, pesticides, chlorinated hydrocarbons, greenhouse gases, noise, etc.) or human health effects (human carcinogens, irritants,

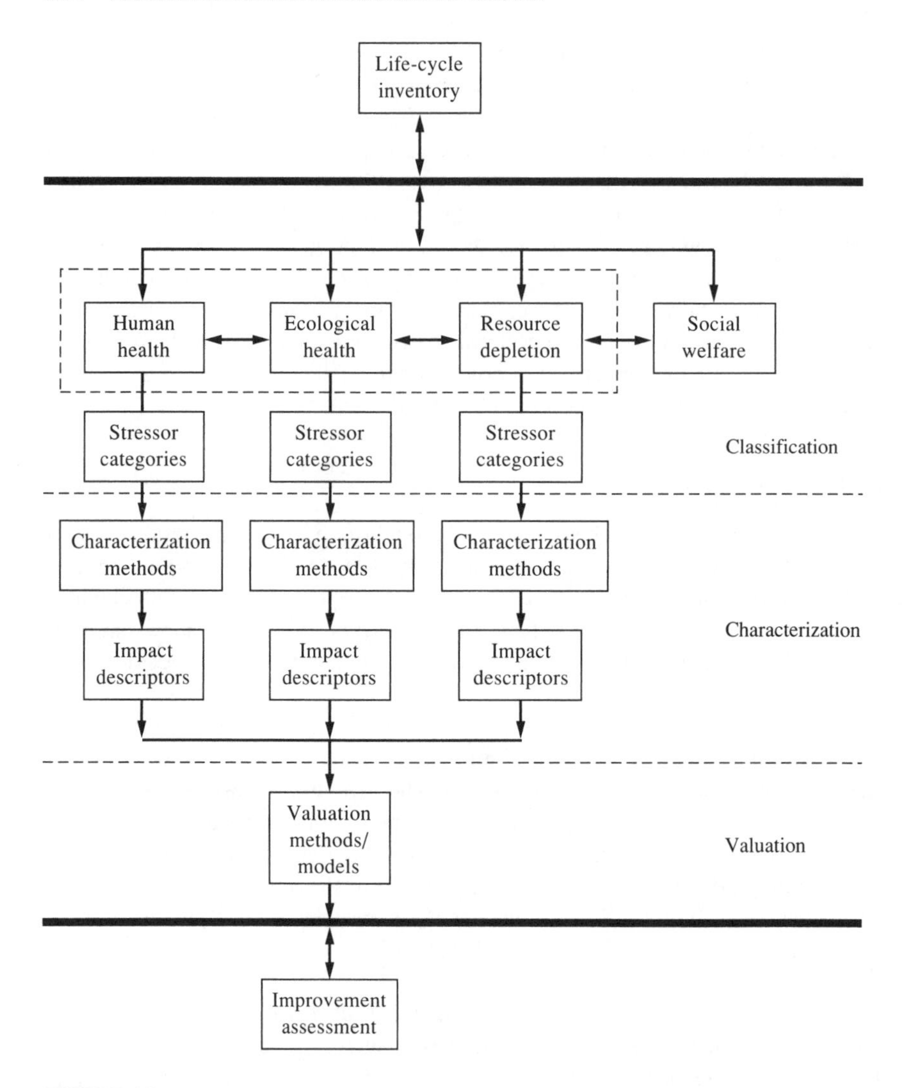

FIGURE 6.7
Conceptual framework for life-cycle impact assessment. (Source: Fava et al., 1993)

odors, behavioral effects). Commonly used environmental issues and stressors are listed in Table 6.1. Thus a material identified in the life-cycle inventory may be assigned to a number of impact groups and to a number of stressor categories within each group. For example, a textile mill may be discharging dyes in its wastewater effluent. These may be included in the ecological health category because of their impacts on fish life and on water quality [i.e., biochemical oxygen demand (BOD) increase]; they may also be included in the human health category because they may be carcinogenic or produce other adverse human health effects. The color imparted to the receiving water may make it aesthetically unpleasant, creating a social impact.

TABLE 6.1
Common environmental issues and stressors

Materials and energy			
Type	**Character**	**Resource base**	**Impacts caused by extraction and use**
Renewable Nonrenewable	Virgin Reused/recycled Reusable/recyclable	Location (local vs. other) Scarcity Quality Management/ restoration practices	Materials/energy use Residuals Ecosystems health Human health
Residuals			
Type	**Characterization**	**Environmental fate**	
Solid waste Air emissions Waterborne	Constituents, amount, concentration, toxicity Nonhazardous Hazardous Radioactive	Containment Bioaccumulation Degradability Mobility/transport	Treatment/disposal impacts
Ecological health			
Ecosystem stressors	**Impact categories**		**Scale**
Physical Chemical Biological	Diversity Sustainability, resilience to stressors	System structure and function Sensitive species	Local Regional Global

Source: "Life Cycle Analysis and Assessment."

CHARACTERIZATION. Characterization is the process of describing the impacts of concern. Usually this is done through the use of models that convert assessment data into impact data. For example, the amount of carcinogenic air pollutants emitted by a factory and reaching a human receptor can be converted into a projected number of new cancers caused by that pollutant loading. The inventory data assigned to each impact category can be analyzed in a variety of ways, from fairly simple to quite complex procedures. Among these, in order of increasing complexity, are:

- Loading.
- Equivalency.
- Inherent chemical properties.
- Generic exposure and effects.
- Site-specific exposure and effects.

Detailed descriptions of each of these procedures can be found elsewhere (Fava et al., 1993; Vigon, 1996).

The *loading technique* sums the data on each input category (i.e., energy usage, BOD contribution, SO_2 emissions) and compares the totals among various options. This is a simple technique that assumes there is a direct relationship between mass loading and environmental or health impact. However, this may not be the case. Moreover, this technique can only be used to compare one process against another based on the one item involved. One process may reduce the amount of particulate emissions into the atmosphere by switching from coal to petroleum as a source of energy and appear to be better based solely on particulate emissions, but this change may result in increased resource depletion of the less abundant petroleum. Thus loading characterization models may be useful as a first cut for comparing the impact of one item for a given process against that of another option, but they are generally not sufficient alone to make good impact analysis decisions.

The *equivalency model* attempts to compare systems on the basis of equivalent impacts. For example, it is difficult to directly compare the impacts of changing solvents so that 1 kg of trichloroethylene is emitted rather than 1 kg of perchloroethylene because they are different chemicals. However, it may be possible to compare them based on equivalency factors. These could include such comparators as acute toxicity values (LD_{50}, or median lethal dose) or cancer potency indexes. In some cases, they could be compared based on regulatory emissions standards, if these were established based on perceived impacts. In some cases, impact potentials for specific impacts have been developed. One example is for global warming chemicals, as described in Section 3.2. Unfortunately, acceptable equivalency factors are not always readily available.

Characterization can also be done based on *inherent chemical properties* associated with the materials emitted. These properties may include toxicity, ignitability, carcinogenicity, and bioaccumulation. This approach is similar to the previously described model, with a different set of conversion factors. Again, sufficient accurate data to make this procedure useful may be lacking. Also, the procedure considers the impacts of the chemicals only in the form and concentration they are in at the time of emission. Some may be biochemically or photochemically degraded to less hazardous materials at a more rapid rate than others and have their impact reduced with time after entry into the environment, but this is not accounted for in the model. The model considers only the inherent properties of the specific compound itself.

Generic exposure and effects models are designed to estimate potential impacts based on generic environmental and human health information. These models are very complex and require the input of large amounts of information; however, much of the required data may not be available or may be of questionable validity. The model needs to describe complex interactions among the stressors, the environment, and humans. Simplistic models cannot do this adequately, and data needed for complex models often are lacking. Much research is under way to improve the use of these models, and they may someday be of value to individual industries, but at present there are none that are generally acceptable for routine use.

Site-specific exposure and effect models are used to determine the actual impacts resulting from process changes based on site-specific fate, transport, and impact information. These are usually more accurate than the generic models because they are based on site-specific information, but they again require large amounts of reliable data.

A method commonly used in Europe to compare environmental impacts from several options is based on the *critical dilution volume*. Developed by the Swiss Federal Ministry for the Environment, the system characterizes each emission in terms of the volume of air or water required to dilute it to the legal limit set for that emission. For example, if the life-cycle environmental loading from a product is 5 kg of chromium and the water quality standard for chromium is 0.05 mg/L, the critical volume, or volume of water that can safely carry that loading, is equal to 5 kg divided by 0.05 mg/L, or 100 million liters of water. Overall results are expressed as a total volume, that is, the sum of the individual dilution volumes. The procedure is simple, but it assumes that good emission standards are available, and it does not consider fate of the contaminant or exposure assessment (Postlethwaite and de Oude, 1996). It is a very conservative approach in that it also assumes that all loadings are cumulative, which often is not the case.

The life-cycle impact assessment characterization process is still in its development stage and there are still gaps in information needed to make good assessments. As more companies undertake impact assessments, the procedures will become more refined and more standardized. This process is needed, as it is the only way to quantifiably correctly decide on the best pollution prevention options to undertake.

VALUATION. The final step in the life-cycle analysis is to assign relative values or weights to different impacts so that the evaluator can compare the importance of the various impacts. For example, if an industry is comparing two possible pollution prevention projects, one that will result in a reduction of 100 kg/day of toluene emitted to the sewer and the other that will decrease air emissions of TCE by 100 kg, how do the evaluators decide which will have the most beneficial environmental impact? They must do a valuation based on the overall environmental impacts of each option. Because the impacts are very different, a weighting system is needed. This step is the least developed of the three phases of an environmental assessment. The major shortcoming is that assigning weights is highly subjective, depending on the values of the assessor and the perceived relative importance of the particular item. One evaluator may consider energy consumption to be the most critical issue, another might favor biodiversity, a third might weight global climate protection more heavily, and so on. Decision making is based on emotional as well as rational valuations. There is no formal scientifically based procedure for doing this at the present time, and there probably shouldn't be. It would eliminate the freedom of people to express their own preferences and apply them in their own evaluation assessment. It does mean, though, that there can be more than one impact assessment outcome.

That said, the valuation step in life-cycle impact assessment can be described as follows:

> Valuation can be understood as the act of using "objective" information and considering it with "subjective" value-based environmental goals with some kind of methodology in order to derive a judgment. (Giegrich and Schmitz, 1996)

Judgment of an issue depends on facts and issues connected with one's values, emotions, previous judgments, and experience. The final impact assessment is a reduction of complex inventory data to impact-related figures and a final judgment of the environmental impacts of these figures.

Various methods have been suggested to conduct the valuation. Among these are the use of ecological scarcity factors, environmental loading factors, human health factors, and environmental acceptability factors. Several of these approaches are described in detail elsewhere (Graedel and Allenby, 1995). Following is a brief discussion of a few of the more widely accepted procedures.

Ecopoint method. This approach was developed in Switzerland in 1984 and later refined. It is a further development of the critical volumes approach described earlier, and it is based on regulatory limits. The fundamental concept is *ecological shortage,* defined as the resilience of an environmental resource to the current pollution level. It measures the relationship between total pollution involved from the process being studied and the maximum permissible pollution. The results are calculated in terms of single dimensionless numbers called *ecopoints.* This consists of the product of the pollution load and an ecological factor for that pollutant based on its contribution to ecological shortage. An overall assessment is obtained by totaling the ecopoints from all of the individual emissions to give a single number, or *ecoscore.*

Environmental effects. This approach is being developed in several European countries. The Dutch are developing a comprehensive system to characterize environmental effects of a variety of chemical substances using environmental indexes (equivalency factors) for various emissions and categories. This requires much research and reams of data, because a suitable indicator must be found for each environmental impact category that can be used when evaluating a pollution load to assess potential impact. For example, the environmental impact of ozone depletion in the stratosphere due to an emitted chemical can be related to the ozone depletion potential of Freon (Giegrich and Schmitz, 1996). The Germans are taking this one step further by incorporating valuation into the equivalency factors. It is likely that ISO will use this system as its preferred valuation approach.

Environmental Priority Strategies. The EPS (Environmental Priority Strategies in Product Design) has been developed in Sweden to assess a wide variety of product types. For each basic material used to produce a product, an *environmental load index* (ELI) is determined. This index assigns values to emissions and resource consumption based on five criteria: biodiversity, human health, ecological health, resources, and aesthetics. Different components of the index account for the environmental impact of a product during three stages of its life-cycle: product manufacture, use, and disposal. These indexes are multiplied by the material's loadings to give environmental load units (ELUs) per kilogram (ELU/kg) of material used. These are then summed to quantify the total environmental load (Graedel and Allenby, 1995; Postlethwaite and de Oude, 1996). Thus, based on the amount of each material required, the environmental indexes are used to calculate the environmental load values at each stage of the product life cycle. Environmental load units derived for different options can be compared to determine the best option. The key to this valuation procedure is developing appropriate ELIs. These can be highly subjective.

A complete life-cycle impact assessment will be invaluable to industries desiring to make proper, informed decisions on environmentally compatible process or product modifications. Studies of this nature have been performed and many more are under way. However, the field is still in its infancy, and there is much room for improvement to the procedures used or for the development of entirely new procedures. The next few years should see a major step forward in refinement of these processes.

6.4.4 Improvement Analysis

The objective of the improvement analysis stage of the life-cycle assessment is to identify opportunities to reduce energy use, raw materials consumption, or environmental emissions throughout the entire life cycle of the product, process, or activity. The analysis relies on the output of the previous two stages. The inventory analysis may identify opportunities to reduce waste emissions, energy consumption, or materials use. The impact analysis identifies those areas that are contributing the greatest environmental impacts and that should be considered for modification. The improvement analysis should use this information to develop strategies for optimizing the improvement of the product or process.

Conducting the improvement analysis is much more difficult than the foregoing description implies. To be effective, there must be ways of adequately equating impacts from various options that may be quite different from one another. The effort that goes into a good LCA that will lead to an effective assessment analysis can be significant. The cost of conducting a comprehensive life-cycle assessment is often enormous; repeating the LCA for each of several options, some of which may be innovative and have little existing data, greatly compounds the cost and the effort required. With this in mind, it can still be stated that an improvement analysis is desirable, even if it must be conducted on a limited basis.

6.5 STREAMLINING LIFE-CYCLE ASSESSMENTS

Because of the complexity and cost of conducting a comprehensive life-cycle assessment, and the extensive time it takes, many firms are reluctant to undertake one. A full-scale life-cycle assessment can cost from $10,000 to several hundred thousand dollars for each product studied (Todd, 1996). This has led researchers to begin developing streamlined procedures for LCAs to produce useful results at lower cost and with less required input. This is often done by truncating the LCA at either or both ends from the actual production steps. Raw material acquisition, final product disposal, or both may be ignored in the streamlined LCA (Curran, 1996a). This must be done very carefully to prevent reaching erroneous conclusions. In other cases, the streamlined LCA is designed to focus on a few critical environmental impacts, such as ozone depletion or acute health hazards, rather than at the "total" life-cycle impacts (Office of Technology Assessment, 1992).

Streamlining refers to various approaches used to reduce the cost and effort required for LCA studies. It is a modified approach to conducting LCAs and should not be confused with screening to determine where additional study is needed.

Assessors using a streamlined LCA approach must select with care the information they will omit from their studies. There is a minimum criterion that should be required of all assessments to ensure validity of the results. One set of minimum standards is as follows (Todd, 1996):

- The study should include some form of inventory; it may also include impact assessment and improvement assessment.
- The study should describe clearly the boundaries defined for the study and the methods used to streamline accepted LCA methodology.
- The study should yield results that are consistent with those produced by a full-scale LCA of the product.

Several approaches have been suggested for streamlining LCAs, including narrowing the boundaries of the study, particularly during the inventory stage; targeting the study on issues of greatest interest; and using more readily available data, including qualitative data (Todd, 1996; U.S. EPA, 1997). Approaches that are currently used are listed in Table 6.2. Each of these approaches has advantages as well as drawbacks. Eliminating particular life-cycle stages may eliminate from consideration important environmental consequences of raw materials extraction or production or ultimate product disposal or reuse. Focusing on one pollutant or one specific environmental impact can also lead to exclusion of other potentially important environmental impacts. Care should be taken to ensure that the results obtained are still accurate and meaningful.

The type of streamlining approach taken should relate to the goals of the study and the intended use of the results. The results then must be evaluated within the context of the study limitations and should not be used to imply something broader than intended (Todd, 1996).

Recently, it has been suggested that the streamlining process be incorporated into the goals and scoping phase of an LCA, rather than considering LCAs and streamlined

TABLE 6.2
Approaches to streamlining LCA methods

Limiting or eliminating life-cycle stages (usually upstream or downstream stages from the main manufacturing stage)
Focusing on specific environmental impacts or issues
Eliminating specific inventory parameters
Limiting or eliminating impact assessment
Using qualitative as well as quantitative data
Using surrogate process data
Establishing criteria to be used as "show stoppers" or "knockouts"
Limiting the constituents studied to those meeting a threshold quantity (e.g., ignore raw materials comprising less than 20% by weight of the life cycle inventory (LCI) total mass)
Combining streamlining approaches

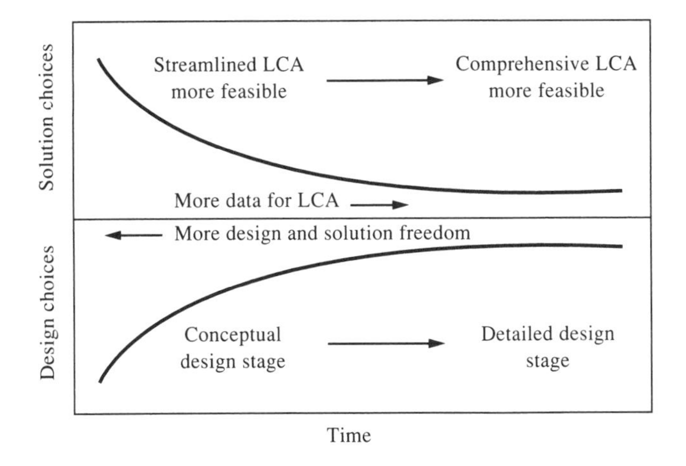

FIGURE 6.8
Design choices and LCA solutions as a function of the design stage. (Adapted from Keoleian, 1994)

LCAs as two separate procedures. Streamlining approaches can be used at the goals and scoping stage to set realistic objectives for the assessment. Streamlining can thus be viewed as "What can be eliminated from a full-scale LCA design and still meet the study goals?" or "What elements of a full-scale LCA must be included to meet the study goals?" (SETAC, 1997).

Streamlined LCAs are often used as a precursor to full life-cycle assessment in order to show where potentially environmentally significant improvements in product design can be made. It is also useful at the conceptual stage of design where limited data may be available. The complete set of life-cycle environmental effects associated with a product system can be evaluated only after the design has been specified in detail. But at this time the opportunities for design change become drastically limited (Keoleian, 1996). Figure 6.8 shows that as the design moves from the conceptual to the detailed stage, both the number of design choices and the number of solutions to environmental problems posed by the design become smaller (Keoleian, 1994). To be useful, comprehensive LCAs require an enormous amount of data that may be available only as the design becomes more detailed, but the opportunities to use the LCA to influence product design are reduced as the design proceeds. Streamlined LCAs may be necessary at these early stages of design.

6.6 POLLUTION PREVENTION FACTORS

The complete life-cycle assessment process is usually very expensive and time-consuming because of the voluminous amount of data required to do a creditable job, and the results may not be totally accurate because of a lack of quality data or the overuse of assumed data. Companies are continually "reinventing the wheel" in their data gathering because there is little sharing of information among various competing companies. In an attempt to overcome this, the U.S. Environmental Protection Agency and Battelle (Tolle et al., 1994) have begun development of a new LCA approach called *P2 factors*.

P2 factors are intended to be an indicator of the general degree of environmental improvement over an entire life cycle that has occurred or might occur as a result of implementing a particular P2 activity (Curran, 1995). A P2 factor is defined as a numerical or semiquantitative ratio between alternative source reduction activities which indicates the magnitude of the resulting environmental effects. It is in the form of a ratio, where the denominator is the summed score for criteria before application of a P2 activity, and the numerator is the summed score of the same criteria after implementation of the P2 criteria. Thus it can be used to compare only two alternative P2 options by one company; it cannot be used to compare products or processes between competing companies. It is not a "full LCA" but rather a simplified LCA methodology involving a mix of life-cycle inventory and impact assessment scoring criteria that can be used to screen candidate P2 activities. It usually is used to study only a few of the life-cycle stages, those that are expected to be affected by implementation of the P2 activity.

The methodology developed for calculating P2 factors includes a matrix of scoring criteria that can be used by different industries. The P2 factors are developed by first identifying which criteria are likely to change as a result of implementing a P2 activity. This is accomplished with the help of the stressor/impact chains, as proposed by SETAC and discussed earlier. Stressor/impact chains can be developed by considering the energy, water, and raw material inputs to each life-cycle stage, as well as the air, water, and solid waste emission impacts. Table 6.3 lists the criteria that may be selected to develop P2 factors for any industry (Tolle et al., 1994). The life-cycle stages where these criteria may be relevant are indicated by an ×. The resulting P2 factors can be used to calculate both an industry average for an entire industry and a site-specific value for an individual company, in order to determine which P2 activities result in the greatest environmental improvement.

Briefly, the P2 factor methodology is as follows (Curran, 1995). Criteria related to environmental impacts that are likely to occur before and those likely to occur after implementation of the P2 activity are selected and scored according to their degrees of impact. The P2 factor is a ratio of the two scores. The environmental impacts are scored individually using a five-number scale (1, 3, 5, 7, and 9) to indicate descending levels of environmental impact likely to be generated by the activity, with 1 indicating the most and 9 indicating the least environmental impact.

To demonstrate the use of this framework for developing P2 factors, the EPA selected the lithographic printing industry for the first case study. Two P2 activities were selected for this case study: (1) solvent substitution for blanket or press wash; and (2) use of waterless versus conventional printing.

The solvent substitution option will be investigated further here. The main reason for possibly instituting this P2 option is to reduce the quantity of volatile organic compounds released to the air. Figure 6.9 shows the stressor/impact chains developed for the solvent substitution P2 options. Based on these, scoring criteria were selected for each of three life-cycle stages: (1) habitat alteration and resource renewability for the raw materials acquisition stage; (2) energy use, airborne emissions, and waterborne

TABLE 6.3
Potential scoring criteria for determining P2 factors

Scoring criteria	RMA	MAN	U/R/M	R/WM
Habitat alteration	×			
Industrial accidents	×	×		
Resource renewability	×			
Energy use	×	×	×	×
Net water consumption	×	×	×	×
Preconsumer waste recycle percentage		×		
Airborne emissions	×	×	×	×
Waterborne emissions	×	×	×	×
Solid waste generation rate		×		
Recycle content		×		
Source reduction potential		×		
Product reuse		×		
Photochemical oxidant creation potential	×	×	×	×
Ozone depletion potential	×		×	×
Global warming potential	×	×	×	×
Surrogate for energy/emissions to transport materials to recycler		×	×	
Recyclability potential (postconsumer)				×
Product disassembly potential				×
Waste-to-energy value				×
Material persistence				×
Toxic material mobility after disposal				×
Toxic content		×		×
Inhalation toxicity		×		×
Landfill leachate (aquatic) toxicity				×
Incineration ash residue				×

Key: × = relevant life-cycle stage; RMA = raw material acquisition; MAN = manufacturing; U/R/M = use/reuse/maintenance; R/WM = recycle/waste management.

effluents for the materials manufacturing (petroleum refining) step of the manufacturing stage; and (3) photochemical oxidant creation potential, ozone depletion potential, global warming potential, and inhalation toxicity for the product fabrication (printing) step of the manufacturing stage. Table 6.4 presents the evaluation criteria for the materials manufacturing portion of the manufacturing life-cycle stage. Similar criteria are developed for the raw material acquisition stage and the product fabrication (printing) stage [see Tolle et al. (1994) for these].

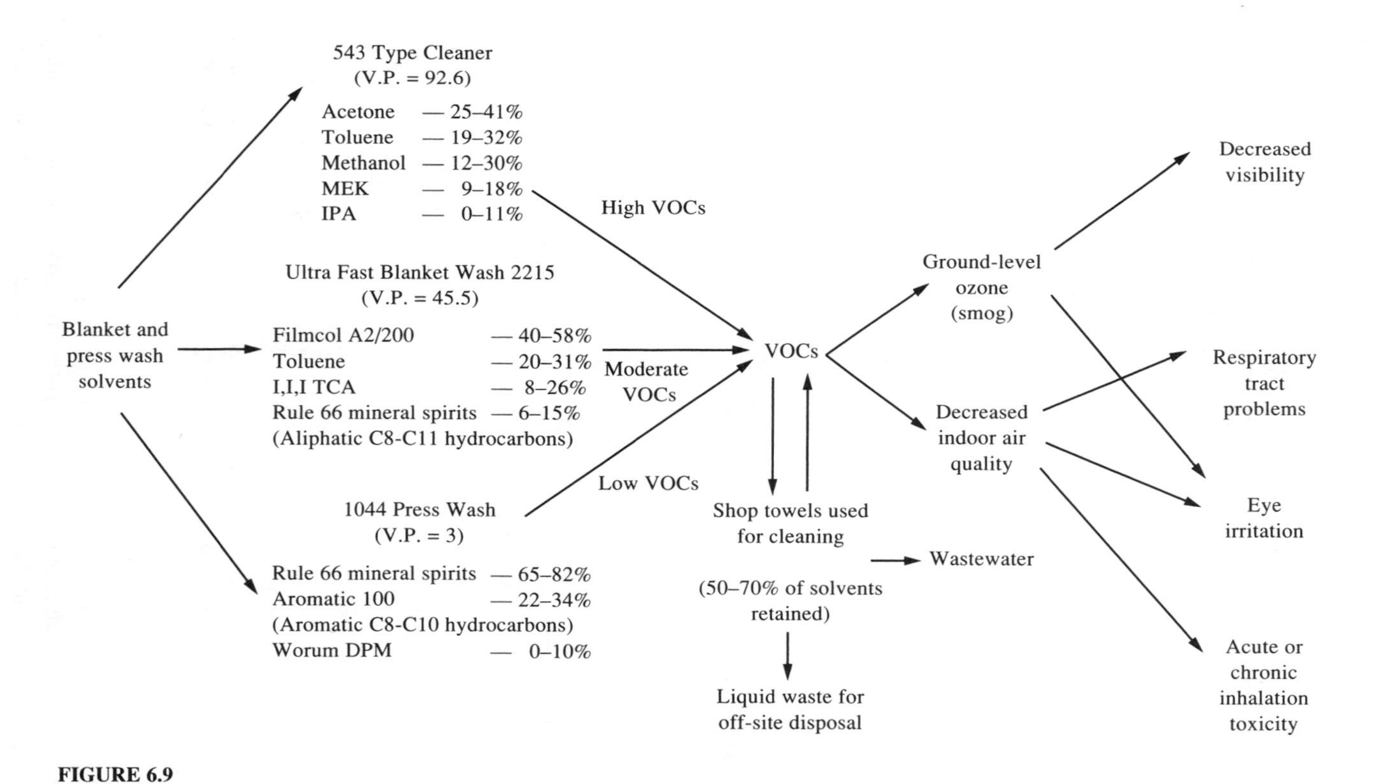

FIGURE 6.9
Stressor/impact diagram for solvent substitution in blanket or press wash in the lithographic printing industry. (Source: Tolle et al., 1994)

TABLE 6.4
Evaluation criteria and scoring ranges for calculation of P2 factors for the material manufacturing life-cycle stage for lithographic printers

	Energy Use
Score	**Criteria ranges for energy used per unit output**
9	<5,000 Btu/lb
7	5,000–10,000 Btu/lb
5	10,000–20,000 Btu/lb
3	20,000–30,000 Btu/lb
1	>30,000 Btu/lb
	Toxic/hazardous airborne emissions
Score	**Criteria ranges based on airborne pollutant regulatory limits**
9	Airborne pollutant emissions consistently > 50% below limits
7	Airborne pollutant emissions consistently > 25–50% below limits
5	Airborne pollutant emissions consistently > 10–24% below limits
3	Airborne pollutant emissions typically at the limits
1	Airborne pollutant emissions often exceed one or more limits
	Waterborne effluents
Score	**Criteria ranges based on waterborne pollutant regulatory limits**
9	Water pollutant emissions consistently > 50% below limits
7	Water pollutant emissions consistently > 25–50% below limits
5	Water pollutant emissions consistently > 10–24% below limits
3	Water pollutant emissions typically at the limits
1	Water pollutant emissions often exceed one or more of the limits

Source: Tolle et al., 1994.

Table 6.5 presents data from three companies that made solvent changes in their printing processes, which were used to calculate P2 factors. When more than one solvent was mixed, scores were calculated for each solvent individually, and then the scores were combined based on the weight percentages used. Company A made two changes in blanket and press wash over a five-year period. The company started with 543 Type Cleaner in 1988, changed to Ultra Fast Blanket Wash 2215 in 1990, and made another switch to 1044 Press Wash in 1993. The first change resulted in a P2 factor of 0.82, indicating a greater environmental impact from the second solvent used. Switching from this solvent to the 1044 Press Wash resulted in a P2 factor of 1.18, indicating a significant improvement in environmental impacts over the use of Ultra Fast Blanket Wash 2215. However, when compared with the original solvent used (543 Type Cleaner), the P2 factor was only 0.97. Thus, from an environmental standpoint, the company may have been better off staying with the original solvent system. Company B initially used 555 Typewash and 70 Press Wash, but switched to 1066 Press Wash.

TABLE 6.5
P2 factors modified scores for solvents used in lithographic printing

Solvent	Energy use	Air-borne emissions	Water-borne emissions	Photo-chemical oxidant creation potential	Inhalation toxicity	Ozone-depleting potential	Global warming potential	Individual solvent mixture score	Average concurrent solvent mixture score	Inter-mediate P2 factor	Overall P2 factor
					Company A						
534 Type Cleaner	4.0	6.5	6.5	4.5	7.7	9.0	7.0	45.1		0.82	
Ultra Fast Blanket Wash	3.0	5.6	5.9	4.6	5.5	6.5	6.0	37.0			0.97
1044 Press Wash	2.9	6.4	5.1	4.4	7.0	9.0	9.0	43.8		1.18	
					Company B						
555 Typewash	2.9	4.9	5.5	4.5	5.4	9.0	6.5	38.7	43.7		
70 Press Wash	3.0	7.0	5.0	5.0	7.0	9.0	9.0	44.9			0.97
1066 Press Wash	2.8	5.7	5.0	4.0	7.0	9.0	9.0	42.5	42.5		
					Company C						
VT3-A	2.7	5.9	5.9	1.9	7.7	9.0	7.0	40.0	39.4		
Power Kleen XF Plus	2.8	5.8	5.1	2.0	4.8	9.0	8.9	38.6			1.05
IP Wash	3.0	6.0	5.0	2.4	7.0	9.0	9.0	41.4	41.4		
											Average P2 ratio = 1.00

Source: Tolle et al., 1994.

The resulting P2 factor was 0.97, indicating a slight increase in environmental stress due to the solvent switch. Company C was initially using VT3-A and Power Kleen XF Plus and switched to IP Wash. The P2 factor was 1.05, indicating a small improvement in environmental impacts by changing the solvent used. These differences may not be significant, though, because the differences found may be within the margin of error of the data available. The average P2 factor for the three companies (four solvent changes) was 1.00, indicating very little overall environmental improvement was achieved, on average, by solvent substitutions.

The second P2 activity evaluated in this study involved switching an offset press used in printing from conventional dampening system printing, using either 2-butoxy ethanol or ethylene glycol in the fountain solution, to a waterless printing system. Waterless printing received a score of 9 for all of the criteria associated with fountain solution solvents, since no fountain solution is required for waterless printing, whereas scores for the fountain solution in the conventional process averaged 6.3. The overall P2 factor for switching from conventional to waterless offset printing was 1.25, indicating decreases in the overall impacts from this P2 activity.

6.7 APPLICATIONS OF LIFE-CYCLE ASSESSMENT

The LCA concept can be used in a number of ways to achieve pollution prevention. Among these are corporate strategic planning, product development, process selection and/or modifications, market claims and advertising (Vigon, 1995), evaluation by government agencies to ensure compliance with pollution prevention requirements, and evaluation and comparison of products by consumers or environmental groups. Thus LCAs may be used by a company as part of its production and sales strategies or by outside groups to evaluate products. A few of these are discussed further here.

6.7.1 Corporate Strategic Planning

Manufacturing organizations are driven by product sales and the resultant net profits. However, many companies in recent years have realized the importance of considering the environmental impacts of their manufacturing processes and products. Consumers, in increasing numbers, are demanding environmental accountability from producers and often are making purchasing decisions based on the "greenness" of a company. By conducting and implementing LCAs and publicizing the results, companies can show that they are concerned about the environment. Since these are *life-cycle* analyses, the management can use the results to make decisions that impact on other industries as well, such as their suppliers, transporters, and potential users of their by-products. By making a corporate decision to improve the product's LCA by requiring suppliers to adhere to strict pollution prevention practices, for example, the producer can affect a much greater sphere than its own operation.

In Chapter 1, the Chemical Manufacturers Association (CMA) Responsible Care Program was described. This program was designed to promote the practice of product stewardship, that is, making safety, health, and environmental protection an integral part of design, manufacture, distribution, product use, recycling, and disposal (Fava and Consoli, 1996).

Many companies are now making pollution prevention a fundamental part of their operating strategy and are using life-cycle assessments to monitor progress in that direction. The LCA results become an intrinsic part of the corporate decision-making process. In some cases, this has meant altering the company's way of deciding what products to produce or processes to modify. Typically, decisions of this nature are made based primarily on financial considerations (i.e., rate of return, benefit-cost ratio, payback period). As will be discussed in the next chapter, these procedures usually do not account well for the benefits of pollution prevention projects. If life-cycle analyses are to be used successfully, accommodations must be made for such factors as reduced long-term liability, extended project implementation times, and improved public image, none of which are usually included in corporate decision making based on financial return.

Scott Paper Company is one example of a company that is incorporating life-cycle considerations as one of the decision criteria for its pulp material and supplier selection for the tissue paper it produces. The company developed a clear set of worldwide positions on environmental management and committed the company to pursue an LCA cradle-to-grave approach, rather than continuing to act on individual issues. Within this program, Scott established a supplier assessment protocol, building environmental criteria into pulp purchasing decisions for its European operations. Suppliers were evaluated using a common set of environmental criteria. Suppliers that did not meet the criteria were given an opportunity to change their practices so that they could comply. Those that could not or would not meet the criteria (about 10 percent of the suppliers) were dropped as suppliers. Thus Scott was able to impact a much larger part of the industrial waste stream than that from its own facilities (Fava and Consoli, 1996).

6.7.2 Product Development

Production of waste materials during manufacturing is a direct sign of an inefficient manufacturing process. Avoidable wastes produced during the use of a product, wastes generated by the disposal of the product after its use, and excessive energy use required to operate the product are also signs of needless inefficiency. These can often be eliminated, or at least reduced, through proper use of life-cycle assessment and implementation of the results. There are many examples of this being successfully accomplished.

Among companies that are incorporating LCA into their decision-making procedures are Motorola, a large electronics company. Motorola is working to integrate environmental considerations into all of its product designs in order to meet customers' environmental demands. The company has found that it can have the biggest pollution prevention impact by focusing on the concept development stage of product design. In an effort to make up for the lack of detailed data available at this stage for making environmental assessments, Motorola has developed a matrix-based streamlined life-cycle assessment procedure for the concept development design stage. The company is beginning to use full-scale life-cycle assessments for the manufacturing stage (Hoffman, 1997).

The most common use of life-cycle assessment is to identify critical areas in which the environmental performance of a product can be improved (Allen, 1996). As discussed earlier, a study of the life-cycle of a woman's polyester blouse indicated that 86 percent of the energy consumed during the life-cycle of the blouse is associated with hot water cleaning and machine drying (see Figure 6.10). It was found that the energy demand associated with cleaning could be reduced by 90 percent if the garment fabric could be redesigned so that it could be washed in cold water and air-dried. Thus, from a life-cycle viewpoint, product improvement should be directed toward development of fabrics, dyes, and detergents that are compatible with cold water washing. Accomplishing this would be a boon to the environment. Any financial benefits from doing this, though, accrue to the consumer and not the garment maker, unless the blouse maker can increase sales by aggressive advertising of the benefits of the new fabric. Thus a beneficial product improvement based on life-cycle assessment may or may not be implemented, depending on whether it benefits the producer.

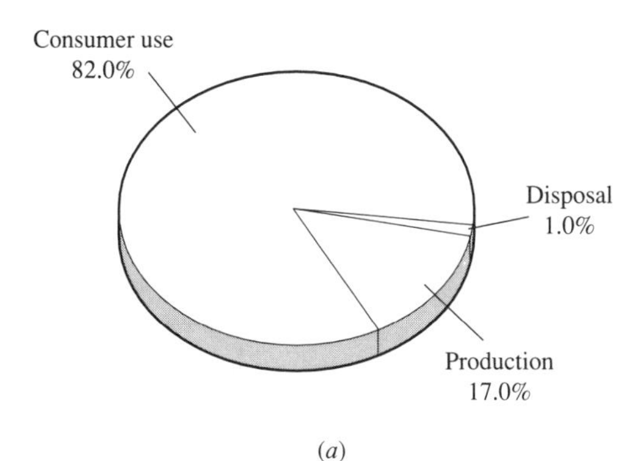

(*a*)

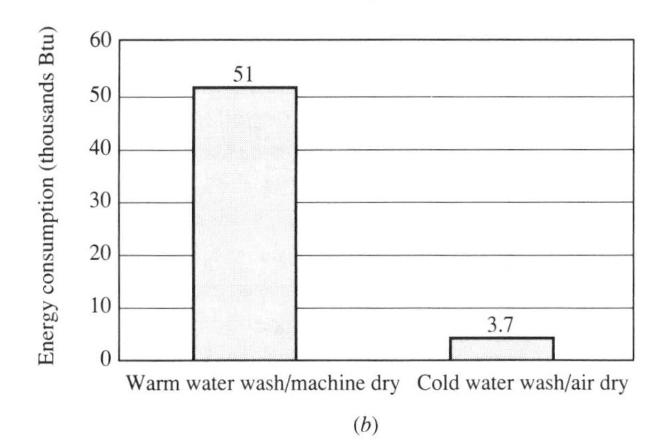

(*b*)

FIGURE 6.10
A life-cycle inventory for a woman's polyester blouse. (*a*) Distribution of life-cycle energy requirements. (*b*) Energy consumption comparisons for one load of home laundry. (Adapted from Franklin Associates, Ltd., 1993)

6.7.3 Process Selection and Modification

Life-cycle assessment has frequently been used to select between several alternative processes to achieve the greatest potential waste prevention or reduction. This can be done as part of an overall LCA for a product or as a focused study to evaluate improved manufacturing processes. An LCA restricted to process modifications may be streamlined because the final product should be the same and should have the same environmental impacts, regardless of the manufacturing process. The study can then be limited to such items as changes in feedstocks, different energy requirements, possibly altered production efficiencies, different by-products, and waste effluents.

IBM is constantly evaluating the environmental burdens associated with the materials used in its information technology equipment (Besnainou and Coulon, 1996; Brinkley et al., 1995). In one study, IBM wanted to know the impacts associated with increasing the amount of recycled polyvinyl chloride (PVC) used in making covers for personal computer monitors. IBM compared a closed-loop recycling option, in which used or surplus monitor equipment was collected at an IBM site and shipped to a qualified recycler who reclaimed the PVC and sent it back to IBM for use as a feedstock in new monitor housings, with two disposal options: (1) landfilling as scrap, and (2) incineration with heat recovery. IBM was mainly interested in studying the recovery/disposal options, so it limited the scope of the LCA to these components, while including all life-cycle phases of that component and material. Since the PVC used in manufacturing the monitor housings was essentially the same whether it came from virgin or recycled materials, the manufacturing process was not studied. The assessment was limited to the three disposal options (see Figure 6.11). Table 6.6 summarizes the life-cycle inventory of materials, energy requirements, air emissions, water effluents, and solid wastes generated by each option. With few exceptions, the recycling option presents the best inventory profile. When coupled with an economic analysis, the ability to recover and use recycled PVC rather than using virgin PVC clearly made this the best option.

The Tennessee Valley Authority (TVA), the major producer of power in the southeastern United States, produces significant quantities of waste materials. In 1993, the TVA began a program to reduce the generation of solid waste by 30 percent by the year 1997. Using an LCA procedure developed by the Electric Power Research Institute (EPRI), the TVA was able to cost-effectively reduce overall solid waste production by 19 percent within two years. In addition to the environmental benefits that resulted from this reduction, the TVA has realized an annual life-cycle cost saving of $41,000 and a one-time saving of $70,000 (West, 1997).

In another study of the way that LCAs can be used to analyze and design manufacturing processes, the process used to produce nitric acid was evaluated (Kniel, Delmarco, and Petrie, 1996). Assessment of the nitric acid process was ideal for this because of the process's simplicity (few materials and energy flows with well-known technologies, one principal waste and product stream; see Figure 6.12) and because necessary data were readily available. This is not often the case. The manufacturing process consists of oxidizing ammonia gas to NO_x with air under high pressure (3.25

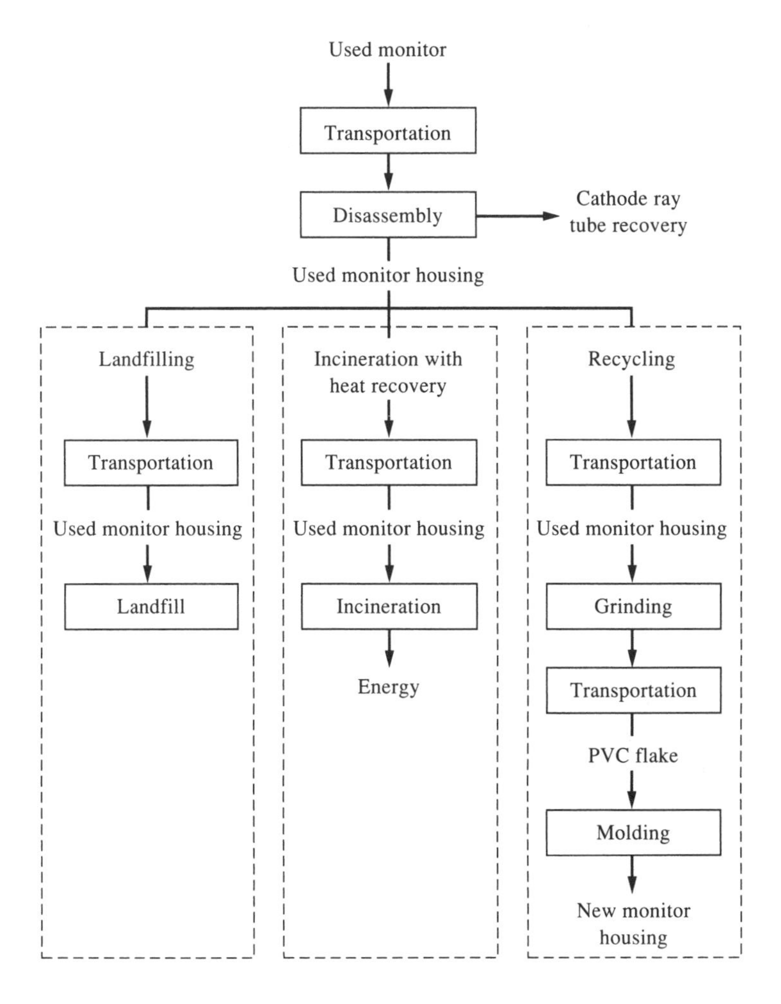

FIGURE 6.11
Functional analysis of the three PVC monitor housing disposal options by IBM. (Adapted from Besnainou and Coulon, 1996)

bar) and then absorbing the NO_x in water in the presence of a catalyst (platinum-rhodium) to form nitric acid (HNO_3). Among the proposed changes that were evaluated were a common end-of-pipe solution; adding a selective catalytic reduction (SCR) unit to the tail gas stream to reduce NO_x levels; and one following waste minimization principles, increasing the pressure in the absorption system to increase the efficiency of NO_x absorption. The nitric acid process modification LCAs and results of economic models for a number of design alternatives aimed at waste reduction were linked and compared to maximize economic returns and minimize environmental impact. The

TABLE 6.6
Life-cycle inventory analysis results for the three disposal options

	Unit	Landfilling	Incineration	Recycling
		Raw materials		
Crude oil	kg	0.036	0.025	−1.07
Coal	kg	0.0002	−0.67	−0.44
Natural gas	kg	0.0001	0.004	−1.28
Limestone	kg		1.50	−0.004
NaCl	kg			−1.5
Water	L	0.007	−0.008	−4.2
		Air emissions		
Particulate matter	g	0.15	33	−8.3
CO_2	g	115.	2400	−4000
CO	g	0.41	1.07	−5.3
SO_x	g	0.16	−13.0	−27
NO_x	g	1.17	−4.17	−33
NH_3	g	0.0007	0.0143	0.0011
Cl_2	g			−0.004
HCl	g		300	−0.48
Hydrocarbons	g	0.31	−13.70	−42.6
Other organics	g	0.00	−0.02	−1.60
		Water effluents		
Biochemical oxygen demand (BOD_5)	g	0.0002	0.0002	−0.18
Chemical oxygen demand (COD)	g	0.0006	0.0007	−2.46
Chlorides	g			−89.4
Dissolved solids	g	0.42	0.48	−2.6
Suspended solids	g	0.0002	−0.004	−5.3
Oil	g	0.005	0.007	−0.10
Sulfates	g			−9.6
Nitrates	g		−0.0004	0.00004
Nitrogen-TKN	g			−0.01
Sodium ions	g			−5.1
Metals	g			−0.45
		Solid wastes		
Hazardous chemicals	g			−0.003
Landfilled PVC	g	2.2	0	0.02
Slags and ash	g		1.7	−0.10
Other	g	0.00005	−0.44	−0.14
		Energy		
Total primary energy	MJ	42		−103
Electricity	kWh	0.0012	−2.1	−2.3

Source: Besnainou and Conlon, 1996.

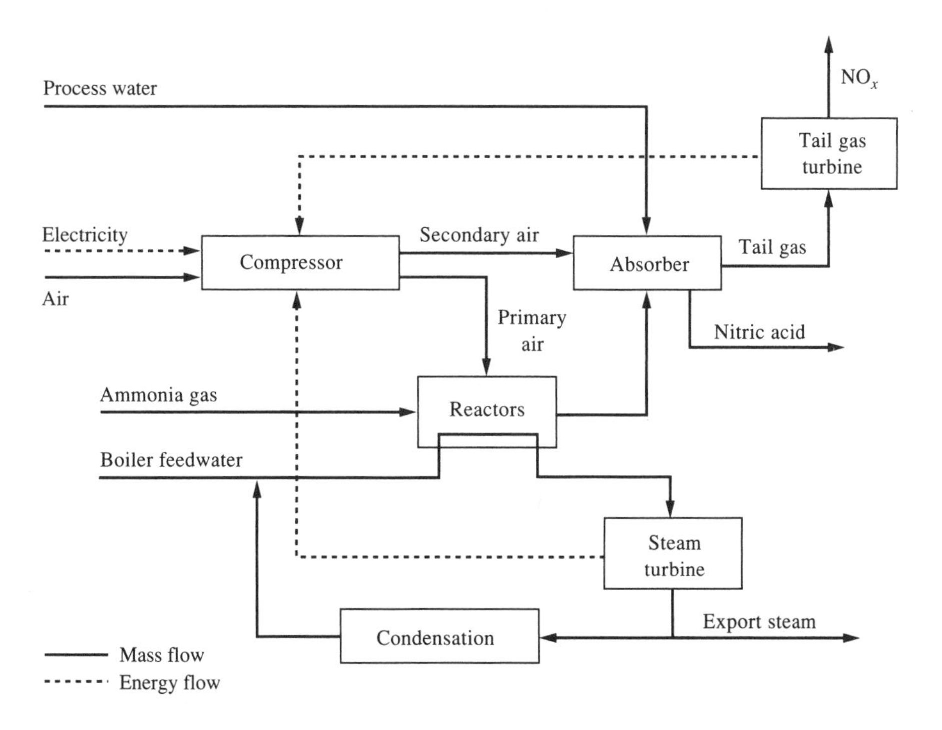

FIGURE 6.12
Process schematic for a nitric acid plant. (Adapted from Kniel, Delmarco, and Petrie, 1996)

comparative study showed that the waste minimization option was clearly superior from an environmental point of view; it also had the highest economic rate of return, because the SCR unit is only a treatment system and provides no additional revenue, whereas the high-pressure modification increases efficiency of absorption and therefore increases production rates.

6.7.4 Marketing Claims and Advertising

Many companies are beginning to see the value of using LCAs to support their claims to product consumers that the company is a good steward of the environment and that its products are environmentally friendly. Typically, though, little information concerning the scope of the LCA or the information derived is provided. Manufacturers generally advertise their products as environmentally friendly based on a single criterion such as recycling potential or the amount of waste generated during one or two stages of the product's life (Bhat, 1996). For example, a manufacturer may label its product as "nontoxic," "biodegradable," "CFC-free," or "phosphate-free," without any other indications as to its "greenness" or the meaning of the terms used. Thus many of these claims must be viewed skeptically. Taking limited data out of context, without all the information from a full life-cycle assessment, can easily lead to misinformation and misleading claims.

6.7.5 Ecolabeling

Industries are not alone in finding a use for life-cycle assessments; both governments and environmental watchdog groups have discovered that LCAs can be used to compare products on an equal basis or to ensure that the products meet minimum environmental standards. These results are used by some governments to ensure regulatory compliance; they can also be used in making product purchase and acquisition decisions by governments, industry, and the general public.

The United States has been slow in adopting LCA techniques as a part of public policy decision making. Progress is being made, though. In 1993, President Bill Clinton signed the Executive Order on Federal Acquisition, Recycling and Waste Prevention, which requires the Environmental Protection Agency to issue guidance that recommends the purchase of "environmentally preferable products and services," based on life-cycle assessment.

In Europe, on the other hand, life-cycle assessment concepts have become a basis for much of the environmental public policy. Several Nordic countries, including Denmark, Norway, and Sweden, are using LCAs of waste management options to determine the best strategies to undertake. A task force in Europe is developing guidelines for *environmental product profiles,* a qualitative description of the environmental impacts of a product for use by commercial and institutional buyers. Under this system, environmental information would be collected for each stage of a product's life cycle. This information will be shared with any user of that product so that the user can add the information from these previous life-cycle stages to its own life-cycle assessment. Thus cumulative impacts can be determined without each manufacturer having to begin the LCA from scratch. Based on an LCA of beverage packaging, Denmark banned the use of nonrefillable bottles and cans and requires the use of refillable bottles. It also has greatly increased the fees for solid waste disposal as an incentive to recycle more waste materials. One of the most far-reaching actions is that taken by Germany, which has passed a law making manufacturers responsible for collecting and recycling various kinds of packaging and requiring manufacturers to take back and recycle many products from consumers after the product's useful life.

Probably the most visible application of LCAs in the public sector is the growing use of environmental labeling. Until recently, environmental labeling claims in the United States have been largely unregulated, sometimes resulting in misleading claims by manufacturers and confusion among consumers. Regulation of ecolabeling is still in its infancy in the United States, but progress is being made. Several states have passed regulations that govern the use of specific terms such as "recyclable" or "biodegradable." The Federal Trade Commission is issuing guidelines on the use of environmental labeling claims on product packaging and advertising. To be effective, evaluation of "greenness" should be performed by a third party, independent of the manufacturers or marketers of the product. A private organization, Green Seal, is using life-cycle assessment to evaluate the environmental impact of various categories of consumer products and to award environmental seals of approval to those products judged to be environmentally preferable (similar to the Underwriters' Laboratories' approval) (Marron, 1995). The expectation is that products receiving the environmental seal of approval will receive a larger share of the market, forcing companies making competing products to change their practices so that they can also receive approval. Recent surveys have shown that four out of five consumers are more likely to purchase a product with a Green Seal label when choosing between products of equal quality and price; the surveys also showed that the Green Seal would have more impact on their purchase decisions than government guidelines.

Ecolabeling in the United States is relatively new, but it is a well-accepted practice in many European countries, Canada, and Japan. Among the G-7 countries, there are six official Eco-Logo Certification programs: Environmental Choice (Canada), Green Seal (United States), Eco Mark (Japan), Ecolabel (European Union, used by the United Kingdom and Italy), NF Environment (France), and Blue Angel (Germany). Each program is somewhat different, but they have some common elements.

ECO-LABELING IN OTHER COUNTRIES. Germany's Blue Angel Program, begun in 1978, is generally regarded as the model for all other ecolabeling programs. The program awards its Blue Angel label to consumer products that are clearly more beneficial to the environment than others in the same product category, as long as the product's primary function and safety are not significantly impaired (Marron, 1995). The assessment program is voluntary, and businesses that are certified pay a fee. The decision-making process for establishing criteria for environmental labeling is a joint effort of German governmental agencies and nongovernmental organizations. The criteria for awarding the Blue Angel includes the efficient use of fossil fuels, alternative products with less of an impact on the climate, reduction of greenhouse gas emission, and conservation of resources. The evaluation procedure consists of using a checklist or matrix to identify important parts of the product's life cycle and the most significant environmental attributes of the product class, supplementing this with the use of expert panels. If there is insufficient information to develop screening criteria, a streamlined LCA will be done using available information with caution (Shen, 1995). Once approved, Blue Angel–labeled products are reviewed every two or three years to reflect state-of-the-art developments in ecological technology and product design. Seventy-one product groups have been identified, with more than four thousand products approved for the Blue Angel seal. The program has been credited with increasing Germany's use of recycled paper and low-VOC paints, lacquers, and varnishes.

In a similar fashion, the Netherlands began awarding the Dutch Ecolabel in 1993 based on a qualitative matrix of environmental criteria. Label criteria are developed by a Board of Experts for product groups based on life-cycle assessments. At present, all consumers see is the label, indicating that the product has passed the "greenness" test, but no background information on the product is provided. In the future, the Dutch plan to expand the labeling program to include simplified descriptions of the product's impact on energy, waste, resources, emissions, and nuisance value (Allen and Rosselot, 1997).

The European Union (EU) is now coordinating the actions of member countries in implementing ecolabeling through its Community Eco Label Award Scheme (Postlethwaite and de Oude, 1996). Member states were requested to develop criteria for a number of product categories, such as shoes and laundry detergents, based on specific ecological criteria. The goal is that roughly 20 percent of the products currently on the market will comply with the initial criteria. There is a period of criteria validity of usually three years, after which the criteria will be reviewed and the standards possibly raised. This will encourage a progressive increase in environmental performance of consumer goods. The intent is to make the standards uniform over all of Europe. Once a set of criteria is approved, it can be used throughout all member countries, avoiding costly and redundant applications. Eventually, all member states will be required to implement the new regulations without any modifications into their national legislation.

The Canadian government began administering its Environmental Choice Program in 1998. The Canadian Standards Association, an independent testing and standards-setting organization, verifies products against guidelines set by the government and licenses companies on behalf of the Environmental Choice Program. The criteria are set such that only 10 percent to 20 percent of the products within a given category can qualify. The program is designed to address long-term environmental issues, including energy efficiency, hazardous by-products, enhanced use of recycled materials, and a design that allows for easier recycling. The criteria are based on a product's full life-cycle assessment. Currently, Canada has more than 1400 approved products.

The ISO 14000 program is also addressing the issue of environmental labeling by developing an international labeling standard. When completed, it will provide a consistent approach and uniform rigor to testing. Currently, labeling programs among various countries are often inconsistent. Major international companies have found that these programs are often confusing and may be based on idiosyncratic, nonscientific objectives. Environmental labeling has the potential to create serious international trade issues through its power to discriminate against products from countries using different environmental criteria. Compliance with multiple labeling requirements is complicated, costly, and administratively burdensome (Cascio, Woodside, and Mitchell, 1996). As a result, many corporations that do business in several countries are reluctant to use such labels. A uniform labeling standard should lead to greater respectability for ecolabeling and a greater rate of participation.

ECOLABELING IN THE UNITED STATES. Green Seal is an independent, nonprofit environmental labeling organization dedicated to protecting America's environment by promoting the manufacture and sale of environmentally preferable consumer products. It awards a "Green Seal of Approval" (see Figure 6.13) to products that cause less harm to the environment than other similar products during the product's manufacture, use, and ultimate disposal. Green Seal standards also consider packaging and product performance. Standards are set on a category-by-category basis, using an "environmental impact evaluation" procedure, based on life-cycle analysis. Categories are generally chosen according to the significance of the associated environmental impacts and the range of products available within the category.

All Green Seal standards are first issued in proposed form for public comment; they may be revised based on information provided before being published in final form. Manufacturers submit products to Green Seal that they would like to have certified. Underwriters' Laboratories, Inc. (UL), is Green Seal's primary testing and factory

FIGURE 6.13
The Green Seal label logo.

inspection contractor. UL evaluates products to determine whether they meet Green Seal's environmental standards. UL is also responsible for conducting follow-up inspections at the manufacturer's facility to monitor continued compliance. To date, Green Seal has awarded its seal of approval to over 250 products. It certifies products in more than 50 categories including paints, water-efficient fixtures, bath and facial tissue, re-refined engine oil, energy-efficient windows, and major household appliances.

6.8 USE OF COMPUTER MODELS IN LIFE-CYCLE ASSESSMENT

Conducting a life-cycle assessment is usually a data-intensive operation. Software packages are beginning to become available to help with the collection and analysis of all the data, primarily from government agencies and university researchers. Some of these include packages to analyze the costs associated with a pollution prevention initiative, as well as the environmental impacts. These programs are designed to provide decision support on process optimization, comparative processes, and environmental needs assessments by looking at purchase, design, installation, start-up, and yearly costs information. They also help the user assess possible effects that the company's processes may have on the environment, such as the generation of hazardous air pollutants (Petty, 1997).

Life-cycle assessment software can be classified into three generic categories: strict LCA tools, product design tools, and engineering tools (Vigon, 1996). Strict LCA tools are intended to supply information to support LCA as a stand-alone activity. Most are intended to be used to collect, organize, and analyze data for life-cycle inventory assessments, but a few are also available for impact assessments; no commercially available software exists as yet for the improvement assessment phase of LCA. Product design–oriented tools are usually directed at design engineers who may not be expert in LCA. They embed LCA computations in a design package, such as that for packaging design. The LCA portion of the software is intended to provide recommendations on materials and process choices based on LCA considerations. Some engineering analysis tools have also been adapted for use in LCA. One example of this is process simulators. In these models, industrial process inputs and transformation functions are described for each process. The model predicts the form and amounts of products, coproducts, and residuals, based on the operating conditions and process rules. These models can provide indications of potential production efficiencies and environmental burdens from a proposed process, and therefore can be very beneficial in a life-cycle assessment. However, these simulations are very complex and require a great deal of data and knowledge about the physical and chemical operations involved. The results will only be as good as the information provided to the model.

Most commercial LCA software is based on a spreadsheet format, using either Lotus 1-2-3 or Excel, but some are much more complex. Unfortunately, most LCA software packages focus on only one or two life-cycle stages, such as packaging or a specific manufacturing process; very few consider the full life-cycle of a product in the assessment. Also, many packages are process or industry specific.

Europe has again taken the lead in developing LCA tools. In addition to computer models, a number of European LCA databases can be used to obtain industry- or product-focused data that can be used in the models (Postlethwaite and de Oude, 1996; SustainAbility Ltd., 1993).

Comprehensive reviews of life-cycle assessment software are provided by Wood, King, and Cheremisinoff (1997) and Vigon (1996). The Environmental Protection Agency also provides useful information on P2 and LCA software (U.S. EPA, 1995; Vigon et al., 1993).

An example of an existing comprehensive LCA software package is TEAM (Tools for Environmental Analysis and Management), a Windows-based C++ program developed to assist with life-cycle cost and life-cycle assessment calculations. The software generates life-cycle inventories and cost results, including sensitivity analysis. It covers raw material acquisition, the manufacturing stage, use/reuse/maintenance aspects, transportation, and recycle/waste management stages. Very complex systems can be analyzed, along with fairly simple estimations (Wood, King, and Cheremisinoff, 1997).

In the future, LCAs and process design will be integrally linked with database systems. An excellent example of how this will occur is the Clean Process Advisory System (CPAS) being developed by a coalition of industry, academia, and government. The CPAS is a computer-based pollution prevention process and product design system that will provide environmental design data to design engineers. It is composed of many separate but integrated software applications containing design information for new and existing clean process or product technology, technology modeling tools, and other design guidance. An overview of the CPAS system is seen in Figure 6.14.

6.9 LIFE-CYCLE ASSESSMENT IN WASTE MANAGEMENT OPERATIONS

The application of life-cycle assessments is not restricted to industrial processes. For example, LCA procedures have been used to evaluate waste management practices by municipalities, particularly in the area of solid waste management. Solid waste disposal is reaching the crisis level in some communities (see Chapter 3), due to lack of acceptable, approved landfill space and a general public opposition to incineration. One solution to this problem that has often been espoused is increased use of recycling of solid waste. Many communities, and some states, have set goals for the percentage of their

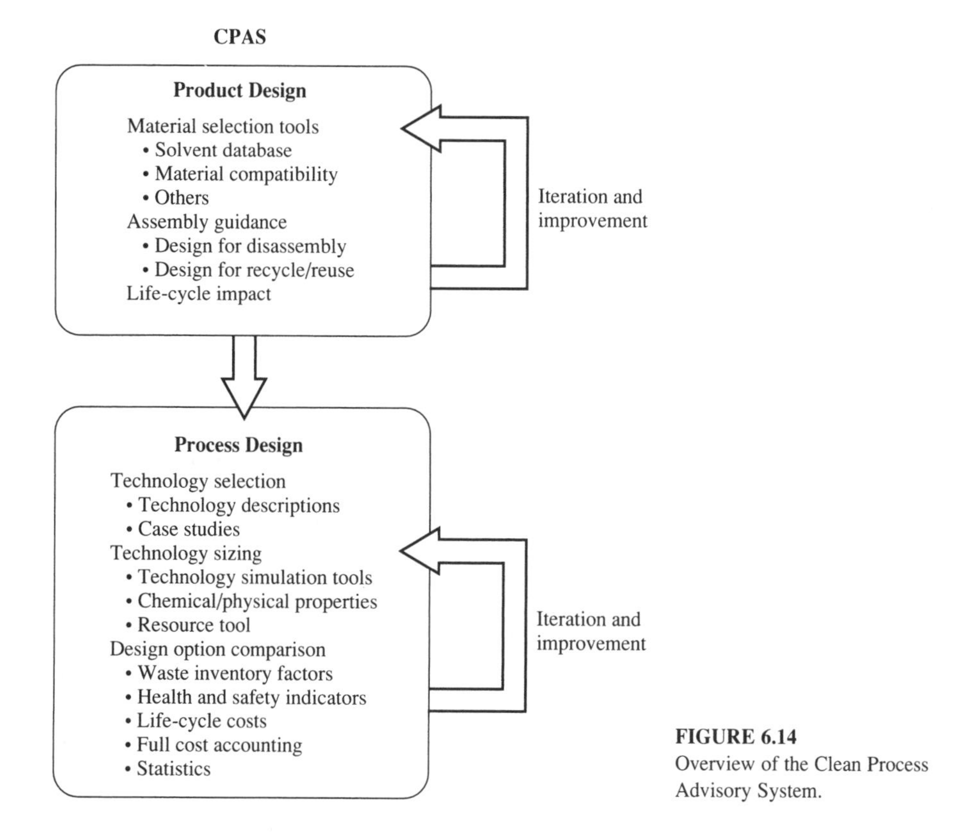

FIGURE 6.14
Overview of the Clean Process Advisory System.

solid waste that must be recycled. In many cases this action is driven by the desire to extend the use of existing landfill space; in other cases the impetus is a desire to conserve resources. In any event, the public perception usually is that recycling will also result in cost savings to the community through sale of the recovered materials. This may or may not be true, depending largely on the high variability in the market prices for these commodities.

As an example, in 1996 the price of recovered cardboard was $130 per ton; in 1997 the price had dropped to $10 per ton. Old newsprint sold for $120 per ton in 1996, while the going price in 1997 was down to $10–$20 per ton. The same trends could be seen for plastic bottles, where the price went from 30 cents per pound to 3 cents per pound in one year. Many communities found that profitable, environmentally correct community recycling programs were losing great amounts of money. It was cheaper to send the separated materials to the landfill than to sell them to a recycler. Recycled materials are commodities subject to the same economic laws of supply and demand as other materials. When supplies are low, prices will be high; these high prices will result in more material being recycled. If too much material is recovered, glutting the market, prices will plunge and recycling will be reduced. Thus the recycling market is highly volatile and cyclical.

A major difficulty in trying to assess the desirability of a solid waste recycling program is determining the true costs, both economic and environmental, associated with recycling. Often a recycling program is considered independent of the community solid waste management program. From an economic standpoint, a community often requires that the recycling program be profitable on its own; that is, profits from sales of recyclables must at least match the cost of recovering the materials in order to break even. However, hidden costs are often overlooked. Removing materials from the waste stream represents an overall negative cost because the recovered material does not need to be landfilled or incinerated. That cost represents a disposal cost savings to the community of $30–$100 per ton of recovered material. If this cost saving is not allocated to the recycling program, it may appear that the recycling program is losing money, even though there is an overall significant cost saving.

The overall (life-cycle) environmental impacts of recycling must also be considered. Removing waste materials from landfills or incinerators and recycling them is generally believed to be beneficial to the environment, but from the life-cycle standpoint this is not always the case. Reclaiming used newsprint greatly reduces the amount of solid wastes going to a landfill. However, during reprocessing, the dyes and fillers found in the newsprint are leached from the paper fibers and sent to wastewater treatment plants, along with chlorine solutions used to bleach the fibers and the fine fibers that are not useful in recycled paper. Thus the solid waste problem has been reduced, but at the expense of increased water pollution. Some of these pollutants removed from the wastewater will be removed as sludge, which may wind up back in the landfill anyway. Another factor that is often overlooked is transportation of recovered materials to the reprocessor. Economic costs for this could be significant and could be the deciding factor in determining whether recycling is economically viable. The environmental impacts (air pollution, fossil fuel consumption) are normally not accounted for. Recycling of low-value renewable materials in one city may be environmentally preferable, but it may not be in a more remote city where there are greater transport impacts ("Life Cycle Analysis," 1995).

To properly assess whether using recycled or virgin materials in a process will result in the least environmental damage, it is essential that a complete LCA be conducted. Life-cycle assessment has begun to be used to evaluate a city or region's future waste management options. The LCA covers the environmental and resource impacts of alternative disposal processes, as well as those other processes that are affected by disposal strategies—different types of collection schemes for recyclables, changed transportation patterns, and so on. This is a very complex analysis, as can be seen in Figure 6.15, a simplified diagram showing some of the different routes that waste might take and some of the environmental impacts incurred along the way.

Life-cycle assessment of waste management operations can be conducted in an analogous way to that for product LCAs. They can consist of the same three components: inventory analysis, impact assessment, and improvement assessment. The only difference is that the functional unit used to define the system is an *input to the system,* rather than an *output from the system.* For solid waste management, this will typically be kilograms of solid waste handled. Waste management LCAs also differ from product LCAs in that waste management LCAs must consider a wider range of materials types (Kirkpatrick, 1996). This complicates the analysis, but the procedures used are the same.

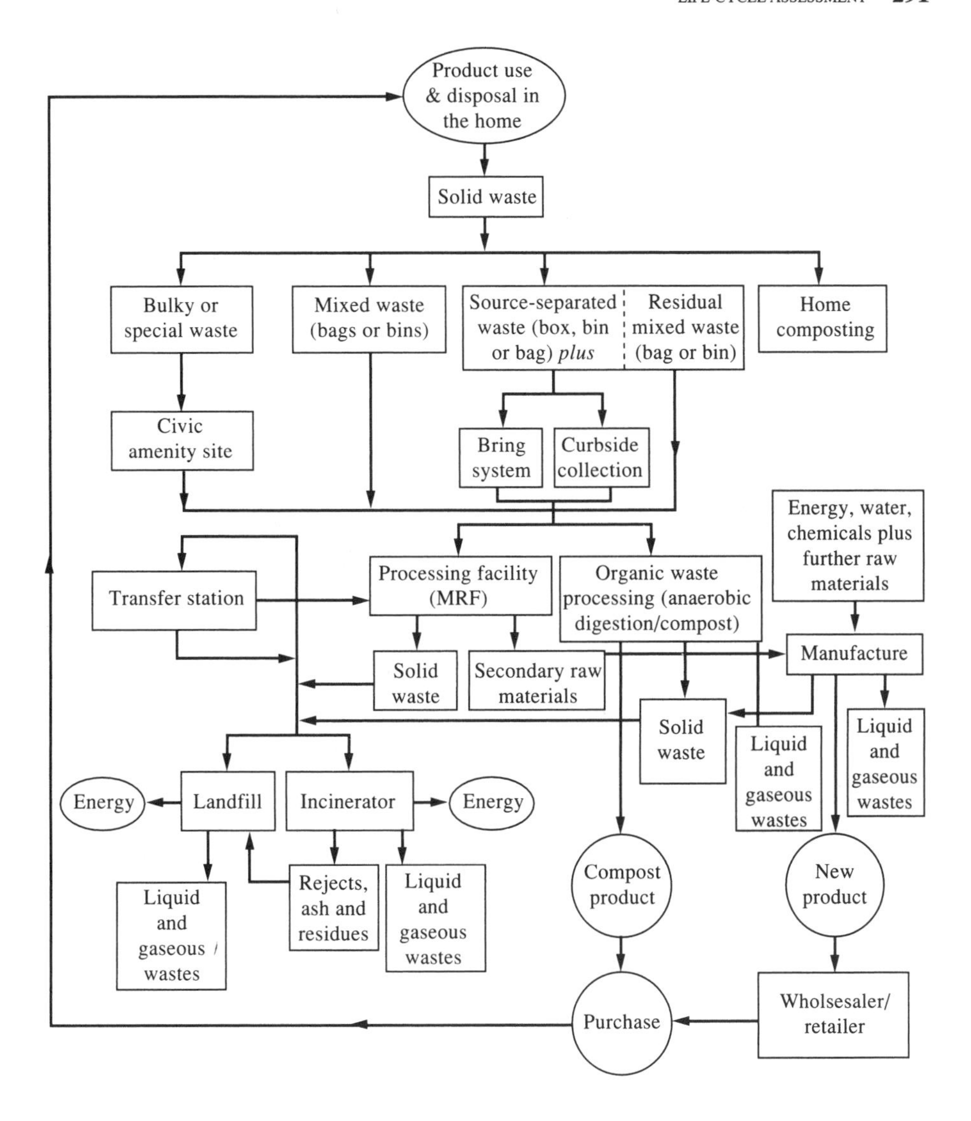

FIGURE 6.15
A simplified flow chart of alternative waste collection and disposal strategies showing the complexity of conducting an LCA. (Adapted from "Life Cycle Analysis," 1995)

The system boundary definition for a waste management LCA can also be complicated. Product LCAs usually consider the life cycle beginning with extraction of the resources from the earth, but where should the solid waste management LCA begin? Often this is defined to be at the point of solid waste pickup. Thus no materials production, manufacturing, transportation, or product use stages prior to disposal are included in the LCA. This means that these LCAs do not include any prior industrial waste minimization or recycling efforts by industry; these activities would be included in the product LCA. The end of the waste management LCA life cycle is typically accepted as occurring when all residues from the system are returned to land. Thus the LCA would include any recycling or energy recovery impacts from the waste management system itself. Once the system boundaries are set, the appropriate data can be gathered for the inventory analysis and impact assessment, analyzed, and used to perform an improvement assessment.

REFERENCES

Allen, D. "Applications of Life-Cycle Assessment." In *Environmental Life-Cycle Assessment,* ed. M. A. Curran. New York: McGraw-Hill, 1996, pp. 5.1–5.18.

Allen, D. T., and Rosselot, K. S. *Pollution Prevention for Chemical Processes.* New York: Wiley-Interscience, 1997.

Besnainou, J., and Coulon, R. "Life-Cycle Assessment: A System Analysis." In *Environmental Life-Cycle Assessment,* ed. M. A. Curran. New York: McGraw-Hill, 1996, pp. 12.1–12.28.

Bhat, V. *The Green Corporation: The Next Competitive Advantage.* Westport, CT: Quorum Books, 1996.

Boguski, T. K., Hunt, R. G., Cholakis, J. M., and Franklin, W. E. "LCA Methodology." In *Environmental Life-Cycle Assessment,* ed. M. A. Curran. New York: McGraw-Hill, 1996, pp. 2.1–2.37.

Breville, B., Gloria, T., O'Connell, M., and Saad, T. *Life Cycle Assessment, Trends, Methodologies and Current Implementation.* Department of Civil and Environmental Engineering, Tufts University, Medford, MA, 1994, pp. 145–55.

Brinckley, A., Kirby, J. R., Wadehra, I.L., Besnainou, J., Coulon, R., and Goybet, S. "Life Cycle Inventory of PVC: Disposal Options for a PVC Monitor Housing." In *Proceedings of the 1995 IEEE International Symposium on Electronics and the Environment.* Orlando, FL: IEEE, 1995, p. 145.

Cascio, J., Woodside, G., and Mitchell, P. *ISO 14000 Guide.* New York: McGraw-Hill, 1996.

Consoli, F., Allen, D., Boustead, I., Fava, J., Franklin, W., Jensen, A., de Oude, N., Parrish, R., Perriman, R., Postlethwaite, D., Quay, B., Seguin, J., and Vigon, B. *Guidelines for Life-Cycle Assessment: A "Code of Practice."* Pensacola, FL: SETAC, 1993.

Curran, M. A. *Environmental Life-Cycle Assessment.* New York: McGraw-Hill, 1996a.

Curran, M. A. "The History of LCA." In *Environmental Life-Cycle Assessment,* ed. M. A. Curran. New York: McGraw-Hill, 1996b, 1.1–1.9.

Curran, M. A. "Using LCA-Based Approaches to Evaluate Pollution Prevention." *Environmental Progress* 14 (1995): 247–53.

Curran, M. A., and Young, S. "Report from the EPA Conference on Streamlining LCA." *International Journal on LCA* 1 (1996): 57–60.

Fava, J. A., and Consoli, F. J. "Application of Life-Cycle Assessment to Business Performance." In *Environmental Life-Cycle Assessment,* ed. M. A. Curran. New York: McGraw-Hill, 1996, pp. 11.1–11.12.

Fava, J. A., Consoli, F., Denison, R., Dickson, K., Mohin, T., and Vigon, B. *Guidelines for Life-Cycle Assessment: A "Code of Practice."* Pensacola, FL: Society of Environmental Toxicology and Chemistry, Workshop Proceedings, N, 1993.

Fava, J., Jensen, A., Lindfors, L., Pomper, S., De Smet, B., Warren, J., and Vigon, B. *Life-Cycle Assessment Data Quality: A Conceptual Approach.* Pensacola, FL: SETAC, 1994.

Foust, T. D., and Gish, D. D. "Future Perspective." In *Environmental Life-Cycle Assessment,* ed. M. A. Curran. New York: McGraw-Hill, 1996, pp. 18.1–18.16.

Franklin Associates, Ltd. *Resources and Environmental Profile Analysis of a Manufactured Apparel Product.* Report to the American Fiber Manufacturer's Association, Washington, DC, 1993.

Giegrich, J., and Schmitz, S. "Valuation as a Step in Impact Assessment: Methods and Case Study." In *Environmental Life-Cycle Assessment,* ed. M. A. Curran. New York: McGraw-Hill, 1996, pp. 13.1–13.14.

Graedel, T. E., and Allenby, B. R. *Industrial Ecology.* Englewood Cliffs, NJ: Prentice Hall, 1995.

Hoffman, W. F. "Recent Advances in Design for Environment at Motorola." *Journal of Industrial Ecology* 1 (1) (1997): 131–40.

Keoleian, G. "The Application of Life Cycle Assessment to Design." *Journal of Cleaner Production* 1 (1994): 143–49.

Keoleian, G. A. "Life-Cycle Design." In *Environmental Life-Cycle Assessment,* ed. M. A. Curran, New York: McGraw-Hill, 1996, pp. 6.1–6.34.

Kirkpatrick, N. "Application of Life-Cycle Assessment to Solid Waste Management Practices." In *Environmental Life-Cycle Assessment,* ed. M. A. Curran. New York: McGraw-Hill, 1996, pp. 15.1–15.14.

Kniel, G. E., Delmarco, K., and Petrie, J. G. "Life Cycle Assessment Applied to Process Design: Environmental and Economic Analysis and Optimization of a Nitric Acid Plant." *Environmental Progress* 15 (1996): 221–28.

Kroschwitz, J., and Howe-Grant, M. *Kirk-Othmer Encyclopedia of Chemical Technology,* 4th ed. New York: Wiley-Interscience, 1991.

"Life Cycle Analysis and Assessment." *Warmer Bulletin* 46 (1995): 1–4.

Marron, J. "Product Labeling." In *Industrial Pollution Prevention Handbook,* ed. H. Freeman. New York: McGraw-Hill, 1995, pp. 313–28.

Meadows, D. H., Meadows, D. L., Randers, J., and Behrens, W. *The Limits to Growth: A Report for the Club of Rome's Project on the Predicament of Mankind.* New York: Universe Books, 1972.

Nash, J., and Stoughton, M. D. "Learning to Live with Life Cycle Assessment." *Environmental Science and Technology* 28 (1994): 236–37.

Office of Technology Assessment. *Green Products by Design: Choices for a Cleaner Environment.* OTA-E-541. Washington, DC: U.S. Government Printing Office, 1992.

Petty, M. "Pollution Prevention in the Information Age." *Environmental Protection* 8 (1997): 34–36.

Postlethwaite, D., and de Oude, N. T. "European Perspective." In *Environmental Life-Cycle Assessment,* ed. M. A. Curran. New York: McGraw-Hill, 1996, 9.1–9.13.

Rudd, D. F., Fathi-Afshar, S., Treviño, A. A., and Stadtherr, M. A. *Petrochemical Technology Assessment.* New York: Wiley, 1981.

SETAC. *Streamlining Environmental Life Cycle Assessment.* Preliminary draft report of the SETAC LCA Streamlining Group. Pensacola, FL: SETAC, 1997.

Shen, T. T. *Industrial Pollution Prevention.* Berlin: Springer-Verlag, 1995.

SustainAbility Ltd. *The LCA Sourcebook: A European Business Guide to Life-Cycle Assessment.* Prepared for the Society for the Promotion of LCA Development and Business in the Environment. London: SustainAbility Ltd., 1993.

Todd, J. A. "Streamlining." In *Environmental Life-Cycle Assessment,* ed. M. A. Curran. New York: McGraw-Hill, 1996, pp. 4.1–4.17.

Tolle, D. A., Vigon, B. W., Becker, J. R., and Salem, M. A. *Development of a Pollution Prevention Factors Methodology Based on Life-Cycle Assessment: Lithographic Printing Case Study.* EPA/600/R-94/157. Cincinnati, OH: U.S. EPA, 1994.

U.S. Environmental Protection Agency. *Life-Cycle Assessment: Public Data Sources for the LCA Practitioner.* EPA/530/R-95/009, Washington, DC: U.S. EPA, 1995.

U.S. Environmental Protection Agency. *Streamlining Life-Cycle Assessment II: A Conference and Workshop.* Proceedings, Cincinnati, OH: U.S. EPA, 1997.

Van den Berg, N. W., Huppes, G., and Dutilh, C. E. "Beginning LCA: A Dutch Guide to Environmental Life-Cycle Assessment." In *Environmental Life-Cycle Assessment,* ed. M. A. Curran. New York: McGraw-Hill, 1996, pp. 17.1–17.41.

Vigon, B. "Life-Cycle Assessment." In *Industrial Pollution Prevention Handbook.* ed. H. Freeman. New York: McGraw-Hill, 1995, 293–312.

Vigon, B. W. "Software Systems and Databases." In *Environmental Life-Cycle Assessment,* ed. M. A. Curran. New York: McGraw-Hill, 1996, pp. 3.1–3.25.

Vigon, B. W., Tolle, D. A., Cornaby, B. W., Latham, H. C., Harrison, C. L., Boguski, T. L., Hunt, R. G., and Sellers, J. D. *Life-Cycle Assessment: Inventory Guidelines and Principles.* EPA/600/R-92/245. Cincinnati, OH: U.S. EPA, 1993.

West, P. *Pollution Prevention Actions Reduce Waste Generation at TVA, Innovators with EPRI Technology.* EPRI Report RP3006, Palo Alto, CA, 1997.

Wood, M. F., King, J. A., and Cheremisinoff, N. P. *Pollution Prevention Software Systems Handbook.* Westwood, NJ: Noyes Publications, 1997.

PROBLEMS

6.1. Draw a flow chart depicting the full life cycle of a peanut butter and jelly sandwich. Be sure to include the feedstocks of the sandwich.

6.2. For the peanut butter and jelly sandwich described in Problem 6.1, conduct an inventory assessment for the sandwich manufacturing stage. Use the Life-Cycle Inventory Checklists reproduced in Figures 6.5 and 6.6 as the basis for your assessment.

6.3. Your local laundromat consumes large quantities of water and energy and transfers dirt from your clothes to the washwater going to the municipal sewer. What are the environmental issues involved with the use of laundromats and what are the stressors?

6.4. An industrial wastewater, flowing at 500 m^3/day, contains 20 mg/L chromium, 40 mg/L cadmium, and 5 mg/L mercury. The receiving water quality standards are, respectively, 50 μg/L, 10 μg/L, and 2 μg/L. What is the critical dilution volume?

6.5. Evaluate Table 6.3 with respect to the manufacture of a peanut butter and jelly sandwich, as described in Problem 6.1. Fill in the boxes marked X with what you think the impact might be.

6.6. Consider the lithographic example described in the text and in Tables 6.4 and 6.5. The company decides that its most pressing issue is reduction of air pollutants. Which of the P2 activities would be the most beneficial, based on P2 factors?

6.7. Three computer monitor housing disposal options are described in Figure 6.11 and Table 6.6. Which option produces the least water effluents? The most water effluents?

6.8. Table 6.6 shows the life-cycle inventory for computer monitor housings. The recycling option presents the best inventory profile. Why isn't it used more often?

6.9. Visit a supermarket and find 10 items for which the manufacturer makes some environmental claims. List the products and their environmental claims, and evaluate whether they are useful to the consumer and whether they are factual or may be misleading.

6.10. Visit a local store and find items that display a Green Seal logo. What are these items? Where is the logo located and how large is it? Does it appear that having the logo makes the products sell better?

6.11. Solid waste recycling programs are generally considered to be environmentally friendly, but paper recycling by a community that incinerates its waste may not be advisable. Why?

6.12. A company is considering imposing its product design by use of LCA. The company produces liquid fabric conditioners. It is considering a variety of product reformulations and packaging alternatives, as listed in the accompanying table. The energy and waste burdens associated with each of these options are also shown. Evaluate the trade-offs and discuss the advantages and disadvantages of each option. Which option is most environmentally acceptable? Which option do you think would be the most acceptable to the public? Why? (Problem developed using data presented in Allen and Rosselot, 1997.)

Strategies for packaging improvement	Decrease in energy needs, %			Decrease in emissions, %		
	Process	Transport	Feedstock	Solid	Aqueous	Air
1. Incorporate 25% recycled paper	3	0	9	9	(+4)	4
2. Encourage 25% consumer recycling	3	2	11	11	(+4)	5
3. Triple concentrate (3×) product	55	53	56	55	54	55
4. Market product in soft pouch	3	18	67	85	(+12)	24
5. Market 3× product in soft pouch	68	73	89	95	63	75
6. Market 3× product in paper carton	53	58	94	91	40	62
7. Encourage 25% composting for strategy 6	53	58	94	92	40	62

CHAPTER

7

POLLUTION PREVENTION ECONOMICS

7.1 OVERVIEW OF ECONOMICS

Frictionless Bearings, Inc., has made the decision to investigate source reduction as a means to minimize the cost of treating its wastes and to reduce its potential long-term liability from disposal of these wastes. After careful analysis of its production processes, the company finds that several alternatives are available, not all of which are mutually exclusive: conversion to a new parts-cleaning process with internal TCE solvent recovery or another process that uses a less toxic solvent that can be discharged to the sewer; installation of new pump seals to reduce the loss of lubricants; recovery of metal shavings, for which there may be a market; and so on. The problem is that Frictionless does not know which of these options will be most effective from an economic standpoint. It cannot invest in all the possible improvements at once but must prioritize the options. To do this, the company must know what the costs and benefits of each option are. The costs used for comparison must be the real overall costs, and not just the immediate investment cost. For example, replacing the solvent recovery system may be more expensive than switching to a new solvent, but if the cleaning effectiveness using the more expensive solvent recovery system is better than use of the alternative less toxic cleaner, resulting in less reject parts, then the overall cost may actually be lower. What is needed is a way to determine the true cost of an alternative process, reagent, or process chemical over the whole life of the product. This can be accomplished using *engineering economics* coupled with *life-cycle assessment.*

As described in the previous chapter, life-cycle assessment is a process designed to evaluate the total life cycle of a product or process from cradle to grave by making detailed measurements of all aspects of a product's manufacture, use, and final disposal. This includes impacts associated with the manufacture of a product, from the mining of the raw materials that go into its production, to transportation of the raw ingredients to the factory, and through its actual production. It also includes environmental impacts associated with its use by the consumer, including power use and repairs, and actions involved with its final disposal after its useful life is completed. The assessment may not even end there. For a potentially toxic material deposited in a landfill, even in a well-designed one, there is always the potential for leakage of the landfill liner, allowing the toxic materials to enter the groundwater and necessitating eventual remediation. Thus there may be a cost associated with the product long into the future.

Life-cycle assessments will be of little value to a company unless economic analyses accompany them. A pollution prevention (P2) proposal may make great sense from an environmental or human health protection viewpoint, but if it is too expensive to implement, it will never be used. Thus the life-cycle costs associated with a P2 option should be included with the life-cycle impacts in a useful, comprehensive life-cycle assessment.

In performing a life-cycle cost analysis it is essential that proper accounting procedures be used in order to be able to correctly compare options. We have discussed life-cycle assessment procedures in some detail; now we review the fundamentals of microeconomics and its impact on the study of engineering economics. Only a brief overview of these subjects is presented here. Those interested in a more detailed discussion of microeconomics can refer to any of the standard texts on the subject, such as the classic text by Samuelson and Nordhaus (1995) or that by Parkin (1993). There are many good texts on engineering economics, including those by Grant, Ireson, and Leavenworth (1990); Newman (1988); Brown and Yanuck (1985); and Steiner (1988). Readers interested in more detail on the impacts of engineering economics on pollution prevention should read *Pollution Prevention Economics* by Aldrich (1996).

Economics involves understanding the way businesses, households, and governments behave. In particular, it is the study of the production and exchange of goods. This includes trends in prices, output, and employment. It also is the study of how people choose to use scarce or limited productive resources, such as land, labor, equipment, and technical knowledge. This vast group of subjects can be condensed to a single definition (Samuelson and Nordhaus, 1995):

> Economics is the study of how societies use scarce resources to produce valuable commodities and distribute them among different groups.

In resolving the quandary posed earlier concerning its parts cleaning process, Frictionless Bearings must ultimately make an economic decision. The company's resources in terms of available cash, personnel, and necessary equipment are limited. The goal is to make the best bearings at the lowest cost possible so that sales can reach a wide market. Installing the new equipment will cost money, create production downtime, and require the use of valuable technical expertise, but the end result may be beneficial in the long run because of lower waste disposal costs. A economic analysis of all of the ramifications is needed.

The field of economics is typically divided into two categories: *macroeconomics,* which studies the functioning of the economy as a whole at the national or international level; and *microeconomics,* which analyzes the behavior of individual components such as industries, firms, and households. Macroeconomics plays a vital role in the decision making of a company, since it controls national monetary policies, unemployment rates, interest rates, and foreign trade policies. However, individual firms have little control over these. They must include macroeconomic factors in their decision-making processes, such as estimating what future interest rates will be or what effect the labor market will have on employee salaries and consequently on the cost of producing a commodity, but they can do little to change these factors. Therefore, we will concentrate on the microeconomics of their decision making.

7.2 MICROECONOMICS

Microeconomics is the study of individual consumers, firms, and markets. How prices for goods are established and the effect of prices on the demand for products are the basic concerns of microeconomics. The interaction between price and demand can take place within the confines of a free-market environment in which supply and demand are interconnected, or in a planned economy, where factors other than demand dictate the supply.

For example, a country with a primarily free-market system, such as the United States, will let the market decide the price for a pair of sneakers, based on the cost for producing those sneakers and the demand for them. If the demand by the public is high, the sneaker manufacturer may be able to set a price well above the production cost and make a good profit. At some point, though, the price becomes too high for many consumers and demand decreases. If the demand remains high, other sneaker manufacturers will produce competing versions, which may again reduce the demand for the original sneaker style and possibly force the price down. If the demand for a product becomes so low that consumers will buy the product only at a cost that is below the production cost, the company will probably cease producing it because it would lose money to continue doing so. However, if the government decided that it was in the national interest to continue producing those sneakers in order to keep the firm in business and maintain high employment, the government might subsidize the manufacture of the product to keep it on the market. This is an example of a planned economy, where supply and demand forces are unlinked.

We will briefly investigate the market mechanisms that have been used in various locations, and then look at the effects of these on supply and demand relationships.

7.2.1 Market Mechanisms

A *market* is an arrangement by which buyers and sellers of a commodity interact to determine its price and quantity (Samuelson and Nordhaus, 1995). This includes all commodities and services, including human resources in the form of wages.

In a free market system, *what* is produced is dictated by what consumers want and will spend money for, assuming that the producer can realize a profit from producing it. This is not always the case. There is a great demand for existing drugs to treat AIDS, but manufacturers are reluctant to manufacture them because the manufacturing cost is very high and the net profit from selling the drugs would be low. They can realize a greater profit by manufacturing other drugs that are in greater demand, can be produced more cheaply, and produce a greater profit. Even though they may garner a small profit from the AIDS drug, manufacturers can earn greater profits from other drugs, and in a capitalist society such as the United States, the company is driven to maximize profits to its stockholders.

How things are produced is determined by the competition among different producers. Profits can be maximized by adopting the most efficient means of production. A producer who discovers a way to produce a commodity at lower cost than competitors will have a decided advantage, either in being able to undercut the prices of the competitors, thus acquiring a greater share of the market, or by keeping the price constant and achieving a greater net profit per unit sold. This is the basis for many pollution prevention activities. If a manufacturer can reduce production costs by recovering and reusing previously lost materials, reducing waste management costs, or switching to more efficient production techniques, its overall production costs will go down and its profits will go up.

The above discussion assumes a free-market economy. Prices are set solely by forces of supply and demand. However, governments often get involved with the market system, artificially setting either the supply or demand side of the picture. They may subsidize the production of goods or services that are not in high demand, such as the railway system in the United States, set prices for certain items to ensure adequate profits and continued production of vital materials such as milk and cotton, or maintain a stable market and stable prices by purchasing unsold materials, as has been done in the past with wheat and butter. They may also provide tax breaks or research resources to certain industries to enhance their functioning. Many of these actions are done with the best interest of the nation's economy at heart, but some are done purely for political reasons. Whatever the reason, government involvement in the economic market can change its operation significantly.

In 1776, Adam Smith wrote *The Wealth of Nations,* in which he stated that any government interference with free competition is almost certain to be injurious. Unfortunately, we seldom have perfect competition among participants in the market, and governments become involved in an attempt to level the playing field. *Imperfect competition* occurs when one firm becomes large enough that it has a monopoly in a segment of the economy. It can then dictate prices, quality of goods, and types of goods available without threat of competition from other firms. Antitrust legislation has been enacted to prevent monopolies from becoming established, since these can have a significant effect on prices and the market system.

Governments also become involved in the market system when they impose regulations for externalities. *Externalities* are spillover effects that occur when firms or people impose costs or benefits on others without those others receiving the proper payment or paying the proper costs. Externalities such as air or water pollution caused by an industry, land destruction through strip mining, and production of unsafe products have placed an unfair burden on the public for the private benefit of the industry. Government regulations are used to minimize these externalities. While the overall objective is worthy, one of their effects is to increase costs to the manufacturer and thus push up prices for the goods produced.

Another example of government involvement in the market system relates to policies for levying taxes. The tax structure in the United States is extremely complicated, and tax loopholes frequently are enacted to benefit a specific sector of business. In addition, tax abatement programs encourage industry to use certain new technologies to increase employment, to locate in certain areas, or to create new businesses. Many of these policies are laudatory and beneficial to the overall economy, but these tax inequities lead to a shift in the supply and demand drivers of the economy.

7.2.2 Supply and Demand

The demand for commodities is in a constant state of flux as businesses bring to the market new products that supplant existing products or as improved technologies allow them to lower the prices of previously expensive products. Personal computers were until recently very rare and expensive; now they are found in essentially every office and home in the country because industry found a way to produce them economically. Computer manufacturers found ways to construct ever better and cheaper computers, responding to consumer demand by supplying a multitude of choices. Consumers' desires to have the latest innovations and continual equipment upgrades now necessitate frequent purchases of new computers.

The supply and demand of a commodity are completely intertwined and serve as the basis for establishing the price for that commodity in a free-market society. The supply and demand functions are governed by empirical laws and parameters (Aldrich, 1996). The *law of demand* states that consumers are willing to buy more of a commodity at lower prices than at higher prices given constant demand pressures. Parameters affecting this law are consumer income, consumer tastes and preferences, relative price of other goods, number of consumers, and expectations for changes in future market conditions. The supply of goods is governed by the *law of supply,* which states that sellers are willing to provide more of a given good at higher prices than at lower prices, given constant supply parameters. Supply parameters include the number of producers, state of technology, size of capital stock, price of inputs, and expectations for future changes to the market.

The assumptions set forth in the laws of supply and demand can be depicted graphically (see Figure 7.1). Table 7.1 shows a hypothetical demand schedule and supply schedule for the purchase of 25-inch television sets, based on the selling price of the set. The demand schedule indicates the number of sets that could be sold for a particular price, while the supply schedule shows the number of units that the manufac-

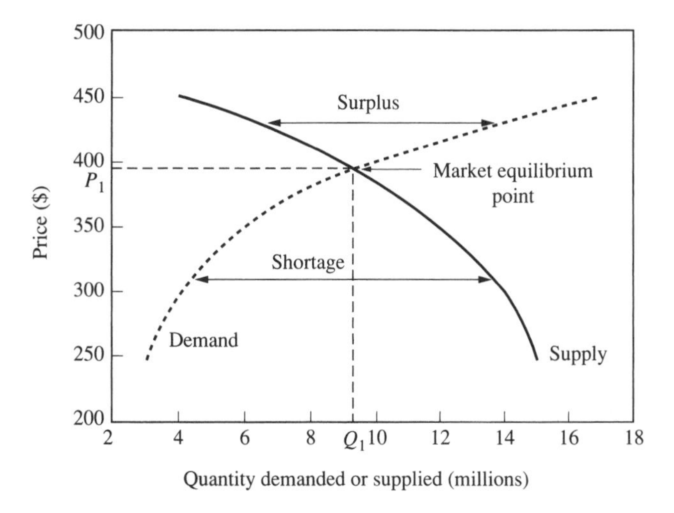

FIGURE 7.1
Market equilibrium for the sale of television sets.

TABLE 7.1
Supply and demand schedules for television sets

Price, $/unit	Quantity buyers willing to purchase (millions/yr)	Quantity sellers willing to supply (millions/yr)	State of market	Pressure on price
250	15	3	Shortage	Upward
300	14	4	Shortage	Upward
350	12	6	Shortage	Upward
390	9.7	9.7	Equilibrium	Neutral
400	9	10	Surplus	Downward
450	4	17	Surplus	Downward

turer would be willing to make and sell at each price. These can be plotted (Figure 7.1) to show trends in potential purchases and sales based on price. The shape of these curves can vary widely. The intersection of these two curves is the *market equilibrium point,* or the point where the supply of goods and the demand for those goods match. Q_1 represents the quantity of goods that would be produced at the equilibrium point, and P_1 is the price that would be assigned to the product at that level of sales. Market pressures should result in a price and supply that match the market equilibrium point. A price higher than this will result in a shift to the left on the demand curve and a reduction in the number of units purchased. This results in the creation of a surplus of unpurchased units. A reduction in price would increase the demand and result in a shortage of units. In a smoothly functioning economy, the price would be adjusted until it produced an equal supply and demand. At the equilibrium price, there are no shortages or surpluses. The economy does not always operate at the equilibrium point, though. At times surpluses or shortages may result from short-term differences in the supply and demand for a commodity. In the long run, though, equilibrium will usually be reached, provided that the demand can be met by the suppliers at a cost they are willing to charge. There may be a minimum price below which manufacturers would be reluctant to produce a product, even if consumers would purchase it at that price.

The preceding analysis assumes that a perfectly competitive market exists. This is rarely the case, however. Often buyers or sellers affect the market price. One producer or a small group of producers who dominate a market (e.g., a computer software producer of a specialized piece of software) can often set the price as they desire because there are not enough competitors to force the price lower. Consumers must either pay the high price or not purchase the product. Conversely, a market dominated by a small number of consumers (e.g., purchase of airplane parts by the aircraft manufacturing industries) can have the price dictated by what that consumer is willing to pay. If the manufacturer is not willing to produce the parts for that price, it may be out of business because there is little other demand for the product that the manufacturer can turn to for sales. In general, though, the basic principles of the supply and demand process hold for most markets.

The supply vs. demand curve shown in Figure 7.1 indicates that the demand curve slopes downward. The quantity demanded decreases as the price rises, and the quantity demanded increases as the price falls. The slope of the curve indicates how responsive the market is to price changes. A steeply sloping curve indicates that small changes in price can have a significant effect on demand, whereas a shallow slope indicates that demand will change little over moderate price increases or decreases. This is a measure of the *price elasticity of demand* for the product. The price elasticity of demand, E_D, can be calculated as

$$E_D = \frac{\text{Percent increase in demand}}{\text{Percent decrease in price}}$$

If E_D is greater than 1, there is elastic demand. If E_D is less than 1, demand is considered to be inelastic, as demand does not increase as fast as the price is lowered. The same analysis can be used to evaluate the elasticity of supply.

Price elasticity is critical to evaluation of the feasibility of potential pollution prevention activities. If there is great price elasticity for a given product, even small increases in the price of the product to pay for the pollution prevention initiative may reduce the market for that product and lead to financial difficulties in its production and sale. This is particularly true if the market for the product is highly competitive and differences between various brands are small. However, if the P2 activity leads to even a small reduction in the price because of cost savings associated with the P2 initiative, there could be a substantial increase in the market demand for the product. Thus it is critical that detailed economic analyses be performed for any product before undertaking a costly pollution prevention activity.

Another factor that could come into play that relates to this discussion is the possibility that the lack of pollution prevention by an industry could also result in an increase in price through pollution taxes or penalties such as fines for discharging contaminants either into the environment or into a municipal sewer. Pollution taxes—such as industrial user fees for discharge into a sewer system, fees to dispose of hazardous wastes in an approved disposal facility, or a tax to discharge wastes into the air or water—are common in most areas. Typically, these costs are rising rapidly. Increases in these costs could also upset the market balance for a manufacturer by forcing it to increase the cost of the product to offset the higher cost of manufacturing due to the pollution penalty. Thus increased prices could be brought about by *not* undertaking a pollution prevention program, as well as by instituting one.

7.2.3 Marginal Cost and Marginal Benefit

The purpose of a manufacturer is to produce goods. The firm must buy raw ingredients and other inputs such as water and energy, use these to produce a product, and then sell the product. The objective is to produce *and sell* the maximum number of products possible at a price that still makes an adequate profit. The cost of producing an item can be broken down into fixed costs and variable costs. Fixed costs, or overhead, are costs for buildings and equipment, interest payments on debt, salaries of permanent employees, and the like. These must be paid even if the firm produces none of the product, and they will not change appreciably if the amount of production increases or decreases. Variable costs vary with the level of output and include such items as materials used to produce the item, the salaries of the production staff, and utilities. The cost of producing an item often falls as the demand increases. Producing one unit of an item may be very expensive because of development and tooling costs, which would have to be included in the cost; producing a thousand of those items would be much cheaper because the initial costs could be spread out over all of the thousand items. Each additional unit produced would cost marginally less. This is not always the case, however. If the plant is working at full capacity, the production of additional units may require the installation of expensive new equipment, resulting in an increase in cost until the demand for the new equipment makes it cost effective. The *marginal cost* is defined as the increase in total cost required to produce one extra unit of output or the reduction in cost to produce one fewer unit of output.

The marginal cost curve can take many shapes. Figure 7.2 depicts the total cost and the marginal cost for one product. The total cost increases with increases in the amount of production. Marginal costs initially decrease as production increases, until a point is reached where a new piece of production equipment or a sizable increase in the work force is needed, or where the cost of the input material increases because of the increased demand. These can result in an increasing marginal cost.

The benefits derived from a particular production decision can be analyzed in a similar fashion. Often the initial benefits derived from a new pollution prevention program can be sizable. As more P2 improvements are made, though, the law of diminishing returns comes into play, and additional improvements may result in only marginally increased benefits. For example, assume we plan to institute a solvent recovery program. Recovery of the first unit of solvent is relatively easy, but as we attempt to increase the recovery efficiency, and more complex recovery systems are needed, the costs escalate rapidly (see Figure 7.3). It may cost X dollars to recover the first 90 percent of the solvent exiting a process. To increase the recovery to 95 percent, only a 5

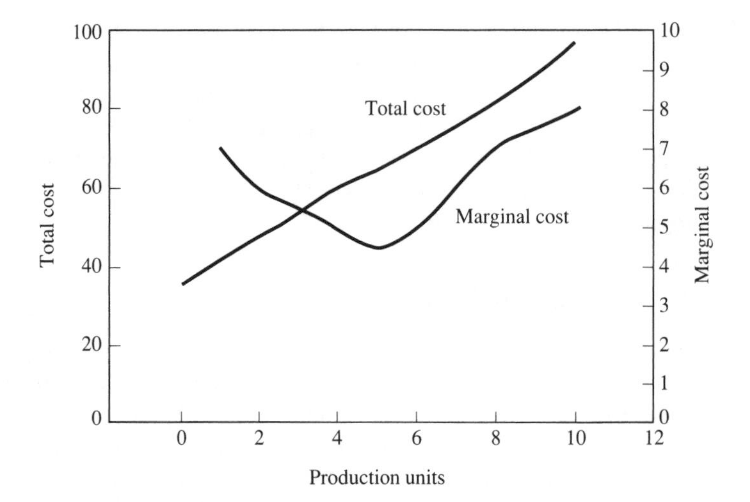

FIGURE 7.2
Total and marginal costs of production.

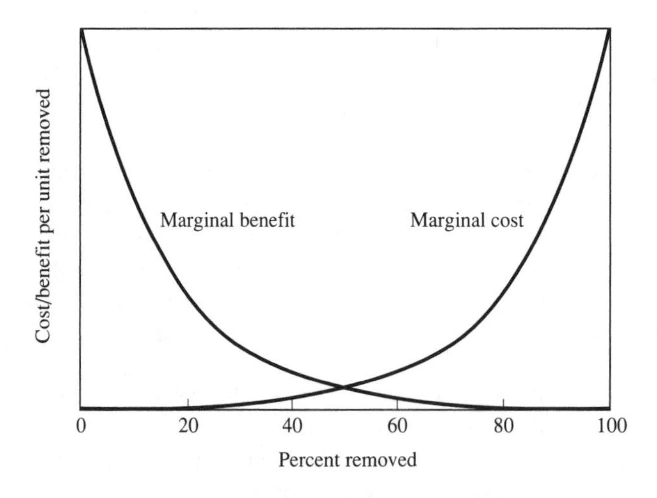

FIGURE 7.3
Marginal cost and marginal benefit associated with increasing percentage removal of a contaminant from a process waste stream.

percent increase in efficiency, may cost another *X* dollars. To increase the recovery to 99 percent may cost 10 times as much as to remove the first 90 percent. The question then becomes, Is it worth the additional cost to get the marginally improved benefit of the higher recovery? The decision must be made using a combination of economic, technical, and legal factors. If the intent is primarily to reduce solvent purchase costs by recycling used solvent and to minimize solvent disposal costs, there will be some break-even point where recovery costs and cost savings from reuse of the solvent are equal. The intersection point between the marginal cost and marginal benefit curves represents the optimal recovery that can be obtained from an economic benefit standpoint. Attempting to increase the recovery beyond this makes no economic sense.

However, because of regulatory requirements or the desire to be rid of the solvent disposal problem, the industry may decide that the additional cost is acceptable.

7.2.4 Market Externalities

Markets do not always operate in a totally open fashion. One problem with a free, open market is that it does not always provide for the common good, only for the good of the immediate producers and consumers or the users of that market. Often this results in shortsighted decisions that maximize immediate profits or benefits at the expense of long-term needs. Burning our petroleum resources to produce energy is currently an inexpensive mode of energy production. But eventually those resources will be gone, and the petroleum will be unavailable for other uses such as manufacture of petrochemicals or pharmaceuticals. To prevent this from occurring, governments could impose limits on the burning of petroleum materials or place user taxes on petroleum to make it uneconomical to burn. Thus government policy can be established to circumvent the open market system in order to provide for the long-term good of the general public, rather than the short-term good of a few. Another example of this is the use of taxes or fines for the discharge of pollutants into the environment. Governments can set the pollution fees at a level that will curb discharges into the environment and thus provide for the good of the general public. This is currently being done with automobiles; special taxes are imposed on large, gas-guzzling automobiles to make their purchase more economically undesirable. Industrial sewer user fees are another example. These costs must be added to the actual manufacturing costs when the industry decides the price to charge for a product. The effect is to disrupt the open market system, because these are charges which have nothing to do with the actual manufacture of the product.

As mentioned, these external effects or impacts created by the market are termed *externalities*. An externality or spillover effect occurs when production or consumption inflicts involuntary costs or benefits on others; that is, costs or benefits are imposed on others but are not paid for by those who impose them or receive them (Samuelson and Nordhaus, 1995). If someone smokes in a confined space, others in the vicinity must involuntarily breathe the smoke. That is a negative externality. The smoker is imposing discomfort on others without paying for the damage done. Likewise, an industry that discharges toxic waste into the air is harmfully impacting others downwind of the plant. In a true free market, this would be acceptable. The company would determine how much pollution control was required based on the marginal damage done to the firm by the pollution versus the marginal cost of abatement. Only the amount of treatment required to be able to efficiently produce and sell the product would be undertaken. But no modern society will allow this. The industry is limited in the amounts of hazardous contaminants it can safely discharge, and fines are levied for discharges beyond these amounts. Thus to overcome the negative effects of these externalities, government imposes treatment requirements and penalties that force the industry to install expensive pollution control equipment, the cost of which is passed on to the consumer through the price of the product. Correcting for externalities can, therefore, have an effect on the market system. From the economist's standpoint, the result is an inefficient economic outcome.

The question, then, is, How is the level of required treatment determined? At one extreme is the laissez-faire approach of letting the industry decide, based on its needs. The other extreme is to allow no pollution. This is usually not achievable, however, and often is not needed to protect the public and the environment. The environment has some capacity to safely assimilate wastes, and people and other biota can safely tolerate some toxins in the environment. What is needed is an analysis of the impacts of various levels of the discharges and the costs necessary to reach acceptable levels.

The most economically efficient system would have *marginal social costs* of abatement equal to *marginal social damage* from the pollution. This occurs when the benefits to health and property of reducing pollution by one unit just equal the costs of that reduction. This can be accomplished by conducting a cost-benefit analysis on the system. Environmental standards would be set by balancing the costs of abatement against the benefits of pollution reduction. Because of the extremely high cost of achieving "zero discharge," this equilibrium point would probably never be at the no-pollution level (see Figure 7.3). Trying to achieve zero pollution would be prohibitively expensive, and often unnecessary, because removing the last tiny amounts of pollution would have little positive impact on health or environmental protection. Treating beyond the point where the marginal benefits exceed the marginal costs is a waste of money, according to economic theory.

Establishing the marginal costs of emission control is relatively straightforward. The costs for control equipment, labor, and operation and maintenance can readily be obtained. Determining the marginal benefits derived from abating the pollution is much more difficult. A dollar figure must be placed on such things as reduced death rates, improved health, a more diverse wildlife, and better aesthetics. In some cases it may mean determining long-term impacts, such as minimization of global warming or reduction in cancers due to the thinning of the stratospheric ozone layer. An entire field of risk analysis study has developed to determine these costs and benefits, and the results are now becoming widely accepted.

There are cases, though, where cost-benefit analysis has led to zero-discharge requirements being imposed. Some materials that are extremely hazardous, such as benzene, PCBs, and some pesticides, have been strictly regulated or banned outright because they are deemed to be too dangerous in the environment at any cost. The allowable discharge limits for dioxins and furans, pollutants produced during combustion, are essentially set at zero. In these cases, it has been determined that marginal benefits of lower discharges will always be greater than the marginal costs.

In theory, a competitive economy should root out pollution over time, as individual businesses strive to increase profits. Pollution is a sign of imperfect efficiency: raw materials and energy that have been imperfectly transformed into products or services. Maximizing profits results from maximizing efficiency, thereby reducing or eliminating wastes and pollution (Shireman and Cobb, 1996).

In most cases, a direct correlation can be made between the profitability of a company and the pollution emissions by that company. In a study of 231 companies from across all industry, it was found that companies with high emissions per revenue dollar were likely to have higher penalties per revenue dollar, more spills per revenue dollar, more NPL sites per revenue dollar, and more RCRA corrective actions per revenue dollar (Bhat, 1996). These all detract heavily from the profit margin, resulting in a negative relationship between emissions and profit margins. High-emissions companies generally have lower profit margins. Thus companies should strive to minimize their pollution emissions as much as possible.

7.2.5 Control Measures

Externalities from business operations can be harmful to some parts of society or the environment. For example, air, water, and land may be polluted; worker safety may be threatened; and the general public may be endangered by unsafe products. The government uses several tactics to control harmful externalities. These include regulations and financial incentives.

REGULATIONS. The government's first line of control is usually a direct form of control through the use of regulations. Emission standards are set for each pollutant or for each pollution source, based on the perceived level that will provide environmental and health safety. Industries are required to meet these standards. If they do not, substantial fines can be levied or those responsible can even be imprisoned. These regulations and fines are described in the pertinent laws (the Clean Air Act, the Clean Water Act, RCRA, SARA, etc.) and their applicable regulations. The result has been a substantial decrease in the number of violations and a general improvement in environmental quality.

Direct controls, in theory, appear to be the best way to reduce pollution because the pollution limits are established through science, and compliance is mandatory. However, the system rarely functions in this ideal fashion. Requirements are not always set at the ideal level because of political lobbying by industry on one side and environmental and health advocates on the other. While complete control of pollution may be desirable from one viewpoint, government must seek to maximize the benefit to all society, by maintaining high employment and fostering a strong economy, for example. Thus the levels must be set at a point where the benefits achieved through pollution control do not outweigh the damage to the economy done by loss of needed businesses. Society wants a vibrant economy as well as a clean environment.

A second problem with the direct approach to control is that enforcement often is not applied equitably across all industry, and in many cases the penalties are not sufficiently high to deter industries from violating the regulations. Only a small number of violations that occur are usually discovered, and the resulting penalties frequently are not severe enough to deter others from repeating the violation. There is a strong incentive to push the edge of the regulations, rather than to readily comply.

Third, it is very difficult to establish what the discharge limits should be. The U.S. EPA spends millions of dollars each year trying to determine the safe levels of many chemicals in the environment. Getting good scientific data to support the regulations in a timely and cost-effective fashion is nearly impossible. Opposition groups, whether industrial or environmental, will question the outcomes of studies and may go to court to try to overturn regulations they do not like. The result is a long, tedious, and expensive process for establishing what the regulations should be.

Fourth, the regulatory process has placed the U.S. EPA in a quandary, for its workers must wear many hats. They are the establishers of environmental regulations, often are the scientists who develop the data on which the regulations are based, are the enforcers of those regulations, and also serve as a repository of environmental technology for the benefit of industry and the general public. Thus they must be both the enforcer of environmental regulations and the provider of assistance to industry on how to best meet environmental quality standards. These are two totally different roles that often conflict with one another. Industry should be able to turn to the EPA for help with environmental problems, but instead they often see EPA as an adversary rather than as a source of technical assistance. This is not the case in many European countries where the two roles of enforcement and environmental assistance are carried out by two separate agencies.

Finally, regulations do not adequately address the key issue, which is reducing overall pollution in a fair and equitable way. Most regulations are set on an across-the-board basis. A Ford Motor Company plant producing tens of thousands of pounds of pollution per year must meet the same discharge requirements as a small local metal plating shop. After acceptable waste treatment to meet discharge standards, which are usually based either on an effluent concentration basis or on a percentage removal basis and not on a total mass discharged basis, the Ford plant may still be discharging many more pollutants than the small metal plater does before he treats, but the local plater must still reduce his waste discharges by the same amount as the Ford plant. From an economic or a fairness standpoint, this makes no sense. The expense imposed on the metal plater is of little value to the overall goal of environmental cleanup, because what it is discharging is minimal in comparison with what the large manufacturer can legally release. This is a clear violation of the marginal cost versus marginal benefit approach. The result is that the general public pays much more to achieve its environmental goals by requiring many small, expensive treatment systems rather than requiring only a few major industries that are the primary causes of the environmental problems to properly clean up their effluents.

To avoid the problems with the direct approach to control based on regulations, some have advocated the use of financial incentives to control the pollution discharges. These can take several forms, some of which may be voluntary.

EMISSIONS FEES. Emissions fees or taxes are an alternative to setting specific emission limits. Under this system, firms pay a fee for each unit of pollutant emitted. In the ideal case, the pollution fee is set to be equal to the environmental damage done by the pollutant. Thus the free-market system—not reliance on mandatory regulations—is the basis for pollution abatement. An industry could continue to pollute, but it would have to pay a hefty fee, the amount of which would have to be passed on to consumers in the form of higher prices for the product. If the fee was high enough that the industry could no longer pollute and still be price competitive, there would be a strong incentive for the industry to install pollution control equipment to avoid the fee. This means that the pollution fees must exceed the cost of treatment. In theory, if a firm decided not to treat and to pay the emissions tax, those funds would be available to pay for the environmental or health damage done by the pollution.

This approach to pollution control has many inadequacies. Establishing a real cost-benefit basis for the emissions fees is very difficult and open to error. Extensive and expensive monitoring would be required to determine the amount of fees due from each industry. This would be particularly difficult for fugitive emissions. High-profit industries would be reticent to go to the trouble to control pollution because they could afford to pay the fees, whereas economically marginal firms would be forced to comply in order to avoid paying them. This could lead to inequities in the market. The biggest problem deals with the concept of choice in pollution control. It is usually much easier and wiser to control pollution at the source than to try to remediate damage caused by pollution. Even though funds would be available to pay for this cleanup, why allow the damage to be done in the first place if it is unnecessary? It may be possible, and in some cases even desirable, to apply the emissions fees concept to industrial wastewaters that are discharged into municipal sewers, as is currently the case in most localities. Some contaminants can be treated more cost-effectively in large volume in one location than at many smaller industrial sites where the equipment and the personnel may be inadequate to do a thorough job. However, this is not the case with air pollutants. If an industry decides to pay the emission fee, for whatever reason, and not treat its emissions, damage to air quality will result. The fees collected will be of little use except to pay for the repair of damaged sectors of the environment or to pay for the medical costs of the affected population.

MARKETABLE PERMITS. Marketable permits offer another approach to increasing flexibility in the pollution control process. This system is already being used for control of certain air and water pollutants. A system of tradeable permits establishes ownership of the right to pollute the air up to a certain limit (Shireman and Cobb, 1996). Permits are issued to allow an industrial facility to emit contaminants up to a certain specified level. The allowable level is established to maintain the desired air quality. Any additional pollutant removal by the industry is desirable, but not necessary to meet the air quality standards. In most cases, the industry would not remove more than is necessary, since this will cost more money. However, if an industry's pollution control facility is very efficient and does release lower than the allowable amount of pollutants, the industry has the option, under the marketable permit system, of using its excess pollution permit capacity (pollution credits) at another pollution source in the company. Alternatively, it may sell its excess capacity credits to any other firm that cannot or does not want to make the investment to reach its permit limits. These credits can be sold on the open market to the highest bidder. Polluters needing additional credits could pay up to the point where the cost of the permit equaled the marginal cost for the additional pollution control required.

The marketable permit system accomplishes the overall objective of maintaining environmental quality while allowing the free-market system to establish who will be responsible for cleanup. Firms with new, more efficient facilities and pollution control equipment can recover some of the cost of their investment by selling excess pollution credits, while older plants that have a hard time meeting the new discharge regulations or have high marginal treatment costs can purchase credits and continue operating without having to make a major, and possibly difficult upgrade to their aging facilities. Both old and new plants can continue functioning, and overall environmental quality is preserved. This also provides an incentive to companies to try to find ways to reduce their pollution, either to gain credits that can be sold or to minimize the amount of credits needed to be purchased. The main objection to this system is the cost to the regulators to monitor the emissions from each source and to keep track of the marketable permits so that they know how much is allowable. Also, a regulatory process is needed to ensure that allowance trading does not result in pollution of previously pristine environments or a heavy buildup of polluting sources in a localized area.

In studies conducted on three waterways—an 86-mile stretch of the Delaware estuary, the lower Fox River in Wisconsin, and the upper Hudson River in New York—the customary mandated approach using across-the-board regulations was found to cost from 50 percent to 200 percent more than the least-cost approach using tradeable permits (Shireman and Cobb, 1996). This system, if properly managed, could be a major boon to both industry and the environment.

Marketable credits need not be limited to industry. The system has also been proposed for countries as a way to deal with expensive reductions in greenhouse gas emissions (Christen, 1997). Most nations are now realizing that their initial commitments to reducing greenhouse gas emissions, made at the 1992 U.N. Framework Convention on Climate Change in Rio de Janeiro, were too ambitious. In the United States, for example, emissions have actually increased by 3 percent since then. Because of the Kyoto Summit in 1997, a proposal is now under discussion that would allow nations that fail to achieve target emissions reductions to borrow against their future emissions rights by purchasing credits from nations that are ahead of schedule. A country that stays under its available permit levels could bank its excess credits for future use or could sell them to countries that exceed the limits. In many cases, the emissions credit sellers would be poor, developing countries that do not have greenhouse gas–emitting infrastructure in place, while credit purchasers would mainly be highly industrialized countries. The result would be a lower cost to developing countries to achieve significant reductions in greenhouse gas emissions. There are problems with this proposal, such as how to assess penalties to those countries that opt to borrow against future credits by delaying action now and then do not meet required future limits, but the discussions are interesting anyway.

A pollution control program based totally on the free-market system will probably never be successful, because no free-market system is truly ideal. There will always be companies that deviate from it. What may be best is a combination of regulations to achieve acceptable environmental quality and a financially based incentive program to enhance the environmental quality.

EMISSION REDUCTION INCENTIVES. The provisions of the Pollution Prevention Act are essentially all voluntary. They are aimed at reducing industrial pollution, but they do not set mandatory limits on pollution. As long as an industry is meeting its NPDES, CWA, or CAA requirements, there is no need for it to do more. Indeed, the free-market system discourages doing more because it will often add to the cost of the final product. Even if a long-term cost savings can be achieved through institution of P2 activities, the industry may be reluctant to undertake it because it has more pressing needs for its resources.

Several localities and states have begun pollution prevention incentive programs to encourage businesses to go beyond what is absolutely necessary in terms of pollution control. These take many forms, including tax abatements, low-interest loans for P2 projects, technical support, and favorable publicity through awards programs. The result of these programs often is a reduction in the cost of the P2 activity or an increase in the benefits derived from the activity. In either case, the impact must be factored into the cost-benefit equation.

The pollution prevention incentives program option will be discussed in more detail in Chapter 13.

7.3 ENGINEERING ECONOMICS

The previous discussion demonstrated that the optimum operating point for a business is where the marginal cost is equal to the marginal benefit. To properly determine this point, though, a thorough analysis of the true costs and benefits, which are often difficult to determine, is required. Engineers are well versed in estimating the costs of installing new processes or using new chemicals, and much research has gone into quantifying the economic impacts of pollution. From the pollution prevention perspective, cost-benefit decisions must be made that will extend over many years. New equipment purchases that will assist in minimizing pollution emissions may mean a sizable initial outlay of capital, but the reduction in pollution control costs and/or the savings in purchase of process chemicals may soon outdistance those initial costs. It is important that the business properly assess these short- and long-term costs and benefits.

Comparing present-day money with the value of that money in the future is difficult because the present-day money, if not spent on the P2 project, could be invested and grow in value. Conversely, savings in the future must be tempered by inflationary changes in the value of money. Thus it is essential that a common basis be used for comparing immediate cash outlays and future returns. To do this, economists convert all costs and benefits to *present worth,* the worth of future expenses and revenues in terms of today's value. These future values are corrected for inflationary increases by applying an appropriate interest rate to them. The rate selected to reduce, or discount, the future value of money to the present value is termed the *discount rate*. It is the annualized rate at which the value of something is reduced to bring it to its present-day value.

7.3.1 Discount Rate

The discount rate is the interest rate that is used to relate the future value of money to its present value. It is the rate of interest or return that a business or person can earn on the best alternative use of the money at the same level of risk. The discount rate is not necessarily the same as the inflation rate. The current inflation rate may be 4 percent. You have the option of investing your money in bank certificates of deposit at 6 percent, in bonds at 7 percent, or in stocks, which have an average return of 10 percent. Thus the discount rate you use can vary and may be different from the inflation rate. It should be noted that the inflation rate usually does significantly affect what the discount rate will be. No one can predict what the inflationary interest rate will be in the future. We have recently seen periods when the inflation rate was as high as 18 percent per year, and other times when it was as low as 2.5 percent per year.

Selecting an appropriate discount rate is critical to conducting a useful economic analysis. If the rate used for the analysis is higher than what actually occurs, future costs and benefits (or revenues) will be reduced to a lower present worth than is correct, thus underestimating the positive value of the project. The reverse is also true. For example, consider the following simple case. Assume that a P2 project requires an initial investment of $10,000. The company determines that investment must pay for itself in future savings within one year. It estimates that the annual marginal benefit from the project after the first year will be $10,500. If it uses a discount rate of 8 percent, the present value of that $10,500 is only $9722 ($10,800/1.08 = $9722), not enough to justify the installation. However, if the company had selected a discount rate of 4 percent, the present value of that projected future marginal benefit would be $10,080 and the project would be worthwhile. The company can't accurately predict what the inflation rate will be over the next year, and either scenario is possible. If the company selected the 8 percent discount rate to make its decision and the inflationary rate was only 4 percent, it would have passed up the savings that would have been realized from going ahead with the project. If the opposite occurred and the company carried out its projections using a 4 percent discount rate while the actual inflation rate was 8 percent, the company would have lost money. Thus it is important to select with great care the discount rate to be used.

The "appropriate discount rate" varies, depending on the economic climate and the amount of risk the company will accept. A conservative firm will tend to use a higher discount rate to err on the conservative side. If the actual inflation rate is lower than the discount rate used, then profits will be greater than predicted. As stated in the definition of "discount rate," the rate selected should be based on the assumption that the investment is at "the same level of risk" as the expected return from the project. A discount rate based on a possible high rate of return from investing in risky junk bonds would not be appropriate if the company usually invests its excess cash in stable corporate bonds that are safer and consequently pay a lower rate of return.

7.3.2 Present Worth

The preceding example assumed that the future marginal costs needed to be discounted by only one year. Since the discount rate was expressed as a certain percentage per year, it could be determined easily by just reducing its value by that percentage. However, what if the company decided that it could tolerate a longer payback period, provided that marginal profits eventually exceeded the initial capital outlay? Could we just multiply the marginal savings by the number of years selected and compare that with the present capital cost? No—in each succeeding year the present value of those revenues is reduced further because of inflation. Each reduction is compounded by the lower value from the preceding year. In the example above, at an 8 percent discount rate, the $10,500 income is worth only $9722 after one year, and this value is reduced by another 8 percent the second year to a present worth value of only $9002. This compounding effect can be significant if a high discount rate is selected.

The present worth of an actual cost or revenue in the future can be determined using the following mathematical relationship:

$$\mathrm{PW} = \frac{F}{(1 + i)^n}$$

where PW = present worth (value) of the revenue or expense
F = actual value of the revenue or expense in the future
i = interest or discount rate being used
n = number of years under consideration

Discounting is essentially the reciprocal of compounding.

Using the previous example, the present value of \$10,500 received two years in the future at an 8 percent discount rate is

$$\mathrm{PW} = \frac{\$10{,}500}{(1 + 0.08)^2} = \$9002$$

If the firm were to receive its benefit of \$10,500 after two years, it would need to value that at a present worth of only \$9002. In essence, if the company had banked \$9002 at 8 percent interest, it would be worth \$10,500 in two years. Thus to spend more than \$9002 up front in capital costs would not make sense if the firm needed to recover that cost in two years and the anticipated revenues would only be \$10,500.

Annual costs, such as operating and maintenance costs, are handled in a slightly different way when determining their present worth. In most cases, it is assumed that all future operating costs will be paid on a uniform basis. Unfortunately, this does not account for inflationary increases in annual operating costs. The present worth of a uniform series of annual payments can be determined using

$$\mathrm{PW} = A \times \mathrm{PWF} = \frac{A \times [(1 + i) - 1]}{i(1 + i)^n}$$

where A = annual cost
PWF = present worth factor

The discounting process is particularly important in determining the potential value of a P2 project because it facilitates the translation of future values to present values. This is especially true for conducting a life-cycle assessment on a product, which will be described later in this chapter.

Example 7.1. A company faces a potential liability 20 years in the future of an estimated \$20 million due to improper disposal of hazardous wastes that may leach into the underlying groundwater. It would like to invest funds now to ensure that it has the needed funds in the future if the contamination does occur. What is the present worth of this \$20 million; that is, what is the amount that should be invested now to ensure that \$20 million will be available in 20 years? Assume that the company can get a 10 percent return on their investment.

Solution. Determine the present worth of \$20 million 20 years from now, using a discount rate of 10 percent.

$$PW = \frac{F}{(1 + i)^n} = \frac{\$20{,}000{,}000}{(1 + 0.10)^{20}} = \$2{,}972{,}873$$

Thus the company needs to invest only $2.97 million now to ensure that it will have the potentially necessary $20 million in 20 years. Alternatively, the company can examine the cost of cleaning up the site now. If the current cleanup cost is less than $2.97 million, and if the firm is confident that a groundwater pollution problem is likely to arise, it may be better off cleaning up now.

The present worth can also be determined using present worth tables, which compile multiplication factors that are based on a range of interest rates and periods of time. One set of data is shown in Appendix E. The present worth can be determined simply by selecting the appropriate discount rate and number of years of interest and finding the multiplier factor from the table. This multiplier is multiplied against the future value to get the present worth.

Example 7.2. Use the present worth table in Appendix E to find the present worth of the investment described in Problem 7.1.

Solution. From the present worth table, for a discount rate of 10 percent and a discount period of 20 years, the appropriate multiplier is 0.14864. Therefore,

$$PW = 0.14864 \times (\$20{,}000{,}000) = \$2{,}972{,}800$$

This is the same value calculated in Problem 7.1 using the present worth equation.

7.3.3 Comparing Investment Alternatives

Progressive companies are constantly facing investment dilemmas. Which of several possible new product lines should be developed? Should an old unreliable piece of equipment be replaced? Should additional personnel be hired in the expectation of increased sales of a product? Each decision is resolved by determining the marginal costs associated with the option and comparing it with the marginal benefits. Several options may have a favorable benefit-cost ratio, but the company may be limited in the number of new ventures it can begin, so some initiatives that make economic sense may still not be undertaken. A project with a net rate of return of 10 percent will be selected over one with a projected 5 percent return, even though the latter may still be a lucrative project.

The same holds true when comparing several pollution prevention options or when comparing a P2 project with a new product line. The P2 project may make sound economic sense, but if its rate of return is lower than that of some other proposed project, the P2 project may not be funded.

Example 7.3. A company conducts a life-cycle assessment for a process that is producing a chemical with a recovery of 70 percent. The remaining 30 percent goes to the sewer. During the LCA, the company finds that the installation of a crystallization step will increase product recovery to 85 percent. The marginal increase in profits from this change is estimated to be $90,000 per year. The crystallizer unit will cost $200,000 to purchase and install, and it will cost $30,000 annually to operate. The company uses an 8 percent discount rate. What will be the company's increase in profits in 5 years if they install and use the crystallizer?

Solution. The present value of the crystallizer and five years of operation can be calculated as follows:

1. Present worth of the crystallizer = $200,000.
2. Present worth of the annual operating costs:

$$\text{PW} = A \times \text{PWF}$$

$$\text{PWF} = \frac{[(1 + i)^n - 1]}{i(1 + i)^n} = \frac{[(1.0 + 0.08)^5 - 1]}{(0.08)(1 + 0.08)^5} = 3.99$$

$$\text{PW} = (\$30{,}000/\text{yr})(3.990) = \$119{,}698$$

3. Total five-year present worth of the crystallizer:

$$\text{PW}_{\text{total}} = \$200{,}000 + \$119{,}698 = \$319{,}698$$

4. The present worth of the increased profits can also be determined:

$$\text{PW} = A \times \text{PWF} = (\$90{,}000/\text{yr})(3.990) = \$359{,}100$$

On a present worth basis, adding the crystallizer will produce a net profit to the company over five years of $359,100 − $319,698 = $39,402. Thus it is profitable to install the crystallizer under the given conditions. However, if the payback period had only been four years, the present worth of the equipment would have been $299,585 and the present worth of the increased profits would have been only $265,360, indicating a loss on investment. Selecting the proper payback period is critical.

Pollution prevention projects may not compete effectively in the capital budgeting process because the conventional cost accounting systems were not specifically designed to evaluate such projects. Conventional cost accounting systems are used to provide information to management, investors, regulators, and other external entities on the financial performance of a business, as well as providing management with information needed to make decisions. Traditional cost categories include direct materials, direct labor, and overhead. Many indirect costs may be hidden in the overhead category (including such items as waste management, regulatory compliance monitoring and reporting, penalties/fines, right-to-know and safety training, personal injury, and insurance) and traditional financial analysts may underestimate the potential savings from a pollution prevention project. Also, conventional cost accounting might not evaluate the project over a long enough time period for the benefits of the pollution prevention project to appear (Ohio EPA, 1995).

There are several ways of evaluating the return on project alternatives, a measure of profitability, to determine which will be the best investment. Among these are the payback period, the benefit-cost ratio, the internal rate of return, and the net present value (Aldrich, 1996).

PAYBACK PERIOD. The payback period approach is used by many companies. The payback period is the length of time it takes for the revenues from a project to equal the initial investment costs. The shorter the payback period, the more attractive the investment. The time it takes for the present value revenues to payback the initial cost is the payback period. The projects with the shortest payback period are selected. Commonly, payback periods of only a few years are necessary for a project to be considered viable. Payback period analysis has two drawbacks: it ignores the time value of money, and it ignores cash flows that occur after the initial investment has been recouped and does not show costs and savings past the point where the project has paid for itself (Ohio EPA, 1995). This approach is often detrimental to pollution prevention projects because they often have long payback periods. They may be expensive up front, but could save large amounts of money in the more distant future from liability expenses for environmental damage done and not detected until later. Payback analysis provides a useful preliminary assessment of a project's attractiveness. However, this initial assessment should be verified by a more detailed analysis.

BENEFIT-COST RATIO. The benefit-cost ratio approach is similar to the payback period, except that it accounts for the time value of money. All costs and revenues are brought back to present value. The benefit-cost ratio represents the ratio of the present value of the project's benefits divided by the present value of the project's costs:

$$\text{Benefit-cost ratio} = \frac{\text{Present value of benefits}}{\text{Present value of costs}}$$

If the resulting ratio is greater than 1.0, the benefits outweigh the costs over the period of time selected, and the project is acceptable. The project still may not be undertaken, though, if another project has a higher benefit-cost ratio, indicating that it would be a better investment of limited resources.

INTERNAL RATE OF RETURN. The internal rate of return (IRR) method also uses the net present value of benefits and costs, but it allows the discount rate to be the variable. The discount rate that makes the present value of the costs and benefits equal is the internal rate of return (sometimes referred to as the return on investment). It represents the actual rate of return of the investment over the life of the project. Projects with the highest rate of return would be selected for implementation, usually provided that they are above some predetermined minimum. It allows the company to compare the project IRR with the company's self-determined discount rate. The discount rate used should be the rate of interest or return that a business can earn on the best alternative use of the money at the same level of risk. Note that the discount rate and the inflation rate are not the same, although the discount rate usually incorporates the predicted inflation rate (Ohio EPA, 1995). If the IRR is greater than the discount rate, the project will generally be accepted; if the IRR is less than the discount rate, the project will be rejected. This approach is often preferable to the benefit-cost ratio, but it requires a large amount of data and predictions of future trends over several time periods.

NET PRESENT VALUE. The net present value (NPV) technique also compares projects based on differences between the present values of their costs and benefits. It is the most powerful tool for assessing profitability over the life of a project. The NPV is the present value of the future cash flows of an investment, minus the investment's current cost. The project with the greatest net profit will be most beneficial to the company. The greater the net present value, the greater the profit to the firm. If the projects have widely varying initial capital costs, though, the firm must decide if it can afford to make a very large investment, even if the projections are for a large return.

The NPV and IRR approaches are essentially the same and should result in the same alternatives being chosen. The IRR shows the rate of return that a project generates, while NPV shows the present-day dollar value of the return that a project generates. However, IRR analysis ignores the impact of the scale of a project. A project that requires an investment of $100 and returns $125 in one year will have the same IRR as a project that requires a $200,000 investment and returns $250,000. The second option has the same IRR as the first, but the net profit is clearly much greater. If project funds are limited, choose the project with the highest NPV, not necessarily the highest IRR.

All of these analytical approaches require making predictions about future conditions, and thus are subject to error. A margin of safety should be employed with all of them. From the standpoint of pollution prevention projects, the biggest unknown is the possibility and extent of future liability. Small changes in these estimates can often lead to big changes in the economic impact analysis.

Another shortcoming with all of these techniques is that the present values are generally calculated over the life of the project, not that of the total product life cycle including eventual product and process waste disposal costs. These costs may not show up in the cost analysis because potential product-associated environmental costs are beyond the lifetime of the *project,* and thus any potential liability savings from the P2 project may not be included. Businesses need to develop a method of viewing P2 projects in a different light, evaluating long-term benefits as well as short-term profits. They also need to develop ways to factor in noneconomic benefits, such as enhanced public approval of their products due to their pro-P2 initiatives.

7.4 ESTIMATING LONG-TERM CLEANUP LIABILITIES

Many inputs used in the financial assessment of a proposed pollution prevention project are relatively easily attainable. The capital costs to purchase the necessary equipment and the costs to maintain and operate them are usually available, as are the projected benefits in terms of reduced chemical wastage, increased productivity, and the like. The costs associated with treatment of wastes resulting from the process can also be fairly accurately estimated. What can't be estimated as easily is long-term liability due to possible future environmental impacts arising from the disposed materials. A company that deposits its wastes in a "secure" chemical landfill may later find that the landfill liner was defective or just wore out (liners have a finite life of maybe only 30 years) and began leaking hazardous constituents into the groundwater, necessitating an expensive remediation action. The company may also be liable for fines or payment of property damage settlements to neighbors of the landfill site. Or the company may discover at some time in the future that materials it used or produced were injurious to its workers or to the general public, making the company susceptible to personal injury settlements. These potential costs should all be part of the cost of doing business, just as salaries and process chemical costs are, and should be factored into the cost analysis for a proposed project, product, or process.

It is often difficult, however, to evaluate potential liability costs because of the uncertainty of what these costs should be or even if they will ever be incurred. As shown earlier in this chapter, the length of time between the present and when a cost may be incurred has a dramatic impact on the present worth of that expense, and consequently on the economic viability of a project. A pollution prevention option that makes good economic sense using typical business accounting procedures may, in actuality, be a poor choice if it results in a major potential long-term liability. By contrast, a project that eliminates or minimizes long-term potential liability may, in the long run, be appropriate, even if the short-term benefits are not as economically acceptable as another option. Long-term liabilities can become the major driving force behind investment decisions, all other things being equal. Unfortunately, because of the uncertainty associated with estimating these costs, they are often not included in the decision-making process.

Example 7.4. Reactive Chemical Company is currently operating a process line that results in 1000 barrels (bbl) per year of a hazardous waste, which it must dispose of at a cost of $250 per barrel in a hazardous waste landfill. It projects that this process line will continue to operate for the next five years, after which it will be shut down. It is estimated that if the landfill liner were to rupture, there would be a remediation cost of approximately $300 per barrel of waste. A proposed pollution prevention project, costing $400,000 to construct and $50,000 per year in operation and maintenance costs, could reduce the amount of wastes produced by half. Should the P2 project investment be made? Assume a discount rate of 8 percent and projected landfill liner failure times of 10 years and 30 years from the present. Neglect the effects of inflation on remediation costs in your calculations.

Solution. First solve for the present worth of the current waste disposal option (without the P2 project). The present worth of the disposal costs for five years of waste generation can be calculated:

$$\text{Annual disposal cost} = 1000\ \text{bbl/yr}(\$250/\text{bbl}) = \$250{,}000/\text{yr}$$

Calculate the present worth of these expenditures for five years with a discount rate of 8 percent:

$$\text{PW} = A \times \text{PWF}$$

$$\text{PWF} = \frac{(1+i)^n - 1}{i(1+i)^n} = \frac{(1+0.08)^5 - 1}{0.08(1+0.08)^5} = 3.99$$

$$\text{PW} = \$250{,}000(3.990) = \$997{,}500$$

The present worth of the future liability can also be calculated:

$$\text{Future liability} = (1000\ \text{bbl/yr})(5\ \text{yr})(\$300/\text{bbl}) = \$1{,}500{,}000$$

For a landfill liner failure after 10 years:

$$\text{PW} = \frac{F}{(1+i)^n} = \frac{\$1{,}500{,}000}{(1+0.08)^{10}} = \$694{,}790$$

For a landfill liner failure after 30 years:

$$\text{PW} = \frac{F}{(1+i)^n} = \frac{\$1{,}500{,}000}{(1+0.08)^{30}} = \$149{,}066$$

The total present worths of the existing system, assuming landfill liner failure after 10 or 30 years, are

$$\text{PW}_{10} = \$997{,}500 + \$694{,}790 = \$1{,}692{,}290$$

$$\text{PW}_{30} = \$997{,}500 + \$149{,}066 = \$1{,}146{,}560$$

Now determine the present worth of the system with the pollution prevention project added. The present worth of the five years of disposal costs after installation of the P2 process will be one-half of that determined above because only 500 barrels of waste will be produced per year.

$$\text{Present worth of disposal costs} = \$125{,}000(3.990) = \$498{,}750$$

The present worth of the operation and maintenance costs can be determined:

$$\text{Present worth of O\&M} = \$50{,}000(3.990) = \$199{,}500$$

The present worth of the capital costs are the actual capital costs, since they are paid for now:

$$\text{Present worth of equipment} = \$400{,}000$$

Total present worth without considering long-term liabilities:

$$\text{PW} = \$498{,}750 + \$199{,}500 + \$400{,}000 = \$1{,}098{,}250$$

The present worth of the future liability can also be calculated:

$$\text{Future liability} = (500\ \text{bbl/yr})(5\ \text{yr})(\$300/\text{bbl}) = \$750{,}000$$

For a landfill liner failure after 10 years:

$$\text{PW} = \frac{F}{(1+i)^n} = \frac{\$750{,}000}{(1+0.08)^{10}} = \$347{,}395$$

For a landfill liner failure after 30 years:

$$\text{PW} = \frac{F}{(1+i)^n} = \frac{\$750{,}000}{(1+0.08)^{30}} = \$74{,}533$$

The total present worths of the existing system, assuming landfill liner failure after 10 or 30 years, are

$$PW_{10} = \$1{,}098{,}250 + \$347{,}395 = \$1{,}445{,}645$$

$$PW_{30} = \$1{,}098{,}250 + \$74{,}533 = \$1{,}172{,}783$$

The present worths can be summarized as follows:

	PW without including liability	PW including liability	
		10-year liner failure	**30-year liner failure**
Without P2 project	$997,500	$1,692,290	$1,146,560
With P2 project	1,098,250	1,445,645	1,172,783

The length of time to the occurrence of the liability is important if future liabilities are included in the cost analysis. Without the inclusion of future liability, the present worth (costs) of the existing system ($997,500) is less than that for the proposed P2 system ($1,098,250), indicating that the proposed pollution prevention project is not economically viable. If a liner failure in 30 years is assumed, and future liabilities are included, the existing system is still less expensive ($1,146,560 present worth for the existing system versus $1,172,783 for the proposed system). However, if the liner might fail in only 10 years, including the P2 project becomes the best option (a present worth of $1,445,645 versus $1,692,290 for the existing system).

7.5 TOTAL COST ASSESSMENT

"Pollution prevention pays!" This slogan used by 3M Corporation equates reducing waste with saving money. As discussed in the last chapter, life-cycle assessments are becoming a significant tool for industries to determine the efficiency of their processes and the environmental friendliness of their products. It is a commonly accepted tenet that a more efficient process (and, therefore, a less polluting one) will be a more profitable process. However, to be able to sell management on a new pollution prevention project, it is usually necessary to show convincingly that the project is economically viable and desirable. Most industry leaders are not engineers or environmentalists; they are more comfortable with economic decision making than with decision making based on the "greenness" of a product. Thus it is important to relate the life-cycle assessment to terms that the business manager can understand—dollars and cents (White, Savage, and Shapiro, 1996). This can be accomplished through a process variously termed total cost assessment, life-cycle costing, or environmental accounting.

Financial analysis of a pollution prevention project will assist a company in determining whether that particular investment will add economic value to the company. This value can be assessed by calculating cash flows over the life of the project and applying measures of profitability.

Environmental accounting has been described as "paying explicit attention, usually in quantitative terms (ideally, in dollars) to environmental considerations in an organization's planning, decision-making, and operations" (Bailey and Soyka, 1997). Environmental accounting can be the assessment of the environmental cost and performance implications of:

- Alternative process or product designs.
- Location options for a new facility.
- Materials use (e.g., raw materials, catalysts, solvents, coatings, cleaners).
- Sources of supplies/inputs.
- Product mix and retention.
- Alternative packaging and delivery systems.
- Return of packaging or discarded products to the producer.
- Waste management and recycling programs.
- Adoption of just-in-time or build-to-order programs.
- Acquisitions and divestitures.

Putting dollar values on the environmental aspects of business operations greatly facilitates their integration into the business planning process.

Pollution prevention projects may not effectively compete in the capital budgeting process because, as described earlier in this chapter, the conventional cost accounting systems are usually not specifically designed to evaluate such projects. *Conventional cost* categories include direct materials, direct labor, overhead, and one-time capital costs such as new equipment or buildings. In addition to these costs, though, there are often many hidden or *indirect costs,* such as environmental permitting and reporting; waste handling, storage, and disposal; natural resource damage; and worker compensation. In many cases, these costs are included in the company's *total costs,* but they are buried in the overhead cost category and applied to all products equally. A pollution prevention project that might result in a dramatic reduction in some of these indirect costs would not receive its proper credit because the cost savings would be spread over all products, unless a new accounting system was used that separated these costs. The long-term, comprehensive analysis of the full range of *internal* costs and savings resulting from pollution prevention projects is termed *total cost assessment* (White, Savage, and Shapiro, 1996).

Another problem with determining the true life-cycle cost of a product, as described earlier, is that most accounting systems do not evaluate the project over a long enough time period. This often causes valid pollution prevention projects to be rejected.

Beyond these costs, which the company can account for directly and over which the company has control, are *external costs.* These are costs that are not normally in the domain of the company, but which the company's decisions might influence when the whole life cycle of the product is considered. For example, if a company decides to switch from buying solvent-based paints to water-based paints, that decision impacts that amount of solvents produced by a petrochemical industry. While the environmental costs associated with the solvent manufacture are not normally considered part of the cost structure of the product being painted, they actually are if the life cycle of the product is considered. Any environmental cost savings due to this switch should be included in the painted product's true life-cycle cost. Also within the category are those indirect costs associated with impacts that the industry might have on the environment that are difficult to assess directly, such as rainforest destruction or greenhouse warming.

When all of the internal costs plus external costs incurred throughout the entire life cycle of a product, process, or activity are collectively analyzed, the process is called *life-cycle costing*. Detailed information on life-cycle costing can be found elsewhere (Brown and Yanuck, 1985). Little information is currently available to properly assess these costs. Therefore, very few organizations have used true life-cycle costing for their products.

7.5.1 Life-Cycle Costing

One industry that has developed procedures for life-cycle cost management is the electric utilities industry. Several states now require pollutant-specific monetary values to be used in energy planning. Some of these states require that impacts from the entire life cycle of the fuel (including coal mining or oil or gas extraction) be included, along with the impacts of emissions from the power plants. These environmental costs are added to the direct costs of producing energy for purposes of resource evaluation and selection only; these costs are not yet reflected in the rates charged to electricity consumers, but they could be in the future (White, Savage, and Shapiro, 1996). By placing monetary values on environmental impacts and including those costs with direct economic costs of energy production, the states ensure that utilities select fuels with the lowest total cost (internal plus external).

As an example, the California Energy Commission (CEC), which oversees utility resource planning, in 1993 developed environmental cost values for each of the state's 13 air quality management districts, reflecting differences in air quality, population density, health costs, and agricultural production among these regions. For the South Coast Air Quality Management District (SCAQMD), which includes Los Angeles and which is known for its poor air quality, the CEC developed the monetary values shown in Table 7.2.

The Electric Power Research Institute (EPRI) developed a Life-Cycle Cost Management System in 1996 for use in these analyses. It can be used to not only determine the life-cycle cost of power production, but also to evaluate pollution prevention

TABLE 7.2
Pollution costs used in Los Angeles to determine the life-cycle environmental cost of energy production

Pollutant	Cost, $/ton
Nitrogen oxides (NO_x)	$14,734
Reactive organic gases (ROGs)	6,911
Sulfur oxides (SO_x)	8,469
Particulate matter	46,479

Source: California Energy Commission, 1993.

measures. ComEd was the first utility to use this software package. ComEd evaluated methods for disposition of small single-phase overhead PCB-contaminated transformers removed from service and found that switching from contracting with an outside firm for disposal to in-house management of the repair, disposal, and salvage resulted in increased net revenues of $635,000 per year. The utility also estimates a life-cycle cost savings of $186,000 per year from using a dehumidification process instead of chemical treatment to control biofouling in the condenser tubes at a generating facility, reducing chemical usage by 250,000 gallons per year. All told, ComEd identified more than $8 million in benefits company-wide during its first year of applying life-cycle cost management (EPRI, 1997).

7.5.2 Life-Cycle Cost Assessment Process

In most cases, sufficient data to conduct a complete life-cycle cost assessment are not available. However, a limited life-cycle cost assessment can be performed, and the results can be invaluable to proper decision making. This may require adjusting the accounting procedures used in the company to accommodate the true costs of a project, rather than lumping many costs into the "overhead" category. Most accounting systems allow for the use of subcategories within the major cost categories. These can be used to achieve the level of detail necessary.

Most large companies have standardized forms, worksheets, or software that they use for data analysis in their decision-making process. Some, such as General Electric, have developed very detailed procedures for evaluating the total costs of a product or process. The U.S. Environmental Protection Agency has developed a template and process for assessing the financial attractiveness of P2 investments that can be used by any company (U.S. EPA, 1989, 1994).

7.5.3 Life-Cycle Cost Assessment Case Study

The use of the life-cycle assessment approach was described in the previous chapter (Section 6.7.3), and its use in the analysis of a nitric acid plant optimization was discussed. That study also evaluated the economics of the pollution prevention alternatives using a life-cycle cost assessment approach. The problem facing the company was how to accommodate both economic and environmental constraints in the design and operation of the processes (Kniel, Delmarco, and Petrie, 1996). Therefore, the company developed a procedure that (1) uses LCA as a means to quantify and compare the environmental performance of a number of alternative process designs for a given system; (2) undertakes an assessment of the economic performance of the selected system; and (3) uses the economic and environmental performance of the system as a basis for a multiobjective optimization, whereby economic returns are maximized and environmental impacts are minimized.

Recall that two P2 options were evaluated; one involved addition of a selective catalytic reduction (SCR) unit to the tailgas stream of the process, and the second consisted of a pressure increase within the absorption system to improve the efficiency of

NO_x absorption. The life-cycle assessment showed that the high-pressure modification was clearly superior to any other alternative, from an environmental impact viewpoint (see Figure 7.4). The SCR modification was also environmentally better than the existing plant, except in terms of fossil fuel use, where its impact was slightly higher, but it scored consistently below the high-pressure modification. From an environmental standpoint, both options would be an improvement over the existing plant, but the high-pressure option would make larger improvements. Depending on the economics of the processes, either or both of the options might be viable.

The P2 options were ranked by calculating an aggregate "environmental index" score for each design option. The weighting factors of environmental significance were based on the marginal changes in each effect score due to a marginal change in mass of product output from the process. For each design, effect scores were multiplied by the weighting factors and the products summed to produce the "environmental index."

The environmental index and the calculated annual rate of return for each design option were plotted as shown in Figure 7.5. The SCR plant has the lowest rate of return because the capital cost of the unit provides no additional revenue, whereas the high-pressure modification increases efficiency of absorption and therefore increases production rates. The high-pressure process clearly has the lowest environmental index value and therefore is the most environmentally desirable option. Its projected rate of return is slightly below that of the existing plant, indicating that from a purely economic viewpoint it is slightly less attractive than the existing plant, but the difference is negligible. An environmentally conscious company could institute the change to the high-pressure process and realize significant environment improvements, with little financial impact.

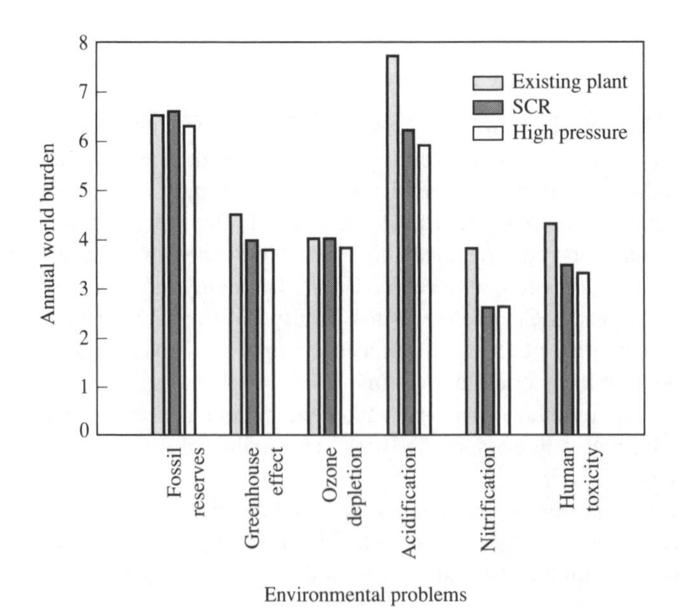

FIGURE 7.4
Normalized environmental profiles for three design alternatives. (Source: Kniel, Delmarco, and Petrie, 1996)

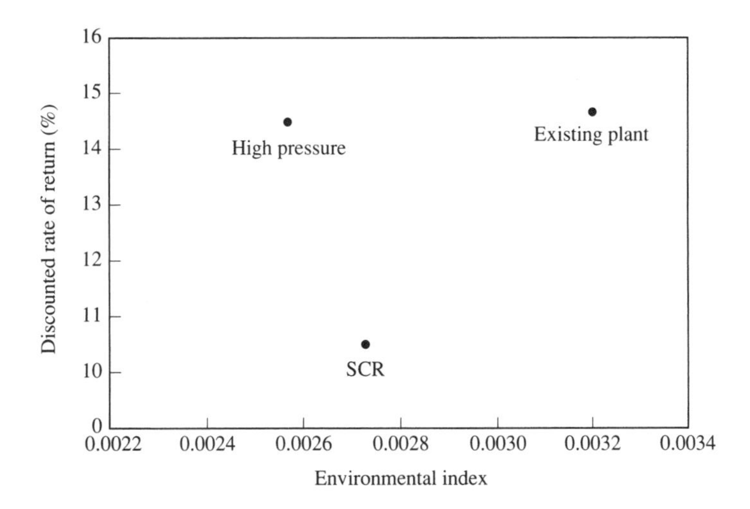

FIGURE 7.5
Economic and environmental performance of three design alternatives. (Source: Kniel, Delmarco, and Petrie, 1996)

7.5.4 Summary

Life-cycle costing is simple in concept. Once a life-cycle assessment is completed for one or more pollution prevention options, the life-cycle costs associated with each option are determined using commonly accepted accounting practices. However, the lack of detailed data on both environmental impacts and the environmental costs associated with these impacts makes effective use of this technique a challenge. These data are being generated by more and more companies now, so that life-cycle costing or environmental accounting may soon become a standard practice for all companies.

REFERENCES

Aldrich, J. R. *Pollution Prevention Economics.* New York: McGraw-Hill, 1996.

Bailey, P. E., and Soyka, P. A. "Introducing Environmental Accounting." *Environmental Engineer* 33, no. 2 (1997): 8–11, 26.

Bhat, V. N. *The Green Corporation: The Next Competitive Advantage.* Westport, CT: Quorum Books, 1996.

Brown, R. J., and Yanuck, R. R. *Introduction to Life Cycle Costing.* Englewood Cliffs, NJ: Prentice Hall, 1985.

California Energy Commission. *Docket No. 90-ER-92S, Appendix F: Air Quality* (1993).

Christen, K. "Banking and Borrowing in Clean Air." *InSites* 5, no. 2 (1997): 6–7.

EPRI. *Life-Cycle Cost Management.* Palo Alto, CA: Electric Power Research Institute, WO3006 (1997).

Grant, E. L., Ireson, W. G., and Leavenworth, R. S. *Principles of Engineering Economy,* 8th ed. New York: Wiley, 1990.

Kniel, G. E., Delmarco, K., and Petrie, J. G. "Life Cycle Assessment Applied to Process Design: Environmental and Economic Analysis and Optimization of a Nitric Acid Plant," *Environmental Progress* 15 (1996): 221–28.

Newman, D. G. *Engineering Economic Analysis.* 3rd ed. San Jose, CA: Engineering Press, 1988.

Ohio EPA. *Financial Analysis of Pollution Prevention Projects.* Ohio Environmental Protection Agency Pollution Prevention Fact Sheet Number 33, Columbus, OH (1995).

Parkin, M. *Microeconomics,* 2nd ed. Reading, MA: Addison-Wesley, 1993.

Samuelson, P. A., and Nordhaus, W. D. *Microeconomics,* 15th ed. New York: McGraw-Hill, 1995.

Shireman, W. K., and Cobb, C. Market-Based Environmental Laws: 100 Ways to Use Prices to Prevent Pollution. fttp://www.quiknet1.quiknet.com/globalff/globmark.html (1996).

Steiner, H. M. *Basic Engineering Economy.* Glen Echo, MD: Books Associates, 1988.

U.S. EPA. *Pollution Prevention Benefits Manual, Vol. I: The Manual; Vol. II: Appendices.* Washington, DC: U.S. EPA, 1989.

U.S. EPA. *P2/Financial User's Manual: Pollution Prevention Financial Analysis and Cost Evaluation System for Lotus 1-2-3 for DOS.* EPA 742-B-94-003. Washington, DC: U.S. EPA, 1994.

White, A. L., Savage, D., and Shapiro, K. "Life-Cycle Costing: Concepts and Applications." In *Environmental Life-Cycle Assessment,* ed. M. A. Curran. New York: McGraw-Hill, 1996, pp. 7.1–7.19.

PROBLEMS

7.1. Externalities, such as sewer user fees, can have an economic impact on industries, forcing them either to raise the prices of their goods to pay for the taxes or to find ways to decrease their pollution discharges. Thus externalities may be beneficial in driving pollution prevention activities. Tax credits for pollution prevention projects can have a similar impact. In your opinion, which system is preferable? Why?

7.2. What are the advantages and disadvantages of marketable permits compared to emission fees?

7.3. For the accompanying supply and demand schedule for ultrawidgets, what is the market equilibrium point? What should happen if the demand for ultrawidgets is 250,000 units per year and the price per unit is $16.00?

Price, $/unit	Quantity buyers willing to purchase (1000s per year)	Quantity producers willing to supply (1000s per year)
10	300	60
12	280	80
14	240	110
16	194	140
18	110	200
20	8	340

7.4. Opponents of marketable permits have argued that this system allows inefficient businesses to stay in business and keep polluting. What arguments can you give to counter this?

7.5. A company is considering a $100,000 P2 project. The current annual inflation rate is 4.0 percent. The company could take out a loan at 8.5 percent interest for the investment. The current production rate of return is 9.5 percent. If the company were to invest the $100,000 rather than carry out the P2 project, the expected rate of return would be 9.25 percent. What should the *discount rate* be for evaluating this P2 option?

7.6. A manufacturer is considering changing its process in order to reduce pollution emissions. Production will remain the same. The current process has an annual operating cost of \$850,000 and an annual waste disposal cost of \$150,000. The new process would have an annual operating cost of \$700,000 and a waste disposal cost of \$85,000. It will cost \$300,000 to modify the existing process line to effect these changes. Based on present worth, and assuming an 8 percent discount rate and a five-year life, does it make economic sense to make the process line changes?

7.7. Repeat Problem 7.6, using the internal rate of return as the basis for your decision.

7.8. The Johnson Chemical Company (JCC) produces an acidic wastewater that requires neutralization before discharge to the sewer. It is currently using 100,000 kg of lime annually as the alkaline agent at a cost of \$65 per metric ton. Lime has a neutralization efficiency (based on an equivalent weight of $CaCO_3$) of only 35 percent, meaning it is not particularly efficient at neutralizing acidic wastes. JCC is considering switching to magnesium hydroxide as its neutralizing agent. It produces $Mg(OH)_2$ as a by-product and can convert it to a usable form for neutralization for \$150 per metric ton. This will reduce the disposal costs for the present $Mg(OH)_2$ by-product waste by \$40 per metric ton. $Mg(OH)_2$ has an additional advantage as a neutralizing agent in that its neutralizing efficiency is 85 percent that of $CaCO_3$. However, its use sometimes produces scale deposits in pipes, which must be removed. It is estimated that this scale removal will cost \$2000 per year. Should the company switch to use of its reprocessed $Mg(OH)_2$ by-product waste as its neutralizing agent or stay with lime?

7.9. A metal plating operator is considering installing an ion exchange unit to recover and reuse metals currently lost in the rinse waters from the plating line. The company presently pays a sewer use fee equivalent to \$4.00 per kilogram of metal sent to the POTW. The ion exchanger can recover 98 percent of this metal. The remainder is small enough that the company will not be subject to the sewer user fee. The ion exchange unit will cost \$50,000 to install and \$12,000 per year to operate. The plater is currently discharging 1000 kg/yr of the metal to the sewer. The metal costs the company \$120 per kilogram to purchase.

(*a*) What is the present worth of the project, assuming a discount rate of 10 percent and an expected equipment life of 10 years?
(*b*) What is the annualized cost for the system?
(*c*) What is the payback period for this project?
(*d*) Should the plater install the ion exchanger?

7.10. A manufacturer receives components, which go into its products, in cardboard boxes. The company is currently paying a fee of \$16,000 per year to have the discarded boxes picked up and hauled away to a landfill for disposal. A local paperboard recycler is willing to pay the company \$4000 per year for the cardboard but will not pick up the cardboard at the plant. The manufacturer estimates that it would cost \$12,000 per year to haul the cardboard to the recycler, which means the firm would only break even, but if it installed a compactor for the boxes, the hauling cost would drop to \$5000 per year. The compactor will cost \$14,000 to install and \$1000 per year to operate. Which option is the best investment? Assume a five-year life for the compactor and a 10 percent discount rate.

7.11. The manufacturer in Problem 7.10 is discussing an arrangement with its parts supplier whereby it would ship the boxes back to the supplier for reuse, rather than disposing of them. The supplier would rebate the company $5000 per year for the returned boxes. The manufacturer estimates that it will cost $14,000 per year to ship back the flattened (noncompacted) boxes for reuse. Now, which is the best option?

7.12. A standard 100-W light bulb costs $1.00 and has a life of 750 h. New energy-efficient light bulbs, costing the same, require only 90 W for the same light output and for the same life. A hotel has 1800 light fixtures that use this type bulb. It is estimated that the average bulb has been used for 375 h. The hotel is planning to switch to the energy-saving bulbs to save on electricity charges, currently running at 8 cents/kWh.

(*a*) If this change is made, how much would the hotel save in electricity charges in the first year, assuming that the light bulbs are on for an average of 6 h/day?

(*b*) Should all of the bulbs be changed now, or should the conversion be done when a bulb burns out? Ignore the cost of bulb installation.

7.13. An equipment maintenance shop uses sixty 55-gal (208-L) drums of trichloroethylene (TCE) per year for parts degreasing. Some of the TCE volatilizes into the atmosphere and some is removed on the cleaned parts and eventually enters the shop's wastewater stream after parts rinsing. However, the shop currently sends 50 drums per year of dirty TCE to a hazardous waste TSDF at a cost of $100 per barrel. Virgin TCE costs the shop $6.00 per gallon. A solvent recovery system supplier reports that it can install a solvent recovery system in the shop for $22,000 that will recover 95 percent of the TCE now being shipped for disposal. It will cost $800 per year to operate the recovery system. Should the shop make the investment? What is the payback period? Assume an 8 percent discount rate.

7.14. The shop in Problem 7.13 is also considering switching to a nonhazardous water-based degreaser that will cost $7.50 per gallon. It is estimated that the equivalent of 20 drums of this degreaser will go to the sewer, either in rinse waters or as discarded dirty degreaser. The sewer user fee for these discharges will be $10,000 per year. Assume the degreaser does not evaporate. Is this a better alternative than installing the solvent recovery system?

7.15. Most cost analyses consider only present capital and operating costs. However, future costs, such as future liability, can sometimes be significant. Consider the following scenario. Clean Chemicals Company (CCC) is setting up a new production facility which will operate for 10 years. CCC will operate an on-site waste treatment plant at an annual cost of $100,000. It can dispose of resulting sludge in an old landfill at a cost of $80,000 per year. CCC officials are concerned, though, that the old landfill may eventually begin leaking into the aquifer below it, requiring expensive remediation. A new lined landfill, with very little potential future liability, is available, but it will cost CCC $150,000 to send its sludge to this landfill. If the company anticipates that the old landfill will fail in 10 years, and the remediation will cost $400,000, should it send its wastes to the less expensive old landfill and pay for cleanup if landfill liner failure occurs, or should CCC pay the higher fee to use the new landfill? If CCC uses the old landfill, how much money should the company put in escrow now to pay for the future liability? Assume a discount rate of 10 percent and an escrow interest rate of 8 percent.

CHAPTER 8

POLLUTION PREVENTION PLANNING

8.1 INTRODUCTION

Successful reduction of waste generation by industry requires careful planning to ensure that the best pollution prevention activities are carried out. However, pollution prevention planning is not enough; the P2 projects must actually be implemented. Too often, P2 planning results in plans that languish on a shelf because other higher priorities for the capital appeared or because the company did not have an overall environmental management system in place that was coordinated with its more common business management system. This chapter discusses ways to set up and implement effective pollution prevention planning programs.

Pollution prevention planning requires a detailed understanding of how a company does business and how it makes its products. The resulting plan should provide a mechanism for a comprehensive and continuous review of the company's activities as they pertain to environmental issues. This task can appear daunting at the beginning, especially to management personnel, who may be unfamiliar with many of the concepts involved. The amount of information and data required for a good pollution prevention plan may also give many pause. However, the task should not be as onerous as it may at first appear. Many of the procedures used are actually analogous to those commonly used to run a business; only the nomenclature may be different. Because pollution prevention activities can have a major impact on the day-to-day operation of a business, pollution prevention planning should be done in parallel with other business planning.

Implementing pollution prevention projects is still voluntary in the United States, but developing a pollution prevention plan is not. Among other things, the Pollution Prevention Act of 1990 amended the Emergency Planning and Community Right-to-Know Act (EPCRA) to require certain industries to describe the steps they are taking to reduce their pollution. This information must be reported annually along with the firm's Toxic Release Inventory (TRI). Unfortunately, the requirement is only to report on pollution prevention plans; it does not require that they be implemented.

In contrast to the voluntary nature of pollution prevention in the United States, many European countries are mandating that companies both plan for and implement waste minimization practices. This will soon become a requirement for all countries within the European Union. The ISO has embarked on a task to standardize environmental quality management. Part of the organization's commitment is the development of ISO 14000, an international series of standards designed to promote a common approach to environmental management, to enhance organizations' abilities to attain and measure improvements in environmental performance, and to facilitate trade and remove trade barriers (Cascio, Woodside, and Mitchell, 1996). These standards will eventually become requirements for essentially any company wishing to conduct international trade. The ISO Environmental Management System will be described in more detail later.

8.2 STRUCTURE OF THE POLLUTION PREVENTION PLANNING PROCESS

The major steps involved in developing a pollution prevention program are outlined in Figure 8.1. The major elements include building support for pollution prevention throughout the company, organizing the program, setting goals and objectives, performing a preliminary assessment of pollution prevention opportunities, and identifying potential problems and solutions (U.S. EPA, 1992).

8.2.1 Organizing the Program

The initiative to develop a company pollution prevention program may come from the upper management level, but more typically it begins in middle management, where the employees are closer to the production line and can more easily see the potential benefits of pollution prevention, or with the environmental control officers in the company. Wherever it begins, it is essential to sell the company executives of its value early on. This may mean having to perform an initial sketchy assessment to show where potential cost savings could occur. It is the executives who make the final decisions on where to invest capital; without their support, even very beneficial P2 projects will not be begun.

Once senior management accepts the concept of pollution prevention and decides to institute a pollution prevention program, they should convey this message to all employees in the company to ensure full compliance. This may take the form of a

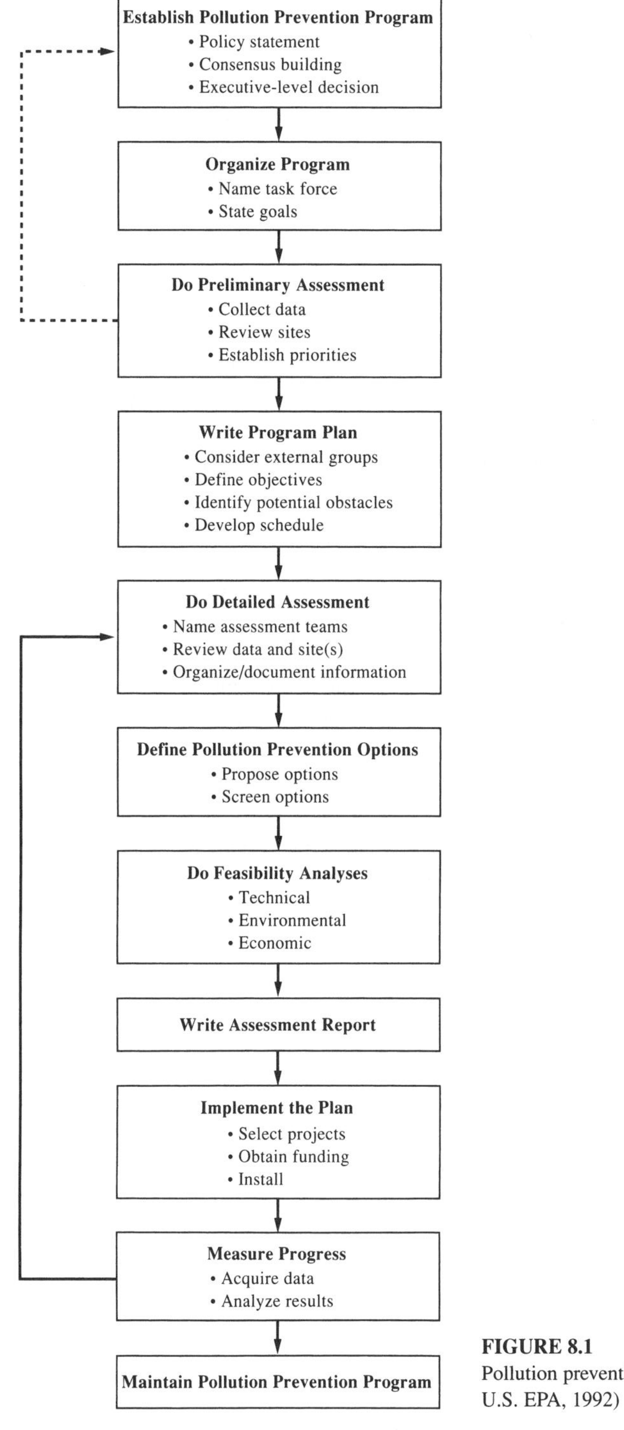

FIGURE 8.1
Pollution prevention program overview. (Source: U.S. EPA, 1992)

formal policy statement. Why the company is implementing the new program, what will be done to implement it, and who will be responsible should all be specified. Whatever method is used, it is important that all employees recognize that the pollution prevention program is not just another piece of paper but a real program that will be enforced. The active cooperation of all employees will be essential for the program to be a success. This is particularly true for the production-level employees, who may have to alter their traditional routines. Upper management must be vigilant to continually show support for the program.

Care should be taken in assembling the team that will develop and direct the implementation of the pollution prevention program. These representatives should collectively be knowledgeable in all facets of the business, both technical and business viewpoints, and represent all levels of employment. This may include executives, environmental and plant process engineers and supervisors, experienced production line workers, purchasing agents, and quality control personnel. The program leader should be someone with authority in the company who has broad-based support and the influence to ensure that the program development and implementation stay on track.

The goals of the program should be established early in order to give direction to the program. The goals should reflect the stated policies of the company, be clearly written so that everyone understands their meaning, and be challenging enough to motivate the employees without being unreasonable or impractical. Some companies begin with a goal of zero discharge, which is probably not practical, and then modify this goal as more information becomes available. The goals should be flexible and adaptable (U.S. EPA, 1992).

8.2.2 Preliminary Assessment

The objective of this phase of the pollution prevention plan development is to review and evaluate existing data and establish priorities and procedures for detailed assessments. Much of the data required for this phase can usually be obtained as part of normal plant operations or in response to existing regulatory reporting requirements. This includes such documents as plant purchasing, accounting, and inventory records; shipping manifests for hazardous wastes shipped from the plant; plant design documents and equipment operating manuals; TRI records; and NPDES, CAA, and SARA Title III reports on the volume, concentration, and degree of toxicity of wastes discharged. The U.S. EPA's *Facility Pollution Prevention Guide* (U.S. EPA, 1992) provides a number of generic and industry-specific worksheets that can be used for gathering these data.

Two primary factors should be considered while conducting this preliminary assessment. The first is that a multimedia approach (air, water, and solid waste) to pollution prevention should be the goal. This includes consideration of all energy and waste streams involved. The second is that the objective of the analysis is pollution prevention, not extensive and expensive data gathering. The amount of data required should be governed by the needs of the study. Detailed data beyond what is absolutely necessary will only add complexity and cost to the exercise.

The assessment team should first become familiar with the targeted processes. This can be accomplished by going into the facility and studying the process in detail. This study should be conducted while the process is in operation and, if possible, during a shutdown/clean-out/start-up period to identify the materials used and wastes generated by this procedure. The assessment team should discuss as much as possible with the line personnel (equipment operators, maintenance and housekeeping staff) and foreman, who have the best working knowledge of the processes being evaluated. When studying the process, the team should note any pollution prevention opportunities and should pay particular attention to the following ("Doing a Detailed Assessment," 1997):

- Procedures of operation followed by line workers.
- Quantities and concentrations of materials (especially wastes).
- Collection and handling of wastes (note if wastes are mixed).
- Record keeping.
- Flow diagram (follow through the actual process).
- Leaking lines or poorly operating equipment.
- Any spill residue.
- Damaged containers.
- Physical and chemical characteristics of the waste or release.

The team should also investigate supplemental operations such as shipping/receiving, purchasing, inventory, vehicle maintenance, waste handling/storage, laboratories, powerhouses/boilers, cooling towers, and maintenance (*Pollution Prevention,* 1993).

Most pollution prevention plans will provide a mechanism for investigating potential P2 initiatives. These studies must be very detailed and focused, as described in the previous two chapters. The level of detail required at the preliminary assessment stage for the program development is much less detailed. The objective here is to assign priorities to processes, operations, and materials that should be addressed later in detail when the plan is being carried out. Table 8.1 lists typical considerations used when prioritizing waste streams for further study. The option rating weighted-sum method, to be described later in this chapter, can be used to determine prioritization.

TABLE 8.1
Typical considerations for prioritizing waste streams for further study

Compliance with current and anticipated regulations
Costs of waste management (pollution control, treatment, and disposal)
Potential environmental and safety liability
Quantity of waste
Hazardous properties of the waste (including toxicity, flammability, corrosivity, and reactivity)
Other safety hazards to employees
Potential for pollution prevention
Potential for removing bottlenecks in production or waste treatment
Potential recovery of valuable by-products
Available budget for the pollution prevention assessment program and projects
Minimizing wastewater discharges
Reducing energy use

Source: U.S. EPA, 1992.

TABLE 8.2
Components of a pollution prevention plan

Components of a pollution prevention plan
Corporate policy statement of support for pollution prevention
Description of the pollution prevention team makeup, authority, and responsibility
Description of how all the groups (production, laboratory, maintenance, shipping, marketing, engineering, and others) will work together to reduce waste production and energy consumption
Plan for publicizing and gaining companywide support for the pollution prevention program
Plan for communicating the successes and failures of pollution prevention programs within the company
Description of processes that produce, use, or release hazardous or toxic materials, including a clear definition of the amounts and types of substances, materials, and products under consideration
List of treatment, disposal, and recycling facilities and transporters currently used
Preliminary review of the cost of pollution control and waste disposal
Description of current and past pollution prevention activities at the facility
Evaluation of the effectiveness of past and ongoing pollution prevention activities
Criteria for prioritizing candidate facilities, processes, and streams for pollution prevention projects

Source: U.S. EPA, 1992.

8.2.3 Pollution Prevention Program Plan Development

The P2 plan task force uses the information collected during the preliminary assessment to write the detailed pollution prevention plan. The plan should define the pollution prevention program objectives, identify potential obstacles and solutions, and define the data collection and analysis procedures that will be used in later detailed P2 studies. Table 8.2 is a detailed list of items that should be included in the plan. Note that this list does not include recommendations for pollution prevention projects; rather, it describes current practices and a methodology for evaluating proposed projects.

It is often useful to consult people outside the company: government officials, local community representatives, or people from similar businesses, for instance. They can provide a broader perspective to the planning process and can help to provide credibility and a sense of community involvement.

A final, critical part of a successful pollution prevention plan is the development of a schedule for implementation. Milestones within each phase of the plan should be detailed and realistic target dates should be assigned. A mechanism for monitoring and ensuring compliance with the schedule should also be included. Many well-conceived plans fail because of lack of proper follow-up to confirm that the plan is being carried out as designed.

8.2.4 Developing and Implementing Pollution Prevention Projects

Writing the pollution prevention plan is only the first step toward achieving a successful pollution prevention program. If continued further action to carry out the provisions of the plan is not performed, the plan will be worthless.

The first step toward implementation of the plan is to conduct detailed assessments of the potential areas of opportunity identified during the preliminary assessment. Assessment teams assigned to each operational area of the facility should review all existing information related to that area and develop more detailed lists of potential waste reduction or energy-saving projects. These may require collection of additional data. Considerable useful information can often come from interviews with the production line workers. Analysis should also include preparation of process flow diagrams and material and energy balances in order to determine pollution sources and opportunities for eliminating them.

Once the sources and nature of wastes generated are determined, the assessment team should propose and then screen pollution prevention options. Their objective should be to generate a comprehensive, prioritized set of options for detailed feasibility assessment. The general categories of pollution prevention actions that should be considered by each group, in approximate order of increasing complexity, include:

- Housekeeping improvements.
- Waste segregation.
- Material substitution.
- Process modification.
- Material recycling and reuse.
- Additional waste treatment.

Some options will be found to have no cost or risk attached; these can be implemented immediately. Others will be found to have marginal value or to be impractical; these will be dropped from further consideration. The remaining options will generally be found to require feasibility assessment.

A number of procedures are available for prioritizing pollution prevention projects. The priority for a project will depend on a number of factors and may vary from facility to facility, depending on the P2 goals established during the planning process. One prioritizing procedure that is commonly used is the *option rating weighted-sum method* developed by the U.S. EPA (U.S. EPA, 1992). This method provides a means of quantifying the important criteria that affect waste management in a particular facility. The first step in this procedure is to determine the important criteria in terms of the program's goals. Sample criteria include:

- Reduction in waste quantity.
- Reduction in waste hazard.
- Reduction in waste treatment/disposal costs.
- Reduction in raw materials costs.
- Reduction in insurance and liability costs.
- Previous successful use within the company.
- Previous successful use within the industry.
- Not detrimental to product quality.

- Low capital cost.
- Low operating and maintenance costs.
- Short implementation period with minimal disruption of plant operations.

The weights (on a scale of 0 to 10, with 0 being of no importance and 10 being highly important) are determined for each criterion in relation to its importance. Thus if reduction in waste treatment and disposal costs is very important, it may be given a weight of 10, and if previous successful use within the company is of only minor importance, it may be given a lower weighting of only 1 or 2. Criteria that are not important are omitted from the analysis. Each option is then rated on each criterion, again using a scale of 1 to 10. Finally, the rating of each option for a particular criterion is multiplied by the weight of the criterion. An option's overall rating is the sum of these products.

Example 8.1 shows how this weighted-sum method can be used to rank various pollution prevention options.

Example 8.1. Goodhealth Pharmaceutical Co. is evaluating various pollution prevention options for its production line. Based on its corporate pollution prevention goals, it has determined that reduction of waste quantity is one of its primary objectives, followed by reductions in treatment costs, liability, and safety hazards. Reductions in the cost of raw ingredients is of interest, but their costs are already rather low and therefore this criterion is weighted lower. Overriding all of these criteria, however, is the need to ensure that there is no loss in product quality since Goodhealth must maintain FDA approval for the medicine. Therefore, this criterion received a weighting of 10. The accompanying table shows the weights applied to the various criteria.

The P2 assessment team developed three potential pollution prevention options for evaluation. One of these (Option A) involved a change in the solvent used in the pharmaceutical separation step. The team estimated that this would significantly reduce safety hazards and future liability while lowering treatment costs somewhat, but the product recovery from the reactor would be impaired by the new solvent and additional purification steps would be needed. The ratings for this P2 option for each of the criteria are also shown in the table, along with the ratings for two other options. Which option should Goodhealth adopt?

		Ratings for each option					
		Option A		**Option B**		**Option C**	
Rating criterion	**Weight (*W*)**	**Rating (*R*)**	$W \times R$	**Rating (*R*)**	$W \times R$	**Rating (*R*)**	$W \times R$
No decrease in product quality	10	5	50	8	80	10	100
Reduce waste quantity	8	8	64	6	64	3	24
Reduce safety hazards	7	9	63	7	49	4	28
Reduce liability	7	8	56	7	49	2	14
Reduce treatment costs	5	6	30	9	45	6	30
Reduce raw material costs	2	3	6	8	16	7	14
Ease of implementation	1	5	5	8	8	9	9
Sum of weight × ratings			274		311		219

Solution. From this screening, Option B rates the highest with a score of 311. Even though Option A results in lower quantities of waste being generated and in reduced safety and liability concerns, the lower product quality makes it a lower rated option than Option B. This does not mean that Option A should not be implemented, however. Assuming the two options are not mutually exclusive, it may be justifiable to implement both options, provided they are both cost effective and the company has sufficient capital to carry out both, and both are improvements over current practices.

The options selected for further analysis should now be examined for technical, environmental, and economic feasibility. The option must first be evaluated from a *technical* standpoint to ensure that it will actually work for a specific application. This may require extensive laboratory or pilot-scale testing and evaluation by a variety of people, possibly including representatives from production, maintenance, quality assurance (or control), and purchasing. It may also require input from the purchasers of the product to verify that any specification changes are acceptable. The option must also be evaluated from the *environmental* viewpoint to ensure that waste is reduced in an environmentally safe fashion. This should include a total life-cycle assessment of the product and materials and processes used, as described in Chapter 6. Energy consumption must also be included in this analysis. Finally, an *economic* evaluation of the option must be performed. No pollution prevention option will be implemented if it is not economically justifiable. Projects that require significant capital costs will need a detailed cost analysis. The life-cycle costing procedures described in Chapter 7 should be used.

8.2.5 Implementing the Pollution Prevention Plan

Once the viable pollution prevention options have been evaluated, prioritized, and tested for worthiness, it is time to sell the best ones to management. This is often the most difficult part of the process. Much of the work to this point has dealt with technical issues, and most of it has probably been done by technical staff. Now these options must be sold to the management side of the business, which may have limited understanding of the environmental issues. These proposed projects may well be competing with other revenue-generating projects for limited resources. This is why accurate life-cycle costing is necessary. Showing that in the long term the pollution prevention project will save the company money, even if it is through reduced liability or reduced treatment costs rather than through increased sales, should convince management of the project's value. A properly structured pollution prevention plan, in which the budgeting decision-making process for pollution prevention is described in detail, will make this exercise easier.

Many pollution prevention projects may require changes in operating procedures, purchasing methods, or materials inventory control. They may also affect employee training procedures. These changes need to be done carefully in order to minimize disruption to normal business activities. A seamless transfer to pollution prevention should mitigate against the common fear of change.

A pollution prevention plan should be a "living document." It should be continually updated as new alternatives become known. A program that includes rewards and recognition to employees who suggest useful pollution prevention ideas or who successfully put those ideas to work is often helpful in garnering the cooperation of the employees and in making them feel that they are truly a part of the program. The suggestions must be properly evaluated and good ones must be implemented, though, or the staff will quickly lose interest.

8.2.6 Measuring Pollution Prevention Progress

Pollution prevention planning does not end with the initiation of a pollution prevention project. It is essential that the new activity be continually monitored to ensure that it is being done correctly and that it is achieving its intended goals. Without continued surveillance, it is often easy to slide back into prior routines. The amount of pollution prevention achieved should be continually measured so that everyone sees the value of the P2 philosophy. Not all projects will be completely successful. It is important to document both successes and failures. The successes can be used as models for future pollution prevention activities. The failures can be studied to see why they failed and what can be changed in the planning process to minimize future failures.

No standardized procedure is available as yet for assessing the success of a pollution prevention project. Wherever possible, pollution prevention progress should be measured on a quantitative basis. The metric used to track pollution prevention should be one that is easily measurable and that reflects the wastes of interest. This may be the quantity of the waste or its toxicity. Measurements may be determined as absolute change in waste quantity, or waste quantity change adjusted for variations in production. Measures of absolute changes in waste may be appropriate to address community concerns or to assess progress toward reduction goals; adjusted measures may give a better picture of chemical use efficiency (Malkin and Baskir, 1995). The unadjusted measure is simply based on waste generation activity over time:

$$\text{Waste reduced or increased} = \text{Waste}_{\text{time2}} - \text{Waste}_{\text{time1}}$$

Not all decreases in waste production may be attributable to pollution prevention, though. For example, reductions in the quantity of product made may also result in less pollution being generated. Therefore, it is important to normalize the pollution data to some other metric, such as the amount of product produced or the number of hours of production over a given time period being used for comparison with previous activities. Measures of P2 normalized for production are often found by calculating an activity index, I, as follows:

$$I = Q_1/Q_0$$

where Q_1 is the production level or activity level in the current year and Q_0 is the quantity in the previous year. The measurement of P2 is then calculated as

$$\text{Normalized P2} = W_1 - W_0 \cdot I$$

where W_1 is the quantity of waste in the current year and W_0 is the quantity of waste in the previous year (Malkin and Baskir, 1995).

The initial challenge is to select the appropriate P2 metric that correlates well with the wastes being measured. Selecting the wrong basis for measuring P2 activity effectiveness or for normalizing the data can lead to erroneous or misleading results. The appropriate factor will be process- and facility-specific. Some factors that could be used include:

- Number of units produced.
- Throughput of a material to be processed, such as barrels of crude oil to be refined.
- Volume of bulk product produced.
- Surface area of product produced.
- Number of production hours required.

Obtaining reliable data is one of the major problems facing environmental managers. The quantification of process emissions is extremely difficult. A standardized methodology for accounting is essential for good decision making. One procedure that can be used to do this is known as the *unit operation system* (UOS). This system prescribes a systematic and unified approach for developing a materials-tracking inventory that can provide plantwide evaluation of pollution prevention cost-effectiveness. The output of the system can be used in performing a life-cycle assessment (Serageldin, 1995). The following steps are among those used in developing the necessary data for the UOS:

- Identify and briefly describe the primary unit operation.
- Draw the boundary of the UOS, which includes all possible points of emission from the unit operation.
- Identify all input/output streams and emission sources crossing the UOS boundaries, and determine the amounts of contaminant in the waste streams.
- Perform a material balance for the UOS to ensure data accuracy.
- Normalize the data based on the important factors that influence the emission rate.

The normalized data can be used to make comparisons between the performances of different plants, to evaluate pollution prevention options, and to determine emission factors for specific unit operation systems.

Reducing the volume of waste without reducing its toxicity may not constitute pollution prevention. Failing to address the issue of waste toxicity assumes that all waste reduction is equivalent; in reality, however, equal reductions of highly toxic and less toxic wastes are not equivalent from either a human health or environmental protection perspective. Better process separation technology, such as through the use of membrane filters rather than gravity separators, may reduce the *volume* of waste generated by concentrating the contaminants into a smaller volume, but this may result in the toxicity remaining the same or even increasing. For toxic wastes, normalizing against toxicity may be necessary. For other types of waste, such as waste packaging, a volume or weight metric may be fine. Care must be taken to ensure that the normalized metric used for a particular waste is suitable for that waste. Unfortunately, no widely accepted method for evaluating toxicity reduction is currently available.

8.3 ENVIRONMENTAL MANAGEMENT SYSTEMS

The procedures described are now being unified into a comprehensive system for managing environmental impacts by industry. The ISO 14000 standards, formally adopted in 1996, establish benchmarks for environmental management performance and describe the measures that industry must take to conform to these standards. The objective is to provide guidance for developing a comprehensive approach to environmental management and for standardizing some key environmental tools of analysis, such as labeling and life-cycle assessment (Cascio, Woodside, and Mitchell, 1996).

The subjects covered by ISO 14000 can be seen in Figure 8.2. Issues dealing with an organization's style of management include environmental management systems, environmental auditing, and environmental performance evaluation. Product evaluation consists of environmental aspects in product standards, environmental labeling, and life-cycle assessment. Standards are being or have been developed for each of these subsections.

The most significant aspect of ISO 14000 is the development of the ISO 14001 document entitled "Environmental Management Systems—Specification with Guidance for Use." This standard lays out the elements of the environmental management system (EMS) that all businesses are required to conform to if they wish to be registered or certified by ISO. A supporting document, ISO 14004, entitled "Environmental Management Systems—General Guidelines on Principles, Systems, and Supporting

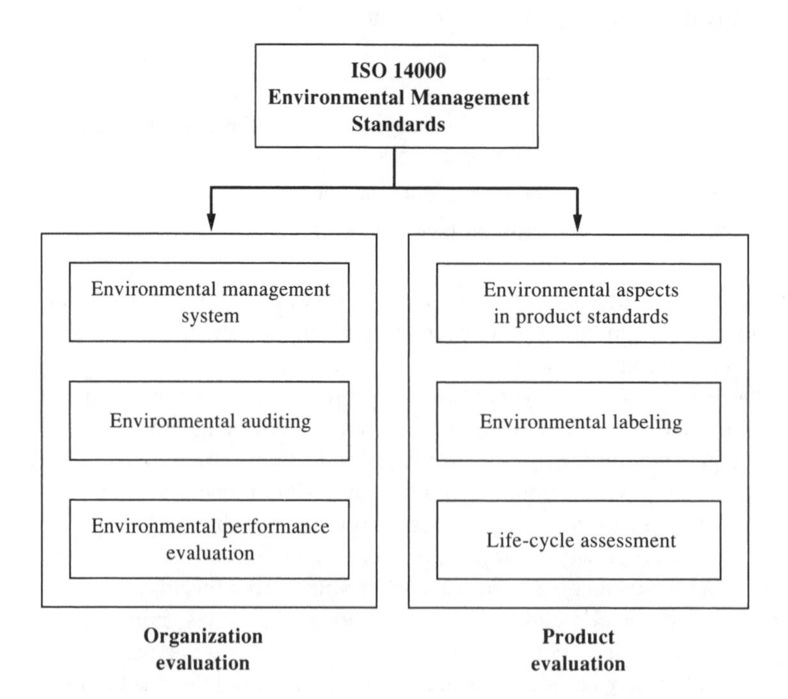

FIGURE 8.2
ISO 14000 family of standards. (Source: Cascio, Woodside, and Mitchell, 1996)

Techniques," is meant to be used as guidance to organizations that are just beginning to implement an EMS. Finally, ISO 14010 provides the general principles on environmental audits that must be used.

The EMS standard described in ISO 14000 prescribes that companies develop an environmental policy; identify environmental aspects and impacts of their activities, products, and services; define the significance of these impacts; prioritize the impacts; identify legal and other requirements governing the organization's operation; establish objectives and targets; implement programs to meet those targets; establish an auditing system and procedures for management review; and implement corrective action, if needed, based on audit findings (Delaney and Schiffman, 1997). An EMS provides a framework to balance and integrate economic and environmental interests. Thus the EMS is essentially the same as the pollution prevention plan described earlier in this chapter; the difference is that the structure of the plan and the procedures to be followed are rigidly prescribed so that an independent third-party reviewer can easily compare plans and results.

An environmental management system is

> that part of the overall management system which includes organizational structure, planning, activities, responsibilities, practices, procedures, processes and resources for developing, implementing, achieving, reviewing and maintaining the environmental policy.

Note that this indicates that the EMS must be an integral part of the overall corporate management structure, and not a stand-alone system. ISO's EMS model has five basic components (see Figure 8.3):

1. *Commitment and environmental policy.* Top management must commit to continual improvement of the EMS, prevention of pollution, and compliance with environmental laws. The policy must be documented, available to the public, and communicated to employees. This serves as the base for all of the components of the EMS, and therefore it is shown at the base of the EMS pyramid.

FIGURE 8.3
Components of an environmental management system.

2. *Environmental management plan.* The plan defines the organization's goals, objectives, and targets for its environmental management program. The content and development of this plan were described earlier in this chapter.
3. *Implementation.* The company specifies how it will ensure that the goals and objectives of the plan are met; describes the responsibilities of each employee; identifies and provides necessary resources; and establishes and implements proper training procedures.
4. *Measurement and evaluation.* A program is required to systematically and regularly measure and monitor the company's environmental performance against its stated objectives and targets. This will also aid in determining what processes or products require improvement.
5. *Continual review and improvement.* Procedures must be developed for continued review and improvement of the EMS in order to improve the overall environmental performance of the company.

ISO 14001 stipulates that an EMS must be established and that it must address all of the topics cited above, but it does allow flexibility in the plan's content and implementation. It encourages companies to *consider* implementation of best available technology, but it does not *mandate* its use. The only requirement in the standard is to consider "prevention of pollution" options when designing new processes or products (Cascio, Woodside, and Mitchell, 1996). Best management practices are often the most efficient and least costly options from a life-cycle standpoint, but it is not necessary for a company to implement them to become accredited.

It is important to keep in mind that ISO 14001, the Environmental Management Standard, is used only to evaluate the EMS. The success of the EMS is measured against its stated objectives to minimize pollution during the process of manufacturing a product. It is not used to measure the environmental performance of the product. A second set of standards deals with an organization's products, processes, and services. ISO 14020 through 14024 deal with environmental labeling (see Chapter 6) requirements, ISO 14040 through ISO 14043 deal with life-cycle assessment procedures, and ISO 14060 (still under development) addresses environmental aspects of products.

Many companies have looked with dread at the implementation of ISO 14000. They see massive amounts of paperwork, copious amounts of time needed to evaluate the company's current environmental status, and the expense of limited resources to make necessary changes to meet the standards. However, the end result should make the effort worthwhile. The Environmental Management System can be used by management to identify manufacturing inefficiencies (losses to waste) and methods to overcome these. And attaining certification indicates to the outside world that the company is interested in preserving the environment, which can provide a significant boost to sales. Any company that plans to conduct international business will soon be required to meet ISO 14000 standards. While this task may now appear onerous, mainly because it is new, ISO environmental certification will eventually become just another part of doing business.

8.4 ENVIRONMENTAL AUDITS

Environmental audits, also known as emission inventories, may be done as part of writing a pollution prevention plan or may be conducted at any time as a stand-alone process. The objective is to identify and characterize the waste streams associated with a process or service so that intelligent decisions can be made concerning pollution reductions. When performed in a systematic, continual fashion, environmental audits are also valuable as a method to verify that a pollution prevention plan is being properly followed, and that environmental regulations are being met.

Environmental auditing is defined by the U.S. EPA as "a systematic, documented, periodic, and objective review by regulated entities of facility operations and practices related to meeting environmental requirements." "Regulated entities" includes private firms and public agencies, as well as federal, state, and local agencies.

Environmental auditing is voluntary, but it is strongly encouraged by the EPA. In 1986, the EPA issued a policy statement that said:

> It is EPA policy to promote the use of environmental auditing by regulated entities to help achieve and maintain compliance with environmental laws and regulations, as well as to help identify and correct unregulated environmental hazards. This policy statement specifically:
>
> - encourages regulated entities to develop, implement and upgrade environmental auditing programs;
> - discusses when the Agency may or may not request audit reports;
> - explains how EPA's inspection and enforcement activities may respond to regulated entities' efforts to assure compliance through auditing;
> - endorses environmental auditing at federal facilities;
> - encourages state and local environmental auditing initiatives; and
> - outlines elements of effective audit programs. (*Federal Register* 51, no. 131 (July 9, 1986): 25004–10)

There are many types of environmental audits and reasons for conducting them. Their use as an integral part of a pollution prevention program was discussed previously. These environmental audits are often termed *management audits* or *operational audits* because they are used by management to establish corporate environmental policy and to direct the management operations to ensure implementation of that policy. Another type of audit is the *compliance audit,* which is used to verify compliance with environmental regulations. These audits differ from management audits in that they are designed mainly to show that regulations are not being violated; their goal is not to determine opportunities for pollution prevention. They usually cover a short time frame and show only a snapshot of the emissions from a process at that time. An *environmental liability audit* is often conducted prior to a property transfer—that is, before the purchase, lease, sale, or financing of buildings or land for commercial, industrial, or multiple-family dwellings. The intent is to determine any potential future impacts to the environment from the site or facility, thus limiting future liability or litigation. Finally, a *waste management contractor audit* is conducted on waste management contractors by waste-generating companies to ensure that the waste management contractors are properly managing the waste being shipped to them. Federal law stipulates that the generator of hazardous wastes is as liable for the proper treatment and disposal of the waste as the waste management contractor. It is important for the waste generator to audit the practices of the waste management contractor to ensure that the wastes are being managed properly.

Environmental audits are conducted in much the same way as previously described for the initiation of a pollution prevention plan, except that the study may be more limited in scope. In many cases, an environmental audit focuses on one process of concern within a company. Thus it does not look at the life cycle of the process or product or all of the materials going into the product. However, to be effective and useful, the audit should receive all of the care that went into the overall pollution prevention plan. It should consist of the same three phases described previously—preaudit, audit, and postaudit activities. It is important that the audit have the full support and encouragement of top management; that a carefully selected and trained audit team be used; that explicit goals and objectives be established and necessary resources be made available; that sufficient high-quality data be collected to achieve the stated objectives; that written reports summarizing the findings be made; and that corrective actions be expeditiously taken when necessary.

An environmental audit should be an accurate, quantifiable depiction of the operation under study. Thus it should be more detailed than the environmental assessment used in the planning process. The environmental assessment allowed for the use of estimates, industry averages, or judgment when necessary data were missing. The objective was to establish a general picture of the operation. An environmental audit should be a true picture of the process. This may necessitate extensive testing to obtain relevant data. Often, samples are taken for analysis specifically for the audit, rather than relying on previous analytical records.

The audit should be designed to provide management with a complete assessment of the wastes being generated by a facility or process and the fate of those wastes. Among the many items that should be addressed are:

- The sources of wastes generated (manufacturing processes and operations, storage facilities).
- The inputs to the process and the process efficiency.
- The types, amounts, and characteristics of waste streams being generated (water, air, and solid waste).
- The frequency of waste generation (continuous or sporadic; if continuous, number of shifts per day).
- Fugitive emissions of wastes.
- Waste handling (are wastes kept segregated or mixed together).
- Energy use.
- Housekeeping procedures.
- Record keeping (both process records and waste generation records).
- Regulatory status of the waste.

TABLE 8.3
Typical data needed in an environmental audit

Design information	
Process flow diagrams	Equipment specifications and data sheets
Materials and heat balances for both production processes and pollution control processes	Piping and instrument diagrams
Operating manuals and process descriptions	Plot and elevation plans
Equipment lists	Equipment layouts and work flow diagrams
Raw material/production information	
Product composition and batch sheets	Operator data logs
Material application diagrams	Operating procedures
Material safety data sheets	Production schedules
Product and raw material inventory records	
Environmental information	
Hazardous waste manifests	Waste analyses
Emission inventories	Previous environmental audit reports
Toxic Release Inventories	Permits and/or permit applications
Biennial hazardous waste reports	
Economic information	
Waste treatment and disposal costs	Operating and maintenance costs
Product, utility, and raw material costs	Departmental cost accounting reports
Other information	
Company pollution prevention plan	Standard procedures
Company environmental policy statements	Organization charts

Source: Adapted from U.S. EPA, 1988.

The types of data typically needed to properly conduct an environmental audit are listed in Table 8.3. As can be seen, acquisition and accumulation of the data will be time-consuming and may be expensive. However, without this information, the audit may result in erroneous findings.

8.4.1 Emissions Inventory

Waste audits should define all wastes being generated in a facility. Not all of these wastes may reach the environment. Some may be recycled or reused, and some may be destroyed during waste treatment processes. *Emissions inventories* are a subset of waste audits. They list the types, amounts, and characteristics of pollutants that are actually being discharged into the environment.

Development of emissions inventories for direct discharges of pollutants (solid, aqueous, or gaseous) into the environment is complex and often requires a major effort, but the release points are usually few in number and fairly well defined. They are usually referred to as *point sources*. They may be an outfall pipe to a river, a connection to a municipal sewerage system, a stack for gaseous emissions, or a landfill that is receiving solid wastes. There are two other potentially significant sources of waste emissions, though, that are not as easy to define or to monitor—fugitive emissions (unintentional releases of liquids or gases from process equipment, storage tanks, or pipelines) and secondary emissions from waste treatment processes or from recycling processes. The procedures for estimating these emissions and processes for their control will be presented in detail in Chapter 12.

8.5 TOXIC RELEASE INVENTORY

The Emergency Planning and Community Right-to-Know Act (EPCRA) has been hailed as one of the most powerful pieces of environmental legislation in the last few decades. The primary goal of EPCRA is to provide information to the public concerning the presence and release of toxic and hazardous chemicals in the community. To fulfill this purpose, EPCRA requires that certain companies that manufacture, process, and use chemicals in specified quantities must file written reports, provide notification of spills/releases, and maintain toxic chemical inventories of chemicals stored on site. Among these is the requirement for certain firms to submit an annual report of releases of listed toxic chemicals, both intentional and unintentional, and a listing of locations and quantities of chemicals stored on site. Facilities are also required to report annually on pollution prevention activities and chemical recycling. This requirement is known as the Toxic Release Inventory (TRI). The intent is to inform communities of the chemicals housed or released in their areas and to help communities prepare to respond to chemical spills and similar emergencies. Consequently, Congress mandated that this information be made public. It is widely available from the U.S. EPA, from state and local environmental and safety agencies, and in many public libraries. It is also available directly on the World Wide Web.

For purposes of the TRI, a *release* is defined as any spilling, leaking, pumping, pouring, emitting, emptying, discharging, injecting, escaping, leaching, dumping, or disposing into the environment (including the abandonment or discarding of barrels, containers, etc., and other closed receptacles) of any "toxic chemical."

8.5.1 Reporting Requirements

All firms that fall within the following classifications must report annually: (1) firms with 10 or more full-time employees; (2) firms included in the Standard Industrial Classification codes 20 through 39 (see Table 8.4); and (3) firms that manufactured or processed a reportable toxic chemical in quantities exceeding 25,000 pounds per year,

TABLE 8.4
Standard Industrial Classification Codes requiring TRI reporting

SIC no.	Industry	SIC no.	Industry
20	Food and kindred products	30	Rubber and miscellaneous plastics
21	Tobacco products	31	Leather and leather products
22	Textile mill products	32	Stone, clay, glass, and concrete products
23	Apparel and other finished fabric products	33	Primary metal industries
24	Lumber and wood products	34	Fabricated metal products
25	Furniture and fixtures	35	Industrial and commercial machinery and computer equipment
26	Paper and allied products	36	Electrical and electronic equipment (except computer equipment)
27	Printing and publishing	37	Transportation equipment
28	Chemicals	38	Measuring, analyzing, and controlling instruments
29	Petroleum refining and coal	39	Miscellaneous manufacturing industries

Source: U.S. EPA, 1997.

or that otherwise used more than 10,000 pounds of any designated chemical (without incorporating it into any product or producing it at the facility, such as using a metal cutting fluid or a tool degreaser). The EPA is planning to soon add a number of nonmanufacturing industry sectors to the TRI program, including metal mining, coal mining, electrical utilities, RCRA Subtitle C hazardous waste treatment and disposal facilities, chemical and allied product wholesale distributors, petroleum bulk stations and terminals, and solvent recovery services. More than 20,000 manufacturing firms are currently required to submit TRIs annually. Because many firms have more than one facility, over 80,000 TRI reports are submitted to the U.S. EPA and state agencies annually.

A wide range of information must be reported by each facility on Form R of the TRI, including the following:

- Amounts of each listed chemical released to the environment at the facility.
- Amounts of each chemical shipped from the facility to other locations for recycling, energy recovery, treatment, or disposal.
- Amounts of each chemical recycled, burned for energy recovery, or treated at the facility.
- The maximum amount of chemical present on site at the facility during the year.
- Types of activities conducted at the facility involving the toxic chemical.
- Source reduction activities.
- The efficiency of waste treatment.

- Environmental permits held.
- Name and telephone number of a contact person.

8.5.2 Toxic Release Inventory Chemicals

The TRI chemical list subject to reporting contained 586 toxic chemicals and 28 chemical categories for 1995 (see Appendix D for the list). The length of this list has been increasing annually and chemicals are continually being added and deleted from the list. Thus comparison of data from one year to the next is difficult. An apparent increase or decrease in total emissions from a firm from one year to the next may merely reflect changes in the chemicals that must be reported. The Pollution Prevention Act (PPA) added new requirements that expanded and made mandatory the source reduction and recycling information that is also reported for the EPCRA list of toxic chemicals. Toxic Release Inventory reports indicate that over 2.2 billion pounds of listed toxic material were discharged into the environment in 1995, most of which (71 percent) went into the air (see Table 8.5).

Most waste materials (3.5 billion pounds) were transferred to off-site locations for recycling, energy recovery, treatment, and disposal in 1995, rather than being directly discharged to the environment. It must be remembered, though, that many of these wastes may still ultimately reach the environment. Table 8.6 shows the amounts of TRI wastes transferred off site in 1995.

Table 8.7 lists the 10 TRI chemicals released in the greatest quantity in 1995. Facilities reported releasing more than 100 million pounds each of four chemicals: methanol, ammonia, toluene, and nitrate. For the first three chemicals, the primary release medium is air; for nitrate compounds it is water. Most of these top 10 compounds are fortunately not acutely toxic or carcinogenic. However, more than 230 million pounds of TRI-listed carcinogens were released to the air, water, and land in 1995. Table 8.8 lists the 10 OSHA carcinogens on the TRI list with the largest quantities of total releases in 1995.

TABLE 8.5
TRI releases in the United States, 1995

Source	Amount, lb
Total releases	2,208,749,411
Fugitive air	85,094,609
Point-source air	1,177,227,504
Surface water	136,315,624
Underground injection	234,979,709
On-site land releases	275,131,965

Source: U.S. EPA, 1997.

TABLE 8.6
TRI transfers in the United States, 1995

Transfer type	Amount, lb
Total transfers	3,534,827,951
Transfers to recycling	2,213,731,389
Transfers to energy recovery	512,029,726
Transfers to treatment	287,576,863
Transfers to POTWs	239,836,516
Transfers to disposal	279,222,397
Other off-site transfers	2,431,060

Source: U.S. EPA, 1997.

TABLE 8.7
Top 10 TRI chemicals released, 1995

Chemical	Total release, lb
Methanol	245,012,356
Ammonia	195,096,446
Toluene	145,096,446
Nitrate compounds	137,743,102
Xylene (mixed isomers)	95,739,943
Zinc compounds	87,648,691
Hydrochloric acid	85,330,532
Carbon disulfide	84,169,763
n-Hexane	77,396,162
Methyl ethyl ketone	70,054,939
Subtotal	1,224,079,403
Total for all TRI chemicals	2,208,749,411

Source: U.S. EPA, 1997.

TABLE 8.8
Top 10 OSHA carcinogens released, 1995

Chemical	Total release, lb
Dichloromethane	57,289,960
Styrene	41,873,608
Trichloroethylene	25,489,839
Formaldehyde	19,426,396
Acetaldehyde	14,410,140
Chloroform	10,600,257
Benzene	9,592,003
Tetrachloroethylene	9,400,811
Acrylonitrile	6,471,484
Acrylamide	6,141,395
Subtotal	200,695,893
Total of all OSHA carcinogens	230,134,414

Source: U.S. EPA, 1997.

8.5.3 Problems with Data

As specified in Section 313 of EPCRA, Form R is used for the Toxic Release Inventory reporting requirement. For listed chemicals, a manufacturer or user must determine if a threshold quantity has been exceeded. If the threshold is exceeded, releases to the various media are determined by a variety of methods, including actual measurements, engineering calculations, mass balances, and published emission factors (Harper, 1991). Releases are recorded as fugitive air emissions, stack emissions, water body releases, releases to land, underground injection, discharges to POTWs, and off-site transfers for disposal, recycling, treatment, and energy recovery.

There definitely are problems associated with the TRI data. An EPA survey several years ago estimated the compliance rate for reporting in 1987 at 66 percent; however, the EPA believes that outreach and enforcement activities in subsequent years have substantially increased the rate of compliance. Harper (1991) found that Form R estimates are frequently prepared by people with little or no technical background, and minimal training in proper estimation techniques.

A TRI requires the reporting of *estimated* data, but it does not mandate that facilities monitor their releases. The company may use readily available data to report the quantities of chemicals used or released to the environment. If no data are available, the law permits the firm to report reasonable estimates. Variations between facilities' estimates can result from the use of different estimation methodologies. A comparison was made between wastewater data sampled and analyzed by POTW personnel in Cincinnati, Ohio, and wastewater emissions data determined by industry and reported as part of the TRI requirements (Dunams, 1997). Data on yearly emissions of TRI chemicals obtained from TRI reports submitted annually by industries, in pounds per year, were evaluated against actual values measured by the POTW to determine if TRI data are representative of the POTW data. It was found that TRI data do not often accurately depict the data collected by the POTW.

Use of existing TRI data to measure P2 progress may be advantageous because no new data collection is required. It also reports on releases to all aspects of the environment, rather than just to one medium, such as air or water. However, the TRI covers only a limited number of substances of the thousands of industrial chemicals in use. Moreover, the accuracy of the data has been questioned (Malkin and Baskir, 1995). Although valuable for evaluating multimedia releases, the estimates may have limited usefulness for determining accurate pollution prevention efforts in the wastewater medium.

Measurement of pollution prevention trends using TRI data is complicated by the fact that many reported decreases in total TRI releases and transfers in fact merely reflect changes in how releases and transfers are estimated or reported and not actual changes in pollution generation patterns (Karam, Craig, and Currey, 1991). Normalized pollution measures (measures of pollution and of pollution reduction per unit of production activity) can help in targeting pollution prevention opportunities and measuring pollution prevention progress. Graham (1993) noted that only 49 percent of her survey respondents normalized measurement procedures for changes in production.

Operators of POTWs need accurate measurements of industrial chemical inputs and must use a sampling strategy that adequately characterizes plant influent and accounts for temporal changes. Wastewater composition will vary depending on diurnal, weekly, and seasonal flows, chemical use patterns, and industrial chemical changes (Epstein and Skavroneck, 1996). As described above, the TRI data submitted are often only estimates and may not adequately mirror the actual discharges to the POTW. Thus TRI data, although relatively easy to evaluate and effective in examining multimedia releases, may not be applicable as a pollution prevention tracking protocol for determining reduction trends in wastewater emissions. Thus rather than relying on TRI data, a program that relies on marginal reductions of wastewater emissions, such as issuance of marketable wastewater permits, should use POTW data that are adjusted to be in a mass-based form to justify percentage reductions or to quantify increasing emission levels.

REFERENCES

Cascio, J., Woodside, G., and Mitchell, P. *ISO 14000 Guide*. New York: McGraw-Hill, 1996.

Delaney, B. T., and Schiffman, R. I. "Organizational Issues Associated with the Implementation of ISO 14000." *Environmental Engineer* 33, no. 1 (1997): 9–11, 23–25.

"Doing a Detailed Pollution Prevention Assessment." *Environmental Regulatory Advisor* 6, no. 10 (1997): 6–8.

Dunams, A. *Guidelines for the Development of a Pollution Prevention Program Pertaining to Industrial Users of the Greater Cincinnati Metropolitan Sewer District.* M.S. thesis. University of Cincinnati, 1997.

Epstein, L. N., and Skavroneck, S. A. "Promoting Pollution Prevention." *Water Environment & Technology* 6 (1996): 55–59.

Graham, A. B. "The Results of PPR's 1993 Survey: Industry's Pollution Prevention Practices." *Pollution Prevention Review* 3 (1993): 369–81.

Harper, P. D. "Application of Systems to Measure Pollution Prevention." *Pollution Prevention Review* 1 (1991): 145–53.

Karam, J. G., Craig, J. W., and Currey, G. W. "Targeting Pollution Prevention Opportunities Using the Toxics Release Inventory." *Pollution Prevention Review* 1 (1991): 131–44.

Malkin, M., and Baskir, J. N. "Issues in Facility-Level Pollution Prevention Measurement." *Environmental Progress* 14 (1995): 240–246.

Pojasek, R. B., and Cali, L. J. "Contrasting Approaches to Pollution Prevention Auditing." *Pollution Prevention Review* 1 (1991): 225–35.

Pollution Prevention: A Guide to Program Implementation. Champaign: Illinois Hazardous Waste Research and Information Center, 1993.

Serageldin, M. A. "Standardized Accounting for a Formal Environmental Management and Auditing System." In *Waste Management Through Process Design,* ed. A. P. Rossiter. New York: McGraw-Hill, 1995, pp. 289–303.

U.S. EPA. *Waste Minimization Opportunity Assessment Manual.* EPA/625/7-88/003. Washington, DC: U.S. EPA, 1988.

U.S. EPA. *Facility Pollution Prevention Guide.* EPA/600/R-92/088. Washington, DC: U.S. EPA, 1992.

U.S. EPA. *1995 Toxics Release Inventory Public Data Release.* Washington, DC: U.S. EPA, 1997.

PROBLEMS

8.1. Visit your university's cafeteria kitchen or that of a local restaurant during serving time and observe the operation. It is much like that at any manufacturing facility. Feedstock (food ingredients) is used along with energy inputs in a manufacturing operation to produce a final product (your meal), which is then sold to and used (eaten) by the consumer. There are opportunities for waste minimization and recycling. Prepare a preliminary assessment of the operation, paying particular attention to those items listed in Section 8.2.2.

8.2. Using the preliminary assessment developed in Problem 8.1, construct a draft P2 program plan for the kitchen. Include the plan components listed in Table 8.2.

8.3. Select several of the potential P2 options identified in Problem 8.2. Using the option rating weighted-sum method, quantify the important criteria that affect these options and prioritize them.

8.4. Select one of the highest rated P2 options from Problem 8.3 and discuss how you would develop this further in order to determine if it is technically and economically feasible and, if it is, how it could be implemented.

8.5. A textile mill has aggressively undertaken a number of P2 projects. One target was a reduction in the amount of BOD in the mill's wastewater effluent. Data in the accompanying table show annual wastewater BOD loadings before (years -2 and -1) and after (years 1 and 2) instituting the P2 projects. What can you say about the P2 program? Has it been effective?

Year	Fabric produced, metric tons/yr	Wastewater volume $\times 10^3$ m^3/yr	BOD concentration, mg/L
-2	800	455	1055
-1	810	465	1097
1	790	372	968
2	870	410	1024

8.6. Using the Internet, find a company that has instituted ISO 14001 Environmental Management Standards and report on its program.

8.7. Locate the major polluting companies in your area, based on total mass releases, using TRI data. What are they, what are their major emissions, and into what media are they discharging?

8.8. Not all chemicals on the TRI list are of equal hazard potential. Referring to your results from Problem 8.7, are the top five companies on your list of major polluters the most harmful in your community, or are there other companies lower on your list that emit less materials, which are potentially more harmful?

CHAPTER 9

DESIGN FOR THE ENVIRONMENT

9.1 INTRODUCTION

Superior Coatings, Inc., specializes in the painting of metal oil filter casings for the automotive industry. Because these filters are located under the hood of the automobile and are not readily visible, the appearance of the coating need not be of high quality. Originally, the company used a dip tank to apply the paint. The filter casings were immersed in a vat of paint, after which the piece was suspended over the tank to allow excess paint to run off. This process was wasteful, however, as paint would accumulate in turnings or depressions in the part, and the paint often drained unevenly from the piece. Later, Superior changed to a spray paint process to minimize the amount of paint being applied and to provide a more uniform coating. The paints used were solvent-borne, with high VOC contents. These coatings were favored because of their fast drying properties. However, they emitted significant quantities of hazardous VOCs as they were applied and as they dried; some of these VOCs are also ozone-depleting compounds.

Because of the potential health danger to their workers and the general public, and to reduce the release of ozone-depleting compounds, Superior Coatings, Inc., an environmentally progressive company, made the decision to switch from the use of solvent-based to water-based paints in 1992. These water-based paints contain only a small amount of organic solvent (the typical water to solvent ratio is 80:20) and thus are much safer to use. However, they do have drawbacks. They still contain some solvents as cosolvents and thus do emit some VOCs, although considerably less than do solvent-based coatings. They require much longer drying times, so that a larger drying room is needed. Also, water-based paints are typically more sensitive to residual oils and greases on the surface of the part. Therefore, more care has to be employed in the initial cleaning of the parts before the painting process. Otherwise, the water-based paint might not adhere properly.

The company is constantly investigating new ways to manufacture their products to provide the highest quality at the lowest cost, both financially and environmentally. Currently, Superior is investigating the use of several new techniques to reduce VOC emissions even further and to improve the quality of the final product. Among these are powder coatings and baked coatings.

Powder technology is the fastest growing coating technology available. Powder coatings are organic coatings that are supplied in extremely fine dry powder form. The most common method of application is to charge the powder in a spray nozzle using a high-voltage electrode. The electrostatically charged particles adhere loosely to the metal substrate being coated. The powder is then cured by heating in a convection oven for 8–20 minutes at a temperature of 325–400°F. Powder coatings are generally the lowest polluting of all coatings, with essentially zero VOC emissions and no hazardous or toxic residues; they produce no air, water, or hazardous waste pollution. They also provide excellent hardness, mar resistance, abrasion resistance, flexibility, and solvent resistance (U.S. EPA, 1996). However, powder coatings are not as versatile as liquid coatings. They cannot be applied to many materials, such as plastics, wood, paper, cloth, masonry, or rubber. They work well with simple parts but are often not suitable for parts with many deep recesses or inaccessible areas. In addition, they require special application equipment and curing temperatures in excess of 325°F.

Baked finishes use water-borne coatings. The main difference between these and regular water-borne coatings is that the baked finishes are cured at 325–350°F, rather than at room temperature. The coatings generally contain polymer resins that cross-link at the high curing temperature, providing the coating with excellent hardness as well as mar and abrasion resistance.

To decide which coating process will be best for the oil filter housings, Superior Coatings, Inc., undertook an extensive testing program to determine the quality of the finished products from each process, the effect of each on the rate of production, the anticipated improvements in environmental quality associated with the plant's emissions, and the financial costs of each process. Based on the results of the life-cycle assessments, the company decided to switch to the powder coating process.

This example is only one of many where a company has decided to alter the design of a product based, at least in part, on its environmental impact. This new paradigm, *Design for the Environment,* is now becoming widely accepted across many manufacturing sectors.

9.1.1 Design for X

Traditionally, engineering design has focused on achieving products of high quality using the best technology available to the company at as low a cost as possible. This "cost" has normally involved only the economic cost of producing the product. Now, though, some companies are going further to include the costs of waste and discarded products along with the costs of the environmental impacts caused by the manufacture, use, and disposal of the product.

The U.S. Congress's Office of Technology Assessment (OTA) coined the phrase "green design" to signify a design process in which environmental attributes of a product are treated as *design opportunities,* rather than design constraints. Green design incorporates environmental objectives with minimum loss to product performance, useful life, or functionality. By this definition, green design involves two general goals: waste prevention and better materials management (see Figure 9.1). *Waste prevention* refers to activities that minimize the generation of the waste at the source. *Better materials management* calls for product design that allows for remanufacturing operations or waste management methods so that the product or its components can be easily recovered and reused after its service life is over (OTA, 1992). The ultimate goal should be "sustainable development."

A company can reduce the pollution resulting from the manufacture or use of a product in a number of ways. For example, it may evaluate the process used to manufacture the product, with the goal of reducing process wastes; it may examine the materials used in manufacture and investigate the efficacy of using less harmful materials or materials that are more sustainable; it may redesign the product so that it is more readily recyclable or more easily disassembled for reuse/recycling; it may choose to redesign the product or the manufacturing process so that its manufacture or use requires less energy; it may redesign the product to extend its useful life; or it may attempt to minimize potential pollution from the final disposal of the product after its use. Each of these options has been given a name, such as Design for the Environment (DfE) or Design for Disassembly (DfD); Table 9.1 lists some of the more commonly used terms. The collective term for all of these options is Design for X (DfX), where X refers to whatever objective has been selected.

Design for the Environment has been defined as "a practice by which environmental considerations are integrated into product and process engineering design procedures" (Allenby, 1991). On analysis of this definition, it readily becomes obvious that DfE and life-cycle assessments are intimately intertwined, both involving the

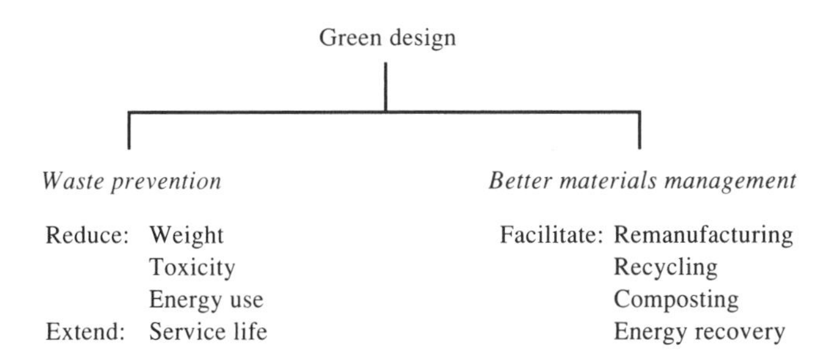

FIGURE 9.1
The dual goals of green design. (Adapted from OTA, 1992)

TABLE 9.1
Design for X categories

Category	Acronym	Description
Environment	DfE	Implement pollution prevention, energy efficiency, and other resource conservation measures to reduce adverse impacts on human health and the environment
Manufacturability	DfM	Integrate a product's manufacturing requirements into the fabrication and assembly processes available in the factory
Disassembly	DfD	Design the product for ease of disassembly and component/material reuse/recycling after the product's useful life is over
Recycle	DfR	Design the product so that it can be easily recycled
Serviceability	DfS	Design the product so that it can be easily installed, serviced, or repaired
Compliance	DfC	Design the product so that it meets all regulatory requirements

refinement of a manufacturing process or of the product itself in order to reduce pollution. As described in Chapter 5, to be truly effective, the process should begin at the earliest conceptual stages of design and continue through preliminary engineering and final process design. All aspects of the product and its manufacturing process, including materials selection, manufacturing process selection, and the product design itself, should be considered. Further, DfE should include an analysis of postconsumer use, including disassembly, recycling and reuse of components and parts, recovery of valuable materials, and environmentally acceptable disposal of the nonreusable fraction (see Figure 9.2).

Figure 9.2 indicates various aspects of a good design. These are not stand-alone design procedures, however. A good product design will incorporate many or all of these design components.

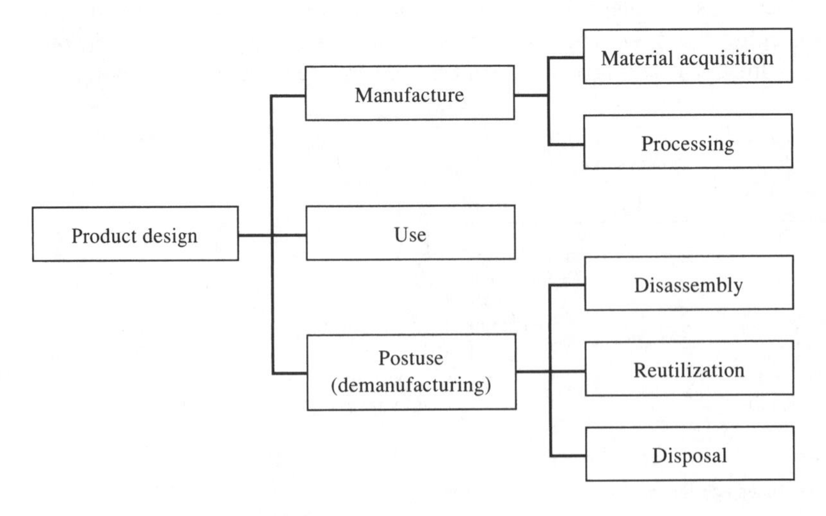

FIGURE 9.2
Major factors that should be considered during product design.

9.1.2 Chapter Focus

This book focuses on Design for the Environment. In particular, Chapter 5 dealt with manufacturing processes and their impact on pollution prevention, while Chapter 6 focused on use of life-cycle assessment to evaluate optimum pollution prevention processes. However, other DfX components also have a significant impact on environmental quality. For example, a product which is designed for ease of disassembly will make reuse/recycling of components of that product easier, thus reducing the amount of material disposed into landfills. Good design will also reduce the amount of raw materials needed, thus reducing resource depletion issues. In many cases, reuse of a material is also more energy efficient than using virgin materials. Later in this chapter we examine Design for Disassembly in greater detail.

The various manufacturing processes and their impacts on waste minimization were described in Chapter 5. This chapter will investigate a number of additional design strategies that can be used to minimize pollution. Among these are the newly emerging field of "green chemistry," which involves the design, manufacture, and use of environmentally benign chemical products and processes; Design for Disassembly and Design for Recycle; packaging; and finally recycling operations. The use of pinch analysis procedures to minimize the use of materials, water, or energy in a manufacturing process is discussed in the next chapter.

9.2 GREEN CHEMISTRY

The primary goal of industry has always been, and will continue to be, to maximize profits from the sale of its products. In recent years, though, other constraints have been placed on industry, including the need to conduct business in an environmentally acceptable fashion. Whether due to regulatory decree or to a desire to decrease environmental management costs or to be perceived by the public as being more environmentally conscious, many industries are exploring the uses of green chemistry.

Green chemistry, also called "benign chemistry" or "clean chemistry," refers to a new field of chemistry dealing with the synthesis, processing, and use of chemicals that reduce risks to humans and the environment (Anastas and Williamson, 1996). It is also defined as synthetic chemistry designed to minimize generation and use of hazardous substances. The ultimate goal is to develop and institute alternative syntheses for important industrial chemicals in order to prevent environmental pollution (Anastas and Farris, 1994). Although no chemical is exactly "benign," much can be done to replace the use of toxic or potentially dangerous chemicals with much more innocuous ones. By using the extensive information available on human health effects and ecological impacts of various chemicals, it is possible to select chemicals that would be more environmentally favorable to use in a particular synthesis or process. These chemicals may be the products themselves, or the feedstocks, solvents, and reagents used in making the products. Further, green chemistry involves a detailed study of the by-products from the synthesis and the effects these by-products have. A synthesis pathway that uses environmentally benign feedstocks and that produces a safe product may still not be a good process if harmful by-products are also produced. Green chemistry concepts can also be used to evaluate the inputs to a synthesis pathway and determine whether it is possible to reduce the use of endangered resources by switching to more plentiful or renewable ones.

Thus industrial chemists can no longer concern themselves only with the chemical they are producing. They must also be cognizant of

1. Hazardous wastes that will be generated during product synthesis.
2. Toxic substances that will need to be handled by the workers making the product.
3. Regulatory compliance issues to be followed in making the product.
4. Liability concerns arising from the manufacture of this product.
5. Waste treatment costs that will be incurred.
6. Alternative product synthesis pathways or processes that may be available.

The first five of these questions were discussed in previous chapters. The last question is addressed next.

9.2.1 Sources of Wastes

The amount of chemicals disposed into the environment is enormous. (For more on this, review Chapter 3.) The chemical manufacturing sector is by far the largest releaser of chemicals into the environment. These wastes arise from the preparation of feedstocks; from the synthesis reaction itself; from product separation; and from the production of energy used in the synthesis. Table 9.2 lists the predominant organic chemicals produced in the United States in 1996.

In the chemical industry, the first step in the synthesis of a chemical is the conversion of raw materials from nature into useful feedstocks and intermediates. These are then reacted to form the desired products, which must be separated from the bulk reaction mixture, purified, packaged, and transported to the consumer. Hydrocarbons from oil or coal are common raw materials for chemical synthesis. The needed feedstocks are obtained by cracking, distilling, pyrolyzing, or otherwise processing these raw materials into the necessary fractions. Figure 9.3 presents a typical process flow chart (adapted from Braithwaite, 1995) for the production of a pharmaceutical from petrochemical feedstocks derived from oil. The intermediates produced along the way may also have value in the production of other chemicals. During the process, the crude oil is converted into propylene, which is converted into acrylonitrile, a commodity chemical used in many processes. The acrylonitrile is converted to 2-diethylaminoethanolamine, a fine chemical. This in turn is converted to metaclopramide, a specialty chemical, and finally to the analgesic drug Paramax, used to combat migraine headaches. At each step, value is added and the chemicals become much more costly. The chemical yields rapidly decrease as the synthesis proceeds. Of the 12 million metric tons (t) per year of propylene processed to produce this drug, only about 50 t/yr of the drug are produced (0.00042 percent of the starting material). Consequently, even minor changes in the price of the raw materials can have a significant effect on the cost of the final product. The same can be said for improvements in yield due to pollution prevention.

All chemical syntheses create pollution. Of particular concern are reactions that produce significant amounts of by-product wastes. Take, for example, the following stoichiometric reaction:

TABLE 9.2
Production of selected organic chemicals and plastics in the United States, 1996

Material	Uses	Amount (millions of pounds)	Change (1986–1996), %
Acrylonitrile	Manufacture of allyl compounds, resins	3,373	44%
Aniline	Manufacture of dyes, resins, medicinals	1,079	31
Benzene	Manufacture of chemicals, polymers, dyes	2,116	55
1,3-Butadiene	Manufacture of polymers, synthetic rubber	3,845	51
Cumene	Manufacture of phenol, acetone, methylstyrene	5,879	57
Ethylbenzene	Conversion to styrene monomer, solvent	10,359	15
Ethylene	Manufacture of alcohols, ethylene oxide, plastics	49,097	49
Ethylene dichloride	Solvent; used as fumigant	11,336	−12
Ethylene oxide	Manufacture of ethylene glycol; fumigant	7,239	33
2-Ethylhexanol	Solvent; mercerizing textiles	760	33
Isopropyl alcohol	Solvent	1,384	6
Polyethylene	Plastics	7,784	−13
Polypropylene	Plastics	11,991	106
Polyamide, nylon type	Synthetic fibers	1,120	141
Polyvinyl chloride	Plastics	13,220	82
Propylene	Plastics	25,111	52
Styrene	Manufacture of plastics, resins, synthetic rubber	11,874	51
Urea	Manufacture of fertilizer, animal feed, plastics	16,258	30
o-Xylene	Solvent, chemical intermediate	887	13
p-Xylene	Solvent, chemical intermediate	6,167	22

Source: Adapted from "Facts and Figures," 1997.

$$A + B \rightarrow C + D$$

where A and B are the reactants, C is the desired product, and D is a waste by-product. For each molecule of C produced, one of D is also produced. This means that there will be an enormous amount of waste to be separated and disposed of. The only way to reduce the amount of waste while maintaining the desired product yield is to use a different synthesis route or to find a use for the by-product (Lester, 1995).

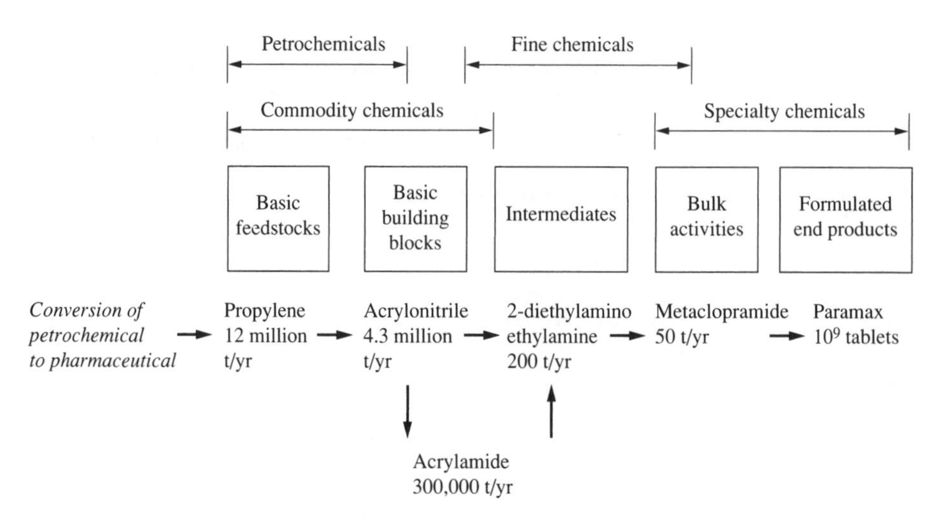

FIGURE 9.3
Chemical synthesis of a pharmaceutical from petroleum.

An example of this is the production of allyl alcohol. The traditional route was the alkaline hydrolysis of allyl chloride, which gave stoichiometric amounts of hydrochloric acid as a by-product:

$$CH_2{=}CHCH_2Cl + H_2O \xrightarrow{\text{Alkali}} CH_2 + CHCH_2OH + HCl$$

A more preferable process involves a two-step reaction using propylene, acetic acid, and oxygen in an aqueous solution:

$$CH_2{=}CHCH_3 + CH_3COOH + {}^1\!/_2\, O_2 \longrightarrow CH_2{=}CHCH_2O\overset{\overset{\displaystyle O}{\|}}{C}CH_3$$

$$CH_2{=}CHCH_2OCCH_3 + H_2O \longrightarrow CH_2{=}CHCH_2OH + CH_3COOH$$

The acetic acid produced in the second reaction can be recovered and used again for the first reaction, leaving no unwanted by-products.

A second source of wastes from chemical synthesis is by-products from secondary reactions, which are undesirable but which occur in conjunction with the desired reaction. These may be represented by

$$A + B \rightarrow C \qquad \text{(desired reaction)}$$

$$A + B \rightarrow D \qquad \text{(undesired side reaction)}$$

By-product D may be a contaminant and of no value, making its presence undesirable. In some cases, the amount of D produced may be reduced by adjusting the operating conditions. A common example of this is chlorination of water for disinfection. The chlorine not only oxidizes cellular constituents in the pathogenic bacteria present, but also oxidizes any organic compounds that may be present, forming harmful chlorinated organic compounds such as chloroform.

A third source of chemical synthesis wastes is the result of a reaction sequence in which the desired product is not the final product, but rather one of the intermediates. In the following sequence, C is the desired product, but C will continue to react to form D unless the reaction is stopped at the correct time:

$$A + B \rightarrow C \rightarrow D$$

Again, it may be possible to adjust reaction conditions to minimize production of D, but this may result in a low conversion rate of reactants A and B to C. An example of this is the production of the very valuable compound ethylene oxide, a much in demand compound that is used as a precursor in the production of ethylene glycol (antifreeze), acrylonitrile (acrylic fibers), and non-ionic surfactants.

$$\underset{\text{Ethylene}}{CH_2{=}CH_2} + \tfrac{1}{2}\,O_2 \longrightarrow \underset{\text{Ethylene oxide}}{H_2C\text{-}CH_2\ (\text{bridged by } O)} \longrightarrow H_2O + CO_2$$

It is difficult to stop the reaction from going all the way to carbon dioxide and water; inhibitors such as halogenated organics (significant pollutants in their own right) can be added, but some 20 percent to 25 percent of the ethylene oxide is still oxidized to carbon dioxide and water (Lester, 1995).

Finally, a host of other sources of wastes from organic compound syntheses may arise from chemical contaminants or impurities in the feedstocks, or the products of reactions these may undergo in parallel with the main reaction; from the solvents, acids, bases, or catalysts used in the reaction; or in some cases from the reactors themselves. Commonly, a significant source of wastes is the unreacted reagents used in the reaction. Most industrial reactions are not allowed to go to completion because of the time involved to get the last part of the material reacted, so that the process mixture at the end of the reaction still contains quantities of unreacted chemicals. To increase the conversion of a valuable reactant, it is common to supply greater than stoichiometric amounts of other reactants to drive the reaction further to the right. This, however, results in large amounts of these excess reactants in the waste. The objective of the synthesis process should be to get the raw materials to as near stoichiometric proportions as possible to minimize unreacted materials, by-products, and necessary purification steps, while maximizing product yield.

Some chemical operations that lead to the production of the desired chemical produce many other chemicals at the same time. Consider the cracking of hydrocarbons such as naphtha in crude oil to produce the very valuable feedstock ethylene. Ethylene is produced, but the yield is low; only about 32 percent of the naphtha is converted to ethylene. Other products include methane (17 percent), propylene (15 percent), butadiene (5 percent), and a mixture of other chemicals (31 percent). These coproducts may be thought of as contaminants of the ethylene, but progressive chemical companies will look for markets or other uses for them if the ethylene yield cannot be increased.

Most or all of these contaminants must be separated from the desired product in order to achieve the desired product quality. Separation is expensive, treatment of the waste is often difficult and can significantly drive up the cost of the product, and inefficient conversion wastes chemicals. Whatever can be done to refine the chemical synthesis processes to minimize wastes will usually result in substantial economic savings and increased profits to the company. Green chemistry provides the best route to this goal.

The U.S. Environmental Protection Agency has taken a lead role in promoting green chemistry. On March 16, 1995, President Bill Clinton announced a voluntary partnership program charged with facilitating this goal. The Presidential Green Chemistry Challenge was established to recognize and promote fundamental and innovative green chemical technologies. The EPA's Office of Pollution Prevention and Toxics is leading this program. For purposes of the program, green chemistry is defined as the use of chemistry for source reduction; it consists of a reduction in, or elimination of, the use or generation of hazardous substances—including feedstocks, reagents, solvents, products, and by-products—from a chemical process. Green chemistry includes all aspects of chemical processes, including reaction conditions. The challenge recognizes and promotes the following green chemistry methodologies:

1. The use of alternative synthetic pathways for green chemistry, such as
   a. Catalysis/biocatalysis.
   b. Natural processes, such as photochemistry and biomimetic synthesis.
   c. Alternative feedstocks that are more innocuous and renewable (e.g., biomass).
2. The use of alternative reaction conditions for green chemistry, such as
   a. Use of solvents that have a reduced impact on human health and the environment.
   b. Increased selectivity and reduced wastes and emissions.
3. The design of chemicals that are, for example,
   a. Less toxic than current alternatives.
   b. Inherently safer with regard to accidental exposure.

These efforts have been summarized into two primary modes of attack: *alternative synthetic pathways,* in which chemists rethink the chemical synthetic steps and pathways used to manufacture industrial and commercial chemicals, and *design of safer chemicals,* in which the chemist designs the molecular structure of the final product so that it is safe or safer than the chemical that it replaces and yet is effective with respect to its intended use (Garrett, 1996).

Each of these categories of potential green chemistry solutions to vexing chemical industry problems will be discussed in more detail in the following sections. It must be kept in mind, though, that the primary criteria for evaluating a synthetic chemical pathway in the manufacture of a chemical product is the yield of the process (Anastas, 1994). *Yield* is the percentage of product obtained versus the theoretical amount one could have obtained for a given amount of starting material. The objective of chemical synthesis is to make the chemical in as high a yield as possible, in an economical fashion. A more benign chemical process that results in a lower yield may be more environmentally acceptable, but it may not be economically viable. On the other hand, a high yield usually correlates well with the greenness of a process, because it indicates that most of the starting materials are converted to product and few will remain in the reacted process stream.

9.2.2 Alternative Synthetic Pathways

Many of the problems encountered during the production of synthetical chemicals arise because of the synthesis pathway selected. By changing the feedstocks or starting materials, it is often possible to minimize, or even eliminate, many undesirable intermediates. Often, the yield of a chemical can be enhanced by use of catalysts or biocatalysts. An increased yield means less starting materials and fewer by-products in the reacted mixture, thus minimizing pollution. Finally, in some cases natural processes can be used to improve the production process.

ALTERNATIVE FEEDSTOCKS. Either more environmentally benign feedstocks can be used or those that are employed can be used in a more efficient manner to reduce the amount needed. This not only reduces the amount of pollution and toxicity of the wastes produced, but also improves worker safety, as workers will be exposed to more benign chemicals or to smaller amounts of toxic materials.

The vast majority of chemical products produced throughout the world are derived from petroleum feedstocks. Many of these, such as benzene, are very hazardous and may be carcinogenic. Much research is under way to find replacements for petroleum feedstocks. One example of this is the manufacture of styrene. The traditional styrene production process consists of two steps. In the first step, benzene is alkylated with ethylene using an acid catalyst. The second step dehydrogenates the ethyl benzene to stryene using a dehydrogenation catalyst:

$$C_6H_6 + H_2C{=}CH_2 \xrightarrow{\text{Catalyst}} C_6H_5{-}CH_2{-}CH_3$$

Benzene | Ethylene | Ethylbenzene

$$C_6H_5{-}CH_2{-}CH_3 \xrightarrow{\text{Catalyst}} C_6H_5{-}CH{=}CH_2$$

Ethylbenzene | Styrene

Both steps give high yields with low production of by-products, but there is significant concern about the process because of the requirement to begin with benzene, a known human liver carcinogen. The process consumes 13 billion pounds of benzene per year. Not only are there concerns about dangers of using benzene in the styrene production process; there are also major dangers with the production and transport of the benzene itself.

Researchers have investigated alternatives to benzene as the starting material. Toluene has been proposed, but no satisfactory pathway has been developed yet. Recently, the use of mixed xylenes has been proposed (Chapman, 1994). Mixed xylenes are the cheapest source of aromatics available and are environmentally safer than benzene. The proposed process uses a single-step high-temperature conversion that does not use a catalyst and gives 40 percent per pass yields of styrene.

Phosgene is an extremely toxic (threshold limit value = 0.1 ppm) and corrosive material that is widely used in chemical synthesis around the world. Many researchers are investigating ways to replace phosgene as a feedstock. One of its primary uses is in the production of polycarbonates and isocyanates. In the production of polycarbonates, bisphenol A is reacted with phosgene and sodium hydroxide in a methylene chloride–water mixture. In addition to problems with the use of phosgene, the methylene chloride is difficult to remove from the polycarbonate and the chlorine impurities have a negative effect on polymer properties.

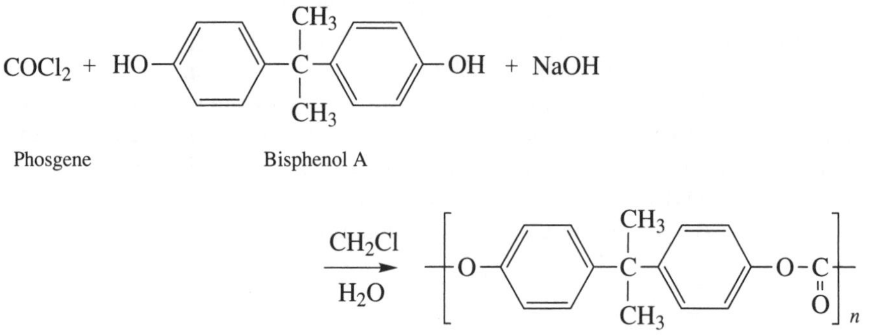

$COCl_2$ + HO–C$_6$H$_4$–C(CH_3)$_2$–C$_6$H$_4$–OH + NaOH

Phosgene Bisphenol A

$\xrightarrow[H_2O]{CH_2Cl}$ [–O–C$_6$H$_4$–C(CH_3)$_2$–C$_6$H$_4$–O–C(=O)–]$_n$

Polycarbonate

Ashai Chemical Industry Co., in Japan, has developed a process to produce polycarbonate without the need for phosgene or methylene chloride (Komiya *et al.,* 1996). In the Ashai process, bisphenol A is reacted with diphenylcarbonate. An amorphous prepolymer is created, crystallized, and then polymerized to give the desired product.

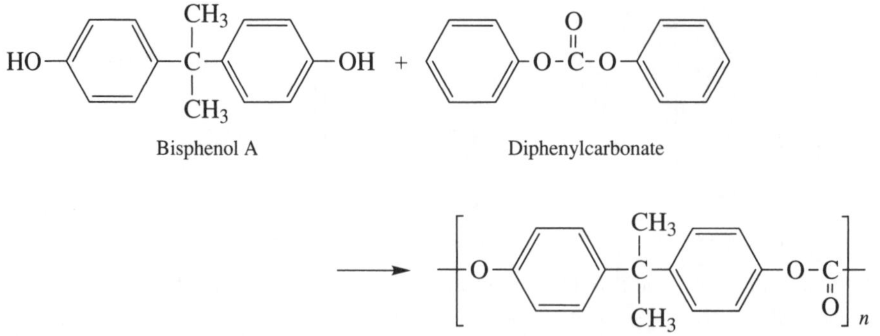

HO–C$_6$H$_4$–C(CH_3)$_2$–C$_6$H$_4$–OH + C$_6$H$_5$–O–C(=O)–O–C$_6$H$_5$

Bisphenol A Diphenylcarbonate

$\longrightarrow$ [–O–C$_6$H$_4$–C(CH_3)$_2$–C$_6$H$_4$–O–C(=O)–]$_n$

Polycarbonate

The company has found that it can produce polycarbonate of higher quality using this process than the conventional process, and the production costs are competitive.

A new and exciting area of research is the replacement of petroleum-based feedstocks with naturally occurring materials such as biomass. In the last century, biomass (i.e., plant matter and, to a lesser extent, animal matter) was the primary source of industrial organic chemicals. Eventually coal, and later petroleum, supplanted these sources and became dominant. It was not until recently that the negative aspects of petrochemical feedstocks caused researchers to again look to biological materials for organic chemical synthesis.

Biological feedstocks offer several advantages:

They are derived from renewable sources.

They are often highly oxidized, which allows for cleaner types of transformations, such as reductions. Most petroleum products are in a highly reduced state and must be oxidized before use. This often requires the use of heavy metals, which can add to the pollution problem.

Most biological feedstocks are less hazardous than petroleum-based feedstocks.

Biomass that is often considered waste materials (e.g., agricultural wastes) can be used as the precursors.

Figure 9.4 depicts examples of potential applications of biological feedstocks.

There are two predominant types of biomass for use in synthetic chemical production: starch and lignocellulosics (Webster, Anastas, and Williamson, 1996). Corn, wheat, sorghum, and potato are representative of the starch class, whereas agricultural wastes (corn cobs, wheat straw, etc.), forestry wastes, and dedicated woody and herbaceous crops comprise the bulk of the available lignocellulosics. Both starch and lignocellulosics contain polymers of sugars that can be broken down into monomers and used in chemical synthesis. The production of fuel ethanol from corn that is mixed with gasoline for automobiles (forming "gasohol") is an example of bioconversion.

Millions of tons of biological waste materials are available to replace petroleum as a feedstock for chemical production (see Table 9.3). The volume is large enough to have a significant impact on the production of industrial chemicals. Because they are generally considered "wastes," their cost to industry would be minimal. For them to be useful, however, a steady supply must be available, and the manufacturer must have the technology available to process this material.

An example of the use of naturally occurring biomass to produce organic compounds is the manufacture of adipic acid. Four billion pounds of adipic acid are produced each year using carcinogenic benzene as the feedstock. Most of this adipic acid is used to produce nylon, although small amounts are used in the preparation of polyurethanes, lubricants, and plasticizers and as a food acidulant. To prepare adipic acid, benzene is hydrogenated to produce cyclohexane, which is subsequently air oxidized in the presence of metal catalysts to yield a mixture of cyclohexanone and cyclohexanol. Nitric acid oxidation of the mixture results in adipic acid in yields of 92 percent to 96 percent (Draths and Frost, 1994).

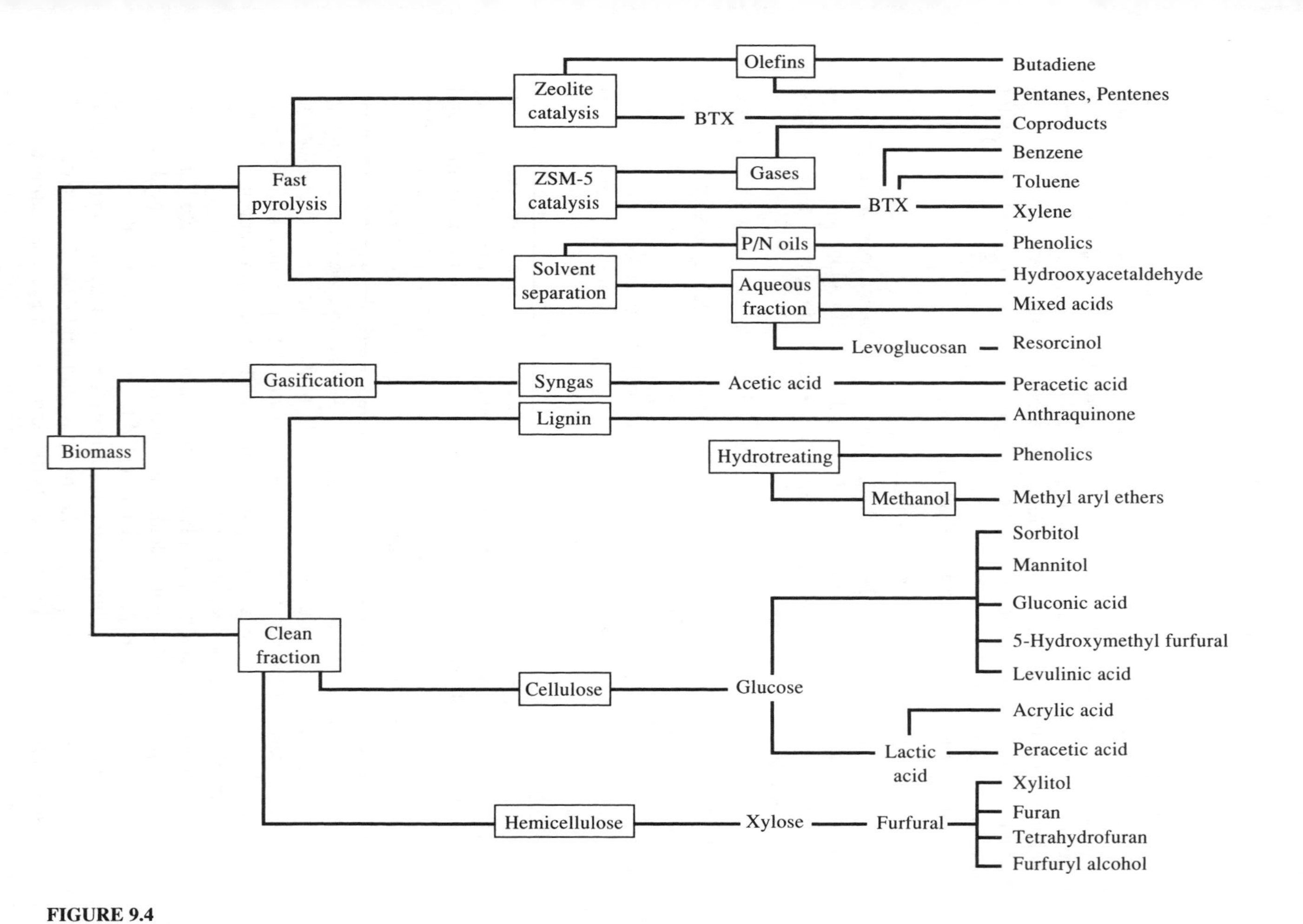

FIGURE 9.4
Flow chart for production of chemicals available from renewable materials. (Adapted from U.S. DOE, 1991)

TABLE 9.3
Current and potential biomass feedstocks in the United States

	Production, millions of dry tons/yr	
Feedstock	**Current**	**Potential**
Crops		
Lignocellulose	391	1565*
Starch crops	103	1575*
Forage grasses	26	414
Waste		
Agricultural	390	390
Livestock	290	325
Industrial	60	60
Municipal	125	160
Forestry	110	140
Total	1495	3064*

* Since some of the same land area may be required by both lignocellulose and starch crops, the quoted potential biomass for each could be mutually exclusive. The total potential feedstock includes only one of these sources.

Source: U.S. DOE, 1991.

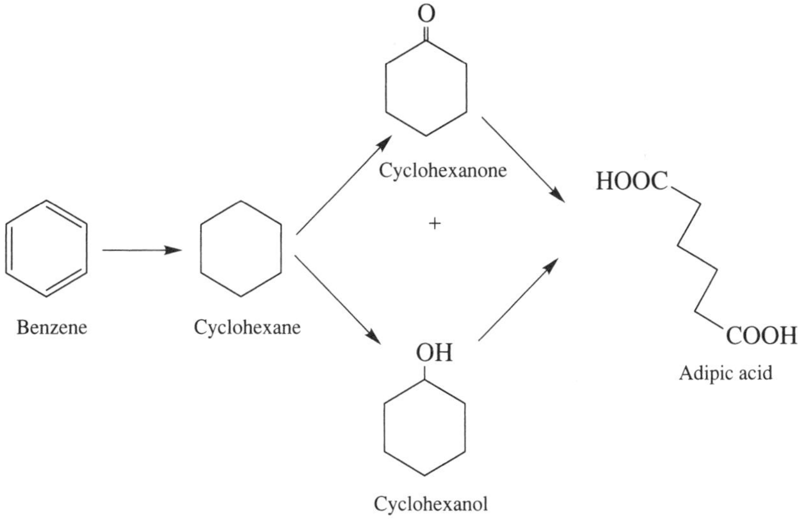

O
Cyclohexanone
+
HOOC
COOH
Adipic acid
Benzene
Cyclohexane
OH
Cyclohexanol

Recently, researchers found new routes to the production of adipic acid, catechol, and hydroquinone using *D*-glucose as the starting material in place of the traditionally used benzene (Draths and Frost, 1994). *D*-Glucose is a nontoxic natural sugar derived from plant starch and cellulose and can be produced in large quantities at little cost.

The two-step synthesis of adipic acid utilizes a biocatalyst to convert *D*-glucose into *cis,cis*-muconic acid, which is subsequently hydrogenated under mild reaction conditions to adipic acid.

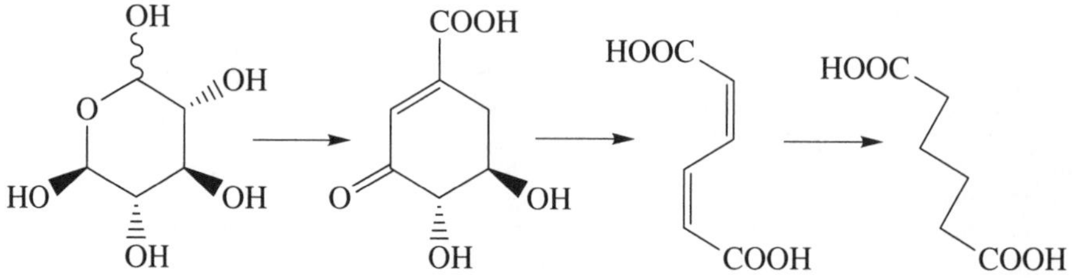

D-glucose → 3-Dehydroshikimate → *cis,cis*-Muconic acid → Adipic acid

Conversion of *D*-glucose to adipic acid does not compare favorably to use of benzene from an economic viewpoint at the present time because of the cost of producing the biocatalyst. However, if full-cost accounting is used, the cost of producing adipic acid from benzene becomes prohibitive because of benzene's toxicity, and synthesis from *D*-glucose becomes the favored pathway.

Green chemistry is not limited to production of chemicals. It can also be used in modifying processes. As an example, consider the oxidation reactions that are used in the bleaching of paper pulp before making paper. Traditionally, these reactions have been carried out using chlorine-based bleaches. Chlorine is a very effective oxidizing agent and produces good results, but in doing so it also produces waste chloride ions and chlorinated organics:

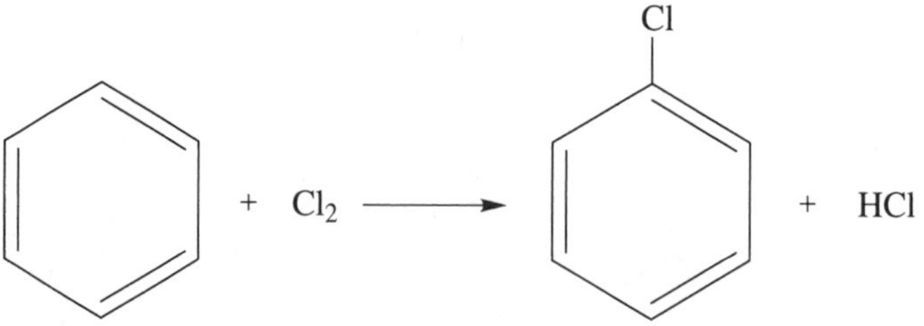

$$C_6H_6 + Cl_2 \rightarrow C_6H_5Cl + HCl$$

These chlorinated organics are significant environmental stressors. The same oxidation-mediated bleaching can be achieved, however, without the undesired toxic by-products by using hydrogen peroxide:

$$H_2O_2 + 2e^- + 2H^+ \rightarrow 2H_2O$$

CATALYSIS. The increasing use of catalysts in chemical manufacturing is producing significant improvements in pollution prevention. Catalysts are materials that modify the mechanism and rate of a chemical reaction. They mediate the reaction but are unchanged while doing so and can be reused. Better than 60 percent of all commercial chemicals and 90 percent of industrial processes are based on use of catalysts. The primary role of catalysis in waste minimization is the development of processes with high yields and therefore minimal waste. More efficient utilization of raw materials and improved yields usually lead to greater profits. The ultimate objective is an industrial process that has 100 percent yield and operates under conditions that require minimal energy. Unfortunately, this is usually impossible to achieve in an economical fashion; we still can try, though, to get as close as economically possible to this goal.

Catalysts can also be used in the treatment technologies used to remove contaminants from process or waste streams. They are commonly used for improving air quality through NO_x removal and by reducing the emissions of volatile organic compounds.

Catalysts find a wide range of applications in industry, including oxidation, hydrogenation, esterification, polymerization, and aromatic substitution. Table 9.4 lists some of the more widely used industrial catalysts, along with some of their advantages and disadvantages (Butterworth, Tavener, and Barlow, 1995). These range from acids used for electrophilic aromatic substitutions or for production of alcohol or methyl methacrylate to zeolites (aluminosilicates) used for cracking of petroleum hydrocarbons and in the production of the styrene precursor ethylbenzene to oxidation of organic substrates using heavy metal catalysts.

Several types of catalysts are available (Simmons, 1996). *Homogeneous catalysis* is a single-phase reaction, typically liquid/liquid. Reactions are usually stoichiometric and use mild reaction conditions. Most use organometallic and coordination complexes,

TABLE 9.4
Major classes of industrial catalysts

Catalyst type	Typical catalysts	Chemical reaction applications	Advantages	Disadvantages
Lewis acids	$AlCl_3$ BF_3 $FeCl_3$	Friedel-Crafts alkylation	Reactive Cheap	Toxic Lots of waste Often not catalytic
Brønsted acids	HF H_2SO_4	Friedel-Crafts alkylation Esterifications	Cheap Effective	Corrosive Handling problems
Zeolites	Zeolite Y ZSM 5 Mordenite	Cracking	Tough Clean Selective Reusable	High temperature Usually gas phase
Enzymes	Yeast Penicillin Acylase Amylases	Natural products Chiral molecules Sugars from starch Peptide synthesis	Enantiopure	Limited application Low stability
Oxidation catalysts and reagents	$Co(OAc)_2$ $K_2Cr_2O_7$ $KMnO_4$	Alkenes Alkyl side chains Alcohols	Active Versatile	Not all catalytic Heavy metal waste Destructive oxidation
Polymerization catalysts	$TiCl_4/AlEt_3$	Ziegler-Natta polymerization of ethylene	Stereoselective	Air-sensitive

Source: Butterworth, 1995.

although a large number of dissolved homogeneous complexes are beginning to find favor. *Heterogeneous catalysis,* which is bi- or multiphased, has dominated industrial operations. Many heterogeneous catalysts are attached to a solid support surface.

During the product development stage, the chemist should pay close attention to the reaction mechanisms involved at each intermediate stage and use the knowledge derived to identify where the process can be enhanced by promoting desirable reactions over all other possible reactions. In some cases, several different catalysts are available for the reaction, but some may be more environmentally benign than others. If possible, the safer catalyst should be selected for use. Following are examples of catalysis that have been found to be effective in pollution prevention.

Transition metals are commonly used as catalysts, particularly in reduction reactions such as hydrogenation. An example of how a reaction can be made more environmentally benign through the careful selection of the catalyst is the synthesis of acetaldehyde (Simmons, 1995):

$$\underset{\text{Ethylene}}{CH_2{=}CH_2} + 1/2\ O_2 \xrightarrow{\text{Catalyst}} \underset{\text{Acetaldehyde}}{CH_3{-}\overset{\overset{\displaystyle O}{\|}}{C}H}$$

The conventional reaction requires a large volume of catalyst, typically a mixture of $PdCl_2$ and $CuCl_2$. During the course of the reaction, $PdCl_2$ is reduced to Pd^0; $PdCl_2$ is regenerated by $CuCl_2$, and the resulting $(CuCl_2)^{-1}$ species is oxidized by molecular oxygen, regenerating $CuCl_2$. The reaction also results in the generation of large quantities of chloride ion; these can subsequently react with organic compounds to form chlorinated organics, which may be toxic or carcinogenic. Research has shown, though, that if $CuCl_2$ is replaced with a vanadium complex, the amount of $PdCl_2$ used and the amount of chloride ion generated is reduced by up to 100 and 400 times, respectively.

A key intermediate in the production of the analgesic ibuprofen is 4-isobutylacetophenone. The conventional manufacturing method is a Friedel-Craft alkylation catalyzed by aluminum trichloride (Braithwaite, 1995):

$$(H_3C)_2CH{\cdot}CH_2{-}C_6H_5 + H_3CCOCl \xrightarrow{AlCl_3} (H_3C)_2CH{\cdot}CH_2{-}C_6H_4{-}C(=O){-}CH_3$$

4-Isobutylacetophenone

$$\downarrow NaCN$$

$$\underset{\text{Ibuprofen}}{(H_3C)_2CH{\cdot}CH_2{-}C_6H_4{-}CH(CH_3){\cdot}COOH} \xleftarrow{HI/P} (H_3C)_2CH{\cdot}CH_2{-}C_6H_4{-}C(CH_3)(OH){-}CN$$

Large quantities of $AlCl_3$ catalyst are required for this reaction; approximately 4 kg for every 5 kg of 4-isobutylacetophenone. The spent catalyst has to be disposed of as aluminum hydroxide. In addition, large amounts of acidic gaseous emissions (HCl) have to be scrubbed from the off-gas stream. Recently, a new process has been developed using hydrogen fluoride in place of the aluminum trichloride. The hydrogen trifluoride catalyst can be readily separated out of the reaction system and recycled back to the process, eliminating the need for spent catalyst disposal.

The second example involves the production of methanol, which traditionally occurs in a two-step process via the steam reforming of methane to produce CO and H_2, which are then reacted to form methanol (Carr, 1996). The process is complex and very costly, requiring 30 kcal/mol of energy. A one-step conversion of methane to methanol by partial oxidation of methane, in the presence of a catalyst, is possible and would be desirable because the reaction is exothermic, thus saving energy costs:

$$CH_4 + \tfrac{1}{2}O_2 \rightarrow CH_3OH \qquad \Delta H = -126\ \text{kJ/mol}$$

However, methanol yields are generally very low and the process is not economical. Recently, researchers at the University of Minnesota have shown that partial oxidation of methane to produce methanol may be viable in a countercurrent moving bed chromatographic reactor, which allows for reaction and separation to occur in one step. In practice, one-step conversion is very difficult, because methanol is further oxidized rather easily to carbon dioxide. However, in this reactor, methanol is rapidly separated from oxygen, minimizing the secondary oxidation of methanol. It also permits contacting the reactants with the catalyst at high CH_4/O_2, where methanol selectivity is high. The catalyst can be mixed with the adsorbent in the reactor, obviating the need to separate and recover the catalyst for reuse.

BIOCATALYSIS. In biocatalysis, enzymes and antibodies are used to mediate reactions. Reactions that use biocatalysts often proceed with exceptionally high selectivity and as a result, these catalysts are becoming increasingly more important (Simmons, 1996). In some cases, they have been shown to increase reaction rates between 9 and 15 orders of magnitude in comparison with uncatalyzed reactions, and a few can increase reaction rates over those using heavy metal catalysts. Biocatalysis may involve the use of whole living microorganisms or of enzymes that are separated from the cell and immobilized in a support medium.

While some enzymes have broad specificity and can catalyze reactions involving a wide range of compounds, probably the most significant aspect of biocatalysts is the very strict specificity that many possess. They may be able to discriminate between different identical groups, within the same molecule based on position in the molecule, between different groups at the same location within a base molecule, or between molecules of different stereospecificity. The latter property promises exciting opportunities and will be described in more detail here.

Stereospecific catalysts can choose between stereoisomers of a racemic mixture or between enantiotropic groups in chiral compounds. The optical activity of some organic compounds was discovered by Jean-Baptist Biot in 1815 and was defined as the ability of a compound to rotate the plane of polarization of light. Louis Pasteur later proposed that this optical property was a consequence of the compound's molecular asymmetry, which produces nonsuperimposable mirror-image structures. A molecule which is not superimposable on its mirror image is called a *chiral* compound (Adger, 1995). The most common type of chiral molecule contains a tetrahedral carbon atom which is attached to four different groups, the carbon atom being the asymmetric center of the molecule. Such a molecule exists as two different compounds, known as *stereoisomers* or *enantiomers*. A mixture of enantiomers is called a *racemic* mixture. An example of a chiral molecule is lactic acid (shown in Figure 9.5), which exists in two forms.

For drug use or pesticide manufacture in particular, enantomerically pure samples of chiral compounds are often necessary, as different stereoisomers can have different, often unwanted, physiological effects. Usually only one of the enantiomers is active, while the other produces no effect or a negative effect. Regulations often require that each enantiomer must be assessed and registered, as well as those of the racemic mixture. Of the 2000 or so drug substances currently marketed, over 70 percent are synthetic. Biological activity almost always produces or recognizes only one enantiomeric form of a substance. Therefore, naturally produced drug compounds are almost all single enantiomer compounds, but synthetic drug compounds are difficult to produce as single enantiomers and nearly all (85 percent to 90 percent) are produced as racemic mixtures. A catalytic process that is specific enough to produce only the desired enantiomer has great benefits.

Chirality is also found in other areas. As mentioned previously, a number of pesticides exist in chiral forms, with one enantiomer being active against the pest in question and the other not. This is not surprising, because the pesticide is acting in a biological way, and in nature usually only one form is produced and would be recognized by the organism. Taste and color perception are also influenced by chirality. For example, *S*-limonene has a lemon flavor, while *R*-limonene tastes like oranges. Similarly, *S*-carvone tastes like caraway, while *R*-carvone tastes like spearmint (Adger, 1995). Thus flavor and fragrance industry chemists also desire to produce enantiomerically pure compounds.

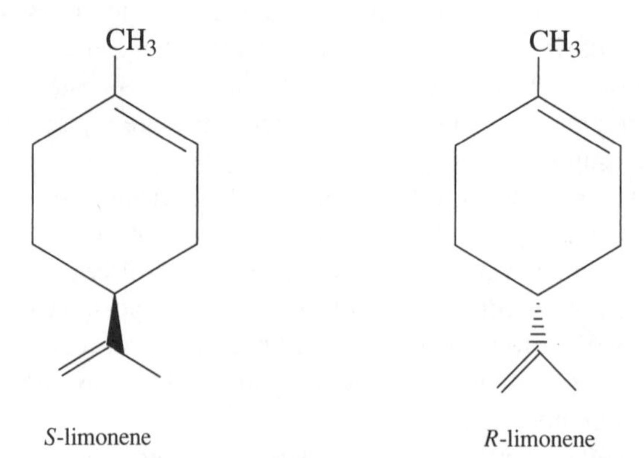

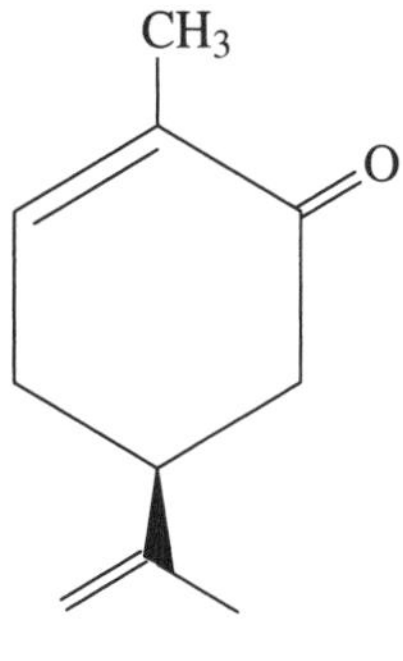

CH_3 O CH_3 O

S-carvone 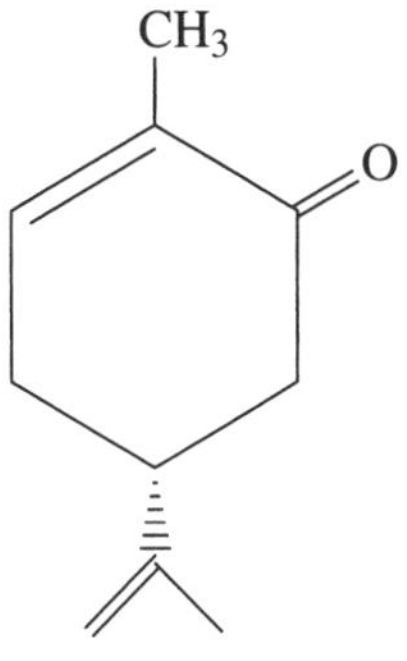 *R*-carvone

Nearly all optically pure compounds are produced catalytically. In some cases, the desired enantiomer is the minor one produced, leading to a large amount of waste as it is separated from other materials in the process stream. To minimize waste production, the asymmetric center should be introduced into the molecule as early as possible in the synthesis, so that the unwanted isomeric species is not carried through the process. Biological catalysts (whole cells or extracted enzymes) are definitely preferred, as they are usually highly specific and often produce only the desired enantiomer. Enzymatic reactions are possible only when producing a naturally occurring substance, though. When a purely synthetic compound is desired, use of enzymes is not possible, and the catalyst used will probably produce a racemic mixture. These must be separated into the desired and the undesired enantiomers, which is rarely an easy task. This is usually done through various crystallization techniques. In some cases, distillation is used.

Another area of intense research activity is the field of catalytic antibodies (Simmons, 1996; Webster, Anastas, and Williamson, 1996). Antibodies are now being engineered to catalyze many types of chemical reactions. The high specificity and diversity of the immune system are attractive for the production of high-selectivity catalysts from specific binding molecules called antibodies. Antibodies can be created against any molecule with a surface structure. They bind with ligands (haptens or antigens), as do enzymes. Animals are injected with a hapten that represents a transition state for the reaction of interest. This triggers an antibody response and the large-scale

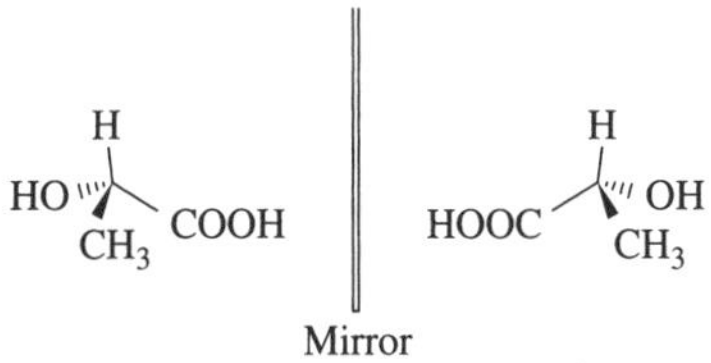

H HO CH_3 COOH | HOOC H OH CH_3

Mirror

FIGURE 9.5
Chiral enantiomers of lactic acid.

production of the desired chemical. In the future, tailored antibody catalysts may have far-reaching potential in organic synthesis and chemical manufacturing.

Biological systems are often capable of mediating highly selective operations such as insertion of oxygen and nitrogen into various organic molecules under mild conditions; performing this nonbiologically is a major challenge. For example, monooxygenase enzymes catalyze monooxygenation reactions under very mild conditions using cytochrome P-450, which is ubiquitous in living organisms. Researchers have developed synthetic catalysts that mimic cytochrome P-450 and that can carry out this reaction without the presence of biological material. Many other synthetic catalysts that mimic the activity of enzymes are now being developed. These *biomimetic catalysts* are efficient and selective, and they offer benign synthesis of chemicals. They have great potential.

In the past, most chemists felt that the complexity of the science of catalysis prevented them from accurately selecting proper catalysts or catalytic conditions for a particular reaction or from predicting the outcome of use of a catalyst. Catalysts were usually selected somewhat arbitrarily in the laboratory based on past experience by the chemist. If one worked, it was used, even if the reason it worked was not known. This scenario is now changing, though, as more becomes known about the science of catalysis. Structural, chemical, and catalytic information about the catalytic sites are being obtained and used to elucidate the catalytic chemistry involved. We are now in the infancy of rational selection of appropriate catalysts and design of catalysts that are specific to a particular reaction. These will be designed to economically achieve high yields of desired products and inhibit the formation of undesirable by-products.

9.2.3 Alternative Reaction Conditions

As we saw in the previous section, the chemical pathway selected and the feedstocks used in making a chemical play a major role in the amount of waste produced. The conditions that are used in synthesizing a chemical can also have a significant effect on the environmental impact of the manufacturing process. They govern the amount of energy required, the types of reactors used, the amounts and types of solvents needed in the reaction or in separation processes, and the separation systems themselves. Several of these parameters were discussed in detail in Chapter 5. In this section, we focus on solvents.

The environmental consequence of using organic solvents in the manufacture of chemical products, the recovery of desired materials during manufacture, the cleaning or degreasing of parts, or the treatment of wastewaters has been a major concern for many years. The question is usually not whether the solvent is needed; usually it is. Rather, the question is: Are there better and more environmentally benign solvents available than the one that is being used? The selection of the appropriate solvent and the use of alternative solvents are active areas of green chemistry.

SOLVENTS. Solvents are used in almost every industry to some extent. In the chemical industry, they are often used to make a material miscible with other materials and to increase the efficiency of the reaction. The rate of a reaction is dependent on the free energy of activation, $\Delta G°$, which is the difference between the energy of the reactants and the transition state (Atherton and Jones, 1995). Figure 9.6 is a schematic free energy profile. Increasing the degree of solvation of the starting materials will reduce their free energy and will increase the energy barrier between reactants and transition state, thus reducing the reaction rate. The converse applies to the transition state: Increasing the degree of solvation will decrease the energy of the transition state and increase the rate of reaction. Thus the rate of reaction can be altered by proper selection of the solvent used.

Most solvents used in organic chemical synthesis have been organic compounds. The most commonly used solvents are volatile organics, such as alcohols, chlorinated hydrocarbons, arenes, and nitriles, which cause many problems in the environment. One approach toward solvent alternatives to VOCs is the increased use of *aqueous reaction systems* in place of organic solvents. Aqueous solvents are usually composed of water, surfactants, and other additives. While many of these show great promise, it is possible that all that is accomplished by using aqueous solvents is trading an air pollution problem for a water pollution problem. The total impacts of the change to an aqueous solvent system must be evaluated before the change is allowed.

Another alternative is to use *solventless chemistry,* in which neat chemicals are used and no solvent waste occurs. The reactions are assisted by benign catalysis, sometimes assisted by microwave activation.

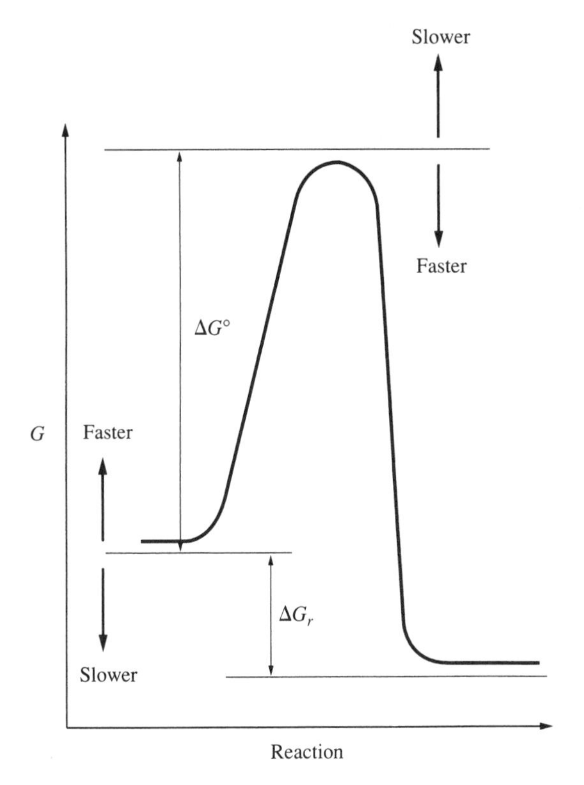

FIGURE 9.6
Effect of solvation of reactants and transition states on reaction rate.

A third option is use of *ionic liquids,* which are liquids that are comprised entirely of ions (Seddon, 1997). For example, molten sodium chloride is an ionic liquid, as is a solution of sodium chloride in water. Ionic solvents that are liquid over a very wide temperature range can be selected, allowing tremendous temperature control. They are excellent solvents for a wide range of inorganic, organic, and polymeric materials; they can be used in small volumes, thus lowering the amount of waste; they have no effective vapor pressure, and therefore do not cause air pollution problems; and they are relatively inexpensive.

One of the most active areas of investigation in alternative solvents, because of their impact on pollution prevention, is focused on the use of *supercritical fluids* (SCFs). The use of supercritical fluids for solvent extraction was described in detail in Chapter 5. They also appear to be excellent for use as a solvent in chemical synthesis and polymerization reactions. Supercritical fluids appear to offer the promise of providing a low-cost, innocuous solvent that can supply "tuneable" properties that can be varied to meet the desired reaction depending on where in the critical region one decides to conduct their chemistry (Anastas and Williamson, 1996). The advantage of supercritical fluids is that separations can be made markedly more efficient and inexpensive by changing the pressure at the end of the reaction to convert the supercritical solvent to a gas.

The hydrolysis of aniline to phenol in supercritical water is the earliest reported study of a simple reaction in a supercritical medium (Clifford, 1995). Supercritical carbon dioxide, which is more commonly used as the supercritical solvent, has recently been shown to be effective in asymmetric catalytic hydrogenation, in which chiral compounds are produced, and in the production of adipic acid from cyclohexene (Morgenstern *et al.,* 1996). In addition, supercritical CO_2 has been shown to be an excellent solvent for free radical brominations, and it can be used to replace chlorofluorocarbons (CFCs, or Freon) in these reactions (Tanko, Blackert, and Sadeghipour, 1994). Table 9.5 summarizes many synthetic reactions that utilize supercritical CO_2. The data show that solubilities are greatly enhanced in the SCF. The yields and selectivities in many cases are equal or superior to those in conventional solvent systems.

TABLE 9.5
Conditions for reactions in supercritical CO_2

Reaction	*P*, psi	*T*, °C	ρ, g/mL	Solubility factor (compared with water)
Hydroformylation	3000, 5000	80	0.60, 0.78	5.15, 6.69
Bromination	2500, 5000	40	0.82, 0.94	6.98, 8.01
Organometallic synthesis	3000	Ambient	0.95	8.08
Polymer synthesis	3000, 5000	40–65	0.70–0.94	6.01–8.01
Hydrogenation	3000	50, 100	0.49, 0.79	4.16, 6.80
Asymmetric hydrogenation	5000	40	0.94	8.01

Source: Adapted from Morgenstern *et al.*, 1996.

Any solvent that is used must be recovered and reused if pollution is to be minimized. The spent or used solvent must be collected from the process stream or off-gas stream to prevent it from entering the environment. It may be reprocessed on site and returned to the same or another process line that is compatible with it, or it may be sent to a third party for reprocessing.

9.2.4 Design of Safer Chemicals

The third prong of the green chemistry paradigm is replacement of hazardous industrial chemicals with new, less hazardous chemicals that have fewer adverse effects on human health and the environment. This concept involves the structural design of chemicals that meet both safety and efficacy of use criteria.

Since the earliest commercial and industrial use of chemicals, the primary emphasis of the chemist has been on producing materials that most efficiently and economically fulfill the needs of their intended use. New structural designs and configurations for chemicals have continually been developed in the pursuit of better performing, lower cost products. However, until recently, little attention was given to the impacts of these chemicals on human health and the environment. This has now changed, though, as federal legislation mandated the toxicity testing and regulation of chemicals and products. The "design of safer chemicals" concept is now widely accepted across all of the chemical industry.

The concept of designing safer chemicals can be defined as

> the employment of structure–activity relationships (SAR) and molecular manipulation to achieve the optimum relationship between toxicological effects and the efficacy of the intended use. (Garrett, 1996)

The key word in this definition is "optimum," since it is seldom possible to achieve zero toxicity or to achieve a maximum level of efficacy; rather, some optimum combination of the two goals is sought.

In many cases the part of a molecule that provides its intended activity or function is separate from the part of the molecule responsible for its hazardous or toxic properties. Therefore, the challenge is to reduce the toxicity of a molecule without sacrificing the efficacy of function. This is a formidable challenge, requiring knowledge of the relationships between chemical structures and biological effects, the pathways of chemicals in biological organisms, and the fates of chemicals in the environment.

There are two main approaches that can be taken to make chemicals safer. The first is to reduce exposure to the chemical by designing chemicals to quickly degrade in the environment to less toxic materials, by making them less volatile or soluble and therefore less likely to disperse in the environment, or by reducing their absorption rates by humans, other animals, or aquatic life. This requires knowledge of physical and chemical properties of the compounds (e.g., volatility, water solubility, biodegradation rates) and uptake by organisms (exposure pathways, molecular size, lipophilicity, excretion and detoxification mechanisms). The second approach is to reduce or eliminate impurities in the chemical intermediates or final product that may cause toxic or hazardous effects.

Today's industrial chemist must be more than a good chemist. He or she must also have a thorough understanding of biochemistry, pharmacology, and the toxicological effects of chemicals and other materials. Many of the decisions concerning development of new chemicals will be based on structure–activity relationships (SARs), a newly developing area that attempts to describe the relationship between the specific structural features of a chemical and its biological activity or biological/toxicological impact on living organisms. These are described in more detail later.

GENERAL PRINCIPLES. To reduce the toxicity of a chemical or to make it safer than a similar chemical substance requires an understanding of the basis of toxicity. Then, structural modifications can be made that attenuate toxicity without reducing the chemical's efficacy. General approaches that can be used to modify molecular structures for more rational design of safer chemicals include the following (DeVito, 1996):

- Reduction of absorption.
- Use of toxic mechanisms.
- Use of structure–activity relationships.
- Use of isosteric replacements.
- Use of retrometabolic ("soft" chemical) design, which allows the chemical to quickly degrade to harmless products.
- Identification of equally efficacious, less toxic chemical substitutes of another class.
- Elimination of the need for associated toxic substances.

A few of these are described more fully here.

There are three fundamental requirements for chemical toxicity: (1) living organism exposure to the chemical; (2) bioavailability of the chemical; and (3) intrinsic toxicity of the chemical. Some substances contain structural features that are not directly toxic but that undergo metabolic conversion to yield toxic compounds. Examples are some azo dyes, which are used widely in the textile industry. These are compounds composed of aromatic moieties linked together through a nitrogen-nitrogen (azo) double bond:

$$\phi - N = N - \phi$$

Many of these dyes are not toxic to humans in their base structure, but within the anaerobic intestinal system they are biodegraded by azo bond cleavage to aromatic amines, which are often carcinogenic. Thus the fates of chemicals in the environment and their resulting products must also be known and evaluated.

Safer chemicals may be designed by keeping the preceding information in mind. Problem chemicals can be modified to decrease absorption, avoiding the incorporation of functional groups that are known to be toxic; or they can be made more readily biodegradable to innocuous intermediates or products.

REDUCING TOXICITY. Toxicity can be reduced by making structural changes that reduce or prevent absorption from the lung, skin, or gastrointestinal tract. When designing a chemical to be absorbed less, one should consider the most likely route of exposure (i.e., dermal, respiratory, oral) and design the molecule accordingly. For example, if the oral route is considered the most critical, it would be helpful to increase the particle size or to keep the substance in an un-ionized form (see Table 9.6). To reduce respiratory exposure to the chemical, the chemical should be designed to have a lower vapor pressure, a low water solubility, or a particle size greater than 5 μm. Skin exposure routes will be minimized if the substance is a solid rather than a liquid, in ionized form, water soluble, and of high molecular weight. A more complete description of these options can be found elsewhere (DeVito, 1996).

Toxicity of a chemical can also be reduced through structural modifications that make the compound less toxic than the unmodified toxic substance. This requires a thorough understanding of toxicology. In general, the chemist may be able to alter the structures of chemicals whose toxic properties are based on their electrophilicity or ability to form free radicals. It has been found that electrophiles are responsible for carcinogenesis in many cases. Chemicals that are electrophilic or are metabolized to electrophilic species are capable of reacting covalently with nucleophilic substituents of cellular macromolecules such as DNA, RNA, enzymes, and proteins, causing irreversible changes to their structures and interfering with their function. Examples of electrophilic substituents commonly found in commercial chemicals that can cause these interactions are seen in Table 9.7. The presence of one of these substituents in a chemical does not always mean that the chemical will be toxic; toxicity also depends

TABLE 9.6
Molecular modifications that reduce absorption

Route	Modification
Gastrointestinal tract	Increase particle size
	Keep substance in un-ionized form (free base, free acid)
	Lower water solubility (log $P > 5$)
	Make molecular size > 500 daltons
	Keep melting point > 150°C
	Make it a solid rather than a liquid
	Make the substance polar by adding substituents such as $-SO_3^-$
Respiratory tract	Make the chemical less volatile by lowering its vapor pressure or increasing its boiling point
	Reduce its water solubility
	Increase its lipophilicity
	Increase particle size to > 5 μm
Skin	Make it a solid rather than a liquid
	Make it polar or un-ionized (*e.g.*, sodium salt of an acid)
	Increase the water solubility or decrease lipophilicity
	Increase the particle size

Source: Adapted from DeVito, 1996.

TABLE 9.7
Electrophilic substituents commonly encountered in commercial substances, their reactions with biological nucleophiles, and possible resulting toxicity

Electrophile	General structure	Nucleophilic reaction	Toxic effect
Alkyl halides	R–X X = Cl, Br, I, F	Substitution	Cancer, granulocytopenia, etc.
α,β-Unsaturated carbonyl and related groups	C=C–C=O C=C–S(=O)(=O)– C≡C–C=O C=C–C≡N	Michael's addition	Cancer, mutations, hepatotoxicity, nephrotoxicity, neurotoxicity, hematotoxicity, etc.
γ-Diketones	R_1–C(=O)–CH_2CH_2–C(=O)–R_2	Schiff base formation	Neurotoxicity
Epoxides (terminal)	–CH–CH_2 (–O– bridging) –O–CH_2–CH–CH_2 (–O– bridging)	Addition	Mutagenicity, testicular lesions
Isocyanates	–N=C=O –N=C=S	Addition	Cancer, mutagenicity, immunotoxicity (e.g., pulmonary sensitization, asthma

on other factors such as molecular size and shape, substituent effects (electronic and steric), bioavailability, and its metabolism in the living system. However, their presence does indicate that the chemical warrants further study.

The electrophilic reactive intermediates believed to be responsible for the carcinogenecity of genotoxic chemicals include carbonium, aziridium, episulfonium, oxonium, nitrenium or arylamidonium ions, free radicals, epoxides, lactones, aldehydes, semiquinones/quinoneimines, and acylating moieties (Lai et al., 1996).

Ideally, electrophilic substituents should never be incorporated into a substance, but this is not always possible as these substituents are often needed for the chemical's intended purpose. Free radicals are highly reactive species that have an unpaired electron. Many chemical substances form free radicals following their absorption in the body. An understanding of free radical–based toxicity is rapidly emerging, and it is becoming apparent that free radicals are a major source of chemical toxicity. The radicals are capable of interacting with and damaging cells and tissues if the body's natural defense mechanisms are overpowered or depleted. Compounds that form free radicals should be replaced if at all possible.

There are many examples of how safer chemicals can be substituted for more toxic ones. Benzene, known to cause hemotoxicity and leukemia in humans, has been widely used in industry. Toluene, the methyl derivative of benzene, is considerably less toxic because it is metabolized in the liver to benzoic acid, which is essentially nontoxic. In many cases, toluene can be effectively used in place of benzene, thus greatly reducing potential toxicity. The development of chlorofluorocarbon (CFC) alternatives—hydrofluorocarbons (HFCs) and hydrochlorofluorocarbons (HCFCs)—for use as refrigerants or aerosol propellants was described in Chapter 3. This is another example of chemical compounds being replaced by less hazardous ones. The main considerations behind the proper choice of HFCs and HCFCs as replacements are described by Webb and Winfield (1995).

STRUCTURE–ACTIVITY RELATIONSHIPS. Structure–activity relationships (SARs) can be used to correlate toxic effects of a compound with its structure. They can be developed qualitatively, based on visual comparison of the structures of the substances in a homologous series and the corresponding effects on the toxicity. From this comparison, the chemist may be able to discern the cause for the toxicity and identify the least toxic members of the group as possible commercial alternatives to the toxic material. An example of qualitative SAR analysis relates to carboxylic acids, chemicals commonly used as synthetic intermediates, plasticizers, catalysts, and preservatives (DeVito, 1996). Some commercial carboxylic acids have medicinal value, whereas others cause liver toxicity or teratogenicity (toxicity to fetuses or offspring of the person receiving the chemical exposure). Through qualitative SAR analysis, it has been found that teratogenicity of carboxylic acids is highly structure-dependent. Teratogenic carboxylic acids contain a free carboxyl group, have only one hydrogen atom at the 2 carbon position, have an alkyl substituent larger than a methyl at C-2, and/or have no double bonds between C-2 and C-3 or between C-3 and C-4. Safer carboxylic acid compounds are those that have none of these properties.

The SAR data can be more rigorously quantified by correlating the data into a regression equation relating biological effects to one or more physicochemical properties of a set of analogous compounds. These *quantitative structure–activity relationships* (QSARs), therefore, describe the effects of structural changes on biological activity and can be used to predict the change in biological potency that may accompany a given change in structure. These physicochemical properties may include such factors as octanol-water partition coefficient ($\log K_{ow}$), water solubility, dissociation constant (pK_a), molecular weight, and percent amine nitrogen (Newsome, Nabholz, and Kim, 1996). An example of a general QSAR equation is

$$\log(1/C) = a(X)^2 + b(X) + c(Y) + d$$

where $1/C$ = biological activity
C = concentration or dose of a substance required to elicit the biological activity
X, Y = physicochemical descriptors of the activity
a, b, c, d = coefficients

The physicochemical properties that best describe the exposure of the chemical in question should be used. For example, K_{ow} or log P (the logarithm of the partition coefficient) is commonly used to quantify the SAR when aquatic toxicity is the major concern, because it governs to a large extent the transport and distribution of a chemical between water and the biological organism.

Once the major physicochemical factor governing toxicity of a given chemical is known, the chemist can evaluate possible structural changes to make it less hazardous while still maintaining its usefulness. The structure could be modified to change the log P value, solubility, and so on. A multitude of possibilities are open to the chemist, once the SAR of the offending chemical is known.

ISOSTERIC REPLACEMENTS. Isosteric substances are substances or substituents that have similar molecular and electronic characteristics—that is, the same charge—caused by having the same number and arrangement of electrons and the same number of atoms. They may not be related structurally, but they often have similar physical or other properties. Some examples are

—H to —F —OH to —NH_2 —CH_3 to —Cl

N to C(H) to S —C(=O)OH to —S(=O)(=O)—N(H)—

By switching from an isostere that exhibits toxic properties to one that does not, it may be possible to carry out the same reaction without the harmful side effects. For example, it may be possible to replace benzene in some reactions with the less toxic thiophene or pyridine, which are all isosteres of each other.

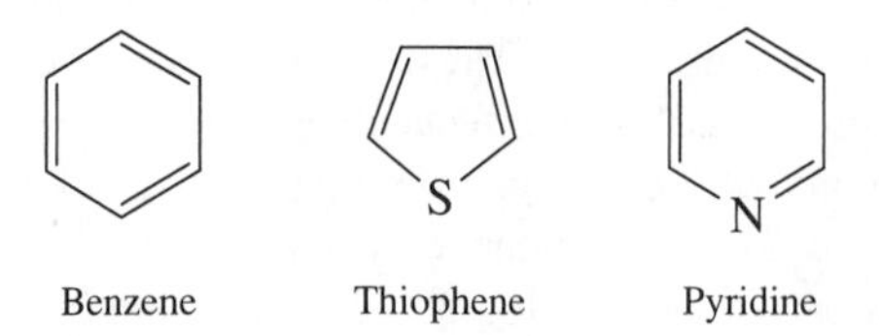

Although they are structurally different, all are aromatic, all are liquid, and all are about equal in molecular size and volume. Many of their chemical properties are similar (DeVito, 1996).

In some cases, slight changes in a carcinogenic molecule can render it noncarcinogenic. An example is the replacement of one hydrogen in the carcinogen 7-methylbenzo[*a*]anthracene with a fluorine, making it noncarcinogenic. The fluorine atom blocks bioactivation and prevents creation of a carcinogenic metabolite.

7-Methylbenzo[*a*]anthracene
(carcinogenic)

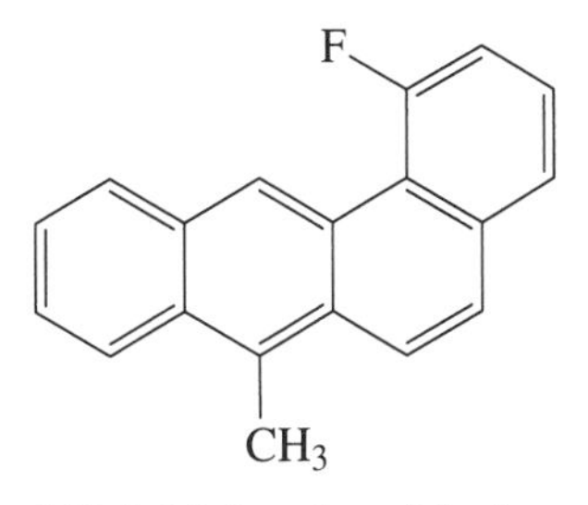

7-Methyl-1-fluorobenzo[*a*]anthracene
(noncarcinogenic)

Another example of how a hazardous isosteric compound can be replaced by a safer one is the use of organosilanes in place of carbon compounds in pesticide manufacture. Organosilanes are inexpensively made, lack intrinsic toxicity, and have many similarities to comparable carbon compounds. Silicon and carbon are both grouped in column IVA of the periodic table, and therefore have similar chemical properties. Silicon is nontoxic, abundant, inexpensive, and available in many forms. Silicon-carbon bonds are not known in nature, suggesting that organosilanes may be short-lived in nature, adding to their desirability as pesticides. Pesticides should last as long as needed to control the target organism, but no longer. The major source of organosilanes is silicone polymer (siloxanes). Most organosilanes can, over time, be degraded by both biotic and abiotic means, rendering them innocuous. A successful example of how safe organosilanes can be used in place of more toxic carbon-based analogs is the insecticide silafluofen (Sieburth, 1996). Natural pyrethrins are effective insecticides that are nontoxic to mammals and birds but are toxic to fish. Silafluofen, an isosteric replacement for naturally occurring pyrethrins, is as effective as natural pyrethrins and is nontoxic to fish. It is now widely used as an insecticide.

Ethofenprox
(a pyrethrin)

Silafluofen

RETROMETABOLIC DESIGN. Retrometabolic design, used primarily in the manufacture of drugs, is also coming into favor for other synthetic chemicals manufacture. This type of design relies on chemicals that are known to undergo detoxification metabolism to harmless intermediates or end products. By designing the substance so that if absorbed, it will undergo rapid metabolism that leads to loss of the toxic properties of the compound, one can avoid potential public health or environmental problems (Bodor, 1996). Either more readily biodegradable substitute chemicals can be used, or biodegradable segments can be built into the desired chemical to render it harmless after it serves its useful purpose or if it comes into contact with biological organisms. This approach has been used commercially to produce a number of safe chemical replacements, including pesticides to replace DDT.

Relatively small changes in molecular structure can often appreciably alter a chemical's susceptibility to biodegradation. The following molecular characteristics generally increase resistance to aerobic biodegradation (Boethling, 1996):

- Halogens, especially chlorine and fluorine.
- Chain branching, especially quaternary carbon and terniary nitrogen.
- Nitro, nitroso, azo, and arylamino groups.
- Polycyclic residues (as in polycyclic aromatic hydrocarbons, or PAHs), especially with more than three fused rings.
- Heterocyclic residues, such as pyridine rings.
- Aliphatic ether (-C-O-C-) bonds.

CHEMICAL DESIGN EXAMPLE. Many dyes used in the textile, leather, plastics, cosmetics, drug, and food industries are based on aromatic amines. Several aromatic amines, such as benzidine, 2-naphthalene, and 4-aminobiphenyl, are known human carcinogens. Many of these dyes, when present in wastewaters, are also not amenable to biological treatment, thus creating potential public health problems from future ingestion of these waters (Jiang and Bishop, 1994). Because of this, many aromatic amine dyes had to be abandoned and replaced with less toxic dyes.

After much SAR testing, the basic requirements for an aromatic amine to be classified as carcinogenic were determined. Structural features important in predicting the carcinogenicity of aromatic amines include (1) the number and nature of aromatic rings; (2) the nature of the amine/amine-generating groups; (3) the position of the amine/amine-generating groups; (4) the number, nature, and position of other ring substituents; and (5) the size, shape, and planarity of the molecule (Lai et al., 1996). By using these data, it is possible to design safer aromatic amine dyes by modifying the chemical structure. Table 9.8 lists several approaches that can be used and their rationale.

9.2.5 Green Chemistry Research Needs

While much has been done in recent years to design products and chemical processes that are more environmentally benign, more still needs to be done. The Council for

TABLE 9.8
Molecular design of aromatic amine dyes with lower carcinogenic potential

Approach	Rationale
Introduce bulky substituent(s) ortho to the amine/amine-generating group(s).	Provide steric hindrance to inhibit bioactivation.
Introduce bulky *N*-substituent(s) to the amine/amine-generating group(s).	Make the dye a poor substrate for the bioactivation enzymes.
Introduce bulky groups ortho to the intercyclic linkages.	Distort the planarity of the molecule, making it less susceptible and a poorer substrate for the bioactivation enzymes.
Alter the position of the amine/aromatic ring(s).	Reduce the length of the conjugation path and thus the force of conjugation which facilitates departure of the acyloxy anion. Nonlinear conjugation path; less resonance stabilization of the electrophilic nitrenium ion. In some cases, distort the planarity of the amine-generating group(s) in the molecule, making it less favorable for DNA intercalation and a poorer substrate for the bioactivation enzymes.
Replace electron-conducting intercyclic links by electron-insulating intercyclic links.	Disrupt the conjugation path and thus reduce the force of conjugation which facilitates the departure of the acyloxy anion. Less resonance stabilization of the electrophilic nitrenium ion.
Ring substitution with hydrophilic groups (e.g., sulfonic acid); especially at the ring(s) bearing the amine/amine-generating group(s).	Render the molecule more water soluble thus reducing absorption and accelerating excretion.

Source: Lai et al., 1996.

Chemical Research has put together a list of the most needy research areas (Hancock and Cavanaugh, 1994). The list includes the following:

- Replace chromium in corrosion protection, which will require development of new redox chemistry.
- Recycle rubber more effectively, which will require new ways to reverse cross-linking and vulcanization.
- Replace traditional acid and base catalysts in bulk processes, perhaps using new zeolites.
- Develop new water-based synthesis and processing methods to minimize use of volatile organic solvents.
- Develop new catalytic processes, based on light or catalytic antibodies, to replace traditional heavy metal catalysts.
- Devise better chelates to separate and recycle heavy metal catalysts.

Because of the complexity of the chemistry involved, significant improvements in development of more benign alternative synthetic pathways will probably require computer assistance. This field is still in its infancy, but progress is being made (Anastas, Nies, and DeVito, 1994; Hendrickson, 1996). Computer programs are available which have the potential for proposing alternative reaction pathways that may subsequently be evaluated for their relative risk and economic viability.

9.3 DESIGN FOR DISASSEMBLY/DEMANUFACTURING

Major environmental impacts of a product are often associated with its ultimate disposal. All of the raw materials, energy, and labor that went into making a product wind up in a landfill or in incinerator residues. This can be a large economic waste and, of more concern, can impose environmental and public health damage if the disposed components are not environmentally benign. This is evident upon examination of our thousands of municipal and hazardous waste landfills that are polluting our groundwaters and our air. In most cases, these are not worthless materials needing disposal; they are instead no longer needed or useful in their present form. The component parts or the materials they are composed of may still be of value if they can be reused or recycled into other products.

Unfortunately, most products are not designed for efficient reuse or recycling. If product design and waste management were coupled more closely, the cost of resources to industry, the loss of valuable resources, and the environmental problems that we all face because of waste disposal could be addressed at the same time. This will require increased employment of material recycling/reuse, which may necessitate changes in the way that products are designed. This means making products that can be, in order of preference, remanufactured, recycled, composted, or safely incinerated with energy recovery (OTA, 1992).

9.3.1 Recycle versus Reuse

Reuse (also called *demanufacturing* or *remanufacturing,* although some distinguish between them) is the additional use of an item after it is retired from a clearly defined duty. It is the process of collecting, dismantling, selling, and reusing the valuable components of products that have reached the end of the intended use. Reformulation of the material is not reuse. However, repair, cleaning, or refurbishing to maintain integrity may be done in transition from one use to the next (Keoleian, 1995). Reusable products may be returned to the same or to less demanding service without major alterations. Minor processing, such as cleaning, may be required. Reuse now extends beyond mere material reuse, though, and requires involvement in original product design, material science, new technologies, and corporate strategy.

Demanufacturing is a segment of reuse in which the product is disassembled for reuse, with the materials or components reused in the same or other products. Remanufacturing is sometimes described as a subcategory of demanufacturing that involves the restoration of the original product to like-new condition. It consists of reuse of obsolescent or obsolete products by retaining serviceable parts, refurbishing usable parts, and introducing replacement components (either identical or upgraded) to create a new product. Such a process is almost always cost-effective and almost always environmentally responsible (Graedel and Allenby, 1995). To be effective, though, remanufacturing systems must rely on:

A sufficient population of old units.

An available trade-in network.

Low collection costs.

Storage and inventory infrastructure.

Returnable bottles that are reused provide a good example of an item that is currently reused a number of times until it is damaged to the point that it can no longer be reused. Items that could easily be reused but that most often are not are the nuts, bolts, screws, and washers used to assemble components. They rarely are damaged during use and could easily be recovered for direct reuse before disposing of a product, but this is seldom done. These reusable items are routinely discarded, even though they may be in good condition. This is often the case because of convenience. It is usually less trouble to discard these usable materials than to recover them because the component is designed in such a way that accessing the parts is troublesome. The owner feels that the value of the parts is too low to justify the time and cost associated with their recovery. However, this reasoning does not consider the time and expense associated with acquiring new parts or the environmental impacts of both disposing of the old parts and manufacturing the new ones. Reusing the parts may actually be less expensive on a life-cycle cost basis. Reuse may also be hampered by the lack of markets for secondary components, because past designs did not consider ease of disassembly, or because new technology makes the component outmoded and of little value. This is the case for many electrical components. The materials in these components may be recycled, however.

Recycling differs from reuse. Recycling involves the reformation or reprocessing of a recovered material (Keoleian, 1995). The recovered and reprocessed material is then used in new products. Not all recycling involves postconsumer products. Most recycling takes place in the factory where unused process materials are returned to the original manufacturing process, such as when recovered pulp fibers from the paper manufacturing process are returned to the paper machine; or when solvents are recovered and reused; or when preconsumer manufactured materials, such as overruns or rejects, are returned for reprocessing. The latter are usually of much higher quality than are postconsumer recycled materials, and are in much higher demand. Nearly all wastepaper produced in a paper mill is recycled because it is less expensive to use these fibers than to process virgin wood chips; however, the market for postconsumer wastepaper is highly variable and often poor because of the low value of this material.

Whether a material will be recycled or reused depends primarily on economic and market factors. A company will not use recycled materials if use of virgin materials is cheaper. Even if recycling is economically feasible, it probably won't be successful unless there is a readily available and steady supply of the material. Lead automobile battery and waste oil recycling programs have been very successful because the

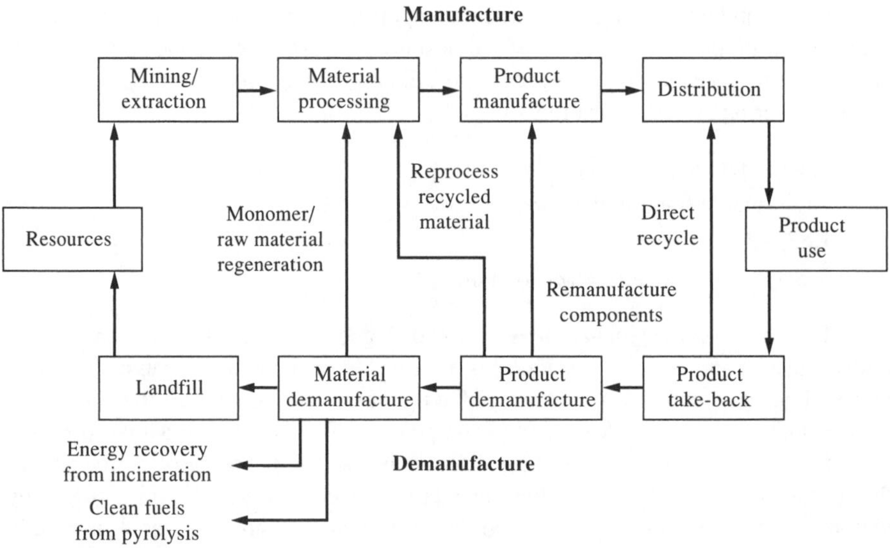

FIGURE 9.7
A product's life cycle from cradle to reincarnation.

infrastructure for collecting, reprocessing, and selling these materials has become universally available. From an economic standpoint, cadmium recycling should be in great demand, but it is almost nonexistent because of a lack of a suitable infrastructure.

Because of the increased pressures to recycle/reuse materials, the typical manufacturing process flow sheet seen in Figure 5.2 should be expanded to include these new options. Figure 9.7 is a flow chart for materials use from extraction through manufacture to use and then back to reuse. This flow chart shows the myriad of alternatives that can be considered when analyzing the life cycle of a product. If successful, very little of the original material will remain in the "disposal" category.

9.3.2 Recycle/Reuse Hierarchy

In a fashion similar to that for the pollution prevention hierarchy pyramid, the recycle/disposal hierarchy can be depicted as seen in Figure 9.8. The highest priority should be to reduce the amount of material in the product through better design. When the product reaches the end of its useful life, as many components as possible should be reused or refurbished for reuse. This preserves all of the resources (material and energy) put into the product during manufacturing. Reuse does require that the product be designed to allow nondestructive disassembly. The products or components that cannot be directly reused should be remanufactured, if possible. Those materials that cannot be reused should be recycled into other products. This is the most common process at this time. An advantage of this step is that destructive disassembly is possible, as only the materials are preserved and not the actual components. Remaining organic

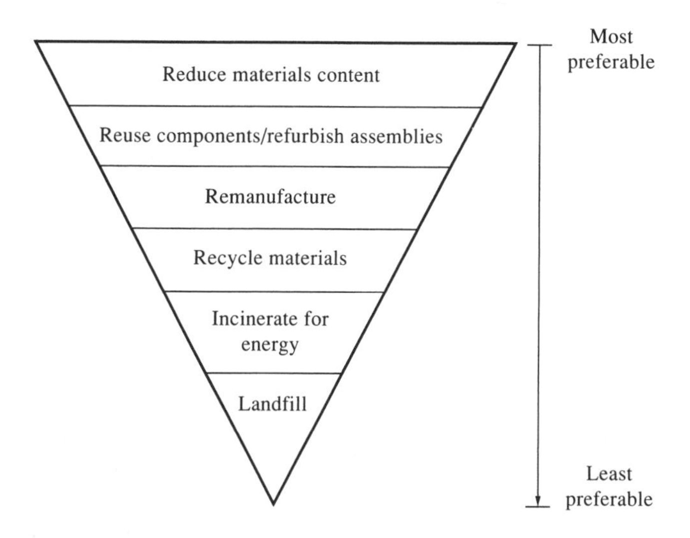

FIGURE 9.8
Hierarchy of recycling options.

materials that can be safely composted should be converted to soil amendment materials, while others could be incinerated to produce energy. Only those components that have no recoverable value should be disposed of as waste.

Instances of successful demanufacturing and remanufacturing abound. For example, more than 25 million personal computers were discarded in 1995. Rather than being dumped into landfills, however, approximately 75 percent of these computers (particularly the computer monitors) are being recycled into new computers or other electronic devices ("End-of-Life Strategies," 1997). Monitors represent about 80 percent of the volume and 60 percent of the mass of a PC (De Fazio, Delchambre, and De Lit, 1997). Over time, computer screens have continually gotten larger and will probably continue to do so in the future. Thus the monitor is usually considered to be the most critical component of a computer from the standpoint of pollution prevention. Television and computer screens have a useful life of around 12 years, which is far longer than that of most computers. Televisions or monitors are typically manually dismantled into nine main components: cables, capacitors, cathode ray tubes, copper wire, aluminum, steel, circuitboards, plastics, and wood. The cathode ray tube, which contributes about 50 percent of the set's total weight, can be refurbished for only about $80. Cables, which are primarily made of copper wire with a plastic casing, can be recovered and recycled. In addition, glass from the screen can be recovered and reprocessed, as can valuable silver, gold, platinum, and lead from the circuitboards, as well as the battery used to maintain the clock during power-off periods. Recycling the monitor may also keep the toxic fine powders of heavy-metal rare-earth salts (e.g., ZnS, Y_2O_2S, Eu_2O_2S), which are attached to the screen with an organic cement, from reentering the environment.

9.3.3 Recycle Legislation

In the United States, recycling/reuse by industry is driven primarily by economics. If an industry feels that it can obtain an economic advantage by using recycled materials or components, it will do so. But there is no legislation requiring companies to use recycled materials or to design their products for easier materials recovery. This is not the case in other countries, however, as in many European countries where legislation mandates recycling. The European Union Directive on Packaging and Packaging Waste, passed in 1994, requires a minimum recycle rate of 15 percent for all plastic, steel, wood, glass, and paper (Payamps, 1997); Germany is already recycling over three-quarters and the Netherlands half of their packaging.

The most sweeping legislation is found in Germany. Not all of the German regulations have gone into effect yet, but they probably will soon. The most significant legislation is known as the "Take-Back Law." It will eventually mandate that manufacturers take back all products from consumers after the product completes its useful life and recycle useful components and materials from these disassembled products. The German Recycling and Waste Management Act of 1991 promulgates a set of principles applying to the entire life cycle of a product, from the moment its materials leave the ground to the time it is recycled, including energy used for transportation. The law gives priority to waste avoidance by requiring the use of low-waste product designs, closed-loop approaches to waste management, and public education to allow for informed purchasing of low-waste and low-pollution products.

Currently, all packaging must be taken back by the manufacturer. By law, all packaging must be designed for waste avoidance. Collected packaging must be reused or recycled to the greatest extent possible. Only those materials that cannot be separated manually or by machine, that are soiled or contaminated by substances other than those that the package originally contained, or that are not integral parts of the packaging are excluded from the recycling mandate. The goal established by the law was to recycle 80 percent of all sales packaging materials by 1995.

Companies soon learned that it was difficult to meet the recycling quotas. Setting up a recovery and reprocessing center on their own was inefficient and very expensive. Consequently, the "Dual-System" (Duales System Deutschland, or DSD) was created. It is a nonprofit organization, funded by business, to collect and recycle packaging and other materials. Qualifying packaging receives a "Green Dot" trademark. Packaging materials with the green dot are placed in separate waste disposal containers by the householder. The material is picked up by the recycling consortium and returned to the manufacturer or is reprocessed.

Automobiles are already among the most recycled consumer products. More than 94 percent of all vehicles that go out of registration in the United States are collected and put into the vehicle recycling process (Eisenstein, 1995). Processes for separation of automobile parts can be very complicated. Figure 9.9 shows schematically how automobile parts can be cut and shredded to liberate materials, which then pass through a magnetic separator to remove ferrous materials, an eddy current separator to remove nonferrous metals, a high-intensity magnetic separator to remove stainless steel, and a foam separator to remove plastics (Hwang, Song, and Tieder, 1996). Once a vehicle

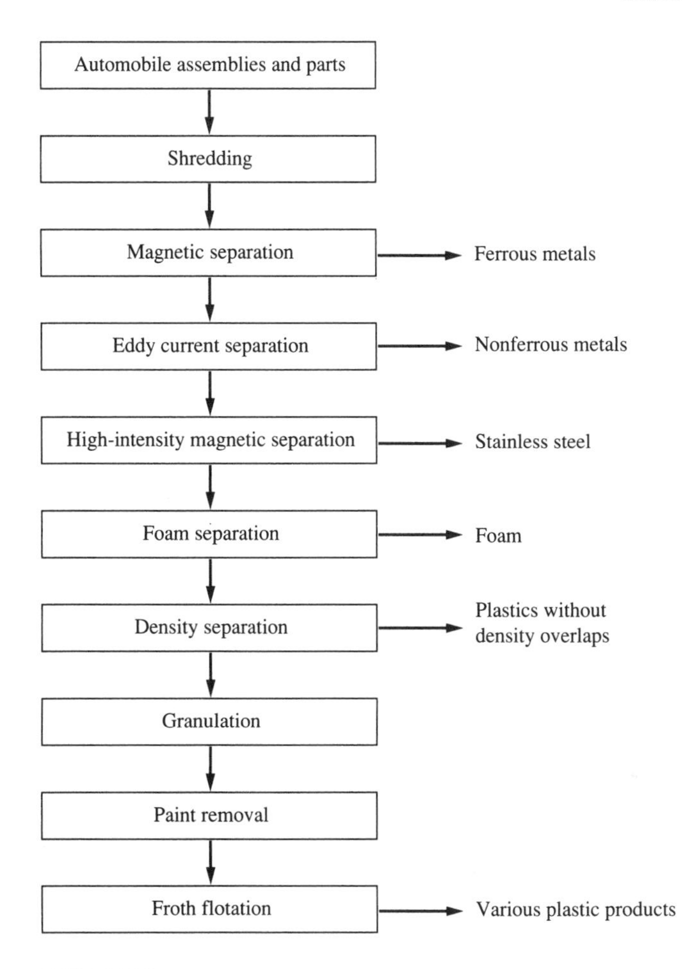

FIGURE 9.9
Possible separation for materials from automobile assemblies and parts. (Adapted from Hwang, Song, and Tieder, 1996)

gets torn up for scrap, about 75 percent of it, by weight, is reused. The rest, which still amounts to an average of 750 pound per vehicle, is known as automobile fluff material and is carted off for disposal.

It is strongly believed, however, that more automobile recycling is achievable. The automotive industry in Germany has agreed to take back and recycle its cars. BMW has built a plant in Munich to recycle automobiles. In less than an hour after entering the plant, a worn-out automobile is reduced to piles of steel, glass, plastic, and other raw materials that will be recycled into new BMWs. As a result of the company's research on how to efficiently dismantle a car, BMWs are now designed to use fewer fasteners, fewer materials, and more easily identified materials. Table 9.9 indicates the recycling targets that have been set for the German automobile industry. As can be

TABLE 9.9
Proposed German reuse and material recycling targets for the automobile industry

	Proposed target, % by weight	
Material	**1996**	**2000**
Steel	100%	100%
Nonferrous metal	88	90
Plastics	20	80
Tires	70	80
Other elastomers	30	50
Glass	30	60
Other fractions	30	60

seen, this will have a profound effect on the way industries function. Other European countries are following suit with mandatory recycling legislation. The European Commission has proposed that beginning in 2002, a maximum of 15 percent of the automobile weight can be landfilled or incinerated without energy recovery; this drops to a maximum of 5 percent in 2015. The automobile industry, as well as many other industries, is intensifying its efforts to develop more easily recyclable products, and products that can function efficiently with recycled materials.

Claims are often made that it is too difficult to increase recycling rates rapidly, but the Germans have disproved this (see Table 9.10). Once the decision was made to recycle in a significant way, the Germans were able to increase their rate of recycling markedly. Between 1993 and 1995, they doubled the amount of paper and tin packaging recycled and increased aluminum recycling by a factor of 10.

An interesting outcome of this enthusiastic acceptance of recycling in Germany is that Germany is now facing a shortage of refuse to feed its solid waste incinerators. A few years ago, over 30 percent of German municipal wastes were incinerated. However, household trash collection decreased by about 50 percent between 1991 and

TABLE 9.10
Germany recycling of household packaging, %

Material	1993	1995
Glass	62	82
Paper/cardboard	55	90
Plastics	29	60
Tinplate	35	64
Composites	26	51
Aluminum	7	70

1995. In many locations, expensive incinerators lie idle because of a lack of waste to burn. These incinerators cost taxpayers hundreds of millions of dollars to construct. The plants must still be paid for, even though they are not operating. To counter this trend, some German cities are now importing solid wastes to feed their incinerators from as far away as Brazil (Sullivan, 1996).

9.3.4 Requirements for Effective Reuse/Recycling

Many large, durable products (e.g., washing machines, refrigerators, automobiles) are simply discarded when their service life is over or when their repair costs exceed their actual value. In other cases, still usable appliances are discarded solely because the owner wants a newer model. Even inoperable units may be inoperable only because of one or a few worn-out parts. Much of the unit may still be usable and have value after proper servicing and remanufacture. For example, most of an electric motor is almost indestructible; rewiring and rewinding the coil may be all that is necessary to make it the equivalent of a new motor. The same holds true for much of the hardware and fasteners used in a product. By dismantling and recovering these usable or refurbishable components, the manufacturer can reduce its need to purchase new ones, thus saving the expense of purchasing virgin parts, conserving valuable resources, and reducing the cost and environmental impacts of landfilling the piece. Manufacturers of laser copier toner cartridges have designed the cartridges for easy refurbishment, and most cartridges are now reused many times. The automobile industry has also been active in this arena for some time (starter motors, carburetors, clutches, etc.); other industrial sectors are slowly beginning to see the merits of reusing usable parts.

For remanufacturing to be economical, though, the products must be designed so that they can be easily and rapidly disassembled into their component parts. The product should be designed with the expectation that it will later be taken apart for component recycling/reuse.

Many factors must be considered in order to make reuse or recycling viable. Some of the more important ones are listed in Table 9.11. Optimization of these may

TABLE 9.11
Factors to consider for design for disassembly

Design for ease of disassembly
Reduce number of assembly operations
Simplify and standardize component fits and interfaces
Make components modular
Select easily recyclable materials for use
Minimize the number of materials used
Use compatible materials
Consolidate parts used
Mark materials to enhance separation
Use water-soluble adhesives when possible

be in conflict with the primary purpose of the product, or with one another, though. For example, it would be much easier to disassemble an automobile and reuse its parts if the parts were bolted together, but this form of assembly is not as strong as welded construction and can lead to squeaks and rattles while driving or to increased damage in the event of an accident. The objective, then, should be to optimize the factors listed above without impairing the primary objective of the product being manufactured or its quality.

EASE OF DISASSEMBLY. Products that can be easily and rapidly disassembled into their component parts are more likely to be reused or remanufactured. Those that are designed so that parts snap together are probably the easiest to disassemble after use. Bolted or screwed components are also easily disassembled. Ease of disassembly also makes repair of the product during use easier, because the part can easily be removed to allow access to other parts that need repair, or the removable part can readily be replaced, if necessary. For example, inner door panels in most automobiles are now assembled with snaps or simple latches, as are computer cases. When metal fasteners are used in a product, carbon or magnetic stainless steel fasteners should be preferred over nonmagnetic stainless steel, aluminum, or brass fasteners to facilitate removal by magnetic separation.

There are two methods of disassembly (Graedel and Allenby, 1995). In *reversible disassembly,* components are disassembled, essentially in the reverse order to which they were assembled, by removing connectors. The second method is *irreversible,* or *destructive, disassembly,* in which parts are broken or cut apart. Destructive disassembly is often quicker than reversible disassembly and is often the method of choice if recovery of the materials, rather than the components, is desired. The product or partial disassembled components are reduced in size by shredders, hammermills, or other size-reduction equipment. The size-reduced materials are separated into various categories, such as magnetic versus nonmagnetic, or high density versus low density. Table 9.12 lists some of the more common size-reduction and materials recovery techniques (de Ron and Penev, 1995). The objective is to get as complete a separation as possible in a cost-effective manner. The more preliminary disassembly that is done and the more material separation steps performed, the purer will be the product and the higher the value. However, each of these steps adds to the processing costs. These must be balanced against the value of the reclaimed materials.

Recycling can be done on site at the manufacturing facility or at a central facility. On-site separation is generally limited to recovery of a few components or materials of use in the plant. Centralized facilities are often more comprehensive since they may be separating a wide variety of materials from several different industries. Consequently, they may incorporate a number of recovery options. Figure 9.10 depicts a polymer recycling facility that employs size reduction, metals and nonmetals removal steps, air classifiers, and wet separation steps (Turk, 1997).

A product that is screwed together or connected with snap fittings can often be easily disassembled by just taking the product apart; the resulting parts could be readily reused. A product that is welded together cannot be so easily disassembled, though; the components must be cut apart, making them unavailable for direct reuse.

TABLE 9.12
Size reduction and materials separation techniques

Process	Description
Shredder/hammermill	The material is destroyed by hammers that beat on the material. In a hammermill, size-reduced materials fall through a grid in the bottom. A shredder does not have a grid.
Cutting mill	Size is reduced by cutting the material using disks mounted on a shaft turning in opposite directions. Often there is a grid on the bottom. Cutting mills are especially suitable to reduce elastic materials.
Cryogrinding	Reduces materials by freezing to a very low temperature with the aid of liquid nitrogen, which increases brittleness and improves the efficiency of the mill and the separation of materials.
Riddle	Particles are classified according to their dimensions, with small-size particles passing through the riddle, while larger particles are retained. By using several sizes of riddles in a cascade fashion, separation into several size ranges is possible.
Wind sifter	Solids are separated by gravity in a wind stream according to density, form, and dimension. Light particles are transported by the airstream, while heavy particles settle.
Gas cyclone	Separates solid particles in an air flow by means of centrifugal forces because of differences of density or dimensions. This is used mainly to separate dust from air.
Float-sink technique	Solid particles are separated according to density by placing the mixture in a liquid having a density value between the density values of the two components to be separated. The heavier particles sink to the bottom, while the lighter particles float to the surface.
Hydroclone	Solid particles in a liquid stream are separated based on differences in density and dimensions by means of centrifugal forces as the particles whirl around in a cylindrical vessel. It is also used to separate contaminants from the liquid.
Ferromagnetic techniques	Ferromagnetic components are separated by directing the waste flow onto a strong magnetic conveyor belt. The ferromagnetic materials are retained by the belt, while nonmagnetic materials fall off.
High-gradient magnetic separation	Separates paramagnetic materials from nonmagnetic materials by directing the waste flow through a strong magnetic field with high gradients.
Eddy current	Ferrous and nonferrous materials, or nonferrous materials alone, can be separated by using the conductivity of the material. An eddy current is generated in conducting materials in a fluctuating magnetic field. Conducting particles are deflected from the flow path, while nonconducting particles are not deflected.
Electrostatic techniques	The size-reduced materials are electrostatically charged and passed between a surface and an electrode of opposite charge. Conducting materials reject their charge and will have a charge equal to that of the surface; they are repulsed. Nonconducting particles are attracted by the surface.

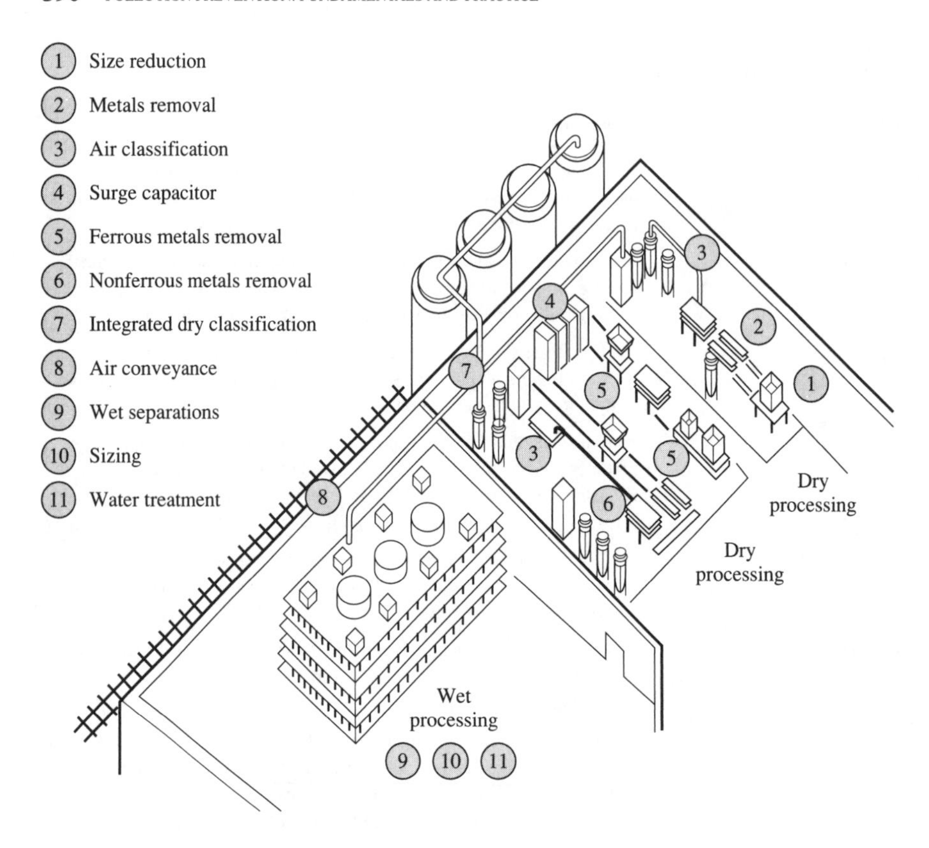

FIGURE 9.10
Polymer recycling facility showing processes for size reducing and separating materials. (Source: Turk, 1997)

An example of an easily disassembled product is a personal computer. Most are designed to be readily accessible so that internal components can be easily replaced or upgraded.

Major factors in the cost of materials or component recovery during reuse/recycling operations is the time and effort required to disassemble the product. Anything that can be done during the design stage to ease the disassembly step can greatly reduce these costs. Connection methods should be simple, and connection points should be apparent. The design should also consider what components are likely to be the most recyclable and have the greatest recycle value. These components should be placed in an easily accessible location and be easy to remove. A potentially valuable item located in an automobile dashboard that can be reached only by laboriously removing and disassembling the entire dashboard will probably never be recovered. Thought must go into the siting and design of these items at the earliest stages of design.

Often products can be made more easy to disassemble by reducing and simplifying the assembly operations. This has a second positive impact in that it usually reduces the assembly costs as well. Another wise step is to make the components modular so that they can be easily disassembled and reused without requiring extensive labor.

SIMPLIFY AND STANDARDIZE COMPONENTS. Simplifying products by minimizing the different types and sizes of components used can also make products easier to recycle. A product that uses many sizes of screws, bolts, washers, nuts, and rivets, requiring the disassembly worker to be constantly changing the tools during disassembly, will greatly slow the process and increase its cost. Often the differences in the size or specification of many parts used in a product are small; in many cases, the parts could be consolidated into one common part. All these different small parts will need to be kept segregated, along with the various larger components that are salvaged. Reducing the number and type of parts used eases this problem. All of these arguments hold true for the manufacturing step as well.

Products should be designed for ease of access to internal components. This usually means a simpler design that is less costly to manufacture. More time may be required during the design stage to arrive at a simplified design that requires fewer parts or one in which all reusable parts are readily accessible, but in the long run this time will be well spent in terms of easier parts procurement, reduced assembly time, and enhanced recycling/recovery later. A modular design will enhance the ease of disassembly.

MATERIALS IDENTIFICATION. A myriad of materials are often used in a product. In many cases, several different similar-appearing materials are difficult to distinguish, leading to additional expense to analyze the material or to further separate mixed materials later. This problem affects secondary recyclers more than original manufacturers. Confusion over the type of material involved is particularly critical with plastics, since there are literally thousands of different polymeric materials in use. Many of these plastic formulations are incompatible with each other and cannot be mixed together. This problem could be alleviated at least somewhat by marking the materials for easy identification.

Uniform standards should be adopted for material identification marking. This standard should include symbols that are easily recognized and logically placed for rapid observation. They should also be specific enough to allow separation, recovery, and recycling systems and operations to properly identify all materials.

Such a universal system does not yet exist, but inroads are being made. Most progress in materials identification systems has involved plastics recycling. Many plastics can be recycled, but, as noted, plastics of different types cannot be mixed. Unfortunately, it is impossible to distinguish one type of plastic from another by sight or touch, and only a small amount of the wrong type of plastic mixed in with the desired material can ruin the melt during reprocessing. Therefore, a universally accepted identification system is needed.

There are currently three major plastics identification systems: ISO 1043-1, ASTM D1972-91, and the Society of Plastics Industry (SPI) Voluntary National Container Material Code System. In addition, the Society of Automotive Engineers (SAE) has a system geared toward plastics used in the automotive industry (Hewlett-Packard, 1992).

The ISO standard calls for an identification of all plastic parts, using the symbols shown in Table 9.13 placed between > and < symbols. If a part consists of a blend of polymers, both are shown; for example, for a blend of polystyrene and polyethylene terephthalate:

>PS+PET<

Other symbols are available to denote the types of fillers included in the blend (see Table 9.13). If the blend above contains 30 percent glass bead as a filler, it would be designated as

>(PS+PET)−GB30<

In the United States, a more commonly used system is one developed by the Society of Plastics Industry. These markings are not as specific as the ISO system, in that they do not account for blends of polymers or for designating the presence of fillers, but they are more easily used by individual consumers. The categories for plastics are listed in Table 9.14, and the identification symbols are as shown in Figure 9.11. Unfortunately, the common plastics polycarbonate (PC) and acrylonitrile-butadiene-styrene (ABS) do not have recycling numbers. Types 1 and 2 are commonly recycled plastics; type 4 is less commonly recycled. The other types are generally not recycled to any great extent. Hewlett-Packard (1992) has proposed a system that incorporates

TABLE 9.13
ISO 1043-1 Plastics Identification System

Plastic		Filler	
>ABS<	Acrylonitrile/butadiene/styrene	GF	Glass fiber
>ABS−FR<	Flame-retardant ABS	GB	Glass bead
>EP<	Epoxy	MP	Mineral powder
>PA<	Nylon (polyamide)	CF	Carbon fiber
>PA6<	Nylon 6		
>PA66<	Nylon 6/6		
>PBT<	Polybutylene terephthalate		
>PC<	Polycarbonate		
>PE<	Polyethylene		
>PE−LLD<	Linear low-density polyethylene		
>PE−LMD<	Low–medium density polyethylene		
>PE−HD<	High-density polyethylene		
>PET<	Polyethylene terephthalate		
>PS<	Polystyrene		
>PS−HI<	High-impact polystyrene		
>PVC<	Polyvinyl chloride		
>SAN<	Styrene/acrylonitrile		
>SI<	Silicone		

TABLE 9.14
Categories of plastics, as defined by the Society of Plastics Industry

Type	Name	Common abbreviation	Uses
1	Polyethylene terephthalate	PET	Soda and water containers, some water-proof packaging
2	High-density polyethylene	HDPE	Milk, detergent and oil bottles, toys, plastic bags
3	Vinyl/polyvinyl chloride	V/PVC	Food wrap, vegetable oil bottles, blister packages
4	Low-density polyethylene	LDPE	Plastic bags, shrink wrap, garment bags
5	Polypropylene	PP	Refrigerated containers, some bags, most bottletops, some carpets, some food wrap
6	Polystyrene	PS	Throwaway utensils, meat packing, protective packing
7	Other		No recycling potential; must be landfilled or incinerated

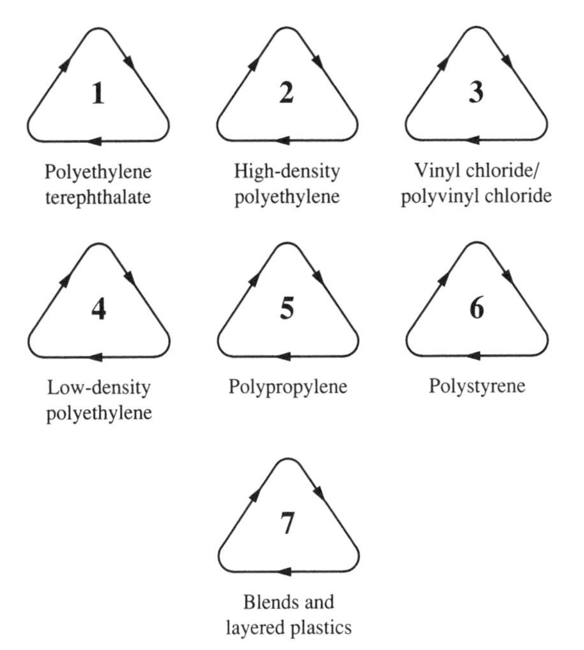

FIGURE 9.11
SPI plastics identification system.

elements of both the ISO and SPI systems (see Figure 9.12). This system allows for the identification of the plastic using both the ISO and SPI designations, the percentage postconsumer and total recycled content, and an indication of other materials incorporated into the plastic. The latter category includes additives that make the plastic biodegradable or photodegradable or that could cause problems during recycling, such as coatings, wax, varnishes, or multilayer nonseparable materials.

Only a few other materials are currently labeled to indicate composition. One is recycled paper; to properly identify paper materials and their recycled content, an identification symbol should be clearly printed on all paper and corrugated packaging. The American Paper Institute (API) uses the recycling symbol (see Figure 9.13*a*) to designate that cartons or other types of paper goods are manufactured using recycled material. However, the standard symbol does not indicate how much recycled fiber is used. It could be as little as 1 percent from postconsumer use and still receive the label. Some manufacturers add text to the symbol indicating the amount of recycled paper and the type of ink used in the paper. Soybean-based inks are gaining favor as a renewable alternative to harsh or toxic petrochemical inks, and their use is now commonly indicated on the symbol. For paper products that can reasonably be expected to be recy-

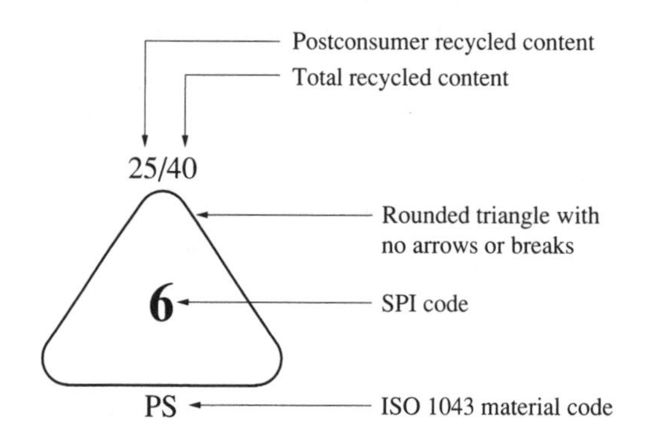

FIGURE 9.12
Proposed Hewlett-Packard plastics identification symbol.

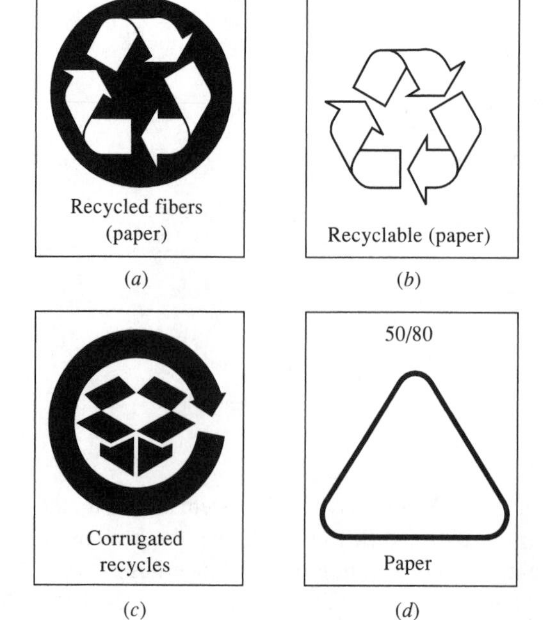

FIGURE 9.13
Recycled materials identification symbols for paper. (*a*) API symbol for recycled paper. (*b*) API symbol for recyclable paper. (*c*) FBA symbol for recyclable corrugated paper. (*d*) Proposed Hewlett-Packard symbol for recycled paper.

clable, the API uses the recyclable symbol (Figure 9.13*b*). The Fiber Box Association (FBA) uses the corrugated recycles symbol (Figure 9.13*c*) to inform users that the corrugated materials can be recycled. Hewlett-Packard (1992) proposes using a variation of its own symbol, developed by the Institute of Packaging Professionals, for plastics to denote the recycled content of paper products (Figure 9.13*d*). The first number indicates the preconsumer recycled paper content and the second number denotes the total recycled content.

USE OF COMPATIBLE MATERIALS. A common problem encountered during recycling of plastics is the mingling of different polymers in the recycle stream. This often causes the entire batch to become unusable because the polymers are incompatible with each other. It is very important to completely separate plastic components into their proper categories based on composition. Table 9.15 indicates those plastics that are compatible and can be processed together. As can be seen, most are not compatible with each other, necessitating a very efficient separation system. The designer should attempt to use as few different types of materials as possible. When it is necessary to use multiple materials in a product, designers should ensure that all materials can be easily separated from the primary plastic. If possible, nonplastic materials should not be attached to plastic materials. If more than one type of plastic is used, they should be compatible with one another so that the recovered materials are not contaminated by a small amount of improperly separated noncompatible plastics.

TABLE 9.15
Material compatibility chart for co-processing of recycled plastics

Matrix Material	Additive											
	PE	PVC	PS	PC	PP	PA	POM	SAN	ABS	PBTP	PETP	PMMA
PE	■	□	□	□	■	□	□	□	□	□	□	□
PVC	□	■	□	□	□	□	□	■	●	□	□	■
PS	□	□	■	□	□	□	□	□	□	□	□	□
PC	□	❍	□	■	□	□	□	■	■	■	■	■
PP	■	□	□	□	■	□	□	□	□	□	□	□
PA	□	□	❍	□	□	■	□	□	□	❍	❍	□
POM	□	□	□	□	□	□	■	□	□	❍	□	□
SAN	□	■	□	■	□	□	□	■	■	□	□	■
ABS	□	●	□	■	□	□	❍	□	■	❍	❍	■
PBTP	□	□	□	■	□	❍	□	□	❍	■	□	□
PETP	□	□	❍	■	□	❍	□	□	❍	□	■	□
PMMA	□	■	❍	■	□	□	❍	■	■	□	□	■

Key: ■ Compatible ● Compatible with limitations ❍ Compatible only in small amounts □ Not compatible

Source: Adapted from Bras and Rosen, 1997.

TABLE 9.16
Classes of supply of the elements

Infinite supply	A, Br, Ca, Cl, Kr, Mg, N, Na, Ne, O, Rn, Si, Xe
Ample supply	Al, Ga, C, Fe, H, K, S, Ti
Adequate supply	I, Li, P, Rb, Sr
Potentially limited supply	Co*, Cr†, Mo, Rh, Ni*, Pb, As, Bi, Pt†, Ir†, Os†, Pa†, Rh†, Rn†, Zr, Hf
Potentially highly limited supply	Ag, Au, Cu, As, Se, Te, He, Hg, Sn, Zn, Cd, Ge, In, Th

* Maintenance of supplies will require mining seafloor nodules.

† Supply is adequate, but virtually all from South Africa and Zimbabwe. This geographical distribution makes supplies potentially subject to cartel control.

Source: Graedel and Allenby, 1995.

MATERIALS SELECTION. Proper selection of materials used in a product can greatly ease the process of recycling or reusing the materials. Materials should be selected during the life-cycle design stage; the objectives should be (1) to minimize environmental impacts caused by the material while maintaining product quality and (2) to select materials that are easily recycled. In some cases, modifications to the design can result in a smaller quantity of a material being needed in the product. The proper selection of materials was discussed earlier in this chapter and will not be repeated here. As described above, the composition of all materials should be identified so that incompatible materials will not be mingled.

When choosing materials, the designer must first ensure that the material is suitable for the purpose intended. In many cases, more than one material may be suitable and other factors should be considered. Cost will certainly be a major factor. Normally, the lowest cost material that is suitable for the task will be selected. Another factor that should be considered, though, is the sustainability of the source of the material. Resource depletion has seldom been a major selection factor in the past, but it may well become one in the future. Resource depletion was discussed in Chapter 3. Table 9.16 indicates the availability of major elements. This listing can be consulted to assess the impact of using a particular material in the product being designed.

The environmental impacts resulting from extracting the natural resource should also be considered, as should the toxicity of the material and the impacts of disposing of the material after the product's use is exhausted, as these may significantly affect the product's life-cycle assessment. Wherever possible, the designer should select abundant, nontoxic, nonregulated natural materials. If suitable recycled materials are available, these should be selected.

9.3.5 Disassembly Strategy

For recovery of useful components, parts, or materials to be effective, a carefully planned disassembly strategy is needed. The objective of the disassembly must be well defined. The following questions must be asked (Johnson and Wang, 1995):

1. How may the recovery process itself generate the highest possible return on investment?
2. Is there a particular disassembly sequence that will maximize the return?
3. Is it better to recover only specific components rather than all components?
4. What design characteristics facilitate ease of disassembly and how are they to be employed?

A three-level disassembly analysis methodology has been developed to answer these questions:

- *Level 1. Feasibility study of materials recovery opportunities.* An initial study is performed to determine:
    1. Percentage of the product (by weight) that can be diverted from disposal.
    2. Preliminary cost estimates: recovery versus disposal.
    3. Other trade-off considerations: learning curve, automation, payback analysis, customer response, trends in legislation, and so on.
- *Level 2. Optimal disassembly sequence generation.* This level generates a preferred sequence of disassembly which will maximize the value to be reclaimed in recovery. The product design characteristics are analyzed, including:
    1. Spatial relationships between components.
    2. Characteristics of individual disassembly operations such as tooling and accessibility.
    3. Clustering of components.
    4. Maximizing the sequence in terms of concurrent disassembly operations and the amount of material recovered.
- *Level 3. Disassembly optimization.* Using the disassembly sequence generated in level 2, the disassembly is optimized in terms of cost, benefit, and degree of disassembly. The analysis results in the following information:
    1. Identification of components which are economically recoverable and those that are not and must go to disposal.
    2. Determination of that point within the disassembly process at which it becomes uneconomical to continue recovery.
    3. Identification of specific component designs which constrain feasible economic recovery of materials.
    4. Identification of components that are most likely to benefit from redesign using Design for Disassembly (DfD) concepts.

The first step in the analysis of a disassembly strategy is to analyze the structure of a product (de Ron and Penev, 1995; Lambert, 1997; Penev and de Ron, 1996). A model of the product's structure is developed using a tree graph in which all relevant components and their physical relationships are displayed. An appropriate type of graph for this purpose is one composed of nodes, representing subassemblies or components, and lines connecting nodes, representing disassembly operations. This diagram (an example is seen in Figure 9.14) shows which components and/or materials

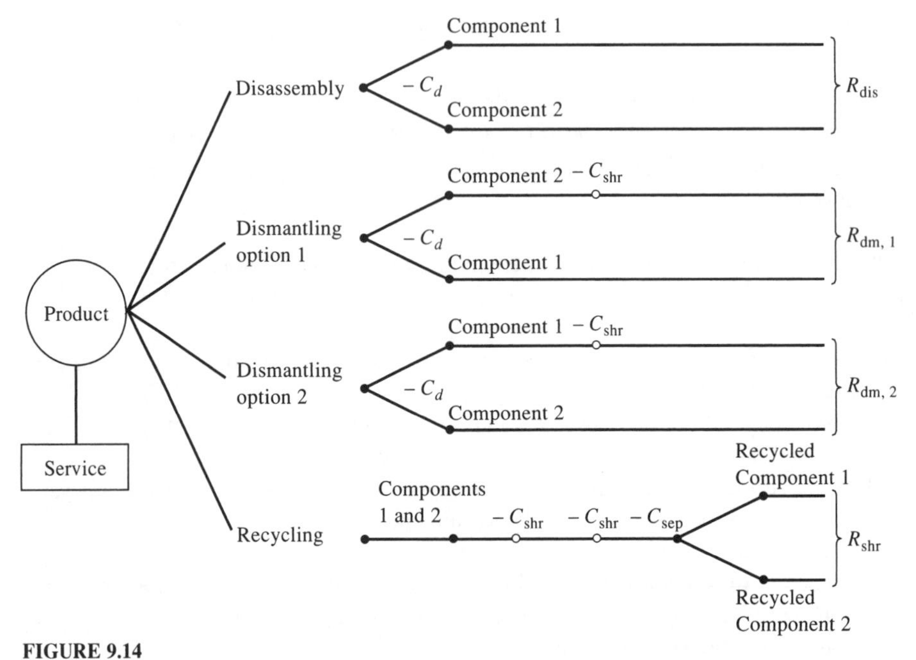

FIGURE 9.14
Typical product disassembly tree diagram, composed of nodes, representing subassemblies or components, and lines connecting nodes, representing disassembly operations. C_d is the cost associated with the disassembly or dismantling, C_{shr} is the cost of shredding the component, and C_{sep} is the cost of separation. R is the revenue derived from the process.

are significant for determination of the disassembly sequence. Components or materials that may contaminate the recovered fraction should be determined, as their removal will determine the number of compulsory disassembly steps.

When the product's structure has been clarified, possible disassembly operations should be evaluated. The objectives are to generate all feasible disassembly operations, determine the optimal level to which the product should be disassembled, and determine the order of execution of these disassembly sequences that will be most economical. Some operations may not be possible because proper separation of materials or components cannot be achieved. If a desired part can be retrieved by destructive methods, these should be used to speed up recovery and lower the cost.

There may be more than one disassembly pathway that can achieve recovery of a particular component or part of interest. In Figure 9.15, product A is being disassembled in order to recover part P. As shown in the figure, this can be achieved (1) by recovering component B and then disassembling this through one of several routes to arrive at part P or (2) by partially destructing product P and then starting disassembly with subassembly H. Each of these will differ in the amount of labor involved, the number and type of tools required, the impact on the recoverability of other materials, and the overall cost. Because the number of theoretical disassembly sequences increases exponentially with the number of parts involved (see Table 9.17), only a few of the possible disassembly sequences can be evaluated. Engineering judgment must be used in deciding which alternatives to study in detail.

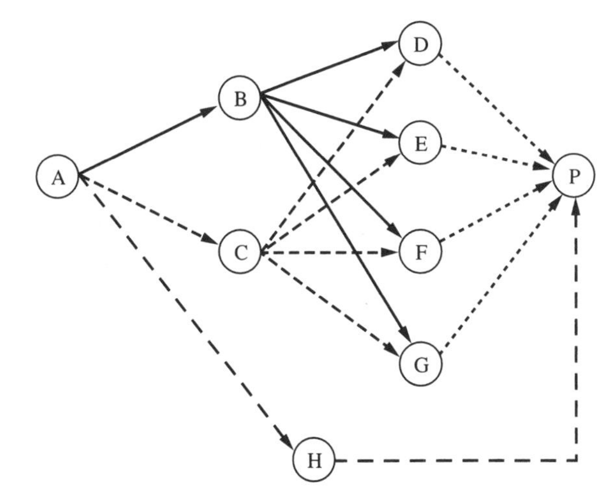

FIGURE 9.15
A product may have multiple disassembly pathways. The designer must pick the optimum path.

TABLE 9.17
Theoretical number of disassembly sequences as a function of the number of parts of the original assembly

Number of parts	Number of subassemblies	Complete disassembly sequences	Total number of disassembly sequences (complete and partial)
1	1	1	1
2	3	1	2
3	7	3	7
4	15	15	41
5	31	105	346
6	63	945	3,797
7	127	10,395	51,157
8	255	135,135	816,357
9	511	2,027,025	15,050,590
10	1,023	34,459,425	314,726,297

To establish an effective plan for optimal disassembly of a product, one should find a balance between various disassembly options. The objective should be to do this in a cost-effective manner. One way to do this is to evaluate the way in which the product was assembled. During the production process, raw materials and energy are used to produce the product. The value added to the product, V_p, increases as it passes through each production step (de Ron and Penev, 1995). This can be expressed as

$$V_p = \sum_{i=1}^{n} (V_{r,i} + V_{m,i} + V_{a,i})$$

where n = number of components
V_p = value added to the product
$V_{r,i}$ = value added to the raw materials of each component
$V_{m,i}$ = value added during the manufacturing process of each component
$V_{a,i}$ = value added during the assembly of each component

The aim of disassembly is to regain the value added to products and materials and to protect the environment. Each of the disassembly operations consists of several sub-processes, and each has a number of economic considerations:

1. The value added to products and materials.
2. The disassembly cost per operation.
3. The revenue per operation.
4. The penalty if noncompatible materials are not completely removed.

The value of the *abandoned product,* V_{ab}, can be determined as follows:

$$V_{ab} = (R_{dis} + R_{shr} - C_{tr} - C_m)$$

where V_{ab} = value of the abandoned products
R_{dis} = revenue from disassembled parts
R_{shr} = revenue from recycling (shredding) the remaining materials
C_{tr} = transport costs
C_m = miscellaneous costs (testing, remanufacturing, quality assurance, etc.)

The revenues obtained by *disassembling* valuable parts of the product and selling them as used parts or selling them to remanufacturers can be expressed by

$$R_{dis} = j = \sum_{j=1}^{n} (P_j - C_{dis,j})$$

where n = number of valuable components which can be extracted from the product
P_j = value of component j
$C_{dis,j}$ = disassembly cost of component j

The revenue from selling the shredded materials remaining after recovery of valuable parts or components is given by

$$R_{shr} = M\left(\sum_{j=1}^{k-1} (r_j \alpha_j - C_{sep,j}) - C_{shr} \right)$$

where k = number of material fractions
M = mass of the shredded part of the product
r_j = revenue of the material j fraction (negative if the material must be dumped)
α_j = mass fraction of material j relative to the total mass of the shredded product
$C_{sep,j}$ = separation cost of a material fraction
C_{shr} = operational cost of the shredder

In most cases, several options will be available for disassembling a product and recovering its valuable products. The objective is to choose among the alternatives and select those that provide the maximum revenue, based on revenues from disassembly and from shredding.

Disassembly, shredding, and sale of components or materials is desirable, but only if the revenues from this operation exceed the associated economic and environmental costs. Within these options, the preferred hierarchy is disassembly and sale of components first, followed by shredding and recovery of materials. However, economic factors will govern which of these is most viable.

9.3.6 Computer-Aided Design

Design for demanufacturing or remanufacturing usually requires that the original product be designed or redesigned to accommodate ease of disassembly. This can be a complex and sometimes daunting task. Recently, some computer-aided design techniques have been modified to include environmental and technical considerations of demanufacturing. This field of study is still in its infancy, but it is growing rapidly. A number of researchers are investigating the design characteristics that allow for easier recovery of components for reuse or recycling.

In one study, Bras and Rosen (1997) are studying the various processes involved in remanufacture that represent the greatest investment in resources, including disassembly, cleaning, inspection, parts replacement, refurbishing, and reassembly. The process analysis procedure consists of a spreadsheet with three components: a parts questionnaire; a worksheet for assembly, disassembly, and testing times; and subassembly data. The resulting output is a single index characterizing the overall remanufacturability of the product.

Boothroyd Dewhurst Inc. (BDI) of Wakefield, Rhode Island, in conjunction with the TNO Institute of Industrial Technology in Delft, Netherlands, has developed a software program that can be used by product designers to analyze and optimize the disassembly sequence of products for end-of-life recovery (Turk, 1997). It is used at the earliest stages of design to ensure easy recycling of the product's materials or components. The resulting data show cost benefits for various options, such as materials recycling, part remanufacture or reuse, and disposal through landfilling or incineration. Digital Equipment Corp. (DEC) helped develop this software and uses it to design product stewardship features into all of its computer equipment. To facilitate recyclability, DEC designs products using modular components for easy and fast disassembly and recovery of components and raw materials after the product's first useful life and for easy reuse of modules in new products.

In another study, Gadh (1997) and his co-workers are developing software that allows assessment of disassembly options during product design development. They have developed two software modules, a CAD-based nondestructive disassembly software package and a module that allows design assessments for components requiring destructive disassembly.

9.3.7 Waste Exchanges

Even after a company has done all it can to reduce the amount of waste it generates by source reduction or internal recycling, there will still be some residual materials (wastes) left. Some of these may have value, although not to the company producing them. These materials are candidates for sale or donation to another company that can use them. The difficulty is in finding that company.

A waste exchange can help businesses find a market or an end user for materials they no longer need. A waste exchange is a service that identifies both producers and markets for by-products, unspent virgin materials, and other forms of waste materials. Typically, a waste exchange acts as an information exchange center, matching producers and users of residual materials. They do not normally handle the materials, but rather provide details about available or wanted waste and surplus materials in the form of catalogs or on an electronic bulletin board service (BBS). The listing usually describes the material, quantity, and form available as well as indicating the geographic location. Names of the suppliers are not given. Potential users may find out who is offering the listed material by contacting the waste exchange. Waste exchanges do not take part in the negotiations or actual shipment of materials.

Waste exchanges are both cost effective and environmentally beneficial. The business generating the waste can avoid paying disposal costs, avoid shipping costs, clear valuable storage space, and minimize regulatory paperwork. The business receiving the materials uses this less expensive feedstock and packaging in place of more expensive new materials. Communities benefit by not having to fill their landfills with this material and by keeping hazardous materials out of the waste. Waste exchanging also reduces resource depletion concerns.

It should be kept in mind that hazardous waste laws pertain to waste exchanges when hazardous wastes are involved. The wastes must be manifested when shipped to the receiver and must be hauled by a licensed hazardous waste transporter. The facility receiving the waste must be a licensed hazardous waste treatment, storage, or disposal facility. Regulatory liability for both companies ends only when the hazardous waste ceases to be a waste, as when it has been beneficially used, reused, or recycled. The waste producer must evaluate the waste receiver carefully, because the generator is liable if the receiver improperly handles or disposes of this waste. Because of this, only small quantities of hazardous wastes have been exchanged. This is not the case for non-hazardous wastes.

The National Materials Exchange Network (NMEN) is an EPA-supported cooperative effort of 36 materials exchanges across North America, offering an on-line database of materials containing more than 5200 listings of materials in 17 categories. There is no charge for the listing service. In addition, more than 50 statewide and local waste exchanges are located across North America. Many of these participate in NMEN. A listing of these can be found elsewhere (U.S. EPA, 1994).

Waste exchanges have been very successful, with about 12 million tons of material exchanged per year. However, this represents only about 10 percent to 30 percent of the materials available for exchange. Typical wastes transferred include plastics, solvents, acids, alkalis, wood and paper, and metals and metal sludges.

An example of a successful waste exchange is that of Triad Energy Resources, Inc., and C&H Sugar Company, both located in California (Decio, 1996). Triad is an agricultural service company that tests and analyzes soils, crops, and water for farmers and then advises landowners about needed nutrients and where to find them. Triad specializes in recommending surplus food and food by-products (e.g., eggshells, mushroom waste) rather than chemical fertilizers. C&H Sugar Company had been unable to find a use for its sugar filtration by-products from the refining process. The company was landfilling these materials at ever-increasing costs. The two companies found each other through CALMAX, the California Integrated Waste Management Board waste exchange. Triad is now converting nearly 100,000 tons per year of C&H Sugar filtration process by-products into an agricultural soil amendment and fertilizer.

9.3.8 Recovery through Composting or Energy Reuse

A final way to recycle resources in a discarded product is to extract the energy present in it or to convert it into compost. Energy recovery through incineration and conversion of the product through composting should be thought of as a recycling option, rather than a disposal option, because the products are not being thrown away, but are converted to a useful commodity. These two options are available only for organically based materials, and only for those that will not be contaminated by other materials present in the product. This means that thought should go into the original design of the product to ensure that energy-recovering incineration or composting will be a viable disposal option. The materials should not contain any toxic or hazardous components that could be emitted to the environment during composting or incineration or that would render any compost produced too dangerous to use.

9.3.9 Barriers to Reuse

Pressure from environmental and regulatory groups to increase the level of demanufacturing and reuse of materials has encouraged many companies to explore ways to better design their products for eventual disassembly and to incorporate remanufactured parts or recycled materials in their products. These companies have also seen the added benefits of public approval of their environmental friendliness and of increased profitability by using recycled materials. Some companies have profited markedly by these efforts. Xerox estimates a saving of almost $200 million in four years as a result of making take-back and stewardship an integral part of its operations. However, widespread use of demanufacturing techniques has been slow to take hold, primarily because of a shortage of scientific knowledge about them. Even companies that are experienced at demanufacturing their own products find it difficult to explain how to do it efficiently and in a methodological fashion. The new software packages described earlier may help to provide this methodology in the future.

A company investigating the economics of demanufacturing and reuse of materials must ask many important questions, including (Spicer and Johnson, 1997)

- What portion of the product is economically feasible to recycle or reuse?
- What is the most profitable or least costly way to retire the product?
- How much of the product will end up in the landfill?
- How long will it take to fully or partially disassemble the product for recycling, and how much will it cost?

- How can the company plan for its demanufacturing requirements?
- How much would it cost to recover a particular selection of components?
- What disassembly plan would optimize the profit of the demanufacturer? That is, what are the economical trade-offs between reuse, remanufacturing, recycling, and disposal?
- How much disassembly is too much?
- How may the recovery process itself generate the highest possible return on investment?
- What would happen if material prices or tipping fees radically changed? How can the economics of demanufacturing account for fluctuating materials prices?
- What will be the effect of design changes on the economics of retirement or remanufacturing?

One of the most difficult questions to answer is that of the economic value of the recovered materials. How much are the components, plastics, metals, glass, and precious metals recovered from discarded consumer products worth, and how stable are those prices? For component parts, are they still useful or refurbishable, or have they become obsolete? Some parts, such as electric motors, change little over the years and can often be refurbished to like-new condition; other parts, such as electronic components, often quickly become obsolete, with little value other than the materials they are composed of. The recovered materials' prices must be compared with the costs of disassembling and recovering/refurbishing the component or material, the prices for virgin materials, and the landfill disposal costs to provide justification for reuse. Unfortunately, all of these prices can fluctuate widely, making accurate predictions difficult. Reuse economics has been likened to a "moving target," driven up or down by the supply and demand forces of the open market (King, 1996). These fears of incorrectly evaluating reuse economics have prevented many companies from enacting product stewardship/demanufacturing programs.

A more stable marketplace for recycled materials and an infrastructure for collecting and redistributing discarded products are needed if remanufacturing/reuse is to become widespread. The recent introduction of recycled materials commodities on the Chicago Board of Trade is a good step in this direction. Another major step has been the creation of waste exchanges across the country. These provide a central point for information on the sources and users of recycled materials, and they are helping to stabilize the markets.

9.4 PACKAGING

One of the largest components of solid wastes going to our landfills—approximately 30 percent of the total—is packaging materials. Much of this is composed of potentially recyclable plastics and paper. Improved packaging of products over the years has resulted in increased consumer safety and reduced damage to the products contained within the package. Until recently, the philosophy was that more packaging meant more safety and therefore was a good objective. Over the last one to two decades, though, concern has arisen about the waste disposal problems associated with this packaging, leading to significant improvements in packaging materials and design in order to reduce the amount of these materials going to our landfills. Figure 9.16 shows that about 30 percent of the wastes going into our municipal landfills is packaging materials, much of which is composed of materials that could be reused or recycled.

In recent years, as the concepts of life-cycle assessment have come to the forefront, an emphasis on minimizing packaging has been renewed. This is particularly true in Europe, where the European Union Directive on Packaging and Packaging Waste has made waste recovery and recycling mandatory. The recycling target for certain kinds of packaging in Europe is between 50 percent and 65 percent by the year 2001.

Reducing the amount of packaging, incorporating recycled materials into the packaging, or changing to a more readily recyclable or biodegradable packaging material not only reduces the environmental impacts associated with disposal, but also results in reduced impacts from producing the packaging and overall reduced costs due to the packaging. Besides the packaging material itself, associated materials such as toxic inks and solvents have impacts that must also be considered.

Packaging should maintain the integrity of the product, ensure consumer safety, and comply with all legal and regulatory requirements while being environmentally responsible (Hewlett-Packard, 1992). Properly designed packaging allows manufacturers to deliver their products to consumers with a minimum of damage, spoilage, or deterioration. Packaging must also be convenient to use by the consumer. Within these requirements, there should also be the objectives of using the minimum amount of material possible while still maintaining the packaging's functionality, and reducing the potential for toxicity entering the environment due to the packaging.

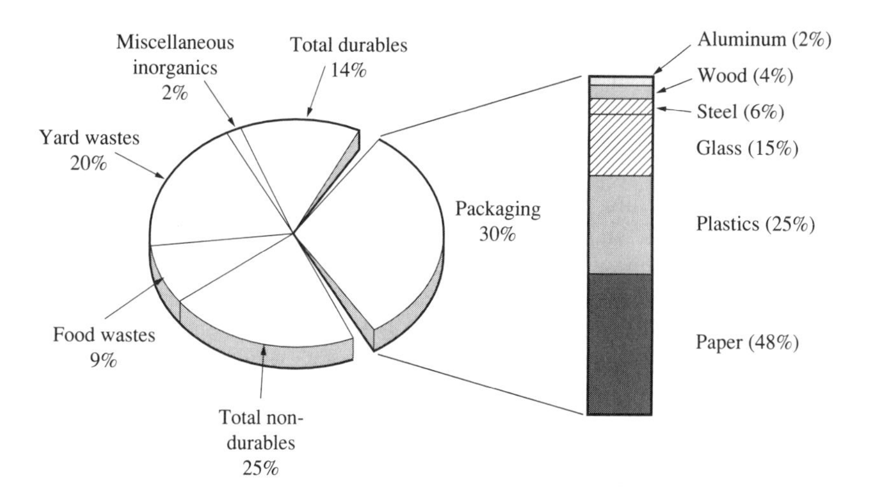

FIGURE 9.16
Components of municipal solid waste and types of packaging waste by weight. (Adapted from Ramsden, 1995)

9.4.1 Minimizing Packaging

Two strategies may be helpful in designing packaging to meet the needs of pollution prevention. The first is to *reduce packaging* by

- Distributing the products unpackaged.
- Using reusable materials.
- Modifying the product so that it requires less packaging.
- Reducing the density of packaging materials.

The second strategy is *material substitution* by

- Using recycled materials.
- Using degradable materials.

The first objective should be to eliminate packaging all together, followed by elimination of excessive secondary or tertiary packaging. Many examples of unneeded packaging abound. When you buy a tube of toothpaste, it often comes inside a cardboard box. A fast-food hamburger may be wrapped in paper, put in a cardboard box, and placed in a paper bag. A computer comes wrapped in plastic, housed in a cardboard box, surrounded by Styrofoam inside another cardboard box. The packaging may weigh as much as the computer. Is all of this packaging necessary, or is it there to impress the consumer? In many cases, much of this packaging can be eliminated.

Not only does this excess material add unnecessarily to the waste stream, but often it complicates potential recycling because several different types of materials (paper, cardboard, plastics, etc.) may be commingled. When plastic bags and cardboard cartons are used, this mixture is obvious. In other cases, it is not. For example, consider a paper bag holding chips. It appears to be a simple thin plastic of minimal weight. On first appearance, it looks like many of the objectives of improved packaging have been met. However, closer examination of the bag shows that it is not as simple as it appears (see Figure 9.17). The bag is actually composed of several extremely thin layers of several different materials, each intended to meet a different packaging need (preserving freshness, indicating tampering, providing product information) (OTA, 1992). The bag does minimize materials use and provide good product shelf life, but the commingled materials make it almost impossible to recycle the bag.

Reducing the amount of material used in containers can significantly impact the amount of packaging that needs to be disposed of. Industry has made major strides in reducing the weight of packaging over the past few years. As shown in Table 9.18, all major soft drink container materials evidenced weight reductions of at least 18 percent during the period 1972–1992.

Changes in product design may also aid in reducing or eliminating packaging requirements. The detergent industry has used this approach by reformulating cleansers into concentrates that will do the same cleaning job with much less cleanser. Therefore,

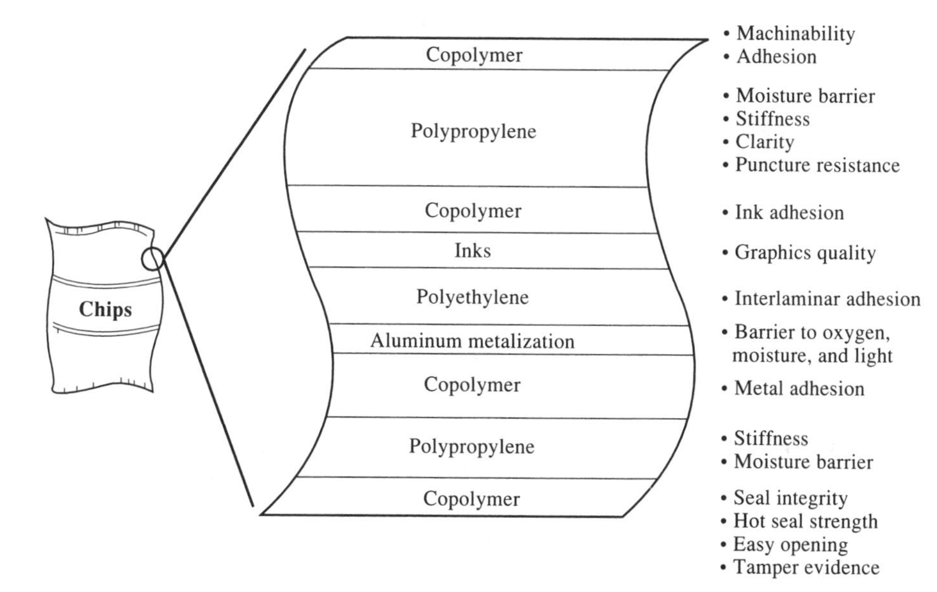

FIGURE 9.17
The cross section of a snack bag illustrating the complexity of modern packaging. The bag is only 0.002 inch thick, but it consists of nine different layers, each with a specific function. (Adapted from Hewlett-Packard, 1992)

TABLE 9.18
Reduction in weights of soft drink containers

	Reduction in pounds per 100 gal of product delivered		
Container	**1972**	**1992**	**Change, %**
One-use glass bottle (16 fl oz)	605.6	384.3	−36.5%
Steel can (12 fl oz)	112.0	76.7	−31.5
Aluminum can (12 fl oz)	48.0	37.5	−22.0
PET bottle (2-L, one-piece)	7.6	22.6	−18.1

Source: Franklin Associates, 1994.

detergent manufacturers can package the material in smaller containers. They are also packaging the cleaners in refillable containers so that the cleaner dispensers do not have to be discarded after being emptied.

Procter & Gamble is one company that has successfully reduced packaging by marketing concentrates. By switching Tide, Dawn, Ivory, and Joy to concentrated forms, P&G has reduced packaging by 16 percent, or 9.7 million pounds per year (5.5

million pounds from plastics, 4.2 million pounds from corrugated fiber in shipping cartons). Further, all bottles are made from high-density polyethylene (HPDE) containing 25 percent postconsumer recycled content.

In general, eliminating or reducing the amount of water in products generally allows for more efficient packaging. Other examples are the reduced packaging requirements for concentrates, powders, and dry foods.

Other examples of packaging changes to promote pollution prevention include changing the configuration of a product to make it easier to protect with less packaging, changing to the use of less toxic inks and pigments for the labels and graphics on the packaging, reducing the thickness of corrugated cardboard containers, and using thinner but stronger plastics. One major study found that flexible pouches and refill containers weigh 75 percent to 90 percent less than traditional rigid containers (*ULS Report,* 1995). These bags, which are commonly used in Europe, consume 73 percent less energy (in the form of material, energy, production, transportation, and disposal costs) than do rigid bottles when the entire life cycles of the two are compared (see Table 9.19). Quaker Oats has come out with a line of bagged cereals, which results in a packaging savings of 85 percent, by weight, over the traditional bag-in-box approach. It is estimated that if all cold cereals were converted to packaging in bags only, roughly 500 million pounds of paperboard would be saved annually (*ULS Report,* 1997).

Evaluation of packaging changes should not be limited to the packaging that goes to the consumer. Packaging used to ship materials and components to the manufacturer should also be analyzed: crates, boxes, drums, stretch film used to bundle materials, pallets, and so on. Where possible, materials should be shipped in bulk, rather than individually.

Many companies have aggressively attacked the packaging issue. Hewlett-Packard, for example, has set a goal of at least 15 percent reduction by weight and/or volume for packaging for new products when compared with the same type of product package two years earlier. HP further states that priority should be given to using packages or packaging components which are reusable without remanufacturing, and that the packaging materials should be capable of being reused safely at least five times (Hewlett-Packard, 1992). As an example of the company's efforts, the HP Deskjet 850 is made with 25 percent recycled plastic and is molded with a new process that allows the use of thinner and lighter plastic. This design will reduce the amount of waste eventually going to landfills by approximately 6 million pounds. For some applications, HP

TABLE 9.19
Life-cycle energy needed to wash 1000 loads of laundry

Container type	Total packaging material, lb	Recycled content, %	Total LCA requirement, Btu
Refillable bag	1.7	25%	62,650
Refillable bottle	7.6	48	228,100

Source: ULS Report, 1996.

uses foam pellets made from wheat instead of traditional Styrofoam packaging materials. In addition, HP has eliminated or significantly reduced heavy metals in packaging inks; vegetable-based inks are now used for some applications.

9.4.2 Degradable Packaging

There are limits to how much reduction in packaging volume can be achieved through redesign of the product or the packaging, as well as in the practice of recycling packaging materials. Eventually, packaging materials must be discarded, and they may then impact the environment. It takes at least 50 years, and usually longer, for plastics to break down in the environment. In recent years, there has been an increasing effort to produce degradable packaging in an attempt to minimize this impact. Degradation may be due to biological or chemical activity or to exposure to sunlight.

Degradable packaging materials have been highly praised. The concept is laudable, but the actual performance does not always meet the claims. Many degradation processes require the presence of sunlight, oxygen, and water; a properly operating landfill is usually deficient in all of these. Thus a material that is potentially biodegradable may not actually degrade after disposal. Newsprint should biodegrade readily in the environment, but numereous studies of decades-old landfills show that in a properly operating landfill where little or no oxygen or water is present, essentially no breakdown of newsprint occurs. The same would occur with "biodegradable" plastics. The main advantage of use of these materials is that packaging indiscriminately discarded into the environment as litter may eventually decompose. Another disadvantage of degradable materials is that they greatly complicate recycling operations by acting as a contaminant in recovered materials. In addition, degradation may enhance pollution by more readily releasing potentially toxic dyes, fillers, and other components from a material that was previously essentially inert (Keoleian, 1995).

The majority of the research into degradable packaging has focused on development of biodegradable plastics. In 1996, some 78,684,000 pounds of plastics were produced in the United States, a 9 percent increase over the previous year. Of this, 27.2 percent was used for packaging. The impact of all this plastic on the environment is considerable. Because of its light weight, the use of plastics in packaging is increasing (see Table 9.20). Fortunately, the rate of plastic bottle recycling is also increasing (see

TABLE 9.20
Materials used in soft drink containers, billions of units

Container type	1989	1991	1993	1995	1996
Aluminum cans	45.7	53.0	58.0	64.6	64.3
Bimetal cans	3.7	2.9	2.1	0.0	0.0
Glass bottles	8.9	8.1	5.9	2.2	1.0
PET bottles	6.9	72.5	77.1	82.8	84.0
Total	65.2	72.5	77.1	82.8	84.0

Source: Taylor, 1997.

Table 9.21). However, the recycling rate is still only 25 percent, leaving a lot of plastic bottles going into landfills or becoming litter.

Biodegradable plastics are defined by ASTM as plastics in which degradation results from the action of naturally occurring microorganisms such as bacteria, fungi, and algae. Most biodegradable packaging is made using a combination of corn starch or cellulose and polyethylene. Some consider the term "biodegradable" a misnomer for these materials. The microorganisms biodegrade the starch, leaving the polyethylene behind as a more dangerous "plastic dust" and releasing toxic additives such as lead, cadmium pigments, and stabilizers. Unfortunately, it has also been found that as the amount of starch used in the plastic is increased, strength of the plastic decreases. Consequently, only a small amount of starch can be used to replace polymer. In addition, these plastics are not heat-stable and cannot be used in injection and blow molding.

Recently, other types of biodegradable plastics have been introduced. Among these are synthetic biodegradable materials, such as poly(ϵ-caprolactone), and processible naturally occurring polymers, such as poly(3-hydroxybutonate) (PHB). The first research on biodegradable plastics focused on PHB, an energy storage material present in a wide variety of bacteria. The bacteria used to produce the PHB, typically *Alcaligenes eutrophus,* can be cultured in biological reactors and the PHB can be readily recovered. The physical properties of PHB are extremely close to that of polypropylene. Unfortunately, the slow degradation rate and the possibility that degradation byproducts may still be problematic have reduced the attractiveness of these plastics. Poly(ϵ-caprolactone) is the material used in the manufacture of seeding pots. It is susceptible to hydrolysis of the ester linkage by microorganisms. More uses for this material are being explored (Bordacs-Irwin, 1995; Ramsden, 1995).

Several photodegradable plastics are available. In some states their use is mandatory for the plastic rings that hold together six-packs of beverage containers, mainly to keep wildlife from becoming entangled in them. Photodegradability is generally achieved by the inclusion of carbonyl groups into the backbone of the polymer; these act as ultraviolet absorbers to break the polymer down. Unfortunately, the rate of breakdown is slow, and degradation is minimal in landfills or other locations where sunlight is absent (Ramsden, 1995).

TABLE 9.21
Plastic bottle recycling, 1995 vs. 1996

	1995		1996	
Container type	Million pounds recycled	Recycling rate, %	Million pounds recycled	Recycling rate, %
All plastic bottles	1,271.7	26.1%	1,307.0	25.0%
PET bottles	644.6	33.1	630.6	29.1
HDPE bottles	587.4	23.4	655.5	24.4

Source: Harler, 1997.

REFERENCES

Adger, B. M. "Industrial Synthesis of Optically Active Compounds." In *Chemistry of Waste Minimization,* ed. J. H. Clark and T. C. Williamson. ACS Symposium Series 626. London: Blackie Academic & Professional, 1995, pp. 201–21.

Allenby, B. R. "Design for Environment: A Tool Whose Time Has Come." *SSA Journal,* September 1991, pp. 6–9.

Anastas, P. T. "Benign by Design." In *Benign by Design: Alternative Synthetic Design for Pollution Prevention,* ed. P. T. Anastas and C. A. Farris. ACS Symposium Series 577. Washington, DC: American Chemical Society, 1994, pp. 2–21.

Anastas, P. T., and Farris, C. A., eds. *Benign by Design: Alternative Synthetic Design for Pollution Prevention.* ACS Symposium Series 577. Washington, DC: American Chemical Society, 1994.

Anastas, P. T., Nies, J. D., and DeVito, S. C. "Computer-Assisted Alternative Synthetic Design for Pollution Prevention at the U.S. Environmental Protection Agency." In *Benign by Design: Alternative Synthetic Design for Pollution Prevention,* ed. P. T. Anastas and C. A. Farris. ACS Symposium Series 577. Washington, DC: American Chemical Society, 1994, pp. 166–84.

Anastas, P. T., and Williamson, T. C. "Green Chemistry: An Overview." In *Green Chemistry: Designing Chemistry for the Environment,* ed. P. T. Anastas and T. C. Williamson. ACS Symposium Series 626. Washington, DC: American Chemical Society, 1996, pp. 1–17.

Atherton, J. H., and Jones, I. K. "Solvent Selection." In *Chemistry of Waste Minimization,* ed. J. H. Clark. London: Blackie Academic & Professional, 1995, pp. 417–40.

Bodor, N. "Design of Biologically Safer Chemicals Based on Retrometabolic Concepts." *Designing Safer Chemicals,* ed. S. C. DeVito and R. L. Garrett. ACS Symposium Series 640. Washington, DC: American Chemical Society, 1996, pp. 84–115.

Boethling, R. S. "Designing Biodegradable Chemicals." *Designing Safer Chemicals,* ed. S. C. DeVito and R. L. Garrett. ACS Symposium Series 640. Washington, DC: American Chemical Society, 1996, pp. 157–71.

Bordacs-Irwin, K. "Biotechnology for Pollution Prevention." In *Industrial Pollution Prevention Handbook,* ed. H. Freeman. New York: McGraw-Hill, 1995, pp. 629–39.

Braithwaite, J. "Waste Minimization: The Industrial Approach." In *Chemistry of Waste Minimization,* ed. J. Clark. London: Blackie Academic & Professional, 1995, pp. 17–65.

Bras, B., and Rosen, D. "Computer-Aided Design for De- and Remanufacturing." Presented at the NSF Design and Manufacturing Grantees Conference, Washington, DC, 1997.

Butterworth, A. J., Tavener, S. J., and Barlow, S. J. "The Use of Catalysis for the Manufacture of Fine Chemicals and Chemical Intermediates." In *Chemistry of Waste Minimization,* ed. J. H. Clark. London: Blackie Academic & Professional, 1995, 522–43.

Carr, R. W. *Partial Oxidation of Methane to Methanol.* Unpublished paper, 1996.

Chapman, O. L. "The University of California–Los Angeles Styrene Process." In *Benign by Design: Alternative Synthetic Design for Pollution Prevention,* ed. P. T. Anastas and C. A. Farris. ACS Symposium Series 577. Washington, DC: American Chemical Society, 1994, pp. 114–20.

Clifford, A. A. "Chemical Destruction Using Supercritical Water." In *Chemistry of Waste Minimization,* ed. J. H. Clark and T. C. Williamson. ACS Symposium Series 626. London: Blackie Academic & Professional, 1995, pp. 505–21.

Decio, K. The 1995 CALMAX Match of the Year, CALMAX (1996). http://www.ciwmb.ca.gov/mrt/calmax/moty95.htm.

De Fazio, T. L., Delchambre, A., and De Lit, P. "Disassembly for Recycling of Office Electronic Equipment." *European Journal of Mechanical and Environmental Engineering* 42 (1997): 25–31.

DeRon, A., and Penev, K. "Disassembly and Recycling of Electronic Consumer Products: An Overview." *Technovation* 15 (1995): 363–74.

DeVito, S. C. "General Principles for the Design of Safer Chemicals: Toxicological Considerations for Chemists." In *Designing Safer Chemicals: Green Chemistry for Pollution Prevention,* ed. S. C. DeVito and R. L. Garrett. ACS Symposium Series 640. Washington, DC: American Chemical Society, 1996, pp. 16–59.

Draths, K. M., and Frost, J. W. "Microbial Biocatalysis Synthesis of Adipic Acid from D-Glucose." In *Benign by Design,* ed. P. T. Anastas and C. A. Farris. ACS Symposium Series 577. Washington, DC: American Chemical Society, 1994, pp. 32–45.

Eisenstein, P. A. "Your Used New Car is Here." *World Traveler,* October 1995: 20–22.

"End-of-Life Strategies." *Pollution Prevention News (USEPA),* EPA 742-N-97-001 (1997): 8–10.

"Facts & Figures for the Chemical Industry." *Chemical & Engineering News* 75, no. 25 (1997): 38–45.

Franklin Associates, Inc. *Characterization of Municipal Solid Waste in the United States: 1994 Update,* EPA 530/R-94/042. Washington, DC: U.S. EPA, 1994.

Gadh, R. "Designing Products for Disassembly for Efficient Solid Waste Recovery: A Focus on Household Appliances." http://www.uwsa.edu/capbud/abstract/gadh.htm, 1997.

Garrett, R. L. "Pollution Prevention, Green Chemistry, and the Design of Safer Chemicals." In *Designing Safer Chemicals: Green Chemistry for Pollution Prevention,* ed. S. C. DeVito and R. L. Garrett. ACS Symposium Series 640. Washington, DC: American Chemical Society, 1996, pp. 2–15.

Graedel, T. E., and Allenby, B. R. *Industrial Ecology.* Englewood Cliffs, NJ: Prentice Hall, 1995.

Hancock, K. G., and Cavanaugh, M. A. "Environmentally Benign Chemical Synthesis and Processing for the Economy and the Environment." *Benign by Design,* ed. P. T. Anastas and C. A. Farris. ACS Symposium Series 577. Washington, DC: American Chemical Society, 1994, pp. 23–30.

Harler, C. "Plastics Markets See Slow Rebound." *Recycling Today* 35, no. 7 (1997): 42–49.

Hendrickson, J. B. "Teaching Alternative Syntheses: The SYNGEN Program." In *Green Chemistry: Designing Chemistry for the Environment,* ed. P. T. Anastas and T. C. Williamson. ACS Symposium Series 626. Washington, DC: American Chemical Society, 1996, pp. 214–31.

Hewlett-Packard. Guidelines for Environmentally Responsible Packaging, 1992. http://www.corp.hp.com/publish/talkpkg/enviro/environm.htm.

Hwang, J., Song, M., and Tieder, R. *Separation of Materials from Retired Automobiles: An Emphasis on Plastics.* Unpublished paper, 1996.

Jiang, H., and Bishop, P. L. "Aerobic Biodegradation of Azo Dyes in Biofilms." *Water Science & Technology* 29, no. 9–10 (1994): 525–30.

Johnson, M. R., and Wang, M. H. "Planning Product Disassembly for Material Recovery Opportunities." *International Journal for Production Research* 33 (1995): 3119–42.

Keoleian, G. A. "Pollution Prevention through Life-Cycle Design." In *Industrial Pollution Prevention Handbook,* ed. H. Freeman. New York: McGraw-Hill, 1995, 253–92.

King, R. "Recycled Material Value: A Moving Target." Speech to the Demanufacturing Partnership Program, Rutgers University, June 20, 1996.

Komiya, K., Fukuoka, S., Aminaka, M., Hasegawa, K., Hachiya, H., Okamoto, H., Watanabe, T., Yoneda, H., Fukawa, I., and Dozono, T. "New Process for Producing Polycarbonate without Phosgene and Methylene Chloride." In *Green Chemistry: Designing Chemistry for the Environment,* ed. P. T. Anastas and T. C. Williamson. ACS Symposium Series 626. Washington, DC: American Chemical Society, 1996, pp. 20–32.

Lai, D. Y., Woo, Y., Argus, M. F., and Arcos, J. C. "Cancer Risk Reduction Through Mechanism-Based Molecular Design of Chemicals." In *Designing Safer Chemicals: Green Chemistry for Pollution Prevention,* ed. S. C. DeVito and R. L. Garrett. ACS Symposium Series 640. Washington, DC: American Chemical Society, 1996, pp. 62–73.

Lambert, A. J. D. "Optimal Disassembly of Complex Products." *International Journal for Production Research* 35 (1997): 2509–23.

Lester, T. "Introduction." In *Chemistry of Waste Minimization,* ed. J. H. Clark. London: Blackie Academic & Professional, 1995, pp. 1–16.

Morgenstern, D. A., LeLacheur, R. M., Morita, D. K., Borkowsky, S. L., Feng, S., Brown, G., Luan, L., Gross, M. F., Burk, M. J., and Tumas, W. "Supercritical Carbon Dioxide as a Substitute Solvent for Chemical Synthesis." In *Green Chemistry: Designing Chemistry for the Environment,* ed. P. T. Anastas and T. C. Williamson. ACS Symposium Series 626. Washington, DC: American Chemical Society, 1996, pp. 133–51.

Newsome, L. D., Nabholz, J. V., and Kim, A. "Designing Aquatically Safer Chemicals." *Designing Safer Chemicals,* ed. S. C. DeVito and R. L. Garrett. ACS Symposium Series 640. Washington, DC: American Chemical Society, 1996, pp. 172–192.

OTA. *Green Products by Design: Choices for a Cleaner Environment,* OTA-E-541. Washington, DC: Office of Technology Assessment, 1992.

Payamps, M. "Changes at Checkout." *Resources: The Magazine of Environmental Management* 19, no. 4 (1997): 6–9.

Penev, K. D., and de Ron, A. J. "Determination of a Disassembly Strategy." *International Journal for Production Research* 34 (1996): 495–506.

Ramsden, M. J. "Polymer Recycling." In *Chemistry for Waste Minimization,* ed. J. H. Clark. London: Blackie Academic & Professional, 1995, pp. 441–61.

Seddon, K. R. "Room-Temperature Ionic Liquids: Neoteric Solvents for Clean Catalysis." Unpublished paper, 1997.

Sieburth, S. M. "Isosteric Replacement of Carbon with Silicon in the Design of Safer Chemicals." In *Designing Safer Chemicals: Green Chemistry for Pollution Prevention,* ed. S. C. DeVito and R. L. Garrett. ACS Symposium Series 640. Washington, DC: American Chemical Society, 1996, pp. 74–83.

Simmons, M. S. "The Role of Catalysts in Environmentally Benign Synthesis of Chemicals." In *Green Chemistry: Designing Chemistry for the Environment,* ed. P. T. Anastas and T. C. Williamson. ACS Symposium Series 626. Washington, DC: American Chemical Society, 1996, pp. 116–30.

Spicer, A., and Johnson, M. "Overcoming Economic Barriers of Demanufacturing and Asset Recovery." Unpublished paper, 1997.

Sullivan, S. "Garbage Gap Alert: There's Not Enough Trash to Go Around." *Newsweek,* October 28, 1996: p. 17.

Tanko, J. M., Blackert, J. F., and Sadeghipour, M. "Supercritical Carbon Dioxide as a Medium for Conducting Free-Radical Reactions." In *Benign by Design: Alternative Synthetic Design for Pollution Prevention,* ed. P. T. Anastas and C. A. Farris. ACS Symposium Series 577. Washington, DC: American Chemical Society, 1994, pp. 99–113.

Taylor, B. "Battles Rage in Packaging War." *Recycling Today* 35, no. 11 (1997): 98–102.

Turk, M. "Software Enables Component Reincarnation." *Design News* 52, no. 13 (1997): 83–84.

ULS Report, "A Study of Packaging Efficiency as It Relates to Waste Reduction." *ULS Report* 2, no. 3 (1995).

ULS Report 3, no. 3 (1996).

ULS Report 4, no. 2 (1997).

ULS Report 4, no. 3 (1997).

U.S. DOE. *Production of Organic Chemicals via Bioconversion.* EGG-2645. Washington, DC: U.S. DOE, 1991.

U.S. EPA. *Pollution Prevention in the Paints and Coatings Industry.* EPA/625/R-96/003. Washington, DC: U.S. EPA, 1996.

U.S. EPA, *Solid Waste and Emergency Response.* EPA-530-K-94-003. Washington, DC: U.S. EPA, 1994.

Webb, G., and Winfield, J. M. "CFC Alternatives and New Catalytic Methods of Synthesis." In *Chemistry of Waste Minimization,* ed. J. H. Clark. London: Blackie Academic & Professional, 1995, pp. 222–46.

Webster, L. C., Anastas, P. T., and Williamson, T. C. "Environmentally Benign Production of Commodity Chemicals through Biotechnology." In *Green Chemistry: Designing Chemistry for the Environment,* ed. P. T. Anastas and T. C. Williamson. ACS Symposium Series 626. Washington, DC: American Chemical Society, 1996, pp. 198–211.

PROBLEMS

9.1. It is common in some industries, such as the manufacture of specialty pharmaceuticals, to have very low yields for the compound of interest. This means that feedstock requirements are high and voluminous amounts of waste result from the manufacturing process. Assume that the reaction yield cannot be easily improved. As the pollution prevention engineer for your company, what steps should you investigate to minimize both the amount of feedstocks consumed and the amount of waste generated?

9.2. Figure 9.3 shows the synthesis pathway for the production of the drug Paramax. Assuming that a typical Paramax tablet weighs 500 mg, of which 50 percent represents propylene precursors, what percentage of the propylene used as the initial feedstock is found in the Paramax?

9.3. Phosgene is a highly toxic chemical that is widely used in industry. Prepare a research paper describing the chemical and physiological properties of phosgene, how it is made, the quantities used in industry, and precautions that are taken in industry to ensure worker safety.

9.4 A classic example of how biomass can be used to replace petroleum-based chemicals is gasohol. Proponents rave about its renewability and its ability to reduce our dependence on imported oil. However, from a life-cycle assessment viewpoint, it also has drawbacks. Prepare a report describing the advantages and disadvantages of using gasohol.

9.5. Chirality is not found only at the molecular level. List several common items found around your home that also exhibit chiral properties.

9.6. A new and exciting technique for carrying out synthesis reactions in an environmentally friendly fashion is called *sonochemistry*. Research this new technique and prepare a short paper describing it, its uses, and its advantages.

9.7. Using the Internet, investigate the Duales System Deutschland and write a short paper summarizing its current status and how successful it has become.

9.8. Examine a battery-operated radio. What impediments do you see to reusing components within it? What about impediments for recycling materials? If you were the designer, how would you change the design to make it more reusable/recyclable?

9.9. Examine several items you have that are made of plastic. Can you easily identify the type of plastic involved? Is the type of plastic identified with a symbol? If so, where is the symbol and is it easily located? Does the item contain more than one type of plastic? If so, are the materials compatible for coprocessing during recycling?

9.10. Much of the packaging that is used is unnecessary. For example, crackers usually come in a plastic bag inside a cardboard box, whereas potato chips usually come in a pressurized bag only, even though the chips are probably more fragile. Why is excess packaging commonly used for goods, even though it adds to the cost of the product, takes up more room, and produces more waste? How can we overcome these obstacles?

CHAPTER 10

WATER, ENERGY, AND REAGENT CONSERVATION

10.1 INTRODUCTION

One major component of the resources used in an industry process—water—is often overlooked. Water is used as a solvent to dissolve reagents and acts as a medium in which reactions occur. It is also used to separate immiscible materials by means of differences in specific gravity and miscible materials by means of differences in affinity of the material between two solvent phases (liquid-liquid, gas-liquid, or solid-liquid). It is commonly used to move materials from one place to another in the manufacturing process. Two major uses are for cleaning and for cooling.

Water is often thought of as an essentially "free" resource because of its relatively low cost. However, it is not free, and when used in large quantities, for cooling or in washing of factory floors or equipment, for example, the cost can add up. It must be remembered that the potable water used in most industrial processes is a resource that should not be squandered. Another significant factor militating against excessive use of water, though, is the cost of treating the resulting contaminated wastewater. Wastewater treatment is expensive, and cost increases as the volume of water to be treated increases. If the water becomes contaminated with hazardous materials, the cost can become prohibitive. Thus, resource conservation aside, it is essential that water use be minimized to avoid costs associated with its treatment.

The implementation of good housekeeping practices to minimize water use was described in Chapter 5, and treatment of industrial wastewaters will be discussed in the next chapter. This chapter examines methods for reducing the amount of water required for some industrial processes, while maintaining good quality control. We begin with a discussion of simple plumbing changes that can greatly minimize water use, and thus wastewater production, in the cleaning or rinsing of materials in the production process. We then describe how a new analytical procedure, known as pinch analysis, can be used to evaluate water recycling schemes that can be used to minimize overall water use. Pinch analysis can also be used to minimize heating and cooling requirements in the production line and to promote conservation of process reagents; these applications will also be presented.

10.2 REDUCTION IN WATER USE FOR CLEANING

As described in Chapter 5, cleaning and degreasing operations are commonly used by industry to remove dirt, oil, and grease from both process input materials and finished products. It is used in the metal finishing industry following machining and after each metal plating or surface finishing step that the part undergoes. Circuit boards and other electronic equipment must also be scrupulously clean, both before and after application of the printed circuit.

As described in Chapter 5, a commonly used cleaning method consists of placing the part to be cleaned in a rinse tank or consecutively in several rinse tanks containing either stagnant or flowing water. The objective is to transfer the contaminants on the surface of the manufactured part to the water by dissolution and dilution. This is the typical process used to remove excess plating solution dragged out of a plating bath. However, it is useful only for removing materials that are water soluble. A single stagnant rinse tank can serve as a cleaning device, but it is not very effective and cleaning ability decreases dramatically with time as the level of contaminants in the rinse tank builds. Large quantities of water are needed to effect cleaning. A series of stagnant rinse tanks improves the situation since each subsequent rinse tank in the progression contains fewer contaminants than the previous one, but this is still not very efficient. By using a flowing tank or series of tanks, cleaning efficiency can be significantly increased and water use decreased. Running the rinse water through the rinse tanks countercurrent to the flow of the pieces being cleaned provides the best and most efficient cleaning possible. Each additional cleaning step in the series contains cleaner water and improves the cleaning operation, but there is a limit to this improvement. Generally, little benefit is achieved by operating more than three countercurrent tanks in series. This can best be seen by examining the operation of various types of rinsing processes.

Figure 10.1 illustrates some of the many types of rinsing configurations. There can be from one to four rinse tanks, with some in series and others in parallel, some stagnant and some flowing, some with concurrent flow and others with countercurrent flow. Some configurations return the concentrated chemicals in the rinse water directly

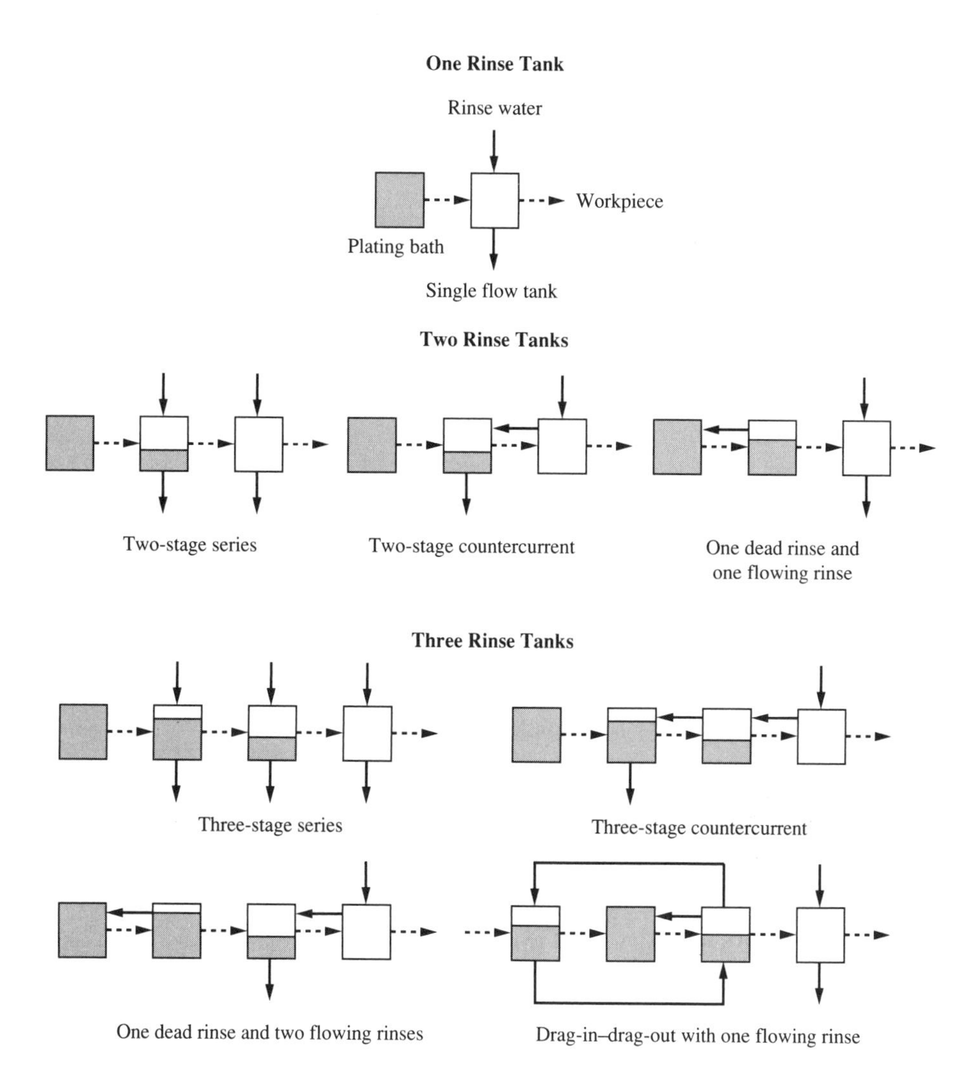

FIGURE 10.1
Rinsing configurations.

to the process tank; in others the rinse water must be processed before reuse of the chemicals contained therein.

Single running rinse tanks require a large volume of water for contaminant removal. Commonly used in the past, these have largely been replaced because of their large water demand and consequent large volume of waste produced, often at a concentration too dilute for efficient recovery. Series rinse tanks reduce the overall waste volume somewhat, and they have an advantage because the tanks can be individually heated or controlled since they have separate feeds. Series rinse tanks are not as efficient as countercurrent rinse systems, though, which provide the most efficient water use and should be used where possible. Stagnant or dead rinse tanks are excellent for the initial rinse because they allow easier recovery of the metal being removed and lower subsequent water use in the flowing tanks. The rinse water in the dead rinse tank can eventually be used to make up dragout and evaporative losses in the plating bath. Spray rinsing may provide the most efficient use of water, but only if the part can be rinsed effectively by the spray. Convoluted turnings that hold dragout will not be cleaned properly. Spray rinses are best for cleaning flat sheets. Water knives that use a high-pressure, low-volume spray applied at an angle to the sheet are often used for this purpose. The drag-in–drag-out system is an interesting innovation. Here, a rinse tank precedes the plating bath. The drag-in tank is connected to the drag-out tank so that their chemical contents are approximately the same. Pieces being processed drag in plating solution to the plating tank, thereby increasing recovery of plating chemicals.

Rinse water requirements needed to achieve a given allowable contaminant concentration on the rinsed workpiece can be determined by use of mass balances. For a single running rinse tank, as shown in Figure 10.2, the mass balance equations can be written as

$$Q_D C_P + Q_R C_R = Q_R C_1 + Q_D C_1$$

$$Q_R = Q_D \frac{(C_1 - C_p)}{(C_R - C_1)}$$

where C_P = constituent concentration in the plating bath
C_1 = constituent concentration in the rinse tank (and effluent)
C_R = constituent concentration in the clean rinse water
Q_D = flow rate of dragout from the plating bath (and from the rinse tank)
Q_R = flow rate of rinse water to the rinse tank

If we assume that C_P is much greater than C_1 and that C_1 is much greater than C_R, the preceding equation can be simplified to

$$Q_R = Q_D \frac{C_P}{C_1}$$

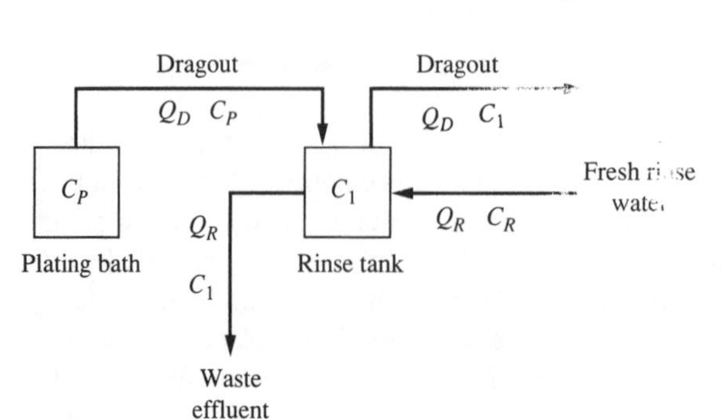

FIGURE 10.2
Mass balance for a single running rinse.

For a series of flowing rinse tanks in which each rinse tank receives a separate inflow of rinse water, the mass balance becomes a little more complicated. A typical flow scheme, which assumes an equal flow rate of rinse water to each rinse tank, is shown in Figure 10.3. The mass balance solution for the first rinse tank is

$$Q_D C_P + Q_R C_R = Q_R C_1 + Q_D C_1$$

where C_1 = the contaminant concentration in the first rinse bath (and in the dragout on the rinsed workpiece)

Each tank in series must be analyzed in the same way. The mass balance for the last rinse tank in the series, tank *n*, is

$$Q_D C_{n-1} + Q_R C_R = Q_D C_n + Q_R C_n$$

where C_{n-1} = contaminant concentration in the next-to-last rinse tank
C_n = contaminant concentration in the last rinse tank

This set of equations must be solved simultaneously to determine the rinse water flow rates. A simplified equation, which assumes that the dragout concentration from the plating bath is much greater than that in the rinsed piece and that the concentration in the rinse tank wastewater is much greater than clean rinse water, can be written as

$$Q_R = \left[C_P \frac{Q_D^n}{C_n} \right]^{1/n}$$

where n = number of rinse tanks in series

Countercurrent rinse tanks are generally more effective than either of the first two examples. The rinsing arrangement shown in Figure 10.4 can be analyzed as follows. The analysis assumes that the rinse water from one tank flows in a countercurrent fashion to the workpiece being rinsed, with flow from one tank to the next. Therefore, Q_D and Q_R are the same for each tank; only the concentrations change. For rinse tank 1, the mass balance equations can be written as

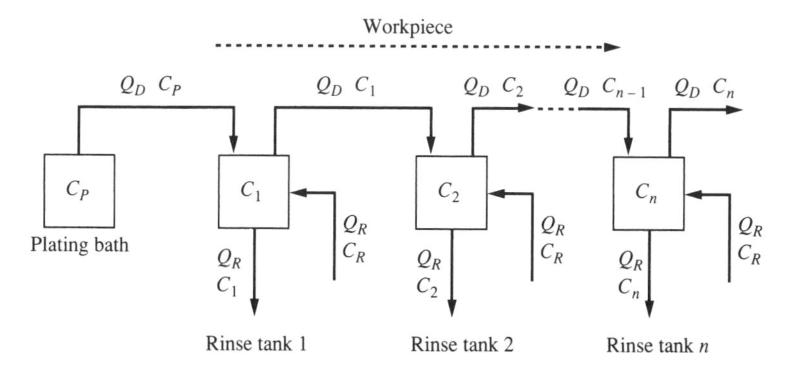

FIGURE 10.3
Mass balance for rinse tanks in series.

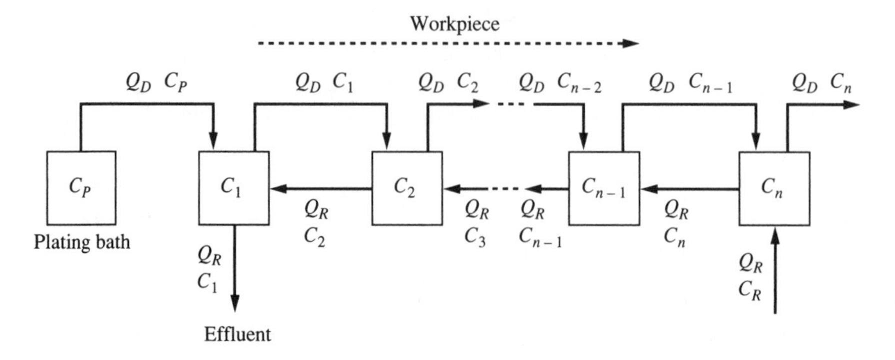

FIGURE 10.4
Mass balance for a countercurrent rinse system.

$$Q_D C_P + Q_R C_2 = Q_D C_1 + Q_R C_1$$

Each rinse tank must be analyzed in a similar way. The mass balance for tank n, the last tank, is

$$Q_D(C_{n-1}) + Q_R C_R = Q_D C_n + Q_R C_n$$

Again, this set of equations must be solved simultaneously to determine the rinse water flow rate. Using the same assumptions as above, however, a simplified expression can be arrived at:

$$Q_R = \left[\left(\frac{C_P}{C_n}\right)^{1/n} + \frac{1}{n}\right] Q_D$$

Finally, a rinsing arrangement in which a dead rinse precedes a series of countercurrent rinse tanks can be examined (see Figure 10.5). The mass balance for the dead rinse can be written as

$$Q_D C_P + Q_M C_M = Q_D C_D + Q_M C_D$$

where Q_M = equivalent flow rate of makeup water to the plating bath (Q_D + evaporative losses)

This equation can be rearranged to solve for C_D:

$$C_D = \frac{Q_D C_P + Q_M C_M}{Q_D Q_M}$$

The live rinses following the dead rinse tank are analyzed as above, with the value for C_D being used as the initial dragout concentration for the rinse tank series.

In all of these analyses, the effluent metal concentration could also be calculated for a given rinse water flow rate.

The effectiveness of using a countercurrent rinse system can be readily seen in Example 10.1, which examines several rinsing options to remove drag-out solution from a metal plating bath.

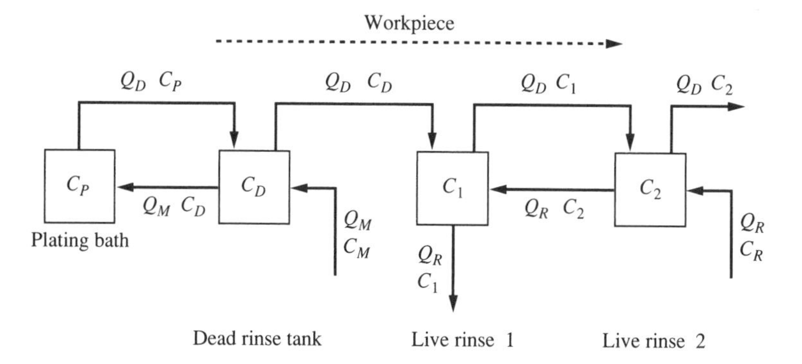

FIGURE 10.5
Mass balance for a dead rinse followed by countercurrent rinse tanks.

Example 10.1. A metal plating firm operates a chromium plating bath that has 85,000 mg/L chromium and a dragout rate of 0.060 L/min. What rinse water flow rate is needed to keep the chromium concentration in the rinse water dragout on the rinsed workpiece at 25 mg/L? Assume that the clean rinse water contains no chromium. Compare (*a*) a single running rinse tank; (*b*) series rinsing with either two or three rinse tanks; (*c*) a countercurrent rinsing process with either two or three rinse tanks; and (*d*) a dead rinse tank followed by two countercurrent live rinse tanks (assume an evaporation rate from the plating bath of 0.01 L/min.).

Solution.

(*a*) Using the simplified mass balance equation for a single flowing rinse tank, we find that the rinse water flow rate must be at least 204 L/min:

$$Q_R = 0.060\ \text{L/min}\ \frac{(25\ \text{mg/L} - 85{,}000\ \text{mg/L})}{(0\ \text{mg/L} - 25\ \text{mg/L})}$$

$$Q_R = 204\ \text{L/min}$$

(*b*) For two tanks in series, the required rinse water flow rate is 7.0 L/min, while for three tanks in series the required flow rate drops to 2.7 L/min. This shows the value of multiple rinse tanks.

Two tanks:

$$Q_{R_2} = \left[\frac{(85{,}000\ \text{mg/L})(0.060\ \text{L/min})^2}{25\ \text{mg/L}}\right]^{1/2}$$

$$Q_{R_2} = 3.5\ \text{L/min}\ \textit{per tank}$$

Three tanks:

$$Q_{R_3} = \left[\frac{(85{,}000\ \text{mg/L})(0.060\ \text{L/min})^3}{25\ \text{mg/L}}\right]^{1/3}$$

$$Q_{R_3} = 0.9\ \text{L/min}\ \textit{per tank}$$

(*c*) For two countercurrent flow rinse tanks in series, the rinse water flow rate is 3.5 L/min (half of that for series rinsing), while for three tanks in series, the total flow rate is only 0.9 L/min. This is only 0.4 percent of the flow rate for a single tank and only one-third the flow rate for a three-tank series rinse.

Two tanks:
$$Q_{R_2} = \left[\left(\frac{85{,}000\ \text{mg/L}}{25\ \text{mg/L}}\right)^{1/2} + \frac{1}{2}\right](0.060\ \text{L/min})$$
$$Q_{R_2} = 3.5\ \text{L/min}$$

Three tanks:
$$Q_{R_3} = \left[\left(\frac{85{,}000\ \text{mg/L}}{25\ \text{mg/L}}\right)^{1/3} + \frac{1}{3}\right](0.060\ \text{L/min})$$
$$Q_{R_3} = 0.9\ \text{L/min}$$

(*d*) To determine the flow rate for a dead rinse followed by two countercurrent rinse tanks, first consider the equivalent dead rinse flow rate, which is equal to the plating bath dragout rate plus the evaporative loss:

$$Q_M = Q_D + \text{Evaporation} = 0.060\ \text{L/min} + 0.2\ \text{L/min} = 0.26\ \text{L/min}$$

The value of C_D can then be determined:

$$C_D = \frac{(0.060\ \text{L/min})(85{,}000\ \text{mg/L}) + (0.26\ \text{L/min})(0\ \text{mg/L})}{0.060\ \text{L/min} + 0.26\ \text{L/min}}$$
$$C_D = 15{,}938\ \text{mg/L}$$

This chromium solution can be returned to the plating bath as makeup water (it is about 20 percent of the concentration of chromium in the plating bath), or it can be concentrated further by procedures described earlier in this chapter. The workpieces, which have now been partially cleaned, are then further rinsed in the countercurrent rinse system. The necessary rinse water flow rate can be calculated as was done above:

$$Q_R = \left[\left(\frac{15{,}938\ \text{mg/L}}{25\ \text{mg/L}}\right)^{1/2} + \frac{1}{2}\right](0.60\ \text{L/min})$$
$$Q_R = 1.5\ \text{L/min}$$

Only 1.5 L/min of rinse water is necessary to achieve the degree of cleaning desired, and essentially all of the chromium in the dragout is recovered for reuse in the plating tank.

Recycling or reuse can be combined with these rinsing systems in order to return excess process chemicals in the rinse waters to the plating bath from which they originated. This is common in the metal plating industry, where the excess metal solution dragged out from a plating bath and that removed from the plated workpiece during the rinse steps are concentrated, possibly using ion exchange or reverse osmosis, and returned to the plating bath. This not only reduces the amount of process chemicals being wasted to the sewer, but also reduces the amount of chemicals needed to be purchased for plating. This can be seen in Example 10.2.

Example 10.2. The metal plating firm described in Example 10.1 has converted to the three-tank countercurrent rinsing process for its chromium plating line. The company is now investigating installation of a reverse osmosis system to concentrate the chromium in the rinse system effluent line for return to the plating bath. Assuming that the reverse osmosis membrane can reject 92 percent of the chromium and pass 90 percent of the water, how much chromium can be recovered for reuse and at what concentration? What will be the chromium concentration in the final waste effluent?

Solution. The rinse water flow rate for a three-tank countercurrent system in Example 10.1 was found to be 0.9 L/min. Assuming the firm operates on two shifts (16 hours), the average daily flow from the chromium plating line is 864 L/day.

The dragout from the plating bath occurred at a rate of 0.060 L/min (57.6 L/day), and it contained 85,000 mg/L chromium. Therefore, the amount of chromium removed from the plating bath as dragout is equal to 4896 g/day:

$$\text{Plating bath dragout Cr} = (85{,}000\ \text{mg/L})(57.6\ \text{L/day}) = 4{,}896{,}000\ \text{mg/day}$$
$$= 4896\ \text{g/day}$$

The design chromium concentration in the final rinse tank was 25 mg/L. The amount of chromium in the dragout from the final rinse tank is therefore equal to 21.6 g/day:

$$\text{Final dragout Cr per day} = (25\ \text{mg/L})(864\ \text{L/day}) = 21{,}600\ \text{mg/day} = 21.6\ \text{g/day}$$

The remainder of the chromium must be in the rinse water effluent. The rinse water chromium concentration is therefore equal to 5640 mg/L:

$$\text{Rinse water Cr} = \frac{(4896\ \text{g/day} - 21.6\ \text{g/day})}{864\ \text{L/day}} = 5.64\ \text{g/L} = 5640\ \text{mg/L}$$

Assuming that 92 percent of the chromium is recovered in 10 percent of the volume, the amount of chromium recovered and returned to the plating bath from the reverse osmosis unit is 4,483 g/day:

$$\text{Cr recovered} = (5.64\ \text{g/L})(0.92)(864\ \text{L/day}) = 4483\ \text{g/day}$$

The concentration of chromium in this stream returning to the plating bath is 51,900 mg/L, as compared with the plating bath concentration of 85,000 mg/L:

$$\text{Recovered Cr concentration} = \frac{4483\ \text{mg/day}}{(864\ \text{L/day})(0.1)} = 51.9\ \text{g/L} = 51{,}900\ \text{mg/L}$$

The final waste effluent chromium concentration is equal to 500 mg/L, as compared with a concentration of 5640 mg/L in the waste without reverse osmosis treatment:

$$\text{Effluent Cr} = \frac{4896\ \text{g/day} - 21.6\ \text{g/day} - 4483\ \text{g/L}}{(864\ \text{L/day})(0.9)} = 0.50\ \text{g/L} = 500\ \text{mg/L}$$

TABLE 10.1
Effect of rinse changes on wastewater effluent quality at a metal plating facility

	Effluent concentration, mg/L	
Constituent	**Before changes**	**After changes**
Cadmium	3.90	0.03
Chromium	4.60	0.07
Copper	0.65	0.01
Cyanide, total	1.35	< 0.02

10.2.1 Case Study

Jewell Electrical Instruments, Inc., is a good example of how rinsing systems can be varied to provide optimum cleaning and chemical recovery. An extensive plating shop was used to electrocoat the parts with copper, cadmium, and other coatings. Parts cleaning was performed using alkaline etch, muriatic acid etch, and trichloroethylene solvent baths, each followed by water rinsing. Initially operated with only a single or a series of once-through flowing water rinse tanks, the plant was converted so that the chromium and Alodine coating lines received a chemical rinse, and the cadmium, copper, and anodizing lines each had a dead rinse followed by countercurrent rinsing. The conversion was so effective that nearly all the plating metals were recovered and returned to the plating baths (see Table 10.1). Rather than having to pay a large municipal sewer user fee for effluent metals, as well as significant costs for hazardous waste disposal, the company's waste management costs were almost nil. In addition, the fee for water use decreased dramatically because rinse water flow rates were greatly reduced. Capital costs were minimal, primarily for plumbing changes since the rinse tanks were already there and only needed to be reconfigured into a dead tank–countercurrent rinse mode.

10.3 PINCH ANALYSIS

In the previous section, we examined the use of mass balances to analyze how materials function in a process and how, by proper reconfiguration of flow patterns, we can greatly reduce pollution from industrial operations. In particular, we examined how we could reduce the flow of rinse water used in metal plating by several orders of magnitude while at the same time concentrating the wasted plating chemicals in the dragout for possible reuse. These simple and inexpensive plumbing changes have greatly reduced the pollution emitted by many plating facilities. Other industries have made similar changes, which have also resulted in significant reductions in wastes reaching the environment while at the same time reducing costs to industry.

Another flow analysis technique that recently came into use—*pinch analysis*—is having an impact on waste minimization by industry. The technique was originally developed by Bodo Linnhoff to optimize energy use in industry, but it has now been adapted for use in optimizing water and chemical use in industrial processes.

Pinch analysis first attracted interest in the 1970s during the energy crisis as industries looked for ways to minimize their consumption of oil. Many processing plants spend as much on energy as on wages, and even small businesses can have energy bills of hundreds of thousands to millions of dollars annually. Much of this expenditure is unnecessary and can readily be saved. The pinch analysis procedure quantifies the potential for reducing energy consumption through improved heat use by design of improved heat exchanger networks (Linnhoff, 1995). Waste heat in one point in the manufacturing process could be used to provide heat at other points where it is needed. An engineering approach was developed to optimize the exchange of this waste heat.

Pinch analysis is widely accepted as a tool for heat optimization. Its use has now expanded to include both energy and mass transfer problems in reaction and separation systems, and it can be used to determine capital and operating cost implications of competing pollution prevention options. It is a holistic design approach based on *process integration,* in which the process is conceived of as an integrated system of interconnected units and streams, rather than as a series of individual and independent steps. In this way, materials or energy that are needed or available in one process step can be supplied by or made available to another process step. Pinch analysis is based on rigorous thermodynamic principles that are used to predict heat and materials flows through the processes making up the manufacturing system. The following section describes how pinch analysis can be used to optimize energy and materials use and to minimize pollution during manufacturing. More detailed descriptions of this procedure can be found elsewhere (El-Halwagi, 1997; Rossiter, 1995; Wang and Smith, 1994).

10.3.1 Thermal Pinch Analysis

Pinch analysis of heat exchange systems grew out of an analytical system for optimizing heat exchangers called heat exchanger network (HEN) synthesis. The objective of HEN synthesis is to maximize the efficiency of energy use in a process by transferring the waste heat from one stage of the process to another stage. HENs have been shown to be useful in producing large cost savings for many industries. Large industrial plants, such as refineries, may have hundreds of heat exchangers to bring process streams to the necessary temperature for the process in question. By integrating the process and intermingling the heat available, the process can often be made much more energy efficient. The rudiments of HEN optimization are presented here as an introduction to a discussion on pinch analysis. More detail can be found in Douglas (1988) and Roosen and Gross (1995).

HEN ANALYSIS. In a HEN analysis, all of the heating and cooling requirements in a process are evaluated to determine the extent to which streams that require heat can be heated by other streams that have excess heat or that need to be cooled. If the heat can be transferred between stages in the process, a procedure known as process integration, fuel needed for heating or cooling will be conserved, and the resultant environmental emissions from that energy use will be minimized.

The energy inputs, and hence the energy costs, may be expressed based on energy losses. Therefore, these losses may be directly coupled to the respective shares of the overall cost for the process. If a proportionality factor between energy loss and

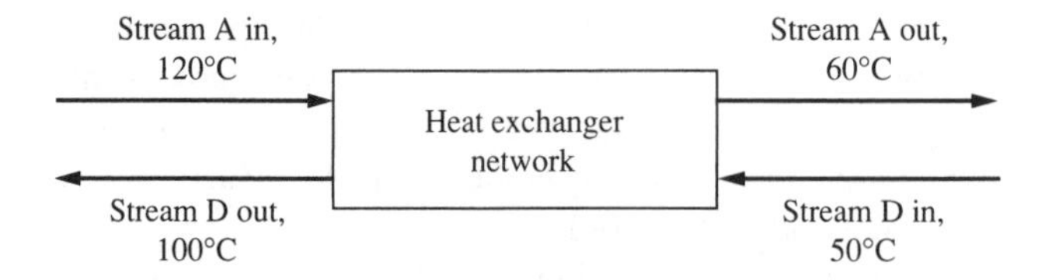

FIGURE 10.6
A simple heat exchange network.

energy costs can be derived, the trade-off between the cost of the heat exchanger and the cost of energy can be determined. For heat exchangers, there is a strong correlation among effective heat exchanger area, temperature drop, construction costs, and energy loss if required temperature ranges and heat loads are known. The rate of heat transfer between two materials is partially governed by the difference in temperatures between the two materials. The larger the difference, the more rapid will be the heat transfer. However, a large ΔT means that available heat is being wasted. For a given choice of heat exchanger material and a known amount of heat to be exchanged, the required exchanger area is a direct function of the allowed minimal temperature gap between the two exchanger sides ΔT_{min}. The required exchanger area can be used to establish the annualized investment cost C_I for the heat exchanger. The entropy production through the process, or the energy losses, can also be coupled to ΔT_{min}. Therefore, energy losses and required investment costs for heat exchangers can be related to each other through ΔT_{min}, plotted, and used as a design characteristic.

Figure 10.6 shows a simple heat exchanger network system. Step A in the process produces a stream at a temperature of 120°C which must be cooled to 60°C before the next processing step. Step D, later in the process, produces a stream at 50°C that must be heated to 100°C before entering Step E. HEN analysis would typically be used to determine whether a heat exchanger could be used economically to transfer excess heat from the Stage A stream to the Stage D stream. Example 10.3 presents an analysis of the available heat balance.

Example 10.3. In the process depicted in Figure 10.6, Stream D needs to be heated from 50°C to 100°C. This stream has an average heat capacity of 2000 J/kg · K and a flow rate of 2 kg/s. Stream A needs to be cooled from 120°C to 60°C. It has an average heat capacity of 1500 J/kg · K and a flow rate of 1 kg/s. The site has steam available at a temperature of 170°C and cooling water at 25°C.

(*a*) What are the utility loadings (steam and cooling water requirements) if steam is used to heat Stream D and cooling water is used to cool Stream A?

(*b*) What will the utility loadings be if a HEN is used to minimize utility use by transferring heat between the two processes?

Solution.

(*a*) Heating requirement (steam):

$$2\ \text{kg/s} \times 2\ \text{kJ/kg} \cdot \text{K} \times (100 - 50)\text{K} = 200\ \text{kJ/s} = 200\ \text{kW}$$

Cooling requirement (cooling water):

$$1 \text{ kg/s} \times 1.5 \text{ kJ/kg} \cdot \text{K} \times (120 - 60)\text{K} = 90 \text{ kJ/s} = 90 \text{ kW}$$

Total energy requirement = 200 kW + 90 kW = 290 kW

(*b*) The stream conditions are shown in Figure 10.7. Assume all of Stream A is contacted with Stream D to bring the temperature of Stream A from 120°C to 60°C.
Exchanger 1 heating balance:

$$1 \text{ kg/s} \times 1.5 \text{ kJ/kg} \cdot \text{K} \times (60 - 120)\text{K} = -90 \text{ kW}$$

Exchanger 1 cooling balance:

$$2 \text{ kg/s} \times 2 \text{ kJ/kg} \cdot \text{K} \times (72.5 - 50)\text{K} = +90 \text{ kW}$$

Note that the 72.5°C temperature of Stream D is the temperature that results to maintain a heat balance through the system. The energy involved in a decrease in temperature of Stream A from 120°C to 50°C will result in an increase of temperature of Stream D from 50°C to 72.5°C.
Exchanger 2 heating balance:

$$2 \text{ kg/s} \times 2 \text{ kJ/kg} \cdot \text{K} \times (72.5 - 100)\text{K} = -110 \text{ kW}$$

Total energy requirement:

$$-90 \text{ kW} + 90 \text{ kW} - 110 \text{ kW} = -110 \text{ kW}$$

With this integrated design, cooling water is no longer necessary, and the amount of steam needed is reduced from 200 kW to 110 kW. The total energy savings is 290 kW − 110 kW = 180 kW. This represents a substantial cost savings and an accompanying environmental improvement due to the reduction in energy use and environmental emissions.

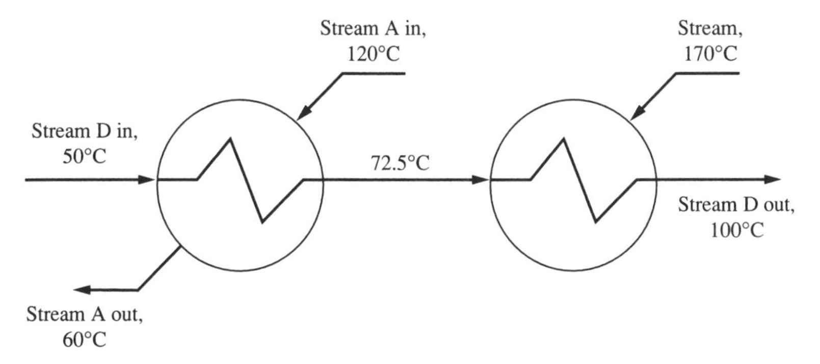

FIGURE 10.7
Stream conditions showing total reuse of Stream A in heat exchanger 1.

PINCH ANALYSIS COMPOSITE CURVES. The pinch analysis procedure for optimizing thermal systems (also known as the Linnhoff method) is based on the rigorous analysis of all sources and sinks of heat within the process in terms of their enthalpy H and temperature T. All hot stream flows and all cold stream flows are put together in an enthalpy–temperature (E-T) diagram. The curves are separated by a temperature selected as the minimum driving force for efficient heat transfer to take place. This minimum temperature is called the *pinch temperature*. The overriding constraints on all heat transfers are that heat may be transferred only from a warmer stream to a colder one, and the amount of heat that can be transferred is limited by the stream's enthalpy. A complete description of this procedure can be found elsewhere (El-Halwagi, 1997; Linnhoff, 1993, 1994, 1995).

The analysis is begun by plotting all heat sinks (cold streams) and all heat sources (hot streams) with respect to their starting and ending temperatures and enthalpies. Figure 10.8 shows the results for a system with two cold streams and two hot streams.

Figure 10.8*a* shows the individual plots for the two cold streams. Stream (a) exists between temperatures T_1 and T_3, while stream (b) goes from T_2 to T_4. The temperature range for the two streams overlaps between T_2 and T_3. Figure 10.8*b* shows a composite of the heat load contributions. The single continuous curve represents all heat sinks of the process as a function of heat load versus temperature. The same analysis is done for the heat sources (hot streams), as shown in Figures 10.8*c* and 10.8*d*. The composite curve in Figure 10.8*d* represents all heat sources in the process as a function of heat load versus temperature.

The two composite curves can be plotted on the same graph using common axes. Figure 10.9 is an example of combined composite curves, with plots from Figures 10.8*b* and 10.8*d* plotted together. The area between the two curves indicates the amount of heat that potentially could be transferred from the hot source to the cold source. The point of minimum vertical separation between the two curves is called the *pinch point*. This represents the minimum difference in temperature between the two curves. In this case, the hot composite curve is always above the cold composite curve, within the region of curve overlap, indicating that heat sources exist within the process to satisfy all heating requirements for the cold streams. There is always a driving force of at least ΔT_{min}. The pinch point also locates the optimum point in the network where the heat exchanger should be placed.

Not all of the heat requirements for the entire process can be met, however, because the two curves do not entirely overlap. The region of potential heat recovery is shown on the figure. The section of the plot to the left of this zone represents the quantity of heat that cannot be recovered. This waste heat must be cooled with cooling water. Similarly, the region to the right of the recovery zone defines the quantity of heat that must be supplied in the form of steam or other heating source. These two numbers represent the minimum amounts of required heating and cooling that can be achieved.

The pinch point represents the minimum amount of temperature difference between the two streams throughout the process. There must be some difference here for practical heat transfer to take place. If the two curves touch, then both streams have the same temperature at this point and there is no temperature difference to provide the

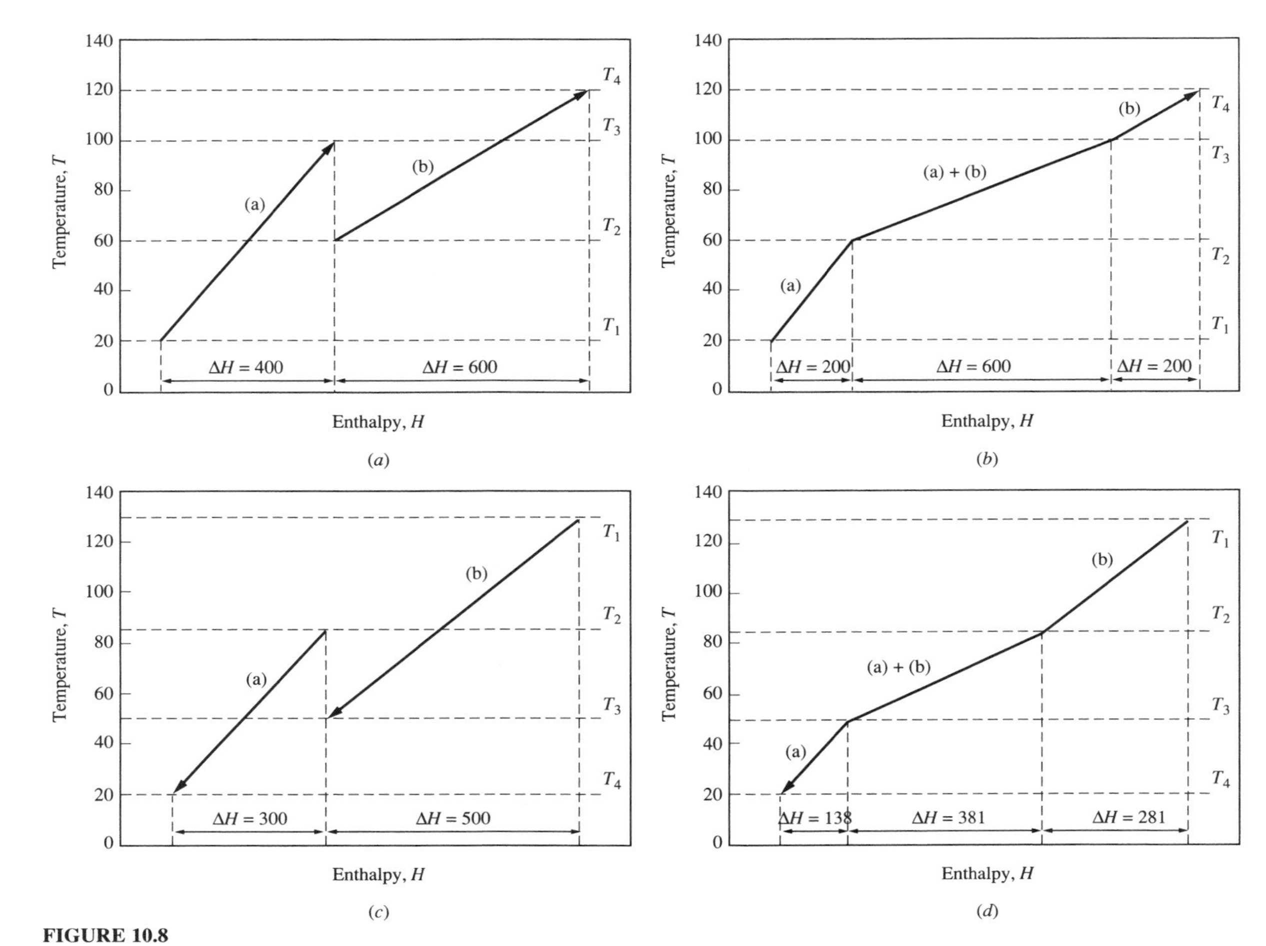

FIGURE 10.8
T-H plots for a system with two hot and two cold streams. (*a*) *T-H* plot for cold stream. (*b*) *T-H* composite plot for cold streams. (*c*) *T-H* plot for hot streams. (*d*) *T-H* composite plot for hot streams.

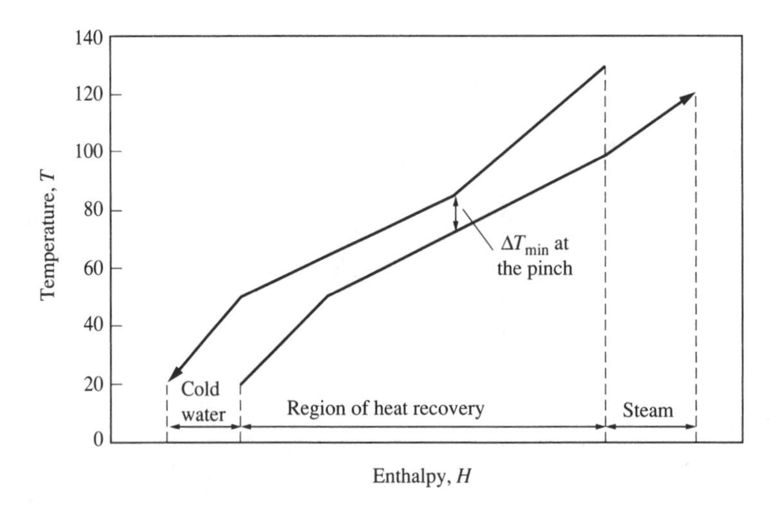

FIGURE 10.9
Combined composite curves.

driving force for heat transfer. If they are a very small difference apart, heat transfer will occur, but the rate might be very small and the size of the required heat exchanger will be huge. Thus the selection of the appropriate ΔT_{min} is an engineering decision, to be based on a trade-off of exchanger size versus allowable temperature gradient.

In the example seen in Figure 10.9, the hot stream always lies above the cold stream, so the hot stream can be used to transfer some of its heat to the cold stream, when needed, throughout the process. This is not always the case, however. For example, in the situation depicted in Figure 10.10, the cold and hot stream composite curves cross each other. It is thermodynamically impossible to transfer heat from the hot stream to the cold stream to obtain the appropriate final temperatures because the temperature of the cold stream must be raised to a value higher than that of the hot stream.

From this, we can draw several conclusions, known as the *three golden rules of HEN design* (Linnhoff, 1995):

- Do not attempt to transfer heat across the pinch.
- Do not use hot utilities to heat any flow stream below the pinch.
- Do not use cold utilities to cool any flow stream above the pinch.

These rules can be used to produce a pinch heat transfer equation that predicts the actual heat consumption of a process:

$$\text{Actual} = \text{Target} + XP_{\text{proc}} + XP_{\text{hot below}} + XP_{\text{cold above}}$$

where Actual = actual heat consumption of process

Target = heat transfer from hot process streams above the pinch to cold process streams below the pinch

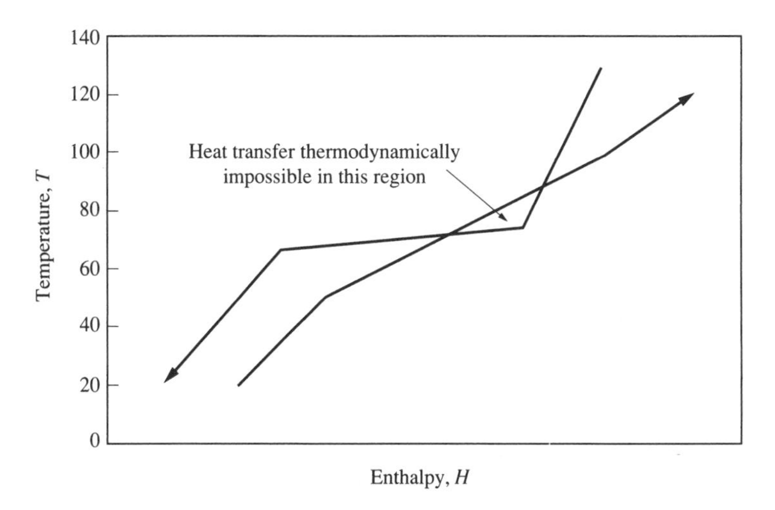

FIGURE 10.10
Example of a thermodynamically impossible heat transfer.

XP_{proc} = heat transfer from hot utilities to cold streams below the pinch
$XP_{cold\ above}$ = heat transfer to cold utilities from hot streams above the pinch

The value for the target is the Δt_{min} established for the process to ensure that there is a sufficient driving force to transfer heat efficiently in the heat exchanger. Enough heat must be supplied to account for that necessary to transfer from the hot to the cold streams in the heat exchanger. Finally, the heat transfers necessary outside the overlap of the combined composite curves must be supplied.

One additional plot can be derived from the *E-T* relationship for the process in question: the *grand composite curve*. In the overlapping portions of the combined composite curves plot, heat is stoichiometrically transferred from one process stream to the other. Thus these two amounts of heat cancel each other out. The amount of heating and cooling required beyond the transferred amount can be depicted by plotting these amounts with the pinch as the origin (see Figure 10.11). The grand composite curve represents the net heating or cooling requirement of the process as a function of temperature. The engineer can use this plot to screen feasible heating or cooling options. For example, low-pressure steam can be used to heat process streams when the required temperature is relatively low, although high-pressure steam is required at higher temperatures. Similarly, cooling water can be used to chill a process stream at lower temperature differentials, but refrigeration may be needed when lower temperatures are called for.

An example of how effective the use of pinch analysis can be is that of BASF's Ludwigshafen (Germany) plant (Buehner and Rossiter, 1996). Between 1970 and 1982, conventional approaches were used to conserve energy. However, from 1979 to 1982, production became limited by the available energy infrastructure at the facility (see Figure 10.12). As part of the overall scheme to reduce energy consumption at the

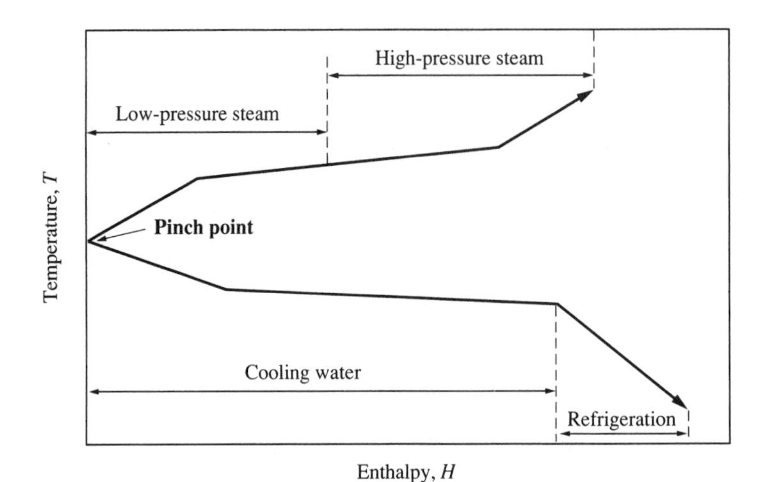

FIGURE 10.11
Grand composite curve showing utility requirements. (Adapted from Linnhoff, 1995)

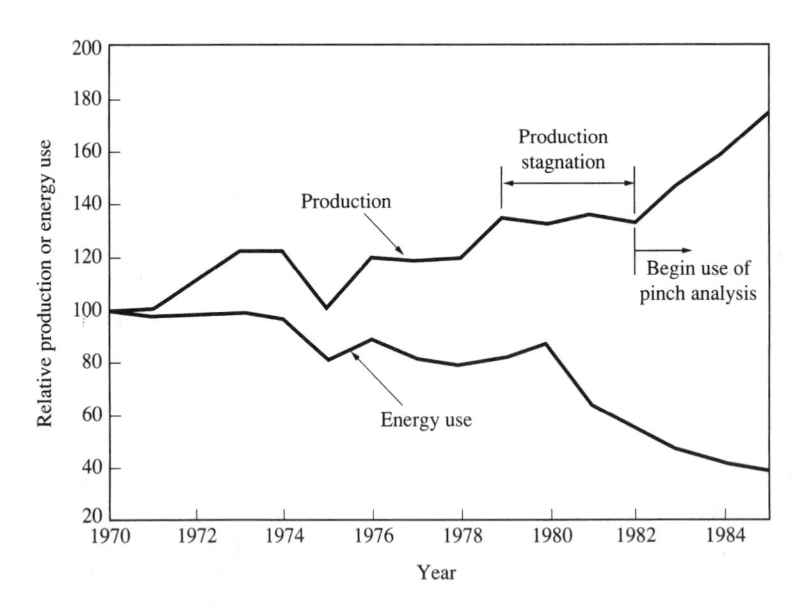

FIGURE 10.12
Effective use of pinch analysis for energy conservation at BASF plant at Ludwigshafen, Germany.

plant, use of pinch analysis was begun in 1982 to improve heat integration of most of the major processes at the plant. A constraint placed on any proposed change was that the change had to have a payback period of less than one year. The plant was able to achieve a total energy savings equal to 790 MW, while actually increasing production. Combustion-related airborne emissions and ash residues were also reduced. Wastewater discharges were also reduced, because less water treatment was needed for steam and cooling water. Emission reductions were substantial (see Table 10.2).

TABLE 10.2
Emission reductions at the BASF plant due to P2 activities

Pollutant	Amount	Pollutant	Amount
Carbon dioxide	240 ton/h	Ash	46 lb/h
Sulfur dioxide	1.4 ton/h	Carbon monoxide	15 lb/h
Nitrogen oxides	0.8 ton/h	Wastewater from water treatment	77 ton/h

As can be seen, utilizing this waste heat from one process step in another process step can greatly reduce overall energy requirements. The heat value that must be supplied to satisfy the two nonoverlapping regions of the combined composite curve is often only a small fraction of that required to cool the hot streams and heat the cold streams without using integrated heat exchangers. It must also be remembered, though, that the optimum heat exchange network established using these procedures may not be the best from an engineering or economic sense. It may not be practical or economical to bring process streams together in a heat exchanger if those process stages are widely separated in the plant. The expense of the additional piping, the maintenance and cleaning costs associated with the extra piping, and the heat losses occurring during flow through the pipe may negate any energy savings. The HEN design should be used as only one of many design inputs that go into the engineering judgment upon which the final design is based.

10.3.2 Pinch Analysis for Water Use

Pinch analysis techniques were developed to optimize energy use, but they can be adapted for other applications. As seen above, substantial energy can be saved through pinch analysis, and concomitant financial savings are significant. Reductions in the amount of energy used also means reductions in the environmental emissions associated with energy production and use, such as thermal pollution from the discharge waters and reductions in NO_x, SO_2, CO, and CO_2 from air emissions. Disposal of coal ash will also be reduced if coal is used as the fuel.

The objective of pinch analysis is not always minimization of energy; it can be applied in minimization of water use and the resultant wastewater production. Clean water is a valuable resource whose cost of production is rapidly spiraling upward. Purchasing water from a public utility is expensive, but most industries that produce their own water by treating raw water sources find that water produced on site is even more expensive. Therefore, most plants will use public water sources where possible. Public water sources are treated to be suitable for human consumption, and they are generally safe from a public health standpoint and palatable, but they may not be of high enough quality for many industrial purposes. Therefore, many industries, such as breweries and electronics firms, must treat this water further. Plants with large boilers for heat or steam production often must treat public water supplies to remove dissolved salts that can cause scale in the boilers and reduce their efficiency. Thus unnecessary use of water can be a major expense, both from the standpoint of the purchase and treatment costs of the water and from the cost of treating and disposing of the resulting wastewater. Pinch analysis can be used to reduce water requirements.

A plant's freshwater demand, and resulting wastewater production, can be reduced by process operation improvements (i.e., replacing water cooling with air cooling, using spray balls [balls sprayed into the reactor that abrade contaminants from the reactor walls] for more effective internal reactor washing, improving energy use, or increasing the number of rinse stages in an operation to reduce water demand) or by increasing water reuse in the process. Process improvement changes were described earlier in this chapter and in Chapter 5. Water reuse is described here.

Water can be reused in several ways. Direct reuse of water from one stage in a process in another stage (or reuse within the same stage) will be suitable only if the water quality is appropriate for the operation to which it is being sent. The level of contamination from the previous step must not interfere with the process in which the water is being reused. This water reuse reduces both freshwater and wastewater volumes, but leaves the contaminant load in the wastewater essentially unchanged.

If water quality is not high enough for direct reuse, the water may be treated to remove the offending contaminants so as to achieve the desired quality, but this comes at a financial cost. Because of the expense of process water, however, it is often more economical to treat the water and reuse it than to discard it (and pay for its treatment before discharge) and purchase new water. The water does not need to be totally reprocessed; only those contaminants that are detrimental to its reuse must be removed. Water renovation may be as simple as pH adjustment or removal of solids by filtration, or may it require much more substantial removal of dissolved materials by such processes as steam stripping, carbon adsorption, reverse osmosis, ultrafiltration, or ion exchange. Water regeneration reduces both freshwater and wastewater volumes and also reduces the contaminant load in the wastewater.

Pinch analysis has recently proven to be effective in reducing water use in industrial plants. The process data are plotted as a contaminant concentration–mass curve (analogous to the temperature-energy curves described earlier). To maximize water reuse, the highest acceptable contaminant *inlet concentration* to the process is needed. This allows the maximum amount of contaminated wastewater to be reused. Specifying the maximum acceptable *outlet concentration* for effective waste management minimizes the water flow rate. The higher the acceptable contaminant concentration in the effluent, the smaller is the amount of dilution water that is needed to make it suitable for reuse.

Consider the case of clean water used to extract contaminants from a process stream (such as washing a chemical precipitate or extracting polar organics from a nonpolar organic product or using a semipermeable membrane to extract contaminants), as shown in Figure 10.13*a* (Smith, 1995; Smith, Petela, and Wang, 1994). As the clean water stream contacts the process water, contaminants in the process water are transferred into the washwater. The washwater contaminant concentration increases and the process stream contaminant concentration decreases. The relationship between these two processes is governed by the partition coefficient of the two streams for the contaminant. The concentration of contaminant versus mass load of that contaminant in the waste stream can be plotted as shown in Figure 10.13*b*. This plot shows the effect of using clean water to extract contaminants from the process stream. However, acceptable results may be achievable using slightly contaminated water as the washwater. As

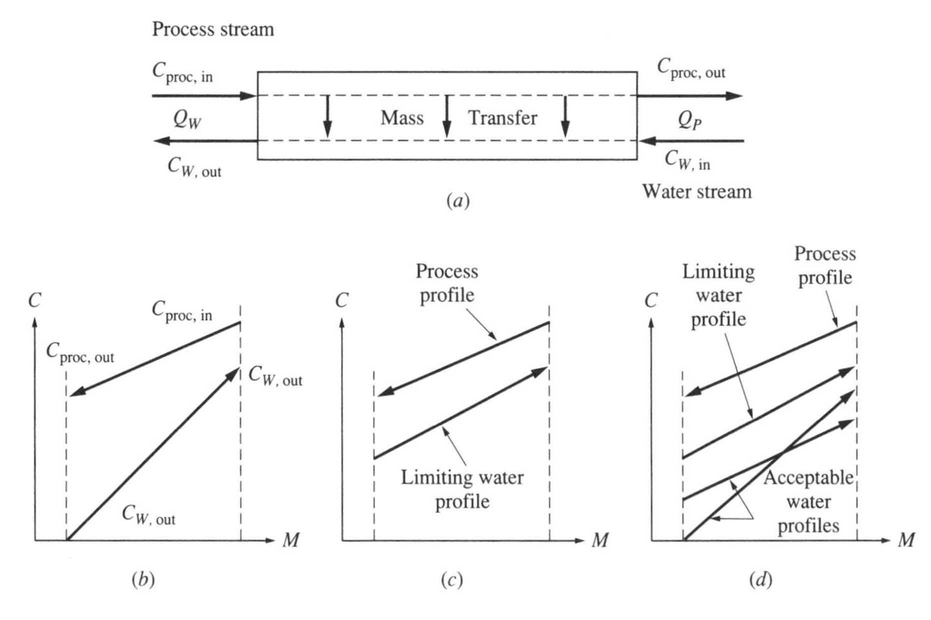

FIGURE 10.13
Water use in industrial processes.

long as the initial contaminant loading is low enough to allow contaminant transfer between the two streams, the desired final process concentration can be achieved (see Figure 10.13*c*). By maximizing the inlet and outlet washwater concentrations to that allowable for adequate process stream treatment, the amount of washwater required is minimized. Maximum inlet and outlet concentrations might result from constraints due to concentration driving forces, solubility, precipitation, fouling, settling, or corrosion (Smith, Petela, and Wang, 1994).

The water profile shown in Figure 10.13*c* represents the one that provides the maximum inlet and outlet concentrations. It shows the limiting conditions and is designated the *limiting water profile*. Any water supply with a combination of flow rate and contaminant concentration in the waste stream below the limiting water profile will satisfy the process requirements (Figure 10.13*d*). The limiting water profile may or may not be the optimum for a particular operation because of other constraints imposed. For example, it may not be possible to push the inlet water concentration to the maximum without elaborate and expensive mixing systems that allow mixing of wastewaters of differing concentration to achieve the desired maximum concentration. It may be more economical to use a cleaner water than necessary, because it is readily available, although some efficiency is thus sacrificed.

The approach described earlier for minimizing energy use through pinch analysis can also be applied here to minimize water use by recycling wastewater to operations where it is still suitable for use. Example 10.4 is modified from an example by Smith, Petela, and Wang (1994).

Example 10.4. Assume that an industrial process has four process steps (see Figure 10.14*b*). The water use and concentration profiles for each step are shown in the first four columns of Table 10.3 (Smith, 1995). Determine the minimum amount of water use required (*a*) if clean water is used for each step and (*b*) if water containing the maximum allowable concentration of contaminants is used.

Solution.

(*a*) The amount of process water needed if clean water is used for each step but the effluent is allowed to go to its maximum allowable concentration is 112.5 t/h (see column 5 in Table 10.3). This can be calculated by determining the amount of water needed per process step as follows:

$$Q = \frac{M}{(C_{\text{out}} - C_{\text{in}})}$$

$$Q = \frac{(2\ \text{kg/h})}{(100\ \text{ppm} - 0\ \text{ppm})(10^{-6})} = 20{,}000\ \text{kg/h} = 20\ \text{t/h}$$

where Q = Water flow rate, t/h
M = Contaminant mass load, t/h
C_{in} = Contaminant concentration in the inlet water, mg/L
C_{out} = Contaminant concentration in the outlet water, mg/L

(*b*) If it was possible to use inlet water containing the maximum allowable inlet concentrations in each process step, the water requirement would actually increase to 170 t/h (see column 6 in Table 10.3) because the driving force for transfer would be reduced by the contaminants in the feedwater. We will see in the next section, though, that this is not the case if water from one process is recycled to another.

WATER REUSE WITH RECYCLE. To allow for the use of the maximum allowable washwater inlet concentrations to minimize water use, it is necessary to recycle some of the wastewater so that it can be blended with the freshwater to achieve the desired acceptable inlet water concentration. This is best done by constructing a composite curve for the process data, as was done for temperature–enthalpy curves for energy conservation. The procedure is described in Example 10.5.

Example 10.5. Using the data from Example 10.4, determine the minimum water requirement if water recycling from one unit to another is allowable. The maximum inlet and outlet concentrations must be maintained for each step.

Solution. Figure 10.15*a* shows the data in Table 10.3 plotted on the same axes. The *x* axis represents the cumulative mass load, while the *y* axis is the contaminant concentration in the wastewater.

The mass loads of the individual streams are combined within each interval in Figure 10.15*b* to produce a limiting composite curve. This indicates how the process would behave if there was a single process step rather than four separate steps.

If the incoming freshwater has no contamination, a water supply line can be plotted from zero, tangent to the limiting composite curve (Figure 10.15*c*). This supply represents the minimum flow rate possible, with the maximum outlet concentration (in this case, approximately 455 ppm). The point of contact between the two curves is the pinch. At the pinch, mass transfer driving forces have gone to a minimum.

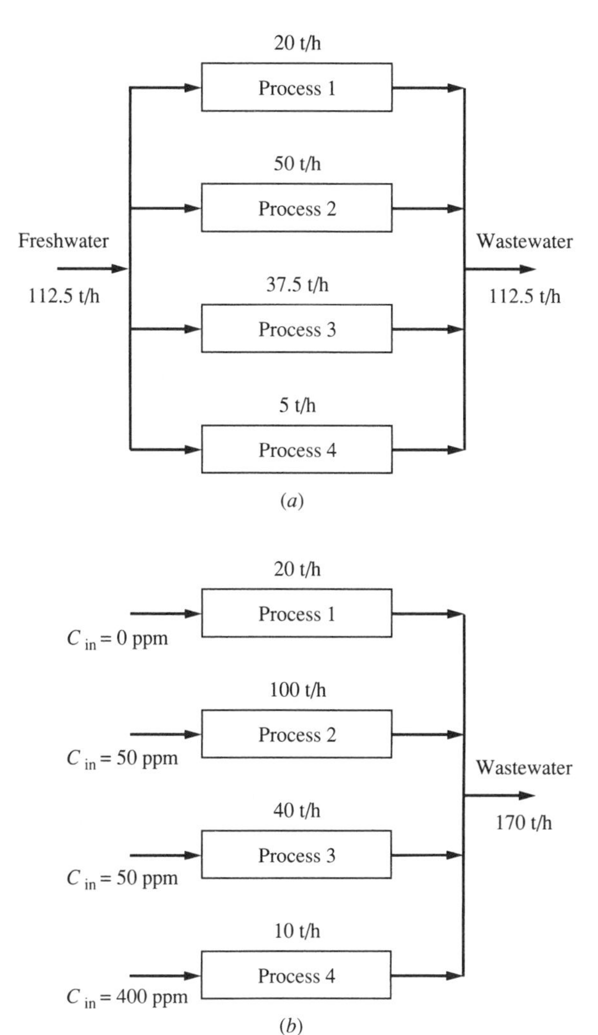

FIGURE 10.14
Process flow sheet for water utilization. (*a*) Water requirements if freshwater is used for each process stage. (*b*) Water requirements if the maximum allowable inlet concentration is used, without recycling wastes.

TABLE 10.3
Limiting process water data

Process Stage	Contaminant mass load, kg/h	Maximum allowable C_{in}, ppm	Maximum allowable C_{out}, ppm	Water flow assuming $C_{in} = 0$, t/h	Water flow assuming maximum allowable C_{in}, t/h
1	2	0	100	20	20
2	5	50	100	50	100
3	30	50	800	37.5	40
4	4	400	800	5	10
Total	41			112.5	170

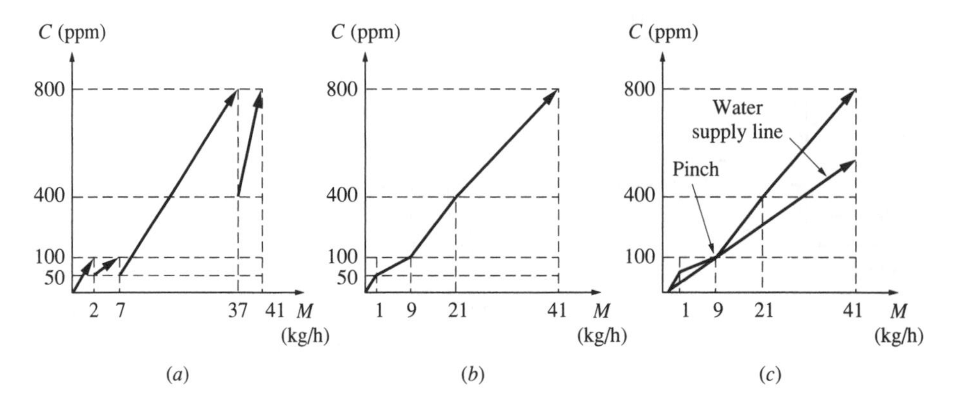

FIGURE 10.15
The limiting composite curve. (*a*) The limiting water profiles. (*b*) Limiting composite curve. (*c*) Matching the water supply line defines the minimum water flow rate. (Adapted from Smith, 1995)

The target value for the minimum water flow rate for this example is calculated as 90 t/h:

$$Q = \frac{41 \text{ kg/h}}{(455 \text{ ppm} - 0 \text{ ppm})(10^{-6})} = 90{,}000 \text{ kg/h} = 90 \text{ t/h}$$

This is 20 percent less than the 112.5 t/h derived without using pinch analysis and 47 percent less than if clean water had been used in all process stages. For this target to be reached, however, it must be possible to economically design a recycle network in which the constraints stipulated in Table 10.3 are satisfied. It is also necessary to keep in mind that there may be more than one contaminant that may interfere. Each contaminant must be checked to ensure that problems with water recycling won't be encountered.

By mixing water streams going to each process, it is possible to achieve the minimum flow rate target. Figure 10.16 shows the final design flow sheet for this process train. Each process step receives a single source of water made up of a mix of freshwater and wastewater from one of the prior stages.

A real-world example of the use of pinch analysis to reduce water use is the Unilever plant in Warrington, England (Buehner and Rossiter, 1996). This factory produces more than 200 products, including paints, glues, and adhesives. Historically, to guarantee product quality, the plant used large quantities of freshwater supplied to each individual reactor. Because of changing perceptions about environmental quality and the need for conservation, and because of the rising costs for raw water and water treatment, the company undertook a campaign to reduce water consumption. Using pinch analysis, Unilever developed a new water flow pattern that used freshwater in some critical processes and a mixture of recycled waters in other steps, such as precleaning and final washes. All that was required was the installation of a storage tank to mix and supply the recycled water to the appropriate units, along with minor plumbing changes. The result was a decrease in water demand of 50 percent and a decrease in wastewater production of 65 percent. This reduced the cost of purchasing water and treating wastewater, but, more important, it greatly reduced the size of the wastewater treatment process required, a substantial savings in capital costs.

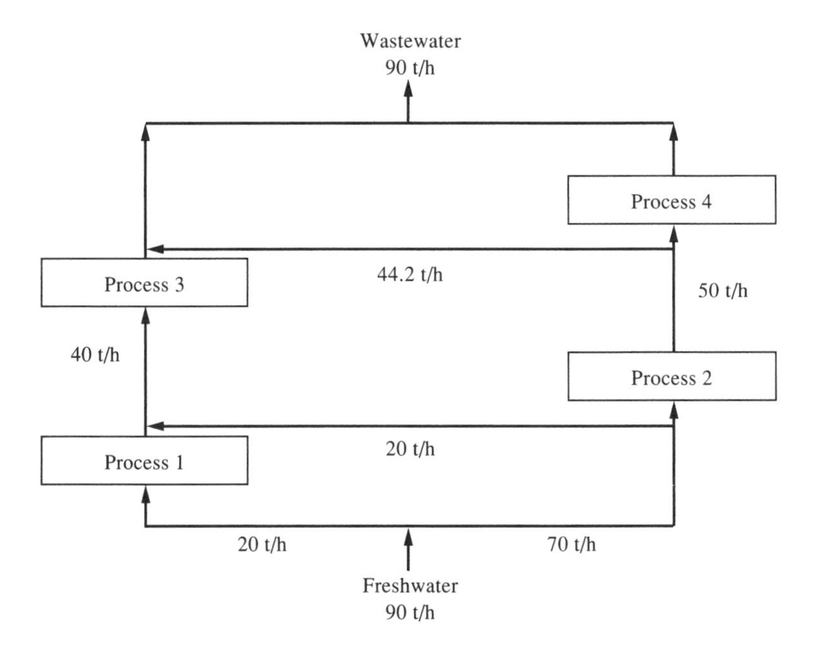

FIGURE 10.16
Final design for the minimum water use.

WATER REUSE WITH REGENERATION. The procedure described above works when wastewater from one process is directly reused by another process. A similar approach can also be used, though, when it is necessary to treat the wastewater before reuse. The maximum allowable contaminant concentration before regeneration of the water is equivalent to the pinch concentration. Allowing the concentration to reach the pinch concentration before regeneration will minimize the amount of water required for the process. During treatment, contaminants in the water will be removed and the concentration will be reduced to C_0 (see Figure 10.17).

Regeneration results in a decrease in contaminant concentration, but the inlet and outlet flow rates are essentially the same (less any losses due to backwash, waste streams, etc., which are usually small). Thus the slopes of the water supply lines in Figure 10.17 before and after the regeneration step are the same. If flow losses during regeneration are appreciable, they can be accounted for by adjusting the slopes before and after regeneration.

For the example above, assuming regeneration lowers the contaminant concentration to 5 ppm, water reuse in the process that includes regeneration reduces the necessary water flow rate to 46.2 t/h, using the design shown in Figure 10.18. Thus intermediate treatment of the water reduces the water requirement to only one-half of that without regeneration and to only one-quarter of that if freshwater had been used in all stages. This results in significant cost savings because less water needs to be purchased; it also means the volume of wastewater that must be treated is greatly reduced.

An evaluation of the Monsanto Chemical plant in Newport, Wales, showed the value of this procedure (Buehner and Rossiter, 1996). Effluent from seven processes at

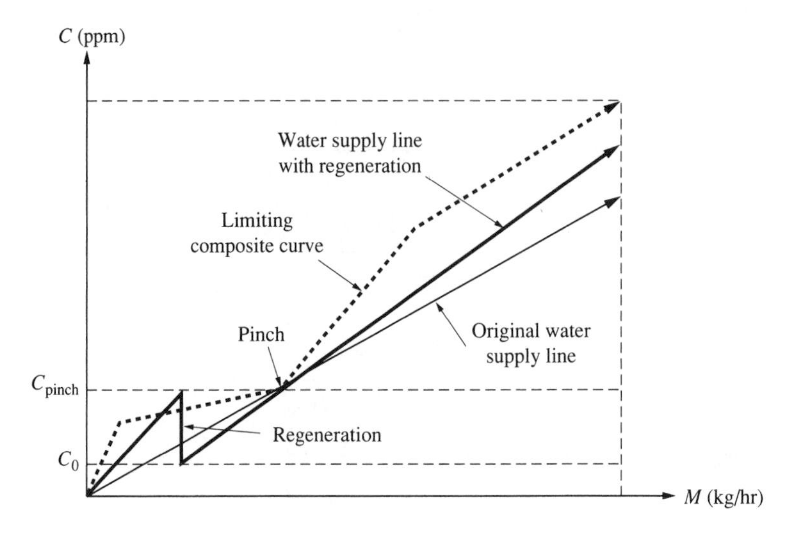

FIGURE 10.17
Water supply requirements when water regeneration is used.

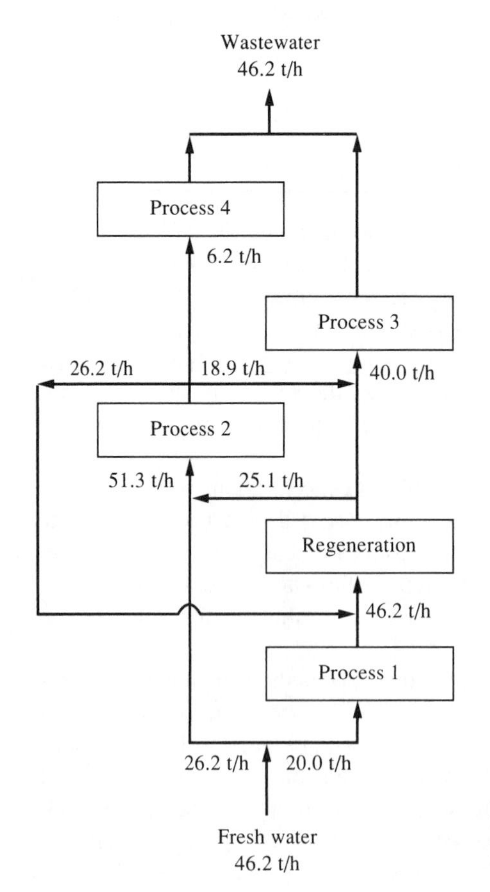

FIGURE 10.18
Process design using water reuse and internal wastewater regeneration.

the plant had been collected, mixed, adjusted for pH, and then discharged to the Severn River estuary. When this no longer became acceptable because of high chemical oxygen demand (COD) loadings (the regulatory authority required a 90 percent reduction in COD before discharge, at a cost of $15 million), Monsanto conducted a pinch analysis of the system. The company found that direct reuse of wastewaters could reduce the freshwater requirement by 30 percent. This would also reduce the size of the wastewater treatment plant required. However, the COD concentration would increase because of the smaller volume, necessitating additional treatment. Monsanto found, though, that if it installed a wastewater treatment unit (in this case, a constructed wetland using reeds), most of the water could be reused, reducing the final effluent by 95 percent. The capital cost for the reed bed was only $500,000. The total investment was $3.5 million, but the annual saving was $1 million in water and other raw materials.

10.3.3 Pinch Analysis for Process Emissions

A relatively new use of pinch analysis is to evaluate the potential for emissions reduction. As described, pinch analysis can be effectively used to minimize water use in industrial processes by reusing wastewater from one process stage in another stage, with or without regeneration of the water quality. Likewise, process streams from one stage can be reused in another to recover or make use of materials being carried in that stream.

The procedure used is identical to pinch analysis for energy or water conservation, as described earlier. The masses of a given material exchanged throughout the various stages of a process are plotted against the respective concentrations or mass fractions of that material for both the material-rich and material-poor streams (see Figure 10.19). Again, as long as the rich stream profile lies below the lean stream profile,

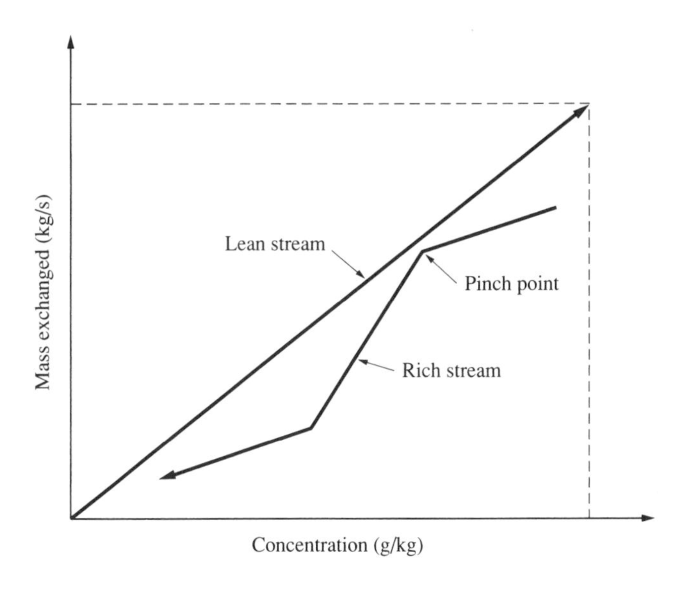

FIGURE 10.19
Pinch analysis for exchange of materials between a rich and lean stream.

material can be exchanged between streams. The efficiency and rate of exchange are governed by a number of factors, including the driving force established by the concentration differences between the two streams. The minimum concentration difference is equal to that at the pinch point.

Example 10.6. A refinery desires to reduce the quantity of water it is using and the volume of waste it produces by reusing water as much as possible. Three process units are heavy consumers of water. All are presently using clean washwaters. The accompanying table shows the flow rates and maximum allowable inlet and outlet concentrations for the primary contaminant in question, in this case, sodium chloride.

(*a*) Determine the current water requirement (i.e., without recycling).

(*b*) Using pinch analysis, determine the minimum allowable water requirement if water recycling is used. What would be the appropriate flow pattern?

(*c*) Assuming that a deionizer capable of treating wastewater to 1 mg/L could be installed after unit 1, what would be the resulting water requirement?

	Process data		
Process unit	**Contaminant mass load, kg/h**	**Max C_{in}, mg/L**	**Max C_{out}, mg/L**
1	5	0	100
2	38	60	300
3	40	130	1,000

Solution.

(*a*) The required water flow rates at $C_{in} = 0$ mg/L and no recycling can be determined:

Unit 1: $Q = (5 \text{ kg/h})/[(100 - 0 \text{ mg/L})(10^{-6} \text{ kg/mg})] = 50{,}000 \text{ kg/h} = 50 \text{ t/h}$

Unit 2: $Q = (38 \text{ kg/h})/[(300 - 0 \text{ mg/L})(10^{-6} \text{ kg/mg})] = 126{,}700 \text{ kg/h} = 127 \text{ t/h}$

Unit 3: $Q = (40 \text{ kg/h})/[(1{,}000 - 0 \text{ mg/L})(10^{-6} \text{ kg/mg})] = 40{,}000 \text{ kg/h} = 40 \text{ t/h}$

Current water requirement $= 50 + 127 + 40 = 217$ t/h.

(*b*) Draw the limiting water profiles and limiting composite curve for the system, and determine the minimum allowable flow rate (see Figure 10.20). First, the limiting water profiles are drawn for each unit (light solid lines in the plot), then the limiting composite curve is drawn (bold solid line) by connecting the points at which limiting concentrations are intersected. Finally, a line is drawn (bold dashed line) from the origin tangent to the limiting composite curve to obtain the effluent concentration. In this case, it is equal to 467 mg/L. The minimum flow rate can then be calculated:

$$Q_{min} = \frac{83 \text{ kg/h}}{(467 \text{ mg/L})(10^{-6} \text{ kg/mg})} = 177{,}700 \text{ kg/h} = 178 \text{ t/h}$$

By recycling flows, the water requirement can be reduced from 217 t/h to 178 t/h, a reduction of 18 percent.

(*c*) The deionizer will reduce the contaminant concentration to 1 ppm after unit 2. Therefore, draw a vertical line down from the limiting composite curve at that point in the graph and then draw a tangent through the pinch point (see Figure 10.21). The

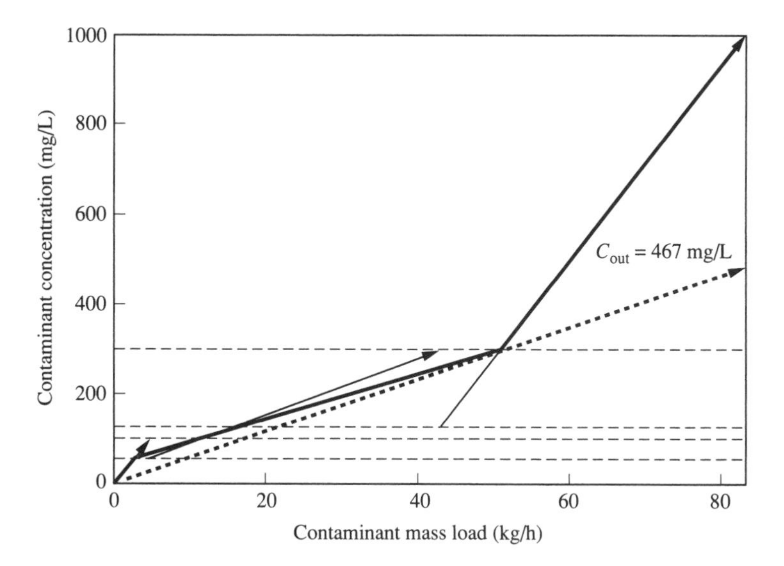

FIGURE 10.20
Construction of the limiting composite curve for Example 10.6.

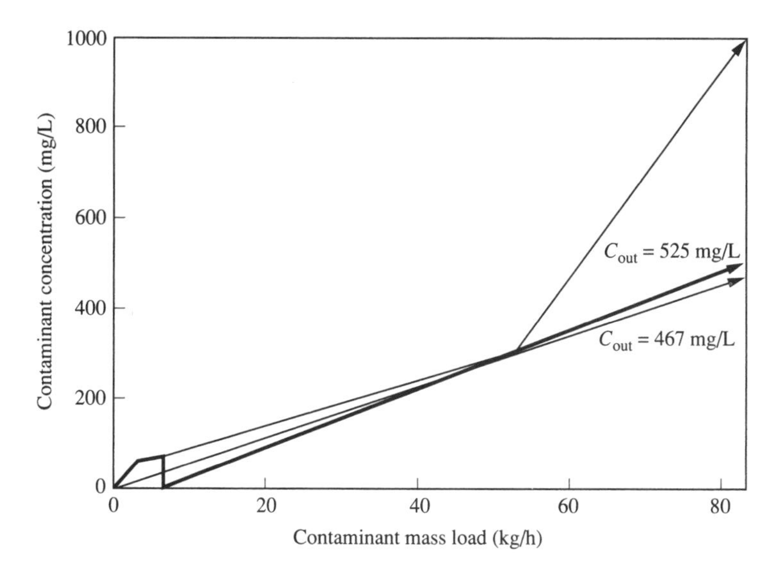

FIGURE 10.21
Construction of the limiting composite curve for Example 10.6, in which wastewater treatment is used for the recycle water.

resulting concentration is 525 mg/L. Therefore, the minimum flow rate with recycle and treatment after unit 1 is 158 t/h. The treatment unit located at this point results in only a small reduction in flow. The savings may not be justified by the cost of the deionizer in this case.

10.3.4 Summary

As was shown, substantial savings in energy, water, and process chemicals can be achieved by applying pinch analysis techniques to industrial processes. These techniques are quite new, however, and still need refinement and practical application procedures. As more examples of their effectiveness become available, they will probably become a major tool for pollution prevention engineers in their quest to minimize resource use and pollution generation.

REFERENCES

Buehner, F. W., and Rossiter, A. P., "Minimize Waste by Managing Process Design." *CHEMTECH* 26, no. 4 (1996): 64–72.

Douglas, J. M. *Design of Chemical Processes.* New York: McGraw-Hill, 1988.

El-Halwagi, M. M. *Pollution Prevention through Process Integration.* San Diego: Academic Press, 1997.

Linnhoff, B. "Pinch Analysis: A State-of-the-Art Overview." *Chemical Engineering Research & Design: Transactions* 71 (1993): 503–22.

Linnhoff, B. "Use Pinch Analysis to Knock Down Capital Costs and Emissions." *Chemical Engineering Progress* 90, no. 8 (1994): 32–57.

Linnhoff, B. "Pinch Analysis in Pollution Prevention." In *Waste Minimization Through Process Design,* ed. A. P. Rossiter. New York: McGraw-Hill, 1995, pp. 53–67.

Roosen, P., and Gross, B. "Optimization Strategies and Their Application to Heat Exchanger Network Synthesis." Presented at Exergoeconomical Analysis and Optimization in Chemical Engineering Seminar, Aachen, Germany, 1995.

Rossiter, A. P. *Waste Minimization through Process Design.* New York: McGraw-Hill, 1995.

Smith, R. "Wastewater Minimization." In *Waste Minimization Through Process Design,* ed. A. P. Rossiter. New York: McGraw-Hill, 1995, pp. 93–108.

Smith, R., Petela, E., and Wang, Y. "Water, Water Everywhere." *Chemical Engineer* 565 (1994): 21–24.

Thom, J., and Higgins, T. "Solvents Used for Cleaning, Refrigeration, Firefighting, and Other Uses." In *Pollution Prevention Handbook,* ed. T. Higgins. Boca Raton, FL: CRC Press, 1995, pp. 199–243.

Wang, Y. P., and Smith, R. "Wastewater Minimisation." *Chemical Engineering Science* 49 (1994): 981–1006.

PROBLEMS

10.1. A nickel plating bath contains 58,000 mg/L nickel. The dragout rate is 0.05 L/min. The parts being plated are placed in a 200-L dead rinse tank, which initially contained nickel-free water, immediately after plating. Draw a plot of the nickel concentration in the dead rinse tank (and of the dead rinse tank dragout) with time. Assume that evaporative losses are minimal.

10.2. For the operation in Problem 10.1, assume that the dead rinse tank is replaced by a single running rinse tank. What rinse water flow rate will be needed to achieve a maximum nickel concentration of 15 mg/L in the dragout? What will be the total mass of nickel lost to the wasted rinse water per day, assuming 24 h/day operation? What will be the nickel concentration in the final rinse water effluent?

10.3. Repeat Problem 10.2, assuming that two countercurrent rinse tanks in series are used. All other parameters remain the same.

10.4. A two-stage manufacturing process has a process stream (10 kg/s) that must be heated from 20 to 60°C before entering the first reactor. After the reaction, the process stream temperature must be increased to 80°C before the next reaction step. The second reaction is exothermic and the product temperature is 90°C. The finished product must then be cooled to 30°C before further processing. Determine the utility loadings (in kilowatts of energy) if steam is used to heat the process stream in the two reactors and cooling water is used to cool the reaction products. Assume the two process streams have a heat capacity of 2500 J/kg·K and the finished product has a heat capacity of 3500 J/kg·K.

10.5. The company described in Problem 10.4 asks you to investigate using a heat exchanger to preheat the two process streams using the reactor effluent. Now what are the utility loads?

10.6. Using the results from Problem 10.4, draw the combined composite curves for the heat exchange network and determine the utility loads. Compare your results with those from Problem 10.5.

10.7. Whey, a by-product of cheese production, is generated in large amounts by the cheese industry. Some of it is processed into other food products and animal feed, but much of it is thrown away because of the processing costs, particularly in developing countries. The whey has a very high BOD and can cause serious environmental problems. Recently, a process was developed in Jordan in which whey is converted to lactic acid by fermentation (see the flow chart in Figure 10.22). Following fermentation, the whey is recovered by filtration to remove coagulated protein and then evaporated to concentrate the lactic

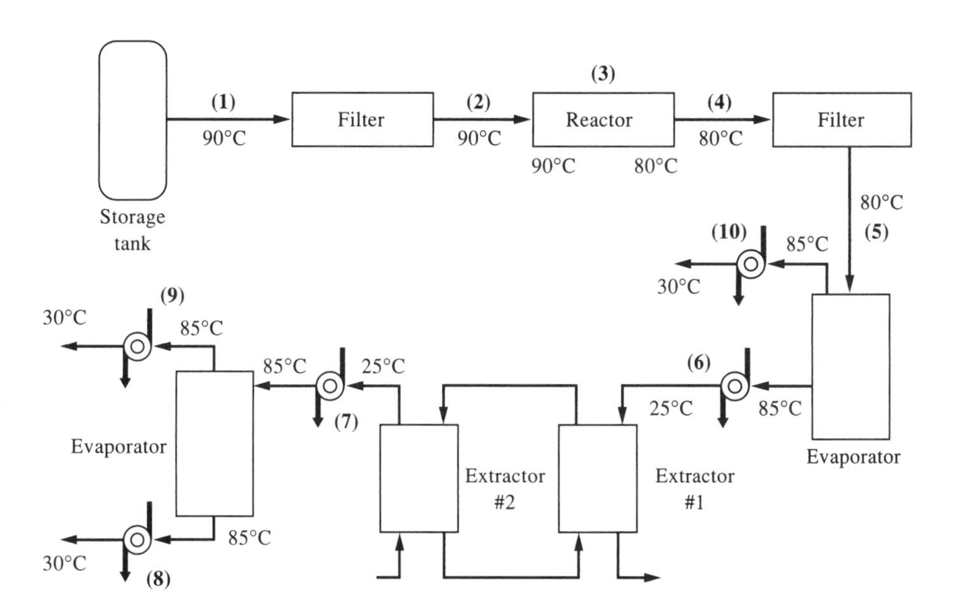

FIGURE 10.22

acid. The concentrate is extracted using liquid-liquid extraction to recover the lactic acid. It then goes through another evaporator to adjust the concentration to the desired value. Material and energy balances were performed on the process to determine the heat requirements for each process step. The accompanying table summarizes the heat requirements.

Stream	Rate of increase in entropy (entropy × flow rate), kJ/h·°C	T_{in}, °C	T_{out}, °C	Rate of heat flow into system, kJ/h
1	—	90	90	0
2	—	90	90	0
3	107.5×10^3	90	80	$+1075 \times 10^3$
4	—	80	80	0
5	—	80	80	0
6	0.771×10^3	85	25	$+46 \times 10^3$
7	1.262×10^3	25	85	-76×10^3
8	0.242×10^3	85	30	$+13 \times 10^3$
9	11.22×10^3	85	30	$+617 \times 10^3$
10	17.32×10^3	85	30	$+952 \times 10^3$

(*a*) Determine the total heating and cooling requirements in kilojoules per hour.
(*b*) If heat exchangers are used to cool stream 6 using heat transfer to stream 7, what will be the heating and cooling requirements? Assume that there must be a minimum temperature difference of 6°C to provide a driving force for heat transfer.
(*c*) Construct a limiting composite curve diagram for the heat exchanger system used.

10.8. Figure 10.23 shows a simplified cattle hide tanning process and the amount of clean water currently being used per day. The hides are salted after they are removed from the

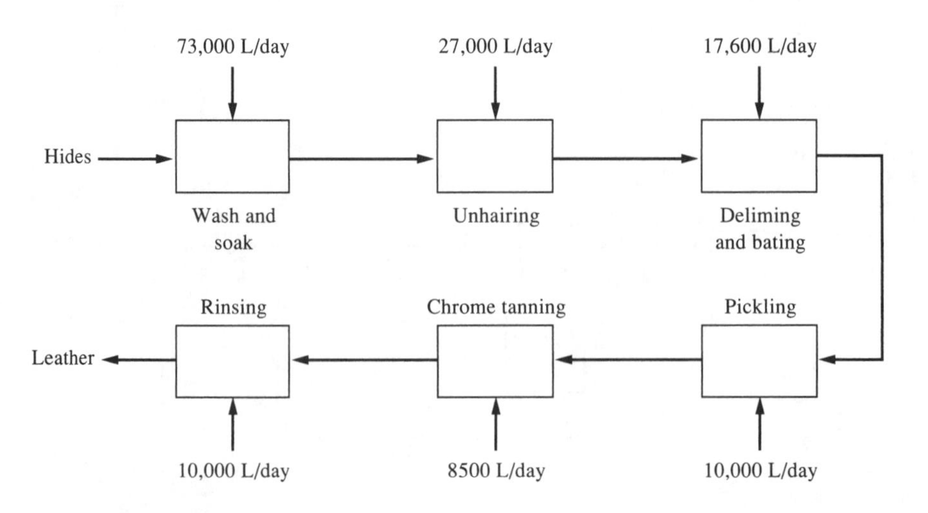

FIGURE 10.23

cows. The first step in the process is to wash the salt out. Hair is removed by using a lime slurry, and then the lime is removed in a soak tank. The hides then go to a pickling bath to lower their pH and then they are chrome tanned. The final step is a rinse to remove the tanning chemicals. The process is quite complex and many factors must be considered when deciding the quality of the water that can be used in the process, but for our purposes let us assume that the volatile solids (VS) concentration (representing organic matter) is the primary constraint. The tannery's lab has determined that the following volatile solids concentrations are the maximum allowable in the process water influents:

		Maximum allowable VS, mg/L	
Process	**Current effluent VS, mg/L**	**Influent**	**Effluent**
Wash and soak	2,200	2,000	3,600
Unhairing	11,000	5,000	18,000
Deliming and bating	6,000	3,000	8,500
Pickling	4,500	2,000	7,000
Chrome tanning	9,000	1,000	13,000
Rinsing	3,000	500	5,000

(*a*) Determine the current water requirements.
(*b*) Determine the amount of water required if clean water is used for each step and the effluent VS is allowed to go to the maximum allowable for each step.
(*c*) Determine the amount of water required if the maximum allowable influent and effluent volatile solids concentrations are used.

10.9. Using the data from Problem 10.8, determine the minimum water requirement using the limiting composite curve approach, assuming that water can be recycled from one step to another.

10.10. There are many ways of laying out a recycle system to minimize water use, and the procedure can be quite complex, requiring computer simulations. However, it is possible to approximate the solution by trial and error. Using the results from Problem 10.9, draw the flow diagram for minimizing water usage in the tannery, including recycle lines. Try to come as close to the answer in Problem 10.9 as you can without violating the maximum allowable influent and effluent volatile solids concentrations for each process step.

CHAPTER 11

RESIDUALS MANAGEMENT

11.1 INTRODUCTION

The application of life-cycle assessment to manufacturing processes and implementation of appropriate pollution prevention techniques can substantially eliminate the majority of wastes from most industries. Except in unusual circumstances, however, no industrial process is 100 percent efficient, and none can lead to zero pollution. Some waste by-products will always be produced during manufacture, and some product will escape capture during recovery operations. Production efficiency can be increased by use of catalysts and optimal operating conditions, and recovery of product can be improved by use of more efficient equipment such as distillation columns, solvent extraction, and high-speed centrifuges. Increasing product recovery, minimizing by-product formation, and recycling waste materials can lead to significant reductions in wastes requiring treatment, but there will always be some residual material that cannot be economically or practically removed from the waste stream. The material may be in the form of liquids, gases, or solids. As stated earlier in this book, a waste is really a resource out of place. However, these wastes will often be mixed with other materials and in low concentrations, making recovery and reuse infeasible. These waste materials must be properly collected, treated, and disposed in a manner that will ensure that they will cause no current or future public health or environmental problems.

A second source of waste materials in industry is leaking equipment and evaporation of volatile organic carbon compounds from storage tanks, open containers, process equipment, waste treatment processes, and other locations. Good pollution prevention practices can minimize but usually cannot completely eliminate these waste emissions. Chapter 12 discusses the source of these wastes, known as fugitive emissions, and what can be done to reduce their severity.

A thorough discussion of the proper disposition of residual wastes would require several textbooks. Here we examine residuals management from a broad perspective, giving the reader an idea of the complexity of the problem and a brief description of some possible solutions. The necessity to practice pollution prevention as rigorously as possible so as to minimize the amount of residuals needing treatment should become readily apparent.

Those interested in more detailed descriptions of treatment options for wastewater should refer to the books by Metcalf & Eddy (1991), the Water Environment Federation (WEF, 1991), Reynolds and Richards (1996), and Davis and Cornwell (1998). Excellent discussions of industrial solid and hazardous waste management can be found in Pfeffer (1992), Tchobanoglous, Theisen, and Vigil (1993), LaGrega, Buckingham, and Evans (1994), and Sincero and Sincero (1996). Air pollution control at industrial facilities is thoroughly discussed in Buonicore and Davis (1992) and Wark, Warner, and Davis (1998). This chapter will present brief summaries of options available for each of these classes of residuals.

11.2 WASTEWATER TREATMENT

Industrial wastewaters vary enormously in composition, strength, and volume from industry to industry and from facility to facility within an industry. Some wastewaters may be acidic or alkaline; contain oxygen-depleting organic materials (BOD_5), nutrients (e.g., nitrogen or phosphorus) that can cause eutrophication, or suspended solids that can settle in receiving waters causing unsightly and oxygen-consuming sludge deposits; be aesthetically unacceptable because of color, taste, or odor; or contain hazardous or toxic components. Therefore, it is impossible to present typical values that would be of use in designing industrial wastewater treatment facilities. Determination of these parameters on a case-by-case basis is essential.

Many older industrial plants mix process, cooling, and sanitary wastewaters in one sewer line. Unfortunately, various components in the wastewater may interfere with the treatment of other components or make their handling more difficult. For example, a wastewater containing easily biodegradable organic constituents can be readily and economically treated by several biological treatment methods, but if a waste stream containing heavy metals is mixed with it, the toxicity of the metals may inhibit the microorganisms from functioning properly. Even after biological treatment, the waste may still be unacceptable for disposal because of the metals. A metal plating waste containing sodium cyanide can also be easily treated, but if an acidic waste is mixed with it, potentially lethal hydrogen cyanide gas can be evolved, creating a major health hazard. Mixing only slightly contaminated clean cooling water with other wastes greatly increases the volume (and cost) of wastes requiring treatment and dilutes the constituents in the other wastes, making recovery or treatment more difficult. Thus segregation of waste streams in a manufacturing facility is essential.

Many industries use a diversity of processes. Some of these may be batch processes; others may be continuous flow processes. Some processes may operate continuously over long periods of time; others are used only sporadically. Some plants operate 24 hours per day, seven days per week; others may operate only 8 hours per day, 5 days per week. These wide variations in operation can lead to significant differences in wastewater flow rates, strength, and composition over time. Most treatment operations require relatively constant flow rates and waste composition. Consequently, many industries find that they must equalize waste streams by holding them in a basin for a period of time to get a stable waste that is easier to treat. In addition to equalizing flow rates, some chemical equalization also occurs due to mingling of wastes over a period of time. If pH in the waste stream fluctuates widely over time, it may also be possible to equalize the pH by addition of acid or base to the equalization basin contents, as necessary.

11.2.1 Treatment Options

Industrial process wastewaters may be disposed of in three ways:

1. They may be treated in an industrial wastewater treatment plant prior to discharge to a receiving stream.
2. They may be discharged directly to a municipal sewer and subsequently treated at the municipal wastewater treatment plant (often referred to as a publicly owned treatment works, or POTW).
3. They may be pretreated at the industrial site to reduce the amount of pollutants present and then be discharged into the municipal sewer for final treatment at the POTW.

Many industries opt to send their wastes to a POTW for processing, provided that the POTW is large enough to adequately process these wastes. This relieves the industry of the responsibility of treating the waste and eliminates much of the onerous paperwork associated with obtaining and complying with the NPDES permit needed to treat and discharge wastes into natural water bodies. The industry pays a fee (known as a user fee) to the municipality for the cost of treatment of their wastewater. Monitoring of waste volumes and compositions discharged to the sewer is still required, but this is usually much less difficult and costly than operating an industrial treatment facility. Often treatment of industrial wastes at a POTW is easier than at the industry because the municipal waste mixed with it dilutes out toxic components to a treatable level. Moreover, the municipal wastewater may add necessary nutrients for biodegradation in biological treatment processes that may be missing in the industrial waste. A major benefit of treatment of industrial wastes at a municipal facility is that treatment is collectively accomplished at one site, rather than at many sites around the municipality. Economies of scale usually make operation of one large facility much less expensive than operating many small ones. Also, a large centralized facility can usually get better trained workers than can one company, where wastewater treatment is considered a drain on cash flows and operating costs must be minimized. Small industrial waste treatment facilities may be operated by the janitorial staff or by the general manufacturing staff, rather than by well-trained waste treatment operators.

Companies that decide to use the POTW for their wastewater treatment may still need to provide equalization or some degree of treatment before discharge to the sewer. All POTWs have established pretreatment programs to control the discharge into the sewer of materials that could be harmful to the sewer or the treatment plant, dangerous to treatment plant workers, impair treatment plant operation, accumulate in the POTW wastewater sludge and make it a hazardous waste, or pass through the treatment plant unchanged and pose a later threat to the environment or public health. For example, acidic industrial wastes could erode sewer lines or POTW process equipment; greasy materials could accumulate on sewer walls, decreasing the carrying capacity of the sewer; excessive heavy metals or toxic organics could impair the microorganisms needed to biodegrade the organics in the municipal wastewater and could pass through the plant untreated; cyanides could be a hazard to POTW workers; and metals could contaminate the sludge, making it unacceptable for land application.

The U.S. EPA has established "prohibited discharge standards" (40 *CFR* § 403.5) that apply to all nondomestic discharges to the POTW and "categorical pretreatment standards" that are applicable to specific industries (40 *CFR* §§ 405–71). The local POTWs are responsible for establishing a pretreatment program and enforcing these standards. The General Pretreatment Regulations prohibit the discharge of the following into a municipal sewer:

1. Pollutants that create a fire or explosion hazard in the POTW, including, but not limited to, waste streams with a closed-cup flashpoint of less than or equal to 60°C using the test methods in 40 *CFR* § 261.21.
2. Pollutants that will cause corrosive structural damage to the POTW (but in no case discharges with a pH lower than 5.0) unless the POTW is specifically designed to accommodate such discharges.
3. Solid or viscous pollutants in amounts that will cause obstruction to the flow in the POTW resulting in interference.
4. Any pollutant, including oxygen-demanding pollutants (such as BOD_5), released in a discharge at a flow rate and/or concentration that will cause interference with the POTW.
5. Heat in amounts that will inhibit biological activity in the POTW resulting in interference, but in no case in quantities that the temperature at the POTW exceeds 40°C unless the approving authority, upon request of the POTW, approves alternative temperature limits.
6. Petroleum oil, nonbiodegradable cutting oil, or products of mineral oil origin in amounts that will cause interference or will pass through the treatment plant.
7. Pollutants that will result in the presence of toxic gases, vapors, or fumes within the POTW in a quantity that may cause acute worker health and safety problems.
8. Any trucked or hauled pollutants, except at discharge points designated by the POTW (Davis and Cornwell, 1998).

Chapter 13 is devoted to a discussion of industrial pretreatment permit programs and how they can be better designed to accommodate and encourage pollution prevention at the industrial facility.

Municipal wastewater treatment operations are not discussed in this book. The reader is referred to the references cited at the beginning of this chapter for in-depth discussions of municipal wastewater treatment. Following are descriptions of a few of the many processes that industry can use for pretreatment of waste before discharge to the municipal sewer or for industrial waste treatment prior to direct discharge to a receiving water in conformance with an NPDES permit. These include physicochemical processes (neutralization, oxidation-reduction processes, sedimentation/precipitation, activated carbon treatment, ion exchange, filtration and membrane processes), biological processes (activated sludge, lagoon and biofilm systems), and sludge treatment and disposal.

11.2.2 Physicochemical Processes

WASTE NEUTRALIZATION. It is common for industrial wastes to have too high or too low a pH for proper treatment or for discharge into a receiving stream or municipal sewer. For these wastes, neutralization is required. Neutralization is a process in which either acid reagents are added to an alkaline waste or alkaline reagents are added to an acidic waste in order to adjust the waste pH to a more acceptable value. It usually is not necessary to achieve neutrality (pH 7.0); the resulting pH only needs to be brought within the acceptable range. The desired pH will depend on the subsequent treatment units used. If the wastes are to be treated biologically, the pH should be adjusted to between 6.5 and 9.0. If the waste is to be discharged into a municipal sewer, the acceptable pH range is often set at 5.0–9.0.

Sometimes it is feasible for an industry to mix an acidic waste stream and an alkaline waste stream together to achieve neutralization of both streams. Care must be taken, though, to ensure that undesirable side reactions do not occur when this is done.

Adjustment of waste pH may also be needed for certain treatment processes to be effective. For example, pH adjustment is often required in oil emulsion breakage processes and in the precipitation of chromium from metal plating or tannery wastes. Cyanide waste treatment also requires careful pH control. Adjusting the pH of these wastes can be accomplished in much the same way as for neutralization; the only difference will be the desired final pH.

The amount of reagent needed to neutralize or adjust the waste pH can be determined by developing a titration curve, in which measured amounts of the reagent are added incrementally to an aliquot of waste and the resulting pH is determined. After plotting the data, the relationship between reagent dosage and desired final pH can be easily determined.

The reaction can take place in batch or continuous-flow reactors. For continuous-feed reactors, the resulting pH is usually monitored using a pH sensor. The sensor is coupled through a feedback system to reagent feed pumps that can vary the reagent flow rate to achieve the desired pH. This can be done using a single tank, but close pH

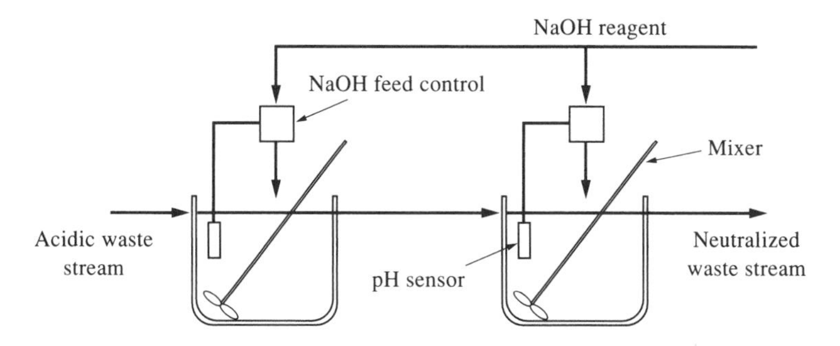

FIGURE 11.1
Alkaline neutralization of an acidic waste.

control with a single tank is difficult because the sensor is detecting the pH in the reactor effluent. If the pH is outside the acceptable range, it is too late to make corrections for that part of the waste. To overcome this problem, multiple neutralization tanks in series can be used, with the major pH control occurring in the first tank and subsequent tanks being used to refine the pH to the acceptable range (see Figure 11.1).

Acidic wastewaters may be neutralized by passing them through a bed of limestone or by the addition of slaked lime, $Ca(OH)_2$; caustic soda, NaOH; or soda ash, Na_2CO_3. Slaked lime is the most commonly used basic material because of its low cost (see Table 11.1). Caustic soda and soda ash are much more expensive than lime, even when the chemical costs are adjusted for the varying neutralizing capacities of the chemicals. However, if the wastewater contains appreciable amounts of sulfuric acid, the sulfates and calcium from the lime will combine in the neutralized wastewater to form calcium sulfate, $CaSO_4$, which has a relatively low solubility and may precipitate, causing scaling and fouling problems. Precipitation of metals (e.g., iron, aluminum, and heavy metals) as metal hydroxides may also occur during neutralization with slaked lime if the metal concentrations are high. This will not occur if a sodium-based reagent is used.

TABLE 11.1
Relative cost of various sources of alkalinity, 1995

Chemical	Equivalent weight	Cost, $/t	Neutralizing capacity, kg alkalinity/kg chemical	Cost/metric ton alkalinity
$Ca(OH)_2$	28	$120	1.8	$ 68
NaOH	40	800	1.25	640
Na_2CO_3	53	400	1.9	210
$NaHCO_3$	84	400	0.6	670
NH_4OH	14	800	3.6	220
MgO	20	360	2.5	140

Source: Speece, 1996.

Typical acidic reagents for neutralizing basic wastes include strong mineral acids, such as sulfuric acid, H_2SO_4, hydrochloric acid, HCl, and carbon dioxide, CO_2. If a source of CO_2 is readily available, such as the factory's boiler flue gas, it is usually the least expensive neutralizing agent. The flue gas is bubbled into the wastewater in the neutralization tank, where the CO_2 reacts with water to form carbonic acid, H_2CO_3, which neutralizes the alkaline waste constituents. If CO_2 is not available, one of the mineral acids can be used. They are more expensive, but their reactions are more rapid.

Example 11.1. A 500-mL sample of an industrial wastewater, with a pH of 2.8, is subjected to a titration test using a 4.0 *N* NaOH titrant solution, resulting in the titration curve seen in Figure 11.2. If 1893 m^3/day (500,000 gal/day) of the wastewater must be neutralized to a pH greater than 7.0, how much sodium hydroxide (4.0 *N*) must be used per day?

Solution. From the titration curve, it can be seen that 6.9 mL of NaOH is needed per 500 mL of waste to bring the pH to 7.0. This is the same as 6.9 L of NaOH per 500 L of waste. Therefore, the amount of sodium hydroxide needed to treat 1893 m^3/day of the waste can be calculated as

$$\text{NaOH required} = \left(\frac{6.9\ \text{L titrant}}{500\ \text{L sample}}\right)(1000\ \text{L/m}^3)(1893\ \text{m}^3\ \text{waste/day})$$

$$= 26{,}123\ \text{L/day}$$

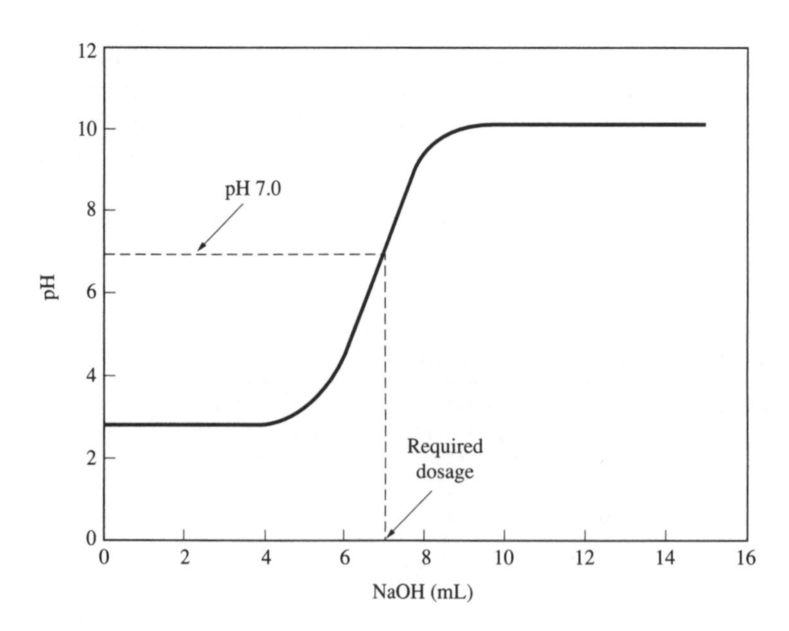

FIGURE 11.2
Titration curve for Example 11.1.

SEDIMENTATION. Many industrial wastewaters contain appreciable amounts of suspended solids. These can cause problems in a sewer system or in subsequent treatment units where they may settle out or cling to pipe or reactor walls. The solids may also contain high concentrations of hydrophobic toxic materials that have sorbed onto them. Proper practice is to remove these suspended solids as quickly after they enter a wastewater as possible.

The settling nature of suspended solids may be classified into one of three types: discrete settling (type 1), flocculent (type 2), and zone settling (type 3). Compression settling (type 4) occurs when solids reach the bottom of a reactor, pile up, and continue compacting as water is squeezed from between the particles.

In discrete settling, the particle maintains its individuality as it settles and does not interact with other settling particles, and therefore does not change in size, shape, or density. It settles due to the competing forces of gravity and buoyancy, with a terminal settling velocity defined by Stokes' law:

$$v = \sqrt{\frac{4g(\rho_p - \rho_l)D}{3C_d\rho_l}}$$

where v = terminal settling velocity of the particle
ρ_p = density of the particle
ρ_l = density of the liquid
D = diameter of the particle
g = acceleration due to gravity
C_d = drag coefficient, which is related to the Reynolds number and particle shape by

$$C_d = \frac{24}{N_{\text{Re}}} + \frac{3}{\sqrt{N_{\text{Re}}}} + 0.34$$

where N_{Re} = Reynolds number = $vD\rho_1/\mu$
μ = liquid viscosity

This type of settling is typical of granular inert materials that do not stick together, such as sand, and that are in dilute suspension. When the liquid flow velocity is small and flow conditions are essentially laminar, as is the case in most settling units, the Reynolds number is typically less than 1.0. Under these conditions, only the first term in the C_d equation is significant and the equation can be reduced to

$$C_d = \frac{24}{N_{\text{Re}}} \qquad N_{\text{Re}} \leq 1.0$$

Using this value for C_d in the settling equation, the equation reduces to

$$v_s = \frac{g(\rho_p - \rho_l)D^2}{18\mu}$$

If N_{Re} is greater than 1×10^4, mixing is turbulent and the equation for C_D can be reduced to

$$C_d = 0.4 \qquad N_{Re} > 10^4$$

For all other Reynolds number values, mixing is considered to be transitional and the full C_d equation must be used. The Reynolds number and the resulting type of mixing should be determined before applying the Stokes' law equation.

The size of an ideal settling tank to remove particles that follow type 1 settling can be determined by assuming that (1) particles entering the tank are initially uniformly distributed throughout the tank depth and (2) to be removed a particle entering the tank at the surface must settle to the bottom before the water exits the tank at the other end. The settling velocity of a particle that settles through a distance equal to the effective depth of the tank in the theoretical detention period of the liquid flowing through the tank can be defined as the *overflow rate* (Eckenfelder, 1989):

$$v_o = \frac{Q}{A}$$

where v_o = tank overflow rate
Q = rate of flow through the tank
A = tank surface area

All particles settling with a settling velocity greater than v_o will be completely removed, and particles with settling velocities less than v_o will be removed in the ratio v/v_o. The removal of discrete particles is independent of tank depth and is a function only of the overflow rate. Actual settling tanks are usually not ideal, due to short-circuiting or turbulence, so settling tank dimensions must be adjusted to account for these nonidealities.

Example 11.2. A foundry produces a wastewater containing metal filings with an effective diameter of 0.6 mm and a density of 2.8. The particles settle by type 1 settling. A horizontal flow velocity in the settling tank of 0.3 m/s will be used to ensure that laminar flow occurs. Assuming that the flow rate is 30,000 L/day, what size settler should be used?

Solution. The terminal settling velocity of the filings can be determined, assuming laminar flow conditions:

$$g = 9.81 \text{ m/s}^2$$

$$\rho_l = 1000 \text{ kg/m}^3$$

$$\rho_s = 2.8\ (1000 \text{ kg/m}^3) = 2800 \text{ kg/m}^3$$

$$D = 0.6 \text{ mm} = 0.6 \times 10^{-3} \text{ m}$$

$$\mu = 9.2 \times 10^{-4} \text{ kg/m} \cdot \text{s}$$

Then

$$v = \frac{\rho_s - \rho_l}{18\mu} gD^2$$

$$v = \frac{2800 - 1000}{18(9.2 \times 10^{-4})}(9.81)(0.6 \times 10^{-3})^2$$

$$v = 0.384 \text{ m/s}$$

Now calculate the required settler surface area to remove a particle with a settling velocity of 0.384 m/s in a wastewater flow of 30,000 L/day:

$$v_o = \frac{Q}{A}$$

$$\therefore A = \frac{Q}{v_o}$$

$$v_o = 0.384 \text{ m/s}$$

$$Q = 30{,}000 \text{ L/day} = 30 \text{ m}^3\text{/day} = 0.347 \text{ m}^3\text{/s}$$

$$A = \frac{0.347 \text{ m}^3\text{/s}}{0.384 \text{ m/s}}$$

$$A = 0.904 \text{ m}^2$$

The required depth of the settler can be determined based on the horizontal flow velocity selected:

$$v_h = \frac{Q}{A_x}$$

$$\therefore A_x = \frac{Q}{v_h}$$

$$v_h = 0.3 \text{ m/s}$$

$$Q = 0.347 \text{ m}^3\text{/s}$$

$$A_x = \text{settler cross-sectional area}$$

$$A_x = \frac{0.347}{0.3} = 1.156 \text{ m}^2$$

The ideal settler should have a surface area of 0.904 m^2 and a cross-sectional area of 1.156 m^2. Assuming a settler width of 0.5 m, the length would be 2.31 m and the depth would be 2.31 m. The actual settler will have to be somewhat larger to allow for possible short-circuiting or turbulence.

Flocculent (type 2) settling occurs when particles do not settle in an independent fashion. Many industrial waste particles follow type 2 settling. As the particles settle, faster settling particles collide with slower settling ones below. The particles may flocculate (i.e., stick together) and become a larger particle with a greater settling velocity. Thus particle settling velocities increase with settling depth. Flocculent particle settling is governed by both overflow rate and tank depth. Settling tank design for flocculent particles requires laboratory settling analyses. Design procedures can be found elsewhere (Davis and Cornwell, 1998; Eckenfelder, 1989; Metcalf & Eddy, 1991; WEF, 1991).

Zone settling occurs with flocculated chemical suspensions and with biological floc when the floc particles adhere and the mass settles as a blanket. The interlocked particles in the sludge blanket settle as a single unit with all particles settling at the same rate. A distinct interface forms between the supernatant and the solids zone. As the solids in the zone continue to settle, the interface lowers, the clarified supernatant zone increases in volume, and solids begin to pile up on the bottom of the settler. Compression, or type 4 settling, occurs at the bottom. Laboratory evaluations are required to design settlers for types 3 and 4 settling.

FLOTATION. Flotation is essentially the reverse of sedimentation. If the wastewater contains solids or immiscible liquids that are lighter than water, they will float to the surface in a flotation tank, where they can be skimmed off. Materials that are heavier than water, such as light solids or grease, may also be removed by flotation if they are made more buoyant by the attachment of air bubbles to their surfaces. In air flotation, air is bubbled into the bottom of the flotation tank or water containing pressurized air is injected into the bottom of the tank so that air bubbles are released from solution when the water pressure is reduced to that in the tank. The rising air microbubbles intercept suspended solids in the water; attach themselves to the particles, making them more buoyant; and carry them to the surface, where they are skimmed off. Clarified liquid is removed from the bottom of the tank. This process is particularly useful for material whose density is close to that of water.

A common use of flotation is in the separation of free oil from waters. The design of gravity separators for oil is specified by the American Petroleum Institute, and the separators are commonly referred to as API separators. These units are designed to remove free oil globules larger than 1.5 mm. Typically, they can reduce the oil content of refinery wastewaters to less than 50 mg/L. The recovered oil can be reprocessed. Other separators have been developed that can remove droplets down to 0.5-mm size, achieving effluents with less than 10 mg/L free oil.

Not all of the oil in water is in the free form, though. Emulsified oils will not readily separate from water; the emulsion must first be broken. This is usually a complex process requiring use of heat, acids, detergents, specialized polymers, alum, or iron salts. The oil released from emulsion can then be recovered in the flotation separator.

COAGULATION. Many industrial wastes contain suspended or colloidal waste materials that are too small to be effectively removed by gravity separation. Removal of colloidal particles is made more difficult by the fact that they usually have a surface electrical charge that causes them to repel other particles, thus preventing agglomeration to a size that could settle. For these small suspended particles and colloids to be removed from suspension, the surface charge must be destabilized and the particles must be brought together so that they can achieve a settlable size. This is the role of coagulation.

Coagulation results from the addition and rapid mixing of a coagulant with the wastewater to neutralize surface charges, collapse the surface layer around the particles, and allow the particles to come together and agglomerate. The resulting formation, called a *floc*, can more readily settle. As the floc settles, it interacts with more particles, enmeshing them in the floc and allowing the floc to grow in size. The mechanisms which take place during coagulation are complex; more detailed descriptions can be found elsewhere (Davis and Cornwell, 1998; Eckenfelder, 1989; Reynolds and Richards, 1996; Viessman and Hammer, 1998).

A number of coagulants are in common use. The most popular coagulants are aluminum sulfate, or alum [$Al_2(SO_4)_3 \cdot 14H_2O$], and iron salts (ferric chloride, ferrous sulfate, ferric sulfate). These work by providing charge destabilization and producing an insoluble hydroxide floc:

$$Al_2(SO_4)_3 \cdot 14H_2O + 3Ca(HCO_3)_2 \rightarrow 2\underline{Al(OH)_3}\downarrow + 3CaSO_4 + 14H_2O + 6CO_2$$
$$2FeCl_3 + 3Ca(HCO_3)_2 \rightarrow 2\underline{Fe(OH)_3}\downarrow + 3CaSO_4 + 6CO_2$$
$$2FeSO_4 \cdot 7H_2O + 2Ca(OH)_2 + \tfrac{1}{2}O_2 \rightarrow 2\underline{Fe(OH)_3}\downarrow + 2CaSO_4 + 13H_2O$$
$$Fe_2(SO_4)_3 + 3Ca(HCO_3)_2 \rightarrow 2\underline{Fe(OH)_3}\downarrow + 3CaSO_4 + 6CO_2$$

As can be seen, each of these reactions consumes a considerable amount of alkalinity. Additional alkalinity, usually in the form of lime, may need to be added to that naturally occurring in the wastewater to make these reactions go and produce a good floc.

Aluminum sulfate is used more frequently than iron salts because it is less expensive, but iron salts have an advantage in that they operate over a wider pH range. The alum floc is least soluble at a pH of 7.0, but it is effective in the pH range of 4.5–8.0. The optimum pH range for iron salts is generally between 4.0 and 12.0. Wastewater pH adjustment may be required if alum is used.

Coagulant aids are also sometimes used to improve coagulation by promoting larger, more rapidly settling flocs. Polyelectrolytes (anionic, cationic, or non-ionic) are large molecular weight polymers that are long molecules and can bridge between particles, bringing them together. Particulate materials, such as clays or activated silica, can be added to bind particles together and to serve as nuclei for floc formation.

Flocculants are usually mixed with the wastewater in a rapid-mix tank, which provides intense mixing to disperse the coagulant uniformly throughout the flow. The detention time in the rapid-mix tank is usually only 20–60 s. Coagulation begins almost immediately, but the high turbulence prevents large flocs from forming. Therefore, the water passes to a flocculation tank where slow mixing is provided using slowly rotating paddles to allow larger floc formation (see Figure 11.3). The flocculated mixture is then settled in a conventional settling tank. Alternatively, the wastewater can be sent to a solids contact or upflow clarifier, which combines the rapid mix, flocculation, and sedimentation steps in one unit (see Figure 11.4). The wastewater and chemicals are mixed in the center cone structure. The coagulated solids flow down under the skirt and

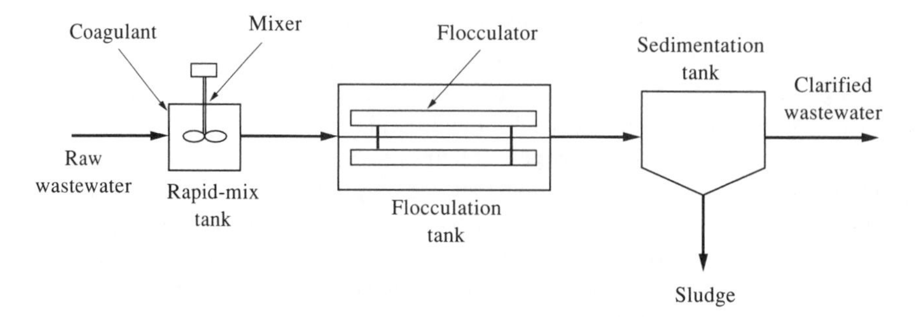

FIGURE 11.3
Flow scheme for coagulation/flocculation of wastewater.

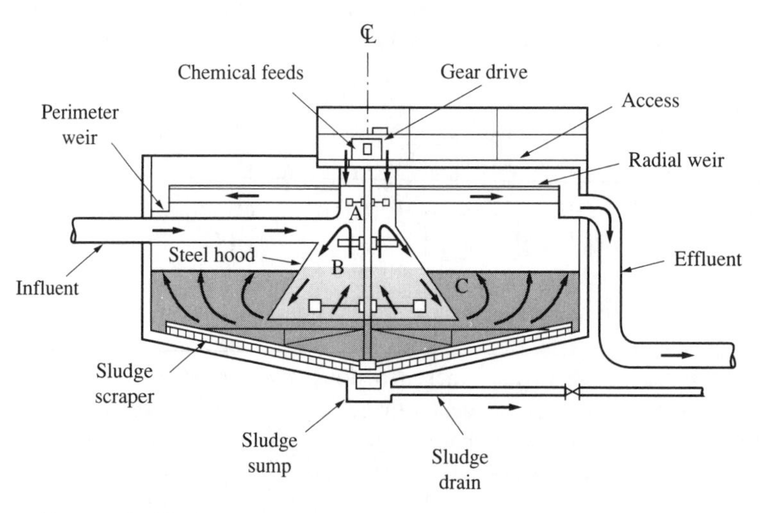

FIGURE 11.4
Typical upflow solids contact clarifier. (Source: American Water Works Association, 1990)

then upward to the effluent trough. As the water flows up, the solids settle into a sludge blanket, through which all additional solids must pass. Solids passing through the blanket are flocculated and removed from the rising water.

PRECIPITATION. Chemical precipitation is the process whereby soluble-phase species are removed from solution by adding a chemical which converts the contaminant to an insoluble form that can be removed by one of the previously described processes. Precipitation is most commonly used for removal of heavy metals from wastewaters from metal plating, the steel and nonferrous industries, the inorganic pigments industry, mining, and the electronics industry. In most cases, this is accomplished by converting the metal to its hydroxide form by the addition of lime or caustic soda, increasing the pH to the point where the metal hydroxide has its minimum solubility.

The minimum solubility pH varies from metal to metal, as can be seen in Figure 11.5. Many metals have a minimum point, rising in soluble concentration both above and below the minimum pH. This is particularly true for metal hydroxides. Wastewaters with these *amphoteric* compounds often require careful pH control during precipitation to ensure effective removal of the metals. Treatment of wastes containing several metals is difficult because of the varying minimum solubility pH values. A system designed to provide a pH of 9.2, the approximate minimum solubility point for lead and zinc, will allow substantial quantities of cadmium to escape as soluble species. Conversely, a pH of 11 will allow most cadmium species to be precipitated, but lead and zinc will have solubilities in excess of 5 mg/L at this pH.

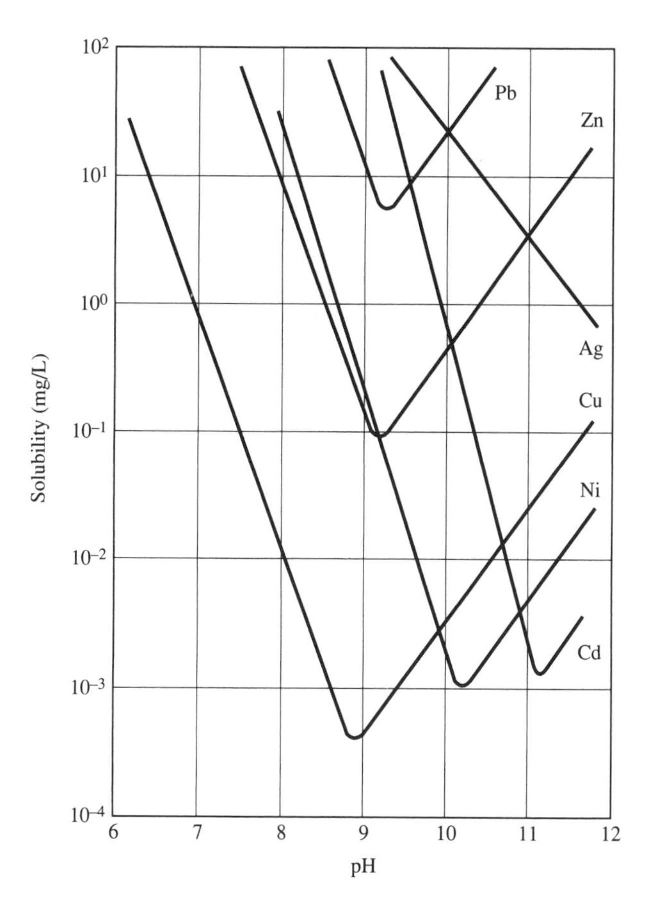

FIGURE 11.5
Solubilities of metal hydroxides in aqueous systems as a function of pH. (Source: U.S. EPA, 1983)

Heavy metals may also be precipitated by reaction with sulfides to form very insoluble metal sulfides. The most commonly used sulfide reagents are sodium sulfide (Na_2S) and sodium bisulfide (NaHS). Sulfide exists in solution as H_2S (aq), HS^- (bisulfide), or as S^{2-} (sulfide), with the dominant species dependent on pH. Below pH 7, H_2S is dominant; above pH 7, bisulfide is dominant until about pH 14, when sulfide becomes the dominant species. Therefore, in most systems, the waste pH is not appropriate for the more efficient reaction with sulfide. Dosages of sulfide greatly in excess of the stoichiometric amount are required, leaving large amounts of unreacted sulfide in the effluent. Sulfides are more expensive than lime but can be quite effective at metals removal; however, they have the drawbacks of being highly odorous and having the potential to generate toxic hydrogen sulfide gas if even slightly acidic conditions are encountered.

CHEMICAL OXIDATION AND REDUCTION PROCESSES. Chemical oxidation or reduction (redox) systems are used to treat several industrial wastewater types. Recent research on *advanced oxidation processes* (AOPs) has been promising and may soon allow oxidation processes to be used on a wider scale.

Chemical oxidation is a process in which one or more electrons are transferred from the chemical being oxidized to the chemical initiating the transfer (the oxidizing agent). Thus *oxidation* of a substance results in the loss of electrons from the substance, whereas *reduction* of a substance results in a gain of electrons in the substance. Oxidation and reduction reactions must be coupled together because free electrons cannot exist in solution. In the case of industrial waste treatment, pollutants are oxidized or reduced to products that are less toxic, more readily biodegradable, or more readily removed by adsorption. They may not be totally oxidized or reduced in all cases, but the intent is that the terminal products of the redox reactions are less toxic or more susceptible to further treatment than the original contaminants.

There are a number of chemical oxidizing and reducing agents and several commonly accepted treatment systems that they can be applied to. Among the more common oxidizing agents are ozone, hydrogen peroxide, chlorine, and potassium permanganate. These are currently used to oxidize cyanides, sulfides, phenols, pesticides, and some other organic compounds. Typical reducing agents are ferrous sulfate, sodium metabisulfite, sodium borohydride, and sulfur dioxide. They are predominantly used to reduce a metal, particularly chromium, to its less soluble and less toxic trivalent state.

Ozone (O_3) is one of the strongest known oxidants, with a standard oxidation potential, E_o, of 2.07 V. Ozone is an unstable compound and decomposes rapidly, so it cannot be shipped or stored and must be made on site, as needed. It is usually produced from dry air or oxygen by use of a high-voltage electric discharge. Electrons within the corona produced by the electric discharge split oxygen-oxygen double bonds, with the resulting oxygen radicals reacting with other oxygen molecules to form ozone gas. Only about 150 g ozone is typically produced per kilowatt-hour of electricity. Ozone is a very selective oxidant, being quite effective in the oxidation of reduced metals or phenol, but the oxidation kinetics are poor for oxidation of such organics as saturated hydrocarbons or chlorinated aliphatics. Ozonation can also be used for the removal of color from wastewaters.

An example of ozone oxidation is the conversion of alcohols to aldehydes and then to organic acids:

$$RCH_2OH \xrightarrow{O_3} RCHO \xrightarrow{O_3} RCOOH$$

Oxidation of unsaturated aliphatic or aromatic compounds causes a reaction with water and oxygen to form acids, ketones, and alcohols. Most of the ozonation products are biodegradable. Thus ozonation usually should not be thought of as a stand-alone treatment process, but rather as a pretreatment step before another treatment process.

A more effective oxidative use of ozone is in advanced oxidation processes. This involves coupling ozonation with the use of ultraviolet light. The uv light–ozone combination promotes the formation of strongly oxidizing free radicals which exhibit oxidation kinetics that are about a million times faster than for molecular ozone alone (Haas and Vamos, 1995). The hydroxyl radical (OH•) is the most important of these. The hydroxyl radical is a stronger oxidant than ozone (E_o = 2.80 V) and is a very nonselective oxidant, oxidizing a wide variety of compounds. Figure 11.6 illustrates the vastly improved oxidations that occur when uv light is used in conjunction with ozone for the oxidation of several types of organics.

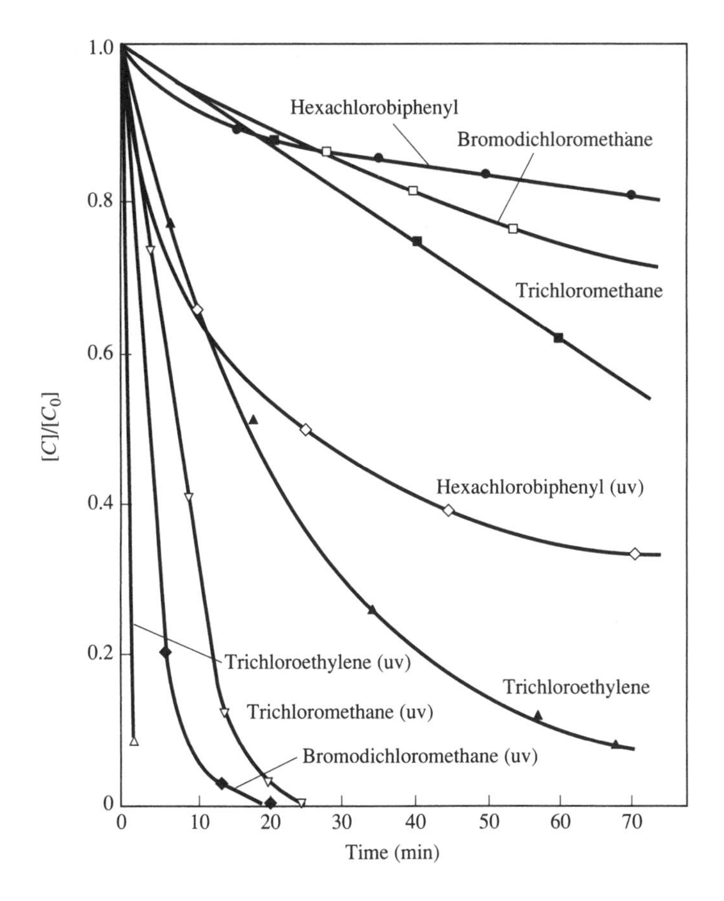

FIGURE 11.6
Destruction of chlorinated organics by ozone and by advanced oxidation processes. At pH 6–7, ozone dose rate = 1.0–1.4 mg/L • min; uv frequency is 254 nm from a low-pressure mercury lamp, with a flux of 0.42 W/L. (Source: Glaze, 1987)

Hydrogen peroxide is an oxidizing agent with a slightly lower oxidation potential than ozone. By itself, it is not very effective in oxidizing organics, but in the presence of a catalyst such as iron it can generate hydroxyl radicals which can react with organics and other reduced compounds in a manner similar to ozone (Eckenfelder, 1989):

$$Fe^{2+} + H_2O_2 \rightarrow Fe^{3+} + OH^- + {}^{\bullet}OH$$

$$Fe^{3+} + H_2O_2 \rightarrow Fe^{2+} + {}^{\bullet}HO_2 + H^+$$

$${}^{\bullet}OH + RH \rightarrow R^{\bullet} + H_2O$$

$$R^{\bullet} + H_2O_2 \rightarrow ROH + {}^{\bullet}OH$$

$${}^{\bullet}OH + Fe^{2+} \rightarrow OH^- + Fe^{3+}$$

As can be seen, the organic compound, RH, is oxidized to the alcohol, ROH. The ferrous iron (Fe^{2+}) acts as a catalyst in the first two steps, oxidizing to ferrous iron and producing the hydroxyl radical (•OH), and then converting back to the ferrous form. Hydrogen peroxide has also been used successfully in combination with uv light to enhance the production of hydroxyl radicals in advanced oxidation processes.

Cyanide can also be oxidized by hydrogen peroxide under alkaline conditions (pH 9.5–11.5) to carbonate and ammonia:

$$CN^- + H_2O_2 \rightarrow CNO^- + H_2O$$

$$CNO^- + OH^- + H_2O \rightarrow CO_3^- + NH_3$$

Cyanides are found in many industrial wastewaters, including those from electroplating, mining, catalytic cracking of petroleum, and coke furnace operations. Oxidation of cyanides is carried out under alkaline conditions and is called *alkaline chlorination*. Care must be taken to ensure that the cyanide waste is segregated and never mixed with acidic wastes, or toxic hydrocyanic acid (HCN) will form. Complete destruction of cyanide is a two-step process. The initial reaction converts cyanide (CN^-) to cyanogen chloride (CNCl):

$$NaCN + Cl_2 \rightarrow CNCl + NaCl$$

The CNCl formed is volatile and toxic. This reaction is independent of pH, but if the pH is maintained at 10 or greater, the sodium cyanate will be quickly converted to the much less toxic sodium cyanate (NaCNO):

$$CNCl + 2NaOH \rightarrow NaCNO + NaCl$$

Sodium cyanate is only about 0.1 percent as toxic as cyanide and readily hydrolyzes in the environment under slightly acidic conditions to ammonia and carbon dioxide. Therefore, many industries discharge the cyanate form of the waste. If it is necessary to oxidize the cyanate further before discharge, it can be reacted under slightly alkaline conditions (pH 8.5) with more chlorine to produce carbon dioxide and nitrogen:

$$2NaCNO + 4NaOH + 3Cl_2 \rightarrow 6NaCl + N_2 + 2H_2O$$

The pH must be reduced from 10 to 8.5 between reaction steps one and two or the reaction kinetics will be very slow. The first reaction consumes from 2 to 8 kg chlorine per kilogram of cyanide, depending on the waste being treated; the second reaction consumes an additional 4 kg chlorine per kilogram cyanide destroyed. An excess of chlorine is used to ensure complete oxidation. Typically this amounts to about 7.5 kg Cl_2/kg CN^-.

Chlorine, as a gas or as the hypochlorite salt, is often used as an oxidant, particularly for the oxidation of cyanide. Chlorine is also commonly used for color removal from wastewaters. It is not normally used for the oxidation of organics because the end products are usually chlorinated organics, which are themselves toxic.

Cyanides may be destroyed using a continuous-flow reactor system with metered chemical flow of chlorine, caustic, and acid to each reactor. Many operations produce only small quantities of cyanide waste. These facilities usually find it easier to treat the wastes on a batch basis.

Supercritical oxidation. Supercritical extraction was described in Chapter 5. Supercritical conditions can also be used to oxidize the waste's constituents in the aqueous state. Oxygenating the water and increasing the temperature and pressure to above the critical point of water (374°C and 218 atm) allows oxidation of organics to quickly occur, producing inorganic salts. Reactor residence times vary from 1 to 20 minutes. The salts are nearly insoluble in supercritical water and precipitate out. The heat produced by the oxidation helps to maintain the reactor temperature. Upon depressurization, cooling, and vapor-liquid separation, the water should be essentially free of contaminants. This process is not yet in commercial operation for waste treatment, but pilot-scale testing indicates that destruction efficiencies greater than 99.99 percent can be achieved for a wide variety of organics. It may soon join our arsenal of industrial waste treatment processes.

Chemical reduction. The principal use of reduction in wastewater treatment is for the removal of chromium from wastes. The reducing agents ferrous sulfate, sodium metabisulfite, and sulfur dioxide are commonly used under acidic conditions ($pH < 3.0$) to reduce toxic and soluble hexavalent chromium to the less toxic and insoluble trivalent chromium. The chromium can then be removed from the wastewater under alkaline conditions as chromic hydroxide by precipitation (see Figure 11.7):

$$Cr^{6+} + Fe^{2+} + H^+ \rightarrow Cr^{3+} + Fe^{3+}$$

$$Cr^{3+} + 3OH^- \rightarrow Cr(OH)_3\downarrow$$

Ferrous sulfate is often used because it is relatively inexpensive and its reaction is very fast; however, a greater amount of sludge is produced which must be disposed of because the resulting ferric ion also reacts with alkalinity to form insoluble ferric hydroxide. Typically, an excess dosage of 2.5 times the stoichiometric requirement for

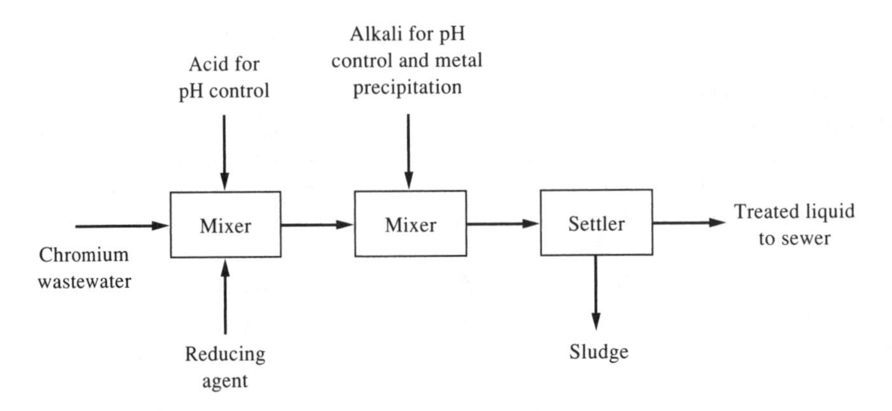

FIGURE 11.7
Schematic of the chromium reduction and precipitation process.

ferrous sulfate is used. The lime requirement for pH adjustment and chromic hydroxide precipitation is about 9.5 kg per kilogram of chromium. The sulfates produced by the other reducing agents will remain soluble and will not add to the volume of sludge generated.

Example 11.3. A metal plater produces 100 m^3/day of a waste containing 50 mg/L Cr^{6+}. The company is considering using ferrous sulfate treatment to precipitate the chromium. What would be the ferrous sulfate requirement and how much sludge would be produced per day if the sludge concentrates to 18,000 mg/L?

Solution. The stoichiometric ferrous sulfate requirement is 1 mol of ferrous sulfate per mole of chromium reduced. Now 50 mg Cr^{6+}/L at 100 m^3/day is

$$\frac{50 \text{ mg/L}}{(1000 \text{ mg/g})(52 \text{ g Cr/mol})}(100 \text{ m}^3\text{/day})(1000 \text{ L/m}^3) = 96.2 \text{ mol Cr}^{6-}\text{/day}$$

Therefore, the theoretical ferrous sulfate requirement is 96.2 mol. However, we need to add it at an excess equal to 2.5 times the stoichiometric amount. The ferrous sulfate requirement, therefore, is

$$\begin{aligned} \text{FeSO}_4 \text{ required} &= 2.5 \times (96.2 \text{ mol/day})(151.8 \text{ g/mol}) \\ &= 36{,}508 \text{ g/day} = 36.5 \text{ kg/day} \end{aligned}$$

Sludge production can be calculated as follows, assuming that all chromic hydroxide and ferric hydroxide produced precipitate:

$$(50 \text{ mg Cr}^{6+}\text{/L})\left(\frac{103 \text{ g Cr(OH)}_3\text{/mol}}{52 \text{ g Cr}^{6+}\text{/mol}}\right) = 99.0 \text{ mg Cr(OH)}_3\text{/L}$$

$$(99.0 \text{ mg/L})(10^{-6} \text{ kg/mg})(100 \text{ m}^3\text{/day})(1000 \text{ L/m}^3) = 9.9 \text{ kg Cr(OH)}_3\text{/day}$$

$$36.5 \text{ kg FeSO}_4\text{/day}\left(\frac{106.8 \text{ g Fe(OH)}_3\text{/mol}}{151.6 \text{ g FeSO}_4\text{/mol}}\right) = 25.7 \text{ kg Fe(OH)}_3\text{/day}$$

The total amount of sludge produced per day (dry weight) is

$$\text{Dry solids} = 9.9\ \text{kg/day} + 25.7\ \text{kg/day} = 35.6\ \text{kg/day}$$

At a concentration of 18,000 mg/L, the total sludge volume to be disposed of will be

$$\frac{35.6\ \text{kg/day}}{(18{,}000\ \text{mg/L})(10^{-6}\ \text{kg/mg})} = 1978\ \text{L/day}$$

ACTIVATED CARBON TREATMENT. Activated carbon adsorption can be used to remove many industrial organic contaminants that are difficult to remove by other processes. Adsorption consists of transferring the contaminant from the aqueous waste phase to the surfaces of an adsorbent material. The most commonly used adsorbent is activated carbon because of its large surface area, ability to sorb a wide variety of compounds, and its low cost relative to other comparable adsorbents. In most cases, the carbon is used in granular form (granular activated carbon, GAC) and is enclosed in a packed bed or fluidized bed reactor. In a few cases, though, it is used in powdered form, mixed with the wastewater and then removed by filtration. Finally, powdered activated carbon (PAC) can be combined with biological treatment; this is described later.

The mechanisms of adsorption were discussed in Chapter 2. In essence, adsorption occurs primarily due to van der Waals' forces, molecular forces of attraction between the solute and the solvent (physical adsorption). When the forces of attraction between the solute and the sorbent are greater than those between the solute and water, the solute will come out of solution and attach itself to the sorbent surface. Activated carbon is an excellent sorbent because of its high porosity and enormous pore surface area (on the order of 600–1000 m^2/g of carbon). Table 2.8 showed the properties of several activated carbons. Most of the adsorption occurs in the interior of the carbon particle due to diffusive transport of the organic contaminant into the pores and adsorption on the pore surfaces. Adsorption can continue until all available adsorption sites are covered. Physical adsorption is a reversible phenomenon, and it is possible to later remove the contaminants from the activated carbon and restore its adsorptive capacity. Some adsorption also occurs because of chemical reactions between the solute and active groups on the activated carbon pore surface (chemical adsorption), but this is usually minor in comparison to physical adsorption. Chemical adsorption is usually irreversible.

The overall rate of adsorption is controlled by the rate of diffusion of the solute molecules within the capillary pores of the carbon particles (Eckenfelder, 1989). The rate varies inversely with the square of the particle diameter; increases with increasing concentration of solute, decreasing pH, and increasing temperature; and decreases with increasing molecular weight of the solute. The adsorptive capacity of a carbon for a solute is dependent on both the carbon used and the sorbent. As described in Chapter 2, the degree to which adsorption will occur and the resulting equilibrium between solute remaining in solution and adsorbing on the carbon can be expressed as an adsorption isotherm. Several isotherm expressions were described there, including the Langmuir and Freundlich isotherms. These isotherms can be used to determine the adsorptive capacity of a given activated carbon for a specified solute at a desired equilibrium contaminant concentration.

Example 11.4. An isotherm study has been conducted for removal of pentachlorophenol (PCP) from water by activated carbon adsorption. The Freundlich isotherm was found to provide the best fit to the experiment data. The Freundlich parameters derived from the study were $K = 150$ mg/g and $1/n = 0.42$. An activated carbon adsorption process is planned to reduce the PCP concentration in the wastewater from 10.0 mg/L to 0.1 mg/L. How much activated carbon will be needed per 1000 m^3 of wastewater treated?

Solution. The Freundlich isotherm is

$$\frac{x}{m} = KC_e^{1/n}$$

where x = mass of sorbent adsorbed
m = mass of adsorbent
C_e = equilibrium concentration of PCP = 0.1 mg/L
K = 150 mg/g
$1/n$ = 0.42

The mass of PCP (per 1000 m^3) to be removed from the wastewater and adsorbed on the carbon can be calculated:

$$x = (10.0\ \text{mg/L} - 0.1\ \text{mg/L})(1000\ \text{L/m}^3)(1000\ \text{m}^3)(10^{-6}\ \text{mg/kg})$$

$$= 9.9\ \text{kg PCP}/1000\ \text{m}^3\ \text{wastewater}$$

Therefore,

$$\frac{9.9 \times 10^6\ \text{mg}}{m} = (150\ \text{mg/g})(0.1\ \text{mg/L})^{0.42}$$

$$m = 173.6 \times 10^3\ \text{g} = 173.6\ \text{kg}$$

The activated carbon requirement is 173.6 kg/1000 m^3 of wastewater treated.

When wastewater is passed through an activated carbon column, contaminants will quickly adsorb to the carbon until equilibrium is reached. Thus in a fresh column of activated carbon, adsorption takes place in a shallow zone at the top of the column (see Figure 11.8). No additional removal takes place below this zone, as the wastewater is already at equilibrium with the carbon. As more wastewater enters the column, though, it encounters carbon at the top that is already saturated with the contaminant and cannot remove more, but the carbon below the saturated adsorption zone can. Thus the adsorption zone front gradually moves down the column. Eventually, the adsorption zone front reaches the bottom of the column; any additional wastewater added will exit the bottom of the column only partially treated. When all of the carbon becomes saturated with contaminants, the effluent concentration will equal the influent concentration.

The result of *breakthrough* of contaminants before all of the carbon is fully loaded with contaminants means that some of the capacity of the carbon cannot be used. When the effluent concentration reaches the maximum allowable discharge concentration, the flow through the carbon column must be stopped, and the carbon must be removed and replaced with fresh or regenerated carbon. An alternative to this is to

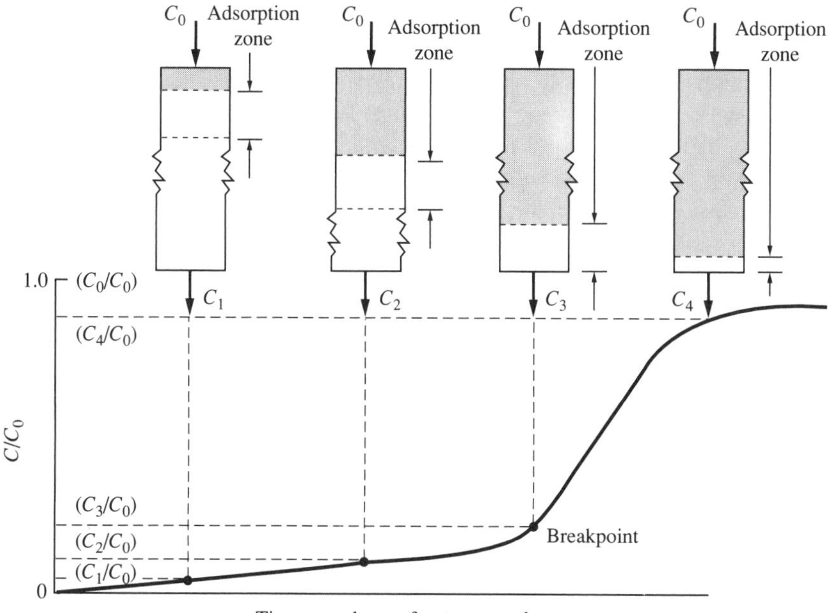

FIGURE 11.8
Typical breakthrough curve for an activated carbon adsorber. (Source: Weber, 1972)

use two columns in series. Contaminants that break through the first column will be trapped by the second column in series. When the first column's capacity becomes totally exhausted, it is removed from service, refreshed with clean carbon, and placed in line after the original second column. In this way, all of the carbon capacity can be used.

ION EXCHANGE. Ion exchange is a chemical treatment process used to remove unwanted ionic species from wastewater. In industrial wastewater treatment, it is predominantly used to remove cations (i.e., heavy metals) from solution, but it can also be used to remove anions such as cyanide, arsenates, and chromate.

As the name implies, ion exchange works by exchanging undesirable cations or anions in solution with less harmful ones from an ion exchange resin. Thus the harmful ions are removed from the aqueous stream and attached to the resin. The ions are not destroyed but rather are removed from the waste stream and concentrated on the resin, where they can be more easily handled. The exchange reaction is reversible, and the exchanged contaminants can later be removed from the resin, making the resin available for reuse.

Ion exchange resins consist of inorganic minerals with a deficit of positive atoms within the crystalline structure, known as *zeolites,* or synthetic organic polymeric materials that have ionizable functional groups, such as sulfonic, phenolic, carboxylic, amine, or quaternary ammonium. Zeolites are essentially all cation exchange materials (they exchange positively charged ions); the negatively charged zeolite surface is counterbalanced by exchangeable cations in solution next to the negative sites. These loosely held cations can be readily replaced by higher valence state cations (i.e., heavy metals) in the wastewater. Synthetic ion exchange resins can produce either cation or anion exchange, depending on the functional groups present. Resins with sulfonic, phenolic, and carboxylic groups are cation exchangers, whereas resins with amine or quaternary ammonium groups are anion exchangers. The carboxyl and phenol functional groups act as a weak acid and the sulfonic group acts as a strong acid, based on their degree of ionization. Cation removal selectivity can be controlled by selection of the proper functional groups. The synthetic resin polymers (typically polystyrene) are usually cross-linked to give them good chemical, mechanical, and thermal stability and to make them insoluble in water.

Industrial cation exchange resins are usually based on either the sodium cycle or the hydrogen cycle, depending on whether sodium or hydrogen is initially attached to the exchanger. The generalized equation for cation exchange by a resin may be represented by

$$\mathrm{Re \cdot M_1 + M_2^+ \rightarrow Re \cdot M_2 + M_1^+}$$

where M_1^+ and M_2^+ are cations of different species and Re is the resin. Ion exchange is a reversible reaction, and therefore the law of mass action may be applied. The equilibrium, or selectivity, constant is represented by

$$K_{M_1^+}^{M_2^+} = \frac{[\mathrm{Re} \cdot M_2][M_1^+]}{[\mathrm{Re} \cdot M_1][M_2^+]}$$

$$= \left[\frac{M_2}{M_1}\right]_{\text{solid}} \times \left[\frac{M_1}{M_2}\right]_{\text{solution}}$$

where M_1 and M_2 are the concentrations of the two cations in solution and on the resin at equilibrium. The greater the magnitude of K, the greater the relative preference of the cation for the exchange resin. Cation exchange resins generally prefer higher valence state ions and more polarized ions. Table 11.2 shows, for both strong and weak cation exchange resins, the cation selectivity preference ranking. Thus it is possible to selectively remove cations from solution by attaching the proper cation to the resin or by choosing between a weak acid or strong acid resin.

In most cases resins contain either an exchangeable sodium or hydrogen, because, being near the bottom of the preference order, they can exchange with any cations above them in the chart. Hydrogen ions can be replaced by essentially all cations except lithium. The result, though, is that the water will have an increased hydrogen ion concentration, and will thus become acidic. Neutralization may be necessary. Sodium-based resins can remove essentially all cations of interest for wastewater treatment; the resulting wastewater will have an increased sodium content, but this is usually of minor or no concern for wastewater discharge. Therefore, sodium-based resins are commonly used. They have the further advantage of being easily regenerated, using common and inexpensive sodium chloride salt.

TABLE 11.2
Selectivity of cation exchange resins to various cations in order of decreasing preference

Strong acid cation exchange resin	Weak acid cation exchange resin
Ba^{2+}	H
Pb^{2+}	Cu^{2+}
Ca^{2+}	Ca^{2+}
Ni^{2+}	Mg^{2+}
Cd^{2+}	K
Cu^{2+}	Na
Zn^{2+}	
Mg^{2+}	
Ce	
Rb	
K	
NH_4	
Na	
H	
Li	

$$Re \cdot M_2 + NaCl \rightarrow Re \cdot Cl + M_2^{+}$$

Hydrogen-based resins are more commonly used to produce high-quality deionized process waters where all cations and anions must be removed (known as demineralization). In this case, the cation exchange–treated water, which is now acidic, passes through an anion exchange resin (which contains exchangeable hydroxyl ions) (see Figure 11.9). The anions in the water are removed to the resin and replaced by hydroxyl ions, which neutralize the acid resulting from the cation exchanger. The result is a pH-neutral, ion-free water. The cation exchange resin is usually regenerated using a strong acid, such as sulfuric acid, whereas the anion exchange resin is regenerated by a strong base, such as sodium hydroxide.

Ion exchange is usually carried out in fixed-bed columns. The equipment required consists of the column containing the exchange resin, a regenerant (i.e., salt or acid) storage tank, and a regeneration solution tank in which the salt is dissolved or the acid is diluted. The resin inside the exchange column is usually supported on a bed of gravel.

The removal mechanisms inside an ion exchange column are similar to those in an activated carbon column. As the water passes through the ion exchange resin, the ions are exchanged within a fairly narrow depth of resin. As the resin exchange sites become exhausted, the exchange zone gradually moves down the column. Eventually,

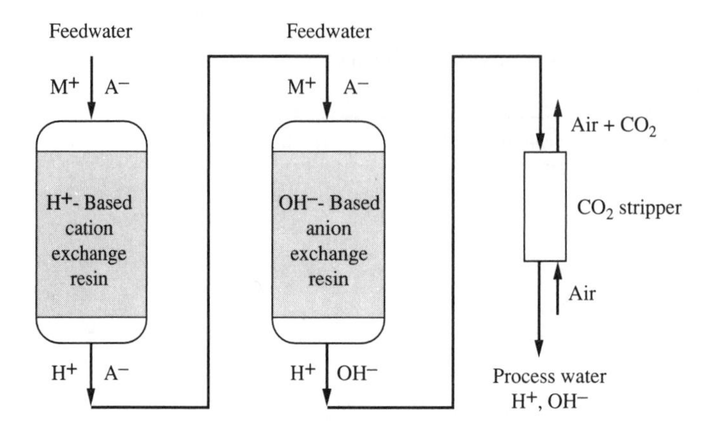

FIGURE 11.9
Demineralization of water by ion exchange.

contaminants will begin to break through the bottom of the bed, signaling the bed needs regeneration. Each resin will have a maximum capacity for exchanging ions, based on the number of exchange sites available on the resin. This is known as the *ion exchange capacity* of the resin (cation exchange capacity, or cec, for cation exchangers). Typical ion exchange capacities range between 200 and 500 meq per 100 g. The ion exchange capacity for a given resin can be used to estimate the amount of resin needed to treat a given volume of wastewater or the time between resin regenerations, as shown in Example 11.5.

Example 11.5. An industrial wastewater with 100 mg/L of zinc is to be treated by a cation exchange column. The column contains 0.4 m^3 resin and has a design flow rate of 200 L/min. The resin has a cation exchange capacity of 5.0 meq/g. How long can the column run before the resin is totally exhausted?

Solution. The resin has a total ion exchange capacity of

$$\text{Resin cec} = (0.4\ \text{m}^3)(1480\ \text{kg/m}^3)(2.0\ \text{eq/kg})$$
$$= 1184\ \text{eq}$$

The zinc loading rate on the column is

$$\text{Zn loading} = (200\ \text{L/min})(100\ \text{mg/L})(10^{-3}\ \text{g/mg})$$
$$= 20.0\ \text{g/min}$$

Convert this to milliequivalents:

$$\text{Zn}^{2+}\ \text{eq. wt.} = 32.7\ \text{g/eq}$$

$$\text{Zn loading} = \frac{20.0\ \text{g/min}}{32.7\ \text{g/eq}}$$
$$= 0.61\ \text{eq/min}$$

The time to exchange resin exhaustion is

$$\text{Exhaustion time} = \frac{1184 \text{ eq}}{0.61 \text{ eq/min}}$$

$$= 1940 \text{ min} = 32.3 \text{ h}$$

FILTRATION. Filtration processes can be used to remove suspended solids from industrial wastewaters, as a stand-alone process to remove larger particles, as a pretreatment device before other treatment processes such as activated carbon adsorption or membrane units, or following chemical treatment. Filters are often used after settling tanks to remove solids that did not settle. There are many processes that fall under the heading of filtration. Among these are screens, granular media filters, vacuum filters, filter presses, and various types of membrane filters. We concentrate here on granular media filters; membrane processes are described in the next section.

Granular media filters consist of beds of granular media, such as sand, through which the wastewater passes. In industry, the filter medium is usually housed in a cylindrical vessel; gravity filters can be used but usually the liquid flows down through the medium due to an applied pressure (see Figure 11.10). The filter medium is supported by a bed of gravel to prevent short-circuiting and ensure uniform flow through the medium and out of the reactor. When buildup of accumulated solids causes pressure losses through the medium to become excessive, flow to the filter is stopped and the filter medium is backwashed by forcing water up through the medium. The medium expands due to the buoyant forces from the rising water, contaminants become dislodged, and they flow out of the filter to waste. When the backwash water is turned off, the fluidized filter medium settles back to the gravel base in a size-segregated fashion with the larger, heavier granules on the bottom and the lighter materials on top.

Granular media filters remove particles on the surface of the filter by straining and throughout the filter depth by a combination of straining, particle adsorption to the filter media's surfaces, and flocculation and settling withing the pores between granules. As suspended solids are removed from the passing water, the bed's porosity and permeability decrease, increasing removal efficiency but also increasing head losses.

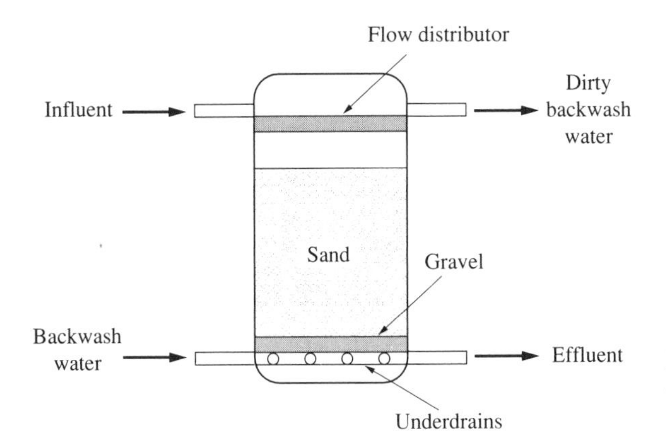

FIGURE 11.10
Cross section through a pressurized granular medium filter.

Graded quartz sand (0.35–0.60 mm diameter) is often used as the filter medium. When the sand is backwashed, it distributes itself with the smaller sand grains at the top and the larger ones on the bottom. This is not very efficient for particle straining. To overcome this, a second medium material, such as anthracite coal with a size range of 1.3–1.7 mm, may be mixed with sand of 0.45–0.60 mm size. These are called *dual media filters*. The coal is less dense than the sand and will settle near the top after backwashing, giving a better medium size gradation. A third option is to use a mixture of several types of media. *Multimedia filters* usually contain garnet (0.25–0.4 mm), sand (0.45–0.55 mm), and anthracite coal (1.0–1.1 mm). On backwashing, these media tend to distribute themselves with the larger sized coal on top, the sand in the middle, and the small garnet on the bottom because of their varying sizes and densities. The top medium provides a coarse filtration step, followed by sand filtration, and then final polishing through the small garnet bottom medium. This provides very effective filtration, with suspended solids generally less than 1 mg/L.

MEMBRANE PROCESSES. Membrane separation of solids from an industrial wastewater is actually a subset of filtration. It can be used to separate colloidal and dissolved solids that are much smaller than those removed by other filtration processes. The membranes used have pores sufficiently small that even small molecules or ions can be removed (see Figure 11.11). The membranes can be made selectively permeable for

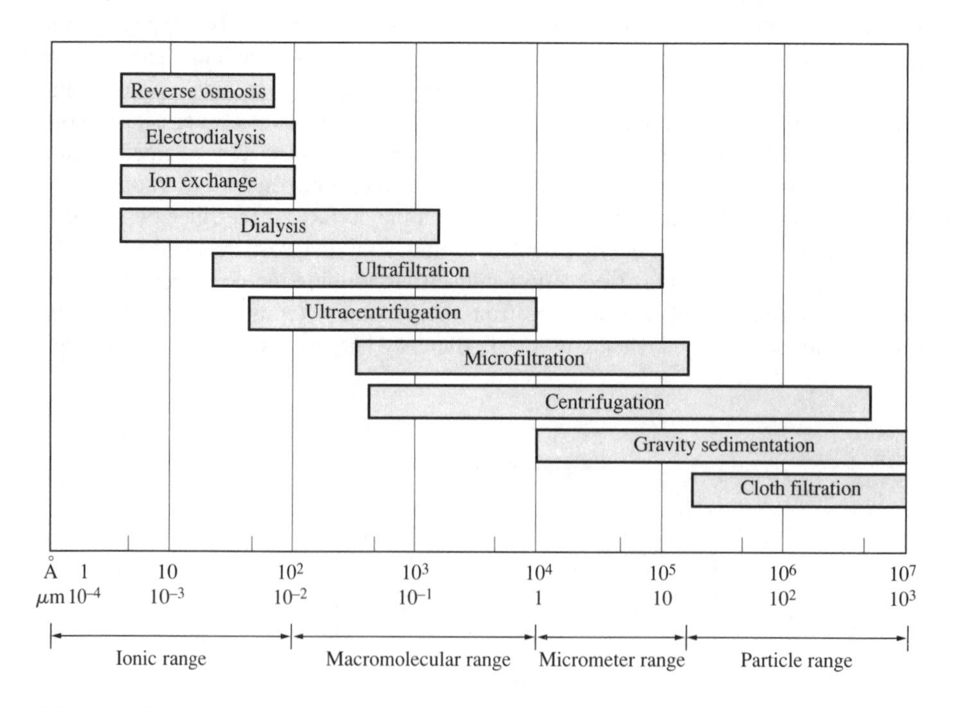

FIGURE 11.11
Effective ranges of various filtration and membrane processes.

specific materials so that separations of these materials can be achieved. Membrane processes are very effective, but the rate of transfer across the membrane is generally slow and pressure drops are high; large membrane areas are generally required.

Membrane filtration includes a broad range of separation processes: ultrafiltration, dialysis, electrodialysis, reverse osmosis, and so on. Ultrafiltration uses membranes with molecular weight cutoffs from several thousand up to one million grams per mole. This allows for separation of proteins and high molecular weight carbohydrates from lower molecular weight materials. These membranes are used in product recovery during material processing, but they are only rarely used in wastewater treatment. Dialysis and electrodialysis can be used to desalinate brackish waters, but they are not typically used for wastewater treatment because of inherent scaling problems and fouling of the membrane. Reverse osmosis, though, has been found to be applicable to a variety of industrial wastes.

Reverse osmosis can remove much smaller molecules than the other membrane processes, including most organic compounds and many inorganic solutes. Reverse osmosis employs a semipermeable membrane and a pressure differential to drive freshwater through the membrane, concentrating the contaminants on the rejection side of the membrane. This phenomenon can best be explained by first examining *osmosis.* In osmosis, if a salt solution is separated from water of lesser salt concentration by a semipermeable membrane, which will allow water to pass through but will reject the passage of salts, the chemical potential generated by the differences in concentration will drive water through the membrane from the purer water side to the concentrated salt side to reduce the potential (see Figure 11.12). The pressure exerted to force the water through the membrane is dependent on the concentration differential between the two sides; it is termed the *osmotic pressure.* As water passes through the membrane, lowering the concentration differential, the osmotic pressure decreases until the concentrations are the same on both sides of the membrane. At that point, the osmotic pressure is zero and no more water passes through the membrane.

This is the reverse of what is wanted in wastewater treatment. Instead of diluting the wastewater with clean water, we want to move water from the wastewater through the membrane, leaving concentrated contaminants behind. This can be accomplished in *reverse osmosis* (RO) by applying a pressure on the wastewater in excess of the osmotic pressure established across the membrane. The higher the pressure exerted above the osmotic pressure, the greater the flux of water that will pass through the membrane. The result is a concentration of the contaminants in a smaller volume of wastewater and the production of clean water. The efficiency of salt rejection depends on the type of membrane selected and the salt concentration gradient, but rejection values of 93 percent to 98 percent are common.

Reverse osmosis membranes typically operate at pressures of 27–41 atm (400–600 psi). Most commonly used membranes are made of cellulose acetate and have a water flux of about 400 L/day $\cdot$ m^2 (10 gal/day $\cdot$ ft^2) at about 40 atm. Thus the flux is very low and large surface areas are required.

Reverse osmosis systems are generally designed based on the rate of water production, the characteristics of the membrane, and the operating conditions. The flux of

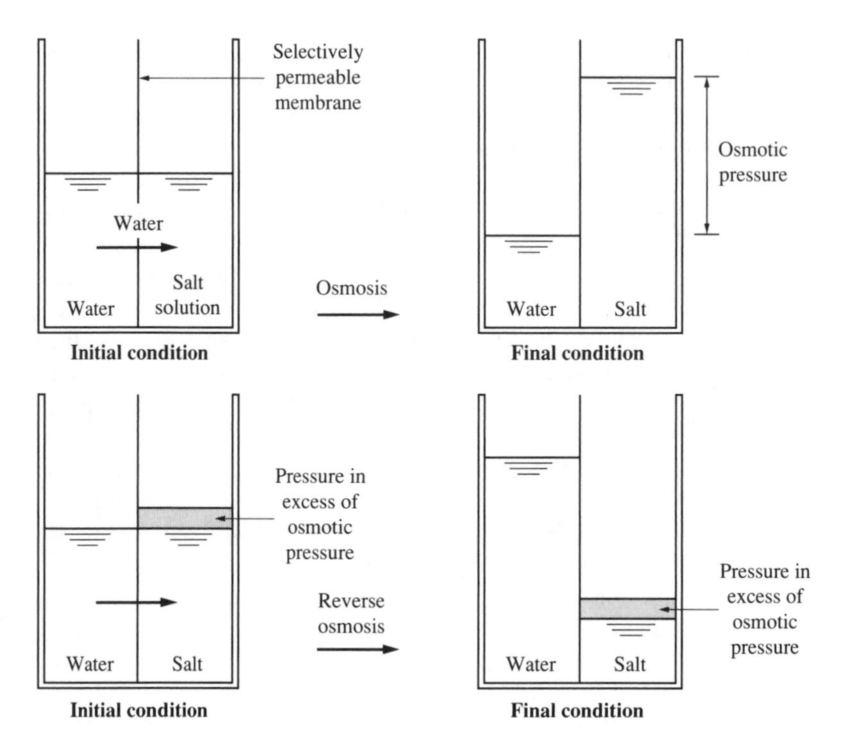

FIGURE 11.12
Osmosis and reverse osmosis mechanisms.

a membrane depends on the membrane characteristics (e.g., membrane thickness and porosity), the pressure applied, and the osmotic pressure of the system, which is governed by the type and concentration of salts in the wastewater:

$$F_W = K(\Delta p - \Delta\pi)$$

where F_W = water flux, $L/m^2 \cdot day$
K = mass transfer coefficient for a unit area of membrane, $L/day \cdot m^2 \cdot kPa$
Δp = pressure differential between feed and product water, kPa
$\Delta\pi$ = osmotic pressure differential between feed and product water, kPa

Wastewater temperature also plays an important role in the water flux. As the temperature increases, the diffusivity will increase and the water viscosity will decrease, leading to an increase in flux. Water production gradually decreases over the life of the membrane because of a gradual densification of the membrane and a decrease in pore sizes, or because of membrane fouling by bacteria or by materials that are difficult to remove during membrane cleaning. Generally, it is best to pretreat the wastewater by filtration to remove large molecules before RO treatment to minimize membrane fouling. Oil and grease are major problems for RO membranes and should definitely be removed.

Because the water flux through an RO membrane is so low, large surface areas are needed for effective treatment. To minimize the space required for these systems, tubular, hollow fiber or spiral-wound membrane modules are used in place of a single sheet of membrane (see Figure 11.13). With tubular membranes, the feed is introduced to the center of the membrane and permeate passes through the membrane and is collected at the outside of the tube. Concentrated effluent exits from the other end of the tube. Hollow fiber membranes are similar to tubular membranes, except that the hollow fibers are much smaller in diameter (usually submillimeter size) and thus provide a greater surface area per unit volume. Flow may be in either direction across the

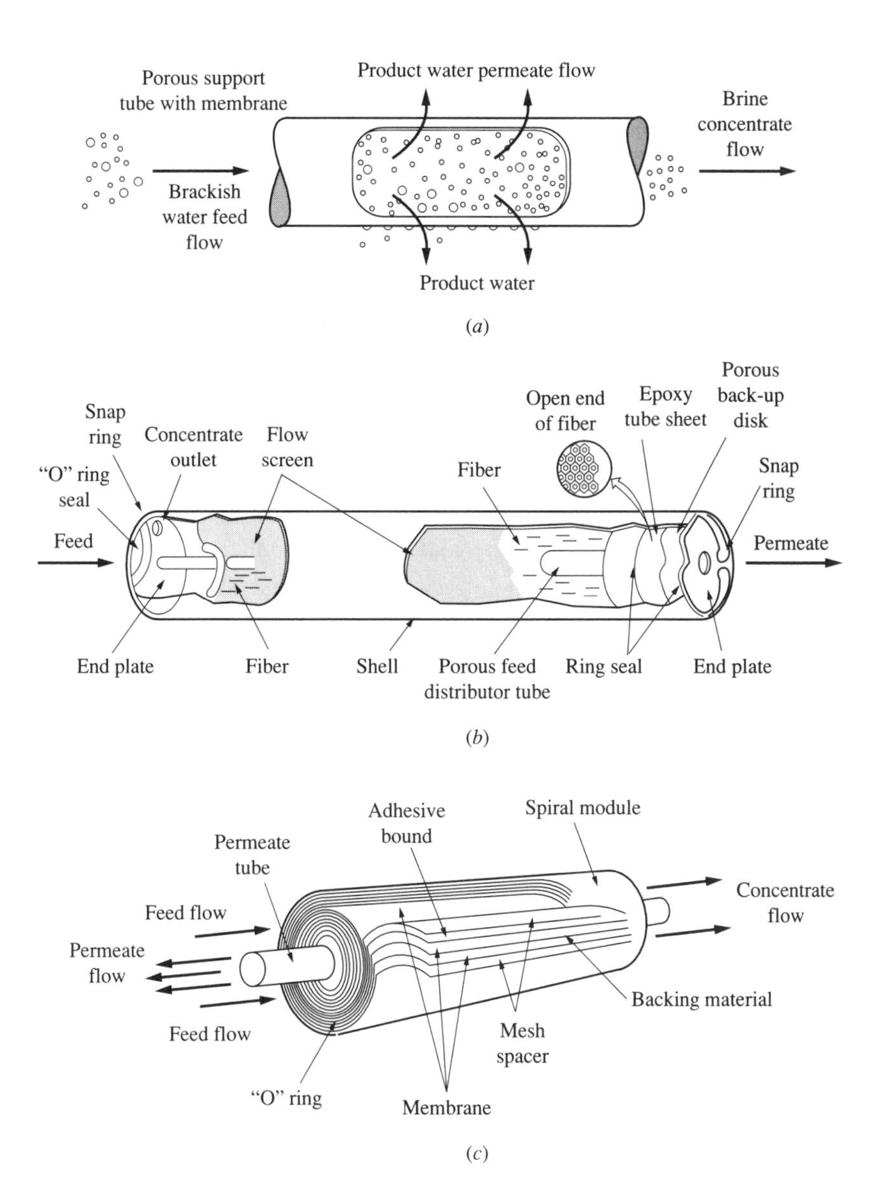

FIGURE 11.13
Schematic diagrams of three types of reverse osmosis membrane systems: (*a*) tubular membrane; (*b*) hollow fiber membrane; (*c*) spiral-wound membrane. (Source: Haas and Vamos, 1995)

TABLE 11.3
Characteristics of different reverse osmosis modules

Module type	Surface area (m^2 membrane/m^3 equipment)	Water flux, L/day · m^2	Water produced, L/s · m^3 of equipment
Tubular	66	1,300	1
Hollow fiber (cellulose acetate)	8,200	400	39
Hollow fiber (nylon)	17,700	41	8
Spiral wound	820	1,300	12

Source: Adapted from Reynolds and Richards, 1996.

membrane, depending on the unit selected. Spiral-wound membranes consist of sheets of membrane material backed by spacer material and rolled up jelly-roll fashion. Water flows across the membrane and is collected in the spacer layer, from which it flows to a central collecting tube. By packing the membranes close together in the modules, a large membrane surface area can be accommodated within a small RO module volume (see Table 11.3).

Reverse osmosis has been applied in industry to produce demineralized process waters and to treat metal plating wastewaters. In addition to producing a high-quality effluent that can often be reused as rinse water in the metal plating operation, the recovered metals in the concentrated reject solution can frequently be returned to the plating bath and be reused, provided that the rinse waters were segregated at the source.

Example 11.6. A metal plating facility wants to install a reverse osmosis system to treat 150,000 L/day of rinse waters containing 20 mg/L of chromium. The effluent requirement is a maximum chromium concentration of 500 μg/L. The mass transfer coefficient for the membrane selected is 0.25 L/day · m^2 · kPa. The manufacturer's recommended pressure differential between the feed and product is 2500 kPa. The osmotic pressure differential between water with 20 mg/L Cr and 50 μg/L Cr is determined to be 280 kPa. The membrane can be purchased in units that provide 800 m^2 surface area per cubic meter of module volume. Can reverse osmosis provide the degree of treatment required? If so, what membrane area is required? What volume will the required modules occupy?

Solution. The required removal efficiency is

$$\text{Fraction removed} = \frac{20\text{ mg/L} - 0.5\text{ mg/L}}{20\text{ mg/L}} = 0.975$$

A chromium removal efficiency of 97.5 percent is required. This is high, but it is within the range for reverse osmosis. The water flux through the selected membrane can be computed as follows:

$$
\begin{aligned}
F_w &= K(\Delta p - \Delta \pi) \\
&= (0.25 \text{ L/day} \cdot \text{m}^2 \cdot \text{kPa})(2500 \text{ kPa} - 280 \text{ kPa}) \\
&= 555 \text{ L/day} \cdot \text{m}^2
\end{aligned}
$$

The membrane surface area needed to treat 150,000 L/d at a flux of 555 L/day $\cdot$ m^2 is

$$\text{Area} = \frac{150{,}000 \text{ L/day}}{555 \text{ L/day} \cdot \text{m}^2} = 270 \text{ m}^2$$

Assuming a membrane packing density of 800 m^2/m^3 in the RO modules, the required volume of the RO module is

$$\text{Volume} = \frac{270 \text{ m}^2}{800 \text{ m}^2/\text{m}^3} = 0.34 \text{ m}^3$$

Even though the membrane surface area required is quite large (270 m^2), the actual volume this RO module occupies is quite small (0.34 m^3) because of the way the membranes are constructed.

11.2.3 Biological Waste Treatment

Except possibly for chemical oxidation-reduction, the previously described processes do not actually destroy the wastes; they only change the volume occupied by the waste or the medium in which it is located. The only destructive technologies for industrial wastes that are commonly used are biodegradation and thermal destruction (incineration). In this section, we briefly discuss biological treatment processes for industrial wastes. The topic is too extensive to go into much detail here. For more information on biological treatment of industrial wastes, the reader is directed to the texts by Metcalf & Eddy (1991), Haas and Vamos (1995), Wentz (1995), and Reynolds and Richards (1996).

Industrial wastewaters containing organic materials can be treated biologically either as the terminal treatment step before discharge to a receiving stream or as a pretreatment step before discharge to a publicly owned treatment works (POTW) for final treatment. Most of the treatment is done by bacteria, which use the organics as substrate for energy and as a source of carbon for new bacterial cell growth. Biological treatment processes mimic the natural biodegradation of organics in the environment but are designed to speed the process so that degradation that may take days or weeks in nature can be accomplished in hours in the treatment plant. Biological treatment is the principal form of treatment for municipal wastewaters. Many of the technologies used there can also be used with industrial wastewaters. Under the right conditions, even toxic organic wastes can be biodegraded to harmless end products.

Microorganisms require a variety of nutrients for growth. These include the major nutrients carbon, hydrogen, oxygen, and nitrogen; the minor nutrients phosphorus, potassium, sulfur, and magnesium; and many trace nutrients. These nutrients are used in the synthesis of new microbial cells or to maintain those that exist. A typical empirical composition of biomass is $C_5H_7O_2NP_{0.2}$. These nutrients must be present in the wastewater in the proper ratios or microbial growth will be retarded. This is normally not a concern with municipal wastewaters because of the commingling of many types of wastes, but it may be in an industrial wastewater. Addition of nutrients may be required for effective biological treatment.

In addition to nutrients for microbial growth, microorganisms also need a source of energy. This may be chemical or photochemical. For chemotrophs, energy is obtained by redox reactions in which electrons (and energy) are transferred from the organic molecule to an electron acceptor. The electron acceptor can be oxygen, in which case the process is considered to be *aerobic;* to an oxygen-containing molecule such as nitrate or sulfate, in which case it is deemed to be an *anoxic* or *facultative* process; or to another more oxidized organic compound, in which case it is called an *anaerobic* process. Much more energy can be obtained by microbes from aerobic processes than from anaerobic processes:

Aerobic	$AH_2 + O_2 \rightarrow CO_2 + H_2O + \text{energy}$	Decreasing energy yield ↓
↓	$AH_2 + NO_3^- \rightarrow N_2 + H_2O + \text{energy}$	
(Facultative)	$AH_2 + SO_4^{2-} \rightarrow H_2S + H_2O + \text{energy}$	
↓	$AH_2 + CO_2 \rightarrow CH_4 + H_2O + \text{energy}$	
Anaerobic	$AH_2 + B \rightarrow BH_2 + A + \text{energy}$	

Anaerobic processes have a low energy yield because some of the organics must serve as the electron acceptor and complete oxidation of the organics cannot occur. The end products (typically methane, CH_4, and organic acids) still contain much of the energy in the original organics:

Aerobic metabolism: $\text{Organics} + O_2 \rightarrow CO_2 + H_2O + \text{energy}$

Anaerobic metabolism: $\text{Organics} \rightarrow \text{Organic acids} + CH_4 + CO_2 + \text{energy}$

Thus microbial growth per unit of substrate in the waste, or *yield,* is lower in processes operating under anaerobic than under aerobic conditions (see Figure 11.14).

Most municipal wastewater treatment systems utilize aerobic biodegradation processes because they are more efficient in degradation of the organic wastes, perform the wastewater renovation more quickly, and are less costly than anaerobic processes. This is also the case for most industrial wastewaters. However, some organic compounds found in industrial wastewaters do not respond well to aerobic degradation and require anaerobic processes (e.g., chlorinated aliphatics such as trichlorethylene), at least for the initial biological treatment. In addition, anaerobic systems can be used to treat industrial wastes that contain high concentrations of oxygen-consuming materials (i.e., high BOD_5) that would make aeration difficult.

Aerobic oxidation kinetics is generally considered to be first order with respect to the substrate concentration in contact with the bacteria:

$$-\frac{dS}{dt} = kS$$

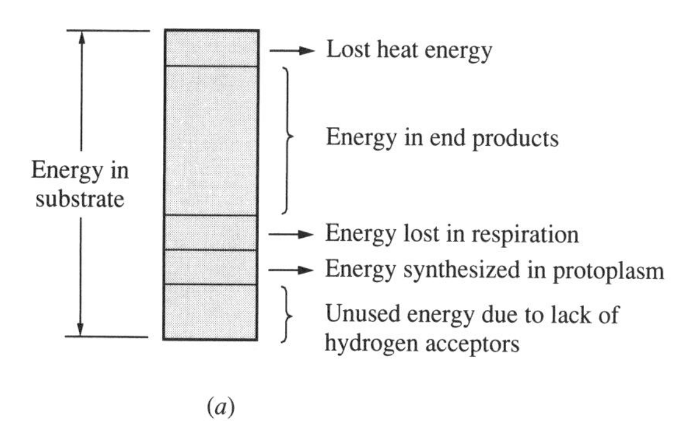

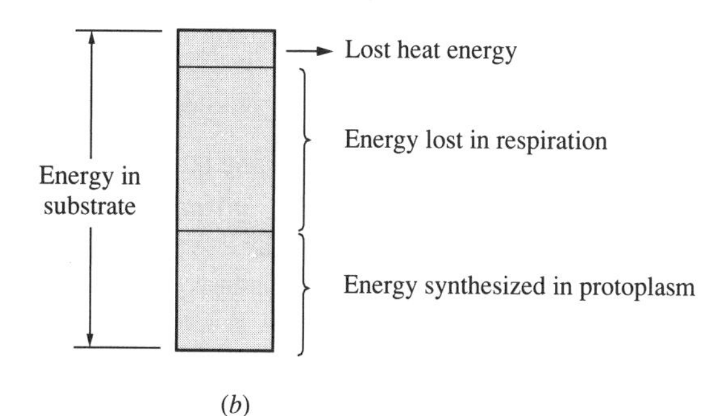

FIGURE 11.14
Energy conversion during aerobic and anaerobic metabolism. (*a*) Anaerobic metabolism; (*b*) aerobic metabolism.

where dS/dt = rate of substrate removal
k = reaction rate constant
S = substrate concentration

Rearranging, we get

$$\frac{S}{S_0} = e^{-kt}$$

where $S = S_0$ at $t = 0$

Similarly, the growth rate of the bacteria can be related to the cell concentration as a first-order process:

$$\frac{dX}{dt} = \mu X$$

where dX/dt = microbial growth rate
X = microorganism population
μ = specific growth rate of the microbes

Growth will be logarithmic, as described above, when substrate for the bacteria is abundant. However, the growth rate is not infinite because there is a finite minimum time needed to replicate all of the enzymes and protoplasm required to produce a new cell. Monod (1949) determined that the growth rate is actually a function of the concentration of the limiting substrate or nutrient (see Figure 11.15). The Monod equation, now used in most biological treatment designs, is

$$\mu = \mu_{max}\frac{S}{K_s + S}$$

where μ_{max} = the maximum specific growth rate
K_s = half-velocity constant

K_s is equal to the substrate concentration at which the specific growth rate is equal to one-half of the maximum specific growth rate. Thus the cell growth rate can be expressed in terms of the substrate concentration:

$$\frac{dX}{dt} = \mu_{max}\frac{S}{K_s + S}X$$

Thus the growth rate is a function of both the substrate concentration and the microbial population, as well as of the maximum specific growth rate, which is governed by the types of microorganisms present and the organics being biodegraded.

There are a large number of biological wastewater treatment systems, which can generally be classified as *suspended growth systems* and *attached growth systems*. Suspended growth systems consist of reactors containing wastewater and bacteria in which the two are intermixed. The bacteria, which usually grow in aggregates known as flocs, are suspended in the wastewater so that there is direct contact between the bacteria and the soluble organics in the wastewater. Examples of suspended growth systems are the activated sludge process and aerobic stabilization ponds. With fixed-film systems, the microorganisms are attached to a stratum or some solid object and the

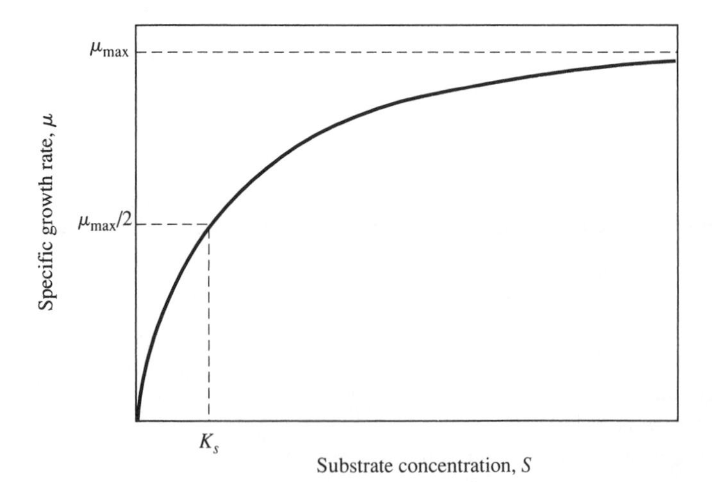

FIGURE 11.15
The effect of substrate concentration on specific growth rate using the Monod equation.

wastewater flows past them. The biomass support surface may be stones or plastic sheets that are essentially fixed in place, or sand or some other particulate material that is fluidized in the wastewater. Examples of fixed-film systems are trickling filters, rotating biological contactors, and fluidized bed reactors.

ACTIVATED SLUDGE SYSTEMS. Activated sludge systems are designed to maintain intimate contact among the wastewater, a large population of bacteria, and oxygen. The large population of bacteria will allow for rapid biodegradation of the organics in the wastewater. An air supply is needed to ensure that there is sufficient oxygen to keep the system aerobic under the high-growth–high-oxygen utilization conditions. Finally, adequate mixing must be supplied to ensure good contact between the bacteria and the wastewater organics.

Figure 11.16 is a schematic of the activated sludge system. In most cases, the wastewater must be pre-settled to remove settleable solids before entering the aeration tank. The bacteria in the aeration tank transform the wastewater organics into new bacterial cells. The microbes usually stick together, forming small floc particles. The microbial mass can be removed from the wastewater in a secondary clarifier, leaving a clarified wastewater with little remaining organics. The high bacterial population necessary in the aeration tank is achieved by recycling biomass removed from the treated wastewater in the final clarifier. The primary settled wastewater usually contains very little bacteria (only about 10 mg/L or less), but bacterial populations of 2000–3000 mg/L are commonly used in the aeration tank. Thus treatment will be much more rapid in the aeration tank than if the organics were degraded in the natural environment. By the time the biomass is returned to the head end of the aeration tank, it has biodegraded essentially all of the available substrate from the wastewater and it is hungry for more substrate. The microorganisms are "activated" for rapid uptake of new substrate, thus the term *activated sludge*.

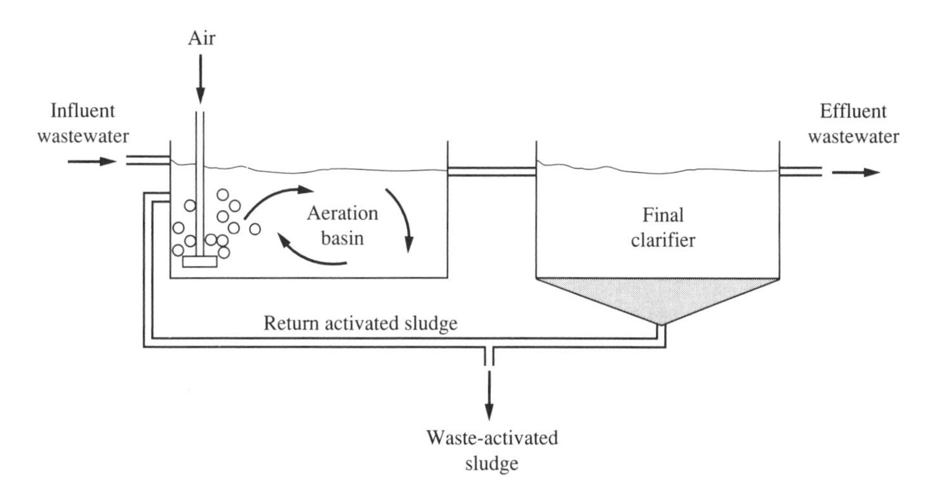

FIGURE 11.16
Schematic diagram of an activated sludge system.

The activated sludge process should be operated with a constant biomass population (termed *mixed liquor suspended solids,* or MLSS) in the aeration tank. Because the microorganisms are growing due to biodegradation of the waste organics, the population will increase unless some of the biomass is continually wasted. This *waste-activated sludge* must be processed and disposed of safely, an expensive and difficult operation. The amount of sludge that must be wasted per day is equal to the amount of biomass produced per day; it can be determined based on the cell yield, or mass of new cells produced per unit of substrate biodegraded. For many aerobic systems, the yield, *Y,* is equal to 0.5 mg biomass/mg BOD_5 biodegraded. This can result in substantial quantities of waste sludge.

A number of variations of the activated sludge process are available, including *step aeration,* in which the wastewater is introduced at intermediate points in a plug flow aeration basin to provide more uniform BOD loading; *contact stabilization,* in which biosorption of organics and their subsequent biodegradation are carried out in separate tanks; the *oxidation ditch,* which uses a closed-loop channel through which the wastewater flows; *high-rate activated sludge,* which uses a high food-to-microorganism (F/M) ratio and a short detention time; and the *extended aeration* process, which provides long aeration times and a low F/M ratio to minimize sludge production. Systems are also available that use pure oxygen, but they are very expensive. Table 11.4 compares the design criteria for these processes.

Another activated sludge variation for industrial waste treatment involves addition of powdered activated carbon (PAC) to the aeration tank. This is known as the

TABLE 11.4
Design parameters for activated sludge treatment processes

Type of process	Mean cell residence time, days	F/M, kg BOD_5/kg MLSS	Loading, kg BOD_5/$m^3 \cdot$ day	Hydraulic retention time, h	MLSS, mg/L	Recycle ratio
Conventional	5–15	0.2–0.4	0.3–0.6	4–8	1,500–3,000	0.25–1.0
Step aeration	5–15	0.2–0.4	0.6–1.0	3–5	2,000–3,500	0.25–0.75
Completely mixed	5–30	0.1–0.6	0.8–2.0	3–6	2,500–4,000	0.25–1.5
Contact stabilization	5–15	0.2–0.6	1.0–1.2	0.5 (contact) 3–6 (stabil.)	1,000–3,000 4,000–10,000	0.50–1.5
High-rate	5–10	0.4–1.5	1.6–16	2–4	4,000–10,000	1.0–5.0
Extended aeration	20–30	0.05–0.15	0.16–0.4	18–36	3,000–6,000	0.75–1.5
Pure oxygen	8–20	0.25–1.0	1.6–3.2	1–3	3,000–8,000	0.25–0.5

Key: Mean cell residence time = average time the bacteria remain in the activated sludge system
F/M = food-to-microorganism ratio
Loading = amount of wastewater BOD supplied per day per unit of bacteria in the system
Hydraulic retention time = average time the wastewater spends in the aeration tank
MLSS = mixed liquor suspended solids concentration; the concentration of biomass in the aeration tank
Recycle ratio = ratio of return sludge flow rate to the aeration tank influent flow rate

Source: Adapted from Metcalf & Eddy, 1991.

PACT process. The PAC becomes enmeshed in the activated sludge floc and adsorbs from the wastewater recalcitrant organics that might otherwise pass untreated through a conventional activated sludge process. In some cases, these adsorbed organics are eventually biodegraded by bacteria growing on the activated carbon surfaces.

BIOFILM SYSTEMS. Fixed-film systems were the first engineered treatment processes developed for wastewater treatment. These original systems consisted of a bed of rocks over which the wastewater was sprinkled. As the wastewater flowed down through the rock bed, bacteria became attached to the rock surfaces and began extracting organics from the passing wastewater. These units, called *trickling filters,* are still in common use today, although their design has been considerably altered. Several types of fixed-film processes are currently used for treating both municipal and industrial wastewaters, including trickling filters, biotowers, rotating biological contactors, and fluidized bed reactors.

All biofilm systems function in essentially the same way. As wastewater passes a solid support surface, bacteria become attached to the surface and grow. As the bacterial colonies grow, the bacteria produce a polysaccharide slime layer around themselves. Thus a biological slime layer, or *biofilm,* is formed. As wastewater passes the biofilm, organic matter and nutrients diffuse into the biofilm, where they are used by the microorganisms (see Figure 11.17). Carbon dioxide and other metabolic by-products

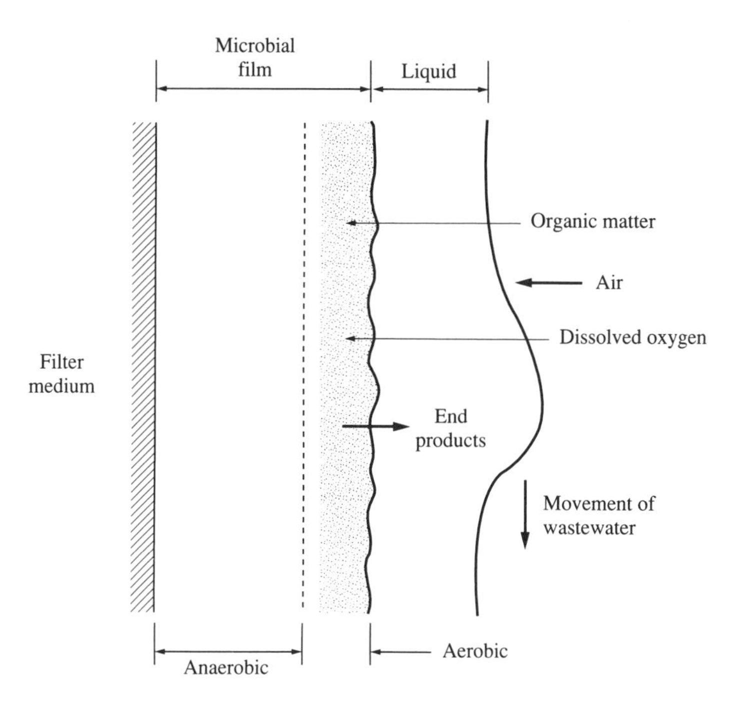

FIGURE 11.17
Schematic diagram showing the operation of a biofilm in a fixed-film wastewater treatment system.

from the microbes diffuse back out into the wastewater and are carried away. The wastewater typically moves through the biofilm reactor in a matter of minutes, but the absorbed organics and nutrients are retained for hours in the biofilm, where they can be metabolized. As the wastewater passes down through a fixed bed of support media, more and more material is removed. Thus there is a gradation of concentrations in the trickling wastewater from the top of the filter to the bottom. This allows for a variation in the types of organisms that can be supported in the biofilter from top to bottom. For example, heterotrophic organisms that can tolerate high organic concentrations can grow at the top of the filter, while nitrifying organisms can grow at the bottom, where most of the organics in the wastewater have been depleted.

In an aerobic biofilm, most of the biofilm will be oxygenated, but the oxygen concentration decreases with depth as it is used by the microorganisms in the biofilm. As the biofilm grows, the thickness of the biofilm increases. Typical biofilm thicknesses range from 100 μm to 2 mm. In a thick biofilm, all of the oxygen may be depleted at some point in the biofilm before the support surface is reached, creating an anoxic or anaerobic layer at the base of the biofilm. At some point, the biofilm may fall, or *slough,* off the support surface when cells at the base of the biofilm starve or because the water flowing by the surface of the biofilm creates shear forces strong enough to pull the biofilm off of the support. This sloughed biofilm must be removed from the wastewater in a final clarifier.

The term "trickling filter" is a misnomer because little or no filtration actually takes place; contaminants are removed by diffusion into the biofilm, where they are biodegraded. A trickling filter consists of a bed of rocks over which the wastewater is sprinkled. Wastewater distribution to the filter bed is usually accomplished using a rotary distributor (see Figure 11.18) that gently distributes the flow over the bed for short periods, then allowing that water to trickle down before the next pass over that

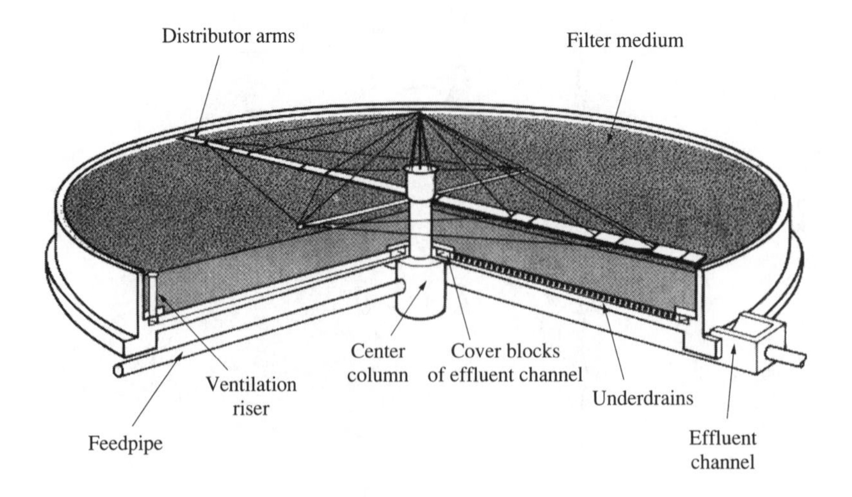

FIGURE 11.18
Cutaway view of a trickling filter with a rotary distributor.

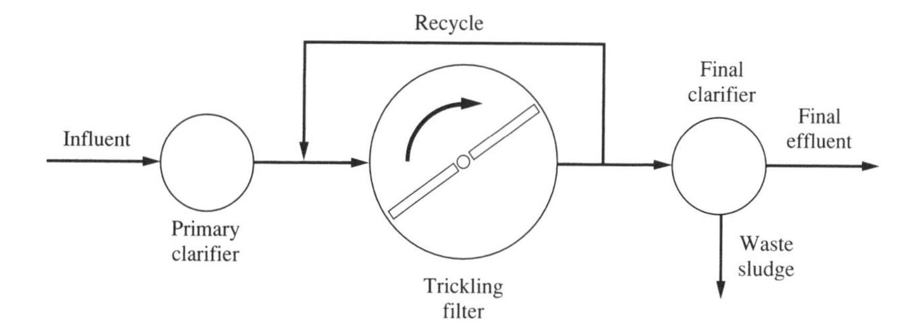

FIGURE 11.19
Flow diagram for a trickling filter system.

portion of the bed. The wastewater trickles over the media in thin films, allowing air for oxygenation of the biofilm to flow through the empty pores between the media. The effluent from the trickling filter passes through a clarifier to remove any sloughed biofilm or other suspended solids before being discharged. Clarified wastewater may be recirculated to the top of the trickling filter to provide a more even flow rate to the filter, to dilute high-strength influent wastes, or to provide additional contact time for removal of recalcitrant organics in the waste. Figure 11.19 is a schematic of a typical trickling filter.

There are several variations on the standard trickling filter design. Typical design criteria for these systems are listed in Table 11.5. *Standard* or *low-rate trickling filters*

TABLE 11.5
Typical design criteria for trickling filters

Item	Low-rate filter	High-rate filter	Superrate filter
Hydraulic loading, $m^3/m^2 \cdot day$	1–4	10–40	40–200
Organic loading, kg $BOD_5/m^3 \cdot day$	0.08–0.32	0.32–1.0	0.8–6.0
Depth, m	1.5–3.0	1.0–2.0	4.5–12.0
Recirculation ratio	0	1–3	1–4
Filter media	Rock, slag, etc.	Rock, slag, synthetics	
Filter flies	Many	Few, larvae are washed away	Few or none
Sloughing	Intermittent	Continuous	Continuous
Dosing intervals	< 5 min	< 15 s	Continuous
Effluent	Usually fully nitrified	Nitrified at low loadings	Nitrified at low loadings

Source: Adapted from Metcalf & Eddy, 1991.

were developed first. They are usually single-stage rock media units with loading rates of 1–4 m^3 wastewater per square meter of filter cross-sectional area per day ($m^3/m^2 \cdot$ day) and a bed depth of 1–3 m. They can produce a high-quality effluent, but they occupy a large land area because of their low loading rate. *High-rate trickling filters* are also rock media filters but use much higher loading rates (10–40 $m^3/m^2 \cdot$ day). They may be single-stage or two-stage systems. To achieve good effluent quality, they must employ recirculation, with a recirculation ratio of 1:3. *Superrate trickling filters* use synthetic plastic media, either as modules or random packed, to increase the available surface area per unit volume of biofilter for biofilm growth (see Figure 11.20). These media have specific surface areas two to five times greater than does rock; they are also much lighter than rock so that they can be stacked higher than can rock. Typical superrate biotowers have loading rates of 40–200 $m^3/m^2 \cdot$ day and plastic media depths of 5–10 m.

A second type of fixed-film treatment system is known as a *rotating biological contactor*. Here, the biofilm growth occurs on disks 2–4 m in diameter closely spaced on a rotating shaft that slowly rotates the disks (1–2 rpm) through a tank containing wastewater (see Figure 11.21). When the disks rotate out of the wastewater, they come into contact with air, allowing the biofilm to become oxygenated. The biosorption and biodegradation that occurs in the biofilm is the same as previously described for trickling filters. The plastic media design of the disks allows for a large specific surface area, thus reducing the required size for the treatment plant. These units are commonly purchased by industries as a complete package treatment plant.

Fixed-film systems can also be operated under anaerobic conditions when the waste strength is too high to maintain aerobic conditions or when anaerobic pathways are needed to biodegrade the waste. In this case, the medium is kept submerged in wastewater. The flow may either be in an upflow or a downflow mode. Recently, *upflow anaerobic biofilters* in which the flow rate is sufficient to expand or fluidize a biofilm-coated sand or granular activated carbon medium have become popular for treating refractory or toxic organics in industrial wastes.

A third type of biofilm system is the *fluidized bed reactor*. In this system, biofilm is attached to a granular support medium, such as sand, which is housed in a vertical column. Wastewater flow enters from the bottom of the column and moves upward at

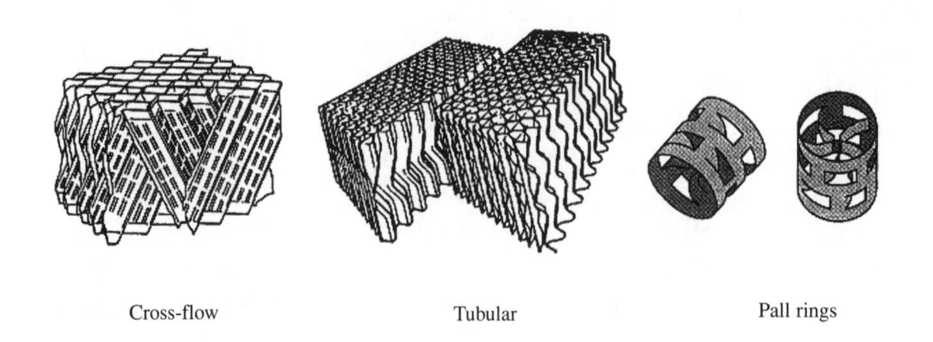

FIGURE 11.20
Schematic diagrams of modular and random packed media used in fixed-film treatment systems. (Source: Bordacs and Young, 1998)

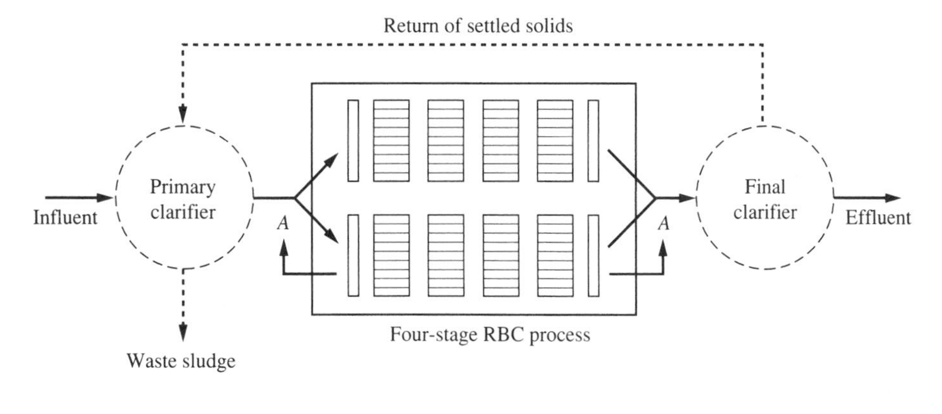

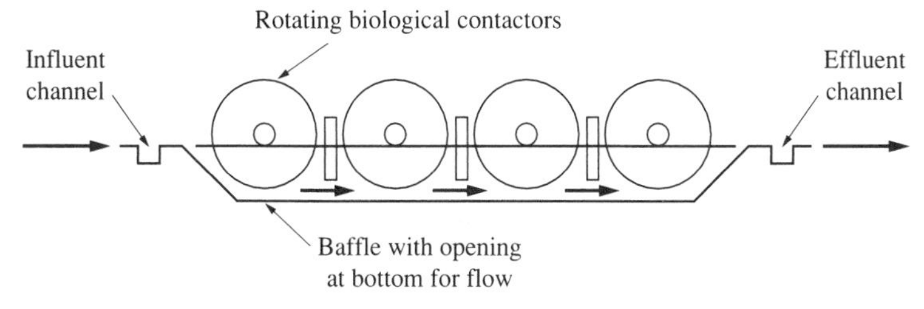

Section *A-A*

FIGURE 11.21
Flow diagram for a rotating biological contactor (RBC) system. (Source: Viessman and Hammer, 1998)

a rate sufficient to slightly lift, or fluidize, the biofilm-coated particles, so that there is good contact between the wastewater constituents and the microbes in the biofilm. Treated effluent exits at the top of the reactor, while the fixed biomass remains inside. These systems provide a large amount of surface area for biofilm attachment on the support medium per unit volume of reactor. Thus small reactors can be built to provide a high degree of treatment. Fluidized bed reactors are commonly used for industrial wastewater treatment.

STABILIZATION PONDS AND LAGOONS. Stabilization ponds and lagoons are commonly used to treat industrial wastes. They are relatively inexpensive to construct and to operate, and they are adaptable to a wide variety of waste types. Ponds and lagoons are usually earthen basins which retain the wastewater for a period of time necessary to achieve the desired degree of treatment. A liner may or may not be required to prevent seepage of wastewater into underlying groundwater. Stabilization ponds can be divided into several categories:

- *Aerobic ponds:* shallow ponds (less than 1 m deep) where dissolved oxygen, mainly due to the action of photosynthesis, is present throughout the depth of the pond.
- *Facultative ponds:* ponds that are 1–2.5 m deep, which have an aerobic upper layer due to photosynthesis or surface reaeration, a facultative middle zone, and an anaerobic lower zone.

TABLE 11.6
Typical design data for industrial wastewater lagoon systems

	Aerobic and facultative lagoons				Anaerobic lagoons			
Industry	Ave. area, m^2	Detention time, days	Loading, $g/m^2 \cdot day$	BOD removal, %	Ave. area, m^2	Detention time, days	Loading, $g/m^2 \cdot day$	BOD removal, %
Meat and poultry	5,260	7	8.1	80	4,050	16	141.2	80
Canning	27,900	38	15.6	98	10,120	15	44.0	51
Chemical	125,460	10	17.6	87	570	65	6.1	89
Paper	339,950	30	11.8	80	287,340	18	38.9	50
Textile	12,550	14	18.5	45	8,900	4	160.6	44
Sugar	80,940	2	9.6	67	141,650	50	26.9	61

Source: Adapted from Eckenfelder, 1989.

- *Anaerobic lagoons:* deep ponds that receive high organic loadings and are anaerobic throughout their depth.
- *Aerated lagoons:* ponds that are oxygenated through the use of surface aerators or diffused aeration systems.

Lagoons are widely used to treat industrial wastes because they are less expensive than activated sludge systems. Typical design data for industrial wastewater lagoons are shown in Table 11.6.

LAND APPLICATION. An alternative to treatment of industrial wastewater using reactors is to use land application, applying the wastewater to land by one of several conventional irrigation techniques. Treatment is provided by natural processes as the wastewater moves through the plant and soil system. Some of the wastewater is lost to the atmosphere through evapotranspiration, but the remainder reenters the groundwater hydrologic system after renovation by filtration and biodegradation. In-depth discussions of the design of land treatment processes can be found elsewhere (Metcalf & Eddy, 1991; Reed, Middlebrooks, and Crites, 1988; Reynolds and Richards, 1996; U.S. EPA, 1973; 1988).

There are essentially four types of land application systems (see Figure 11.22). The *slow-rate process* is the most widely used land treatment system and is achieved by applying the wastewater to vegetated land as irrigation water for the crops. The wastewater can be applied by sprinklers or by flooding the land using ridge-and-furrow techniques. Some of the applied wastewater is evapotranspirated and some is used by the plants, while the remainder percolates through the soil, where soil bacteria decompose the organic matter. The system is designed to minimize surface runoff. *Rapid-infiltration systems* are designed to apply the wastewater to relatively permeable soil within a basin at a much higher rate than in the slow-rate system. Plants may be

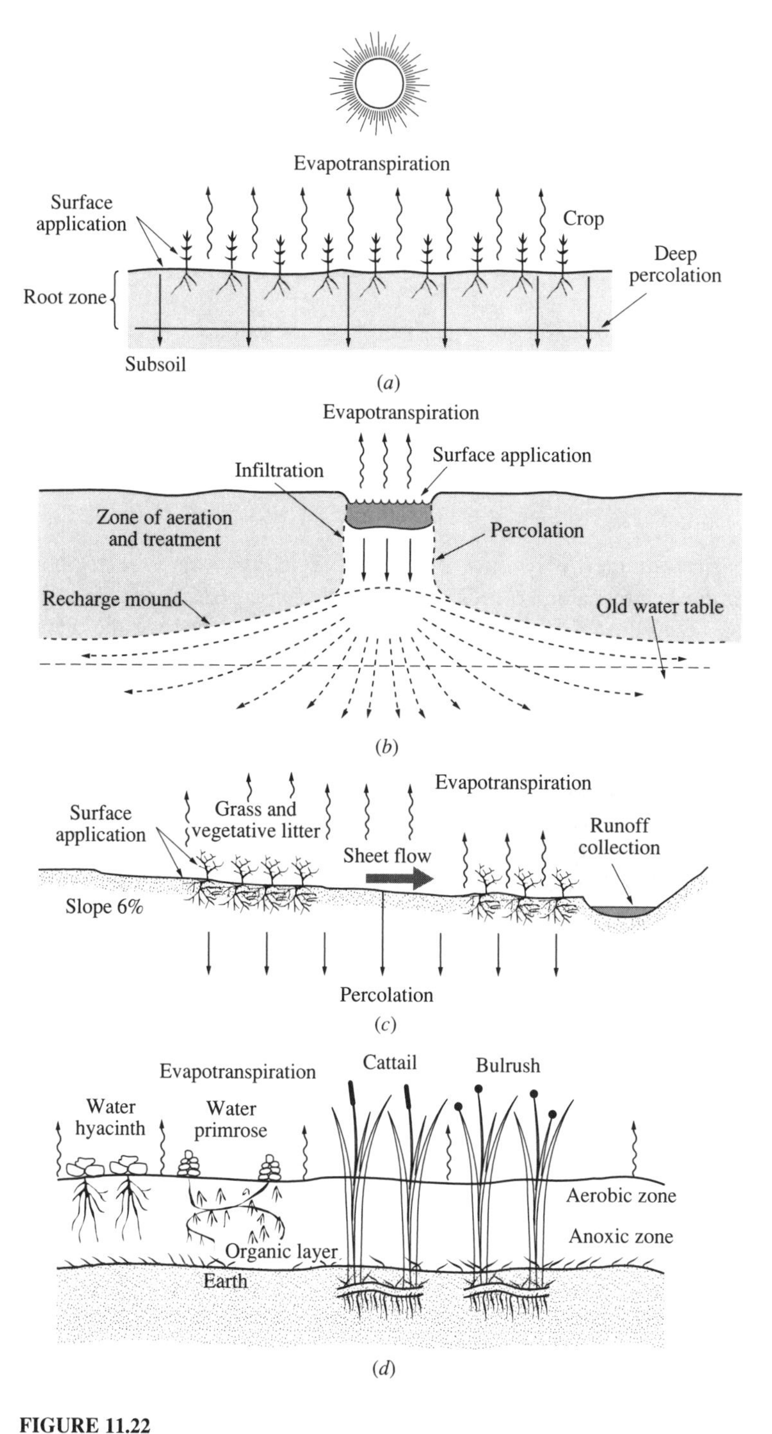

FIGURE 11.22
Four land treatment systems: (*a*) slow-rate irrigation; (*b*) rapid infiltration; (*c*) overland flow; (*d*) wetland. (Adapted from Sincero and Sincero, 1996)

TABLE 11.7
Comparative characteristics of land treatment systems

Feature	Slow-rate irrigation	Rapid infiltration	Overland flow	Artificial wetland
Hydraulic loading rate, cm/day	0.2–1.5	1.5–30	0.6–3.6	1.4–4.7
Land required for 1000 m^3/day (1000 × m^2)	63.4–396	3.2–63.4	26.4–159	21.3–71.4
Soil type	Loamy sand to clay	Sands	Clay to clay loam	Loamy sand to clay
Soil permeability	Moderately slow to moderately rapid	Rapid	Slow	Slow

present in the basin, but they play only a minor role in wastewater renovation. Treatment is accomplished by natural processes in the soil. These systems require much smaller land areas because of the higher application rates, and they have the added advantage of recharging the groundwater in areas where water is scarce. The *overland flow method* is designed to minimize percolation into the soil. Treatment is provided by the plants and the microbial biomass at the soil surface as the wastewater flows down a sloping field. The renovated wastewater is collected at the bottom of the slope and discharged into a surface water body. The fourth land application system is *wetland treatment,* also known as *aquatic treatment.* Wastewater is applied to natural and artificial wetlands, and as it flows through these systems, renovation is accomplished by sedimentation and biological activity by the plants and microorganisms present. Typical design characteristics for the first three systems are listed in Table 11.7.

Land application systems are commonly used to treat some types of industrial wastes. They have found wide acceptance for cannery, pulp and paper, dairy, and tannery wastes. Because of the large land areas required and the potential for odors from some wastes, they are usually applicable only in rural areas.

11.2.4 Sludge Management

One of the most vexing problems associated with on-site industrial waste treatment, and a major reason why many industries opt to send their wastes to a POTW for treatment, is management of the sludge that results from almost all industrial wastewater treatment processes. These sludges are generated by solids-liquid separation processes (sedimentation, filtration), by chemical treatment processes (precipitation), or from biological treatment systems (waste biological sludges). Many of these sludges are high-volume, gelatinous, water-holding materials that are difficult to process further. Most of the sludge is still water, often as much as 90 percent to 99 percent. Landfilling of this large volume of material that contains only a small percentage of contaminants is generally not economically viable. Thus additional processing is needed to dewater these sludges to a more manageable volume. Dewatering is often a difficult and expensive process, though.

Organic sludges often still contain large amounts of biodegradable organic material; left untreated, this will result in highly odorous conditions and oxygen depletion in the environment. The biodegradable organic content can be reduced by use of aerobic or anaerobic treatment systems or by composting. These stabilized sludges, as well as chemical sludges, still require dewatering before they can be disposed of. Dewatering techniques include vacuum filtration, filter presses, drying beds, centrifugation, lagooning, and thermal processing. The resulting dewatered sludge may be landfilled or, if primarily organic, incinerated.

11.3 AIR POLLUTION CONTROL

Many industrial processes create air contaminants that cannot be discharged into the atmosphere. The easiest way to control air pollution is to eliminate it at the source. The techniques described in Chapters 5 and 9 should greatly minimize the amount of air pollutants generated. There will probably still be some air contaminants remaining, however, that must be removed or destroyed. Industries do not have the option of sending these to a municipal treatment facility as they do with liquid or solid wastes. These emissions must be controlled at the industrial facility. This chapter provides only a brief overview of air pollution control options. More complete discussions can be found elsewhere (Buonicore and Davis, 1992; Wark, Warner, and Davis, 1998).

Typically, air pollution control devices are divided into those that are used to control particulates and those that are intended for control of gaseous pollutants. Some processes may be capable of removing both types of contaminants.

11.3.1 Particulate Control

SETTLING CHAMBER. Several devices are available for control of particulate matter. The simplest is a settling chamber, which relies simply on gravitational settling of large particulates. The chamber consists of an increase in the size of the exhaust flue cross-sectional dimensions so that the air flow velocity is reduced, allowing larger particles to settle. Stokes' law can be used to determine the settling velocity of a particle:

$$v = \sqrt{\frac{4}{3} \cdot g \cdot \frac{\rho_p D}{C_d \rho_a}}$$

where v = terminal settling velocity
g = gravitational acceleration
C_d = drag coefficient
ρ_p = density of the particulate
ρ_a = density of air
D = particulate diameter

If we set L equal to the chamber length, W to the width, and Q to the air flow rate, v is equal to Q/WL. Substituting this into the equation above, we can solve for the particulate size that will be completely removed in the settling chamber:

$$D = \frac{0.75 C_d Q^2 \rho_a}{g L^2 W^2 \rho_p}$$

Particles of smaller diameter will also be removed, in proportion to the ratio of their settling velocity to that of the terminal settling velocity of the particle defined above. The practical lower particle size limit for settling chambers is only about 50–100 μm, about the diameter of a human hair. For smaller particles, other devices are needed.

CYCLONE. For particle diameters down to about 10 μm, the collection device of choice is usually a cyclone. This is a simple, economical unit with no moving parts, that relies on inertial effects for particulate removal. Particulate-laden air is sent into a conical cylinder, where it is forced into a spiral flow path and accelerates (see Figure 11.23*a*). The centrifugal force imparted on the particulates forces them to move to the wall of the chamber, where they then slide down to the bottom of the cone and are removed. The clean air exits up through the center of the cyclone.

BAGHOUSE FILTER. When particles are smaller than 10 μm or a higher collection efficiency is required, a baghouse filter can be used. These are widely used in industry. A baghouse filter is similar to a conventional home vacuum cleaner. It consists of a chamber housing natural or synthetic cloth bags through which the dirty air is pumped (Figure 11.23*b*). Particulates larger than the openings between the fibers are filtered out; smaller particles are removed by interception on the fibers themselves and by electrostatic attraction between the particles and the fibers.

Once particles begin to accumulate, the openings become smaller and the importance of sieving increases. The cleaned air passes through the bag fabric and exits through an opening in the baghouse chamber. Particulates collect on the inside surfaces of the bags. Eventually, the head loss through the fabric will become excessive, necessitating removal of the particulates. The bags are periodically shaken to remove the accumulated dust, or the bag is isolated and air is blown into the bag from outside to dislodge particles. The released dust falls into a hopper below.

Baghouse filters are very efficient and can remove even submicrometer size particles. However, they cannot be used for wet air streams, because the particulates may cake on the filter surface, or the gases may be corrosive to the filter fabric. They also cannot be used for treatment of high-temperature gas streams (> 90–100°C for cotton or wool bags or > 260°C for glass fiber bags) because damage to the filter material will result.

SCRUBBER. A scrubber is another device that can be used to remove particulates from air. Scrubbers are of particular value where the contaminated air is wet, corrosive, or hot, applications where baghouses cannot be used. Simple spray chambers can be used for removal of larger particle sizes. Dirty air flows through a chamber into which atomized water droplets are sprayed. The water droplets accumulate on the particulates in the air, increase their size and weight, and cause them to settle more rapidly and efficiently than in a settling chamber. The removed particulates in the collected spray water at the bottom of the spray chamber are drawn off to a settling basin, where the particulates are settled. The clarified water is usually recycled back to the scrubber.

For high-efficiency removal of fine particles, a combination venturi scrubber and cyclone can be used (see Figure 11.23*c*). A fine mist is sprayed into the dirty air as it flows through the venturi. The air then enters a cyclone, where the now large and heavy

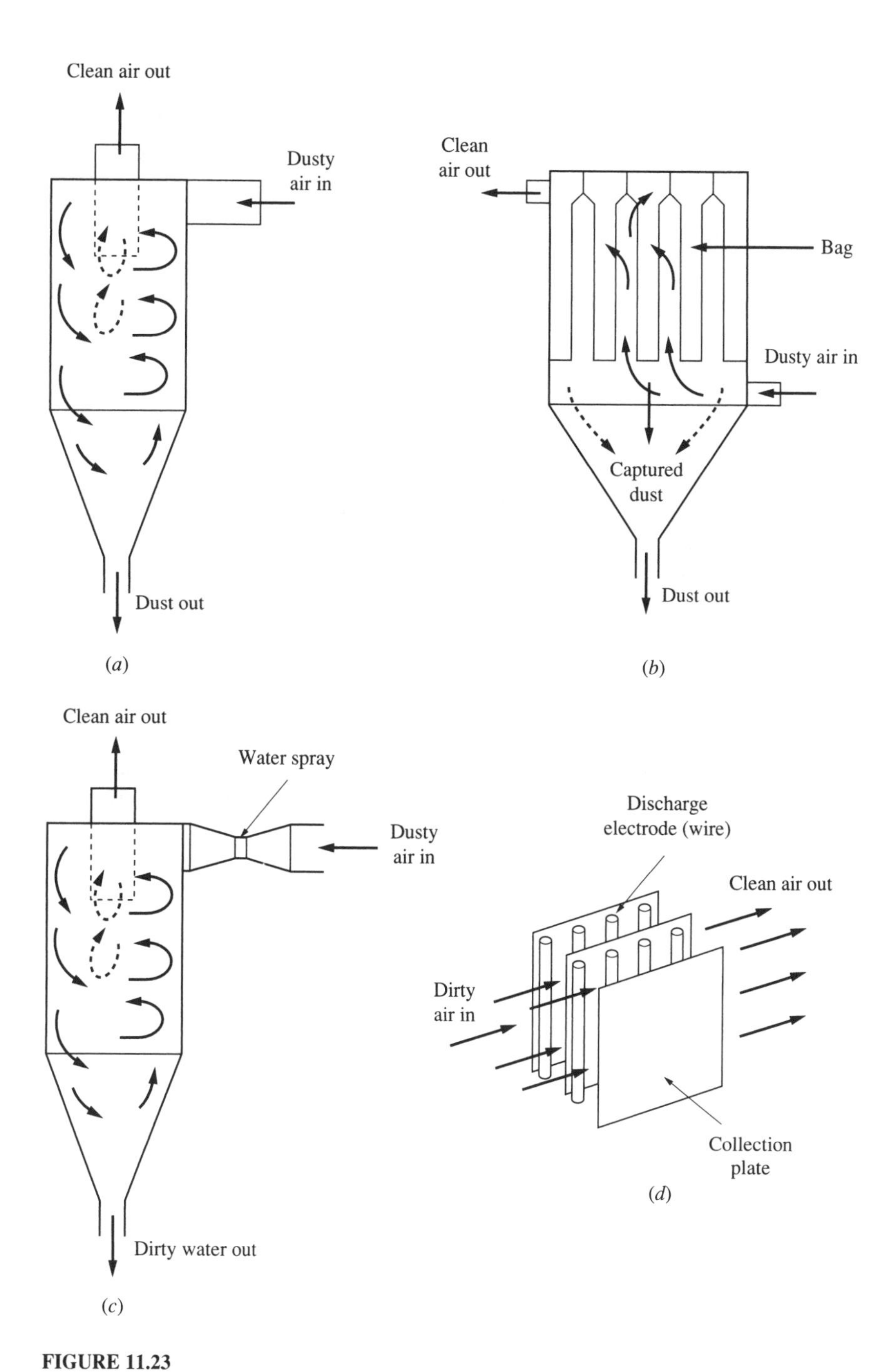

FIGURE 11.23
Schematic diagrams of particulate control devices: (*a*) cyclone; (*b*) baghouse; (*c*) venturi cyclone wet scrubber; (*d*) electrostatic precipitator.

water-laden particulates are removed by cyclonic action. Particulate removal efficiencies can be as high as 99 percent with a well-designed scrubber system. However, the air exiting the scrubber is saturated with water vapor and the stack plume may be very visible, creating an aesthetic problem. In addition, the waste is still present, but now it is in liquid form, requiring further treatment, rather than in a solid form as it is from one of the previously described particulate removal processes.

ELECTROSTATIC PRECIPITATOR. A fourth type of particulate control device is the electrostatic precipitator (ESP). The ESP is a high-efficiency dry collector of particulates from air. The particulate matter is removed by applying a high electrical direct current potential (30–75 kV) between alternating plates and wires (see Figure 11.23*d*). A full-scale ESP may have hundreds of parallel plates, with very large surface areas. As the particle-laden gas stream passes through this ion field, ions attach to the particulates, giving them a net negative charge. The particulates then migrate to the positively charged plates, where they are neutralized and stick. They are periodically removed from the ESP plate surfaces by rapping the plates. Efficiencies can be very high, even for small particulates. Electrostatic precipitators are commonly used in electric power plants. They are also used to control air pollution from blast furnaces, cement kilns, metal roasters, and acid production facilities.

In general, the efficiencies of various particulate control devices depend on the size and composition of the particles. Figure 11.24 plots approximate collection efficiencies, as a function of particle size, for several control devices. As can be seen, bag filters are the most efficient devices available, followed closely by scrubbers. Figure 11.25 shows the range of particle sizes over which various control devices are effective.

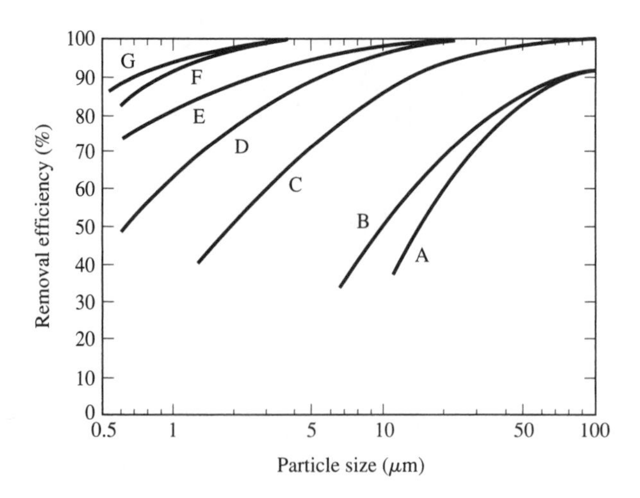

FIGURE 11.24
Comparison of approximate removal efficiencies of several air particulate control devices. A = settling chamber, B = simple cyclone, C = high-efficiency cyclone, D = electrostatic precipitator, E = spray tower wet scrubber, F = venturi scrubber, G = bag filter. (Adapted from Lappe, 1951)

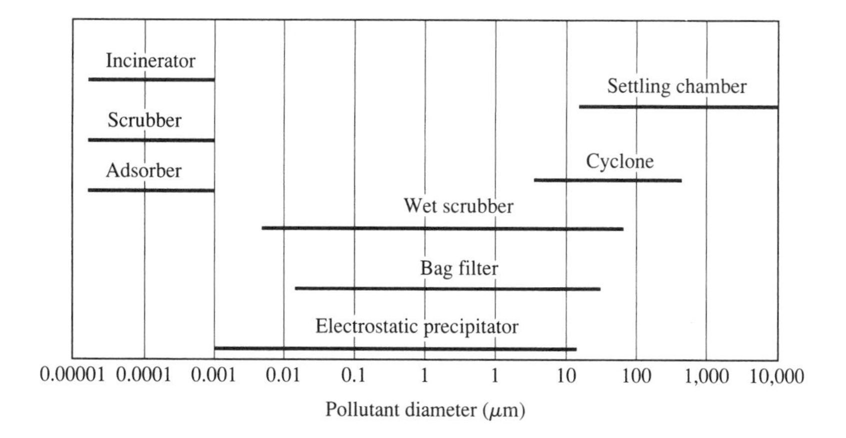

FIGURE 11.25
The effectiveness of various air pollution control devices. (Adapted from Vesilind, 1997)

11.3.2 Gas Removal

The primary gaseous contaminants of concern as air pollutants are SO_x, NO_x, CO, CO_2, hydrocarbons, and other organic and inorganic gases. The principal methods for control of gaseous emissions include absorption, adsorption, and incineration. Some of these processes merely move the contaminants to another medium; others change them chemically or even destroy them. In the process of destroying the contaminants, though, as with incineration, new pollutants may be created (e.g., CO_2 or gaseous HCl from the incineration of chlorinated hydrocarbons).

Wet scrubbers, described earlier, can remove gaseous contaminants by absorbing them into water droplets. These devices transfer the pollutant from the air phase to the water phase, where further treatment may be required. Only pollutants that are highly soluble in water can be treated effectively. For less soluble materials, scrubbers may be employed in which a chemical injected into the scrubber water reacts with the pollutants, either chemically altering them or making them easier to remove. A common example of this is flue gas desulfurization, in which SO_2 in the flue gas may be removed by reacting it with a solution of lime or limestone in water:

$$SO_2 + CaO \rightarrow CaSO_3$$

or

$$SO_2 + CaCO_3 \rightarrow CaSO_4 + CO_2$$

Both the calcium sulfite and calcium sulfate (gypsum) can be separated from the absorption water in clarifiers.

Adsorption systems for removal of air pollutants usually consist of a bed of adsorbing material, such as activated carbon, housed within a pressure vessel through which the contaminated air passes. Contaminants are transferred from the air to the adsorber. Activated carbon is the most common adsorber used for air pollutants, but other adsorbents can be used, such as activated alumina, silica gel, or molecular sieves. When the adsorber becomes saturated with contaminants, breakthrough will occur and the adsorbent must be replaced or regenerated.

Incineration or *flaring* can be used to oxidize carbon monoxide or organic air pollutants (hydrocarbons, organic acids, aldehydes, and ketones, as well as organics containing chlorine, sulfur, and nitrogen) to carbon dioxide and water. Both direct flame combustion and catalytic combustion can be used. Catalytic incineration employs the use of catalysts to speed up the oxidation reaction and allow it to proceed at a much lower temperature than is necessary with direct flame combustion. Combustion processes cannot be used for all organic wastes; some organics are oxidized to materials that are more toxic than the original material. For example, the combustion of trichloroethylene produces phosgene, which was used as a poison gas in World War I. Combustion of other materials can lead to the production of small, but harmful, quantities of dioxins.

11.4 SOLID WASTE DISPOSAL

Industries create refuse, trash, and garbage just as municipalities do. Many industries must also contend with disposal of mountains of packaging materials and process wastes. These must be disposed of in a safe and sanitary fashion, usually in landfills or by incineration. Although management of these solid wastes is often complex and expensive, a more serious problem is disposal of toxic and hazardous wastes. Most of these wastes are now banned from disposal in landfills or municipal incinerators. The industry may process the toxic wastes to destroy, detoxify, or reduce in volume the waste as much as possible, but ultimate disposal of the residual waste is usually left to dedicated hazardous waste disposal companies, with which the industry contracts. Improper disposal in the past, as at Love Canal, led to the hazardous waste sites now on the Superfund list and to stringent waste management requirements.

Hazardous waste landfills and incinerators are usually much too expensive to construct and operate for an individual industry, and the paperwork required to operate an industry-owned hazardous waste facility is overwhelming. The proper management of industrial solid wastes is not discussed in this book. Rather, the reader is directed to other sources of information on this topic, including books by Pfeffer (1992), Tchobanoglous, Theisen, and Vigil (1993), Sincero and Sincero (1996), and LaGrega, Buckingham, and Evans (1994).

REFERENCES

Allen, D. T., and Rosselot, K. S. *Pollution Prevention for Chemical Processes.* New York: John Wiley & Sons, 1997.

American Water Works Association. *Water Treatment Plant Design.* New York: McGraw-Hill, 1990.

Bordacs, K., and Young, J. C. "Biological Processes for Industrial Wastewater Treatment." In *Standard Handbook of Industrial Hazardous Waste Treatment and Disposal,* 2nd ed., ed. H. F. Freeman. New York: McGraw-Hill, 1998, pp. 9.3–9.22.

Buonicore, A. J., and Davis, W. T. *Air Pollution Engineering Manual.* Pittsburgh: Air and Waste Management Association, 1992.
Davis, M., and Cornwell, D. D. *Introduction to Environmental Engineering,* 3rd ed., New York: McGraw-Hill, 1998.
Eckenfelder, W. W. *Industrial Water Pollution Control,* 2nd ed. New York: McGraw-Hill, 1989.
Glaze, W. H. "Drinking Water Treatment with Ozone." *Environmental Science and Technology* 21 (1987): 224–30.
Haas, C. N., and Vamos, R. J. *Hazardous and Industrial Waste Treatment.* Upper Saddle River, NJ: Prentice Hall, 1995.
LaGrega, M. D., Buckingham, P. L., and Evans, J. C. *Hazardous Waste Management.* New York: McGraw-Hill, 1994.
Lappe, C. E. "Processes Use Many Collection Types." *Chemical Engineering* 58 (1951): 145.
Metcalf & Eddy. *Wastewater Engineering: Treatment, Disposal and Reuse,* 3rd ed. New York: McGraw-Hill, 1991.
Monod, J. "The Growth of Bacterial Cultures." *Annual Review of Microbiology* 3 (1949): 371–93.
Nemerow, N. L., and Agardy, F. J. *Strategies of Industrial and Hazardous Waste Management.* New York: Van Nostrand Reinhold, 1998.
Pfeffer, J. T. *Solid Waste Management Engineering.* Englewood Cliffs, NJ: Prentice Hall, 1992.
Reed, S. C., Middlebrooks, E. J., and Crites, R. W. *Natural Systems for Waste Management and Treatment.* New York: McGraw-Hill, 1988.
Reynolds, T. D., and Richards, P. A. *Unit Operations and Processes in Environmental Engineering,* 2nd ed. Boston: PWS Publishing, 1996.
Sincero, A. P., and Sincero, G. A. *Environmental Engineering: A Design Approach.* Upper Saddle River, NJ: Prentice Hall, 1996.
Speece, R. E. *Anaerobic Biotechnology for Industrial Wastewaters.* Nashville, TN: Archae Press, 1996.
Tchobanoglous, G., Theisen, H., and Vigil, S. *Integrated Solid Waste Management.* New York: McGraw-Hill, 1993.
U.S. EPA. *Constructed Wetlands and Aquatic Plant Systems for Municipal Wastewater Treatment.* Washington, DC: U.S. EPA, 1988.
U.S. EPA. *Development Document for Effluent Limitations Guidelines and Standards for the Metal Finishing Point Source Category.* EPA 440/1-83-091. Washington, DC: U.S. EPA, 1983.
U.S. EPA. *Hazardous Waste Treatment, Storage, and Disposal Facilities (TSDF): Air Emission Models.* EPA-450/3-87-026. Research Triangle Park, NC: U.S. EPA, 1987.
U.S. EPA. *National Air Quality and Emissions Trend Report, 1996.* EPA 454/R-97-010. Research Triangle Park, NC: U.S. EPA, 1998.
U.S. EPA. *Wastewater Treatment and Reuse by Land Application.* EPA-660/2-73/006. Washington, DC: U.S. EPA, 1973.
Vesilind, P. A. *Introduction to Environmental Engineering.* Boston: PWS Publishing, 1997.
Viessman, W., and Hammer, M. J. *Water Supply and Pollution Control,* 6th ed. Menlo Park, CA: Addison-Wesley, 1998.
Wark, K., Warner, C. F., and Davis, W. T. *Air Pollution: Its Origin and Control,* 3rd ed. Menlo Park, CA: Addison Wesley Longman, 1998.
WEF. *Design of Municipal Wastewater Treatment Plants.* Alexandria, VA: Joint Task Force of the Water Environment Federation and the American Society of Engineers, 1991.
Weber, W. J. *Physicochemical Processes for Water Quality Control.* New York: Wiley-Interscience, 1972.
Wentz, C. A. *Hazardous Waste Management.* New York: McGraw-Hill, 1995.

PROBLEMS

11.1. A wide variety of alkaline agents are available for neutralizing acidic wastes (e.g., sodium hydroxide, lime, sodium carbonate, limestone). Lime is the least expensive of these per pound. Explain why a higher priced reagent may be more beneficial and actually more economical than lime.

11.2. A pH titration was performed on a 100-mL sample of an acidic industrial waste using 2.0 *N* NaOH, yielding the following results:

Amount, mL 2.0 *N* NaOH	pH	Amount, mL 2.0 *N* NaOH	pH
0.0	2.0	6.0	6.9
1.0	2.1	6.5	8.0
2.0	2.2	7.0	8.8
3.0	2.3	7.5	9.4
4.0	2.5	8.0	9.7
4.5	3.5	9.0	10.0
5.0	4.7	10.0	10.1
5.5	5.8	11.0	10.1

If 380 m^3/day of the wastewater is to be neutralized to pH 6.0 using 4.0 *N* NaOH, how much NaOH must be used per day? How difficult will it be to maintain the pH at exactly 6.0? Why?

11.3. Determine the terminal settling velocity in water of a spherical particle having a diameter of 1.0 mm and a specific gravity of 2.60. Assume settling follows Stokes' law, and that (*a*) the Reynolds number is 0.2, and (*b*) the Reynolds number is 100.

11.4. An industry produces a wastewater containing suspended sand particles (specific gravity 2.65). Three sand particle sizes are present: 0.7 mm (300 mg/L), 1.0 mm (500 mg/L), and 1.5 mm (200 mg/L). The company plans to use an existing tank as a settling basin to remove as much of the sand from the wastewater as possible. The tank is 5 m long, 2 m wide, and 2 m deep. The flow rate into the tank will be 6.0 m^3/s. What fraction of each particle size will be removed, assuming that the particles entering the tank are uniformly distributed vertically?

11.5. Ferric sulfate is to be used as a coagulant for removal of fine particles from an industrial wastewater. It has been determined that a ferric sulfate dosage of 10 mg/L as Fe^{3+} will be required.

(*a*) How much ferric sulfate as $Fe_2(SO_4)_3$ must be used?

(*b*) If the wastewater contains 45 mg/L alkalinity as $CaCO_3$, will the coagulation process be successful? Why?

11.6. An industry plans to use coagulation/sedimentation to treat its wastewater. It has found that an iron dosage of 10 mg/L as Fe is suitable. The company can use ferric chloride, ferric sulfate, or ferrous sulfate as the source of iron. The waste has sufficient alkalinity for any of the reagents. The company would like to minimize the amount of sludge to be disposed of. Assuming that the sludge produced by all three reagents will concentrate to 3 percent solids, which reagent would be the best choice? Ignore differences in commercial reagent strength; assume they are all 100 percent active reagent.

11.7. Chlorine is an oxidizing agent that can be used to chemically degrade large, complex organic molecules into smaller ones. Some researchers have expressed concern about this technique, though, because of the chlorinated organic by-products that are formed.

Research this topic and prepare a short paper describing the effectiveness of this process and any limitations it may have.

11.8. Recently, Fenton's reagent has become increasingly popular for chemical oxidation. What is it and how does it work?

11.9. An activated carbon breakthrough curve was developed for treatment of an industrial wastewater using a specific carbon column. The curve is shown in Figure 11.26.

(*a*) Assuming that at least 85 percent of the target chemical must be removed from the wastewater before discharge, how much wastewater can be treated in a single column before the activated carbon must be replaced or regenerated?

(*b*) If two columns are used in series, how much water can be treated before carbon replacement or regeneration is needed?

11.10. A reverse osmosis unit is to demineralize 200,000 L/day of industrial wastewater for reuse in the manufacturing process. The unit has a mass transfer coefficient of 0.18 $L/day \cdot m^2 \cdot kPa$ and a membrane area of 600 m^2. To achieve the desired product water quality, the osmotic pressure differential between the feed and product water should be maintained at 300 kPa. The RO unit manufacturer's recommended maximum pressure difference between the feed and product water for this unit is 2500 kPa.

(*a*) Is the RO unit suitable for this application?

(*b*) If in the future the wastewater flow requiring treatment increases to 250,000 L/day, will this RO unit still be capable of handling the flow?

11.11. A synthetic rubber manufacturer produces a wastewater that contains butadiene and styrene. The company wants to treat this waste biologically. Research the biodegradability of these compounds and make recommendations to the company as to which biological process would be best and how it should be operated.

11.12. Determine the dimensions for an air pollution settling chamber to remove 60 μm diameter particles that have a specific gravity of 2.3 from a gas stream flowing at 30 m^3/s. The

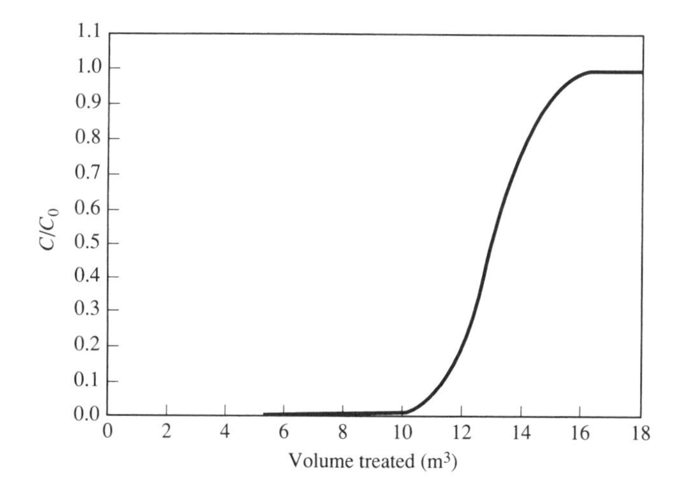

FIGURE 11.26

air temperature is 20°C. Assume a gas velocity of 0.3 m/s is to be used, and the particle drag coefficient is equal to 0.4.

11.13. An industry's stack emissions contain dust with a particle diameter range of 0.1–10 μm, 20 percent SO_2, and 10 percent methyl isobutyl ketone. What treatment device or devices would you recommend for this gas stream?

CHAPTER 12

FUGITIVE EMISSIONS

12.1 INTRODUCTION

By applying the principles described in the previous eleven chapters, a company can usually greatly decrease the pollution emanating from its facilities. However, those in some industries, particularly large industries such as refineries and petrochemical facilities, are unlikely to be able to reduce emissions to zero. An average-sized manufacturing plant may have 3000–30,000 components such as pumps, valves, compressor seals, and pipe flanges that may leak. Larger facilities may have as many as 100,000 of these connection points. Large refineries may have more than 250,000 pieces of potentially leaking equipment. Well maintained, nonleaking equipment emits very little process fluids. However, in a large facility, even a well-maintained one, some equipment seals, packing materials, or gaskets commonly leak, resulting in some unintentional releases of process liquids and gases. Such unintentional releases, referred to as *fugitive emissions,* may be continual leakage of small amounts of process fluids due to faulty process equipment or sudden, major leaks due to equipment failure. Continual monitoring for possible leaks or potential equipment failure and rectifying any leaks that are found present an enormous task for the company. Compounding this problem is the fact that various components will have different environmental compliance schedules and may be evaluated on different monitoring schedules. This can quickly become a logistical nightmare.

Essentially all industries, from large petrochemical plants to local gasoline filling stations, must comply with environmental regulations. Consider your local dry cleaning establishment. Dry cleaners are a major source of perchloroethylene (PCE), one of the toxic air pollutants regulated under the Clean Air Act Amendments (CAAA; U.S. EPA, 1994). There are two sources of PCE emissions at dry cleaning facilities: process vent emissions (i.e., the dry cleaning machine vent) and fugitive emissions (emissions from the wet clothing in the clothing transfer between the clothing washer and dryer, equipment leaks, open containers, etc.). In 1996 about 25,000 commercial and industrial dry cleaning facilities emitted 95,700 tons of PCE into the atmosphere (U.S. EPA, 1997). Most of this was in the form of fugitive emissions. Proper control techniques can largely eliminate these emissions. Process vent emissions must be controlled through the use of refrigerated condensers or activated carbon adsorbers. Fugitive emissions from clothing transfer are controlled through room enclosures, while other fugitive emissions are controlled through leak detection and repair, along with pollution prevention activities, such as good housekeeping.

There are many federal, state, and sometimes local regulations that apply to leak monitoring. These standards normally define

- Chemical streams that must be monitored.
- Types of components (pumps, valves, connections, etc.) to be monitored.
- Measured concentration that indicates a leak.
- Frequency of monitoring.
- Actions to be taken if a leak is discovered.
- Length of time in which an initial attempt to repair the leak must be performed.
- Length of time in which an effective repair of the leak must be made.
- Actions that must be taken if a leak cannot be repaired within guidelines.
- Record-keeping and reporting requirements (What Is LDAR?, 1997).

The CAAA specify the allowable leak rate and monitoring frequency within each industry segment. Typically, a leak is defined as emission to the atmosphere as a result of the dripping of liquid VOCs, and/or the detection of 10,000 parts per million by volume (ppmv) of a contaminant in the air at the surface of the potential source. Leaking sources commonly must be repaired so that measured concentrations will be less than 500 ppmv. As the Hazardous Organic National Emission Standards (the HON Rule) are specified for each industry, such as refining or manufacture of synthetic organic chemicals, the magnitude of the leak detection and repair (LDAR) requirements has blossomed. The HON Rule defines emission standards, along with monitoring, inspection, record-keeping and reporting requirements for process vents, storage vessels, transfer racks, and wastewater operations. In addition to increasing the number of locations that must be monitored, these new regulations are reducing the allowable fugitive emissions limits, in some cases drastically. For example, the previous standard of 10,000 ppmv for some valves used in chemical processing was reduced in 1997 to 500 ppmv (Brown, 1997). This will have a significant impact on industry.

In addition to emissions from stationary sources such as stacks and process vents, CAAA requires determination of fugitive emissions. Facilities must include fugitive emissions in the "potential to emit" value used in determining major source status for hazardous air pollutants for the facility (Maisel, 1996). A *major stationary source* is defined by the EPA as one which directly emits, or has the potential to emit, 100 tons per year or more of any air pollutant, including fugitive emissions. Facilities designated as major sources must comply with much more stringent regulations and increased compliance is required. Any facility that is close to or over the limit as a major source should minimize fugitive emissions so as to remove that facility from the major source category.

To overcome the often monumental record-keeping and reporting requirements, a number of companies have developed computerized leak detection and repair programs to assist in this effort.

12.2 SOURCES AND AMOUNTS

Fugitive emissions can originate at any place where equipment leaks may occur. The primary sources are pumps, flanges, compressor seals, valves (particularly pressure-relief valves), and any other point where pipe connections are made. Table 12.1 lists potential sources of fugitive emissions that should be tested. Fugitive emissions can also arise from evaporation of hazardous compounds from open-topped tanks or reservoirs. We usually think of fugitive emissions as fluids (liquids or gases), but another source of fugitive emissions is dust from such activities as construction and demolition, mining and quarrying, traffic over roads, waste collection, and agriculture (Sullivan, 1997).

Although fugitive emissions from a single valve, connector, or pump may be small, the cumulative impact from the thousands of such components in a facility can be staggering. Fugitive emissions account for about one-third of all organic emissions at synthetic organic chemical manufacturing industries (SOCMI). Figure 12.1 shows a typical profile of fugitive-emitting equipment in a facility; the actual percentages vary from industry to industry. While there are many sources of fugitive emissions, the major culprit is usually leaking valves. Thus much of the activity to reduce fugitive emissions centers around reducing losses from valves. Methods for doing this are described later, along with control strategies for reducing other fugitive emissions.

TABLE 12.1
Equipment leak emission sources

Agitator/mixer shaft seals	Meters
Compressor seals	Open-ended lines
Connectors	Pipe connectors
Diaphragms	Pressure-relief devices
Drains	Pump seals
Flanges	Stuffing boxes
Hatches	Valves
Instrument insertion points	Vents

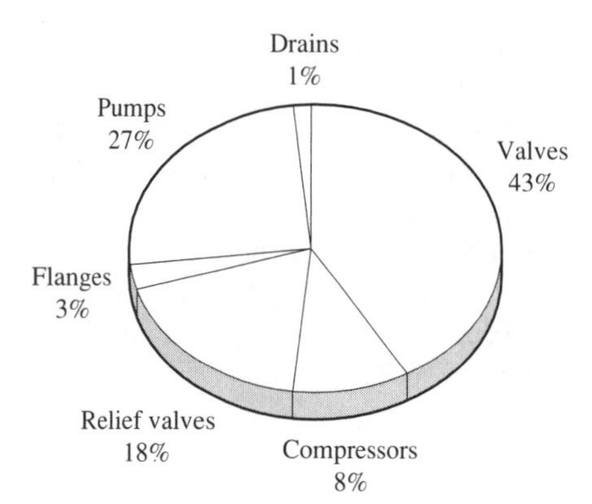

FIGURE 12.1
Typical distribution of fugitive emissions in process plants. (Source: Siegell, 1997)

The cumulative mass of hydrocarbon emissions from fugitive sources is of serious concern. As described in Chapter 3, hydrocarbons are precursors of smog formation and are also important constituents in ozone depletion and global warming. Thus it is essential that fugitive emissions be controlled.

12.3 MEASURING FUGITIVE EMISSIONS

A major activity that many companies must undertake on a regular basis is measuring the amounts of fugitive emissions emanating from their plants. To fulfill this CAAA requirement, typically a portable gas detector is used to measure emissions from each fugitive emission source. Most organic vapor analyzers are based on gas chromatography principles, usually using a flame ionization detector, because these analyzers have a fast response (< 3 s to as much as 30 s), a broad detection range up to 50,000 ppm, and are accurate (U.S. EPA, 1995). Photoionization detectors are also available. Recently, other types of monitoring equipment have come onto the market, including catalytic bead, nondispersive infrared, and combustion analyzers, but the standard flame ionization detector is still the most frequently used device. Some of these detectors have built-in data logging capabilities, but with others the data must be logged manually, a slow and often error-prone process. In general, these portable VOC monitors are equipped with a probe that is placed at the possible leak interface of a piece of equipment. A pump within the instrument draws a continuous sample of gas from the leak interface area to the instrument detector. The instrument response is a relative value expressed in parts per million by volume (ppmv). Actual concentrations are not measured because each compound present will respond in a different way. To measure concentrations of individual organics would be essentially impossible with this equipment. The instrument is usually calibrated using methane or isobutylene as the reference compound, so the measured concentration at the leaking device is only an approximate concentration of the total carbon content of the vapor being sampled.

In some cases, it is necessary to measure mass emissions of specific organic compounds from a piece of equipment, rather than measuring only total organics. This requires that the interface of the piece of equipment be "bagged." A tent or bag of material impermeable to the compounds of interest is constructed around the leak interface of the piece of equipment. A known rate of carrier gas is flowed through the bag, and a sample of gas exiting the bag is collected and analyzed using laboratory instruments.

The goal of the VOC screening program is to measure VOC concentrations at all potential leak points. All equipment to be included in the screening survey needs to be identified before the screening program begins. Each source should be uniquely identified, using tags or, more preferably if automatic data logging is to be used, bar codes. A flow diagram of the process and the process streams is needed so that an efficient survey route can be established. When sampling, the component should be identified for the records and then the probe inlet should be placed directly on the leak interface; for equipment with moving parts, the probe should be placed 1 cm from the leak interface. The probe should be moved around the interface until the maximum reading is shown on the meter and then left at that point for approximately two times the instrument response time. The maximum reading is recorded as the emission value.

To ensure that data quality is maintained, the data should be recorded on prepared data sheets. The collected data should include the following:

1. Monitoring instrument type and model number.
2. Operator's name.
3. Date.
4. Component identification number.
5. Component type (valve, connector, open-ended line, etc.).
6. Location/stream (a brief description of where the screened component is located and the composition of material in the equipment).
7. Service (gas, light liquid, or heavy liquid).
8. Number of hours per year the component is in service.
9. Screening value (ppmv).
10. Background concentration (ppmv).
11. Comments, if any (U.S. EPA, 1995).

The service categories are defined as gas/vapor, light liquid, and heavy liquid. A *gas/vapor* is a material in the gaseous state at operating conditions. A *light liquid* is a material in a liquid state in which the sum of the concentrations of individual components with a vapor pressure over 0.003 atm (2.3 mm Hg) at 20°C is greater than or equal to 20 weight percent. A *heavy liquid* is a liquid other than a light liquid.

The task of monitoring all potential leakage sites in a large facility is staggering. As stated previously, these facilities may have hundreds of thousands of components, many of which require monitoring more than once per year. It is just not economically feasible to monitor all of these devices on a regular basis using a hand-held monitor. Assuming that a person could take one reading every two minutes (which is not feasible because of the time needed to log the identification information on the device, make the measurement, and then move from that device to the next), it would take about 2 worker-years to make 100,000 measurements. Thus the U.S. Environmental Protection Agency has established alternative methods for estimating the fugitive emissions.

There are four approaches for estimating equipment leak emissions that are approved by the U.S. EPA. These are, in order of increasing refinement:

1. Average emission factor approach.
2. Screening ranges approach.
3. EPA correlation approach.
4. Unit-specific correlation approach.

These approaches are commonly used in the petroleum industry and the synthetic organic chemical manufacturing industries (SOCMI). All require an accurate count of equipment components by type of equipment and by type of service (i.e., gas, light liquid, or heavy liquid). Except for the first approach, they all also require screening data using portable analyzers.

12.3.1 Average Emission Factor Approach

Since about 1980, the U.S. EPA and the American Petroleum Institute have conducted several studies to determine typical fugitive emissions emanating from equipment and components at gas plants, petroleum refineries, synthetic chemical plants, and other industrial operations. Table 12.2 is a compilation of some of these emission factors. The average emission factor approach for estimating emissions is a relatively simple procedure that is based on knowledge of the number and type of each component

TABLE 12.2
Average emission factors for estimating fugitive emissions

		TOC emission factor, kg/h · source		
Equipment type	**Service**	**SOCMI**	**Refinery**	**Marketing terminal**
Valves	Gas	0.00597	0.0268	1.3×10^{-5}
	Light liquid	0.00403	0.0109	4.3×10^{-5}
	Heavy liquid	0.00023	0.00023	—
Pump seals	Gas	—	—	6.5×10^{-5}
	Light liquid	0.0199	0.144	5.4×10^{-4}
	Heavy liquid	0.00862	0.021	—
Compressor seals	Gas	0.228	0.636	1.2×10^{-4}
	Light liquid	—	—	1.3×10^{-4}
Pressure-relief valves	Gas	0.104	0.16	1.2×10^{-4}
Fittings (connectors and flanges)	Gas	0.00183	0.00025	4.2×10^{-5}
	Light liquid	0.00183	0.00025	8.0×10^{-6}
	Heavy liquid	0.00183	0.00025	—
Open-ended lines	All	0.0017	0.0023	—
Sampling connections	All	0.0150	0.0150	—

Source: U.S. EPA, 1995

(flange, valve, etc.), the service each component is in (gas, light liquid, or heavy liquid), the total organic concentration of the stream, and the time period each component was in service.

The following equation is used to determine the estimated total organic carbon (TOC) mass emissions from each component (U.S. EPA, 1995):

$$E_{\mathrm{TOC}} = F_A \cdot WF_{\mathrm{TOC}}$$

where E_{TOC} = TOC emission rate from a component (kg/h)
F_A = applicable average emission factor for the component (kg/h)
WF_{TOC} = average mass fraction of TOC in the stream serviced by the component

Not all components in the system will be leaking at the rates shown. The range of possible leak rates from individual pieces of equipment spans several orders of magnitude (U.S. EPA, 1995). Typically, most of the leakage will come from a few components of a given type in a system, while others of that type will leak very little. The emissions factors are based on the average emissions from a number of units in a system. Thus the calculation should be used only to determine whether the aggregate of units is emitting more VOCs than allowable; it cannot be used to pinpoint specific units that need repair. Further, the estimate is based on national averages of many facilities. It does not account for specific differences at an individual facility, such as age of equipment or adequacy of equipment maintenance, which could have a significant effect on leakage rates.

Example 12.1. A small segment of a refinery operation contains 45 valves, 550 flanges and other pipe connections, 6 pumps, 10 open-ended lines, and 4 sampling connections. The light liquid process fluid contains 90 wt % TOC and 10 percent water. Determine the fugitive emissions from the plant segment.

Solution. Using the equation

$$E_{\mathrm{TOC}} = F_A \cdot WF_{\mathrm{TOC}}$$

calculate the fugitive TOC emission from each component, multiply by the number of components, and sum the results. This can best be done in a spreadsheet.

Sample calculation:

$$E_{\mathrm{TOC}} = (0.0109 \text{ kg/h} \cdot \text{unit})(0.90 \text{ wt }\%)(45 \text{ units})$$

$$E_{\mathrm{TOC}} = 0.442 \text{ kg/h}$$

Equipment type	Emission factor, kg/h · unit	TOC, wt %	Number of units	Estimated fugitive emissions, kg/h
Valves	0.0109	0.90	45	0.442
Connections	0.00025	0.90	550	0.124
Pump seals	0.144	0.90	6	0.778
Open-ended lines	0.0023	0.90	10	0.021
Sampling connections	0.0150	0.90	4	0.054

As can be seen, even though there are only four pumps present in this system, leakage from them represents 55 percent of the total losses. Total fugitive emissions from this segment of the facility is estimated to be 1.42 kg/h as TOC. On an annual basis, this is equivalent to 12,439 kg/yr, a significant loss; it could also cause a significant deleterious impact on the environment.

12.3.2 Screening Ranges Approach

This approach, formerly referred to as the leak/no-leak approach, should be more exact than the average emissions approach because it relies on screening data from the facility, rather than on industrywide average values. When applying this approach, it is assumed that components having screening values greater than 10,000 ppmv have a different average emission rate than components with screening values less than 10,000 ppmv. These two ranges were formerly referred to as leaking and nonleaking components, respectively. The emission factors to be applied to components in these two categories are shown in Table 12.3.

The screening ranges approach is applied in a similar manner to the average emission factor approach, except that the number of components leaking less than and more than 10,000 ppmv are calculated separately. The following equation is used:

TABLE 12.3
Screening range emission factors for estimating fugitive emissions

		TOC emission factor, kg/h · source					
		SOCMI		Refinery		Marketing terminal	
Equipment type	Service	≥10,000 ppmv	<10,000 ppmv	≥10,000 ppmv	<10,000 ppmv	≥10,000 ppmv	<10,000 ppmv
Valves	Gas	0.0782	1.31×10^{-4}	0.2626	0.0006	—	1.3×10^{-5}
	Light liquid	0.0892	1.65×10^{-4}	0.0852	0.0017	2.3×10^{-2}	1.5×10^{-5}
	Heavy liquid	0.00023	2.10×10^{-3}	0.00023	0.00023	—	—
Pump seals	Gas	—	—	—	—	—	—
	Light liquid	0.243	1.87×10^{-3}	0.144	0.0120	7.7×10^{-2}	2.4×10^{-4}
	Heavy liquid	0.216	2.1×10^{-3}	0.021	0.0135	—	—
Compressor seals	Gas	1.608	0.0894	0.636	0.0894	—	1.2×10^{-4}
	Light liquid	—	—	—	—	3.4×10^{-2}	2.4×10^{-5}
Pressure-relief valves	Gas	1.691	0.0447	0.16	0.0447	3.4×10^{-2}	1.2×10^{-4}
Fittings (connectors and flanges)	Gas	0.113	8.1×10^{-5}	0.00025	0.00006	3.4×10^{-2}	5.9×10^{-6}
	Light liquid	0.113	8.1×10^{-5}	0.00025	0.00006	6.5×10^{-3}	7.2×10^{-6}
	Heavy liquid	0.113	8.1×10^{-5}	0.00025	0.00006	—	—
Open-ended lines	All	0.01195	0.0015	0.0023	0.0015	—	—

Source: U.S. EPA, 1995.

$$E_{\text{TOC}} = (F_G \cdot N_G) + (F_L \cdot N_L)$$

where E_{TOC} = TOC emission rate for an equipment type
F_G = applicable emission factor for sources with screening values ≥ 10,000 ppmv, kg/h · source
N_G = equipment count for sources with screening values ≥ 10,000 ppmv
F_L = applicable emission factor for sources with screening values < 10,000 ppmv, kg/h · source
N_L = equipment count for sources with screening values < 10,000 ppmv

The screening-range emission factors are a better indication of the actual leak rate from individual equipment than the average emission factors, but a complete analysis of existing equipment is needed to ascertain which fall above or below the 10,000 ppmv limit. Because average values are still used with this procedure, actual releases may vary from those predicted. Further, the results of this approach give only a snapshot of the conditions at the time that the components were screened. If components begin leaking after the screening, they will not be accounted for in the estimate.

Example 12.2. For the industrial process described in Example 12.1, assume that a screening survey of the components was conducted and 25 percent of each component type was found to emit more than 10,000 ppmv TOC. Estimate the fugitive emissions from the system, using the screening ranges approach.

Solution. Again, it is best to use a spreadsheet approach to the solution, as shown below. To calculate E_{TOC}, use the equation

$$E_{\text{TOC}} = (F_G \cdot N_G) + (F_L \cdot N_L)$$

Sample calculation:

$$E_{\text{TOC}} = (0.0852 \text{ kg/h} \cdot \text{unit})(11 \text{ valves}) + (0.0017 \text{ kg/h} \cdot \text{unit})(34 \text{ valves})$$

$$E_{\text{TOC}} = 0.995 \text{ kg/h}$$

Equipment type	**F_G, kg/h · unit**	**N_G**	**F_L, kg/h · unit**	**N_L**	**E_{TOC}, kg/h**
Valves	0.0852	11	0.0017	34	0.995
Connections	0.00025	138	0.00006	412	0.059
Pump seals	0.144	2	0.0120	4	0.336
Open-ended lines	0.0023	2	0.0015	8	0.017

The estimated total fugitive emissions from this system using the screening ranges approach is 1.407 kg/h, or 12,325 kg/yr. This is very close to the value obtained using the average emission factor approach (12,439 kg/yr). However, if other percentages of highly leaking equipment were used, the result could be quite different.

12.3.3 EPA Correlation Approach

The EPA correlation approach is a further refinement of the estimating procedures that predicts mass emission rates as a function of screening values for a particular equipment type. Correlations relating screening values to mass emission rates for SOCMI process units and for petroleum industry process units (including refineries, marketing terminals, and oil and gas production operations) are listed in Table 12.4. The EPA correlation approach is preferred when actual screening values are available.

To obtain the total fugitive emissions from a process, the sum of the emissions associated with each of the screening values is determined. Equipment pieces with a screening value of zero are assigned the default-zero leak rate. It is generally assumed that all components leak to some extent. The default-zero leak rate is the mass emission rate associated with a screening value of zero. This provides an emission rate for components where the screening rate was below the detection limit of the organic vapor analyzer.

12.3.4 Unit-Specific Correlation Approach

This is the most exact, but also the most expensive, method for determining fugitive emissions. It requires that screening values and corresponding mass emissions data (i.e., bagging data) be collected for a statistically significant number of each piece of process unit equipment. The mass emission rates determined by bagging, and the associated screening value, can be used to develop a leak rate–screening value relationship, or correlation, for that specific equipment type in that process unit.

TABLE 12.4
EPA correlations for estimating fugitive emissions

	TOC leak rate from correlation,* kg/h · unit		
Equipment type	**SOCMI**	**Refinery**	**Default-zero emission rate, kg/h · unit**
Gas valves	1.8×10^{-6} SV$^{0.873}$	—	6.6×10^{-7}
Liquid valves	6.41×10^{-6} SV$^{0.797}$	—	4.9×10^{-7}
Valves (all)	—	2.29×10^{-6} SV$^{0.746}$	7.8×10^{-6}
Light liquid pumps	1.90×10^{-5} SV$^{0.824}$	—	7.5×10^{-6}
Pump seals (all)	—	5.03×10^{-5} SV$^{0.610}$	2.4×10^{-5}
Connectors	3.05×10^{-6} SV$^{0.885}$	—	6.1×10^{-7}
Connectors	—	1.53×10^{-6} SV$^{0.735}$	7.5×10^{-6}
Flanges	—	4.61×10^{-6} SV$^{0.703}$	3.1×10^{-7}
Open-ended lines	—	2.20×10^{-6} SV$^{0.704}$	2.0×10^{-6}

*SV = screening value (ppmv).
Source: U.S. EPA, 1995.

A minimum number of leak rate measurements and screening value pairs must be obtained to develop the correlations. Mass emissions data must be collected from individual sources that have screening values distributed over the entire emissions range. In general, for each equipment type and service type, a random sample of a minimum of six components should be chosen for bagging from each of the following screening ranges: 1–100 ppmv; 101–1000 ppmv; 1001–10,000 ppmv; 10,001–100,000 ppmv; and >100,000 ppmv. This will result in analysis of a minimum of 30 bags. In a few cases, a smaller number of bag analyses is allowable (see U.S. EPA, 1995). With the mass emissions data and screening values, leak rate–screening value correlations can be generated using least-squares regression analyses.

12.4 CONTROLLING FUGITIVE EMISSIONS

Two primary techniques are used for reducing fugitive emissions from equipment: (1) modifying or replacing existing equipment, and (2) implementing a leak detection and repair (LDAR) program. Examples of equipment modifications include installing a cap on an open-ended line, replacing an existing pump with a sealless type, and installing on a compressor a closed-vent system that collects possible leaks and routes them to a control device. An LDAR program is a structured program to detect and repair equipment that is identified as leaking (U.S. EPA, 1995).

12.4.1 Equipment Modification

Equipment modification to control emissions is accomplished either by installing additional equipment that eliminates or reduces emissions or by replacing existing equipment with sealless types. It may also include redesign and reconstruction of a process to eliminate leakage points; for example, a pipeline may be shortened to minimize joints and flanges. Some methods of equipment modification are more effective than others. Table 12.5 summarizes potential equipment modifications and their approximate control efficiencies.

Most fugitive emissions from a typical industrial facility, on a total mass basis, come from leaking valves. A packing material is commonly used around the valve stem to form a seal that allows the valve stem to move while keeping process fluids from leaking (see Figure 12.2). Deterioration of the packing, or a valve stem that is bent or nicked, can allow process fluids to escape between the valve stem and the housing (Allen and Rosselot, 1997). Three areas must be addressed in controlling fugitive emissions from valves: component monitoring, stem sealing, and mechanical condition.

Component monitoring requires periodic inspections of each valve to determine the emissions level. Those valves found to be leaking above a specified level are either repaired or replaced. Monitoring programs are the most cost-effective means for controlling fugitive valve emissions, with reductions of 24 percent to 73 percent being typical, depending on the monitoring frequency and the level established for repair or replacement of the valve (Bello and Seigell, 1997).

TABLE 12.5
Equipment modifications to reduce fugitive emissions and their control efficiencies

Equipment type	Modification	Approximate control efficiency, %
Pumps	Sealless design	100%
	Closed-vent system	90
	Dual mechanical seal with barrier fluid maintained at a higher pressure than the pumped fluid	100
Compressors	Closed-vent system	90
	Dual mechanical seal with barrier fluid maintained at a higher pressure than the pumped fluid	100
Pressure-relief devices	Closed-vent system	Varies
	Rupture disk assembly	100
Valves	Sealless design	100
Connectors	Weld together	100
Open-ended lines	Blind, cap, plug, or second valve	100
Sampling connections	Closed-loop sampling	100

Source: U.S. EPA, 1995.

Initial valve repairs include tightening the packing gland to further compress the packing and to seal the path. However, overtightening the packing can be detrimental because the valve may become difficult or impossible to operate. Attempts to operate the valve can also damage the overcompressed packing ("Valve Packings," 1995). If tightening the packing is not successful, the packing may be replaced with "low emission" packing the next time the equipment is out of service. Several quality packing materials are available today, including virgin polytetrafluoroethylene (PTFE) packing, filled PTFE packing, graphite and carbon-based packing, and perfluoroelastomer (PFE) packing; the latter is becoming the packing of choice for complying with fugitive emission standards (Wolz, Winkel, and Paul, 1995). These packings generally claim to meet fugitive emissions requirements of 500–1000 ppmv.

Live load springs can be used to maintain a relatively constant load on the packing, reducing the need for frequent bolt tightening. Live-loading involves installing a series of disk springs that will provide a constant load to the packing when properly torqued. As the packing consolidates over time, the spring action of the washers continues to apply a load, eliminating the need for retorquing. There are now valves that combine a duplex valve packing employing two sets of packing rings (one is a graphite composite and the other is a PTFE-carbon ring) and live-loading springs to maintain a consistent packing ring stress. The PTFE-carbon ring set is designed to provide a sealing capability below 500 ppmv emissions, while the composite was selected for smooth valve operation over an extended period, without maintenance (Blickley, 1995). Poor valve mechanical condition can cause the packing to improperly compress, which will

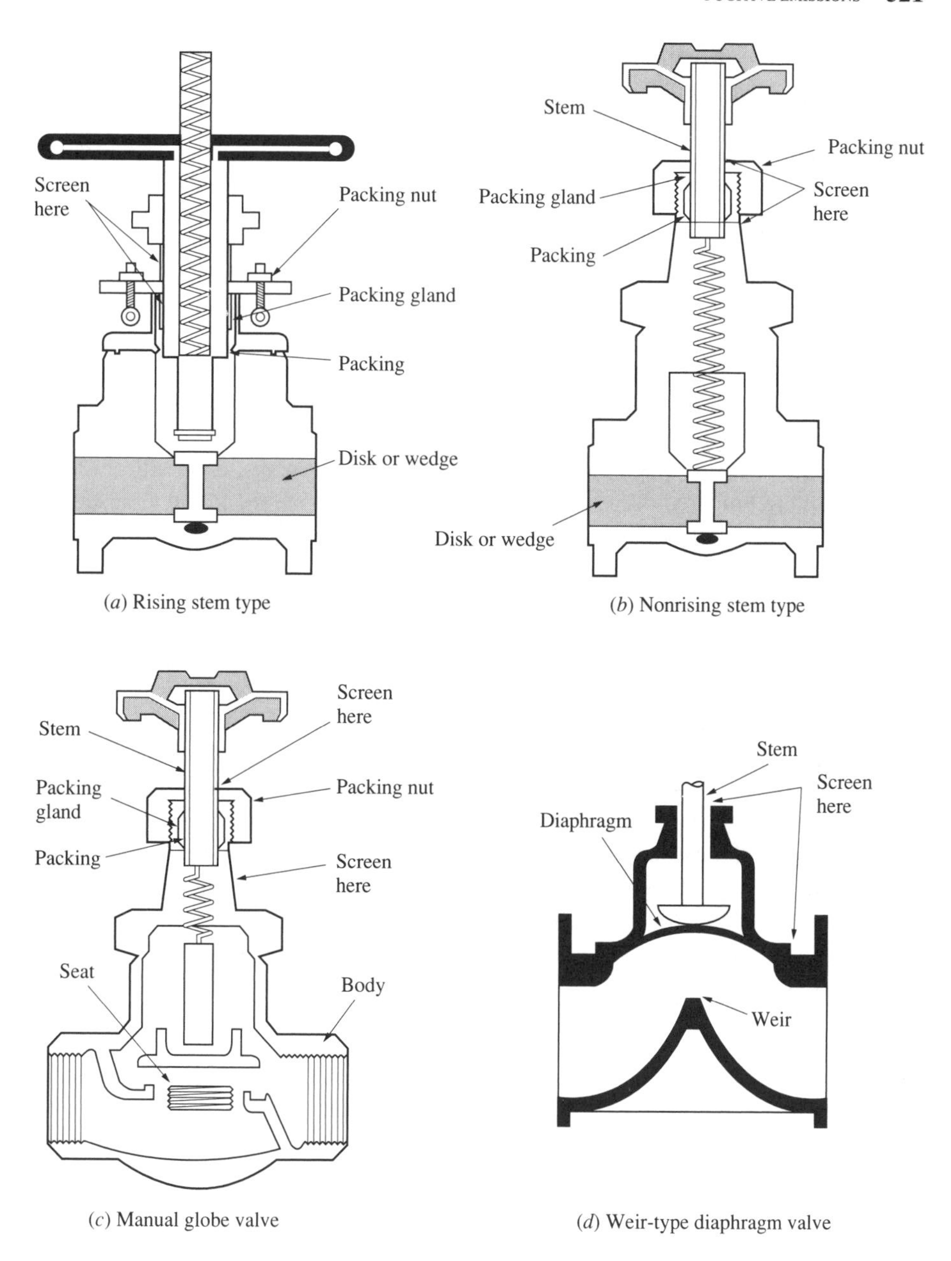

FIGURE 12.2
Typical valves used in industry. Points where the valve should be screened for leaks are denoted on the figures. (Source: U.S. EPA, 1995)

also lead to valve leakage. If the valve stem is corroded or bent, it may be repaired or replaced the next time the equipment is out of service (Siegell, 1997).

Emissions from process valves can be eliminated if the valve stem can be isolated from the process fluid. Sealless diaphragm valves serve this purpose. The diaphragm is used to regulate flow through the valve; it also serves as a barrier between the process fluid and the valve stem. As long as the integrity of the diaphragm is maintained, there will be virtually no fugitive emissions. However, if the diaphragm ruptures, large emissions may result.

Pumps and compressors, which are similar to pumps except that they pump gases, are generally the second largest emitters of fugitive emissions. Emissions usually occur when there is a faulty or failed seal between the rotating pump shaft and the pump housing. Equipment modifications to prevent leaks can include routing leaking vapors to a closed-vent system that captures the vapors or incinerates them in a flare, installing a dual mechanical seal containing a barrier fluid, or replacing the pump with a sealless type. A dual mechanical seal contains two seals between which a barrier fluid is circulated at a pressure higher than the pumped fluid. Because of this higher barrier fluid pressure, the pumped fluid cannot leak and control efficiency is essentially 100 percent. However, if both the inner and outer seals fail, large releases of the pumped fluid could result. Sealless pumps (diaphragm pumps, magnetic drive pumps, etc.) will not leak, when operating properly, because the process fluid cannot escape to the atmosphere. However, a catastrophic failure of the sealless pump will lead to large emissions.

Pressure relief valves are designed to reduce pressure by opening when the operating pressure exceeds a given limit; they close again when the pressure returns to a safe level. Emissions occurring as a result of these pressure-reducing discharges are not considered to be fugitive emissions, but rather are included in the plant's normal atmospheric contaminant discharges. Fugitive emissions from pressure-relief valves usually occur as a result of improper seating of the valve after a release or faulty maintenance of the seal if the process is operating too close to the set pressure of the valve. These vapors can be controlled by process design that provides a larger safety margin between the fluid operating pressure and the relief valve "pop-off" pressure, by good housekeeping of the valve condition, by use of a closed-vent system and a flare, or by use of a rupture disk–pressure relief valve combination. A rupture disk, which is leakless under normal operation, bursts when the operating pressure exceeds its limit, allowing process fluid to escape until a new disk is installed. They can be installed upstream of a pressure-relief valve to prevent fugitive emissions through the pressure-relief valve seat.

Flanges and other types of pipe connectors are the most numerous components that can cause fugitive emissions in most industrial facilities. EPA regulations state that no more than 500 ppmv of any volatile hazardous air pollutant can be emitted from a bolted flange connection. Fortunately, the emissions rate per connector is usually low. Flanges consist of units bolted together at the end of pipes that contain a gasket to prevent leakage. Leakage that does occur usually is due to a faulty gasket or to poor assembly. A study of fugitive emissions from flanges found that emissions could be reduced by increasing the compressive load on the gasket, by using thinner gaskets, and by using a gasket material that conforms well to flange surface imperfections such as those made from filled restructured PTFE sheet. More important, the smoother the surface finish of the flange surface, the lower the emission rate (Drago, 1995). Other types of connectors include threaded connections and nut-and-ferrule connectors. In cases where connectors are not required for safety, maintenance, process modification, or periodic equipment removal, fugitive emissions can be eliminated by welding the connectors together.

12.4.2 Leak Detection and Repair Programs

The most cost-effective control measure to reduce fugitive emissions is to initiate a leak detection and repair program. An LDAR program is designed to identify pieces of equipment that are emitting sufficient amounts of material to warrant reduction of the emissions through repair. Individual components are sampled to determine VOC emissions; if the hydrocarbon concentration at a particular component is above a predetermined "leak level," the component is repaired. These inspection and maintenance programs often result in reduced losses that exceed the program's cost (Siegell, 1997).

These programs are best applied to equipment types that can be repaired on-line or to equipment for which equipment modification is not feasible. They are best suited to valves, pumps, and connectors. LDAR is not readily applicable to compressors, because spare compressors that could be used in a bypass mode during repairs are not usually available, or to open-ended lines, pressure-relief valves, or sampling connections, where equipment modification is more useful. Heavy liquid services are usually not included in an LDAR program because the emission rates from these leaks are usually very small.

The U.S. Environmental Protection Agency has developed an approved monitoring procedure, *U.S. EPA Method 21* (U.S. EPA, 1993) and a suggested method for conducting an LDAR (U.S. EPA, 1995). This approach can be used to estimate the control effectiveness of an LDAR program for light liquid pumps, gas valves, light liquid valves, and connectors. The approach is based on the relationship between the percentage of equipment pieces that are leaking and the corresponding average leak rate for all of the equipment.

It is necessary to define the frequency of component sampling and the screening value at which a "leak" is indicated for the LDAR program. Table 12.6 summarizes the estimated average control effectiveness for several SOCMI facilities and refineries that have instituted LDAR programs. A 100 percent control effectiveness cannot be obtained using this procedure because components that are emitting fugitive emissions at a screening value less than 10,000 ppmv are not repaired or controlled. Two control strategies are compared: (1) monthly LDAR with a leak definition of 10,000 ppmv, and (2) quarterly

TABLE 12.6
Control effectiveness for an LDAR program at SOCMI and refinery process units

	Control effectiveness, %			
	SOCMI process units		Refinery process units	
Equipment type and service	Monthly monitoring	Quarterly monitoring	Monthly monitoring	Quarterly monitoring
Valves, gas	87	67	88	70
Valves, light liquid	84	61	76	61
Pumps, light liquid	69	45	68	45

Source: U.S. EPA, 1995.

LDAR with a leak definition of 10,000 ppmv. As should be expected, more frequent monitoring results in more effective control, but the expense of monitoring is greatly increased. If the more stringent proposed National Emission Standards for Hazardous Air Pollutants (NESHAP) standard of 500 ppmv standard is applied as the screening value, the LDAR control effectiveness would be between 75 percent and 95 percent.

Leak detection and repair programs are data intensive and usually require sophisticated data management software. With the screening limit dropping to 500 ppmv, even more components will need to be tracked and more analyses will need to be performed. Several commercially available LDAR software packages are now on the market.

12.5 FUGITIVE EMISSIONS FROM STORAGE TANKS

Another source of fugitive emissions is storage tanks containing organic liquids. These can be found in many industries, including petroleum producing and refining, petrochemical and chemical manufacturing, bulk storage and transfer operations, and other industries consuming or producing organic liquids. Nearly all industries have storage tanks for fuel oil. In addition, many industries store gases in large storage containers. All of these tanks can give off fugitive emissions.

Six basic tank designs are used for organic liquid storage vessels: fixed-roof (vertical and horizontal), external floating roof, domed external floating roof, internal floating roof, variable vapor space, and pressure (low and high) tanks. The first four are depicted in Figure 12.3 (U.S. EPA, 1997).

Fixed-roof tanks (Figure 12.3*a*) can be installed vertically or horizontally and can be constructed above or below ground of steel or fiberglass-reinforced polyester. Fixed-roof tanks are either freely vented to the atmosphere or are equipped with a pressure/vacuum vent. Fugitive emissions from fixed-roof tanks are caused by changes in temperature, pressure, and liquid level. Fixed-roof tanks are the least expensive storage tanks to construct, but they are generally considered to be the minimum acceptable equipment for storing liquids because of their potential to release fugitive emissions whenever the liquid level in the tank increases.

External floating roof tanks (Figure 12.3*b*) consist of an open-topped cylindrical steel shell equipped with a steel-plate roof that floats on the surface of the liquid. The roof rises and falls with the liquid level in the tank. The floating roof is equipped with a rim seal system, which contacts the tank wall and reduces evaporative losses of the stored liquid. Fugitive emissions should be limited to evaporative losses from an imperfect rim seal system, from fittings in the floating deck, and from any exposed liquid on the tank wall when liquid is withdrawn and the roof lowers.

An *internal floating roof tank* (Figure 12.3*c*) has both a permanent fixed roof and a floating roof inside. Evaporative losses are minimized by installing a floating roof under the existing fixed roof. Evaporative losses from internal floating roof tanks may come from deck fittings, nonwelded deck seams, and the annular space between the floating deck and the wall. The space between the fixed and floating roof is generally freely vented, so any vapors that move into this space will be vented to the atmosphere.

A *domed external floating roof tank* (Figure 12.3*d*) is similar to the internal floating roof tank in that it has a fixed roof over a floating roof. They are usually the result of retrofitting an existing floating roof tank with a fixed roof to block the wind and minimize evaporative losses.

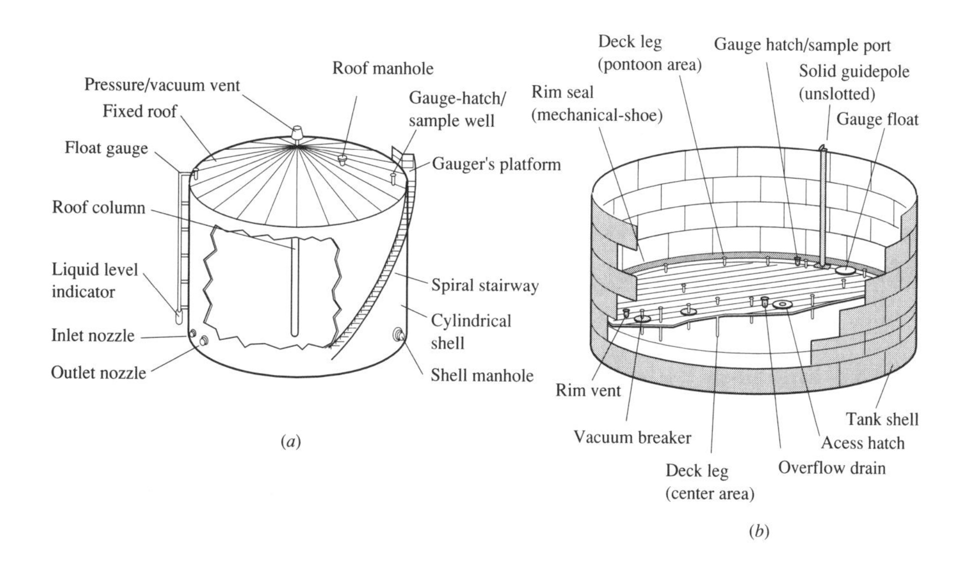

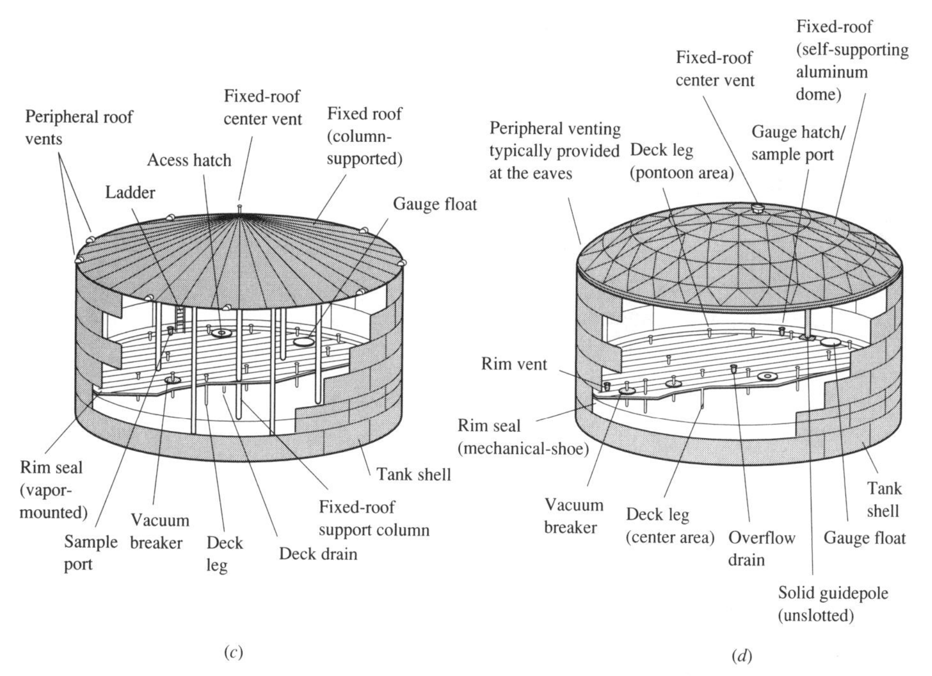

FIGURE 12.3
Schematics of four types of liquid storage tanks: (*a*) typical fixed-roof tank; (*b*) external floating roof tank (pontoon type); (*c*) internal floating roof tank; (*d*) domed external floating roof tank. (U.S. EPA, 1997)

Variable vapor space tanks are equipped with expandable vapor reservoirs to accommodate vapor volume fluctuations attributable to temperature and barometric pressure changes. Many of these use a flexible diaphragm membrane to provide expendable volume. They may be either separate gasholder units or integral units mounted atop fixed roof tanks. Losses from these tanks are generally limited to tank filling times when vapor is displaced by liquid and the tank's vapor storage capacity is exceeded.

Pressure tanks generally are used for storing organic liquids and gases with high vapor pressures. Pressure tanks are equipped with a pressure/vacuum vent that is set to prevent venting loss from boiling and breathing loss from temperature and barometric pressure changes. Losses from these tanks should be minimal, provided that the vent is well maintained and the tanks are not overpressurized, leading to releases from the pressure-relief valve.

12.5.1 Emissions Estimation

The U.S. EPA, in conjunction with the American Petroleum Institute, has developed procedures for estimating emissions from storage tanks. These procedures are compiled in the publication *Compilation of Air Pollutant Emission Factors* (U.S. EPA, 1997). These procedures have also been used as the basis for TANKS, a software program to estimate emissions from storage tanks.

As an example of the estimating procedures, consider tank losses from fixed-roof tanks. Losses can occur continually while the liquid is standing in the tank, and working losses can occur when liquid is being added or withdrawn from the storage tank. The model assumes that the tanks are substantially liquid- and vapor-tight and operate at atmospheric pressure. Total losses from these fixed-roof tanks are equal to the sum of the standing storage loss and working loss:

$$L_T = L_S + L_W$$

where L_T = total losses, lb/yr
L_S = standing storage losses, lb/yr
L_W = working losses, lb/yr

The standing storage losses are due to breathing of the vapors above the liquid in the storage tank. They can be estimated as follows:

$$L_S = 365 \cdot V_V W_V K_E K_S$$

where V_V = vapor space volume, ft^3
W_V = vapor density, lb/ft^3
K_E = vapor space expansion factor, dimensionless
K_S = vented space saturation factor, dimensionless

and 365 represents days per year. The vapor density, W_V, is calculated using the equation

$$W_V = \frac{M_V P_{VA}}{R T_{LA}}$$

where W_V = vapor density, lb/ft^3
M_V = vapor molecular weight, lb/lb mol
R = universal gas constant, 10.731 psi-ft^3/°R-lb mol

P_{VA} = vapor pressure at daily average liquid surface temperature, psi
T_{LA} = daily average liquid surface temperature, °R

The vapor phase expansion factor, K_E, is calculated as follows:

$$K_E = \frac{\Delta T_V}{T_{LA}} + \frac{\Delta P_V - \Delta P_B}{P_A - P_{VA}}$$

where ΔT_V = daily temperature range, °R
ΔP_V = daily pressure range, psi
ΔP_B = breather vent pressure setting range, psi
P_A = atmospheric pressure, psi

The vented vapor saturation factor, K_S, is calculated using the formula

$$K_S = \frac{1}{1 + 0.053 P_{VA} H_{VO}}$$

where H_{VO} = vapor space outage, ft (= height of a cylinder of tank diameter D whose volume is equivalent to the vapor space volume of the tank).

The working losses, L_W, can be estimated from:

$$L_W = 0.0010 M_V P_{VA} Q K_N K_P$$

where Q = annual net throughput (tank capacity (bbl) times annual turnover rate), bbl/yr
K_N = turnover factor, dimensionless
for turnovers > 36/year, $K_N = (180 + N)/6N$
for turnovers ≤ 36, $K_N = 1$
N = number of tank volume turnovers per year
K_P = working loss product factor, dimensionless
for crude oils = 0.75
for all other liquids = 1.0

Example 12.3. Determine the yearly fugitive emission rate for chloroform stored in a vertical fixed-cone roof tank. The tank is 20.0 ft in diameter, 30.0 ft high, and usually contains product to a depth of approximately 25.0 ft. The tank working volume is 58,758 gal. The number of turnovers for the tank is 10 (i.e., the throughput of the tank is 587,580 gal/yr, or 13,990 bbl/yr). Assume the average liquid surface temperature is 55°F (515°R), with a minimum liquid temperature of 45°F (505°R) and a maximum liquid temperature of 60°F (520°R). The vapor temperature in the tank ranges from 36°F (496°R) to 64°F (524°R). At these temperatures, the vapor pressure of chloroform ranges from 1.72 psia at 45°F to 2.48 psia at 60°F. The breather vent is set to actuate at 0.03 psia above or below the atmospheric pressure.

Solution. The emissions losses are made up of standing storage losses and working losses:

$$L_T = L_S + L_W$$

To determine the chloroform vapor density, assume the following properties for chloroform:

mol wt = 119.39

liquid density = 12.49 lb/gal

vapor pressure at 55°F = 2.20 psia

Then

$$W_V = \frac{M_V P_{VA}}{RT_{LA}}$$

$$= \frac{(119.39 \text{ lb/lb-mol})(2.20 \text{ psia})}{(10.731 \text{ psi} \cdot \text{ft}^3/°\text{R} \cdot \text{lb-mol})(515°\text{R})} = 0.0475 \text{ lb/ft}^3 = 6.84 \text{ psi}$$

$$V_V = \frac{\pi}{4D^2 H_{VO}}$$

For a cone roof, the vapor space outage, H_{VO}, is calculated by

$$H_{VO} = H_S - H_L + H_{RO}$$

where H_S = tank shell height = 30.0 ft
H_L = liquid height = 25.0 ft
H_{RO} = roof outage = 1/3 $(S_R)(R_S)$ and
S_R = roof slope = 0.0625 ft/ft
R_S = tank shell radius = 1/2 D = 10.0 ft
H_{RO} = 1/3 (0.0625)(10) = 0.208 ft

Thus

$$H_{VO} = 30 - 25 + 0.208 = 5.208 \text{ ft}$$

Therefore,

$$V_V = \frac{\pi}{4(20.0)^2(5.208)} = 1635 \text{ ft}^3$$

The vapor space expansion factor, K_E, can be determined as follows:

$$K_E = \frac{\Delta T_V}{T_{LA}} + \frac{\Delta P_V - \Delta P_B}{P_A - P_{VA}}$$

where ΔT_V = 524 − 496 = 28°R
ΔP_V = 2.48 − 1.72 = 0.76 psia
ΔP_B = breather vent pressure range = 0.03 + 0.03 = 0.06 psia
T_{LA} = 515°R
P_A = 14.7 psia
P_{VA} = 1.20 psia

Therefore,

$$K_E = \frac{28}{515} + \frac{0.76 - 0.06}{17.7 - 2.20} = 0.0092$$

The vented vapor space expansion factor, K_S, is found as follows:

$$K_S = \frac{1}{1 + 0.053P_{VA}H_{VO}}$$

$$= \frac{1}{1 + 0.053(2.20\text{ psia})(5.028\text{ ft})} = 0.652$$

We can now calculate the *standing storage losses:*

$$L_S = 365W_V V_V K_E K_S$$

$$= (365\text{ days/yr})(0.0475\text{ lb/ft}^3)(1{,}635\text{ ft}^3)(0.0092)(0.652) = 170.0\text{ lb/yr}$$

Now we turn to *working losses:*

$$L_W = 0.0010M_V P_{VA} Q K_N K_P$$

$$= (0.0010)(119.39)(2.20)(13{,}990)(1)(1) = 3{,}674\text{ lb/yr}$$

We can now calculate the total annual losses from the storage tank:

$$L_T = L_S + L_W$$

$$= 170 + 3674 = 3844\text{ lb/yr}$$

Thus nearly 2 tons of chloroform are lost per year due to fugitive emissions from this tank.

12.5.2 Emissions Control

Emissions from organic liquids in storage occur because of evaporative losses of the liquid during its storage (known as breathing losses or standing storage losses) and as a result of changes in the liquid level during filling and emptying operations (known as working losses). Several methods are available to control these losses (U.S. EPA, 1997).

Emissions from fixed-roof tanks can be controlled by installing an internal floating roof and seals to minimize evaporation of the product being stored. This will control releases by 60 percent to 99 percent. Vapor balancing can also be used; this is commonly done at gasoline filling stations. As the storage tank is being filled, the vapors expelled from the tank are directed to the emptying gasoline tanker truck. The truck then takes the vapors to a centralized station for vapor recovery. This system can control atmospheric releases during tank filling by 90 percent to 98 percent. Vapor recovery systems can also be installed on the storage tank to collect emissions and convert them to a liquid product, using such processes as vapor/liquid absorption or vapor cooling. Alternatively, the vapors can be destroyed by thermal oxidation, using a flame or flameless incinerator to combust the vapors. Control efficiencies for these systems are on the order of 96 percent to 99 percent.

Losses from floating roof tanks include losses from evaporation of product remaining on the walls during product withdrawal from the tank and roof lowering and evaporative standing storage losses. The majority of rim seal vapor losses result through wind action. Improved seals or installation of secondary seals can greatly cut down on these evaporative losses. Wiper seals can be installed to wipe the walls as the roof moves down during emptying, thus minimizing the amount of material remaining on the walls. Deck fittings and seams should be sealed to prevent losses from these.

12.6 FUGITIVE EMISSIONS FROM WASTE TREATMENT AND DISPOSAL

Many industries have wastewater treatment facilities either for treating the wastes sufficiently for direct discharge to a receiving stream or for pretreatment before waste discharge to a municipal sewer. Because of the nature of these pollution control devices, they are often significant sources of fugitive emissions to the atmosphere. Many units require vigorous mixing and turbulence, as in chemical mixing/flocculation, diffused air flotation, or the activated sludge process; or aeration, as again in the activated sludge process or diffused air flotation, or in chemical oxidation using air. Other units contain more quiescent liquids but require large expanses of surface area exposed to the air, as in settling/clarification basins, biofilm treatment systems, lagoons, and sunlight-catalyzed oxidation systems. Land application of wastewaters is another significant source of fugitive emissions. All of these processes promote the loss of volatile contaminants to the atmosphere, and all must be controlled.

Computer models are now available to estimate the emissions from a given treatment unit or process train. These must be able to account for VOC disappearance via a variety of pathways, including volatilization, biological decomposition, adsorption, oxidation-reduction, photochemical reaction, and hydrolysis. The importance of each of these pathways varies, depending on the specific process being analyzed (see Table 12.7). Among the VOC fate models available, only a few are currently in wide use

TABLE 12.7
Pathways for hazardous waste area emission sources

Pathway	Surface impoundments	Wastewater treatment plants		Land treatment	Landfill
		Aerated	Nonaerated		
Volatilization	I	I	I	I	I
Biodegradation	I	I	I	I	S
Photodecomposition	S	N	N	N	N
Hydrolysis	S	S	S	N	N
Oxidation-reduction	N	N	N	N	N
Adsorption	N	S	S	N	N
Hydroxyl radical reaction	N	N	N	N	N

Key: I = important; S = secondary; N = negligible or not applicable.
Source: U.S. EPA, 1987.

(Melcer, 1994; Melcer et al., 1994; Zhu et al., 1996). These include Water8 and TOX-CHEM. In addition, the process simulation package SuperPro Designer can be used to estimate VOC fates.

The causes of emissions from these treatment units are inherent in their design or operation. One of the only ways to minimize fugitive emissions from these units is to cover them, collect the escaping VOCs, and treat or destroy them. This can be done using the procedures described in the preceding section. Other control strategies might include minimizing turbulence at points where it is not needed (e.g., weirs or drop structures), using powdered activated carbon in the activated sludge process to trap VOCs so that they cannot volatilize, or switching to treatment processes that are less likely to have fugitive emissions, such as activated carbon adsorption or membrane processes (Shen, Schmidt, and Card, 1993).

REFERENCES

Allen, D. T., and Rosselot, K. S. *Pollution Prevention for Chemical Processes.* New York: John Wiley & Sons, 1997.

Bello, C. E., and Siegell, J. H. "Why Valves Leak: A Search for the Cause of Fugitive Emissions." *Environmental Progress* 16 (1997): 13–15.

Blickley, G. J. "Valve Users Get Help on Fugitive Emissions." *Control Engineering* 42, no. 10 (1995): 117–22.

Brown, K. H. "Fugitive Emissions Monitoring Trends." *Environmental Protection* 8, no. 2 (1997): 15–16.

Drago, J. "Limiting Fugitive Emissions from Flanges." *Plant Engineering* 49, no. 4 (1995): 79–80.

Maisel, B. "Making the Most of Air Emissions Measurement Requirements." *Pollution Engineering: Online,* November (1996), pp. 1–5.

Melcer, H. "Monitoring and Modeling VOCs in Wastewater Facilities." *Environmental Science & Technology* 28 (1994): 328A–35A.

Melcer, H., Bell, J. P., Thompson, D. J., Yendt, C. M., Kemp, J., and Steel, P. "Modeling Volatile Organic Contaminants' Fate in Wastewater Treatment Plants." *Journal of Environmental Engineering* 120 (1994): 588–609.

Shen, T. T., Schmidt, C. E., and Card, T. R. *Assessment and Control of VOC Emissions from Waste Treatment and Disposal Facilities.* New York: Van Nostrand Reinhold, 1993.

Siegell, J. H. "Control Valve Fugitive Emissions." *Hydrocarbon Processing* 76, no. 8 (1997): 45–47.

Sullivan, R. A. "From Dirt to Toxic Dust." *Environmental Protection* 8, no. 5 (1997): 12–13.

U.S. EPA. *Compilation of Air Pollutant Emission Factors,* 5th ed. AP-42. Washington, DC: U.S. EPA, 1997.

U.S. EPA. *Control Techniques for Fugitive VOC Emissions from Chemical Processing Facilities.* EPA-625/R-93-005. Washington, DC: U.S. EPA, 1993.

U.S. EPA. *Hazardous Waste Treatment, Storage and Disposal Facilities (TSDF): Air Emission Models.* EPA-450/3-87-026. Washington, DC: U.S. EPA, 1987.

U.S. EPA. *New Regulations Controlling Emissions from Dry Cleaners.* EPA-453-F-94-025. Washington, DC: U.S. EPA, 1994.

U.S. EPA. *Protocol for Equipment Leak Emission Estimates.* EPA-453/R-95-017. Research Triangle Park, NC: U.S. EPA, 1995.

"Valve Packings Conquer Fugitive Emissions." *Chemical Engineering* 102, no. 11 (1995): 151–52.

What Is LDAR? http://www.orrsafety.com/ldarlow.htm (1997).

Wolz, D., Winkel, L., and Paul, B. O. "Controlling Fugitive Emissions by Retrofitting Valves." *Chemical Processing,* April 1995.

Zhu, H., Orton, T. C., Keener, T. C., Bishop, P. L., and Khang, S. J. "Characterizing VOC Emissions from Aeration Units." *Proceedings of the Water Environment Federation WEFTEC '96,* Alexandria, VA: Water Environment Federation, Vol. 6 (1996): 363–74.

PROBLEMS

12.1. A medium-sized refinery has 80,000 components that must be evaluated for fugitive emissions. Of these, 60,000 are connectors, 9000 are light liquid valves, 6000 are heavy liquid valves, 4000 are gas valves, 340 are pressure relief valves, 400 are open-ended lines, 100 are light liquid pump seals, 60 are heavy liquid pump seals, and 100 are compressor seals. The light liquid process fluid contains 90 percent by weight organic compounds and 10 percent water. Determine the fugitive emissions from this refinery using the average emissions approach.

12.2. For the refinery in Problem 12.1, assume that a screening survey of the components found that all of the open-ended lines, 30 percent of the valves, 25 percent of the connectors, and 20 percent of all of the other components emitted hydrocarbons in excess of 10,000 ppmv TOC. Estimate the refinery's fugitive emissions using the screening ranges approach.

12.3. The refinery described in Problem 12.1 decides to check its fugitive emissions using the EPA correlation approach. It determines that the screening values are as follows: valves, 50,000 ppmv; pump seals, 70,000 ppmv; connectors, 65,000 ppmv; flanges, 30,000 ppmv; and open-ended lines, 450,000 ppmv. Assume all components leak. What is the estimated fugitive emissions rate using this procedure?

12.4. An SOCMI facility has 2000 connectors, of which only 50 (2.5 percent) exceed 10,000 ppmv. The screening value is determined to be 220,000 ppmv. What percentage of the plant's total fugitive emissions is contributed by these 50 connectors?

12.5. The U.S. EPA's TANKS 3.1 model for simulating fugitive emissions losses from tanks is available on the Internet at http://www.epa.gov/ttn/chief/ap42c7.html. Using this model, estimate the amount of chloroform that would be released from the tank described in Example 12.3 and compare the results with the numerical solution determined there. State all other assumptions you make.

CHAPTER 13

MUNICIPAL POLLUTION PREVENTION PROGRAMS

13.1 INTRODUCTION

Industry-initiated pollution prevention activities are increasing, particularly among larger companies which have experienced the advantages of preventing pollution at the source and have the monetary capabilities to see projects to completion. Pollution prevention is becoming more desirable than pollution production followed by cleanup at a later period. Many industries, though, have not yet initiated P2 in a significant way and continue to discharge unnecessary wastes into municipal sewers. As a result, publicly owned treatment works (POTWs) are increasingly experiencing system overloads, resulting in inadequate wastewater treatment before discharge to receiving streams.

An option available to the POTW is to expand the facility to meet this increased demand, but this is usually very expensive. Industrial pretreatment programs are required at all POTWs. In a pretreatment program, before using the municipal sewer, industries must receive a permit, which restricts discharges of certain contaminants. These restrictions usually consist of concentration-based limits, although in some cases, mass-based limits or volume limits are also imposed to prevent industries from reaching the concentration limit by merely diluting the wastewater with clean water. In almost all cases, pretreatment programs are based on end-of-pipe treatment. A second option available to municipalities is to institute a pollution prevention program as part of the pretreatment program. A successful municipal P2 program would benefit both the industry and the municipality, because the industrial process would be made more efficient and less waste would be generated that requires treatment.

To date, only a few municipalities have fully implemented pollution prevention programs, but many more are beginning to put them in place or to study how to structure them. This chapter discusses elements of effective municipal pollution prevention programs and how to properly implement them. The goal of such a P2 program is to develop a well-structured waste reduction protocol that is not a burden to industry and that benefits all participants, monetarily and environmentally.

The remainder of this chapter will lead to the development of a pollution prevention protocol that can be enacted by sewerage agencies. This goal is achieved by reviewing pollution prevention regulations, the essence of pretreatment programs, elements of pollution prevention programs, preferences of POTWs and industries, and the pollution prevention program development flowchart.

13.2 REGULATORY BASIS FOR POLLUTION PREVENTION PROGRAMS

Pollution prevention has been an afterthought of previous regulatory efforts which focused on controlling and cleaning up the most acute environmental problems. Hence the development of regulations and performance standards has proceeded with a technology orientation and resulted in controls that are based on the treatment of pollutants rather than the prevention or elimination of pollutants.

The impetus for pollution prevention has been an indirect message in much of the nation's environmental legislation. Major statutes that address pollution control and management in a specific medium acknowledge some role for prevention as a national policy or a goal for public sector programs; they and other statutes provide a strong incentive for pollution prevention by including expenses and responsibilities for properly managing existing pollution. Congress first emphatically endorsed hazardous waste minimization in the Hazardous and Solid Waste Amendments of 1984, a reauthorization of the Resource Conservation and Recovery Act. This act introduced pollution prevention into the hazardous waste regulatory scheme by stipulating that hazardous waste generators must have a waste minimization program in place and that they must report the results of their waste minimization efforts (Ross and Konrad, 1993).

The following sections discuss relevant environmental statutes; note that not all of the major environmental statutes are discussed in the following section. This is because many environmental statutes focus on the control, management, and disposal of pollutants, rather than their prevention. Although many of these environmental statutes do not directly authorize or mandate pollution prevention, many indirectly promote pollution prevention by establishing regulatory programs that increase the cost, potential liability, and public scrutiny associated with managing hazardous materials. A good example is the Comprehensive Environmental Response, Compensation, and Liability Act (CERCLA). Thurber and Sherman (1995) note that CERCLA is not a pollution prevention statute, but it does promote pollution prevention through its pervasive liability scheme.

13.2.1 Resource Conservation and Recovery Act and Hazardous and Solid Waste Amendments

The Resource Conservation and Recovery Act (RCRA) addresses the management of solid waste, hazardous waste, and underground storage tanks that contain petroleum or hazardous substances. The act establishes a comprehensive regulatory scheme applicable to hazardous wastes. However, the RCRA's hazardous waste provisions regulate wastes after they are generated; they generally do not authorize the EPA to regulate in-process materials. Hence, the RCRA does not provide extensive authority to mandate pollution prevention—but the statute does provide some authority for addressing pollution prevention.

The Hazardous and Solid Waste Amendments (HSWA) added several new provisions to the RCRA, some of which do require pollution prevention. These provisions make it clear that pollution prevention is a fundamental element of the United States hazardous waste management policy. This statute states that "wherever feasible, the generation of hazardous waste is to be reduced or eliminated as expeditiously as possible." These amendments mandate that hazardous waste generators (i.e., industrial users) and treatment, storage, and disposal facilities have waste minimization programs in place. Such programs must exist to the extent that they are economically practical.

With the passage of the HSWA, the reduction or elimination of hazardous waste generation at the source was given legal priority over the management of hazardous wastes after they are generated. The RCRA legislation applied to hazardous waste generators that transport their wastes off site. It declares that the generators are required to certify on their hazardous waste manifests that they have programs in place to reduce the volume or quantity and toxicity of hazardous waste generated to the extent economically practicable. These programs are preventive in nature.

13.2.2 Emergency Planning and Community Right-to-Know Act

The primary goal of the Emergency Planning and Community Right-to-Know Act (EPCRA) is to provide information to the public concerning the presence and release of toxic and hazardous chemicals in the community. To fulfill this purpose, the EPCRA requires that certain companies that manufacture, process, and use chemicals in specified quantities must file written reports, provide notification of spills/releases, and maintain toxic chemical inventories.

Companies that meet certain requirements must submit an annual report of releases of listed toxic chemicals pursuant to EPCRA Section 313, known as the Toxic Release Inventory (Form R). These requirements for reporting apply to companies (1) with 10 or more full-time employees; (2) that fall under the Standard Industrial Classification codes 20 through 39; and (3) that manufactured or processed a reportable toxic chemical in quantities exceeding 10,000 pounds.

In 1994, more than 300 toxic chemicals and 20 chemical categories were subject to reporting (Ohio EPA, 1995). With the passage of the Pollution Prevention Act (PPA), new requirements were added to expand and make mandatory the source reduction and recycling information that is reported for the EPCRA list of toxic chemicals. Dunams

(1997) found in his analysis of Hamilton County in Ohio that the TRI indicates that more than 5.5 million pounds of toxic material was discharged to Hamilton County's POTWs (nearly 60 percent of Ohio's overall discharges to POTWs).

The EPCRA is a reporting and public right-to-know law and not necessarily a pollution prevention statute. It is nevertheless one of the primary drivers of pollution prevention through the publicity it generates in this current era of environmentalism.

13.2.3 Clean Water Act

The Clean Water Act (CWA) establishes legislation to restore and maintain the chemical, physical, and biological integrity of the nation's water. The act contains five main components that work to accomplish this goal:

1. Technology-based, industry-specific minimum national effluent (i.e., wastewater discharge) standards.
2. Water quality standards.
3. A permit program for discharges to U.S. bodies of water.
4. Specific provisions applicable to certain toxic and other pollutant discharges (oil, hazardous chemical, etc.).
5. A revolving POTW construction loan program.

The primary focus of this act is to ensure that toxic levels of pollutants are not discharged to the nation's waters. This is achieved by restricting the types and amounts of pollutants that are discharged. Such restrictions are imposed through the use of enforceable effluent standards specified in National Pollution Discharge Elimination System (NPDES) permits. By nature, the NPDES permit program relies primarily on treatment (i.e., pollution control) as the principal means of achieving compliance with discharge restrictions. In this sense, the CWA does not focus on pollution prevention. However, CWA programs do include significant pollution prevention components, such as the development of effluent standards and best management practices (BMPs), which are inherently considered pollution prevention practices.

Recently, pollution prevention practices have become part of the NPDES program, working in conjunction with BMPs to reduce potential pollutant releases. Best management practices have become recognized as an important part of the CWA's NPDES permitting process in preventing the release of toxic and hazardous chemicals (Dennison, 1996). Over the years, as BMPs for many different types of facilities have been developed, numerous case studies have demonstrated not only the successes but the flexibility of the BMP approach in controlling releases of pollutants to receiving waters. Based on the authority granted by the CWA regulations, BMPs span the universe of pollution prevention activities encompassing production modifications, operational changes, materials substitution, materials and water conservation, and other such measures.

Best management practices may be divided into general BMPs, applicable to a wide range of industrial operations, and facility-specific BMPs, tailored to the requirements of an individual process. General BMPs are widely practiced measures that are independent of chemical compound, source of pollutant, or industrial category; they are typically low in cost and easy to implement. Common general BMPs include facility housekeeping, preventive maintenance, inspections, security, employee training, and record keeping and reporting.

Facility-specific BMPs are measures used to control releases associated with individually identified toxic and hazardous substances and/or one or more particular ancillary sources. Facility-specific BMPs will vary from site to site depending on chemical toxicity, proximity to water bodies, proximity to populace, climate, age of facility or equipment, process complexity, engineering design, employee safety, and environmental release history.

13.2.4 Pollution Prevention Act

Since 1984, the direct references to pollution prevention, albeit minor, have been expanding in federal legislation. The Pollution Prevention Act (PPA) of 1990 is notable among federal statutes in this regard; it overtly supports pollution prevention, focuses on reducing the volume and toxicity of wastes at the source, and directs the EPA to implement pollution prevention activities as the national environmental management policy of the United States. In its findings, Congress stated that source reduction opportunities often went unexploited because of a variety of factors: existing regulations and industrial resources were focused on treatment and disposal; the applicable regulations did not require or address a multimedia approach to pollution prevention; and there was a lack of essential information on source reduction technologies that industry needed to overcome institutional barriers to source reduction (Thurber and Sherman, 1995). The institutional barriers will be discussed later. With the passage of the PPA, the first real opportunity to change the pollutant production culture arrived.

In particular, the act advances the following four approaches (Ross and Konrad, 1993):

- Funding of state technical assistance programs.
- Integration of pollution prevention into EPA activities.
- Establishment of a pollution prevention information clearinghouse.
- Reporting on pollution prevention activities as an addition to the Toxics Release Inventory requirements.

Note that these approaches are largely nonregulatory, with the exception of the reporting requirement.

The framework for a preventive approach in the United States was laid out in law as early as 1984 in HSWA and RCRA, but it was not until passage of the Pollution Prevention Act that the goals of pollution prevention were seriously addressed. The HWSA introduced the term *waste minimization,* which includes source reduction and recycling. Interestingly, the subsequent Pollution Prevention Act of 1990 did not specifically define pollution prevention, but it did define source reduction. Source reduction covers any practices that reduce or eliminate the creation of pollutants; it includes equipment or technology modifications; process or procedure modifications; reformulation or redesign of products; substitution of raw materials; and improvements in housekeeping, maintenance, training, or inventory control.

Specific provisions of the PPA require the U.S. EPA to establish source reduction programs and other practices that reduce or eliminate the creation of pollutants and provide states with information and technical assistance on source reduction. Thus source reduction, recycling, and other waste minimization efforts are fast becoming a significant environmental regulatory compliance issue. The EPA has also used this act to reinforce its environmental management options hierarchy. This hierarchy of pollution management, in order of priority, is source reduction, recycling, treatment, and disposal. The EPA defines pollution prevention as source reduction, while some states also include recycling (in-process or on-site recycling). The latter two elements in the P2 hierarchy should be employed only as a last resort, and then only in an environmentally safe manner.

Prior to the PPA, the effluent guideline development process was focused on the water medium alone and was treatment oriented. After the PPA was enacted, the EPA began to emphasize multimedia effects, waste reduction, raw material substitution, and in-process recycling in the development process. The future will undoubtedly hold less reliance on treatment technology as the primary means to achieve the goals of the CWA.

The PPA is basically an enabling act which states congressional commitment to waste reduction and recycling activities and which mandates that the EPA implement pollution prevention strategies and regulations. The PPA also requires that companies report their pollution prevention and recycling practices for each facility and each toxic chemical under SARA Title III, also known as EPCRA. Moreover, the PPA specifies that facilities are required to report all multimedia releases.

13.2.5 Regional Pollution Prevention Initiatives: Great Lakes Water Quality Initiative

In addition to national legislation that deals with pollution prevention, there are also a number of regional initiatives. One regional program is the Great Lakes Water Quality Initiative (GLI), an agreement between the United States and Canada to prohibit discharge of toxic substances in toxic amounts and to eliminate discharge of persistent toxic substances into the Great Lakes Basin. This program differs from the current national water quality program by using bioaccumulation factors that account for direct uptake from the waters of the Great Lakes system, plus uptake from the food chain (WEF, 1996). This initiative will affect all Great Lakes states and will provide for lower limits, especially on metals, even for wastewater treatment plants (WWTPs) that do not discharge directly into the Great Lakes Basin.

The CWA requires that the Great Lakes states adopt water quality standards that are "consistent with" the Great Lakes Water Quality Initiative by March 23, 1997. The following benchmarks are to be used in setting standards (National Wildlife Federation, 1997):

- Are the state's environmental programs sufficient to control the most important sources of bioaccumulative toxics that are fouling the Great Lakes and inland lakes and rivers?
- Do the state's environmental programs properly prioritize the most dangerous chemicals—those that bioaccumulate and persist in the environment?
- Are the state's environmental programs consistent with the U.S.–Canada Great Lakes Water Quality Agreement and the Great Lakes Governors Toxic Substance Control Agreement?

Specifically, the GLI urges the Great Lakes states to control diffuse sources of pollution and implement pollution prevention programs. Within the GLI, pollution prevention is clearly defined and required for all dischargers seeking variances or lengthened compliance schedules to meet permit limits.

13.3 SOURCE CONTROL AND PRETREATMENT PERMIT PROGRAMS

13.3.1 Mandatory Programs

The Clean Water Act sets the framework for the imposition of industrial wastewater control programs on municipalities and the subsequent setting of regulations by municipalities on industrial users. Specifically, Sections 307(b) and (c) of the CWA set forth the authority for the EPA to establish pretreatment standards for existing and new sources discharging industrial wastewater to POTWs. The CWA also states that it is unlawful for any industrial user to operate a facility in violation of the pretreatment standards established by the EPA.

The CWA sets requirements for the development of local pretreatment programs for all POTWs that have a total design flow greater than 5 Mgal/day and that receive toxic discharges from industrial users that may pass through or interfere with the operation of the POTW or otherwise be subjected to pretreatment standards. This is provided for by an NPDES permit issued to a POTW. This permit includes conditions that require, among other things, identification of the character and volume of industrial pollutants and a program to ensure compliance with EPA pretreatment standards. The NPDES permit, at a minimum, needs to contain a requirement that a POTW complies with its approved pretreatment program. The NPDES permit issued to a POTW becomes an independent source of legal obligations imposed on the POTW for implementing and/or enforcing pretreatment requirements on industrial users (IUs) permitted through the pretreatment program. Permitted industrial users are typically classified as categorical or significant industrial users. Categorical IUs are industrial sectors categorized by the CWA as being heavy and chronic dischargers. Significant IUs discharge over 25,000 gal/day of process wastewater into the sewer system and have violated or have the potential to violate pretreatment standards or adversely affect the wastewater treatment plant.

Industrial pretreatment programs, an initial solution to escalating wastewater problems, were designed to shift the burden of handling pollutants back to the generators. However, Lindsey, Neese, and Thomas (1996) found that traditional pretreatment programs have been flawed by ineffective enforcement and suffer the theoretical shortcoming of dealing with a problem after it exists, rather than identifying the problem at the source and modifying the system to prevent it from occurring again. The U.S. EPA (1993) states that even with full compliance with categorical pretreatment standards, categorical industrial users would continue to discharge 14 million pounds of toxic metals and 51 million pounds of toxic organic pollutants to POTWs each year. Therefore, there is a need to incorporate pollution prevention within the treatment plant pretreatment infrastructure to deal with the root of the problem.

More than any other public authority, POTW pretreatment program personnel maintain close contact with local industrial dischargers and have an understanding of their specific industrial process operations and waste streams. Hence pretreatment personnel have many opportunities to encourage and to educate industries on adopting pollution prevention measures. Epstein and Skavroneck (1996) found, based on a 1992 survey, that pollution prevention has been successfully integrated into some existing pretreatment programs. The extent and successes encountered with such implementation were not documented.

In implementing the federal pretreatment regulations, POTWs should have the authority to promote pollution prevention in a number of capacities, such as requiring spill control plans and toxic organic management plans. To incorporate pollution prevention planning or other pollution prevention requirements into permitting and enforcement actions, however, POTWs might need to expand their authority; this is discussed later in this chapter. Obtaining the legal authority to mandate pollution prevention is where POTW staff may have problems with incorporating such measures. POTWs need to address the fact that activities promoting pollution prevention are integral to meeting the goals of the local pretreatment program. They can use enforcement discretion, which is inherent in a pretreatment program, to provide incentives to pursue pollution prevention projects. In addition, POTWs have the authority to prohibit the discharge of chemicals that interfere with permit compliance, solids disposal, or marketing of biosolids products. This authority can be used to ensure that sewer dischargers implement pollution prevention strategies to meet the performance-based requirements of the pretreatment program. The most appropriate time to require pollution prevention planning and analysis is when industrial dischargers are seeking or renewing pretreatment permits. Within the procedure of regulating industrial waste, the particulars of pollution prevention should be incorporated in the WWTP's established pretreatment programs. The benefits of pollution prevention in, and to, pretreatment programs is twofold: (1) to assist in addressing current and anticipated compliance problems, and (2) to generally try to encourage opportunities to reduce toxic loadings to the sewers.

Nonregulated industries, which are usually small-quantity emitters, are not administered by sewerage agencies under the pretreatment program but are interesting candidates for pollution prevention education and other voluntary programs. These voluntary programs and the previously mentioned mandatory programs are discussed further later in this chapter.

13.3.2 Local Discharge Standards, Limits, and Authority

As more stringent requirements are imposed on POTW discharges, municipalities must increase their control over what is discharged to the POTW system. The EPA has determined that WWTPs with pretreatment programs must develop local discharge standards or prove that none are required. Local discharge standards are required to enforce the general pretreatment prohibitions and to prohibit the discharge of chemicals that interfere with WWTP operations, that pass through and violate the NPDES permit, that affect the disposal or marketability of biosolids, and that may affect the health and safety of WWTP workers and the general public. These local discharge standards often drive regulated facilities to examine pollution prevention as a means of achieving compliance with the regulations and as an alternative to end-of-pipe pretreatment technologies.

Local authority for regulation of industrial users is set forth in ordinances, rules, and regulations or other such enactments of a local government agency or municipality. The authority for a municipality to enact local requirements is driven by state and federal law. The CWA provides POTWs with the legal authority to restrict industrial and commercial pollutants from the wastewater they receive. To protect a WWTP from such problems, the CWA endorses local action beyond federal pretreatment regulations. However, unless a sewerage agency can document a problem with pass-through, interference, or threats, the CWA does not endorse going beyond federal pretreatment regulations (Sherry, 1988). Therefore, although federal law or regulations may require POTWs to have certain authority under certain conditions, they do not provide the POTW with the authority to enact local requirements. The federal regulations set forth minimum requirements for implementation of a pretreatment program. These regulations, however, can be and have been improved to provide greater control of industrial user activities or to facilitate the POTW's implementation activities. A number of policy and program decisions must be made to require industrial users to undertake pretreatment activities beyond the federal minimum. The municipality, however, should keep in mind the potential liability associated with a commitment to implement requirements beyond the federal minimum (WEF, 1996).

A broad authorizing authority provides the municipal entity the power to enact local laws, regulations, or other enforceable requirements, setting forth specific requirements. Municipal action setting forth the particulars of an ordinance and regulations would be necessary for the municipal entity to establish enforceable industrial user requirements. This authority has been shown in places such as Palo Alto, California, and New Hampshire, where sewer ordinances have been modified to include applicable pollution prevention requirements.

Hence industrial waste control can be effective only if there is legal authority for the municipality to develop waste control programs, inspect industries, monitor industrial waste characteristics, and enforce compliance with these programs (WEF, 1996). The ability of a POTW to enforce a municipal industrial waste program is strictly controlled by the power granted to the municipality by the state and by the enabling legislation adopted by the state legislature.

13.4 TYPES OF POTW-ADMINISTERED POLLUTION PREVENTION PROGRAMS

Publicly owned treatment works were the recipients of wastewater from an estimated 30,000 significant industrial users (SIUs) in 1991 (Knight, 1995; Lindsey, Neese, and Thomas, 1996). The EPA further estimates that even if all the categorical pretreatment standards were met by these SIUs, discharges would still exceed 6400 tons of toxic metals and 23,000 tons of organics each year (Lindsey, Neese, and Thomas, 1996). By reducing the quantity and toxicity of IU discharges beyond pretreatment standards, pollution prevention can provide a means for POTWs to meet increasingly stringent federal and state environmental quality standards, reduce the transfer of influent contaminants from wastewater to other environmental media, improve POTW worker safety, reduce sewer collection system hazards from toxic or hazardous gases, and reduce POTWs' high sludge management costs.

In response to continued strengthening of wastewater discharge permit requirements, WWTPs need to conduct aggressive pollution prevention programs to comply with such stringent limits and to proactively reduce emissions. POTW programs should initially focus on reducing discharges of pollutants of concern, which have the most potential to impact the environment. Other factors that should be considered in setting pollution prevention program priorities include violations of environmental standards (by both POTW and IU), compliance with permit requirements, previous actions to reduce discharges of specific pollutants, and the ability and authority of the treatment plant to achieve discharge reductions of the pollutant (PARWQCP, 1996). Each pollutant of concern should be monitored to identify the cause and sources, and then potential pollution prevention measures should be investigated.

Once pollutants of concern and users of concern are targeted, a pollution prevention strategy can be developed. Incorporation of pollution prevention strategies into POTW pretreatment programs offers many opportunities for improving efficiency of both the POTW's operations and those of its industrial users. As stated previously, POTWs do have the authority to require permitted industrial users to meet discharge limits through pollution prevention activities. Hence POTWs have an enforcement mechanism in place (i.e., the pretreatment permit program) to regulate permitted IUs. However, wastewater from small industrial and commercial sources is largely unregulated. The approach aimed at nonregulated industries is typically voluntary. In this section, mandatory and voluntary programs and their development are discussed.

13.4.1 Voluntary Programs

The EPA has issued strong policy initiatives to advance nationwide pollution prevention activities. However, regulatory, economic, and cultural barriers as well as disparity in the availability of information essentially combine to form an impediment to pollution prevention opportunities. They thus create a business climate that focuses company resources and corporate policy on end-of-pipe solutions rather than solutions embedded in the raw materials or processes of industry (Atcheson, 1995). Voluntary programs became one of the early, primary strategies used to unleash and stimulate the market potential (by overcoming economic barriers) for pollution prevention programs. Moreover, voluntary programs are seen as a tool to foster innovation, stimulate information exchange, and, ultimately, change the culture surrounding environmental compliance. Therefore, voluntary programs incorporated the first wave of pollution prevention implementing devices, because the main barrier to overcome was a change in the culture of industrial decision making—a culture that would not be subdued by mandatory programs.

In the recent past, POTWs administered a limited number of pollution prevention programs. Of these, very few have mandatory aspects; rather, the EPA and state organizations have tried to encourage voluntary commitments to pollution prevention. These voluntary efforts have been promoted largely through participation in various programs, such as the 33/50 Program.

The 33/50 Program is one of the earliest voluntary programs. This program was based on a challenge to industry to reduce emissions of 17 of the most ubiquitous and toxic chemicals by 33 percent by 1992 and 50 percent by 1995. Epitomizing a flexible and voluntary approach, this program allows participating industries to choose their preferred methods for reducing releases, which need not be pollution prevention, as long as chemical discharges are reduced. In exchange, participants would receive assistance, be given recognition, and be eligible for awards. Although effective, this program did elicit its share of criticism. Atcheson (1995) states that the primary objections were that the program merely rewarded ongoing efforts and did not encourage new activities. Also, the expedience pushed participating industries to achieve results quickly, often foisting pollution control technology over pollution prevention, since pollution control programs were proven and were considered the standard. Voluntary programs described in this chapter, however, do stress the benefits of pollution prevention.

The emphases of voluntary programs are twofold: (1) to promote pollution prevention among permitted industrial users in a nonregulatory manner and (2) to address nonregulated industries that have neither mandatory requirements nor pretreatment limits to comply with. Therefore, voluntary programs are applicable to both regulated and nonregulated industries.

Voluntary provisions are set forth to encourage source reduction, waste minimization, and recycling activities; these provisions attempt to achieve pollution reduction goals through elective practices and programs. Traditionally, voluntary programs have been initiated through educational outreach and technical assistance mechanisms. Newly developed programs, however, are initially voluntary; subsequently, guidelines are followed to mandate goals for reduction efforts. These types of programs are discussed in this section.

TYPES OF VOLUNTARY PROGRAMS.

Educational Outreach and Technical Assistance. Voluntary programs by POTW pretreatment staff have generally focused on educational outreach and technical assistance mechanisms with their industrial users to promote use of pollution prevention technologies and techniques. Educational outreach programs provide pollution prevention information to local industries through workshops, advisory groups, on-site inspections, and routine correspondence. Educational outreach can be executed by sponsoring workshops and training sessions, convening local pollution prevention forums, publicly recognizing pollution prevention achievements, compiling and distributing pollution prevention information, and publicizing industrial waste exchanges. These efforts employ all of the traditional tools that a POTW has at its disposal to communicate information to the business community and also allow the POTW to act as a medium for information transfer. This act of educating and convincing industry of the importance of pollution prevention may be the most effective way to make pollution prevention viable, rather than just a compliance program (Boyer, 1993).

Technical assistance programs provide industries with a technical service to identify and evaluate site-specific opportunities for pollution prevention efforts. Technical assistance programs (TAPs) are usually more costly than educational outreach, since TAPs require trained staff who can provide specific pollution prevention advice. For POTWs to institute TAPs, pretreatment inspectors need training that will help them understand specific pollution prevention techniques and the industry-specific options that exist for eliminating pollutants.

In addition, POTWs can encourage their industrial dischargers to adopt best management practices such as inventory controls, employee training, and basic maintenance and inspection activities. In general, BMPs can be implemented at little or no cost and can often achieve significant reductions in toxic discharges. Usually, small industrial and commercial users may not be aware of simple BMP steps to cleaner, more efficient operations and could benefit from the POTW's guidance. Typically, BMPs are a component of TAPs.

One of the most effective ways that POTWs can identify potential pollution prevention measures and promote pollution prevention, in general, is to explore opportunities during routine facility inspections. During inspections of permitted industries, industrial investigators can discuss pollution prevention opportunities and highlight specific pollution prevention program elements applicable to each industry. Visits and follow-up discussions provide significant opportunities to identify specific pollution prevention opportunities, to distribute pollution reduction literature related to each individual industrial process, and to educate staff at each industry about pollution prevention. Pollution prevention opportunities can be enhanced because POTW officials and pretreatment inspectors can use their longstanding relationships with industrial users to facilitate the spread of pollution prevention practices.

The program of waste minimization audits for industrial users by the City of San Francisco (1996) incorporates technical assistance and educational outreach. The city's Bureau of Environmental Regulation and Management (BERM) instituted waste minimization audits for industrial and commercial businesses to identify and reduce pollutants in wastewater that may exceed discharge limits or impact the San Francisco Bay without initiating labor-intensive long-term permitting programs. The BERM's waste management audit program focuses on specific industrial sectors.

The San Francisco program's approach includes the following components:

- Developing educational materials about pollution prevention and best management practices.
- Visiting each industry.
- Assessing each industry's current pollution generation and pollution prevention activities.
- Encouraging voluntary adoption of BMPs through correspondence.
- Revisiting selected industries to assess their efforts toward preventing pollution.

Each industrial sector approached (automotive service facilities, commercial printing, metal finishing, etc.) has a pollution prevention checklist and an information package that targets specific and proven industrial pollution prevention techniques.

The City of Albuquerque (1993) has a similar program that is a nonregulatory technical assistance program for industries which focuses on the latest waste reduction methods and technologies. This program differs by including a reporting requirement, which asks each industry to include a narrative statement with each semiannual report describing any source reduction, waste minimization, or pretreatment efforts undertaken during the period in review. If no such pollution prevention efforts are undertaken, a statement to that effect must be included in the report. Hence the city extends information on pollution prevention technologies and practices through educational outreach and TAPs, and the industries voluntarily report on actions taken.

The City of Palo Alto has developed a technical assistance program called the Pollution Prevention Technology Access Program. The objectives of this program are to (1) demonstrate successful pollution prevention techniques; (2) assist industry in identifying environmental and economic benefits of pollution prevention; (3) provide companies with technical resources to identify and implement pollution prevention projects; and (4) recognize companies that implement pollution prevention projects (PARWQCP, 1996). The first objective, P2 demonstrations, looks at the complete process of a cooperative pollution prevention implementation project between the city and a chosen industrial user. The sewerage agency puts out requests-for-proposals (RFPs) aimed at specific sectors of industry to initiate a comprehensive pollution prevention project. After selection and during project implementation, the POTW provides technical oversight and equipment for testing wastewater samples; it also partially funds the project. The demonstration project encourages rapport between industry and the POTW. The industry works with the POTW in development of a pollution prevention technique that provides future compliance and cost savings for the industry. Best of all, it gives similar industries ideas about possible pollution prevention opportunities and reaffirms the ideal of working with a regulatory agency to obtain a common goal—the reduction of pollutants in the waste stream.

Voluntary Reduction Programs. These programs deal with reductions in toxic effluents achieved through a voluntary approach. The City of Albuquerque (1993) has a voluntary reduction program geared to nonpermitted industries that also includes the educational and technical assistance aspects (i.e., BMPs are established), as noted above, but adds a certification requirement and recognizes cooperating businesses through sampling and analysis of wastewater discharges. This program, the 5 Parts Per Million (PPM) Silver Program, has established constructive working relationships with businesses to achieve voluntary compliance of silver wastewater discharges. Industries may participate in this voluntary program as an alternative to typical pretreatment and monitoring programs. Components of this program include (1) sewer agency distribution of silver reduction information and assistance in developing reduction techniques; (2) IU implementation and maintenance of a silver management plan; (3) IU provision of records detailing the fate of silver; and (4) sewer agency visits to the facility yearly and provision of sampling services for the IU. The program approach taken is proactive and is not promoted as being enforcement oriented. For example, an industry that does not meet the 5 PPM standard must work with the city's Pollution Prevention Program to find an appropriate method to reduce silver discharges within a certain time limit. Therefore, noncompliance leads to a working relationship with the Pollution Prevention Program and not an enforcement action such as a fine. As noted in the previous City of Albuquerque program, a narrative statement describing the silver reduction program and sampling results or evidence of participation in the 5 PPM Silver Program is a requirement. This program requires annual recertification and awards certificates to individual businesses.

The Buffalo Sewer Authority (BSA) has a similar voluntary and nonregulatory program that was initiated to counteract problems encountered during heavy rains for combined sewer overflow (CSO) systems (BSA, 1997). During significant rains, toxic contaminants are discharged directly into receiving waters through CSOs. In this situation, pretreatment programs do not have the opportunity to provide the protection that a pollution prevention approach would provide, since pollution prevention activities are aimed at reducing or eliminating toxic discharges. Pollution control activities are aimed at minimizing the effects of industrial pollutants on WWTPs, but for plants experiencing CSO problems, portions of the incoming wastewater stream will not even enter the wastewater treatment system. Hence pollution prevention can reduce the effects of toxic discharges during overloading.

Unlike the City of Albuquerque program, participating facilities in Buffalo are both permitted and nonpermitted industrial users. The Buffalo program consists of a nonregulatory pollution prevention technical assistance and educational outreach program, a sampling and analysis program, and encouragement from the sewer authority to industrial users to adopt multimedia source reduction techniques for eliminating both direct and stormwater discharges. Since the Buffalo Sewer Authority is in a Great Lakes state, additional emphasis must be placed on actual discharges and potential discharges involving bioaccumulative chemicals of concern. One important aspect of this program deals with the voluntary status. This program is currently voluntary due to sewer use ordinances that do not incorporate pollution prevention efforts as a viable alternative or addendum to the pretreatment program. This situation can be remedied as discussed in Section 13.3.

PROBLEMS WITH VOLUNTARY PROGRAMS. Voluntary programs, in reaching their goals, may employ mechanisms that hinder the achievement of these goals. These hindrances can affect the industry's involvement and can affect the implementation of voluntary programs by the POTWs.

With respect to industry, providing information and assistance can be valuable, and the free provision of information concerning pollution prevention can serve as an incentive that will induce firms to reduce their use of toxic substances. However, merely providing information and training may not be sufficient to overcome barriers or disincentives to pollution prevention, especially among smaller industries (Boyer, 1993).

Not only are there barriers to overcome, but competitiveness becomes an issue. Perception is heightened in cases where a company may be aware that its competitors are not involved in a voluntary program and therefore may have gained an unfair advantage over those involved in voluntary reduction programs. With a mandatory program, company X knows that company Y has the same requirements to be met; the same assurances cannot be said for voluntary programs.

With respect to the POTW, technical assistance programs may pose a liability problem. This concern is primarily related to a POTW assuming two potentially divergent roles: enforcer and adviser. Specifically, the following question arises: Does a POTW assume any responsibility for a wastewater discharge violation if its advice does not help an industry with a preexisting compliance problem or causes the violation (Sherry, 1988)? To protect against liability problems, POTWs need to clearly state that the technical assistance program is voluntary. Outside the legalities, POTW staff may find that their offerings of outreach and technical assistance activities conflict with their roles as enforcement authorities. This may cause the industry to be gun-shy about working with a municipality. This inconsistency can be remedied by locating the pollution prevention effort in a separate organization unit [e.g., the City of Albuquerque and the State of North Carolina house their pollution prevention efforts in dedicated Offices of Pollution Prevention (U.S. EPA, 1994)], not in the same line of supervision as the enforcement function, or by providing the outreach and technical assistance through a third party.

13.4.2 Regulatory and Enforcement Programs

As pollution prevention has gained prominence in recent years, the EPA and other regulatory bodies have sought to incorporate pollution prevention requirements into rules, inspections, adjudications, consent orders, and other aspects of their regulatory programs. Specifically, POTWs can promote pollution prevention among industrial and commercial dischargers by expanding or more aggressively enforcing existing requirements and/or by establishing new mandates. Hence the development of regulatory pollution prevention programs.

Regulatory and enforcement programs provide enormous opportunity to foster the development of industrial pollution prevention programs. These programs require industrial users to conduct pollution prevention opportunity assessments, to conduct pollution prevention training, to incorporate pollution prevention into management policy, and to possibly alleviate the severity of a violation. These activities can include developing and issuing user permits and responding to user noncompliance, among other activities. Before regulatory programs get to the development stage, there must be a regulatory mechanism to initiate these changes to pretreatment permits. These changes are initiated by modifying sewer use ordinances. This section looks at how sewer use ordinances can be modified and then subsequently lead to the development of mandatory pollution prevention programs.

MODIFIED SEWER USE ORDINANCES AND USER PERMITS. Mandatory pollution prevention programs applied to industrial users are the direct effects of modifications to the sewer use ordinance. Once pollution prevention ideals are incorporated in the regulatory backbone of the sewerage agency, pollution prevention elements can be adapted within industrial user permits. This modified permit envelops the whole realm of mandatory pollution prevention programs that can be applied to industrial users.

Publicly owned treatment works, with the appropriate authority established in the sewer use ordinances, can use the permitting process as an effective mechanism for instituting pollution prevention as a local requirement for industrial and commercial users. For example, the City of Palo Alto (1992) revised its sewer use ordinance to establish an authority to require waste minimization studies, new local limits for silver, and requirements for photoprocessors. Because existing effluent silver concentrations were 400 percent greater than the new effluent discharge limit and were causing bioaccumulation of silver in San Francisco Bay clams, the City of Palo Alto used its ability to revise the sewer use ordinance. Since most of the silver came from one source—the photoprocessing industry—the ordinances could be revised to grant the sewerage agency the authority to go after this nonregulated commercial category.

A suitable example for a model sewer use ordinance was developed by the New Hampshire Department of Environmental Services (1996). This model ordinance specifically includes language intended to promote pollution prevention activities as well as making a clear statement that pollution prevention is the primary purpose of the sewer use ordinance. Relevant areas of pollution prevention inclusion are found in sections that discuss local discharge restrictions, mass-based limits, pollution prevention plans, industrial discharge permit contents, and publication of pollution prevention measures. With regard to local limits and mandatory pollution prevention actions, the model ordinance states:

> Pollutants that exceed fifty percent of their allowable headworks loading at the wastewater treatment facility are considered to be of concern and have resulted in development of local limits. Pollutants that exceed twenty percent of their allowable headworks loading at the wastewater treatment facility are targeted for mandatory pollution prevention action.

This description is indicative of the pollution prevention aspects contained within this model ordinance and sets the tone for the inclusion of pollution prevention requirements.

The model sewer use ordinance outlines strategies that either directly require facilities to adopt certain pollution prevention practices or to create incentive structures that directly promote pollution prevention. These strategies include (1) requiring pollution prevention plans; (2) setting new local limits; (3) controlling discharges from small nonregulated industrial and commercial users; and (4) employing mass-based local limits.

Future modifications to industrial user permits could alleviate the focus of single-medium reductions. As described earlier, the single-medium focus of most federal regulations (Clean Water Act, Clean Air Act, etc.) indirectly means that pollutants can be regulated out of one medium and into another. Reducing pollution in this manner is ineffective for obtaining absolute reductions. However, permit modifications can incorporate multimedia or integrated permitting schemes. This incorporation will be significant for the overall development of an environmental management scheme. Incorporating the benefits of the foregoing strategies with multimedia applications, a modified permit would minimize cross-media transfers of pollutants and overall risk, as well as encouraging facilities to incorporate pollution prevention objectives.

POLLUTION PREVENTION PLANS. Currently, many POTWs require industrial dischargers to develop and implement solvent management plans and spill contingency plans to control the emissions of hazardous pollutants to the sewers (Sherry, 1988). A pollution prevention planning requirement would be a similar type of mandate, only with a more comprehensive goal. The goal is to eliminate or reduce hazardous pollutant loads to the sewers, without transferring the pollutants to another medium. POTWs can require industrial and commercial users to develop and submit pollution prevention plans as part of the permitting process. As discussed previously, POTWs may need to amend sewer use ordinances to provide them with the authority to require submission of pollution prevention plans. For example, the City of Palo Alto modified its sewer use ordinance in 1995 to provide the sewerage agency with the ability to require that new industrial facilities complete a waste minimization/pollution prevention plan (PARWQCP, 1996). This study strengthens a prior requirement by mandating that specific steps, included in the pollution prevention plan, be taken to minimize toxic constituents in discharged wastewater.

Pollution prevention plans, therefore, can require the industrial user to implement activities recommended in the plan. Also, as encountered in New Jersey, pollution prevention plans can include reporting requirements. This requirement compels industry to provide government officials with quantitative data pertaining to pollution prevention reduction goals and to also provide information on why failures occurred (Beardsley, Davies, and Hersh, 1997).

Typically, pollution prevention plans contain detailed and systematic assessments of a facility's ability to reduce the volume and toxicity of discharges through pollution prevention activities. A pollution prevention assessment or audit conducted by facility owners and operators or an outside group can be the single most effective means for identifying technically and economically feasible pollution prevention opportunities capable of achieving long-term reductions in the generation of toxic waste streams (U.S. EPA, 1993).

If industrial users have not already developed pollution prevention plans that address the waste streams destined for the sewers, pretreatment programs should consider exploring the possibility of incorporating a pollution prevention planning provision into the permitting process. Such a provision would require that a facility interested in renewing an existing permit or obtaining a new permit submit a detailed pollution prevention plan. Pollution prevention plans should consist of the following elements (U.S. EPA, 1993; Dennison, 1996):

- Specific performance goals for the prevention of pollution.
- A commitment to implement the plan to achieve plan goals.
- Employee awareness and training programs to involve employees in pollution prevention planning and implementation.
- A process flow diagram showing a mass balance of toxic constituents.
- An estimate of the amount of regulated waste generated by each process.
- Assessment of current and past pollution prevention activities, including an estimate of the reduction in amount and toxicity of regulated waste achieved by the identified actions.
- A listing of pollution prevention opportunities applicable to the facility's operations.
- Development of technically, economically, and environmentally feasible pollution prevention opportunities, including an assessment of the cost, benefits, and cross-media impacts of the identified opportunities.
- An implementation timetable.

Personnel can assist their industrial and commercial dischargers in developing pollution prevention plans by identifying pollution prevention opportunities (such as BMPs) during inspections, coordinating meetings specifically targeted at pollution prevention, and providing technical materials and contacts. The most direct means for achieving widespread implementation of BMPs is to require pollution prevention planning as a condition for obtaining or renewing a discharge permit. Industry-specific BMPs should be inherent in a pollution prevention plan, although typically they are developed in TAPs.

NEW LOCAL LIMITS. A POTW has the authority to require users to meet discharge limits set by their pretreatment permit and to set other requirements to prevent pass-through of toxic contaminants and disruptions of normal wastewater treatment operations. In general, setting local limits covering a wide range of contaminants, or just one pollutant of concern, for industrial and commercial sources provides a strong incentive for implementing pollution prevention measures. The cost of sewer user fees or pretreatment generally rises with the stringency of local limits; as this occurs, pollution prevention becomes a more desirable means to assist industrial and commercial users in meeting local limits.

New limits are typically set based on conclusions made from maximum headworks loading analyses that indicate that certain chemicals are reaching the maximum allowable limits. The effects of setting local limits based on allowable headworks loadings, as proposed in the New Hampshire model sewer use ordinance, were discussed earlier. Also, new regulations, such as the Great Lakes Initiative, will tighten the limits for some metals and for compounds that bioaccumulate.

As discussed previously, the City of Palo Alto (1992) developed local silver limits due to stringent POTW effluent requirements and the bioaccumulation of silver in clams. The new upstream limits were determined based on effluent limits set by the Regional Board and the expected treatment process removal rates. The upstream local limits were reduced from 5.0 mg/L to 0.25 mg/L for industries and 0.50 mg/L for commercial users (photoprocessors). In addition to developing local limits, the City of Palo Alto's Silver Pilot Program also consisted of enhanced enforcement (an administrative penalties program), waste minimization studies, and metals mass discharge fees. However, commercial users were required to adhere to the discharge limit, treat the spent fixer or haul it off site, self-monitor for silver, and submit an annual report. This is a significant contrast to the City of Albuquerque's voluntary silver reduction program.

MASS-BASED LIMITS. Most POTWs issue pretreatment permits specifying the allowable *concentrations,* typically in milligrams per liter, of certain contaminants in wastewater discharged to sanitary sewers. By increasing the volume of water discharged, a pollutant's concentration can obviously be easily lowered. Although prohibited by POTWs, dilution can be and is used by firms to comply with discharge limits. Enforcing dilution prohibition requirements can be difficult, but dilution activities are usually associated with noticeable increases in water consumption.

In contrast, a *mass-based limit* is a set mass of pollutant, expressed usually in pounds or kilograms per day, that is allowed to be released to the sewer over a specified period of time. Since mass-based limits are based on the actual pollutant load discharged, and not on the pollutant's concentration, there is no incentive to use dilution as a strategy to remedy a violation. Hence mass-based limits tend to encourage the conservation of water by industrial users.

In terms of pollution prevention, mass-based limits offer an excellent alternative to the traditional concentration-based limits. For example, eliminating one part of a waste stream through pollution prevention or reducing water consumption might cause a facility to increase pollutant concentrations, even though the total mass of the pollutants does not increase and might even decrease. In switching to mass-based limits, POTWs must have reliable data on industrial flow, along with the concentrations of pollutants in the wastewater, to monitor facilities accurately for compliance with the mass-based limits. Whereas reliable concentration data may be relatively easy to collect, accurate flow data may be more difficult to obtain (U.S. EPA, 1993). Maintaining concentration-based limits in conjunction with mass-based limits is an option that a POTW may follow after instituting mass-based limits (Sherry, 1988). This provides a POTW with a significant degree of enforcement flexibility in case mass-based data are unattainable due to flow measurement difficulties. With this type of regulatory arrangement, a POTW does not give up its ability to enforce its traditional concentration-based limits.

The City of Palo Alto devised a program of compliance options that industries must initiate, based on different types of monitoring. This program was specifically aimed at permitted metal finishers and printed circuitboard manufacturers to reduce the discharge of copper (PARWQCP, 1996). In 1994, service area cities of the sewerage agency adopted new sewer use ordinance requirements specific to these two industries. The first compliance option is the *reasonable control measure* (RCM) option, which involves the installation of proven reasonable control measures while meeting an annual average copper concentration limit of 0.4 mg/L in the discharge. Typically, RCMs are considered to be pollution control techniques, but in this context the RCMs are proven pollution prevention activities. The second option, called the *mass limit option,* involves performing a pollution prevention study [conducted by a regional water quality control plant (RWQCP)] and using the study to set a facility-specific annual copper discharge mass limit consistent with implementing identified pollution prevention measures. Both the RCM and the facility-specific measures are designed to have a simple payback period of five years or less and to meet safety and product quality criteria. The City of Palo Alto found that industry preference was split between the two compliance options (PARWQCP, 1996).

CONTROLLING DISCHARGES FROM NONPERMITTED INDUSTRIAL USERS. Commercial and small industrial dischargers (dental offices, hospitals, universities, photoprocessors, etc.) are typically not required to obtain discharge permits. These nonregulated industries, however, may represent a significant portion of the total loadings of a toxic pollutant entering a POTW. Recall the silver problems encountered in Palo Alto that were primarily due to photoprocessors, a nonregulated industry. In this situation, a POTW could benefit from imposing local limits on and/or promoting pollution prevention with commercial and small industrial users. A sewer use ordinance modification can provide the necessary control over small industrial and commercial users, but this general ordinance does not allow a POTW to set user-specific requirements that can be incorporated into individual discharge permits. Therefore, a pollution prevention program can be developed that is aimed at nonpermitted industrial users. This program would incorporate a regulatory element for a user of concern or for the reduction of a pollutant of concern.

In dealing with nonpermitted industries, some sewerage districts develop another class of permits to regulate these typically nonregulated industries. For example, the City of San Francisco (1996) issued Class II discharge permits to more than 600 nonsignificant industrial users in three years in an effort to reduce the level of problem constituents. These permits differ from SIU permits by not requiring a mandatory waste minimization plan, unless a waste minimization checklist confirms that a Class II industry should be required to submit a mandatory plan. The permits also specify compliance requirements with pretreatment local limits and a zero-discharge option (verified through inspection of hazardous materials inventories and manifests). Sherry (1988) notes that other cities in California have initiated small generator permits for all businesses that discharge nondomestic sewage to the sewers.

The City of Palo Alto also has a permitting program for nonindustrial facilities; currently vehicle service facilities, machine shops, and photoprocessors are permitted. At these permitted, nonindustrial facilities, modified ordinance requirements include them within the realm of RWQCPs' enforcement protocol, but only after educational and technical assistance methods have been tried and failed (PARWQCP, 1996).

TYPES OF PROGRAMS FOR NONCOMPLIANT INDUSTRIES. By taking full advantage of their authority to deal with users who are in noncompliance with pretreatment requirements, POTWs can encourage or mandate pollution prevention. Under normal program activities (nonmodified sewer use ordinance), POTWs can mandate pollution prevention, but they cannot require specific measures beyond BMPs (U.S. EPA, 1993). In response to user noncompliance, however, a POTW can require specific pollution prevention measures as part of a mutually agreed upon compliance schedule with the user.

In requiring the development of a corrective action plan, POTWs can require facilities in noncompliance to conduct pollution prevention planning, to identify cost-effective pollution prevention measures, to develop an implementation schedule with interim and final milestones, and to initiate a supplemental environmental project (SEP). These initial activities were discussed earlier; now the focus turns to SEPs.

Companies confronted with fines for violating environmental regulations (i.e., violating discharge limits) may consider looking into SEPs. Typically, the violating industry and the EPA agree to include commitments by the industry to implement specific pollution prevention activities in the settlement as partial relief for the original violations and as a supplemental project to mitigate related environmental problems (Ross and Konrad, 1993). The incentive for an industry to agree to an SEP is generally a settlement that reduces a fine below the normal maximum potential fine. Typically, costs for implementing the SEP are more expensive than the reduction in the potential fine, but the future benefit (savings through process changes) should make up the difference. Therefore, SEPs have the potential to improve the relationship between industry and municipalities, advance pollution prevention efforts, and provide cost advantages, while the companies can still be held liable for their actions.

The SEP program is a modification of the EPA's enforcement activities which demonstrates a quasi-voluntary approach to pollution prevention. In the end, SEPs provide "extra" environmental benefits to the public while ensuring that violators continue to pay penalties which both reflect the seriousness of the violation and eliminate the economic advantages of noncompliance (Rosenberg et al., 1995). Supplemental environmental projects also increase the prospects that the violator will be able to remain in compliance in the future. There are seven allowable categories, defined by the EPA for SEP projects: pollution prevention, public health, pollution reduction, environmental restoration and protection, assessments and audits, environmental compliance promotion, and emergency planning and preparedness.

Some states are looking at enforcement settlements as a way to fuel the implementation of pollution prevention projects by industries (Ross and Konrad, 1993). For example, in Rhode Island the Woonsocket Regional Wastewater Commission began incorporating pollution prevention into its Notices of Violation (NOV). Noncompliant industries may allow University of Rhode Island/Department of Environmental Management personnel to carry out an on-site pollution prevention assessment and to recommend modifications (RIDEM, 1994). However, the activity is undertaken in lieu of a fine or penalty. This pollution prevention opportunity is seen as a cooperative effort between the regulators and industry, and it encourages goodwill between the two entities.

PROBLEMS WITH MANDATORY PROGRAMS. The main obstacle industries face from mandatory policies is accepting the notion that a municipality can tell an industry how to run its business. Pollution control efforts mandate a method for minimizing what the industry is discharging, but pollution prevention mandates a retooling of the processes and materials a business utilizes. Teeter (1993) clearly develops a picture of the problems encountered by stating that early public sector pollution prevention programs were largely built on technical assistance and educational outreach. These programs not only addressed legitimate needs, they also reflected attitudes about how much municipality-administered pollution prevention efforts can or should intrude into the private sector—especially into industry's decision making about the production process.

To promulgate effective pollution prevention efforts, a sewerage agency must focus on providing either information or incentives (i.e., voluntary programs) to facilities through agency programs sufficiently distant from its regulatory arms; regulatory approaches would accomplish little more than creating resistance. The facility owners, not government, know best whether and how to integrate pollution prevention into their environmental programs and strategies. However, arguments can be and have been made to dispel the last statement, concerning whether industries have the know-how and the impetus to implement pollution prevention activities on their own.

Mass-based limits also pose problems that need to be overcome. For example, when a facility increases its production capabilities, it is also most likely increasing its pollution potential. Under a mass-based situation, the discharge amount increases. Under concentration-based limits, concentration measures will probably remain the same. So, from the business perspective, mass-based limits can be a detriment to the possible expansion of an industrial facility. This could also provide a problem for the local Chamber of Commerce, which may be interested in facilities expanding their potential to provide jobs for local residents. Environmental groups and regulators will indicate that, if an industry wants to expand, the industry will probably have to evaluate pollution prevention technologies to maintain or reduce toxic mass loadings. To remedy this situation, normalized mass loadings may be used to avoid thwarting expansion by industries. Normalized loadings will be based on pounds emitted per day divided by pounds of product produced/manufactured per day. This method is very accurate as an indicator of true pollution prevention trends. Typically, when outputs of a pollutant of concern decrease, you may initially think there has been a reduction in pollutants emitted or substitution of a safer chemical for the more toxic one, but in actuality more information is needed to establish pollution prevention reductions. This apparent reduction could merely reflect a drop in production; less production means less waste.

13.4.3 Market-Based Programs and Pollution Prevention Incentives

Businesses are not likely to implement pollution prevention activities to achieve acceptable compliance rates without incentives. Voluntary and mandatory pollution prevention programs have significant shortcomings from the standpoint of local governments that may want to create incentives to induce small business participation in pollution prevention programs (Boyer, 1993). Therefore, other options must be found to encourage pollution prevention: market-based programs and positive incentives are two ideals that are rooted in economic value. Regulated industries could be convinced to invest and support pollution prevention activities through positive economic incentives. If industry has the flexibility to discover the least expensive method for compliance, enhanced environmental pollution prevention benefits could be achieved.

Pollution prevention grants and subsidized pollution prevention loans, however, may not be effective as economic incentives to pollution prevention implementation (Laughlin and Corson, 1995). Outcome-based tax credits would better help manufacturers link pollution prevention with profitability, thus effectively promulgating pollution prevention implementation, and would significantly extend the reach of a pollution prevention program. For example, the State of Indiana initially proposed a program of tax credits and deductions. This solution (bill HB 1441) would have authorized up to $10 million of annual corporate income tax credits to encourage industries to adopt and implement pollution prevention measures (Laughlin and Corson, 1995). Although fair, this program was considered by politicians as "corporate welfare." The bill was never introduced, but there are nearly a dozen states that are interested in similar market-based programs. If implemented, this bill would award manufacturers a 30 percent credit for the incremental cost of replacing a toxic raw material with a less toxic or non-toxic substitute. The legislation would also establish a 30 percent credit for the cost of making process improvements needed to reduce waste. The credit would be adjusted based on the percentage reduction of waste that a firm actually achieves. For example, if a firm makes a $100,000 investment to improve production equipment, the maximum one-time credit that it could receive would be $30,000. However, if the firm achieves only a 50 percent reduction in waste, the credit would be reduced to $15,000. This outcome-based determination promotes accountability and effectiveness among industries affected and is a positive incentive option to consider.

To provide incentives for companies to implement waste minimization studies and pollution prevention measures, the City of Palo Alto (1992) adjusted the billing structure for sewer use charges. The purpose was not to raise new revenue but to shift the current sewer fee schedule to reflect removal of metals as well as conventional parameters.

Because small industries are especially sensitive to the impacts of compelling regulations, and because negative incentives are very difficult to apply, incentives for pollution prevention should, to the maximum extent possible, be based on positive inducements for voluntary participation. Boyer (1993) found that industry representatives on Erie County's (New York) advisory committee agreed that recent changes in regulatory standards and in civil and criminal liability, along with the resulting escalation in costs of waste handling and disposal, created ample negative incentives for firms to minimize the amounts of waste they generate; therefore, additional negative incentives would not remedy the situation. Besides, Boyer (1993) states that negative incentives may simply inspire resentment and resistance by affected industries.

MARKETABLE WASTEWATER PERMITS. Like pretreatment programs, marketable wastewater (MWW) permits are an option with real possibilities of pollution prevention and economic benefit (Aldrich, 1996). In essence, the permit is a license to pollute: the holder of an MWW permit may emit any amount of pollutant up to the level specified in the marketable permit. If only the permissible amount of pollution is discharged, water quality will be maintained.

Initially the POTW needs to establish the total amount of pollution that can be discharged into a given watershed from its WWTP, based on the NPDES permit limits. Then, marketable wastewater permits, which allow that pollution loading per toxic chemical, would be made available to the polluters. The polluter (industrial users) would then obtain any number of permits through the market system, purchasing them from the POTW or from other industries that are discharging at less than their allowable limit.

The financial advantage of marketable wastewater permits accrues when the industry reduces toxic discharges by implementing pollution prevention. Pollution prevention, which would lower or eliminate emissions, would mean one of two things to a business. The industrial user could continue operations at the current level at a lower cost gained by environmental savings or could reduce its emissions and sell its excess permits to other industrial users or back to the sewerage agency.

13.4.4 Measuring Pollution Prevention Progress

Once pollution prevention is implemented, requirements for reporting pollution prevention trends differ, depending on the intended use of the data. Industries are pressured to prove the success of their environmental efforts. Sewerage agencies are interested in gauging the benefits of pollution prevention activities. The public demands information on what industries are doing to reduce the quantities and toxicities of harmful chemicals in their communities. Environmental groups have an interest in tracking environmental performance and pollutant reductions.

To encourage industries to provide viable and accurate results to POTWs and to report to the public and environmental groups that reductions are being made, an effective tool that assesses progress in pollution prevention is essential (Warren and Craig, 1995). For industry, knowing and tracking the components of a pollution prevention program that are most successful allows management to save time and resources by directing efforts toward the more efficient activities. For municipalities, incentive programs may be tied to the reduction of pollutants (i.e., marketable wastewater permits), thus they need an efficient and accurate method for tracking pollution trends. Accurate measurements allow municipalities to evaluate the successes and failures of pollution prevention programs and activities.

The procedure used most frequently for measuring chemical releases is Form R of Section 313 of EPCRA (used for the Toxic Release Inventory reporting requirement). For listed chemicals, a manufacturer or user must determine if a threshold quantity has been exceeded. If the threshold is exceeded, releases to the various media are determined by a variety of methods, including actual measurements, engineering calculations, mass balances, and published emission factors. Releases are recorded as fugitive air emissions, stack emissions, water body releases, releases to land, underground injection, discharges to POTWs, and off-site transfers for disposal, recycling, treatment, and energy recovery. Harper (1991) found that Form R estimates are frequently prepared by persons with little or no technical background, and with minimal training in proper estimation techniques. Although valuable for evaluating multimedia releases, the estimates have limited usefulness for determining accurate pollution prevention efforts in the wastewater medium.

In addition to the information furnished by Harper (1991), Karam, Craig, and Currey (1991) note that measuring pollution trends is complicated by the fact that many reported decreases in total TRI releases and transfers reflect changes in how releases and transfers are estimated or reported and not actual changes in pollution generation patterns. Normalized pollution measures (measures of pollution and of pollution reduction per unit of production activity) can help in targeting pollution prevention opportunities and measuring pollution prevention progress. Graham (1993) noted that 49 percent of survey respondents normalized measurement procedures for changes in production.

With respect to industry needs, the means by which reductions are calculated and the releases being measured should be clearly stated, as should the method of tracking and accounting for each release (Pojasek and Cali, 1991). A key element in any viable pollution prevention program is a loss tracking system, which enables management to identify in a timely manner the location and circumstances for each process loss. In addition, a tracking system should allow an industry to monitor its loss reduction (pollution prevention) performance as a function of production. Harper (1991) further adds that targeting "zero defects" and making waste minimization a quality control parameter are valuable only if management already knows (1) where waste is generated in the production process, (2) how to control that generation without having a negative impact on product quality or production rate, and (3) how to measure the parameter in a straightforward manner that allows on-line adjustments to reduce waste generation. With improvements in areas such as inventory control and production rates, measurements can often be made to provide more timely or more detailed information concerning generation and reduction of waste.

To fulfill the POTW's responsibilities for accurate measurements and tracking, plant staff must use a sampling strategy that adequately characterizes plant influent and accounts for temporal changes, because wastewater composition will vary depending on diurnal, weekly, and seasonal flows, chemical use patterns, and industrial chemical changes (Epstein and Skavroneck, 1996). Dunams (1997) made a comparison between wastewater data sampled and analyzed by the Greater Cincinnati Metropolitan Sewer District (GCMSD) and wastewater emissions data determined by industry and reported as part of their TRI requirements. Yearly emissions of TRI chemicals, in pounds per year, were evaluated against GCMSD values to determine whether TRI data are representative of GCMSD's data. In general, the findings indicate that TRI data do not accurately depict the data collected by the GCMSD. The inaccuracies are mostly due to industry inexperience in performing wastewater analysis rather than lack of integrity. The analysis was executed to indicate that TRI data, although easier to evaluate and more encompassing in examining multimedia releases, may not be applicable as a pollution prevention tracking protocol for determining reduction trends in wastewater emissions. In developing a program, such as marketable wastewater permits, which relies on marginal reductions of wastewater emissions, any sewerage agency's data that are in a mass-based form should be used in justifying percentage reductions or to quantify increasing emission levels.

13.5 DESIRABLE QUALITIES OF A MUNICIPAL P2 PROGRAM

13.5.1 Industry Requests

A survey was recently conducted of industry environmental compliance coordinators in Cincinnati to determine what municipal P2 programs would be desirable for industry (Dunams, 1997). Eleven possible pollution prevention program options were presented to those surveyed:

1. Educational outreach and technical assistance programs.
2. Waste minimization audits.
3. Demonstration projects.
4. Voluntary reduction programs.
5. Marketable wastewater permits.
6. New local limits.
7. Pollution prevention plans.
8. Mass-based limits.
9. Reasonable control measures.
10. New permitting class and requirements.
11. Supplemental environmental projects.

The programs range from voluntary to mandatory and include market-based and modified pollution control programs. Table 13.1 is a summary of the rating results by the industrial users. Each program was rated from 1 to 5, with 5 being indicative of a desirable program.

The total score indicates the degree to which programs are accepted by industry. The four voluntary programs (rows 1–4) are rated higher (average rating over 18) than most of the mandatory programs (average rating of 15.4, not including the marketable permits program), except for the supplemental environmental projects program, the marketable permits program, and the new permit classification. Educational outreach and technical assistance programs were ranked the highest. Voluntary programs may be acceptable as a control measure, as long as pollution prevention projects are being initiated and leading to effluent reductions from industrial users and the POTW can remain in compliance with its NPDES permit.

The lowest ranked program is the setting of new local limits. This program is generally perceived to be more an aspect of pollution control than a pollution prevention program, since setting new limits in the past has necessitated that IUs develop control programs to meet them. Hence industry is indicating that pollution control is less desirable than real pollution prevention programs; IUs want to avoid mandatory programs that they may associate with pollution control programs.

13.5.2 Publicly Owned Treatment Works Requests

To evaluate the effectiveness of pollution prevention programs administered by POTWs, a survey was sent to wastewater treatment plant operators in the Great Lakes region:

TABLE 13.1
Industrial user evaluation of pollution prevention programs

	Criteria					
Program	**Imple-mentability of program**	**Admini-strative impact**	**Cost of imple-mentation**	**Encourages pollution prevention**	**Alternative compliance**	**Total score**
Education outreach/ technical assistance	3.8	3.5	3.2	4.3	4.0	18.8
Waste minimization audits	4.0	3.0	3.5	4.0	3.7	18.2
Demonstration projects	3.7	2.7	2.8	3.8	4.0	17.0
Voluntary reduction	4.0	2.7	3.2	4.2	4.0	18.1
Marketable wastewater permits	3.5	3.0	3.2	3.8	3.7	17.2
New local limits	2.8	2.3	2.5	3.3	2.5	13.4
Pollution prevention plans	3.0	2.3	2.7	3.5	3.3	14.8
Mass-based limits	3.3	2.5	2.7	3.8	3.8	16.1
Reasonable control measures	3.0	2.2	2.2	3.3	2.4	13.1
New permitting class and requirements	4.0	3.3	2.7	3.8	3.3	17.1
Supplemental environ-mental projects	4.0	3.0	3.2	3.8	3.8	17.8

Source: Dunams, 1997.

Illinois, Indiana, Michigan, Minnesota, New York, Ohio, Pennsylvania, and Wisconsin (Dunams, 1997). The survey was designed to indicate the positive and negative factors involved in implementing a pollution prevention program and to determine which programs encourage industry to move toward pollution prevention as a viable waste management technique.

The range of average daily flows treated by the responding POTWs is 15–220 Mgal/day, with the average being 71.6 Mgal/day. All of the WWTPs have pretreatment programs, which is a necessity since they treat more than 5 Mgal/day and they receive pollutants from industrial users that may pass through or interfere with the POTW operations. A few of the plants experience problems with chlorine control, zinc, or ammonia, and one had a significant silver problem that was alleviated in 1994. Mercury seems to be the future chemical of concern.

Regulatory compliance, environmental issues, quality of biosolids, and local issues all influenced POTWs to initiate pollution prevention activities with industrial users. Over half the respondents indicated that regulatory requirements or conditions of noncompliance prompted pollution prevention actions. One POTW noted that pollution prevention programs with industrial users were approached as "preventive medicine" for long-term compliance objectives. Two WWTPs initiated pollution prevention activities to promote goodwill and cooperative efforts that benefit the environment. One WWTP's impetus was involvement in an EPA grant program that evaluated the viability of using P2 approaches to promote source reduction of toxic substances among industrial users.

The POTWs were asked to indicate which programs are aimed at nonpermitted industrial users, permitted industrial users, and internal POTW activities and who administers the program. Administration of POTW pollution prevention programs was indicated to be done by POTW staff, except for one program where an outside organization administered the program. Only one program was reported to have a dedicated pollution prevention staff. The rest of the programs are administered by pretreatment staff.

Targeting nonpermitted users takes a little more imagination from POTW staff since they have no mechanism, like a permitting program, to authorize pollution prevention activities. Programs aimed at nonpermitted users include educational outreach (pollution prevention conferences, telephone consultations, and library development), technical assistance, and voluntary reduction programs (development of BMPs and inclusion of educational programs). Most of the voluntary reduction program efforts are aimed at reducing mercury levels in the wastewater.

Programs aimed at permitted industrial users include educational outreach, technical assistance programs, setting of local limits, levying of fines and enforcement actions, beginnings of a mass-based program, and voluntary reduction programs. Local limits seem to be the preferred method of dealing with permitted industries. One POTW noted that changes to sewer use ordinances, which led to new local limits, were necessary for the POTW to meet new NPDES permit limits. Another POTW stated that sewer use ordinances and local limits are programs consistent with pollution control. However, the lowering of limits has the potential to modify the pollution management scheme of industries, and with education outreach and technical assistance, POTWs will encourage industries to look into pollution prevention as a viable solution over pollution control.

The survey evaluated the same 11 programs described previously as well as an additional program (12) that establishes mandatory monitoring requirements for nonpermitted industrial users. Program 12 targets hospitals that are required to monitor for mercury once every reporting period. Table 13.2 summarizes the results of the survey of POTW operators. Again, the programs were rated from 1 to 5, with 5 being indicative of a desirable program.

These rankings are very subjective but they provide an interesting look into several pollution prevention activities being conducted by POTWs. The main conclusion drawn is that most POTWs are involved only in voluntary aspects of pollution prevention, if any; two-thirds of the respondents have educational and technical assistance programs ongoing, and a few have voluntary reduction programs. Many POTW operators expressed the feelings that P2 programs require much staff time to oversee and that they are costly. Some also had concerns about the time and energy required to develop and implement mandatory programs as well as the burdensome amount of paperwork. However, as POTW operators gain knowledge of pollution prevention and apply pollution prevention solutions to industrial problems, the tendency shifts to the inclusion of mandatory programs to the already established voluntary programs.

TABLE 13.2
POTW evaluation of pollution prevention programs

	Criteria					
Program	**Imple-mentability of program**	**Enforce-ability of program**	**Admini-strative impact**	**Cost of imple-mentation**	**Program satisfaction**	**Total score**
Education outreach/ technical assistance	4.0	2.0	3.9	4.6	3.3	17.8
Waste minimization audits	3.0	5.0	3.0	3.0	4.0	18.0
Demonstration projects	2.0	1.0	2.0	4.0	5.0	14.0
Voluntary reduction	3.2	2.0	3.2	3.7	4.3	16.4
Marketable wastewater permits*						
New local limits	3.0	5.0	2.7	3.0	3.7	17.4
Pollution prevention plans	3.5	3.0	3.0	4.0	3.0	16.5
Mass-based limits	1.0	5.0	1.0	1.0	5.0	13.0
Reasonable control measures	2.0	1.0	2.0	2.0	5.0	12.0
New permitting class and requirements†						
Supplemental environ-mental projects	3.0	5.0	2.0	5.0	5.0	20.0
Mandatory monitoring for nonpermitted users	3.0	1.0	3.0	3.0	3.0	13.0

*New program under development and will not have been applied by POTWs.

†No Reclassification Permits developed by the responding POTWs.

Source: Dunams, 1997.

13.6 DEVELOPMENT OF PUBLICLY ADMINISTERED POLLUTION PREVENTION PROGRAMS

13.6.1 Program Recommendation

A process for the development of a municipal wastewater pollution prevention program and an implementation approach for POTWs has been proposed to the GCMSD and is now being implemented. The overall program recommended is a combination of voluntary and mandatory programs to be implemented in two phases.

Sewer use ordinances should provide a regulatory basis for these programs. Modifications to the sewer use ordinance may be required if POTWs encounter obstacles to enacting pollution prevention requirements. These barriers can be overcome by employing logical reasoning and the vigor of prior regulations. The Clean Water Act grants POTWs the legal authority to restrict industrial and commercial pollutants beyond federal pretreatment standards. In effect, the CWA provides POTWs with a regulatory floor, but not a ceiling, in their efforts to prevent pass-through of pollutants, interference with plant operations, and other problems resulting from nondomestic discharges (Sherry, 1988). For example, the City of Palo Alto modified its sewer use ordinance (SUO) in response to the effects of pollutants passing through treatment and bioaccumulating in clams and to new proposed stringent limits.

Modifications to the sewer use ordinance will not directly impede or inspire the implementation of voluntary programs, but these modifications will be necessary if and when voluntary programs do not meet reduction goals. They will grant the POTW the necessary mechanisms to require permitted users to implement pollution prevention measures and to conceivably reclassify nonpermitted industrial users to a special permit class (i.e., a small-contributor permit).

Voluntary programs should be initiated to deal with specific pollutants of concern by nonpermitted industrial users and for industries in compliance (permitted users who have developed a pollution prevention plan) which desire to be more economically and environmentally efficient. The voluntary route will foster good will, fairness, and cooperation between industry and the POTW; it will provide an easy, introductory, administrative route for encouraging compliance. Voluntary programs would consist of educational programs, technical assistance programs (TAPs), and a voluntary reduction program. These were all described earlier.

Under the proposed municipal pollution prevention program, if an industry cannot effectively initiate or carry out pollutant reductions through the voluntary programs, mandatory programs will be triggered. Nonpermitted industries could be reclassified into a new permitted class that will be subject to a regulatory agenda. With regard to programs aimed at specific pollutants of concern, permitted users will directly follow a course of mandatory requirements. Mandatory programs consist of facility pollution prevention plans and recommended methods of implementation, an option of a mass-based limit program or a reasonable control measure program, and a system of fines and supplemental environmental projects.

The municipal pollution prevention program could be implemented in phases. Phase I could consist of the entire voluntary program and components of the mandatory program, such as pollution prevention plan development and implementation, and the levying of fines and development of supplemental environmental projects. Phase II could be enacted a few years later, after initiation of Phase I. This phase would include an additional mandatory compliance endeavor. This endeavor is composed of two options: the reasonable control measure option and the mass-based limit option.

This may seem like a lot to achieve, but the best pollution prevention programs provide similar exhaustive aspects and continue to add relevant features to their pollution prevention plans as needed. Again, the POTW will probably not tackle all of these tasks at once, but will gradually phase them in to accomplish a complete pollution prevention program. This phasing approach will benefit both the industrial users and the POTW staff by fostering good will and providing pollution prevention staff an opportunity to realize and integrate more advanced pollution reduction techniques.

13.6.2 Program Implementation

A flowchart can be used as a guideline for the POTW to follow in implementing pollution prevention programs for permitted and nonpermitted industrial users under

Phase I and Phase II conditions (see Figure 13.1). Note that this program can also be aimed toward commercial users; they would be considered small nonpermitted industrial users.

In addition to establishing viable solutions for alternative compliance methods (i.e., pollution prevention instead of pretreatment) and meeting stringent requirements

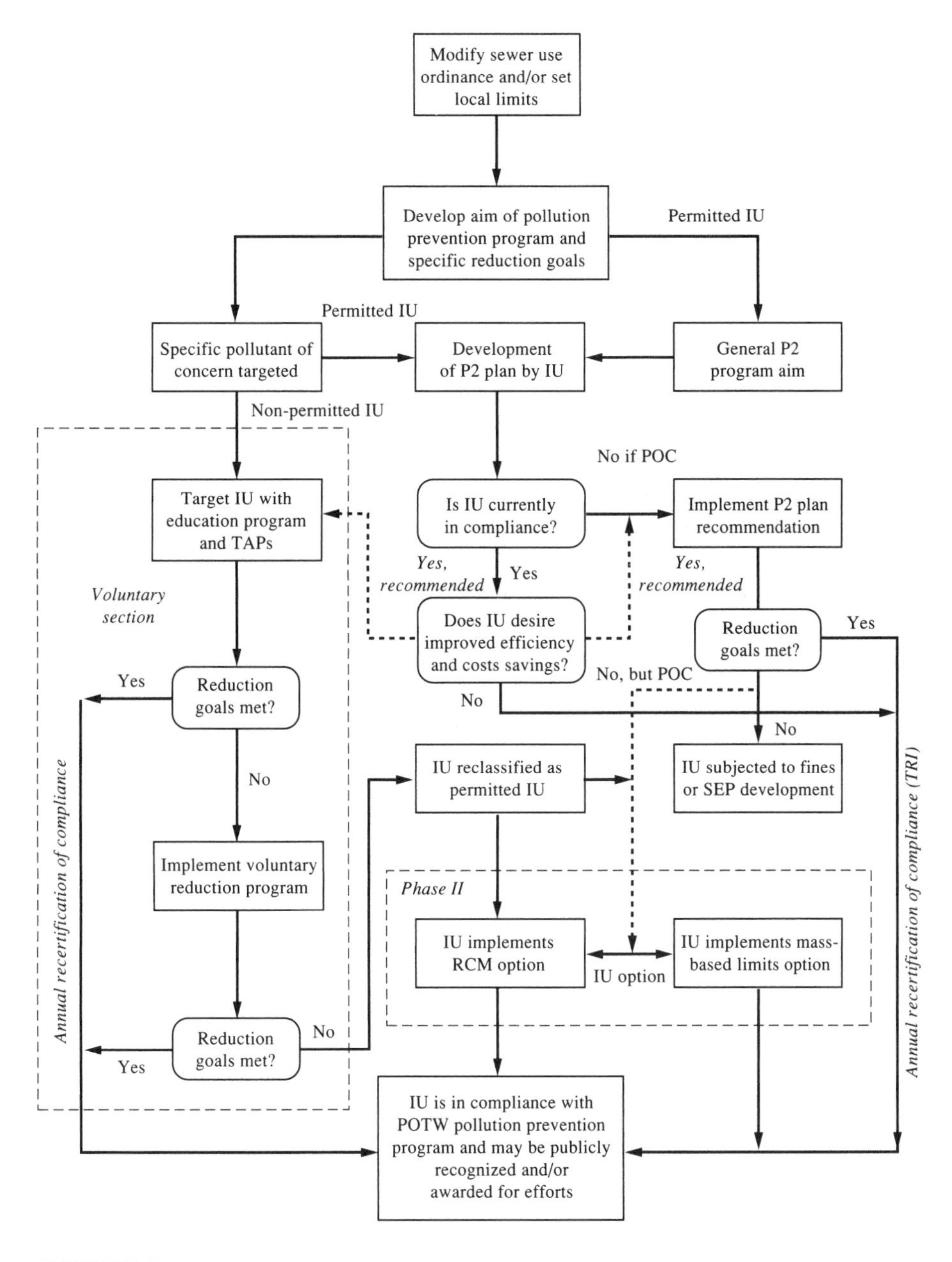

FIGURE 13.1
Proposed municipal P2 program flowchart. (Source: Dunams, 1997)

from upcoming regulations (e.g., the Great Lakes Initiative), this program flowchart provides for all aspects of implementation. As recommended by Sherry (1988), it instructs POTW staff to:

- Identify specific users of concern (industrial and commercial sectors) and/or pollutants of concern for priority attention.
- Set reduction goals for those waterborne pollutants of concern.
- Evaluate the effectiveness and implementation of voluntary or mandatory programs or an infusion of both for possible inclusion into a pollution prevention effort.
- Submit a proposed program to the POTW governing board that identifies the pollution prevention activities selected for implementation, along with timetables and financial support required.
- Recommend any changes to the existing local sewer use ordinance necessary to implement the pollution prevention program proposed.

This flowchart was developed to advise POTWs of different and flexible options that industries desire in a pollution prevention program. The first step is the evaluation of the current sewer use ordinance, followed by its modification to reflect desired pollution prevention language. New language will include the reclassification of nonpermitted users, development of pollution prevention plans, development of new local limits, development of mass-based limits, and local supplemental environmental projects (an option added to minimize the levying of fines, which is currently the only penalty available).

The next step of the flowchart requires a decision by the POTW concerning the aim of a specific pollution prevention program and the reduction goals anticipated. The POTW may implement pollution prevention programs specifically targeted to problematic pollutants emitted by industrial users, it can develop a general strategy for all of their permitted industrial users, or it can do both. As stated earlier, voluntary programs would provide a more viable and less demanding starting point for the POTW to foster a cooperative relationship with industries and therefore would be initiated for nonpermitted users that discharge pollutants of concern.

When focusing on pollutants of concern (POC), the POTW needs to determine the industries to be targeted. Pollutants of concern are usually determined based on maximum headworks loading analysis, and then respective upstream industries and commercial users are targeted with respect to a specific pollutant of concern. The emitting industries can be identified through the POTW's laboratory information management system (LIMS) data, the knowledge of industrial inspectors, and the pretreatment permit, which lists chemicals permitted industrial users have on site and pollutants they are discharging.

The pollution prevention program described here is aimed at individual industrial users, but if the POTW ascertains that a sector of industry (e.g., photoprocessors) is typically noncompliant, specific ordinance requirements can be developed for that industrial sector (see the City of Palo Alto, 1992).

PHASE I. The activation of Phase I (Figure 13.1) will provide the POTW with an almost complete pollution prevention program. Aspects of this phase will also set a cooperative platform for fostering a working relationship with industry. The rest of this section describes the voluntary and mandatory aspects of the proposed pollution prevention program.

Voluntary Program Components. The POTW should initiate a voluntary reduction program that will include disseminating information on currently tested reduction techniques, on-site inspections, and sampling and analysis of discharges. Voluntary program requirements will be the initial aspects applied to nonpermitted users that discharge pollutants of concern. A regimen of technical assistance and educational outreach will assist and inform industries of general, cursory pollution prevention opportunities. Industries will be given an opportunity to develop their own individual solutions to achieve reduction goals. To achieve success in implementing pollution prevention technology, they must demonstrate compliance with the POTW's goals, and they will then be publicly recognized. Incentives will be indirectly provided through financial benefits inherent in the exchanging of emissions control objectives for pollution prevention goals; further, they will no longer need to continue the POTW's increasingly stringent pollution prevention program. Industrial users will be required to be recertified yearly to ensure compliance with constantly changing laws and requirements.

Industries that are unsuccessful with self-initiated pollution prevention programs may be given a longer time to achieve compliance if they display a "good-faith effort" in their desire to reduce the discharge of pollutants. If compliance is still not met, the industrial user will be required to employ a voluntary reduction program. Under the voluntary reduction program, the POTW will perform a waste minimization audit for and with the company and will subsequently encourage it to adopt pollution prevention measures and plans that are revealed during the audit. Mandates for reduction will be established, but the company only needs to provide documentation that it implemented and maintained a reduction plan and that reductions are met. The POTW's sampling and analysis will confirm reductions. No mandatory pollution prevention protocols are established; the industrial user will select its own preferred prevention option. If the industrial user is successful in meeting pollutant reductions, annual recertification will be required to ensure future compliance. However, if the industrial user does not comply, a longer time line may be established to allow it to comply or the IU may be required to develop a mandatory pollution prevention plan after being reclassified as a permitted small contributor, as discussed above. This reclassification of typically nonpermitted IUs is not a permanent label, but it provides POTWs with a mechanism for achieving reductions of pollutants of concern from an established nonpermitted industry. Afterwards, if a reclassified permittee is in compliance and not problematic with respect to a POC, the permittee status can be dropped.

Mandatory Program Components. The components of the mandatory program in Phase I consist of developing pollution prevention plans, implementing P2 measures recommended by the plan, and the levying of fines or supplemental environmental projects. This program is directly aimed at permitted IUs and specific users of concern (UOC) discharging pollutants of concern. This latter category includes a reclassified permitted UOC that could not achieve reductions after two voluntary attempts.

Permitted users of concern will have pollution prevention requirements imposed whenever a pretreatment permit is renewed or during the initial application. Note that the inclusion of pollution prevention requirements in the pretreatment permit does not supersede any of the typical monitoring and inspection activities by pretreatment staff or the industry. The goal in modifying sewer use ordinances and industrial permits is to promote and require pollution prevention activities as the preferred waste management option.

A major requirement of the permitted UOC and the reclassified UOC will be the development of a pollution prevention plan that recommends possible areas of reduction. These plans will document the benefits of switching to pollution prevention technology and the anticipated waste savings resulting from increased efficiency. The importance of market-based programs is evident at this point. The industrial user learns that pollution prevention is not just a compliance option but a way of improving company efficiency, which in turn can reduce waste management costs or even derive profits from waste management. Under a marketable wastewater permit program, an industry's adoption of pollution prevention will lead to reduced emissions and the subsequent sale of a permit no longer needed.

The permitted UOC and the reclassified UOC, under the correct conditions, will be required to implement pollution prevention measures determined during the development of the pollution prevention plan. If implementation is not a requirement (i.e., the IU has compliance status) and the user does not heed the incentives to implement pollution prevention measures, the industry will be considered in compliance and will adhere to other aspects of its permit (self- and POTW-monitoring, etc.). However, users of concern discharging pollutants of concern will be required to implement specific measures aimed at the reduction or elimination of POC.

If the industrial user's pollution prevention plan implementation effort was futile (reduction goals were not met), the levying of fines or the option of a reduced fine with a supplemental environmental project will be initiated. It must be noted that an agreement between the POTW and the industry must occur for the industry to initiate the development of a local supplemental environmental project. The purpose of a POTW's pollution prevention program is to avoid the levying of fines while providing industry with the flexibility and options to achieve compliance.

PHASE II. After the POTW has had sufficient time to implement and efficiently manage Phase I of this pollution prevention program, Phase II (Figure 13.1) can begin, as needed. Phase II will add a mandatory compliance option, aimed mostly at pollutants of concern, which will provide one more mechanism for industrial users to implement pollution prevention and reduce/eliminate the POC. With respect to the pollutants of concern, a system of fines is not an option; reductions must be made. All opportunities for IUs to implement pollution prevention are for their own benefit, economically and environmentally, and also have significant benefits to the POTW. Therefore, these pollution prevention plans are providing industry with options for future successes.

Phase II adds a compliance option that provides the industrial user with two routes for bringing POC into compliance: the mass-based option and the reasonable control measures option. Industries are given a date to meet their new discharge limits, and the POTW will provide technical assistance and will vigorously pursue available methods to ensure that industries will succeed in achieving compliance.

This compliance option will necessitate additional duties for the POTW's staff. Not only will the staff need to develop listings of proven reasonable control measures for various industries, but they will also be involved in the task of implementing an exhaustive pollution prevention plan under the mass-based option.

This Phase II option is the last line of pollution prevention programs. It will require that all resources of the POTW and the industry are maximized to minimize or eliminate the quantity of POC. This compliance option was developed to ensure pollution prevention results.

This compliance option is triggered by one of two processes. The first process is activated when an industry is in noncompliance with the voluntary reduction program (i.e., does not meet reduction requirements in minimizing POC). The noncompliant industry, which has had two voluntary opportunities to come into compliance, will be reclassified as a permitted contributor and will be required to implement one of the two compliance options. The thought of being permitted is sometimes enough to intimidate an IU into using pollution prevention technology. The second process is activated when a permitted industrial user continues to be noncompliant, under the general P2 program, after implementing a pollution prevention measure that was recommended in its pollution prevention plan to deal with a specific pollutant of concern. This noncompliant permitted industry then has the option of implementing either compliance method or being subjected to a fine or an SEP. Again, the overall goal is to avoid fining the industrial user by providing significant opportunities to comply with the pollution prevention program.

13.7 INTERNAL POLLUTION PREVENTION IN PUBLICLY OWNED TREATMENT WORKS

In addition to requiring industries to modify their operations, POTWs should also take a close look at their internal operations to determine if pollution prevention efforts will be worthwhile. They should practice pollution prevention in their own operations by reducing or eliminating chlorination of wastewater; consulting with suppliers to ensure that the most environmentally friendly supplies (including process additives) are purchased; and using spill prevention, inventory control, and other good housekeeping measures to generate fewer waste chemicals (Epstein and Skavroneck, 1996).

Currently, the approach taken by most POTWs is to transfer the bulk of the pollutants they receive that are not biologically degradable to other, more easily managed media (Lindsey, Neese, and Thomas, 1996). Materials such as toxic heavy metals are precipitated and incorporated into sludges that are most often buried or applied to land. Problems may also be encountered with operations employed to maintain aerobic conditions (aeration, spraying, agitation) which tend to promote the volatilization of organic solvents. Lindsey and colleagues found that significant expenses will need to be incurred by POTWs to modify operations or to install covers on existing aeration, spraying, and filtration units to confine vapors. These additional compliance costs add to the POTW's overhead without improving treatment efficiency or capacity. Moreover, Andrews (1994) notes that many POTWs are overtreating during periods of routine flow and, as a result, are wasting money.

Several other activities have been undertaken to minimize pollution at POTWs. Internal activities are another means for POTWs to reduce pollutants at their own institution. One POTW provided pollution prevention technical assistance to its on-site auto repair shop. Another POTW used waste pickle liquor for phosphate removal from its wastewater, rather than purchasing chemicals. Water conservation was stressed by another POTW. Some POTWs have a P2 committee that identifies problems and seeks solutions, such as recycling, use of long-life light bulbs, and product substitutions. A few POTWs list recovery of chemicals from the laboratory as an on-site P2 activity.

13.8 IMPLEMENTATION OF PUBLICLY OWNED TREATMENT WORKS POLLUTION PREVENTION PROGRAMS

Current regulations for industrial users within municipal wastewater treatment plant pretreatment programs are replete with outdated pollution control methods. These programs provide for the control and reduction of waste after it is produced. With regulations becoming more stringent and discharge limits being tightened, pollution control measures often can no longer effectively ensure compliance. Environmental management should accomplish more than just being a partial, short-term, single-medium solution, especially since the technology exists that justifies change.

A viable pollution prevention program can be incorporated within a POTW's pretreatment program. Figure 13.1 outlines a strategy for treatment plants to select the appropriate elements for their community for a complete pollution prevention approach to environmental management. The incorporation of pollution prevention in the pretreatment program can be initiated by modifying sewer use ordinances, which will in turn modify the industrial pretreatment program. Pollution control elements would remain, for the time being, but it should be made abundantly clear that pollution prevention is the preferred option for complying with regulations and reducing discharges. The goal of the POTW should be for industries to spend money attaining efficient operations and compliance rather than noncompliance issues and fines.

In conclusion, municipal pollution prevention programs can provide benefits to both the POTW and to the industrial users that cannot be provided by pollution control.

REFERENCES

Aldrich, J. R. *Pollution Prevention Economics: Financial Impacts of Business and Industry.* New York: McGraw-Hill, 1996.

Andrews, J. F. "Dynamic Control of Wastewater Treatment Plants." In *Environmental Science and Technology* 28 (1994): 434A–40A.

Atcheson, J. "Voluntary Pollution Prevention Programs." In *Industrial Pollution Prevention Handbook,* ed. H. M. Freeman. New York: McGraw-Hill, 1995.

Beardsley, D., Davies, T., and Hersh, R. "Improving Environmental Management: What Works, What Doesn't." *Environment* 39 (1997): 6–9.

Boyer, B. *Creating Pollution Prevention Incentives for Small Businesses: The Erie County Program.* Buffalo, NY: Erie County Department of Environment and Planning, Office of Pollution Prevention, 1993.

Buffalo Sewer Authority (BSA). *Buffalo River Combined Sewer Overflow Pollution Prevention Project.* Buffalo, NY: BSA, 1997.

City and County of San Francisco. *Water Pollution Prevention Program: Annual Report Fiscal Year 1995–96.* San Francisco: Department of Public Works, Bureau of Environmental Regulation and Management, 1996.

City of Albuquerque. *5 PPM Silver Program and Pollution Prevention Program.* Albuquerque, NM: Public Works Department, 1993.

City of Palo Alto and James M. Montgomery Consulting Engineers, Inc. *Silver Reduction Pilot Program: Palo Alto Regional Water Quality Control Plant.* Palo Alto, CA, 1992.

Corson, L. A. Director, Indiana Clean Manufacturing Technical Institute, personal conversation, 1997.

Dennison, M. S. *Pollution Prevention Strategies and Technologies.* Rockville, MD: Government Institutes, Inc., 1996.

Dunams, A. W. *Guidelines for the Development of a Pollution Prevention Program Pertaining to Industrial Users of the Greater Cincinnati Metropolitan Sewer District.* M.S. thesis, University of Cincinnati, 1997.

Epstein, L. N., and Skavroneck, S. A. "Promoting Pollution Prevention." *Water Environment & Technology* 8, no. 11 (1996): pp. 55–59.

Graham, A. B. "The Results of PPR's 1993 Survey: Industry's Pollution Prevention Practices." *Pollution Prevention Review* 3, Autumn 1993, pp. 369–81.

Harper, P. D. "Application of Systems to Measure Pollution Prevention." *Pollution Prevention Review* 1, Spring 1991, pp. 145–53.

Karam, J. G., Craig, J. W., and Currey, G. W. "Targeting Pollution Prevention Opportunities Using the Toxics Release Inventory." *Pollution Prevention Review* 1, Spring 1991, pp. 131–44.

Knight, L. "Encouraging Pollution Prevention through Publicly Owned Treatment Works Activities." In *Industrial Pollution Prevention Handbook,* ed. H. M. Freeman. New York: McGraw-Hill, 1995.

Laughlin, J. D., and Corson, L. A. "A Market-Based Approach to Fostering P2." *Pollution Prevention Review* 5, Autumn 1995, pp. 11–16.

Lindsey, T., Neese, S., and Thomas, D. "Implications of Pollution Prevention for Water Pollution Control." *Water Quality International,* March/April 1996, pp. 32–36.

Marchetti, J. A., Poston, B., McPherson, E., and Webb, J. R. "Overcoming the Barriers to Pollution Prevention." *Pollution Prevention Review,* Winter 1996, pp. 41–50.

National Wildlife Federation. *Bringing the GLI Home: The Status of States' Compliance with the Great Lakes Water Quality Initiative.* Washington, DC: NWF, 1997.

New Hampshire Department of Environmental Services (NHDES). *Model (Example) Sewer Use Ordinance.* Concord: State of New Hampshire, 1996.

Ohio Environmental Protection Agency. *1994 Toxic Release Inventory Annual Report.* Columbus, OH: OEPA, 1995.

Palo Alto Regional Water Quality Control Plant. *1996 Clean Bay Plan.* Palo Alto: PARWQCP, 1996.

Pojasek, R. B., and Cali, L. J. "Measuring Pollution Prevention Progress." *Pollution Prevention Review,* Spring 1991, pp. 119–30.

Rhode Island Department of Environmental Management and the University of Rhode Island (RIDEM). *Assessment of Regulatory and Non-regulatory Approaches to Source Reduction in the Rhode Island Textile Industry.* EPA Pollution Prevention Incentives Grant No. NP820607, Progress Report No. 4. Kingston, RI: URI, 1994.

Rosenberg, P., Hindin, D., Packman, J., and Kraus, G. "EPA's Revised SEP Policy and the Negotiation of P2 SEPs." *Pollution Prevention Review,* Autumn 1995, pp. 1–9.

Ross, B., and Konrad, C. "The Federal Agenda: Impact on State and Local Pollution Prevention Efforts." In *Pollution Prevention: A Practical Guide for State and Local Government,* ed. D. T. Wigglesworth. Chelsea, MI: Lewis Publishers, 1993.

Sherry, S. *Reducing Industrial Toxic Wastes and Discharges: The Role of POTWs.* Sacramento, CA: Local Government Commission, 1988.

Skavroneck, S. *Survey of Sewerage Districts on Pollution Prevention Activities.* Milwaukee: Milwaukee Metropolitan Sewerage District, 1992.

Teeter, D. "Incorporating Pollution Prevention into Facility Permits." In *Pollution Prevention: A Practical Guide for State and Local Government,* ed. D. T. Wigglesworth. Chelsea, MI: Lewis Publishers, 1993.

Thurber, J., and Sherman, P. "Pollution Prevention Requirements in United States Environmental Laws." In *Industrial Pollution Prevention Handbook,* ed. H. M. Freeman. New York: McGraw-Hill, 1995.

U.S. EPA. *Guides to Pollution Prevention: Municipal Pretreatment Programs.* EPA/625/R-93/006. Washington, DC: U.S. EPA, Office of Research and Development, 1993.

U.S. EPA. *Pollution Prevention at POTWs: Case Studies.* EPA/742/F-94/001. Washington, DC: U.S. EPA, Office of Pollution Prevention and Toxics, 1994.

Warren, J., and Craig, J. "Measuring Pollution Prevention Progress." In *Industrial Pollution Prevention Handbook,* ed. H. M. Freeman. New York: McGraw-Hill, 1995.

Water Environment Federation. *Developing Source Control Programs for Commercial and Industrial Wastewater: Manual of Practice No. OM-4.* Alexandria, VA: WEF, 1996.

PROBLEMS

13.1. Locate and describe the best management practices that have been established for the textile industry. These can be found on the Internet or in your university library government documents section.

13.2. Research the Great Lakes Water Quality Initiative. What standards and timetables have been set? Are states on schedule to meet the GLI goals? Will an industry in Indianapolis, 200 miles from the Great Lakes, have to meet these standards?

13.3. It has been stated that even with full compliance with categorical pretreatment standards, POTWs in the United States would still continue to receive in excess of 14 million pounds of toxic metals and 51 million pounds of toxic organic pollutants each year from industries. Why is this so?

13.4. The 33/50 Program was considered to be a successful demonstration of a voluntary pollution prevention program. What were the 17 toxic chemicals targeted for reduction, how many industries participated, and how successful was the program?

13.5. One difficulty with voluntary educational programs and waste audits conducted by POTW personnel for industry is that the POTW is often viewed as an adversary by industry. The POTW administers the pretreatment program, monitors for compliance, and levies fines and other penalties for violations. Industries often will not welcome POTW employees viewing the inner workings of the facility, where they may see other possible infractions. If you were in charge of setting up a POTW-sponsored P2 program for industries, what would you do to mitigate these fears?

13.6. An industry is currently discharging 25,000 kg/yr of a toxic chemical to a POTW. Under new local limits, the company will have to reduce this to 5000 kg/yr or pay a new fee of $5.00/kg of the chemical above the limit. It can install a pretreatment system that will provide the required discharge reduction at a capital cost of $60,000 and an annual operating cost of $15,000. A second alternative is to implement certain pollution prevention measures that will cost $130,000 to install and $18,000 per year to operate but will result in savings of $25,000 per year in process chemicals. Assuming a 10 percent discount rate and a five-year period for the project, should the industry pay the sewer user fee, install the pretreatment system, or initiate the P2 projects? Ignore the effects of inflation.

13.7. Dental offices, which most people consider environmentally benign and which usually are not regulated under POTW pretreatment programs, are increasingly coming under scrutiny. Why? What contaminants or practices are of concern?

CHAPTER 14

TOWARD A SUSTAINABLE SOCIETY

Mahatma Gandhi [when asked if, after independence, India would attain British standards of living]: "It took Britain half the resources of the planet to achieve its prosperity; how many planets will a country like India require?"

14.1 INTRODUCTION

As the municipalities of the world grow and develop, they contribute to the depletion of natural resources and addition of pollution to the air, water, and land. As if an enormous debt were suddenly falling due, the United States and the rest of the world are discovering that unfettered consumption of Earth's natural resources and nonrenewable energy sources in support of economic growth has polluted air, water, and land and is visibly damaging the environment. This damage is exacting the heavy price of global climate deterioration, loss of biodiversity, and dangerous accumulation of hazardous and solid waste.

In the book *Silent Spring,* written in 1965, Rachel Carson said:

> For the first time in the history of the world, every human being is now subjected to contact with dangerous chemicals, from the moment of conception until death. In the less than

> two decades of their use, the synthetic pesticides have been so thoroughly distributed throughout the animate and inanimate world that they occur everywhere. Residues of these chemicals linger in soil to which they have been applied a dozen years before. They occur in the mother's milk, and probably in the tissues of the unborn child.

The book painted an accurate picture of environmental conditions at that time. More than three decades later, very little has changed. Indeed, in some places it is now worse. Carson said that nothing but reform can change this, but reforms must be radical in approach to change fundamental aspects of society. We have to change the way we do business to integrate the concept of sustainability of the planet. Only then can we say that business operations are environmentally friendly. But does this mean we cannot have economic growth or development? No, but we must find new ways of development to ensure that it doesn't degrade the environment. The conventional view of competition between economic growth and environmental quality must be changed through public policy, management, and investment practices. To maintain the critical balance between a healthy economy and a healthy environment, economic growth and environmental quality must be treated as complementary objectives. This balanced perspective is essential to "sustain" our communities.

Our understanding of the relationship between people and the landscape is complex and intricate, often blurred, and at times contradictory. Business as usual is no longer an option—for government, the private sector, or individual citizens. Our soils, waters, forests, and minerals are not inexhaustible. Farms, industries, homes, and lifestyles must become more sustainable, in every community on our planet.

Sustainable development is not a new concept. It is the latest expression of a longstanding ethic involving people's relationship with the environment and the current generation's responsibilities to future generations. For a community to be truly sustainable, it must adopt a three-pronged approach that considers economic, environmental, and cultural resources. Communities must consider these needs not only in the short term, but also in the long term. To be sustainable, development must improve economic efficiency, protect and restore ecological systems, and enhance the well-being of all peoples. Ours is a finite global system, and in some cases the limit has been reached beyond which no more resources can be taken out from the system, hence the need for conservation and sustainability.

There is growing evidence that many current global trends in the use of resources or sinks for wastes are not sustainable. This is not unique to the late twentieth century, but the magnitude of the problem has certainly increased during this period. Throughout the history of mankind, man has destroyed or damaged the natural resources. Now this destruction has reached a point where it is seriously affecting the future availability of natural resources and biodiversity in the global environment. Different regions may have different problems and focal points (e.g., one area may have problems with sustainability of renewable resources such as forests, whereas another area may be gripped with problems of waste, pollution, or the greenhouse effect). But all of these problems point to one main problem, and that is unsustainability.

We continually interact with the environment around us. This environment encompasses the entire natural world: air, land, oceans, plants, and animals. We have gained increasing control over the environment, leading to increases in our ability to alter the environment in which we live. But the capacity of the environment to absorb those alterations is limited. Invariably, the "carrying capacity" of every natural resource is either overstretched or exceeded. Carrying capacity for human beings can be defined as "the maximum rate of resource consumption and waste discharge that can be sustained indefinitely in a given region without progressively impairing the functional integrity and productivity of the relevant ecosystems" (Rees, 1992).

Environmental degradation is caused primarily by people who have the means or money to enjoy all the comforts of material life, such as automobiles and electronics, which contribute a great deal to pollution. Poor people, on the other hand, reuse most of the goods which wealthier people normally throw away. They might overuse some natural resources, such as fresh water and wood or forests for fire, but the contribution per person to such problems as greenhouse gases and water pollution is often much less than those of their wealthier counterparts. Sustainability depends on people using as few resources as possible.

14.2 DEFINING THE PROBLEM

To sustain means to support without collapse. Sustenance is that which supports life. Currently, humans are unequally provided with sustenance, and many suffer actual deprivation. One way to address the problem is by increasing their economic activity, but unfortunately this is usually accompanied by environmental degradation. In addition, we know that the present economic activity level is already unsustainable.

14.2.1 Biodiversity

Probably the worst thing facing civilization is not energy depletion, economic collapse, limited nuclear war, or conquest by a totalitarian government. As terrible as these catastrophes would be for us, they can be repaired within a few generations. The one ongoing process that will take millions of years to correct is the loss of genetic and species diversity by the destruction of natural habitats. This is the folly for which our descendants are least likely to forgive us (Wilson, 1992). The world is losing about 150 species per day because of human activities such as deforestation, pollution, application of pesticides, and urbanization (Reid and Miller, 1989). Some will say that we have taken steps to prevent all this by creating national reserves and parks, placing limits on hunting, and other means, but despite our concentrated recent efforts to protect our natural resources and landscapes, results are not very encouraging. How then do we reconcile the unrelenting need to protect natural systems with the impulse to transform them into human systems? Perhaps we can achieve this through an inclusive view that nature and culture are, in fact, not merely "two sides of the same coin." Rather, we need to engage in nonlinear and cyclical modes of thinking about nature, culture, and landscape. This is a complex relationship, one which is best understood through clarification, rather than through simplification.

Biodiversity can be taken as an indicator of environmental health, as it is assumed to be essential for the resilience of ecosystems; only flexible systems can bounce back to steady state after being subjected to a shock. Almost everything humans do—such as agricultural practices, trade, and regional development—affects biodiversity, and every effort must be undertaken to preserve this diversity by making sure that no species are on the brink of extinction.

14.2.2 Impediments to Achieving Sustainability

It is clear that we cannot continue to consume resources as we have in the past. The main reasons for our inability to achieve sustainability are (Pearce, Markandya, and Barbier, 1989):

Resource scarcity. This is one of the most important reasons for society's lack of sustainability at present. We are using our natural resources indiscriminately and at a very rapid rate. If we continue to consume at such a rate and do not find new renewable resources, some resources will soon become scarce and our standard of living will deteriorate. Sustainability hurdles are discussed further later in this chapter.

Environmental problems. Overproduction and overconsumption in recent years are the fundamental causes of the accelerating destruction of the global ecosystems, which in turn is making society unsustainable. The 20 metric tons of waste which every American generates each year must initially come from the environment and will later be dumped back into it as pollution and waste. The greenhouse effect is another major environmental problem. Scientists agree that fuel use must be cut by at least 60 percent to achieve a sustainability goal. For a future population of 11 billion people, this means that the average per capita fuel use would have to be cut to 6 percent of the present industrialized nations' average. Continued growth and development is the top priority of all nations, and unfortunately even present levels of production and consumption are not sustainable.

Poverty and the Third World. There are enough resources to feed everyone on the planet, provided there is an equal and just distribution of the resources. But rich and developed nations control most of the world's resources, and even within Third World countries there is a wide gap between rich and poor. Recent economic development has usually resulted in indiscriminate growth, with a very small minority garnering profits and benefits. Unless all people have basic amenities and comforts, Third World countries and their people will remain apathetic toward the fate of the environment, a situation that will clearly prevent world sustainability.

Conflicts of interest between nations. If all nations continue the quest to raise production, consumption, and living standards there can be no reduction in conflicts of interest between nations and the concominant overuse of world resources.

Falling living standards. Despite the increase in real gross national product (GNP) of rich, developed nations, the quality of life has remained stagnant or has actually gone down in many countries. For example, the real GNP per capita in

Australia has tripled since World War II, but some surveys indicate that there has been a decline in the standard of living rather than improvement. Indexes of social breakdown (e.g., doubling of suicide rates, increased number of divorces) point to a significant reduction in quality of life. This might be due to a single-minded pursuit of affluence, growth, commercialization, and so on, which in turn brings destruction of the community and a production of social wreckage in the form of unemployment and poverty.

Inadequate economic systems. Neither socialist nor capitalist economies allow us to reduce production to the minimum levels sufficient to give all people comfortable living standards. If we were to cease production of unnecessary goods and waste, there would be a jump in unemployment and bankruptcy. Present economic systems require an enormous amount of unnecessary production and waste. They are not suited for distributing goods according to need or for accomplishing critical policies.

Lack of technology. Despite large technological leaps during the last few decades, all major global problems continue to proliferate. Costs of acquiring resources escalate daily; in fact, they have been rising at almost 4 percent per year. The effects of the law of diminishing returns are already visible. Most people would like to provide the same living standards to people in Third World countries as we have in rich countries. However, to achieve this goal by 2060, world output has to increase by at least a factor of 10. Thus we must assume that within the next 60 years we must make the world sustainable, even though the present level of production and growth is unsustainable. Such a change cannot be acccomplished through technology alone; we will probably have to change our life-style and our attitudes to make this world sustainable.

Now, before we define sustainability more completely, let us go back in time and study how all this started and how the concept of sustainability became the buzzword of today.

14.3 THE HISTORY OF SUSTAINABILITY

One might be wondering, How could all this environmental damage have happened? In this section we discuss briefly some of the major factors that have contributed to our current situation. We also discuss some of the major international events that laid the foundation for the concept of sustainability.

The history of technological development and progress can be summarized as follows (Hatcher, 1996):

Environmental disconnection. As we moved from a hunter-gatherer to an agrarian society, our connection to environmental processes and ecological cycles gradually diminished. Hunter-gatherers generally used only what they needed and then moved on, allowing local species and the ecosystem to rebound from their harvest. But this changed as the first agriculturalists began to harvest and cultivate edible plants over other species, clearing land and planting areas with a single plant species. This disruption increased as agricultural technology improved, until almost continuous cultivation became the norm. Ecosystems no longer rebounded to replenish the soil. By disrupting the ecosystem, agrarian practices upset the ecosystem's robust nature and its ability to endure as a viable food-producing system, even in unfavorable conditions.

Social stratification. As agriculture developed, the social structure supporting it grew in parallel. As humanity proceeded from agriculture to industry, social stratification became increasingly intense and widespread. Stratification became international as industry sought cheap labor in the mostly agricultural countries of the Southern Hemisphere. All of this led to a class-based society where the powerful took ownership of the land, and the weak and poor became laborers to provide for their food and shelter (Sanderson, 1988).

Standard of living. As measured by health and the amount of time spent on food production, the standard of living for the overall population declined as a result of agricultural development. Increased agricultural intensity and production did not bring tangible benefits to the individual. However, the surplus did promote population growth, which in turn encouraged increased agricultural production. The same or more labor was needed to maintain increased demand for production, resulting in a poor standard of living for the laboring class, who now had to spend more time working on the land. On the other hand, the landowners spent much less time working, and their standard of living exceeded that of everyone else (Sanderson, 1988). Soon the Industrial Revolution, medical advances, safe drinking water, and sanitary disposal of wastes increased the average life span, but low-paying jobs in industry kept the average standard of living down. The gulf between the rich and the poor kept widening.

Population and technology growth. Technological development and population growth became locked into an upward spiral, each supporting the other. The spiraling growth continued, with the industrial revolution accelerating the problem in the mid-eighteenth century. New and more powerful machines made life easier, resulting in a population explosion, which in turn created more demand. The increase in demand for goods led to increased extraction of natural resources at the expense of the environment.

Momentum, lag, and uncertainty. The solution to environmental and ecological problems is exacerbated by our limited perception and understanding. Environmental effects build up slowly, gaining momentum as they build. The problem goes unnoticed at first, and by the time it is identified and detected, the problem is often so bad that even an immediate response cannot solve it. Even when we are able to measure the problem, our techniques and predictive models are inadequate or too inexact to give the degree of certainty we expect in order to make sound decisions. Good examples of this are damage to the ozone layer and global warming.

These and many other reasons have led us to our present state. If we don't take steps to alleviate this grave environmental situation, we are likely to see increased problems in the future.

14.3.1 Origin of the Sustainability Concept

The International Development Strategy of the First United Nations Development Decade of the 1960s was based on the belief that the fruits of accelerated economic growth would trickle down to the low-income population strata. This trickle-down effect did not occur and social justice became one of objectives of the Second Development Decade in the hope that it would lead to equitable distribution of the results of economic growth. The Third Development Decade of the 1980s realized that there were inequities and imbalances in international relations. The strategy, therefore, included the goal of establishing a "New International Economic Order." Unfortunately, this also failed. The latest attempts at an International Development Strategy for the 1990s (i.e., in the Fourth United Nations Development Decade) again calls for the acceleration of economic growth (U.N. General Assembly, resolution 45/199). According to the U.N. resolution, the 1980s were characterized as a decade of falling growth rates, declining living standards, and deepening poverty, with a widening gap between rich and poor countries. Although the Fourth United Nations Development Decade calls for fast economic development, it also stresses the need to simultaneously eradicate poverty and hunger, and develop and protect the environment. It led to the United Nations Conference of Environment and Development (UNCED), which was held in Rio de Janeiro, Brazil, on June 3–14, 1992.

The Rio conference was a watershed in the worldwide development of the concept of sustainable development. The outcomes of Rio were:

- The Rio Declaration on Environment and Development.
- The Convention of Climate Change.
- Convention on Biological Diversity.
- Conservation and Sustainable Development of All Types of Forests (Forests Principles Program) and *Agenda 21*.

Initiatives by international organizations which have contributed to the concept of sustainable development were synthesized in the Brundtland Report (Reid, 1993), which is summarized in Table 14.1.

14.3.2 United Nations Conference on Environment and Development

On June 3, 1992, the United Nations Conference on Environment and Development (UNCED) began in Rio de Janeiro, Brazil. It brought together diplomats, politicians,

TABLE 14.1
Summary of activities leading to our current understanding of sustainability

Year	Activity	Description
1972	*The Stockholm Conference.*	The agenda of the United Nations Conference held in Stockholm was the human environment, and it expressed concern with the global spread of environmental damage. This led to the establishment of Environmental Protection Agencies in a number of countries. The remedial steps taken were aimed at controlling the extent of environmental damage by setting limits or by requiring restoration of environmental quality. However, such an approach neither took a holistic approach to the environment nor integrated the environment and development.
1980	*The World Conservation Strategy.*	The International Union of Nature and Natural Resources (IUCN) published its World Conservation Strategy (WCS). The strategy defined development as "the modification of the biosphere and the application of human, financial, living and non-living resources to satisfy human needs and improve the quality of human life." The WCS said that conservation is a process which must be applied "cross-sectorally," and not be seen as a separate "activity sector in its own right," if the fullest sustainable benefits are to be derived from the resource base. The WCS also calls for anticipatory environmental policies and national accounting systems which will include nonmonetary indicators of success in conservation.
1980	*A Programme for Survival.*	The independent Commission on International Development Issues published a report, *A Programme for Survival,* calling for a reassessment of the notion of development as well as a new economic relationship between the richer North and the poorer South.
1982	*The United Nations World Charter for Nature.*	The United Nations published its *World Charter for Nature,* which adopted the principle that every form of life is unique and should be respected, regardless of its value to humankind. It also called for respect for our dependence on natural resources and control of exploitation of them: "Ecosystems and organisms shall be managed to achieve and maintain optimal sustainable productivity."
1986	*The IUCN Ottawa Conference on Environment and Development.*	The IUCN followed up the World Conservation Strategy with the Ottawa Conference on Conservation and Development. Sustainable development, the "emerging paradigm," is derived from two closely related paradigms of conservation: that nature should be conserved, which is "reaction against the *laissez-faire* economic theory that considered living resources as free goods, external to the development process, essentially infinite and inexhaustible"; and a second, derived from the moral injunction to act as steward, and responding to warnings expressed in publications such as *Silent Spring* and *Limits to Growth.*

and experts on environment and development from 172 of the 178 member states of the United Nations. The Rio Conference was not only a conference of governments. More than 1100 nongovernmental organizations (NGOs) were officially accredited to the conference and some 200 of them substantially influenced the documents which resulted from Rio. Moreover, in parallel with this conference, an environmental forum was held with 3738 NGOs from 153 countries participating. In general, the NGOs were able to reach more conclusive agreements than the diplomats at UNCED. The outcome of the forum was indicative of the trends that countries will have to follow in the future. During the last two days of the conference, the Earth Summit was organized.

The Rio conference was the third main conference on the environment organized by the United Nations. The first, the Conference on the Human Environment, was held in 1972 in Stockholm, Sweden. The dominant idea was that environmental problems are essentially by-products of intense industrialization and use of technology by the society, and a scientific-technical approach would therefore be able to solve them. The second conference was held in 1982 in Nairobi, Kenya. It was marked by a growing awareness that environmental problems in fact have a much wider reach than their technical-scientific scope. Socioeconomic factors were already seen as essential co-determinants of environmental issues. All of this led to the publication of the report of the World Commission of Environment and Development in 1987. This commission, which was chaired by Norwegian Prime Minister Gro Harlem Brundtland, focused on sustainability as the main benchmark of environmental policy. Sustainability was defined as

> the rearrangement of technological, scientific, environmental, economic, and social resources in such a way that the resulting heterogenous system can be maintained in a state of temporal and spatial equilibrium.

Thus the Brundtland Report represented both a deepening and an elaboration of the ideas discussed in Nairobi in 1982.

We next discuss in detail some of the main documents that came out of the Rio Conference.

EARTH SUMMIT AND AGENDA 21. Agenda 21 (U.N., 1992a; UNCED, 1992) is a plan of action for the world's governments and citizens. It sets forth strategies and measures aimed at halting and reversing the effects of environmental degradation and promoting environmentally sound and sustainable development throughout the world. The agenda comprises some 40 chapters and totals more than 800 pages. UNCED has grouped Agenda 21's priority activities under seven social themes:

1. The Prospering World (revitalizing growth with sustainability).
2. The Just World (sustainable living).
3. The Habitable World (human settlement).
4. The Fertile World (efficient resource use).

5. The Shared World (global and regional resources).
6. The Clean World (managing chemicals and waste).
7. The Peoples' World (public participation and responsibility).

RIO DECLARATION ON EARTH AND ENVIRONMENT. At the Earth Summit in Rio de Janeiro in 1992, over 118 countries took up the battle cry for sustainability, calling for local government changes that would tie economic growth to environmental protection. A set of principles was adopted to guide future development. These principles define people's rights to development and their responsibilities to safeguard the common global environment. To a large extent, they build on ideas from the Stockholm Declaration at the 1972 United Nations Conference on the Human Environment.

The Rio Declaration states that the only way to achieve long-term economic progress is to link it with environmental protection. This will happen only if nations establish a new and equitable global partnership involving governments, their people, and key sectors of societies. They must build international agreements that protect the integrity of the global environment and the development system.

The Rio Declaration consists of a preamble and 27 articles, reflecting the general principles of Agenda 21, the conventions, and a deforestation statement. The principles included in the Rio Declaration are presented in Table 14.2.

It was at Rio that the concept of sustainable development as integrating concerns for economic, ecological, and human well-being was propelled into the arena of global decision making, after the Brundtland Commission had brought the issues to world attention five years earlier. The means of achieving sustainable development was further emphasized at the World Summit for Social Development at Copenhagen in March 1995. This summit acknowledged that people are at the center of our concerns for sustainable development and that they are entitled to a healthy and productive life in harmony with the environment. It was concluded that social and economic development cannot be secured in a sustainable way without the full participation of women, and that equality and equity between women and men is a priority for the international community and as such must be at the center of economic and social development. The Copenhagen Declaration on Social Development stated that economic development, social development, and environmental protection are interdependent and mutually reinforcing components of sustainable development, and that democracy is an indispensable foundation for the realization of social and people-centered sustainable development. Sustainable social development is an integrated process of building human capacity to fight poverty, create productive employment, and promote social integration.

14.4 SUSTAINABILITY AND WHAT IT MEANS

Definitions of sustainability have been based on both weak and strong concepts (Pearce, Markandya, and Barbier, 1989); they are most easily distinguished from one

TABLE 14.2
Principles of the Rio Declaration

People are entitled to a healthy and productive life in harmony with nature.
Development today must not undermine the development and environment needs of present and future generations.
Nations have the sovereign right to exploit their own resources, but without causing environmental damage beyond their borders.
Nations shall develop international laws to provide compensation for damage that activities under their control cause to areas beyond their borders.
Nations shall use the precautionary approach to protect the environment. Where there are threats of serious or irreversible damage, scientific uncertainty will not be used to postpone cost-effective measures to prevent environmental degradation.
To achieve sustainable development, environmental protection shall constitute an integral part of the development process, and cannot be considered in isolation from it.
Eradicating poverty and reducing disparities in living standards in different parts of the world are essential to achieve sustainable development and meet the needs of the majority of people.
Nations shall cooperate to conserve, protect, and restore the health and integrity of Earth's ecosystem. The developed countries acknowledge the responsibility that they bear in the international pursuit of sustainable development in view of the pressures their societies place on the global environment and of the technologies and financial resources they command.
Nations should reduce and eliminate unsustainable patterns of production and consumption and promote appropriate demographic policies.
Environmental issues are best handled with the participation of all concerned citizens. Nations shall facilitate and encourage public awareness and participation by making environmental information widely available.
Nations shall enact effective environmental laws and develop national law regarding liability for the victims of pollution and other environmental impacts of proposed activities that are likely to have a significant adverse impact.
Nations should cooperate to promote an open international economic system that will lead to economic growth and sustainable development in all countries. Environmental policies should not be used as an unjustifiable means of restricting international trade.
The polluter should, in principle, bear the cost of pollution.
Nations shall warn one another of natural disasters or activities that may have harmful transboundary impacts.
Sustainable development requires better scientific understanding of the shared global problems. Nations should exchange knowledge and innovative technologies to achieve the goal of sustainability.
The full participation of women is essential to achieve sustainable development. The creativity, ideals, and courage of youth and the knowledge of indigenous peoples are needed, too. Nations should recognize and support the identity, culture, and interests of indigenous peoples.
Warfare is inherently destructive of sustainable development, and nations shall respect international laws protecting the environment in times of armed conflict and shall cooperate in their further establishment.
Peace, development, and environmental protection are interdependent and indivisible.

another with reference to often unstated assumptions about how technology and human ingenuity can be assumed to substitute for natural resources and ecological services. Strong definitions of sustainability assume that the possibility for such substitution is limited enough, or at least uncertain enough, to make continued industrial growth ecologically precarious. Weak definitions tend to assume that efficiency in use of resources, reflecting the substitution of ingenuity for resources inputs, will continue to increase as in the past (Cairncross, 1993; Daly and Cobb, 1989). Historical patterns of technological change tend to support the second point of view (Ayres, 1989a), one that has been referred to as "techno-optimism." A weak and explicitly economic definition of sustainability would be organized around an industry's or a region's ability to continue providing income, either through employment or indirectly through the multiplier effects of local spending of that income. Stronger definitions of sustainability are less likely to consider the imperatives that sustainability and competitiveness be compatible or reconcilable. This is at least partly because strong definitions of sustainability tend to assume limits on carrying capacity, which in turn imply constraints on continued increases in economic output in the rich countries.

14.4.1 Definitions of Sustainability

Many definitions of sustainability have been proposed. Following is a sampling of a few of them.

> The word sustainable has roots in the Latin "subtenir," meaning "to hold up" or "to support from below." A community must be supported from below—by its inhabitants, present and future. Certain places, through the peculiar combination of physical, cultural, and, perhaps, spiritual characteristics, inspire people to care for their community. These are the places where sustainability has the best chance of taking hold. (Muscoe Martin, 1995)

> Sustainability refers to the ability of a society, ecosystem, or any such ongoing system to continue functioning into the indefinite future without being forced into decline through exhaustion. . . of key resources. (Robert Gilman, president of Context Institute)

> Sustainability is the [emerging] doctrine that economic growth and development must take place, and be maintained over time, within the limits set by ecology in the broadest sense—by the interrelations of human beings and their works, the biosphere and the physical and chemical laws that govern it . . . It follows that environmental protection and economic development are complementary rather than antagonistic processes. (William D. Ruckelshaus, former U.S. EPA administrator, 1989)

> Sustainable development meets the needs of the present without compromising the ability of future generations to meet their own needs. (Gro Harlem Brundtland, former prime minister of Norway, 1987)

> Then I say the earth belongs to each . . . generation during its course, fully and in its own right, no generation can contract debts greater than may be paid during the course of its own existence. (Thomas Jefferson, September 6, 1789)

Another way of describing sustainability can be seen in Figure 14.1.

Sustainability is:

Safe
Universally accepted
Stable
Technology that benefits all
Antipollution
Improvement in quality of life
Nontoxic
Awareness
Beautiful
Indigenous knowledge
Least-cost production
Income
Total quality
Youth

FIGURE 14.1
Factors influencing sustainability. (Source: Olaitan Ojuroye, Nigeria)

Sustainable development most commonly refers to ecological sustainability, but terms like social, economic, community, and cultural sustainability have slowly come into use. All of these should be combined when we talk about sustainable development. In this book, sustainable development and sustainability are used interchangeably.

14.4.2 What Is Sustainable Development?

Sustainability is basically made up of three closely connected precepts:

- The environment is an integral part of the economy; it is not a free resource.
- Equity between the developing and developed world is essential. The developing world gives more importance to the pace of development and wants to reach a high standard of living as soon as possible. This poverty issue must be addressed, and there has to some equity between these worlds.
- Every entity (from countries to individuals) should have long-term futuristic goals in mind and should not operate on the basis of short-term benefit. Longer planning horizons are needed, and policies need to be proactive rather than reactive.

In terms of the Brundtland Commission definition, sustainable development is development that meets the needs of the present without compromising the ability of future generations to meet their own needs. This definition contains within it two key concepts:

- The concept of needs, particularly the essential needs of the world's poor, to which overriding priority should be given.
- The concept of limitations imposed by the state of technology and social organization on the environment's ability to meet present and future needs.

To better understand the concept of sustainable development, the term can be broken down into its individual components:

Sustain: Maintain; supply with necessities or nourishment; support.

Develop: Expand or realize the potentialities of growth; bring gradually to a fuller, greater state.

Thus the goals of economic and social development must be defined in terms of sustainability in all countries—developed or developing, market-oriented or centrally planned. Interpretations will vary, but they must share certain general features and must flow from a consensus on the basic concept of sustainable development and on a broad strategic framework for achieving it.

For the business enterprise, sustainable development means adopting business strategies and activities that meet the needs of the enterprise and its stakeholders today while protecting, sustaining, and enhancing the human and natural resources that will be needed in the future. The sustainable business has interdependent economic, environmental, and social objectives; its management understands that long-term viability depends on integrating all three objectives in decision making. Rather than regarding social and environmental objectives as costs, a sustainable enterprise seeks opportunities for profit in achieving these goals.

14.4.3 Conceptualization of Sustainability

ECONOMIC CONCEPT. Sustainability can be best conceptualized economically by the Hartwick's "cake-eating model" (Common, 1995). The origin of this appellation is the analogy with the problem of dividing up a cake between a large number of would-be cake eaters. Suppose that the only natural resource available is a finite stock of a nonrenewable resource, and recycling is impossible. Next suppose that humans directly live off this resource, which is essential to life. Also, for the sake of simplicity, suppose that the human population size is constant across generations. The question then is, What is the largest constant rate of per capita consumption that can be maintained indefinitely? What is the maximum sustainable rate of consumption? The answer, of course, is clearly zero. There is no positive use rate for a finite nonrenewable resource stock that can be maintained indefinitely. In a cake-eating world, sustainability is impossible.

ECOLOGICAL CONCEPTS. Before we look further into an economic conceptualization of sustainability, it is necessary to define some important terms. *Stability* refers to a propensity for return to an equilibrium level following a disturbance. According to Conway (1985), stability is the degree to which productivity is constant in the face of a small disturbance caused by the normal fluctuations of climate and other environmental variables. *Resilience* is a property of an ecosystem, rather than of a population within an ecosystem. It is the ability of a system to maintain its structure and patterns of behavior in the face of a disturbance. Note that resilience means the system remains unchanged. An *ecosystem* is an interconnected biotic assembly of plants, animals, and microbes, together with its abiotic physiochemical environment. A system is sustainable in the ecological sense if it is resilient. Complex ecosystems aren't necessarily more stable, and a low stability system can demonstrate high resilience. Basically, the concept of resilience is what characterizes an ecological approach to sustainability.

Sustainability, according to Conway (1985), is the ability of the system to maintain productivity in spite of a major disturbance. Clearly, this concept of sustainability is a resilience concept.

The transition to a sustainable society requires a careful balance between long- and short-term goals and an emphasis on efficiency, equity, and quality of life, rather than on quantity of output. It requires more than productivity and more than technology. It also requires maturity, compassion, and wisdom (Meadows, Meadows, and Randers, 1992).

Key characteristics of sustainability are as follows:

- Sustainability is a normative, ethical principle. It has both necessary and desirable characteristics. Because opinions may differ over what is desirable, there is no single version of a sustainable society.
- Both environmental-ecological and sociopolitical sustainability are required for a sustainable society.
- No one can or wants to guarantee the persistence of any particular system. We want to preserve the capacity for the system to change. Thus sustainability is never achieved once and for all, but is only approached. It is a process, not a state. It will often be easier to identify unsustainability than sustainability.

Common components of a sustainable society are listed in Table 14.3.

In the United States, an environmental impact statement (EIS) is commonly required by the U.S. EPA for many construction projects or other initiatives in order to evaluate the impacts of the project on the environment before approval for the project is granted. The project may be rejected if significant negative impacts are predicted. Thus an EIS can be thought of as a device that could be used to ensure that a project meets the objectives of sustainability. However, an EIS usually is site-specific, takes into account only negative impacts of particular projects on a case-by-base basis, and does not consider alternative options. The EIS may not provide a clear picture of the

TABLE 14.3
Common components of a sustainable society

What sustainability is	What sustainability is not
Integrated decision-making process	Justification for business as usual
Research and information	Growth at all costs
Democratic values	Heavier command and control systems
Community participation	All things to all people
Collaboration	Static or declining economy
Equity, justice, and shared progress	Quick fixes and ad hoc solutions
Obligations to future generations	
Leadership in all sectors far beyond compliance	
Long-term solutions	

long-term impacts on the environment or on our natural resources. Sustainability, however, must also relate to equity between and within generations, biodiversity, and population issues. Trying to improve the basic efficiency of the EIS process should not be confused with the broader objective of integrating considerations of environmental impact with the basic principles of sustainability. The EIS process can't adequately assess the cumulative intergenerational effects of individual projects. More detailed and more complex evaluation processes are needed.

14.4.4 Hurdles to Sustainability

Essentially all sustainability efforts so far have failed because they lacked a sense of urgency and commitment or because sustainability is not commonly thought of as a long-term philosophy. Along with these difficulties are several other hurdles that a sustainability movement must face. Some of these are described here (Welford, 1995):

Level of consumption. Mass consumption is not possible indefinitely, and if we continue to consume the resources indiscriminately, there will be nothing left for the future. The way of life in the developed countries is often extremely expensive in terms of per capita resource and energy consumption. Every year each American consumes, on average, 20 metric tons of new materials, including energy equal to 12 tons of coal. If all people likely to be living on Earth during the latter part of the twenty-first century were to consume energy at such a rate, the world energy production would have to be 14 times its present level, and all potentially recoverable (as distinct from currently known) energy resources (excluding breeder and fusion reactors) would be exhausted in about 14 years (Trainer, 1985). Just to provide the present world population with the affluent world's diet would require eight times the present world cropland (which is likely to decrease in the future due to increased population and housing demands), or more than the world's entire land area (Rees, 1992).

Apathy. Wealthy people are often apathetic to the need for conserving, and affluent living styles commonly lead to mass consumption, as stated above.

Developing nations. Third World people aspire to higher standards of living and often disregard the importance of conserving the environment. Only one-fifth of the world's population now lives affluently. Half the world's population averages a per capita income one-sixtieth of that found in the rich countries, and more than 1 billion people live in desperate poverty. Deprivation takes the lives of more than 40,000 Third World children every day. There is much development going on in Third World countries, but the development is often of little benefit to the inhabitants. Throughout the 1980s, living conditions for many, if not most, of the Third World people deteriorated (Kakwani, 1988). Much of the time, the production capacity is being used for the benefit of rich countries, and as a result, the environmental status of the developing countries keeps on deteriorating because they have more urgent needs than environmental improvement. According to Trainer (1995), "The rich must live more simply so that the poor may simply live."

Lack of public awareness. Everyday choices people make have a bearing on the environment, and small steps will go a long way toward helping and saving the environment. The public needs to be provided with suitable information and needs to be made aware of the importance of achieving a sustainable world.

Lack of knowledge. Developed countries do not fully know how to achieve sustainability, and even less is known in the developing countries. More research is needed to overcome this deficiency.

Magnitude and number of uncertainties. There are so many uncertainties associated with the concept of sustainability that it is difficult to estimate and forecast "what if" scenarios. Human behavior is unpredictable, as are people's reactions under new, and possibly more stressful, conditions. These, coupled with inaccuracies of data and uncertainties and assumptions associated with predictive models, make the job all the more difficult.

14.5 ACHIEVING SUSTAINABLE DEVELOPMENT

Some of the critical objectives for improving the environment and for instituting better development policies result from the concept of sustainable development. These objectives include (ECWD, 1987):

- Reviving growth.
- Changing the quality of growth.
- Meeting essential needs for jobs, food, energy, water, and sanitation.
- Conserving and enhancing the resource base.
- Reorienting technology and managing risk.
- Merging environment and economics in decision making.

To achieve sustainable development, we must consider sustainable development in all its dimensions—ecological, social, economic, and political. Policies on wilderness preservation or pollution prevention instituted in isolation are not enough to achieve sustainable development. We need to consider poverty concerns, gender issues, institutional organization, and decision making under one umbrella and address them in unison in an integrated way (Dale and Robinson, 1996). A strategy or framework is required to guide us to the path of sustainability.

14.5.1 Sustainable Development Framework

The incorporation of environmental concerns in development planning and policy formulation requires the introduction of space as an explicit dimension of an integrative framework (Bartlemus, 1994). This spatial effect is introduced by the fundamental concept of ecosystems. Ecosystems represent the area of interaction of a biotic community with its nonliving environment (Odum, 1971).

The sustainable development framework shown in Table 14.4 introduces four strategic functions: assessment, research and analysis, planning and policies, and support. These four functions are then analyzed in terms of (1) sustainable economic growth and (2) development at the local, national, and international levels. Table 14.4

TABLE 14.4
A framework for sustainable growth and development

Strategic function	Development paradigm					
	Local Development		National Development		International Development	
	Regionally sustainable economic growth	Ecodevelopment	Sustainable economic growth	Sustainable development	Development of a supportive economic environment	Global sustainable development
Assessment	Integrated regional (monetary) accounting	Environment statistics	Integrated environmental economic accounting	Environment statistics	International (comparative) environmental accounting	International (integrated) databases, statistical compendia, and reporting
		Regional (physical) resource accounting Statistical ecology		Natural (physical) resource accounting National development reports		
Research and analysis	Modeling spatial disparities in income and growth	Modeling ecodevelopment	Integrated micro-, meso- and macroeconomic analysis and modeling	Models of integrated (physical) planning and development	International and global economic/environmental analysis and integrated modeling	International (physical) modeling of global environmental concerns and their socioeconomic implications
		Modeling carrying capacity Development of ecotechniques				
Planning and politics	Regional planning and policies of sustainable economic growth	Ecodevelopment planning and administration (decentralized planning and strategies)	Reorientation of macroeconomic policies toward sustainability	Integrated (physical) planning and project formulation	International strategies of sustainable economic growth	International standards for sustainable development
			Policies of structural change	Demographic policies on population, resources, environment, and development		International strategies and conventions (e.g., Agenda 21)

(continued)

TABLE 14.4 (*cont.*)
A framework for sustainable growth and development

	Development paradigm					
	Local Development		National Development		International Development	
Strategic function	**Regionally sustainable economic growth**	**Ecodevelopment**	**Sustainable economic growth**	**Sustainable development**	**Development of a supportive economic environment**	**Global sustainable development**
Planning and politics (*cont.*)			Economic (dis)incentives for microeconomic planning and management	Programs of human needs satisfaction		
Support	Extension service		Public awareness building and participation		Promotion of sustainable growth and development at international and global levels	
	Technical assistance (for local-level projects)		Programs and projects of education, training, and public information		Multilateral support for international (global) sustainable development	
	Public awareness building and participation (support to grassroots movements/organizations)		Institution building and environmental law/regulations		International institution building (e.g., UNCSD)	
			Technical cooperation, including transfer (import) of environmentally sound technology and capacity building			

Source: Bartlemus, 1994.

provides information about particular strategy functions for all regional levels or, if read vertically, on comprehensive approaches to sustainable growth or development at national, subnational, and international levels.

Sadler and Fenge (1993) described seven basic principles that must be considered when establishing a strategy for sustainable development. They are listed in Table 14.5.

This approach to creating a sustainable development strategy at the national level has been applied. Table 14.6 outlines the core elements of a Canadian national sustainable development strategy (Sadler and Fenge, 1993).

Once a sustainability framework is in place and strategies to achieve the goals of sustainability are formulated, it then becomes necessary to develop ways to implement these strategies.

TABLE 14.5
Seven principles of sustainable development strategy making

When developing a strategy for sustainable development, one should consider the following seven principles:

Integrative approach. Strategy should be integrative, forward-looking, cross-sectoral processes for linking and balancing environmental, social, and economic policy objectives.

Focus on issues. Strategy should directly address the major structural issues and constraints on achieving an economically viable, socially desirable, and ecologically maintainable future. This involves addressing current problems within a longer term policy horizon.

Goal orientation. Strategy should be based on clearly defined objectives and priorities with measurable targets and time frames for meeting them. A long-term vision of a sustainable society is a useful starting point, because it can be framed in terms of broad, shared values and hopes for the future.

Compatibility with policy processes. Strategies must be adjusted to the policy cycle and institutional culture and must initiate change in the direction and process of decision making.

Consensus building. Strategies should invite wide public involvement and consultation. To establish strategies regarding societal values and ethics, ensuring that they incorporate the visions and aspirations of citizens and facilitate life-style and behaviorial changes, is indispensable.

Action orientation. Strategies should lead to immediate, practical steps that lay the ground for a longer term, systemic transition in patterns of production and consumption.

Capacity enhancement. Strategies should be capacity-building processes that strengthen institutions, sharpen concepts and tools of sustainability, improve skills and competencies, and promote public awareness.

Source: Sadler and Fenge, 1993.

TABLE 14.6
Components of a Canadian national sustainable development strategy

Introduction
  The rationale and benefits of an NSDS
State of the nation
  Assessment of global trends, national issues, regional prospects
  Canada in a global context
Vision for tomorrow
  Values and ethics of a sustainable society
  Images and aspirations for the future
Goals, objectives, and targets
  Overall goal
    To improve quality of life consistent with obligations to future generations

(continued)

TABLE 14.6 (*cont.*)

Components of a Canadian national sustainable development strategy

- Goals, objectives, and targets (*cont.*)
  - Objectives
    - To improve social welfare and individual well-being
    - To provide an equitable distribution of opportunities
    - To maintain natural capital at or above current levels
  - Targets for meeting economic, social, and environmental objectives
    - To adopt minimum standards to guarantee environmental sustainability
    - To pay down the environmental deficit through rehabilitation measures
- Guiding principles
  - To integrate decision making
  - To harness markets
  - To share responsibility
  - To build consensus
  - To increase public awareness
  - To enforce accountability of decision makers
  - To ensure open decision making
  - To empower people
- Cross-cutting elements
  - Institutional reform
    - Harmonize environmental regulations
  - Greening the economy
    - Clean industry strategies
  - Resource and environmental stewardship
    - Biodiversity strategy
  - Healthy communities
    - Urban renewal; promotion of rural and traditional livelihoods
  - Individual action and initiative
    - Environmental citizenship
  - Population and immigration
    - Replacement and growth options
  - Scientific and technical innovation
    - Raw material substitution
  - Global commitments
    - Response to convention on climate change
- Sectoral plans and initiatives
  - Direction and dimensions for achieving sustainability in major policy and economic sectors
  - Detailed plans to be drawn up by each industry sector
    - Energy, transportation, agricultural, forestry, fisheries, tourism, chemicals, etc.
- Regional dimensions
  - Links to provincial and local sustainability strategies
  - Responses to regional issues and concerns
  - The North and the national interest
- Indigenous peoples
  - Use of traditional knowledge and life-styles
  - Land and resource use
  - Community and economic development
- Means of implementation
  - Communication and outreach
  - Policy dialogue on cross-cutting sectoral issues
  - Monitoring progress on implementing the strategy
  - Supporting measures
    - Research agenda, information tools and technologies
  - Policy tools and instruments

Source: Adapted from Sadler and Fenge, 1993.

14.5.2 Application of Sustainability Strategies

The concepts of sustainability can be applied universally. We now briefly examine how they can be applied to various aspects of our environment, including land, water, air, housing, and energy.

LAND. Land is a major resource for agriculture, environment, society, and so on. Over time, land has been primarily used for grain and food production, but during the past few decades increased pressure has been applied to land for recreational uses and urban housing. Expansion of land use for agriculture, recreation, and housing resulted in a loss of wildlife habitat, ecological imbalance, and changes in watershed patterns. All of these land uses have resulted in soil degradation and a loss or removal of the top layer of soil, with a concomitant loss of soil fertility and reduction in grain production. Organic matter in soil, which is important for soil fertility, is also affected by different land uses, including cultivation. Diverting some of the land back for wildlife habitats can complement the sustainability of agriculture.

WATER. Water is the most important resource for human survival, and much of the world is gripped by a short supply of usable water, be it potable water or water for agricultural or industrial use. Even though Earth is almost 80 percent water, usable water is only a small fraction of the total supply; hence the need for sustainability. Precious water resources are contaminated through seepage from agricultural lands, landfills, and indiscriminate pumping of groundwater. Greater scrutiny of water use and pollution is required.

AIR. Like water, air has been polluted by the activities of society. Indeed, the probability of air becoming polluted is great, because misuse of free goods is more likely than that of expensive resources. If, somehow, a charge could be applied on the use of this free resource, we would come closer to solving the problem. The atmosphere now has a relatively large concentration of greenhouse gases, which are causing global warming, resulting in the uncertainty of climate as more and more fossil fuels are being burnt. Diverting more land to forest cover would fulfill the needs for sustainability of both land and air resources.

HOUSING. Buildings designed, renovated, constructed, operated, and demolished in an environmentally sound and energy-efficient manner benefit the environment, building owners, and the community. For example, increased energy and water efficiency will enhance resource conservation and reduce the need for utilities to build new facilities. Increased allowance for daylighting in building design can improve productivity and occupant health. Implementing source reduction and waste recycling policies can reduce disposal costs and prolong local waste disposal capacity. Carefully considering public transportation access to buildings will reduce traffic congestion and improve air quality. *The Sustainable Building Technical Manual: Green Building Design, Construction and Operations,* produced jointly by the U.S. Department of Energy and Public Technology, Inc. (USDOE, 1996), shows how to design, operate, and maintain environmentally friendly buildings.

ENERGY. Although many of us only think about reducing our energy use during an energy shortage, the choices we make about our everyday energy activities—how we get to work and school, what kind of lighting we buy, and what kind of appliances we purchase—have a significant impact on both the economy and our environment. In fact, many organizations such as Worldwatch believe that energy will be the most important subject of the next decade. If we are to help ensure a sustainable future for ourselves and our global neighbors, we must work together. We need to discuss problems and share examples of successful environmentally sustainable energy initiatives in both rural and urban areas and demonstrate how communities, environmentalists, businesses, and government agencies can become partners in promoting sustainable energy choices (e.g., where conventional grid supply systems cannot be extended economically to remote villages, renewable energy sources offer the greatest promise).

14.5.3 Indicators of Sustainability

Once we take steps to achieve sustainability, we need some indicators to tell us if we are on the right track. Indicators are required to make any necessary midcourse corrections.

> Trying to run a complex society on a single indicator like the Gross National Product is literally like trying to fly a 747 with only one gauge on the instrument panel . . . imagine if your doctor, when giving you a checkup, did no more than check your blood pressure. (Hazel Henderson, 1991)

What are the indicators by which health, well-being, and improvement of a community can be measured? Traditional measurements often analyze an individual issue, for example, the number of new jobs in a particular community. But such a one-dimensional approach does not reveal the quality of those jobs or their impact on the local economy. Other indicators measure the number of children living in poverty, indicating the relationship between social health and local economic performance. It is evident that many indicators can be chosen, but few of them will provide an overall picture of our multidimensional world.

Nontraditional methods, such as "indicators of sustainability," are designed to provide information for understanding and enhancing the relationships between the economic, energy use, environmental, and social elements inherent in long-term sustainability. For example, sustainability indicators can serve as valuable tools for profiling local energy consumption patterns as a sustainability benchmark. Communities such as Seattle, San Francisco, and Toronto are using indicators to gather and evaluate information on both current energy use and future alternatives for the residential, commercial, industrial, and transportation sectors. This information is vital in planning for and managing the energy resources that will support sustainable development.

The concept of sustainability indicators has been derived from the concept of natural ecosystem indicators. The sustainability indicators should

1. Consistently represent a critical ecosystem component.
2. Be amenable to isolation in the environment.
3. Be measured accurately and repeatedly.

4. Be understood in terms of the health of the ecosystem.
5. Be well understood and accepted by the community.
6. Have the potential to be linked to other sustainability indicators.
7. Represent and relate to important community values.

Sustainability indicators identify key characteristics of the existing human and natural ecosystems. The role of an indicator is to make complex systems understandable or perceptible. An effective indicator or set of indicators helps a community determine where it is, where it is going, and how far it is from chosen goals. They provide perspective on a community's progress and guideposts for changes in its activities. Indicators of sustainability examine a community's long-term viability based on the degree to which its economic, environmental, and social systems are efficient and integrated.

To measure the degree of efficiency and integration, a set of indicators is often required. These indicators can incorporate several broad categories such as economy, environment, society and culture, government and politics, resource consumption, education, health, housing, quality of life, population, public safety, recreation, and transportation. Examples of indicators currently in use from several of these categories are described in Table 14.7.

TABLE 14.7
Sustainability indicators

Indicator	Description
	Economy
Income	Distribution of jobs and income, gross domestic product (GDP), gross national product (GNP), stock market averages
Business	Percentage of wages earned within a community and spent within the community
Training	Employer payroll dedicated to continuing training/education
Human development/ quality of life	Education, health care, cost of living, cultural diversity
	Environment
Air	CO_2 emissions from transportation sources
Drinking water	Percentage reduction in drinking water supplies from 1990
Land use	Percentage of development occurring annually within an urban area
	Resource use
Energy	Percentage of energy used that is derived from renewable sources
Hazardous materials	Consumption of pesticides
Water	Number of gallons of water saved through leak repair
	Society and culture
Abuse	Child abuse/neglect/abandonment
Diversity	Racism perception
Volunteerism	Volunteer rate for sustainability activities

Source: Hart, 1995.

The usefulness and accuracy of indicators of sustainability depend on their ability to create a snapshot of the community's economic, environmental, and social systems. Choosing the appropriate indicators and developing a program is a large-scale process requiring collaboration between many sectors, including government agencies, the public, research institutions, civic and environmental groups, and business. The indicator programs profiled by Hart (1995) offer a wealth of information on the process of indicator program development, rationale for specific indicator selection, and ongoing challenges communities face in using indicators.

For sustainability to be successful, economic, environmental, and social equity issues must be considered together, since they are inextricably linked. Government must provide direct and meaningful interaction among those affected; information delivery must undergo vast technological changes to enable citizens and institutions to participate more fully. New methods of governance that are more collaborative, such as private–public partnerships, must be instituted. Sustainability indicators help us in measuring the effects of our sustainability efforts.

14.5.4 What Is Being Done to Achieve Sustainability?

Let us now examine what has been done to achieve sustainability. Following are brief summaries of steps taken or being taken by several groups to achieve sustainable development.

The U.S. Environmental Protection Agency, under Project XL, is giving selected companies the regulatory flexibility needed to streamline their manufacturing processes to allow them to both reduce costs and produce superior results. Project XL is a national pilot program that tests innovative ways of achieving better and more cost-effective public health and environmental protection. Through site-specific agreements with project sponsors, EPA is gathering data and project experience that will help the agency redesign current approaches to public health and environmental protection. Under Project XL, sponsors—private facilities, industry sectors, federal facilities, and communities—can implement innovative strategies that produce superior environmental performance, replace specific regulatory requirements, and promote greater accountability to stakeholders. XL projects are real-world tests of innovative strategies that achieve cleaner and cheaper results than conventional regulatory approaches would achieve. The EPA will grant regulatory flexibility in exchange for commitments to achieve better environmental results than would have been attained through full compliance with regulations. The EPA has set a goal of implementing 50 pilot projects in four categories: XL projects for facilities, sectors, government agencies, and communities.

The Nature Conservancy and Georgia-Pacific Corp. agreed in 1994 to implement a unique partnership to manage 21,000 acres of wetlands along North Carolina's lower Roanoke River that teems with animal life. Georgia-Pacific owns the land, but a joint committee—including representatives of the U.S. Fish and Wildlife Service, the Nature Conservancy, and Georgia-Pacific—decides where and under what conditions timber harvesting can occur.

One of the most successful pregnancy prevention programs in the country is Teens Teaching Teens. Started in Atlanta, Georgia, by the public schools and the Grady

Health System, the program trains high school juniors and seniors to encourage eighth graders to postpone sex. A study has shown that after participating in this peer counseling, students are less likely to be sexually active. Being sexually inactive is probably the best way to achieve population control, and this program means fewer births and a slower population growth. Most of the world's social and environmental problems are linked to excessive population; programs of this type are models of how population growth can be brought under control through education.

Several initiatives have been undertaken by states and communities to achieve sustainable development. Denver, Colorado's, Environmental Program is a comprehensive effort to protect the health and welfare of Denver citizens and the region's economy through protection and enhancement of environmental quality. Citywide environmental planning and implementation efforts are coordinated with the Denver Regional Council of Governments, the Regional Air Quality Council, and the Clean Air Colorado program.

The Minnesota Sustainable Development Initiative brought together 105 citizen-leaders representing local agencies and environmental, business, and civic organizations to prepare a sustainable development plan for Minnesota that would reconcile the economic and environmental goals of the state, ensuring environmental protection while allowing for economic and job growth. The initiative came up with 400 specific strategies for protecting the environment and developing the economy in the areas of agriculture, energy, forestry, manufacturing, minerals, recreation, and settlement.

Sustainable Urban/Rural Enterprise (SURE) is a civic nonprofit corporation promoting the dual goals of economic development and environmental stewardship for the City of Richmond and Wayne County, Indiana. SURE promotes enterprises that are consistent with its principles of sustainability: that the only development-related activities that are pursued are those that can be perpetuated continuously by future generations; that economic and agricultural systems should be both adaptable and resilient; and that no development activity should deplete the natural resource base. One of SURE's initiatives is directed toward agriculture: to reduce soil erosion and chemical use, maintain greenery and trees on farmland, study hydroponics in greenhouses, and maintain a diverse agriculture. Other SURE initiatives include recycling, recycled product manufacturing, and the development of neighborhood gardens.

14.6 SUSTAINABILITY IN THE UNITED STATES

14.6.1 President's Council on Sustainable Development

The President's Council on Sustainable Development (PCSD) was established by President Bill Clinton on June 29, 1993, by Executive Order 12852. The council adopted the definition of sustainable development as stated in the original Brundtland Commission report: "development that meets the needs of the present without compromising the ability of future generations to meet their own needs." The 25-member council is a partnership, drawing leaders from industry and government and from environmental, labor, and civil rights organizations. It is charged with developing bold new approaches to integrate economic and environmental policies.

The mission of the PCSD is:

1. To develop and recommend to the president a national sustainable development action strategy that will foster economic vitality.
2. To develop an annual Presidential Honors Program, recognizing outstanding achievements in sustainable development.
3. To raise public awareness of sustainable development issues and participation in opportunities for sustainable development.

Council members serve on eight task forces. The task forces and their duties are listed in Table 14.8.

The council has adopted 10 national goals for a sustainable future. They are interdependent and must be achieved in unison, considering economic, environmental, and social equity issues:

1. Ensure every person the benefits of a healthy environment.
2. Sustain a healthy economy that affords the opportunity for a high quality of life.
3. Ensure equity and opportunity for economic, social, and environmental well-being.
4. Protect and restore natural resources for current and future generations.
5. Encourage stewardship.
6. Encourage people to work together to create healthy communities.

TABLE 14.8
The task forces and duties of the President's Council on Sustainable Development

Task force	Duty
Eco-Efficiency	Identifies models of sustainable manufacturing, pollution prevention, and product stewardship that will enhance recommendations for policy change.
Energy and Transportation	Develops long- and short-term policies to contribute to a more sustainable energy future.
Natural Resources Management and Protection	Develops guidelines to better manage and protect the nation's natural resources.
Principles, Goals, and Definitions	Articulates sustainable development principles and goals.
Population and Consumption	Identifies the impact of population and consumption patterns on sustainable development and recommends actions to address these issues.
Public Linkage, Dialogue, and Education	Works to foster public dialogue and develop educational outreach activities.
Sustainable Agriculture	Examines and makes recommendations relating to sustainable agriculture production, practices, and systems.
Sustainable Communities	Explores the obstacles and opportunities for sustainable development at the community level.

7. Create full opportunity for citizens, businesses, and communities to participate in and influence the natural resource, environmental, and economic decisions that affect them.
8. Move toward stabilization of the U.S. population.
9. Lead in developing and carrying out sustainable development policies globally.
10. Ensure access to formal education and lifelong learning that will prepare citizens for meaningful work and a high quality of life and give them an understanding of concepts involved in sustainable development.

The PCSD has provided many recommendations for local government, industries, and individuals that will allow them to better do their part toward the goal of achieving sustainability. A few of these are listed in Table 14.9.

TABLE 14.9
PCSD recommendations for furthering sustainability

Build a "New Framework for a New Century."
Increase the cost-effectiveness of the current environmental management system by creating opportunities for attaining environmental goals at lower costs.
Create a new, flexible, and performance-based regulatory management system to achieve superior results and cost savings through innovation.
Adopt a voluntary system of extended responsibility for products through their entire life cycle where designers, suppliers, producers, users, and disposers work together to exercise environmental stewardship from procurement of raw materials, through manufacture and distribution, to use, disposal, and reuse.
Expand market-driven pollution control programs, such as emissions trading.
Establish a national commission to review the effect of federal tax and subsidy policy on the goals of sustainable development and recommend changes.
Change tax policies—without increasing the overall tax burden—to encourage employment and economic opportunity while discouraging environmentally damaging production and consumption decisions. Tax reform should not place a disproportionate burden on lower income individuals and families.
Eliminate government subsidies that are inconsistent with economic, environmental, and social goals.
Revamp the federal government's method of collecting, organizing, and disseminating data on economic, environmental, and social conditions to improve its quality and accessibility.
Improve the collection, coordination, and dissemination of scientific and health information available via computers.
Promote widespread access to information through computers, by offering computer skills training, making information formats more consistent within and among government agencies, and improving computer networks.
Develop better natural resource and quality-of-life baseline information and methods to measure the quality and quantity of renewable and nonrenewable resources, such as forests, lakes, minerals, and fish, and develop indicators of progress toward sustainability goals.
Change the education system to teach students at all levels the interdependence of the environment, social equity, and the economy.
Revise business accounting practices to link products with their environmental costs.

(continued)

TABLE 14.9 (*cont.*)
PCSD recommendations for furthering sustainability

Identify key issues, create a vision for the future, and set goals and measurable benchmarks. This strategic planning process by a diverse group should identify unique local advantages and set goals to utilize them.
Create federal incentives to spur communities to deal with issues that transcend jurisdictions. Activities could include pooling local property taxes to increase equity in public services, improving education, and reducing economic incentives for sprawl.
Encourage builders, architects, developers, contractors, and community groups to design and rehabilitate buildings to use energy and natural resources efficiently, enhance health and the environment, preserve history and natural settings, and contribute to a sense of community.
Design new communities and improve existing ones by using land efficiently, promoting mixed-use and mixed-income development, retaining public open space, and providing diverse transportation options.
Manage the geographic growth of communities and create a plan for decreasing sprawl, conserving open space, respecting nature's carrying capacity, and protecting against natural hazards.
Revitalize "brownfield" sites (contaminated, abandoned, or underused land), making them more attractive for redevelopment by providing regulatory flexibility and incentives.
Issue executive orders (at federal and state levels) for agencies to use voluntary, multistakeholder approaches to manage natural resources and resolve natural resource conflicts.
Protect water quality, biodiversity, and other natural resources in unison through cooperative efforts across entire ecosystems, such as watersheds.
Create incentives to promote stewardship among landowners, corporations, government, and resource users to pursue stewardship or protection of natural resources.
Require commercial users of natural resources to pay the full cost of resource depletion.
Restore habitat and eliminate overfishing to rebuild and sustain depleted wild stocks of fish in U.S. waters.
Move toward stabilization of the U.S. population.
Create partnerships to enhance opportunities for women, with particular interest in curbing unintended pregnancy among teens and the disadvantaged.
Encourage the Commission on Immigration Reform to continue its work and support research to promote the implementation and fair enforcement of responsible immigration policies.

14.6.2 Role of Local Governments

Local governments hold a unique position as leaders in sustainability: They have an interest in promoting sustainable development, the facilities to act as "sustainability laboratories," the authority to initiate positive change, and the flexibility to tailor programs to specific local circumstances. Consumption practices cannot be reversed overnight; sustainability planning is an ongoing, dynamic activity requiring careful development, nurturing, implementation, and review. Sustainable development is a trial-and-error proposition. Local governments can be, and in many areas are, leaders in refining the vision for a sustainable community. This leadership requires finding tools and creating working programs with clear benefits at local, national, and global levels. A mayor of a large urban city remarked over four years ago that although the term "sustainability" is not entirely clear or without controversy, the concept of an effective and sustainable integration of economics and the environment is one that a local government can both understand and practice.

Cities and counties own, operate, and manage large numbers of buildings and facilities, as well as vehicle fleets and mass-transit systems; they design, plan, finance, and operate major water, wastewater, and solid and hazardous waste management systems; they enact local plans and policies that affect residential, commercial, and industrial development as well as land-use and transportation choices. Local governments have vital interests in improving and maintaining the quality of the air, water, and land resources essential to our economic and environmental well-being and quality of life.

Most cities and counties will implement sustainability measures incrementally and promote them in meaningful ways. The vision of sustainability is not limited to unique places: Communities across the country are involving residents, government officials, and businesspeople in dialogue and action toward the goal of fostering sustainable development. Citizens from all social and economic groups should take part in decision making. It is critical to their healthy survival.

Numerous cities and counties are designing and implementing sustainable plans. Others are in the midst of identifying indicators of progress or benchmarks of sustainability; many other programs are not specifically labeled as "sustainable" but are consistent with the vision of sustainable communities.

14.7 SUSTAINABILITY IN THE THIRD WORLD

The concepts of local and global sustainability sometimes are in direct opposition. For example, some of the world's wealthiest nations have been relatively successful in sustaining their economies, but they often rely heavily on another nation's capital (e.g., natural resources, labor) or use another nation as a global sink (e.g., disposing of hazardous wastes there because the host country is willing to accept the waste in exchange for needed capital). The consumption and production patterns more or less dictate exploitation of resources of poor nations. So, even though these practices may lead to local "economic" improvement, viewed holistically (including environmental sustainability), they may not really lead to sustainability. This implies that we may need some kind of international law which limits one nation's use of another nation's resources; however, most of the actions still need to be taken at the local level to achieve sustainability.

14.7.1 Barriers to Sustainability in the Third World

Development efforts in the Third World can be classified into three categories—indigenous, Western, and a hybrid of indigenous and Western (James, 1996). These categories are prominently visible in all fields of life, including agricultural, economic, and social systems. Each of these systems has its own environmental benefits and disadvantages; the environmental state of the country depends, to a certain degree, on the categories in use there. Indigenous systems were not historically very harmful to the environment, but lately they have increasingly resulted in deforestation due to increased land clearing and farming. Western farming methods often result in more production per hectare, thus avoiding the need to bring more and more land area under agriculture to feed the increasing population, but they often result in loss of soil fertility and eutrophication due to excessive application of artificial fertilizers. Eutrophication, caused by the accumulation of nutrients from fertilizers in bodies of water, has destroyed many lakes around the world. Water pollution, soil erosion, and air pollution from increased agricultural and economic development activities have led to a decrease in biodiversity. Unplanned development is another major cause of loss of biodiversity.

It is often argued that resource exploitation is a necessary and unavoidable step for a Third World country striving to become prosperous and industrialized. However, it is also claimed that reduction of the level of poverty would reduce pressures on the exploitation of our natural resources because the forests would no longer be the primary source of food and income for the indigenous people of the developing world. Greater wealth would allow them the ability to more wisely manage their natural resources. Some of the developing countries are now tempted to import hazardous wastes from the developed nations to achieve quick economic gains, but the environmental ramifications of such endeavors are enormous, and the repercussions are unlimited. A wealthier nation would not need to allow its environment to be despoiled in exchange for needed capital.

Tourism is the basis of the economy of some countries. Many countries promote tourism because it attracts foreign investments, adds to the country's infrastructure (roads, hotels, etc.), and increases job potential. But the end result of this tourism development is often degradation of the environment, as the natural carrying capacity of resources is exceeded. Trash problems on Mt. Everest and a shrinking of wild animal habitats in Africa, which attract many tourists, are significant problems now.

Corruption is another major bottleneck for sustainable development in some Third World countries. Corrupt practices create an atmosphere of unaccountability and a distorted decision-making process, as well as being destructive to social, economic, and environmental conditions.

Another major problem facing developing nations is population growth. Increases in population translate into an increase in demand, which results in increased pressure on social and economic structure, ultimately resulting in degradation of the environment. The United Nations Population Fund (UNFP, 1991) blames population growth for two-thirds of the increase of carbon-dioxide emissions, 80 percent of tropical forest depletion, the dwindling and degrading of freshwater resources, and the degradation of coastal areas. To make matters worse, developing nations lack the resources and knowledge to tackle all these problems. Therefore, sustainable development collides head on with the idea of economic development. To make the world sustainable, harsh but necessary decisions must be made. This will be a very daunting task; whether humanity can achieve it remains to be seen.

14.7.2 Models of Macroeconomic Management for Third-World Countries

Several models that can be used for macroeconomic management of Third World countries have been proposed. These are discussed in this section.

The *revised minimum standard model* was developed by the World Bank and focuses on external debt. Based on a national accounts and balance of payments framework, it contains behavioral and technical relations among income, expenditure, investment, saving, and the domestic and foreign credit markets. The model has been tested and implemented in several developing countries.

Computable general equilibrium models are generally the extensions of social accounting matrixes in which behavioral equations are simulating the interactions of different economic agents. Due to their complexity, they have been the focus of research, rather than practical application.

A *public sector planning and management information system* is conceived as a mixture of a database, modeling, and management system, consisting of different interlinked modules. Among these modules are:

A policy analysis and planning system (for modeling at the macro-, meso-, and project level).

An economic monitoring system (database).

A financial management system (budget preparing and monitoring).

A debt management system (debt monitoring system).

A investment project bank (project monitoring).

A resource mobilization system (sources of project and program funding).

14.8 A FRAMEWORK FOR SUSTAINABILITY

14.8.1 The Role of Individuals

A sustainable society requires an open, accessible political process that puts effective decision-making power at the level of government closest to the situation and lives of the people affected by a decision (Dale and Robinson, 1996). In a sustainable society, all persons should have freedom from extreme want and from vulnerability to economic coercion, as well as the positive ability to participate creatively and directly in the political and economic system, thereby maintaining a minimum level of equality and social justice in a society.

Individuals can play a major role in achieving sustainability. Following are some actions that individuals can take categorized by the areas that they will influence.

Unemployment. To achieve a sustainable society, full employment is a requirement. This may mean reducing hours worked per week to ensure that jobs are available for everyone.

Life-style. We could have an impact on sustainability by making the following changes in our life-styles: sharing goods rather than owning them individually; increasing the energy efficiency of homes; landscaping with native vegetation, which requires little watering or chemicals, and which would contribute to energy conservation by reducing the exposure of houses to both heat and cold. We should start thinking "less is better," insist on efficient appliances, and accustom ourselves to smaller living spaces and increased use of common areas.

Education. Education for sustainability should be firmly entrenched in schools, colleges, and universities. Elementary and secondary schools should start acting as centers for developing sustainable society skills. All schools should adopt a "green school" program, which teaches students how to implement methods by which schools can operate more sustainably. In this way, the concept of sustainability will become instilled in the minds of students, who constitute a major part of the population.

Health and pharmaceutical companies. To achieve a sustainable society, we should hospitalize patients only for treatment of acute problems. People should be encouraged to recover at home and realize that home care is often the best opportunity for the individual to recuperate successfully. More leisurely lifestyles and healthier environments might help reduce the need for hospitalization. Reduced consumption of meat reduces burden on our livestock as well as promoting health. People who are healthy, both mentally and physically, have less demand for prescription and nonprescription pharmaceutical products. As a result, less waste and better utilization of resources by pharmaceutical manufacturers will be achieved.

Transportation. We could cut down on need for travel by using advancements in science and technology. Much business can be conducted electronically using videoconferencing, telephones, and computer links. Reducing long-distance travel for recreational purposes and requiring full-cost pricing of fuels would result in increased and more efficient use of public transportation. Using bicycles or walking wherever possible, and eventually using efficient small electric cars or cars powered by fuel cells, could also greatly reduce the environmental impacts of transportation.

Farming. Farmers can do their share by going back to natural pest control and nutrient recycling techniques, instead of relying on modern methods, thus eliminating the dependency on synthetic chemicals. Where possible, farmers should harness wind, solar, and biomass energy as a power source for the farm and use appropriate land preparation techniques for tillage and irrigation to reduce soil erosion.

Finally, individuals should be conservative and thoughtful in using all resources, even those commonly thought of as "free," such as water. In the end, no resource is free, because we all pay for environmental remediation through increased consumer goods prices or through taxation.

14.8.2 The Role of Industry

Industry can play a major role in achieving a sustainable society by shifting to cleaner manufacturing methods and instituting green marketing and ecolabeling, for instance.

Clean manufacturing and clean production have been the focus of this book. These terms must become a part of every engineer's lexicon because they provide for a combined environmental-economic benefit. Cleaner production focuses on the methods and processes which prevent pollution and emphasize waste minimization, and which work in concert to provide greater efficiency and energy conservation. This will require significant participation by all engineers, who can and must provide the critical nexus to achieve effective cleaner production implementation. The emphasis behind the cleaner production philosophy must be based on cradle-to-grave concepts. Industrial production systems require resources: *materials* from which products are made, *energy* which is used to transport and process materials, as well as *water* and *air*. Present production systems are usually *linear* (see Figure 14.2), often using hazardous substances and finite resources in vast quantities and at high rates.

The goal of clean production is to fulfill our need for products in a sustainable way; that is, using renewable, nonhazardous materials and energy efficiently while conserving biodiversity. Clean production systems are *circular,* rather than linear, and use fewer materials and less water and energy (see Figure 14.3). Resources flow through the production-consumption cycle at slower rates. The clean production approach questions the very need for the product or looks at how else that need could be satisfied or reduced.

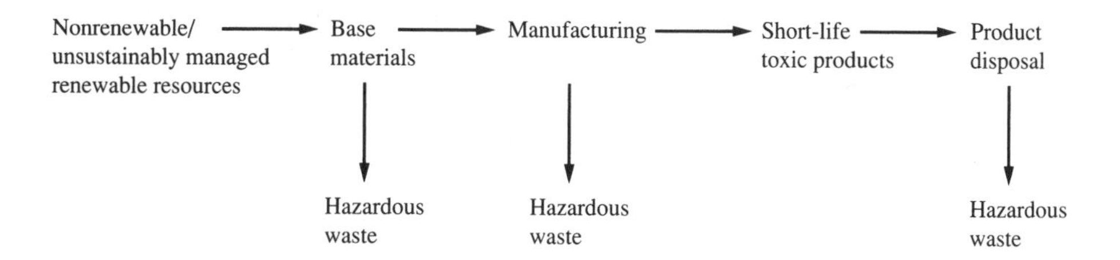

FIGURE 14.2
Linear structure of the industrial economy.

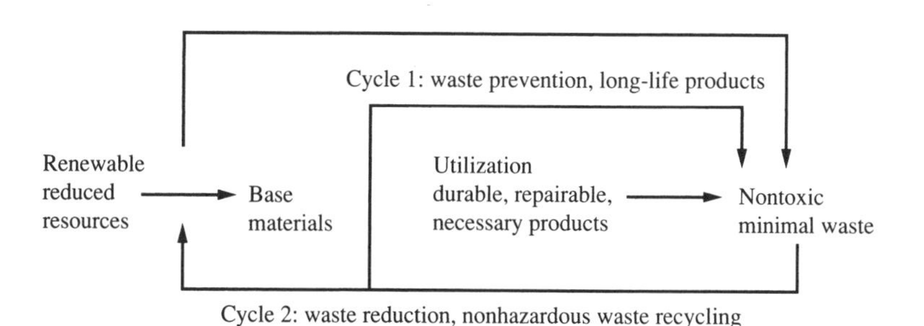

FIGURE 14.3
Circular structure of a sustainable economy.

14.8.3 The Four Elements of Clean Manufacturing

The clean production philosophy can be approached from four perspectives (Green-Peace International, 1995):

The precautionary approach. The precautionary approach puts the burden of proof on the potential polluter. Rather than communities having to prove harm, the potential polluter must prove that a substance or activity will do no environmental harm. This approach rejects the sole use of quantitative risk assessment in decision making because it recognizes the limitations of scientific knowledge in determining whether the use of a chemical or an industrial activity should proceed. This approach does not ignore science, but rather acknowledges that since industrial production also has social impacts, other public decision makers (not just scientists) must be involved.

The preventative approach. The preventative approach uses as its primary emphasis the fact that it is cheaper and more effective to prevent environmental damage than to attempt to manage or "cure" it. Prevention requires going upstream in the production process to forestall the problem at the source instead of attempting damage control downstream. Pollution prevention replaces pollution control. For example, prevention requires process and product changes to avoid the generation of incinerable waste streams, instead of more sophisticated incinerator design. Similarly, demand- and supply-side energy efficiency practices replace the current overemphasis on the development of new fossil fuel energy sources.

Democratic control. The democratic control approach says that access to information and involvement in decision making ensures more sound environmental decisions. Clean production should involve all those affected by industrial activities, including workers, consumers, and communities. Hence communities must have information on industrial emissions and access to pollution registers, such as the Toxics Release Inventory (TRI), toxic use reduction plans, and data on product ingredients.

Integrated and holistic approach. This philosophy says that society must adopt an integrated approach to environmental resource use and consumption. Currently, environmental management is fragmented, which results in the transfer of pollutants from one medium to another (i.e., between air, water, and soil). Reductions in polluting emissions from production processes lead to the hazard being transferred to the product. These dangers can be minimized by addressing all material, water, and energy flows; the whole life cycle of the product; and the economic impact of the change to clean production. The tool used to assist in maintaining a holistic approach is the life-cycle assessment, described in Chapter 6. An integrated approach is essential to ensure that as hazardous materials like chlorine are phased out, they will not be replaced by materials that pose new environmental threats.

Clean production systems are energy efficient and nonpolluting, and they make efficient use of both renewable and nonrenewable materials. As a result, products manufactured using these systems are durable and reusable; easy to dismantle, repair, and rebuild; and minimally and appropriately packaged for distribution using reusable or recycled and recyclable materials. Table 14.10 compares the traditional way of manufacturing goods with cleaner production approaches.

Clean manufacturing is both a process and a goal. Clean production processes can be broadly divided in two steps: changing the production process and changing the product. These topics were covered in Chapters 5 and 9.

Since the advent of the industrial age, new mass production techniques have increased the supply of goods to the point where, in some instances, they have overtaken the demand. Therefore, industries had to find new ways of selling their goods to keep their mass production systems at full capacity (Peattie, 1989). Marketing is a social process by which individuals and groups obtain what they need or want through creating and exchanging products and values with others (Kotler, 1984). Marketing is so basic that it cannot be considered a separate function. Green marketing and ecolabeling were described in Chapter 6.

TABLE 14.10
A comparison of pollution control and cleaner production attitudes

Pollution control approach	Cleaner production approach
Pollutants are controlled by filters and waste treatment methods.	Pollutants are prevented at their sources through integrated measures.
Pollution control is evaluated when processes and products have been developed and when problems arise.	Pollution prevention is an integrated part of product and process development.
Pollution controls and environmental improvements are always considered cost factors for the company.	Pollutants and waste are considered to be potential resources and may be transformed into useful product and by-products, providing they are not hazardous.
Environmental challenges are to be addressed by environmental experts such as waste managers.	Environmental improvement challenges should be the responsibility of people throughout the company, including workers and process and design engineers.
Environmental improvements are to be accomplished with techniques and technology.	Environmental improvements include nontechnical and technical approaches.
Environmental improvement measures should fulfill standards set by the authorities.	Environmental improvement measures should be a process of working continuously to achieve higher standards.
Quality is defined as meeting the customers' requirements.	Total quality means the production of goods that meet customers' needs and have minimal impacts on human health and the environment.

Source: Stahel, 1992.

14.9 INDUSTRIAL ECOLOGY

The concept of industrial ecology is still evolving and can be best explained with the help of the following example (Cote and Plunkett, 1996). In Kalundborg, Denmark, 10 industries are engaged in a system of mutually beneficial symbiosis. From the oil refinery and the power plant to the fish farm and the pharmaceutical company, one firm's waste is another's feedstock and one firm's by-product is another's raw material. The Kalundborg case, as seen in Figure 14.4, is a model of industrial ecology, because as a system it mimics, albeit in a limited manner, the cycling of materials and energy that occurs in a natural ecosystem (Knight, 1992).

Ecology is the study of the interrelationships of biota with their physicochemical environment. An ecosystem is a bounded system of dynamic, interdependent relationships between living organisms and their physical, chemical, and biological environment. The system has mechanisms by which nutrients are disseminated and replenished. Through the co-evolution of species, ecosystems acquire self-stabilizing mechanisms and a dynamic internal balance (McMichael, 1993). The goal of industrial ecology is to integrate production systems and product cycles with natural ecosystems and material cycles. The processes of the natural ecosystems have evolved over long

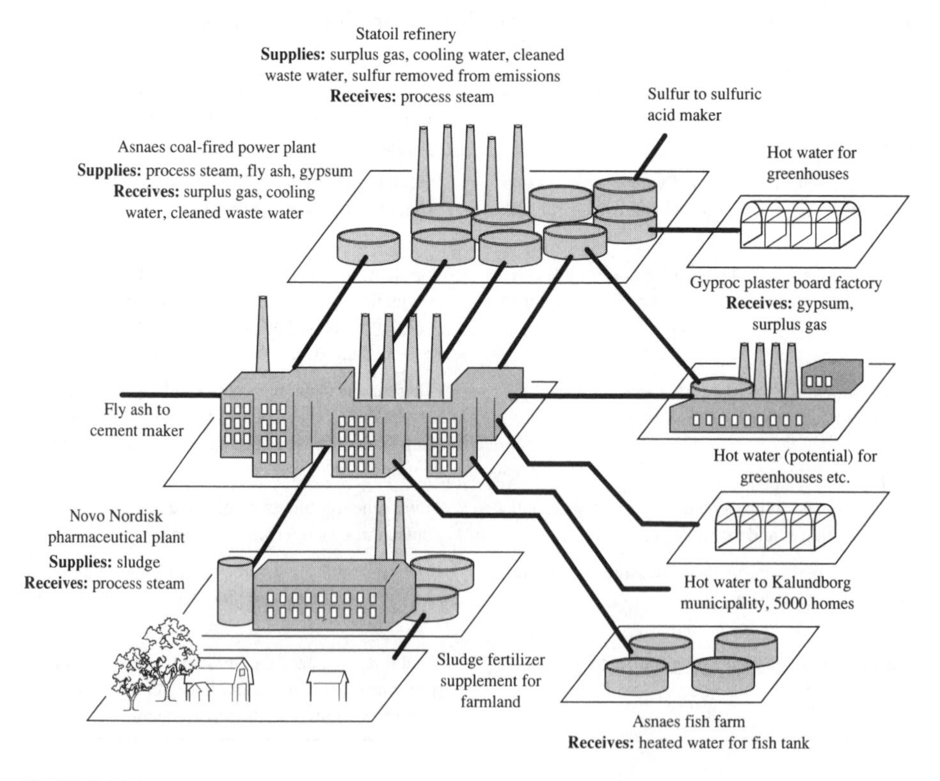

FIGURE 14.4
The Kalundborg, Denmark, industrial park.

periods into more or less stable communities because they are efficient, meaning that materials are not wasted but rather are conserved and reused in various forms (Odum, 1993). This is in sharp contrast to the increasing burden of waste that the industrial economy has imposed on the natural environment in the air, soil, and water through its linear, once-through pattern of material use.

As in natural systems, the cycling of materials must become the underpinning of industrial ecosystems. The various cycles should be maximized for recycling material in new production, optimizing use of materials and energy, minimizing waste generation, and reevaluating wastes as raw materials for other processes (Tibbs, 1992). Materials webs must be established linking producers, consumers, and scavengers; and multimaterial, multidimensional recycling must begin to emulate the complex webs in nature. Wherever possible, symbiotic relationships should be encouraged.

Another important feature of an ecosystem is diversity, which allows an ecosystem to recover after a stress is imposed. Industries that rely on a single source of supply or option for waste disposal are at greater risk of collapse than those that have diverse sources of materials or waste disposal options.

Industrial ecology can be defined as the study of industrial development policies and practices and the interrelationship of industrial and natural systems. It is a framework for designing and operating industrial systems as living systems interdependent with natural systems (Graedel and Allenby, 1995).

Tibbs (1992) outlines six principal elements of industrial ecology:

1. Fostering cooperation among various industries, so that the waste of one production process becomes the feedstock for another.
2. Balancing industrial input and output with the constraints of natural systems by identifying ways that industry can safely interface with nature, in terms of location, intensity, and timing, and developing indicators for real-time monitoring.
3. Striving to decrease materials use and energy intensity in industrial production.
4. Improving the efficiency of industrial processes by redesigning production processes and patterns for maximum conservation of resources.
5. Developing renewable energy supplies for industrial production by creating a worldwide energy system that functions as an integral part of industrial ecosystems.
6. Adopting new national and international economic development policies by integrating economic and environmental accounting in policy options.

14.9.1 Ecoindustrial Parks

Ecoindustrial parks are being developed to test and implement industrial ecology. Ecoindustrial parks, like standard industrial parks, are designed to allow firms to share infrastructure as a strategy for enhancing production and minimizing costs. The distinguishing feature of ecoindustrial parks is their use of ecological design to foster collaboration among firms in managing environmental and energy issues. Company production patterns, as well as overall park maintenance, work together to follow the principles of natural systems through cycling of resources, working within the constraints of local and global ecosystems, and optimizing energy use. Ecoindustrial parks offer firms the opportunity to cooperatively enhance both economic and environmental performance through increased efficiency, waste minimization, innovation and technology development, access to new markets, strategic planning, and attraction of financing and investment.

14.9.2 Industrial Ecology Principles

An ecological industrial park is primarily based on two principles: to drive down pollution and waste while simultaneously increasing business success. Other ecological and economic green manufacturing concepts are related to industrial ecology. These principles are discussed next, with respect to their influence on industrial ecology and pollution prevention.

THE PRECAUTIONARY PRINCIPLE. During the 1990s, the precautionary principle has emerged as an increasingly popular theory. It has been applied to the areas of environmental law and resource management on both a national and international level. Basically, the precautionary principle holds that the existence of scientific uncertainty regarding the precise effects of human activities on the natural environment constitutes legitimate grounds for constraining such activities, rather than for pursuing them. In essence, this principle calls for a reduction of all inputs to the environment. The precautionary principle was first referred to in an official setting at the Second International Conference on the Protection of the North Sea, held in London in 1987. Regulation of marine pollution was the subject, and the precautionary principle was advanced in an attempt to shift the burden of proof from the regulatory authority to the polluter.

The precautionary principle is also the basis of Principle 15 of the June 1992 Rio Declaration:

> In order to protect the environment, the precautionary approach shall be widely applied by states according to their capabilities. Where there are threats of serious or irreversible damage, lack of scientific certainty shall not be used as a reason for postponing cost-effective measures to prevent environmental degradation.

In 1992, many governments signed the Intergovernmental Agreement on the Environment, which also adopted the precautionary principle. It defines the precautionary principle as follows:

> Where there are threats of serious or irreversible damage, lack of full scale scientific certainty should not be used as a reason for postponing measures to prevent environmental degradation.

In the application of the precautionary principle, public and private decisions should be guided by (1) careful evaluation to avoid, wherever practicable, serious or irreversible damage to the environment and (2) assessment of the risk-weighted consequences of various options.

The precautionary principle addresses a problem central to sustainability: the inability to predict all the future consequences for human interests of current actions with environmental impacts. The precautionary principle describes one theory of how the environmental versus regulatory communities should deal with the problem of true uncertainty. The principle states that rather than waiting for certainty, regulators should act in anticipation of any potential environmental harm to prevent it. The precautionary principle states that where there are threats of serious or irreversible damage, lack of scientific certainty will not be used as a reason for postponing cost-effective measures to prevent environmental degradation.

THE INTERGENERATIONAL EQUITY PRINCIPLE. This principle requires that the needs of the present generation be met without compromising the ability of future generations to meet their own needs. This depends on the combined and effective application of the other principles for sustainable development.

THE INTRAGENERATIONAL EQUITY PRINCIPLE. This requires that *all* people within the present generation have the right to benefit equally from the exploitation of resources, and that they have an equal right to a clean and healthy environment. This principle applies to the relationship between groups of people within a country and between countries.

THE SUBSIDIARITY PRINCIPLE. This requires that decisions be made by the communities affected, or on their behalf, by the authorities closest to them. This principle encourages local ownership of resources and responsibility for environmental problems and their solutions.

THE POLLUTER-PAYS PRINCIPLE. This principle, also known as PPP or the 3Ps principle, requires that the polluter bear the cost of preventing and controlling pollution. It forces polluters to internalize all the environmental costs of their activities so that these are fully reflected in the costs of the goods and services they provide.

THE USER-PAYS PRINCIPLE. This is an extension of the polluter pays principle. The user-pays principle (UPP) requires that the cost of a resource to a user include all the environmental costs associated with its extraction, transformation, and use.

ENVIRONMENTAL PERFORMANCE BONDS. This is a variation of the deposit-refund system. It is designed to incorporate both known and uncertain environmental costs into the incentive system, as well as inducing positive environmental technological innovation.

Assume there is a firm that wants to proceed with a project, but it needs the permission of a regulatory agency, such as the EPA, in order to proceed. Under this system, the EPA would charge the firm directly for any known environmental damages and levy an assurance or performance bond requirement to the firm equal to the current best estimate of the worst conceivable environmental consequences of the project. The bond would be kept in an interest-bearing account for a predetermined length of time, defined as the longest lasting conceived consequence of the project.

In keeping with the precautionary principle, this system requires a commitment of resources up front to offset the potentially catastrophic future effects of current activity. Portions of the bond (plus some proportion of the interest) would be returned if and when the firm could demonstrate that the suspected worst-case damages had not occurred or would be less than those originally assessed. The withheld portion of the interest is to cover the EPA's administrative costs and to finance EPA research. If damages did occur, portions of the bond would be used to rehabilitate or repair the environment and possibly to compensate injured parties. The firm would get back the remainder of the bond, with appropriate interest adjustments, at the end of the bond's life. If all the funds are required for remediation, the firm would get nothing back, forfeiting the full value of the bond. Funds tied up in bonds could still be used for other economic activities. The only cost would be the difference (plus or minus) between the interest on the bond and the return that could be earned by the business had it invested in other activities. On average, this difference should be minimal. In addition, the "forced savings" that the bond would require could actually improve overall economic performance in economies that chronically undersave, like that of the United States.

By requiring the users of environmental resources to post a bond adequate to cover uncertain future environmental damages (with the possibility for refunds), the burden of proof and the cost of the uncertainty are shifted from the public to the resource user. The advantages claimed for such an instrument are in terms of the incentives it creates for the firm to undertake research to investigate environmental impact and means to reduce it, as well in terms of stopping potentially damaging projects. This plan also promotes innovation, which is required for betterment of human life and sustainable development.

CONSERVATION STRATEGIES. The industrial ecology approach is acknowledged in the 1991 update of the World Conservation Strategy (WCS). The revision calls on industry and government to identify as one of their priorities the task of committing business to sustainability and environmental excellence, expressed in high performance standards and advanced by economic instruments. In satisfying this goal, there will be a need to consider the occupational health and safety of workers; energy, material, and water efficiency of practices, processes, and products; control over the life cycle of manufacturing; and integrated approaches to pollution prevention and control (IUCN, 1991).

SOFT MATERIAL PATHS. The operating principle of this philosophy is efficiency: meeting people's needs with as little as possible of the most appropriate materials available.

INDUSTRIAL METABOLISM. This principle views an industrial process as being similar to the metabolic process of a living organism. The industrial process has inputs, such as materials and energy, and outputs, such as biomass and work. This term was coined by Ayres (1989b), who has great concern for the fact that the inputs and outputs of living organisms are in balance with the ecosystem and the inputs and outputs of the industrial processes are not. He also introduces the materials balance concept, a means of identifying the dissipative elements within an industrial process in order to reduce that dissipation.

E-FACTOR. This term was coined by Makower (1993) to describe a bottom-line approach to environmentally responsible business. The e-factor encompasses economics, enforcement, empowerment, education, efficiency, and excellence. In Makower's view, emphasis should be placed on excellence, thus linking this philosophy to total quality management (TQM). The common themes between the e-factor and TQM are (1) improvement in productivity and profits; (2) new corporate culture and leadership; (3) emphasis on long-range planning; (4) more flexibility within organizations; (5) improvement in information exchange, training, and accountability; and (6) continuous self-auditing.

3RS. This stands for reduce, reuse, and recycle; practicing the 3Rs is a precondition for industrial ecology. The 3Rs represent a move toward nature's hallmarks of efficiency and material cycles, and they represent the most obvious means of achieving economic benefits for industry. Contributing to the implementation of the 3Rs and the creation of industrial symbioses are a number of tools and concepts, such as life-cycle assessment (see Chapter 6) or ecobalancing, ecoauditing, and total environmental quality management. Ecoauditing requires a holistic approach, rather than the traditional mechanistic one, to evaluating the impact of an industrial process on the environment. Total environmental quality management refers to the expansion of traditional approaches, emulated in TQM and ISO 9000 standards of quality assurance, with implementation of ISO 14000 standards, in order to incorporate environmental issues.

14.10 MEASURES OF ECONOMIC GROWTH

To progress toward the goal of sustainability, we need an effective way to measure economic growth that is in line with the concepts of sustainable development. But before we discuss some of the green accounting and economic activity measurement techniques, let us examine some of the traditional economic indicators and problems associated with business-as-usual measurement methods.

14.10.1 Gross Domestic Product and Gross National Product

The most widely used measures of national income are gross national product and gross domestic product (U.N., 1992b). *Gross domestic product* (GDP) measures total domestic demand for goods; it also measures the output produced to meet that demand of the country's population. Therefore, GDP is seen as a measure of economic performance. An increase in GDP means that more is being produced; it commonly is used as an economic status indicator. Three methods are used to measure GDP:

1. The total output sold by firms, measured by value-added techniques. It is not the total value of sales by firms, because this would result in double-counting.
2. The sum of the incomes earned by persons in the economy. This is the most obvious rationale for calling GDP the "national income." The sum of incomes is equal to the value of total output produced by firms by virtue of the convention that output is measured in terms of value added.

3. Total expenditure by individuals on consumption plus expenditure by firms on capital equipment (investment). Note that firms' expenditures on intermediate goods is not included here; only their expenditure on items of durable capital equipment is considered.

Theoretically, each of the ways of measuring the GDP should produce the same numerical result; in reality, errors arise in the collection of data from the very large number of firms and individuals in an actual economy. Thus a residual error term usually is included to account for these errors and differences.

Gross national product (GNP), which measures a nation's total production including exports, is often considered to be an obsolete measure of progress in a society striving to meet a nation's needs as effectively as possible and with the least damage to the environment. What counts is not growth in total output, which may benefit primarily people in other countries, but rather the quality of the services rendered to the population in question. Bicycles and light rail, for example, are less resource-intensive forms of transportation than automobiles and contribute less to GNP, but a shift to mass transit and cycling for most passenger trips would enhance urban life by eliminating traffic jams, reducing smog, and making cities safer for pedestrians. Switching to these modes of transportation would cause the GNP to decrease, but overall they would increase the well-being of the public (Anderson, 1989). Gross national product is further distorted by the fact that it absorbs, as positive adjustments, the costs of environmental disasters, natural disasters, and medical expenditures associated with them. The costs associated with remediation of environmental disasters, such as the Bhopal gas leak disaster in India, or with natural disasters such as the earthquake at Kobe City, Japan, will result in a sharp increase in GNP for these countries. The only losses registered by the GNP are lost production and lost working hours. Thus striving to boost GNP is often inappropriate and counterproductive. As ecologist and philosopher Garrett Hardin puts it, "For a statesman to try to maximize the GNP is about as sensible as for a composer of music to try to maximize the number of notes in a symphony" (Hardin, 1991).

From the standpoint of macroeconomic stabilization, GDP is a reasonable measure of economic performance, but it neglects environmental considerations. Hence it also is not satisfactory. The GDP measures neither sustainable income nor maximum consumption possible during a period which would leave the society with the same wealth at the end of the period as at the beginning of it. The society's wealth is its stock of capital and its productive assets. Capital equipment wears out as it is used in production processes; the extent to which it wears out in a period is known as *depreciation*. A measure of sustainable income for a period would therefore better be described as output less depreciation. Thus GDP overstates sustainable income.

Gross national product measures the value of the output produced by domestically owned factors of production, regardless of the physical location of production. Like GDP, GNP does not take environmental factors into account, and hence is not suitable for measurement of sustainable development. There are many other factors which make either GDP or GNP unsuitable for this purpose; a detailed discussion of these can be found elsewhere (Common, 1995).

It is now nearly universally agreed that the proper measure of national income, for the purposes of monitoring national economic performance, is *net domestic product* (NDP), equivalent to GDP less depreciation. This can be further refined to account for effects on the environment. To do this, we must accept the following two points:

1. Defensive expenditures to offset environmental degradation actually increase national income.
2. National income measures do not reflect the depletion and degradation of environmental assets caused by economic activity.

We should modify the GDP in the following ways in order to make it more usable and acceptable for measuring sustainable income:

- Defensive expenditures should be subtracted from the conventional GDP (this is known as the *environmentally adjusted GDP*).
- Environmental cost (the reduction over a given period in the value of the total stock of environmental assets, measured by multiplying the physical damage in the size of each environmental asset by a corresponding unit price and then summing the values arising across the different assets) should be subtracted from the conventional GDP (this is known as the *sustainable GDP*).
- Depreciation of man-made assets should be subtracted from the sustainable GDP (this is known as the *proper net domestic product,* or PNDP).

14.10.2 Green Accounting

As has been discussed, GDP is not a good indicator of a country's level of sustainability. "A country could exhaust its mineral resources, cut down its forests, erode its soils, pollute its aquifers, and hunt its wildlife to extinction, but measured income would not be affected as these assets disappeared" (Uno, 1995). Here, "measured income" is national income calculated on the basis of existing accounting conventions, such as the GDP.

"Only if the basic measures of economic performance . . . are brought into conformity with a valid definition of income will economic policies be influenced towards sustainability" (Repetto et al., 1989). This quotation describes the national income estimating procedures in Indonesia. Similar critiques of current national income accounting conventions have also been reported by Daly and Cobb (1989), Jacobs (1991), Pearce, Markandya, and Barbier (1989), and Anderson (1991).

Hence sustainable income should be expressed as the maximum consumption possible during a period, such that the society has the same wealth at the end of the period as at the start of it. Wealth is the total value of society's assets, including both assets produced by economic activity and environmental assets. This sustainable income is also called proper net domestic product (PNDP) (Common, 1995). Green accounting, more properly *natural resource accounting* or *environmental accounting,* is the means by which PNDP data would be produced.

ENVIRONMENTAL ACCOUNTING. In order to take steps toward sustainability, we need a framework to measure or account for the environment which is consistent with the principles of sustainability. The accounting framework describes the state of the environment for a particular point in time, that is, environmentally related activities within a year. The dynamic change of environment and related activities can be described by time series comparisons using such a framework. One example of indicators which can be derived from the accounting framework is what is called "green GNP."

The relationship among various factors or components of the environment can be formulated as follows:

$$\frac{Z}{A} = \frac{N}{A} \times \frac{Y}{N} \times \frac{E}{Y} \times \frac{Z}{E}$$

where $A =$ geographical area
$E =$ resources (e.g., energy)
$N =$ population
$Y =$ income (product)
$Z =$ pollutants

Thus environmental limits are expressed as the multiple of population density, per capita income, resources intensity, and efficiency of resource use.

A NEW PRODUCE, CONSUME, AND RECYCLE ERA. Throughout history, man has largely indulged in a produce-consume-forget cycle, and as a result has made our growth unsustainable. We need to look beyond this concept and start working toward a produce-consume-recycle era. We should realize that exhaustion of natural resources and uncontrolled emissions of hazardous materials have created problems of global scope; one of the best ways to alleviate these problems is by recycling. The accounting framework should have provisions by which we can properly trace the flow of material inputs to our goods being produced and the amount of materials being recycled.

There is much disagreement within society as to what needs to be sustained and what does not. One point of view is that all natural resources should be sustained, and the commitment toward making sure that people's needs are met should be strengthened. Other groups say that it is more important to sustain different aspects of human development and activities, such as economic growth, human development, and social and political stability. This is often a social, moral, technical, and ethical debate, but the final decision as to what resources will be sustained will be a political one.

14.11 STEPS FOR ADOPTING A SUSTAINABILITY APPROACH

Achieving sustainability will not be easy, but even minimal progress toward sustainability will not occur unless we begin soon. Following are some suggestions for the first steps of this new journey:

1. Define what sustainability means for you and your organization.
2. Begin research and conduct an internal survey to determine where corporate practices have an impact on sustainability. Establish the descriptors for these practices.

3. Develop a sustainability framework, followed by a strategy for achieving sustainability.
4. Decide on goals and set a time limit to achieve these goals.
5. Decide on sustainability indicators and their weights, if any.
6. Develop a method for analysis of the evaluation results.
7. Decide on tools for sustainability planning and practice, keeping in mind the following points:
   Safeguard biodiversity at all costs. Remember that biodiversity is not a conflicting claim for resources by creatures other than humans but rather is the basis for future life. Try not to reshape the existing ecosystem by intensively growing or raising only a few crops or animals, as this will result in biodiversity reductions.
   Minimize interference with the ecosystem. All sustainability tools and plans should ensure that they minimize interference with ecosystems and respect their carrying capacity. Human-induced waste should be avoided.
8. Evaluate your needs periodically. Make a balanced judgment on whether the consumption of certain goods or services adds happiness, and learn to say *enough*.

Sustainability initiatives should be pursued using a combination of tools, rather than a single tool or a one-dimensional strategy. Sustainable development is inherently cross-disciplinary, involving the integration of previously distinct fields. Sustainable development requires policies and management approaches that coordinate a wide variety of disciplines. It should be remembered that sustainability initiatives and programs require us to think "outside of the box" when it comes to implementation and finance. There is no blueprint for a sustainable society waiting to be discovered. The problem itself changes over time as the result of the economy-environment linkages and their repercussions in human societies. Any solution to the sustainability problem will require successful adaptation to changing circumstances.

14.12 THE OUTLOOK FOR SUSTAINABILITY

Change is inevitable and necessary for the sake of future generations and for ourselves. Beneficial change is that which leads to the mutually reinforcing goals of economic growth, environmental protection, and social equity. Economic growth based on technological innovation, improved efficiency, and expanding global markets is essential for progress toward greater prosperity, equity, and environmental quality. To improve environmental quality, the current regulatory system needs to provide enhanced flexibility in return for superior environmental performance and low costs. Environmental progress, in turn, depends on individual, institutional, and corporate responsibility, commitment, and stewardship. Hence a new collaborative decision-making process that leads to better decisions; more rapid change; and more sensible use of human, natural, and financial resources in achieving our goals is required.

Since economic growth, environmental protection, and social equity are linked, integrated policies to achieve these national goals should include programs that stabilize global population. This objective is critical to maintain the resources needed to ensure a high quality of life for future generations. Steady advances in science and technology should be directed toward helping to improve economic efficiency, protect and restore natural systems, and modify consumption patterns.

Above all, a knowledgeable public, the free flow of information, and opportunities for review and redress are critical to open, equitable, and effective decision making, and hence to sustainable development. Citizens must have access to high-quality and lifelong formal and nonformal education that enables them to understand the interdependence of economic prosperity, environmental quality, and social equity and that prepares them to take action to support all three. Finally, we can state that we know what we have to do and we know how to do it. But will we do it? If we fail to convert our self-destructive economy into one that is environmentally sustainable, future generations will be overwhelmed by environmental degradation and social disintegration (Brown, 1993).

REFERENCES

Anderson, V. *Alternative Economic Indicators.* London: Routledge, 1991.

Anderson, H. "Moving beyond Economism: New Indicators of Culturally Specific Sustainable Development." In *Redefining Wealth and Progress: New Ways to Measure Economic, Social and Environmental Change.* New York: Bootstrap Press, 1989.

Ayres, R. U. *Technological Transformations and Long Waves.* RR-89-1. Laxenburg, Austria: International Institute for Applied Systems Analysis, 1989a.

Ayres, R. U. "Industrial Metabolism." In *Technology and Environment,* ed. J. H. Ausubel and H. E. Sladorich. Washington, DC: National Academy Press, 1989b, pp. 23–49.

Bartelmus, P. *The Concepts and Strategies of Sustainability, Environment, Growth and Development.* London: Routledge, 1994.

Brown, L. R. "A New Era Unfolds." In *State of the World 1993: A Worldwatch Institute Report of Progress toward a Sustainable Society.* ed. L. Starke. London: Norton, 1993.

Brundtland, G. H. *Our Common Future.* Oxford: Oxford University Press, 1987.

Cairncross, F. *Costing the Earth: The Challenges of Governments, the Opportunities for Business.* Boston: Harvard Business School Press, 1993.

Callenbach, E., Capra, F., Goldman, I., Lutz, R., and Marburg, S. *EcoManagement: The Elmwood Guide to Ecological Auditing and Sustainable Business.* San Francisco: Berett-Koechler Publishers, 1993.

Carson, R. *Silent Spring.* Boston: Houghton Mifflin, 1994.

Coddington, W. *Environmental Marketing: Positive Strategies for Reaching the Green Consumers.* New York: McGraw-Hill, 1993.

Common, M. *Sustainability and Policy: Limits to Economics.* Cambridge: Cambridge University Press, 1995.

Conway, G. R. "Agroecosystem Analysis." *Agricultural Administration,* 20 (1985): 31–55.

Costanza, R., and Cornwell, L. "The 4P Approach to Dealing with Scientific Uncertainty." *Environment,* 34, no. 9 (1992): 12.

Cote, R., and Plunkett, T. "Industrial Ecology: Efficient and Excellent Production." In *Achieving Sustainable Development,* ed. A. Dale and J. B. Robinson. Vancouver: University of British Columbia Press, 1996.

Dale, A., and Robinson, J. B. *Achieving Sustainable Development.* Vancouver: University of British Columbia Press, 1996.

Daly, H. E., and Cobb, J. B. *For the Common Goal: Redirecting the Economy toward Community, the Environment and a Sustainable Future.* Boston: Beacon, 1989.

Drucker, P. F. *Management: Tasks, Responsibilities, Practices.* New York: Harper & Row, 1973.

ECWD, *Our Common Future,* World Commission on Environment and Development (ECWD). Oxford: Oxford University Press, 1987.

Graedel, T. E., and Allenby, B. R. *Industrial Ecology.* Englewood Cliffs, NJ: Prentice Hall, 1995.

GreenPeace International. "What Is Clean Production." http://www.rec.org/poland/wpa/cpbl.htm,1995.

Hardin, G. "Paramount Portions in Ecological Economics." In *Ecological Economics: The Science and Management of Sustainability,* ed. R. Costanza. New York: Columbia University Press, 1991.

Hart, M. *Guide to Sustainable Development.* Ipswich, MA: Atlantic Center for the Environment, 1995.

Hatcher, R. L. *Sustainable Development: The Pre-Brundtland Commission Era.* Brussels: VUB Press, 1996.

Henderson, H. Paradigms of Progress. http://www.subjectmatters.com/indicators/htmlsrc/WhatIsAnIndicator.html.

International Union for the Conservation of Nature. *Caring for the Earth: A Strategy for Sustainable Living.* Gland, Switzerland: IUCN, 1991.

Jacobs, M. *The Green Economy: Environment Sustainable Development and the Politics of the Future.* London: Pluto, 1991.

James, V. *Sustainable Development in Third World Countries: Applied and Theoretical Perspectives.* Westport, CT: Praeger, 1996.

Kakwani, N. *The Economic Crisis of the 1980s and Living Standards in Eighty Developing Countries.* CAER Paper, University of New South Wales, 1988.

Knight, P. "Closing the Loop." *Tomorrow* 2, no. 2 (1992): 40–43.

Kotler, P. *Marketing Management: Analysis, Planning and Control.* Englewood Cliffs, NJ: Prentice-Hall, 1984.

Lowe, F. *Discovering Industrial Ecology: An Overview and Strategies for Implementation.* Oakland, CA: Change Management Center, 1992.

Makower, J. *The e-factor: The Bottom Line Approach to Environmentally Responsible Business.* Toronto: Random House, 1993.

Martin, M. "A Sustainable Community Profile," *Places* 9, no.3 (1995): 30.

McMichael, A. J. *Planetry Overload: Global Environmental Change and the Health of the Human Species.* Cambridge: Cambridge University Press, 1993.

Meadows, D. H., Meadows, D. L., and Randers, J. *Beyond the Limits: Global Collapse or a Sustainable Future.* London: Earthscan, 1992.

Morrison, W. I. *Data and Modeling in the United Nations Integrated Macroeconomics Development Planning and Management for Sustainable Development: Guidelines for Island Developing Countries.* New York: United Nations, 1993.

Odum, E. P. *Ecology and Our Endangered Life Support Systems,* 2nd ed. Sunderland, MA: Sinauer Associates, 1993.

Odum, E. P. *Fundamentals of Ecology,* 3rd ed. Philadelphia: W.B. Saunders, 1971.

Pearce, D., Markandya, A., and Barbier, E.B. *Blueprint for a Green Economy.* London: Earthscan, 1989.

Peattie, R. C. *Green Marketing.* M & E Handbooks. London: Pitman, 1989.

Rees, C. "The Ecologist's Approach to Sustainable Development." *Finance and Development,* December 1993.

Rees, W. E. "Ecological Footprints and Appropriate Carrying Capacity: What Urban Economics Leaves Out." *Environment and Urbanization* 4, no. 2. (1992): 121–36.

Reid, D. *From Nature Conservation to National Sustainable Development Strategies. A Review of the Literature of Nature Conservation, Sustainable Development and Sustainability—1966–93.* Centre for Human Ecology, University of Edinburgh, 1993.

Reid, W. V., and Miller, K. R. *Keeping Options Alive: The Scientific Basis for Conserving Biodiversity.* Washington, DC: U.S. Department of Agriculture, 1989.

Repetto, R., Magrath, W., Wells, M., Beer, C., and Rossini, F. *Wasting Assets: Natural Resources in the National Income Accounts.* Washington, DC: World Resources Institute, 1989.

Ruckelshaus, W. D. "Toward a Sustainable World." *Scientific American* 261 (1989): 166.

Sadler, B., and Fenge, T. "A National Sustainable Development Strategy and the Territorial North." *Northern Perspectives* 21, no. 4 (1993).

Sanderson, S. K. *Macrosociology.* New York: Harper & Row, 1988.

Stahel, W. "Product Design and Waste Minimization," In *Waste Minimization and Clean Technology: Management Strategies for the Future,* ed. W. Forester and J. Skinner. San Diego: Academic Press, 1992.

Tibbs, H. B. C. "Industrial Ecology: An Environmental Agenda for Industry." *Whole Earth Review,* Winter 1992, pp. 4–19.

Trainer, F. E. *Abandon Affluence!* London: Zed Books, 1985.
Trainer, T. *The Conservative Society: Alternatives for Sustainability.* London: Ted Books, 1995.
United Nations. *Agenda 21: The United Nations Programme of Action from Rio.* New York: UN, 1992a.
United Nations. *Handbook of National Accounting, Integrated Environmental and Economic Accounting.* New York: United Nations Department of Economic and Social Development, 1992b.
United Nations Conference on Environment and Development. *The Global Partnership for Environment and Development, A Guide to Agenda 21.* New York: UNCED, 1992.
United Nations Environment Programme (UNEP). *Population and the Environment: The Challenges Ahead.* London: Banson, 1991.
Uno, K. *Environmental Options: Accounting for Sustainability.* Norwell, MA: Kluwer Academic Publishers, 1995.
U.S. DOE, *Sustainable Building Technical Manual: Green Building Design, Construction, and Operations.* Washington, DC: U.S. DOE, 1996.
U.S. EPA. *Status Report on the Use of Environmental Labels World-Wide.* Washington, DC: Abt Associates for the U.S. EPA, 1993.
Welford, R. *Environmental Strategy and Sustainable Development.* New York: Routledge, 1995.
Welford, R., and Gouldson, A. *Environmental Management and Business Strategy.* London: Pitman, 1993.
Wilson, E. O. *The Diversity of Life.* Cambridge, MA: Belknap Press, 1992.
World Commission on Environment and Development (ECWD). *Our Common Future.* Oxford: Oxford University Press, 1987.

PROBLEMS

14.1. Section 14.2.2 lists seven major impediments to achieving sustainability. Which of these do you think is the most significant? What could be done to alleviate this impediment?

14.2. The UNCED conference in Rio de Janeiro in 1992 was the last major international conference on sustainability. Research what has been done since the Rio conference to implement its recommendations, and prepare a paper summarizing your findings.

14.3. Can an open capitalistic market ever be truly sustainable? State and justify your views.

14.4. It is often argued that it is not important to do without some luxuries now for the sake of conservation, because future generations will find replacements for any resources that become depleted. Can you cite any examples of this for the college-age generation? Do you agree or disagree with this philosophy? Why?

14.5. "Underdeveloped countries should not be held to the same stringent sustainability standards as developed countries because they need to, at least temporarily, consume resources (e.g., timber, high sulfur coal, etc.) at a higher rate and use more polluting lower technology industrial processes in order to catch up with the developed countries." State and support your views on this philosophy. What could be done to alleviate this impediment?

APPENDIX A

THE ELEMENTS

TABLE A.1
Atomic Weights of Elements

Element	Symbol	Atomic number	Atomic weight	Element	Symbol	Atomic number	Atomic weight
Actinium	Ac	89	[227]	Chromium	Cr	24	51.996
Aluminum	Al	13	26.9815	Cobalt	Co	27	58.9332
Americium	Am	95	[243]	Copper	Cu	29	63.54
Antimony (stibium)	Sb	51	121.75	Curium	Cm	96	[247]
Argon	Ar	18	39.948	Dysprosium	Dy	66	162.50
Arsenic	As	33	74.9216	Einsteinium	Es	99	[254]
Astatine	At	85	[210]	Erbium	Er	68	167.26
Barium	Ba	56	137.34	Europium	Eu	63	151.96
Berkelium	Bk	97	[249]	Fermium	Fm	100	[253]
Beryllium	Be	4	9.0122	Fluorine	F	9	18.9984
Bismuth	Bi	83	208.980	Francium	Fr	87	[223]
Boron	B	5	10.81	Gadolinium	Gd	64	157.25
Bromine	Br	35	79.904	Gallium	Ga	31	69.72
Cadmium	Cd	48	112.40	Germanium	Ge	32	72.59
Calcium	Ca	20	40.08	Gold (aurum)	Au	79	196.967
Californium	Cf	98	[251]	Hafnium	Hf	72	178.49
Carbon	C	6	12.01	Hahnium	Ha	105	[260]
Cerium	Ce	58	140.12	Helium	He	2	4.0026
Cesium	Cs	55	132.905	Holmium	Ho	67	164.930
Chlorine	Cl	17	35.453	Hydrogen	H	1	1.0079

TABLE A.1 (*cont.*)

Atomic Weights of Elements

Element	Symbol	Atomic number	Atomic weight	Element	Symbol	Atomic number	Atomic weight
Indium	In	49	114.82	Radon	Rn	86	[222]
Iodine	I	53	126.904	Rhenium	Re	75	186.2
Iridium	Ir	77	192.22	Rhodium	Rh	45	102.905
Iron (ferrum)	Fe	26	55.847	Rubidium	Rb	37	85.4678
Krypton	Kr	36	83.80	Ruthenium	Ru	44	101.07
Lanthanum	La	57	138.91	Rutherfodium	Rf	104	[257]
Lawrencium	Lr	103	[257]	Samarium	Sm	62	150.35
Lead (plumbum)	Pb	82	207.2	Scandium	Sc	21	44.956
Lithium	Li	3	6.941	Selenium	Se	34	78.96
Lutetium	Lu	71	174.97	Silicon	Si	14	28.086
Magnesium	Mg	12	24.305	Silver	Ag	47	107.868
Manganese	Mn	25	54.9380	(argentum)			
Mendelevium	Md	101	[256]	Sodium	Na	11	22.9898
Mercury	Hg	80	200.59	(natrium)			
(hydroargyrus)				Strontium	Sr	38	87.62
Molybdenum	Mo	42	95.94	Sulfur	S	16	32.06
Neodymium	Nd	60	114.24	Tantalum	Ta	73	180.948
Neon	Ne	10	20.179	Technetium	Tc	43	98.9062
Neptunium	Np	93	237.0482	Tellurium	Te	52	127.60
Nickel	Ni	28	58.7	Terbium	Tb	65	158.9254
Niobium	Nb	41	92.906	Thallium	Tl	81	204.37
Nitrogen	N	7	14.0067	Thorium	Th	90	232.038
Nobelium	No	102	[254]	Tin (stannum)	Sn	50	118.69
Osmium	Os	76	190.2	Titanium	Ti	22	47.90
Oxygen	O	8	15.9994	Tungsten	W	74	183.85
Palladium	Pd	46	106.4	(wolfram)			
Phosphorus	P	15	30.9738	Uranium	U	92	238.03
Platinum	Pt	78	195.09	Vanadium	V	23	50.94
Plutonium	Pu	94	[242]	Xenon	Xe	54	131.30
Polonium	Po	84	[210]	Ytterbium	Yb	70	173.04
Potassium (kalium)	K	19	39.096	Yttrium	Y	39	88.9059
Praseodymium	Pr	59	140.9077	Zinc	Zn	30	65.38
Promethium	Pm	61	[145]	Zirconium	Zr	40	91.22
Protactinium	Pa	91	[2311231.0359]				
Radium	Ra	88	[2261226.0254]				

Note: Based on ^{12}C = 12.0000 amu. A value given in brackets denotes the mass number of the longest-lived or best-known isotope.

TABLE A.2
Periodic table of the elements

Key: ATOMIC NUMBER → 30; ATOMIC WEIGHT (1) → 65.38; OXIDATION STATES → 2; BOILING POINT, °C → 906; MELTING POINT, °C → 419.5; DENSITY (g/mL) (2) → 7.14; SYMBOL → Zn; NAME → Zinc

Notes:
(1) Atomic weights are 1971 values. Parentheses indicated most stable or best known isotope.
(2) Density values for gaseous elements are for liquids at the boiling point.

Group IA	IIA	IIIB	IVB	VB	VIB	VIIB	VIII	VIII	VIII	IB	IIB	IIIA	IVA	VA	VIA	VIIA	VIIIA
1 1.0079 −252.7 1 −259.2 0.071 H Hydrogen																	2 4.0026 −268.9 −269.7 0.126 He Helium
3 6.939 1330 1 180.5 0.53 Li Lithium	4 9.0122 2770 2 1277 1.85 Be Beryllium											5 10.811 — 3 (2030) 2.34 B Boron	6 12.01115 4830 ±4,2 3727 2.26 C Carbon	7 14.0067 −195.8 ±3,5,4,2 −210 0.81 N Nitrogen	8 15.9994 −183 −2 −218.6 1.14 O Oxygen	9 18.9994 −188.2 −1 −219.6 1.505 F Fluorine	10 20.183 −246 −248.5 1.20 Ne Neon
11 22.9898 892 1 97.8 0.97 Na Sodium	12 24.312 1107 2 650 1.74 Mg Magnesium											13 26.9815 2450 3 880 2.70 Al Aluminum	14 28.086 2680 4 1410 2.33 Si Silicon	15 30.9738 280 ±3,5,4 44.2 1.82 P Phosphorus	16 32.064 444.6 ±2,4,6 119.0 2.07 S Sulfur	17 35.453 −34.7 ±1,3,5,7 −101.0 1.56 Cl Chlorine	18 39.948 −185.8 −189.4 1.40 Ar Argon
19 39.098 760 1 63.7 0.86 K Potassium	20 40.08 1440 2 838 1.55 Ca Calcium	21 44.956 2730 3 1539 3.0 Sc Scandium	22 47.90 3260 4,3 1668 4.51 Ti Titanium	23 50.942 3450 5,4,3,2 1900 6.1 Va Vanadium	24 51.996 2665 6,3,2 1875 7.19 Cr Chromium	25 54.938 2150 7,6,4,2,3 1245 7.43 Mn Manganese	26 55.847 3000 2,3 1536 7.86 Fe Iron	27 58.933 2900 2,3 1495 8.9 Co Cobalt	28 58.71 2730 2,3 1453 8.9 Ni Nickel	29 63.54 2595 2,1 1083 8.96 Cu Copper	30 65.38 906 2 419.5 7.14 Zn Zinc	31 69.72 2237 3 29.8 5.91 Ga Gallium	32 72.59 2830 4 937.4 5.32 Ge Germanium	33 74.922 613 ±3,5 817 5.72 As Arsenic	34 78.96 685 −2,4,6 217 4.79 Se Selenium	35 79.909 58 ±1,5 −7.2 3.12 Br Bromine	36 83.80 −152 −157.3 2.6 Kr Krypton
37 85.47 588 1 38.9 1.53 Rb Rubidium	38 87.62 1380 2 768 2.6 Sr Strontium	39 88.905 2927 3 1509 4.47 Y Yttrium	40 91.22 3580 4 1852 6.49 Zr Zirconium	41 92.906 3300 5,3 2468 8.4 Nb Niobium	42 95.94 5560 6,5,4,3,2 2610 10.2 Mo Molybdenum	43 (98) — 7 2140 11.5 Tc Technetium	44 101.07 4900 2,3,4 2500 12.2 Ru Ruthenium	45 102.906 4500 2,3,4,6,8 1966 12.4 Rh Rhodium	46 106.4 3980 2,4 1552 12.0 Pd Palladium	47 107.870 2210 1 960.8 10.5 Ag Silver	48 112.40 765 2 320.9 8.65 Cd Cadmium	49 114.82 2000 3 156.2 7.31 In Indium	50 118.69 2270 4,2 231.9 7.30 Sn Tin	51 121.75 1380 ±3,5 630.5 6.62 Sb Antimony	52 127.60 969.8 −2,4,6 449.5 6.24 Te Tellurium	53 126.904 183 ±1,5,7 113.7 4.94 I Iodine	54 131.30 −108.0 −111.9 3.06 Xe Xenon
55 132.9064 690 1 28.7 1.90 Cs Cesium	56 137.34 1640 2 714 3.5 Ba Barium	57 138.91 3470 3 920 ★ 6.17 La Lanthanum	72 178.49 5400 4 2222 13.1 Hf Hafnium	73 180.948 5425 5 2996 16.6 Ta Tantalum	74 183.85 5930 6,5,4,3,2 3410 19.3 W Wolfram	75 186.2 5900 7,6,4,2,−1 3180 21.0 Re Rhenium	76 190.2 5500 2,3,4,6,8 3000 22.6 Os Osmium	77 192.2 5300 2,3,4,6 2454 22.5 Ir Iridium	78 195.09 4530 2,4 1769 21.4 Pt Platinum	79 196.967 2970 3,1 1063 19.3 Au Gold	80 200.59 357 2,1 −38.4 13.6 Hg Mercury	81 204.37 1457 3,1 303 11.85 Tl Thallium	82 207.19 1725 4,2 327.4 11.4 Pb Lead	83 208.980 1560 3,5 271.3 9.6 Bi Bismuth	84 (210) — 2,4 254 (9.2) Po Polonium	85 (210) — ±1,3,5,7 (302) — At Astatine	86 (222) (−61.8) (−71) — Rn Radon
87 (223) — 1 (27) — Fr Francium	88 (226) — 2 700 5.0 Ra Radium	89 (227) — 3 1050 ★★ — Ac Actinium	104 — — — — [Rf] — [Ku] Rutherfordium	105 — — — — — Ha Hahnium													

★	58 140.12 3468 3,4 795 6.67 Ce Cerium	59 140.907 3127 3,4 935 6.77 Pr Praseodymium	60 144.24 3027 3 1024 7.00 Nd Neodymium	61 (147) — 3 (1027) — Pm Promethium	62 150.35 1900 3,2 1072 7.54 Sm Samarium	63 151.96 1439 3,2 826 5.26 Eu Europium	64 157.25 3000 3 1312 7.89 Gd Gadolinium	65 158.924 2800 3,4 1356 8.27 Tb Terbium	66 162.50 2600 3 1407 8.54 Dy Dysprosium	67 164.930 2600 3 1461 8.80 Ho Holmium	68 167.26 2900 3 1497 9.05 Er Erbium	69 168.934 1727 3 1545 9.33 Tm Thulium	70 173.04 1427 3,2 824 6.98 Yb Ytterbium	71 174.97 3327 3 1652 9.84 Lu Lutetium
★★	90 232.04 3850 4 1750 11.7 Th Thorium	91 (231) — 5,4 (1230) 15.4 Pa Protactinium	92 238.036 3818 6,5,4,3 1132 19.07 U Uranium	93 (237) — 6,5,4,3 637 19.5 Np Neptunium	94 (242) 3235 6,5,4,3 640 — Pu Plutonium	95 (243) — 6,5,4,3 — 11.7 Am Americium	96 (247) — 3 — — Cm Curium	97 (247) — 4,3 — — Bk Berkelium	98 (249) — 3 — — Cf Californium	99 (254) — — — — Es Einsteinium	100 (253) — — — — Fm Fermium	101 (256) — — — — Md Mendelevium	102 (254) — — — — No Nobelium	103 (257) — — — — Lr Lawrencium

APPENDIX B

PROPERTIES OF SELECTED COMPOUNDS

Name	Typical uses	Formula	Molecular weight, g/mol	List*	Water solubility, mg/L	Vapor pressure, mm Hg	Henry's constant, atm · m³/mol	K_{oc}, mL/g	log K_{ow}
Acenaphthene	Dye intermediate; used in plastics	$C_{12}H_{10}$	154.21	P C N	3.42	1.55×10^{-3}	7.92×10^{-5}	4.5×10^{3}	4.00
Acenaphthylene	Product of incomplete combustion	$C_{12}H_8$	152.2	P C N	3.93	2.90×10^{-2}	2.80×10^{-4}	2.5×10^{3}	3.70
Acetaldehyde	Manufacture of acetic acid, butanol, flavors	C_2H_4O	44.05	N H	Miscible	750	6.61×10^{-5}		0.43
Acetic acid	Manufacture of acetates, cellulose acetate	C_2H_4O	60.05		Miscible	11	1.23×10^{-3}		−0.34
Acetone	Solvent; manufacture of MIBK, acetic acid	C_3H_6O	58.09	C N	Miscible	270	3.97×10^{-5}	2.2	−0.24
Acetonitrile	Solvent; synthesis of many compounds	C_2H_3N	41.06	P N H	Miscible	74	3.46×10^{-6}	2.2	−0.34
Acrolein	Manufacture of plastics	C_3H_4O	56.06	P C N H	2.0×10^{5}	220	4.4×10^{-6}	0.8	−0.09
Acrylamide	Monomer for plastics; used in dyes	C_3H_5NO	71.08	N H	2.1×10^{6}	7.0×10^{-3}	3.03×10^{-9}	−0.67	
Acrylic acid	Manufacture of plastics	$C_3H_4O_2$	72.07	N H	Miscible	3.2		0.16	0.31
Acrylonitrile	Manufacture of acrylic fibers	C_3H_3N	53.07	N H	7.35×10^{4}	100	1.1×10^{-4}	0.85	0.25
Aldicarb	Insecticide	$C_7H_{14}N_2O_2S$	190.29	N	7.8×10^{3}	1.0×10^{-4}	1.0×10^{-10}		
Aldrin	Insecticide; now banned	$C_{12}H_8Cl_6$	364.9	P C N	0.18	6.0×10^{-6}	1.4×10^{-6}	9.6×10^{4}	5.3
Allyl alcohol	Manufacture of allyl compounds, resins, plastics	C_3H_6O	58.09	N	Miscible	20	5.0×10^{-6}	1.3	−0.03
Allyl chloride	Manufacture of allyl compounds	C_3H_5Cl	76.53	H	3.8×10^{3}	340	1.08×10^{-2}		1.79

* P = priority pollutant (designated under the Clean Water Act); C = hazardous substance list (designated under CERCLA); N = Appendix IX chemicals (designated under RCRA); H = hazardous air pollutant (designated under the Clean Air Act Amendments).

(cont.)

Name	Typical uses	Formula	Molecular weight, g/mol	List*	Water solubility, mg/L	Vapor pressure, mm Hg	Henry's constant, atm · m³/mol	K_{oc}, mL/g	log K_{ow}
4-Aminobiphenyl	Detection of sulfates	$C_{12}H_{11}N$	169.24	N H	842	6.5×10^{-5}	3.83×10^{-6}	107	2.78
Ammonia	Manufacture of nitric acid, synthetic fibers, fertilizers	NH_3	17.04	H	5.3×10^{5}	8.7	2.9×10^{-4}	3.1	0.0
Aniline	Manufacture of dyes, resins, medicinals	C_6H_7N	93.1	N H	35.4×10^{3}	0.6	0.14	3.9×10^{3}	0.93
Anthracene	Source of dyestuffs	$C_{14}H_{10}$	178.24	P C N H	4.5×10^{-2}	1.7×10^{-5}	1.77×10^{-5}	1.4×10^{4}	4.45
Antimony	Flame retardant, catalyst, pigments	Sb	121.75	P H	Insoluble	1.0			
Arsenic	Metallurgy; semiconductors	As	74.92	P H	Insoluble	1.0			
Atrazine	Herbicide	$C_8H_{14}ClN_5$	215.7		33	2.8×10^{-7}	2.6×10^{-9}	148	2.68
Benzene	Manufacture of chemicals, polymers, dyes	C_6H_6	78.12	P C N H	1.78×10^{3}	95.2	5.5×10^{-5}	83.0	2.12
Benzidine	Manufacture of direct azo dyes	$C_{12}H_{12}N_2$	184.26	P N H	400	10	3.88×10^{-11}	10.5	1.3
Benzo[*a*]anthracene	Carcinogen in coal tar	$C_{18}H_{12}$	228.3	P C N	5.7×10^{-3}	2.2×10^{-8}	1.2×10^{-5}	1.4×10^{6}	5.6
Benzo[*a*]pyrene	Carcinogen in coal tar	$C_{20}H_{12}$	252.32	P H	3.8×10^{-3}	5.6×10^{-9}	2.4×10^{-6}	5.5×10^{6}	6.1
Beryllium	Source of neutrons when bombarded	Be	9.01	P N H	Insoluble	10			
1,1′-Biphenyl	Heat transfer agent; fumigant; used in organic synthesis	$C_{12}H_{10}$	154.22	H	7.5	9.6×10^{-3}	4.15×10^{-4}	1.4×10^{3}	4.09
Bis(2-chloroethyl) ether	Reagent for organic synthesis; solvent	$C_4H_8Cl_2O$	143.02	P C N	1.2×10^{4}	0.71	1.3×10^{-5}	13.9	1.5

Bis(2-chloroiso-propyl) ether		$C_6H_{12}Cl_2O$	171.08	P C N	1.7×10^3	0.85	1.1×10^{-4}	61.0	2.1
Bis(chloromethyl) ether	Manufacture of plastics and ion-exchange resins	$C_2H_4Cl_2O$	114.96	P H	2.2×10^4	30.0		1.2	3.8
Bis(2-ethylhexyl) phthalate	Plasticizer	$C_{24}H_{38}O_4$	390.62	P C N H	40	2.0×10^{-7}	1.1×10^{-5}		4.66
1,3-Butadiene	Manufacture of polymers, synthetic rubber	C_4H_6	54.1	H	735	2.5×10^3	0.063	120	1.99
n-Butane	Gas/fuel, manufacture of synthetic rubber	C_4H_{10}	58.12		61.4	1.8×10^3	0.93		2.89
Cadmium	Batteries, electroplating, pigments	Cd	112.4	P H	Insoluble	1.0			
Camphor	Plasticizer	$C_{10}H_{16}O$	152.24		590	0.18	3.0×10^{-5}		2.42
Captan	Fungicide, bactericide	$C_9H_8Cl_3NO_2S$	300.59	H	3.3	6.0×10^{-5}		6.4×10^3	2.35
Carbaryl	Contact insecticide	$C_{12}H_{11}NO_2$	201.24	H	40	5.0×10^{-3}	1.27×10^{-5}	229	2.36
Carbon disulfide	Manufacture of rayon, carbon tetrichloride, solvent	CS_2	76.13	C N H	2.3×10^3	260	0.013	54	2.00
Carbon tetrachloride	Solvent, manufacture of organic compounds	CCl_4	153.81	P C N H	800	90	0.030	110	2.64
Chlordane	Insecticide	$C_{10}H_6Cl_8$	409.76	P C N H	0.56	1.0×10^{-5}	4.8×10^{-5}	1.4×10^5	3.32
Chlorobenzene	Manufacture of phenol, aniline; solvent	C_6H_5Cl	112.56	P C N H	500	11.7	1.07×10^{-5}	330	2.84
Chloroethane	Refrigerant, solvent, alkylating agent	C_2H_5Cl	64.51	P	4.7×10^3	1.0×10^3	1.78×10^{-6}	35	1.43
Chloroform	Solvent, cleansing agent	$CHCl_3$	119.39	P C N H	8.0×10^3	160	2.5×10^{-4}	31.0	1.97
Chloromethane	Refrigerant	CH_3Cl	50.49	P	0.8	3.75×10^3	6.5×10^{-4}		0.0
2-Chlorophenol	Antiseptic	C_6H_5ClO	128.56	P N	1.9×10^4	1.42	8.28×10^{-6}		2.17

* P = priority pollutant (designated under the Clean Water Act); C = hazardous substance list (designated under CERCLA); N = Appendix IX chemicals (designated under RCRA); H = hazardous air pollutant (designated under the Clean Air Act Amendments).

(cont.)

Name	Typical uses	Formula	Molecular weight, g/mol	List*	Water solubility, mg/L	Vapor pressure, mm Hg	Henry's constant, atm · m^3/mol	K_{oc}, mL/g	log K_{ow}
Chromium	Manufacture of chrome-steel, alloys; in metalplating	Cr	52.0	P H	Insoluble	1.0			
Chrysene	Found in coal tar	$C_{18}H_{12}$	228.3	P C N H	1.8×10^{-3}	6.3×10^{-9}	7.26×10^{-20}	2.0×10^5	5.61
Copper	Manufacture of bronze, brass, alloys, electrical conductors	Cu	63.55	P	Insoluble	1.0			
o-Cresol	Manufacture of synthetic resins, disinfectant, solvent	C_7H_8O	108.15	N H	2.6×10^4	6.3×10^{-7}	1.23×10^{-6}		1.94
p-Cresol	Manufacture of synthetic resin, disinfectant, solvent	C_7H_8O	108.15	N H	1.8×10^4	0.24	1.92×10^{-7}		1.93
Crotonaldehyde	Manufacture of butanol, butyraldehyde, maleic acid	C_4H_6O	70.09	N	1.33×10^5	19	1.96×10^{-5}		
Cyclohexane	Solvent, paint remover, manufacture of adipic acid	C_6H_{12}	84.18		30	95	0.194		3.44
4,4′-DDD	Insecticide, DDT degradation by-product	$C_{14}H_{10}Cl_4$	320.04	P C N	0.02	1.89×10^{-6}	2.16×10^{-5}	7.7×10^5	6.2
4,4′-DDE	DDT degradation by-product	$C_{14}H_8Cl_4$	318.02	P C N H	4.0×10^{-2}	6.5×10^{-6}	2.34×10^{-5}	4.4×10^6	7.0
4,4′-DDT	Insecticide, now banned	$C_{14}H_9Cl_5$	354.48	P C N	5.5×10^{-3}	1.9×10^{-7}	5.20×10^{-5}	2.4×10^5	6.19
Dibenzo[*a,h*] anthracene	Carcinogen from coal tar	$C_{22}H_{14}$	278.36	P C N	5.0×10^{-4}	1.0×10^{-10}	7.33×10^{-9}	3.3×10^6	6.8
1,2-Dichlorobenzene	Solvent, insecticide, fumigant	$C_6H_4Cl_2$	147	P C N	100	1.0	1.9×10^{-3}	1.7×10^3	3.6
1,4-Dichlorobenzene	Insecticide, room deodorant	$C_6H_4Cl_2$	147	P C N H	79	1.18	4.5×10^{-3}	1.4×10^3	3.6
1,1-Dichloroethane	Intermediate in manufacture of vinyl chloride, TCA	$C_2H_4Cl_2$	98.96	P C N	5.5×10^3	180	5.9×10^{-3}	30	1.79

1,2-Dichloroethane	Vinyl chloride monomer, manufacture of solvents	$C_2H_4Cl_2$	98.96	P C N	8.7×10^3	61	9.1×10^{-4}	14	1.48
1,1-Dichloroethylene	Solvent; manufacture of acetyl cellulose	$C_2H_2Cl_2$	96.94	P C N	2.25×10^3	600	0.021	65	1.84
2,4-Dichlorophenol	Intermediate in herbicide 2,4-D synthesis	$C_6H_4Cl_2O$	163.0	P C N	9.8×10^3	0.12	6.66×10^{-6}	380	3.08
1,2-Dichloropropane	Solvent, in dry cleaning fluids, degreasing	$C_3H_6Cl_2$	112.99	P C N	2.7×10^3	38	2.9×10^{-3}	48	2.00
Dieldrin	Insecticide, now banned	$C_{12}H_8Cl_6O$	380.9	P C N	0.19	1.8×10^{-7}	5.8×10^{-5}	160	2.97
Diethyl-*o*-phthalate	Solvent, plasticizer	$C_{12}H_{14}O_4$	222.26	P C N	896	0.05	8.46×10^{-7}	142	2.50
1,1-Dimethylhydra-zine	Rocket fuels	$C_2H_8N_2$	60.12	N H	Miscible	157	2.45×10^{-9}		−2.42
Dimethylphthalate	Solvent, plasticizer	$C_{10}H_{10}O_4$	194.2	N H	4.0×10^3	< 0.01	4.2×10^{-7}		2.28
4,6-Dinitro-*o*-cresol	Herbicide, insecticide	$C_7H_6N_2O_5$	198.15	P C N H	290	3.2×10^{-4}	4.3×10^{-4}	240	2.70
2,4-Dinitrophenol	Manufacture of dyes; wood preservative	$C_6H_4N_2O_5$	184.12	P C N H	5.6×10^3	3.9×10^{-4}	1.6×10^{-9}	53	2.29
2,4-Dinitrotoluene		$C_7H_6N_2O_4$	182.15	P C N H	270	5.1×10^{-3}	8.67×10^{-7}	45	2.00
2,6-Dinitrotoluene		$C_7H_6N_2O_4$	182.15	P C N	180	3.5×10^{-4}	2.17×10^{-7}	92	2.00
1,4-Dioxane	Solvent, stabilizer in chlorinated solvents	$C_4H_8O_2$	88.11	N H	4.31×10^5	37	4.88×10^{-6}	3.5	0.01
Endosulfan	Insecticide	$C_9H_6Cl_6O_3S$	406.91	N	0.53	1.0×10^{-5}	1.01×10^{-4}	2.2×10^3	3.55
Endrin	Insecticide, now banned	$C_{12}H_8Cl_6O$	380.91	N		1.0×10^{-5}	1.91×10^{-5}		5.02
Ethanol	Alcoholic beverages, solvent, antiseptic	C_2H_6O	46.08		6.52×10^4	75		2.20	−0.32
2-Ethoxyethanol	Solvent; in varnishes and paints	$C_4H_{10}O_2$	90.14	N	1.0×10^6	3.80			−0.54
Ethyl acetate	Flavor, essence, solvent	$C_4H_8O_2$	88.12	N	7.4×10^4	72.8	1.34×10^{-4}	1.1×10^3	3.15

* P = priority pollutant (designated under the Clean Water Act); C = hazardous substance list (designated under CERCLA); N = Appendix IX chemicals (designated under RCRA); H = hazardous air pollutant (designated under the Clean Air Act Amendments).

(*cont.*)

Name	Typical uses	Formula	Molecular weight, g/mol	List*	Water solubility, mg/L	Vapor pressure, mm Hg	Henry's constant, atm · m³/mol	K_{oc}, mL/g	log K_{ow}
Ethylbenzene	Conversion to styrene monomer; resin solvent	C_8H_{10}	106.18	P C N H	152	7.1	8.68×10^{-3}		3.11
Ethyl ether	Solvent	$C_4H_{10}O$	74.12	N		440	1.28×10^{-3}		0.83
Fluoranthene	Product of incomplete combustion	$C_{16}H_{10}$	202.26	P C N	0.21	0.01	0.017	3.8×10^4	4.90
Formaldehyde	Manufacture of resins, plastics, foam insulation, fertilizers	CH_2O	30.03	N H	4.0×10^5	3.88×10^3	1.67×10^{-7}	3.6	0.35
Formic acid	Used in dyeing, tanning, chemical synthesis	CH_2O_2	46.03	N	1.0×10^6	35.0	1.67×10^{-7}		−0.54
Furan	Occurs in wood oils	C_4H_4O	68.08	N	9.9×10^3	602			1.34
Heptachlor	Insecticide, largely banned	$C_{10}H_5Cl_7$	373.3	P C N H	0.18	3.0×10^{-4}	2.3×10^{-3}	1.2×10^4	4.40
Hexachlorobenzene	Organic synthesis	C_6Cl_6	284.76	P C N H	6.0×10^{-3}	1.9×10^{-5}	1.7×10^{-3}	3.9×10^3	5.23
Hexachlorobutadiene		C_4Cl_6	260.74	P C N H	2.0	0.15	0.026	2.9×10^4	4.78
Hexachloroethane	Used in metal refining	C_2Cl_6	236.72	P C N H	4.2×10^4	0.40	2.5×10^{-3}	2.0×10^4	4.60
n-Hexane	Used in place of Hg in thermometers	C_6H_{14}	86.17	H		124	1.18		4.0
Hydrazine	Manufacture of agricultural chemicals, rocket fuel; antioxidant	H_4N_2	32.06	N H	3.41×10^8	16.0		0.10	−3.08
Hydrogen cyanide	Exterminating rodents	HCN	27.03	N H	1.0×10^6	620			0.35
Hydrogen sulfide	Production of elemental sulfur, sulfuric acid	H_2S	34.08	N H	4.13×10^3	1.52×10^4			0.96
Indeno[1,2,3-*c,d*] pyrene	Product of incomplete combustion	$C_{22}H_{12}$	276.34	P C N	5.3×10^{-4}	1.0×10^{-10}	2.96×10^{-20}	1.6×10^6	6.50

Isobutanol	Manufacture of flavors; solvent	$C_4H_{10}O$	74.14	N	7.3×10^4	10.4	9.25×10^{-6}		0.76
Isophorone	Solvent	$C_9H_{14}O$	138.23	P C H	1.04×10^4	0.38	0.01	41	2.22
Kepone	Insecticide, now banned	$C_{10}Cl_{10}O$	490.6	N	7.60	2.3×10^{-7}	0.031	5.5×10^4	2.00
Lead	Construction material, batteries; pigments	Pb	207.19	P H	Insoluble	0.05			
Lindane	Insecticide	$C_6H_6Cl_6$	290.83	N H		9.4×10^{-6}	3.25×10^{-6}	2.0×10^3	3.76
Malathion	Insecticide	$C_{10}H_{19}O_6PS_2$	330.38		145	4.0×10^{-5}	4.9×10^{-9}	1.8×10^3	2.89
Mercury	In instruments; arc lamps; pharmaceuticals	Hg	200.59	P N	0.056	1.2×10^{-3}	0.011		
Methanol	Solvent; manufacture of formaldehyde; antifreeze	CH_4O	32.05	N H	Miscible	126	4.66×10^{-6}		−0.77
Methylene chloride	Solvent, paint remover, degreaser	CH_2Cl_2	84.93	N H	2.0×10^4	349	2.7×10^{-3}	8.80	1.30
Methyl ethyl ketone (MEK)	Solvent, manufacture of synthetic resins	C_4H_8O	72.12	C N H	3.5×10^5	77.5	4.7×10^{-5}	4.50	0.26
Methyl isobutyl ketone (MIBK)	Solvent	$C_6H_{12}O$	100.18	N H	1.72×10^4	15.6	1.5×10^{-5}	19	1.19
Methyl methacrylate	Manufacture of resins, plastics	$C_5H_8O_2$	100.13	N H	1.28×10^4	28.0	2.46×10^{-4}	1.8	0.79
1-Naphthalamine	Manufacture of dyes	$C_{10}H_9N$	143.2	N	1.46×10^3	6.5×10^{-5}	1.3×10^{-10}	61	2.07
2-Naphthalamine	Manufacture of dyes; antioxidant in rubber	$C_{10}H_9N$	143.2	N	586	2.6×10^{-4}	2.0×10^{-9}	130	2.07
Naphthalene	Manufacture of resins, phthalic acids used in dyes	$C_{10}H_8$	128.17	P N H	3×10^3	0.054	4.6×10^{-4}	1.3×10^3	3.51
Nickel	Metal plating; alloys	Ni	58.71	P H	Insoluble	1.0			
Nitrobenzene	Manufacture of aniline; in soaps	$C_6H_5NO_2$	123.12	P C N H	1.9×10^3	0.15	2.45×10^{-5}	36.0	1.85

* P = priority pollutant (designated under the Clean Water Act); C = hazardous substance list (designated under CERCLA); N = Appendix IX chemicals (designated under RCRA); H = hazardous air pollutant (designated under the Clean Air Act Amendments).

(cont.)

Name	Typical uses	Formula	Molecular weight, g/mol	List*	Water solubility, mg/L	Vapor pressure, mm Hg	Henry's constant, atm · m^3/mol	K_{oc}, mL/g	log K_{ow}
2-Nitrotoluene	Manufacture of dyes, toluidines	$C_7H_7NO_2$	137.14			0.15	4.5×10^{-5}		2.30
n-Octane	Found in petroleum	C_8H_{18}	114.23		Insoluble	10.37	3.23		5.18
Oxalic acid	Analytical reagent; dyeing; cleaning wood	$C_2H_2O_4$	90.04		1.3×10^5	17.4	0.95		−0.62
Parathion	Insecticide	$C_{10}H_{14}NO_5PS$	291.26	N	Insoluble	4.0×10^{-4}	8.6×10^{-8}	3.2×10^3	3.43
Pentachlorobenzene		C_6HCl_5	250.32	N	0.24	6.0×10^{-3}	7.1×10^{-3}	1.3×10^4	5.19
Pentachlorophenol	Insecticide; wood preservative	C_6HCl_5O	266.32	P C N H	14.0	1.1×10^{-4}	3.4×10^{-6}	5.3×10^4	5.00
Pentane	Found in petroleum	C_5H_{12}	72.15			426	1.26		3.62
Perchloroethylene (PCE)	Dry cleaning; textile processing; degreasing	C_2Cl_4	165.82	P H	< 1.0	18.5	0.015	665	3.4
Phenanthrene	Found in coal tar	$C_{14}H_{10}$	178.24	P C N H	1.9×10^3	2.1×10^{-4}	2.56×10^{-4}	7.3×10^4	−0.19
Phenol	Disinfectant; manufacture of resins, dyes, organic compounds	$C_6H_6O_6$	94.12	P N H	8.2×10^4	0.20	3.97×10^{-7}	14.2	1.46
Propane	Fuel gas, refrigerant	C_3H_5	44.10		Insoluble	6.54×10^3	0.71	27	2.36
n-Propyl alcohol	Solvent	C_3H_8O	60.10			14.5	6.74×10^{-6}		0.30
Pyrene	Found in coal tar	$C_{16}H_{10}$	202.26	P C N H	0.13	2.5×10^{-6}	1.87×10^{-5}	3.8×10^4	4.88
Pyridine	Solvent; synthetic intermediate	C_5H_5N	79.11	N H	1.0×10^6	20	1.2×10^{-5}		0.66
p-Quinone	Oxidizing agent; manufacture of dyes; tanning	C_6H_4O	108.10	H		0.1	9.5×10^{-7}		0.20

Styrene	Manufacture of plastics, resins, synthetic rubber	C_8H_8	104.16	C N H	300	5.12	2.6×10^{-3}	920	2.95
1,2,4,5-Tetrachloro-benzene		$C_6H_2Cl_4$	215.88	N	0.30	0.10	0.010	1.6×10^{3}	4.67
2,3,7,8-Tetrachloro-dibenzo-*p*-dioxin (TCDD)	Carcinogen; product of incomplete combustion	$C_{12}H_4Cl_4O_2$	321.96	P N H	2.0×10^{-4}	1.0×10^{-6}	5.4×10^{-2}	3.3×10^{6}	6.72
1,1,2,2-Tetrachloro-ethane	Solvent; manufacture of paint and resins	$C_2H_2Cl_4$	167.84	P N H	2.9×10^{3}	5.0	4.6×10^{-4}	54	3.03
Tetrachlorethylene (PERC)	Dry cleaning; textile processing; degreasing	C_2Cl_4	165.82	P H	150	14	0.015	98	4.10
Tetraethyl lead	Gasoline additive	$C_8H_{20}Pb$	323.47	N	0.80	0.15		4.9×10^{3}	
Tetrahydrofuran	Solvent for PVC; used in organic synthesis	C_4H_8O	72.11	N	Miscible	145	7.1×10^{-5}		0.46
Thiourea	Photo fixative; manufacture of resins	CH_4N_2S	76.13	N	9.2×10^{5}			1.60	−1.14
Toluene	Manufacture of benzoic acid, dyes, benzaldehyde	C_7H_8	92.15	P C N H	515	22.0	6.7×10^{-3}	300	2.73
Toxaphene	Insecticide	$C_{10}H_{10}Cl_8$	413.8	P C N H	0.50	1.0×10^{-6}	0.063	964	3.30
1,2,4-Trichloro-benzene	Insecticide	$C_6H_3Cl_3$	181.44	P C N H	30.0	0.29	2.3×10^{-3}	9.2×10^{3}	4.30
1,1,1-Trichloro-ethane	Solvent	$C_2H_3Cl_3$	133.4	P C N	4.4×10^{3}	100	0.018	152	2.50
1,1,2-Trichloro-ethane	Solvent	$C_2H_3Cl_3$	133.4	P C N H	4.4×10^{3}	32	9.1×10^{-4}	56	2.47
Trichloroethylene (TCE)	Solvent, degreaser; manufacture of organic chemicals	C_2HCl_3	131.38	P N H	1.10	60	9.1×10^{-3}	126	2.38

* P = priority pollutant (designated under the Clean Water Act); C = hazardous substance list (designated under CERCLA); N = Appendix IX chemicals (designated under RCRA); H = hazardous air pollutant (designated under the Clean Air Act Amendments).

(*cont.*)

Name	Typical uses	Formula	Molecular weight, g/mol	List*	Water solubility, mg/L	Vapor pressure, mm Hg	Henry's constant, atm · m^3/mol	K_{oc}, mL/g	log K_{ow}
2,4,5-Trichloro-phenol	Fungicide, bacteriacide	$C_6H_3Cl_3O$	197.44	C N H	1.19×10^3	0.022	1.8×10^{-7}	159	2.53
2,4,6-Trichloro-phenol	Fungicide, bacteriacide, preservative	$C_6H_3Cl_3O$	197.44	P C N H	800	0.017	9.1×10^{-8}	89	3.72
2,4,6-Trinitrotoluene (TNT)	Explosives	$C_7H_5N_3O_6$	227.15		200	4.3×10^{-3}			2.25
Vinyl acetate	Manufacture of plastics, lacquers	$C_4H_6O_2$	86.09	H	2×10^4	83	4.8×10^{-4}		0.73
Vinyl chloride	Manufacture of plastics; refrigerant	C_2H_3Cl	62.5	P C N H	4.27×10^3	2.6×10^3	0.056	3.9	0.60
m-Xylene	Solvent, chemical intermediate	C_8H_{10}	106.18	N H	175	8.3	6.3×10^{-3}	57	1.38
o-Xylene	Solvent, chemical intermediate	C_8H_{10}	106.18	N H	130	10.0	5.4×10^{-3}	2.38	2.95
p-Xylene	Solvent, chemical intermediate	C_8H_{10}	106.18	N H	198	8.8	6.3×10^{-3}		3.26
Zinc	Galvanizing, alloys, batteries	Zn	65.37	P	Insoluble	1.0			

* P = priority pollutant (designated under the Clean Water Act); C = hazardous substance list (designated under CERCLA); N = Appendix IX chemicals (designated under RCRA); H = hazardous air pollutant (designated under the Clean Air Act Amendments).

APPENDIX C

HAZARDOUS WASTE LISTS

TABLE C.1
Hazardous wastes from nonspecific sources (F list)

Industry and EPA hazardous waste number	Hazardous waste from nonspecific sources	Hazard code*
F001	The following spent halogenated solvents used in degreasing: tetrachloroethylene, trichloroethylene, methylene chloride, 1,1,1-trichloroethane, carbon tetrachloride, and chlorinated fluorocarbons; all spent solvent mixtures/blends used in degreasing containing, before use, a total of 10% or more (by volume) of one or more of the above halogenated solvents or those solvents listed in F002, F004, and F005; and still bottoms from the recovery of these spent solvents and spent solvent mixtures.	T
F002	The following spent halogenated solvents: tetrachloroethylene, methylene chloride, trichloroethylene,1,1,1-trichloroethane, chlorobenzene, 1,1,2-trichloro-1,2,2-trifluoroethane, orthodichlorobenzene, trichlorofluoromethane, and 1,1,2-trichloroethane; all spent solvent mixtures/blends containing, before use, a total of 10% or more (by volume) of one or more of the above halogenated solvents or those listed in F00l, F004, or F005; and still bottoms from the recovery of these spent solvents and spent solvent mixtures.	T

*Hazard code definitions: C = corrosivity; E = EP toxicity; H = acute hazardous waste; I = ignitability; R = reactivity; T = toxic waste.

TABLE C.1 (*cont.*)
Hazardous wastes from nonspecific sources (F list)

Industry and EPA hazardous waste number	Hazardous waste from nonspecific sources	Hazard code*
F003	The following spent nonhalogenated solvents: xylene, acetone, ethyl acetate, ethylbenzene, ethyl ether, methyl isobutyl ketone, *n*-butyl alcohol, cyclohexanone, and methanol; all spent solvent mixtures/blends containing, before use, a total of 10% or more (by volume) of one or more of the above nonhalogenated solvents, listed in F00l, F002, F004, and F005; and still bottoms from the recovery of these spent solvents and spent solvent mixtures.	I
F004	Wastes, including but not limited to, distillation residues, heavy ends, tars, and reactor cleanout wastes from the production of chlorinated aliphatic hydrocarbons, having carbon content from one to five, utilizing free radical catalyzed processes. (This listing does not include light ends, spent filters and filter aids, spent dessicants, wastewater, wastewater treatment sludges, spent catalysts, and wastes.)	T
F005	The following spent nonhalogenated solvents: toluene, methyl ethyl ketone, carbon disulfide, isobutanol, pryridine, benzene, 2-ethyoxyethanol, and 2-nitropropane; all spent solvent mixtures/ blends containing, before use, a total of 10% or more (by volume) or one or more of the above nonhalogenated solvents or those solvents listed in F001, F002, or F004; and still bottoms from the recovery of these spent solvents and spent solvent mixtures.	I,T
F006	Wastewater treatment sludges from electroplating operations except from the following processes: (1) sulfuric acid anodizing or aluminum; (2) tin plating on carbon steel; (3) zinc plating (segregated basis) on carbon steel; (4) aluminum or zinc-and-aluminum plating on carbon steel; (5) cleaning/stripping associated with tin, zinc, or aluminum plating on carbon steel; and (6) chemical etching and milling of aluminum.	T
F007	Spent cyanide plating bath solutions from electroplating operations.	R,T
F008	Plating bath residues from the bottom of plating baths from electroplating operations where cyanides are used in the process.	R,T
F009	Spent stripping and cleaning bath solutions from electroplating operations where cyanides are used in the process.	R,T
F010	Quenching bath residues from oil baths from metal heat-treating operations where cyanides are used in the process.	R,T
F011	Spent cyanide solutions from salt bath pot cleaning from metal heat-treating operations.	R,T
F012	Quenching wastewater treatment sludges from metal heat-treating operations where cyanides are used in the process.	T
F019	Waste treatment sludges from the chemical conversion coating of aluminum.	T
F020	Wastes (except wastewater and spent carbon from hydrogen chloride purification) from the production or manufacturing use	H

TABLE C.1 (*cont.*)
Hazardous wastes from nonspecific sources (F list)

Industry and EPA hazardous waste number	Hazardous waste from nonspecific sources	Hazard code*
F020 (*cont.*)	(as a reactant, chemical intermediate, or component in a formulating process) or tri- or tetrachlorophenol, or of intermediates used to produce their pesticide derivatives. (This listing does not include wastes from the production of hexachlorophene from highly purified 2,4,5-trichlorophenol.)	
F021	Wastes (except wastewater and spent carbon from hydrogen chloride purification) from the production or manufacturing use (as a reactant, chemical intermediate, or component in a formulating process) of pentachlorophenol, or of intermediates used to produce its derivatives.	H
F022	Wastes (except wastewater and spent carbon from hydrogen chloride purification) from the manufacturing use (as a reactant, chemical intermediate, or component in a formulating process) of tetra-, penta-, or hexachlorobenzenes under alkaline conditions.	H
F023	Wastes (except wastewater and spent carbon from hydrogen chloride purification) from the production of materials on equipment previously used for the production or manufacturing use (as a reactant, chemical intermediate, or component in a formulating process) of tri- and tetrachlorophenols. (This listing does not include wastes from equipment used only for the production or use of hexachlorophene from highly purified 2,4,5-trichlorophenol.)	H
F024	Wastes, including but not limited to, distillation residues, heavy ends, tars, and reactor cleanout wastes from the production of chlorinated aliphatic hydrocarbons, having carbon content from one to five, utilizing free radical-catalyzed processes. (This listing does not include light ends, spent filters and filter aids, spent dessicants, wastewater, wastewater treatment sludges, spent catalysts, and wastes.)	T
F026	Wastes (except wastewater and spent carbon from hydrogen chloride purification) from the production of materials on equipment previously used for the manufacturing (as a reactant, chemical intermediate, or component in a formulating process) of tetra-, penta-, or hexachlorobenzene under alkaline conditions.	H
F027	Discarded unused formulations containing tri-, tetra-, or penta-chlorophenol or discarded unused formulations containing compounds derived from these chlorophenols. (This listing does not include formulations containing hexachlorophene synthesized from prepurified 2,4,5-trichlorophenol as the sole component.)	H
F028	Residues resulting from the incineration or thermal treatment of soil contaminated with EPA hazardous waste numbers F020, F021, F022, F023, F026, and F027.	T

*Hazard code definitions: C = corrosivity; E = EP toxicity; H = acute hazardous waste; I = ignitability; R = reactivity; T = toxic waste.

TABLE C.2
Hazardous wastes from specific sources (K list)

Industry/EPA hazardous waste number	Hazardous wastes	Hazard code*
	Wood preservation	
K001	Bottom sediment sludge from the treatment of wastewater from wood-preserving processes that use creosote and/or pentachlorophenol.	T
	Inorganic pigments	
K002	Wastewater treatment sludge from the production of chrome yellow and orange pigments.	T
K003	Wastewater treatment sludge from the production of molybdate orange pigments.	T
K004	Wastewater treatment sludge from the production of zinc yellow pigments.	T
K005	Wastewater treatment sludge from the production of chrome green pigments.	T
K006	Wastewater treatment sludge from the production of chrome oxide green pigments (anhydrous and hydrated).	T
K007	Wastewater treatment sludge from the production of iron blue pigments.	T
K008	Oven residue from the production of chrome oxide green pigments.	T
	Organic chemicals	
K009	Distillation bottoms from the production of acetaldehyde from ethylene.	T
K010	Distillation side cuts from the production of acetaldehyde from ethylene.	T
K011	Bottom stream from the wastewater stripper in the production of acrylonitrile.	R,T
K013	Bottom stream from the acetonitrile column in the production of acrylonitrile.	R,T
K014	Bottoms from the acetonitrile purification column in the production of acrylonitrile.	T
K015	Still bottoms from the distillation of benzyl chloride.	T
K016	Heavy ends or distillation residues from the production of carbon tetrachloride.	T
K017	Heavy ends (still bottoms) from the purification column in the production of epichlorohydrin.	T
K018	Heavy ends from the fractionation column in ethyl chloride production.	T
K019	Heavy ends from the distillation of ethylene dichloride in ethylene dichloride production.	T
K020	Heavy ends from the distillation of vinyl chloride in vinyl chloride monomer production.	T
K021	Aqueous spent antimony catalyst waste from fluoromethanes production.	T
K022	Distillation bottom tars from the production of phenol/acetone from cumene.	T
K023	Distillation light ends from the production of phthalic anhydride from naphthalene.	T

*Hazard code definitions: C = corrosivity; E = EP toxicity; R = acute hazardous waste; I = ignitability; R = reactivity; T = toxic waste.

TABLE C.2 (*cont.*)

Hazardous wastes from specific sources (K list)

Industry/EPA hazardous waste number	Hazardous wastes	Hazard code*
	Organic chemicals (*cont.*)	
K024	Distillation bottoms from the production of phthalic anhydride from naphthalene.	T
K093	Distillation light ends from the production of phthalic anhydride from *o*-xylene.	T
K094	Distillation bottoms from the production of phthalic anhydride from *o*-xylene.	T
K025	Distillation bottoms from the production of nitrobenzene by the nitration of benzene.	T
K026	Stripping still tails from the production of methyl ethyl pyridines.	T
K027	Centrifuge and distillation residues from toluene diisocyanate production.	R,T
K028	Spent catalyst from the hydrochlorinator reactor in the production of 1,1,1-trichloroethane.	T
K029	Waste from the product steam stripper in the production of 1,1,1-trichloroethane.	T
K095	Distillation bottoms from the production of 1,1,1-trichloroethane.	T
K096	Heavy ends from the heavy ends column from the production of 1,1,1-trichloroethane.	T
K030	Column bottoms or heavy ends from the combined production of trichloroethylene and perchloroethylene.	T
K083	Distillation bottoms from aniline productions.	T
K103	Process residues from aniline extraction from the production of aniline.	T
K104	Combined wastewater streams generated from nitrobenzene/aniline production.	T
K035	Wastewater treatment sludges generated in the production of creosote.	T
K036	Still bottoms from toluene reclamation distillation in the production of disulfoton.	T
K037	Wastewater treatment sludges from the production of disulfoton.	T
K038	Wastewater from the washing and stripping of phorate production.	T
K039	Filter cake from the filtration of diethylphosphorodithioic acid in the production of phorate.	T
K040	Wastewater treatment sludge from the production of phorate.	T
K041	Wastewater treatment sludge from the production of toxaphene.	T
K098	Untreated process wastewater from the production of toxaphene.	T
K042	Heavy ends or distillation residues from the distillation of tetrachlorobenzene in the production of 2,4,5-T.	T
K043	2,6-Dichlorophenol waste from the production of 2,4-D.	T
K099	Untreated wastewater from the production of 2,4-D.	T
K123	Process wastewater (including supernates, filtrates, and washwaters) from the production of ethylenebis (dithiocarbamic acid) and its salt.	T

TABLE C.2 (*cont.*)

Hazardous wastes from specific sources (K list)

Industry/EPA hazardous waste number	Hazardous wastes	Hazard code*
	Organic chemicals (*cont.*)	
K124	Reactor vent scrubber water from the production of ethylenebis (dithiocarbamic acid) and its salts.	C,T
K125	Filtration, evaporation, and centrifugation solids from the production of ethylenebis (dithiocarbamic acid) and its solids.	T
K126	Baghouse dust and floor sweepings in milling and packaging operations from the production or formulation of ethylenebis (dithiocarbamic acid) and its salts.	T
	Explosives	
K044	Wastewater treatment sludges from the manufacturing and processing of explosives.	R
K045	Spent carbon from the treatment of wastewater containing explosives.	R
K046	Wastewater treatment sludges from the manufacturing, formulation, and loading of lead-based initiating compounds.	T
K047	Pink/red water from TNT operations.	R
	Petroleum refining	
K048	Dissolved air flotation (DAF) float from the petroleum refining industry.	T
K049	Slop oil emulsion solids from the petroleum refining industry.	T
K050	Heat exchanger bundle cleaning sludge from the petroleum refining industry.	T
K051	API separator sludge from the petroleum refining industry.	T
K052	Tank bottoms (leaded) from the petroleum refining industry.	T
K085	Distillation or fractionation column bottoms from the production of chlorobenzenes.	T
K105	Separated aqueous stream from the reactor product washing step in the production of chlorobenzenes.	T
K111	Product washwaters from the production of dinitrotoluene via nitration of toluene.	C,T
K112	Reaction by-product water from the drying column in the production of toluenediamine via hydrogenation of dinitrotoluene.	T
K113	Condensed liquid light ends from the purification of toluenediamine in the production of toluenediamine via hydrogenation of dinitrotoluene.	T
K114	Vicinals from the purification of toluenediamine in the production of toluenediamine via hydrogenation of dinitrotoluene.	T
K115	Heavy ends from the purification of toluenediamine in the production of toluenediamine via hydrogenation of dinitrotoluene.	T
K116	Organic condensate from the solvent recovery column in the production of toluene diisocyanate via phosgenation of toluenediamine.	T
K117	Wastewater from the reactor vent gas scrubber in the production of ethylene dibromide via bromination of ethene.	T

*Hazard code definitions: C = corrosivity; E = EP toxicity; R = acute hazardous waste; I = ignitability; R = reactivity; T = toxic waste.

TABLE C.2 (*cont.*)
Hazardous wastes from specific sources (K list)

Industry/EPA hazardous waste number	Hazardous wastes	Hazard code*
	Petroleum refining (*cont.*)	
K118	Spent adsorbent solids from purification of ethylene dibromide in the production of ethylene dibromide via bromination of ethene.	T
K136	Still bottoms from the purification of ethylene dibromide in the production of ethylene dibromide via bromination of ethene.	T
	Inorganic chemicals	
K071	Brine purification muds from the mercury cell process in chlorine production, where separately prepurified brine is not used.	T
K073	Chlorinated hydrocarbon waste from the purification step of the diaphragm cell process using graphite anodes in chlorine production.	T
K106	Wastewater treatment sludge from the mercury cell process in chlorine production.	I
	Pesticides	
K031	By-product salts generated in the production of MSMA and cacodylic acid.	T
K032	Wastewater treatment sludge from the production of chlordane.	T
K033	Wastewater and scrub water from the chlorination of cyclopentadiene in the production of chlordane.	T
K034	Filter solids from the filtration of hexachlorocyclopentadiene in the production of chlordane.	T
K097	Vacuum stripper discharge from the chlordane chlorinator in the production of chlordane.	T
	Iron and steel	
K061	Emission control dust/sludge from the primary production of steel in electric furnaces.	T
K062	Spent pickle liquor generated by steel finishing operations of facilities within the iron and steel industry (SIC Codes 331 and 332).	C,T
	Secondary lead	
K069	Emission control dust/sludge from secondary lead smelting.	T
K100	Waste leaching solution from acid leaching of emission control dust/sludge from secondary lead smelting.	T
	Veterinary pharmaceuticals	
K084	Wastewater treatment sludges generated during the production of veterinary pharmaceuticals from arsenic or organoarsenic compounds.	T
K101	Distillation tar residues from the distillation of aniline-based compounds in the production of veterinary pharmaceuticals from arsenic or organoarsenic compounds.	T
K102	Residue from the use of activated carbon for decolorization in the production of veterinary pharmaceuticals from arsenic or organoarsenic compounds.	T

TABLE C.2 (*cont.*)

Hazardous wastes from specific sources (K list)

Industry/EPA hazardous waste number	Hazardous wastes	Hazard code*
	Ink formulation	
K086	Solvent washes and sludges, caustic washes and sludges, or water washes and sludges from cleaning tubs and equipment used in the formulation of ink from pigments, driers, soaps, and stabilizers containing chromium and lead.	T
	Coking	
K060	Ammonia still lime sludge from coking operations.	T
K087	Decanter tank tar sludge from coking operations.	T

TABLE C.3

Hazardous wastes from commercial products, intermediates, and residues (P and U lists)

Hazardous waste number	Chemical abstracts number	Hazardous substance	Hazard code*
		P list	
P023	107-20-0	Acetaldehyde, chloro-	
P002	591-08-2	Acetamide, *n*-(aminothioxomethyl)-	
P057	640-19-7	Acetamide, 2-fluoro-	
P058	62-74-8	Acetic acid, fluoro-, sodium salt	
P066	16752-77-5	Acetimidic acid, *n*-[(methylcarbamoy1)oxy]thio-, methyl ester	
P002	591-08-2	1-Acetyl-2-thiourea	
P003	107-02-8	Acrolein	
P070	116-06-3	Aldicarb	
P004	309-00-2	Aldrin	
P005	107-18-6	Allyl alcohol	
P006	20859-73-8	Aluminum phosphide	R,T
P007	2763-96-4	5-(Aminomethyl)-3-isoxazolol	
P008	504-24-5	4-α-Aminopyridine	
P009	131-74-8	Ammonium picrate	R
P119	7803-55-6	Ammonium vanadate	
P010	7778-39-4	Arsenic acid	
P011	1303-28-2	Arsenic oxide As_2	
P011	1303-28-2	Arsenic pentoxide	
P012	1327-53-3	Arsenic trioxide	

*Hazard code definitions: C = corrosivity; E = EP toxicity; H = acute hazardous waste; I = ignitability; R = reactivity; T = toxic waste.

TABLE C.3 (*cont.*)
Hazardous wastes from commercial products, intermediates, and residues (P and U lists)

Hazardous waste number	Chemical abstracts number	Hazardous substance	Hazard code*
		P list (*cont.*)	
P038	692-42-2	Arsine, diethyl-	
P036	696-28-6	Arsonous dichloride, phenyl	
P054	151-56-4	Aziridine	
P013	542-62-1	Barium cyanide	
P024	106-47-8	Benzenamine, 4-chloro-	
P077	100-01-6	Benzenamine, 4-nitro-	
P028	100-44-7	Benzene, (chloromethyl)	
P042	51-43-4	1,2-Benzenediol, 4-[1-hydroxy-2-(methylamino)ethyl]-,	R
P046	122-09-8	Benzeneethanamine, α,α-dimethyl	
P014	108-98-5	Benzenethiol	
P001	81-81-2	2*H*-1-Benzopyran-2-one, 4-hydroxy-3-(3-oxo-l-phenylbutyl)-, and salts	
P028	100-44-7	Benzyl chloride	
P015	7440-41-7	Beryllium dust	
P016	542-88-1	bis(chloromethyl) Ether	
P017	598-31-2	Bromoacetone	
P018	357-57-3	Brucine	
P021	592-01-8	Calcium cyanide	
P022	75-15-0	Carbon bisulfide	
P095	75-44-5	Carbonic dichloride	
P023	107-20-0	Chloroacetaldehyde	
P024	106-47-8	*p*-Chloroaniline	
P029	544-92-3	Copper cyanide	
P030		Cyanides (soluble cyanide salts), not otherwise specified	
P031	460-19-5	Cyanogen	
P033	506-77-4	Cyanogen chloride	
P034	131-89-5	2-Cyclohexyl-4,6-dinitrophenol	
P036	696-28-6	Dichlorophenylarsine	
P037	60-57-1	Dieldrin	
P038	692-42-2	Diethylarsine	
P041	311-45-5	Diethyl-*p*-nitrophenyl phosphate	
P040	297-97-2	*O,O*-Diethyl *O*-pyrazinyl phosphorothioate	
P043	55-91-4	Diisopropyl fluorophosphate (DEP)	
P004	309-00-2	1,4:5,8-Dimethanonaphthalene, 1,2,3,4,10,10-hexachloro-1,4,4*a*,5,8,8*a*-hexahydro-, (1α,4α,4β,5α,8α,8*a*β)	
P060	465-73-6	1,4:5,8-Dimethanonaphthalene, 1,2,3,4,10,10-hexachloro-1,4,4*a*,5,8,8*a*-hexahydro-, (1α,4α,4β,5β,8β,8*a*β)-	

TABLE C.3 (*cont.*)

Hazardous wastes from commercial products, intermediates, and residues (P and U lists)

Hazardous waste number	Chemical abstracts number	Hazardous substance	Hazard code*
		P list (*cont.*)	
P037	60-57-1	2,7:3,6-Dimethanonaphth[2,3*b*]oxirane, 3,4,5,6,9,9-hexachloro-l*a*,2,2*a*,3,6,6*a*,7,7*a*-octahydro-, (l*a*α,2β,2*a*α,3β,6*a*α,6*a*β,7β,7*a*α)-	
P051	72-20-8	2,7:3,6-Dimethanonaphth[2,3*b*]oxirane, octahydro-, 1*a*α,2β,2*a*α,3α,6α,6*a*β,7β,7*a*α)-	
P044	60-51-5	Dimethoate	
P045	39196-18-4	3,3-Dimethyl-1-(methylthio)-2-butanone, *O*-[(methylamino) carbonyl]oxime	
P046	122-09-8	α,α-Dimethylphenethylamine	
P047	534-52-1	4,6-Dinitro-*o*-cresol and salts	
P048	51-28-5	2,4-Dinitrophenol	
P020	88-85-7	Dinoseb	
P085	152-16-9	Diphosphoramide, octamethyl-	
P039	298-04-4	Disulfoton	
P049	541-53-7	2,4-Dithiobiuret	
P050	115-29-7	Endosulfan	
P088	145-73-3	Endothal	
P051	72-20-8	Endrin	
P042	51-43-4	Epinephrine	
P101	107-12-0	Ethyl cyanide	
P054	151-56-4	Ethyleneimine	
P097	52-85-7	Famphur	
P056	7782-41-4	Fluorine	
P057	640-19-7	Fluoroacetamide	
P058	62-74-8	Fluoroacetic acid, sodium salt	
P065	628-86-4	Fulminic acid, mercury (2+) salt	R,T
P059	76-44-8	Heptachlor	
P062	757-58-4	Hexaethyltetraphosphate	
P116	79-19-6	Hydrazinecarbothioamide	
P068	60-34-4	Hydrazine, methyl-	
P063	74-90-8	Hydrocyanic acid	
P063	74-90-8	Hydrogen cyanide	
P096	7803-51-2	Hydrogen phosphide	
P064	624-83-9	Isocyanic acid, methyl ester	
P060	465-73-6	Isodrin	

*Hazard code definitions: C = corrosivity; E = EP toxicity; H = acute hazardous waste; I = ignitability; R = reactivity; T = toxic waste.

TABLE C.3 (*cont.*)

Hazardous wastes from commercial products, intermediates, and residues (P and U lists)

Hazardous waste number	Chemical abstracts number	Hazardous substance	Hazard code*
		P list (*cont.*)	
P007	2763-96-4	3(2*H*)-Isoxazolone, 5-(aminomethyl)-	
P092	62-38-4	Mercury, (acetato)phenyl	
P065	628-86-4	Mercury fulminate	R,T
P082	62-75-9	Methamine, *n*-methyl-*n*-nitroso	
P016	542-88-1	Methane, oxybis[chloro-]	
P112	509-14-8	Methane, tetranitro-	R
P118	75-70-7	Methanethiol, trichloro-	
P050	115-29-7	6,9-Methano-2,4,3-benzodioxathiepen, 6,7,8,9,10,10-hexachloro-1,5,5*a*,6,9,9*a*-hexahydro-, 3-oxide	
P059	76-44-8	4,7-Methano-1*H*-indene, 1,4,5,6,7,8,8-heptachloro-3*a*,4,7,7*a*-tetrahydro-	
P066	16752-77-5	Methomyl	
P067	75-55-8	2-Methyl aziridine	
P068	60-34-4	Methyl hydrazine	
P064	624-83-9	Methyl isocyanate	
P069	75-86-5	2-Methyllactonitrile	
P071	298-00-0	Methyl parathion	
P072	86-88-4	α-Naphthylthiourea	
P073	13463-39-3	Nickel carbonyl	T
P074	557-19-7	Nickel cyanide	
P075	54-11-5	Nicotine and salts	
P076	10102-43-9	Nitric oxide	
P077	100-01-6	*p*-Nitroaniline	
P078	10102-44-0	Nitrogen dioxide	
P081	55-63-0	Nitroglycerine	R
P082	62-75-9	*N*-Nitrosodimethylamine	
P084	4549-40-0	*N*-Nitrosomethylvinylamine	
P085	152-16-9	Octamethylpyrophosphoramide	
P087	20816-12-0	Osmium tetroxide	
P088	145-73-3	7-Oxabicyclo[2.2.1]heptane-2,3-dicarboxylic acid	
P089	56-38-2	Parathion	
P034	131-89-5	Phenol, 2-cyclohexyl-4,6-dinitro-	
P048	51-28-5	Phenol, 2-4-dinitro-	
P047	534-52-1	Phenol, 2-methyl-4,6-dinitro-, and salts	
P020	88-85-7	Phenol, 2-(1-methylpropyl)-4,6-dinitro	

TABLE C.3 (*cont.*)

Hazardous wastes from commercial products, intermediates, and residues (P and U lists)

Hazardous waste number	Chemical abstracts number	Hazardous substance	Hazard code*
		P list (*cont.*)	
P009	131-74-8	Phenol, 2,4,6-trinitro-, ammonium salt	R
P092	62-38-4	Phenylmercury acetate	
P093	103-85-5	Phenylthiourea	
P094	298-02-2	Phorate	
P095	75-44-5	Phosgene	
P096	7803-51-2	Phosphine	
P041	311-45-5	Phosphoric acid, diethyl 4-nitrophenyl ester	
P039	298-04-4	Phosphorodithioic acid, *O,O*-diethyl *S*-[2-(ethylthio)ethyl] ester	
P094	298-02-2	Phosphorodithioic acid, *O,O*-diethyl *S*-[(ethylthio)methyl] ester	
P044	60-51-5	Phosphorodithioic acid, *O,O*-dimethyl *S*-[2-(methylamino)-2-oxoethyl] ester	
P043	55-91-4	Phosphorofluoric acid, bis(1-methylethyl)-ester	
P089	56-38-2	Phosphorothioic acid, *O,O*-diethyl *O*-(4-nitrophenyl) ester	
P040	297-97-2	Phosphorothioic acid, *O,O*-diethyl *O*-pyrazinyl ester	
P097	52-85-7	Phosphorothioic acid, *O*-[4-[(dimethylamino)sulfonyl] phenyl] *O,O*-dimethyl ester	
P071	298-00-0	Phosphorothioic acid, *O,O*-dimethyl *O*-(4-nitrophenyl) ester	
P110	78-00-2	Plumbane, tetraethyl-	
P098	151-50-8	Potassium cyanide	
P099	506-61-6	Potassium silver cyanide	
P070	116-06-3	Propanal, 2-methyl-2-(methylthio)-, *O*-[(meyhylamino)carbonyl] oxime	
P101	107-12-0	Propanenitrile	
P027	542-76-7	Propanenitrile, 3-chloro-	
P069	75-86-5	Propanenitrile, 2-hydroxy-2-methyl-	
P081	55-63-0	1,2,3-Propanetriol, trinitrate	R
P017	598-31-2	2-Propanone, 1-bromo	
P102	107-19-7	Propargyl alcohol	
P003	107-02-8	2-Propenal	
P005	107-18-6	2-Propen-1-ol	
P067	75-55-8	1,2-Propylenimine	
P102	591-08-2	2-Propyn-1-ol	
P008	504-24-5	Pyridinamine	

*Hazard code definitions: C = corrosivity; E = EP toxicity; H = acute hazardous waste; I = ignitability; R = reactivity; T = toxic waste.

TABLE C.3 (*cont.*)

Hazardous wastes from commercial products, intermediates, and residues (P and U lists)

Hazardous waste number	Chemical abstracts number	Hazardous substance	Hazard code*
		P list (*cont.*)	
P075	54-11-5	Pyridine, (*S*)-3-(1-methyl-2-pyrrolidinyl)-, and salts	
P111	107-49-3	Pyrophosphoric acid, tetraethyl ester	
P103	630-10-4	Selenourea	
P104	506-64-9	Silver cyanide	
P105	26628-22-8	Sodium azide	
P106	143-33-9	Sodium cyanide	
P107	1314-96-1	Strontium sulfide	
P108	57-24-9	Strychnidin-l0-one and salts	
P018	357-57-3	Strychnidin-10-one, 2,3-dimethoxy-	
P108	57-24-9	Strychnine and salts	
P115	10031-59-1	Sulfuric acid, thallium(I) salt	
P109	3689-24-5	Tetraethyldithiopyrophosphate	
P110	78-00-2	Tetraethyl lead	
P111	107-49-3	Tetraethylpyrophosphate	
P112	509-14-8	Tetranitromethane	R
P062	757-58-4	Tetraphosphoric acid, hexaethyl ester	
P113	1314-32-5	Thallic oxide	
P113	1314-32-5	Thallium(III) oxide	
P114	12039-52-0	Thallium(I) selenite	
P115	10031-59-1	Thallium(I) sulfate	
P109	3689-24-5	Thiodiphosphoric acid, tetraethyl ester	
P045	39196-18-4	Thiofanox	
P049	541-53-7	Thioimidodicarbonic diamide	
P014	108-98-5	Thiophenol	
P116	79-19-6	Thiosemicarbazide	
P026	5344-82-1	Thiourea, (2-chlorophenyl)-	
P072	86-88-4	Thiourea, 1-naphthalenyl-	
P093	103-85-5	Thiourea, phenyl	
P123	8001-35-2	Toxaphene	
P118	75-70-7	Trichloromethanethiol	
P119	7803-55-6	Vanadic acid, ammonium salt	
P120	1314-62-1	Vanadium (V) oxide	
P084	4549-40-0	Vinylamine, N-methyl-N-nitroso	
P001	81-81-2	Warfarin	
P121	557-21-1	Zinc cyanide	
P122	1314-84-7	Zinc phosphide	R,T

TABLE C.3 (*cont.*)

Hazardous wastes from commercial products, intermediates, and residues (P and U lists)

Hazardous waste number	Chemical abstracts number	Hazardous substance	Hazard code*
		U list	
U001	75-07-0	Acetaldehyde	I
U034	75-87-6	Acetaldehyde, trichloro	
U187	62-44-2	Acetamide, *N*-(4-ethoxyphenyl)-	
U005	53-96-3	Acetamide, *N*-9*H*-fluoren-2-yl	
U112	141-78-6	Acetic acid, ethyl ester	I
U144	301-04-2	Acetic acid, lead salt	
U214	563-68-8	Acetic acid, thallium (1+) salt	
U232	93-76-5	Acetic acid, (2,4,5-trichlorophenoxyl)-	
U002	67-64-1	Acetone	I
U003	75-05-8	Acetonitrile	I,T
U004	98-86-2	Acetophenone	
U005	53-96-3	2-Acetylaminofluorene	
U006	75-36-5	Acetyl chloride	C,R,T
U007	79-06-1	Acrylamide	
U008	79-10-7	Acrylic acid	I
U009	107-13-1	Acrylonitrile	
U011	61-82-5	Amitrole	
U012	62-53-3	Aniline	I,T
U014	492-80-8	Auramine	
U015	115-02-6	Azaserine	
U010	50-07-7	Azirino (2′,3′:3,4)pyrrolo[1,2-*a*[indole-4,7-dione, 6-amino-8-[((aminocarbonyl)oxy)methyl]-1,l*a*,2,8,8*a*,8*b*-hexahydro-8*a*-methoxy-5-methyl-	
U016	225-51-4	3,4-Benzacridine	
U017	98-87-3	Benzal chloride	
U157	50-49-5	Benzo[*j*]aceanthrylene, 1,2-dihydro-3-methyl-	
U047	91-58-7	*β*-Chloronaphthalene	
U048	95-57-8	*o*-Chlorophenol	
U049	3165-93-3	4-Chloro-*o*-toluidine, hydrochloride	
U032	13765-19-0	Chromic acid, calcium salt	
U050	218-01-9	Chrysene	
U051	8021-39-4	Creosote	
U052	1319-77-3	Cresols (cresylic acid)	
U053	4170-30-3	Crotonaldehyde	
U055	98-82-8	Cumene	I
U246	506-68-3	Cyanogen bromide	

*Hazard code definitions: C = corrosivity; E = EP toxicity; H = acute hazardous waste; I = ignitability; R = reactivity; T = toxic waste.

TABLE C.3 (*cont.*)

Hazardous wastes from commercial products, intermediates, and residues (P and U lists)

Hazardous waste number	Chemical abstracts number	Hazardous substance	Hazard code*
		U list (*cont.*)	
U197	106-51-4	2,5-Cyclohexadiene-1,4-dione	
U056	110-82-7	Cyclohexane (1)	
U057	108-94-1	Cyclohexanone (1)	
U130	77-47-4	1,3-Cyclopentadiene, 1,2,3,4,5,5-hexachloro-	
U058	50-18-0	Cyclophosphamide	
U240	94-75-7	2,4-D, salts and esters	
U059	20830-81-3	Daunomycin	
U060	72-54-8	DDD	
U061	50-29-3	DDT	
U062	2303-16-4	Diallate	
U063	53-70-3	Dibenz[*a,h*]anthracene	
U064	189-55-9	Dibenzo[*a,i*]pyrene	
U066	96-12-8	1,2-Dibromo-3-chloropropane	
U069	84-74-2	Dibutyl phthalate	
U070	95-50-1	*o*-Dichlorobenzene	
U071	541-73-1	*m*-Dichlorobenzene	
U072	106-46-7	*p*-Dichlorobenzene	
U073	91-94-1	3,3′-Dichlorobenzidine	
U074	764-41-0	1,4-Dichloro-2-butene	I,T
U075	75-71-8	Dichlorodifluoromethane	
U078	75-35-4	1,1-Dichloroethylene	
U079	156-60-5	1,2-Dichloroethylene	
U025	111-44-1	Dichloroethyl ether	
U081	120-83-2	2,4-Dichlorophenol	
U082	87-65-0	2,6-Dichlorophenol	
U240	94-75-7	2,4-Dichlorophenoxyacetic acid, salts and esters	
U083	78-87-5	1,2-Dichloropropane	
U084	542-75-6	1,3-Dichloropropene	
U085	1464-53-5	1,2:3,4-Diepoxybutane	I,T
U108	123-91-1	1,4-Diethyleneoxide	
U086	1615-80-1	*N,N*-Diethylhydrazine	
U088	84-66-2	Diethyl phthalate	
U087	3288-58-2	*O-O*-Diethyl-*S*-methyl-dithiophosphate	
U089	56-53-1	Diethylstilbestrol	
U090	94-58-6	Dihydrosafrole	
U091	119-90-4	3,3′-Dimethoxybenzidine	
U092	124-40-3	Dimethylamine	I
U093	60-11-7	Dimethylaminoazobenzene	

TABLE C.3 (*cont.*)

Hazardous wastes from commercial products, intermediates, and residues (P and U lists)

Hazardous waste number	Chemical abstracts number	Hazardous substance	Hazard code*
		U list (*cont.*)	
U094	57-97-6	6,12-Dimethylbenz[*a*]anthracene	
U095	119-93-7	3,3′-Dimethylbenzidine	
U096	80-15-9	α,α-Dimethylbenzylhydroperoxide	R
U097	79-44-7	Dimethylcarbamoyl chloride	
U098	57-14-7	1,1-Dimethylhydrazine	
U099	540-73-8	1,2-Dimethylhydrazine	
U101	105-67-9	2,4-Dimethylphenol	
U102	131-11-3	Dimethyl phthalate	
U103	77-78-1	Dimethyl sulfate	
U105	121-14-2	2,4-Dinitrotoluene	
U106	606-20-2	2,6-Dinitrotoluene	
U107	117-84-0	Di-*n*-octyl phthalate	
U108	123-91-1	1,4-Dioxane	
U109	122-66-7	1,2-Diphenylhydrazine	
U110	142-84-7	Dipropylamine	I
U111	621-64-7	Di-*n*-propylnitrosamine	
U001	75-07-0	Ethanal	I
U174	55-18-5	Ethanamine, *N*-ethyl-*N*-nitroso	
U155	91-80-5	1,2-Ethanediamine, *N,N*-dimethyl-*N*′-2-pyridinyl-*N*′-(2-thienylmethyl)-	
U067	106-93-4	Ethane, 1,2-dibromo-	
U076	75-34-3	Ethane, 1,1-dichloro-	
U077	107-06-2	Ethane, 1,2-dichloro-	
U131	67-72-1	Ethane, hexachloro-	
U024	111-91-1	Ethane, 1,1′-[methylenebis(oxy)]bis[2-chloro]-	
U117	60-29-7	Ethane, 1,1′-oxybis-	I
U025	111-44-4	Ethane, 1,1′-oxybis[2-chloro]-	
U184	76-01-7	Ethane, pentachloro-	
U208	630-20-6	Ethane, 1,1,1,2-tetrachloro-	
U209	79-34-5	Ethane, 1,1,2,2-tetrachloro-	
U218	62-55-5	Ethanethioamide	
U359	79-00-5	Ethane, 1,1,2-trichloro-	
U227	110-80-5	Ethanol, 2-ethoxy-	
U173	1116-54-7	Ethanol, 2,2′-(nitrosoimino)bis-	
U004	98-86-2	Ethanone, 1-phenyl	
U043	75-01-4	Ethene, chloro-	

*Hazard code definitions: C = corrosivity; E = EP toxicity; H = acute hazardous waste; I = ignitability; R = reactivity; T = toxic waste.

TABLE C.3 (*cont.*)
Hazardous wastes from commercial products, intermediates, and residues (P and U lists)

Hazardous waste number	Chemical abstracts number	Hazardous substance	Hazard code*
		U list (*cont.*)	
U042	110-75-8	Ethene, (2-chloroethyoxy)-	
U078	75-35-4	Ethene, 1,1-dichloro-	
U079	156-60-5	Ethene, 1,2-dichloro-	E
U210	127-18-4	Ethene, tetrachloro-	
U228	79-01-6	Ethene, trichloro-	
U112	141-78-6	Ethyl acetate	I
U113	140-88-5	Ethyl acrylate	I
U238	51-79-6	Ethyl carbamate	
U038	510-15-6	Ethyl 4,4′-dichlorobenzilate	
U114	111-54-6	Ethylenebis[dithiocarbamic acid], salts and esters	
U067	106-93-4	Ethylene dibromide	
U077	107-06-2	Ethylene dichloride	
U359	110-80-5	Ethylene glycol monoethyl ether	
U115	75-21-8	Ethylene oxide	I,T
U116	96-45-7	Ethylene thiourea	
U117	60-29-7	Ethyl ether	I
U076	75-34-3	Ethylidene dichloride	
U118	97-63-2	Ethyl methacrylate	
U119	62-50-0	Ethylmethanesulfonate	
U120	206-44-0	Fluoranthene	
U122	50-00-0	Formaldehyde	
U123	64-18-6	Formic acid	C,T
U124	110-00-9	Furan	I
U125	98-01-1	2-Furancarboxyaldehyde	I
U147	108-31-6	2,4-Furandione	
U213	109-99-9	Furan, tetrahydro-	I
U125	98-01-1	Furfural	I
U124	10-00-9	Furfuran	I
U206	18883-66-4	*o*-Glucopyranose, 2-deoxy-2(3-methyl-3-nitrosoureido)-	
U126	765-34-4	Glycidylaldehyde	
U163	70-25-7	Guanidine, *N*-methyl-*N*′-nitro-*N*-nitroso-	
U127	118-74-1	Hexachlorobenzene	
U128	87-68-3	Hexachlorobutadiene	
U129	58-88-9	Hexachlorocyclohexane (gamma isomer)	
U130	77-47-4	Hexachlorocyclopentadiene	
U131	67-72-1	Hexachloroethane	
U132	70-30-4	Hexachlorphene	
U243	1888-71-7	Hexachloropropene	

TABLE C.3 (*cont.*)
Hazardous wastes from commercial products, intermediates, and residues (P and U lists)

Hazardous waste number	Chemical abstracts number	Hazardous substance	Hazard code*
		U list (*cont.*)	
U133	302-01-2	Hydrazine	R,T
U086	1615-80-1	Hydrazine, 1,2-diethyl-	
U098	57-14-7	Hydrazine, 1,1-dimethyl-	
U099	540-73-8	Hydrazine, 1,2-dimethyl-	
U109	122-66-7	Hydrazine, 1,2-diphenyl-	
U134	7664-39-3	Hydrofluoric acid	C,T
U134	7664-39-3	Hydrogen fluoride	C,T
U135	7783-06-4	Hydrogen sulfide	
U096	80-15-9	Hydroperoxide, 1-methyl-1-phenylethyl-	R
U136	75-60-5	Hydroxydimethylarsine oxide	
U116	96-45-7	2-Imidazolidinethione	
U137	193-39-5	Indeno[1,2,3,-*c*,*d*]pyrene	
U139	9004-66-4	Iron dextran	
U190	85-44-9	1,3-Isobenzofurandione	
U140	78-83-1	1sobutyl alcohol	I,T
U141	120-58-f	Isosafrole	
U142	143-50-0	Kepone	
U143	303-34-4	Lasiocarpine	
U144	301-04-2	Lead acetate	
U146	1335-32-6	Lead, bis(acetato-*O*)tetrahydroxytri-	
U145	7446-27-7	Lead phosphate	
U146	1335-32-6	Lead subacetate	
U129	58-89-9	Lindane	
U147	108-31-6	Maleic anhydride	
U148	123-33-1	Maleic hydrazide	
U149	109-77-3	Malononitrile	
U150	148-82-3	Melphalan	
U151	7439-97-6	Mercury	
U152	126-98-7	Methacrylonitrile	I,T
U092	124-40-3	Methanamine, *N*-methyl-	I
U029	74-83-9	Methane, bromo-	
U045	74-87-3	Methane, chloro-	I,T
U046	107-30-2	Methane, chloromethoxy-	
U068	74-95-3	Methane, dibromo-	
U080	75-09-2	Methane, dichloro-	
U075	75-71-8	Methane, dichlorodifluoro-	

*Hazard code definitions: C = corrosivity; E = EP toxicity; H = acute hazardous waste; I = ignitability; R = reactivity; T = toxic waste.

TABLE C.3 *(cont.)*

Hazardous wastes from commercial products, intermediates, and residues (P and U lists)

Hazardous waste number	Chemical abstracts number	Hazardous substance	Hazard code*
		U list *(cont.)*	
U138	74-88-4	Methane, iodo-	
U119	62-50-0	Methanesulfonic acid, ethyl ester	
U211	56-23-5	Methane, tetrachloro-	
U153	74-93-1	Methanethiol	I,T
U225	75-25-2	Methane, tribromo-	
U044	67-66-3	Methane, trichloro-	
U121	75-69-4	Methane, trichlorofluoro-	
U123	64-18-6	Methanoic acid	C,T
U154	67-56-1	Methanol	I
U155	91-80-5	Methapyrilene	
U142	143-50-0	1,3,4-Metheno-2*H*-cyclobuta[*c,d*]pentalen-2-one, 1,1*a*,3,3*a*,4,5,5,5*a*,5*b*,6- decachlorooctahydro-	
U247	72-43-5	Methoxychlor	
U154	67-56-1	Methyl alcohol	I
U029	74-83-9	Methyl bromide	
U186	504-60-9	1-Methylbutadiene	I
U045	74-87-3	Methyl chloride	I,T
U156	79-22-1	Methylchlorocarbonate	I,T
U226	71-55-6	Methylchloroform	
U157	56-49-5	3-Methylcholanthrene	
U158	101-14-4	4,4′-Methylenebis(2-chloroaniline)	
U068	74-95-3	Methylene bromide	
U080	75-09-2	Methylene chloride	
U159	78-93-3	Methyl ethyl ketone (MEK)	I,T
U160	1338-23-4	Methyl ethyl ketone peroxide	R,T
U138	74-88-4	Methyl iodide	
U161	108-10-1	Methyl isobutyl ketone	I
U162	80-62-6	Methyl methacrylate	I,T
U163	70-25-7	*N*-Methyl-*N*′-nitro-*N*-nitrosoguanidine	
U161	108-10-I	4-Methyl-2-pentanone	I
U164	56-04-2	Methylthiouracil	
U010	50-07-7	Mitomycin C	
U059	20830-81-3	5,12-Naphthacenedione, (8-*S*-*cis*)-8-acetyl-10-[(3-amino-2,3,6-tridexoy)-α-L-lyxo-hexo(pyranosyl)oxy]-7,8,9,10-tetrahydro-6,8,11-trihydroxy-1-methoxy-	
U165	91-20-3	Naphthalene	
U047	91-58-7	Naphthalene, 2-chloro-	
U166	130-15-4	1,4-Naphthalenedione	

TABLE C.3 (*cont.*)
Hazardous wastes from commercial products, intermediates, and residues (P and U lists)

Hazardous waste number	Chemical abstracts number	Hazardous substance	Hazard code*
		U list (*cont.*)	
U236	72-57-1	2,7-Naphthalenedisulfonic acid, 3,3′-[(3,3′dimethyl-(1,1′-biphenyl)-4,4′-dyl)]-bis(azo)bis(5-amino-4-4-hydroxy)-, tetrasodium salt	
U166	130-15-4	1,4-Naphthoquinone	
U167	134-32-7	α-Naphthylamine	
U168	91-59-8	β-Naphthylamine	
U026	494-03-1	2-Naphthylamine, *N,N′*-bis(2-chloromethyl)-	
U167	134-32-7	l-Naphthylenamine	
U168	91-59-8	2-Naphthylenamine	
U217	10102-45-1	Nitric acid, thallium (1+) salt	
U169	98-95-3	Nitrobenzene	I,T
U170	100-02-7	*p*-Nitrophenol	
U171	79-46-9	2-Nitropropane	I,T
U172	924-16-3	*N*-Nitrosodi-*n*-butylamine	
U173	1116-54-7	*N*-Nitrosodiethanolamine	
U174	55-18-5	*N*-Nitrosodiethylamine	
U176	759-73-9	*N*-Nitroso-*N*-ethylurea	
U177	684-93-5	*N*-Nitroso-*N*-methylurea	
U178	615-53-2	*N*-Nitroso-*N*-methylurethane	
U179	100-75-4	*N*-Nitrosopiperidine	
U180	930-55-2	*N*-Nitrosopyrrolidine	
U181	99-55-8	5-Nitro-*o*-toluidine	
U193	1120-71-4	1,2-Oxathiolane, 2,2-dioxide	
U058	50-18-0	2*H*-1,3,2-Oxazaphosphorin-2-amine, *N,N*-bis(2-chloroethyl)tetrahydro-2-oxide	
U115	75-21-8	Oxirane	I,T
U126	765-34-4	Oxiranecarboxyaldehyde	
U041	106-89-8	Oxirane, (chloromethyl)-	
U182	123-63-7	Paraldehyde	
U183	608-93-5	Pentachlorobenzene	
U184	76-01-7	Pentachloroethane	
U185	82-68-8	Pentachloronitrobenzene (PCNB)	
U242	87-86-5	Pentachlorophenol	
U186	504-60-9	1,3-Pentadiene	I
U187	62-44-2	Phenacetin	
U188	108-95-2	Phenol	
U048	95-57-8	Phenol, 2-chloro-	

*Hazard code definitions: C = corrosivity; E = EP toxicity; H = acute hazardous waste; I = ignitability; R = reactivity; T = toxic waste.

TABLE C.3 (*cont.*)

Hazardous wastes from commercial products, intermediates, and residues (P and U lists)

Hazardous waste number	Chemical abstracts number	Hazardous substance	Hazard code*
		U list (*cont.*)	
U039	59-50-7	Phenol, 4-chloro-3-methyl-	
U081	120-83-2	Phenol, 2,4-dichloro-	
U082	87-65-0	Phenol, 2,6-dichloro-	
U089	56-53-1	Phenol, 4,4′-(1,2-diethyl-1,2-ethenediyl)bis-	
U101	105-67-9	Phenol, 2,4-dimethyl-	
U052	1319-77-3	Phenol, methyl-	
U132	70-30-4	Phenol, 2,2′-methylenebis[3,4,6-trichloro-]	
U170	100-02-7	Phenol, 4-nitro-	
U242	87-86-5	Phenol, pentachloro-	
U212	58-90-2	Phenol, 2,3,4,6-tetrachloro-	
U230	95-94-4	Phenol, 2,4,5-trichloro-	
U231	88-06-2	Phenol, 2,4,6-trichloro-	
U150	148-82-3	L-Phenylalanine, 4-[bis(2-chloroethyl)amino]-	
U145	7446-27-7	Phosphoric acid, lead salt	
U087	3288-58-2	Phosphorodithioic acid, *O,O*-diethyl-*S*-methyl-, ester	
U189	108-95-2	Phosphorous sulfide	R
U190	85-44-9	Phthalic anhydride	
U191	109-06-8	2-Picoline	
U179	100-75-4	Piperidine, 1-nitroso-	
U192	23950-58-5	Pronamide	
U194	107-10-8	1-Propanamine	I,T
U111	621-64-7	1-Propanamine, *N*-nitroso-*N*-propyl-	
U110	142-84-7	1-Propanamine, *N*-propyl-	I
U066	96-12-8	Propane, 1,2-dibromo-3-chloro-	
U084	542-75-6	1-Propane, 1,3-dichloro-	
U149	109-77-3	Propanedinitrile	
U152	126-98-7	2-Propanenitrile, 2-methyl-	I,T
U171	79-46-9	Propane, 2-nitro-	I,T
U027	39638-32-9	Propane, 2,2′-oxyhis[2-chloro]-	
U193	1120-71-4	1,3-Propane sulfone	
U235	126-72-7	1-Propanol, 2,3-dibromo-, phosphate (3:1)	
U140	78-83-1	1-Propanol, 2-methyl-	I,T
U002	67-64-1	2-Propanone	I
U007	79-06-1	2-Propenamide	
U243	1888-71-7	1-Propene, hexachloro-	
U009	107-13-1	2-Propenenitrile	
U008	79-10-7	2-Propenoic acid	I
U113	140-88-5	2-Propenoic acid, ethyl ester	I

TABLE C.3 (*cont.*)

Hazardous wastes from commercial products, intermediates, and residues (P and U lists)

Hazardous waste number	Chemical abstracts number	Hazardous substance	Hazard code*
		U list (*cont.*)	
U118	97-63-2	2-Propenoic acid, methyl-, ethyl ester	
U162	80-66-2	2-Propenoic acid, 2-methyl-, methyl ester	I,T
U233	93-72-1	Propionic acid, 2-(2,4,5-trichlorophenoxy)-	
U194	107-10-8	*n*-Propylamine	I,T
U083	78-87-5	Propylene dichloride	
U148	123-33-1	3,6-Pyridazinedione, 1,2-dihydro-	
U196	110-86-1	Pyridine	
U191	109-06-8	Pyridine, 2-methyl-	
U237	66-75-1	2,4(1*H*,3*H*)-Pyrimidinedione, 5-[bis(2-chloroethyl)amino]-	
U164	56-04-2	4-(1*H*)-Pyrimidinone, 2,3-dihydro-6-methyl-2-thioxo-	
U180	930-55-22	Pyrrolidine, 1-nitroso	
U200	50-55-5	Reserpine	
U201	108-46-3	Resorcinol	
U202	81-07-2	Saccharin and salts	
U203	94-59-7	Safrole	
U204	7783-00-8	Selenious acid	
U204	7783-00-8	Selenium dioxide	
U205	7446-34-6	Selenium sulfide	R,T
U015	115-02-6	L-Serine, diazoacetate (ester)	
U233	93-72-1	Silvex	
U206	18883-66-4	Streptozotocin	
U103	77-78-1	Sulfuric acid, dimethyl ester	
U189	1314-80-3	Sulfur phosphide	R
U232	93-76-5	2,4,5-T	
U207	95-94-3	1,2,4,5-Tetrachlorobenzene	
U208	630-20-6	1,1,1,2-Tetrachloroethane	
U209	79-34-5	1,1,2,2-Tetrachloroethane	
U210	127-18-4	Tetrachloroethylene	
U212	58-90-2	2,3,4,6-Tetrachlorophenol	
U213	109-99-9	Tetrahydrofuran	I
U214	15843-14-8	Thallium(I) acetate	
U215	6533-73-9	Thallium(I) carbonate	
U216	7791-12-0	Thallium chloride	
U217	10102-45-1	Thallium(I) nitrate	
U218	62-55-5	Thioacetamide	
U153	74-93-1	Thiomethanol	I,T

*Hazard code definitions: C = corrosivity; E = EP toxicity; H = acute hazardous waste; I = ignitability; R = reactivity; T = toxic waste.

TABLE C.3 (*cont.*)

Hazardous wastes from commercial products, intermediates, and residues (P and U lists)

Hazardous waste number	Chemical abstracts number	Hazardous substance	Hazard code*
		U list (*cont.*)	
U244	137-26-8	Thioperoxydicarbonic diamide, tetramethyl	
U219	62-56-6	Thiourea	
U244	137-26-8	Thiuram	
U220	108-88-3	Toluene	
U221	25376-45-8	Toluenediamine	
U223	26471-62-5	Toluene diisocyanate	R,T
U328	95-53-4	*o*-Toluidine	
U353	106-49-0	*p*-Toluidine	
U222	636-21-5	*o*-Toluidine hydrochloride	
U011	61-82-5	1-*H*-1,2,4-Triazol-3-amine	
U226	71-55-6	1,1,1-Trichloroethane	
U227	79-00-5	1,1,2-Trichloroethane	
U228	79-01-6	Trichloroethylene	
U121	75-69-4	Trichloromonofluoromethane	
U230	95-95-4	2,4,5-Trichlorophenol	
U231	88-06-2	2,4,6-Trichlorophenol	
U234	99-35-4	1,3,5-Trinitrobenzene	R,T
U182	123-63-7	1,3,5-Trioxane, 2,4,6-trimethyl-	
U235	126-72-7	Tris(2,3-dibromopropyl) phosphate	
U236	72-57-1	Trypan blue	
U237	66-75-1	Uracil mustard	
U176	759-73-9	Urea, *N*-ethyl-*N*-nitroso	
U177	684-93-5	Urea, *N*-methyl-*N*-nitroso-	
U043	75-01-4	Vinyl chloride	
U248	81-81-2	Warfarin, when present at concentrations of 0.3% or less	
U239	1330-20-7	Xylene	I
U200	50-55-5	Yohimban-16-carboxylic acid, 11,17-dimethoxy-18-[(3,4,5-trimethoxybenzoyl)oxy]-, methyl ester	
U249	1314-84-7	Zinc phosphide, when present at concentrations of 10% or less	

TABLE C.4
Hazardous air pollutants

Compound	Compound
1,1,2,2-Tetrachloroethane	4-Aminobiphenyl
1,1,2-Trichloroethane	4-Nitrobiphenyl
1,1-Dimethylhydrazine	4-Nitrophenol
1,2,4-Trichlorobenzene	4-Picoline*
1,2-Dibromo-3-chloropropane	Acetaldehyde
1,2-Diphenylhydrazine	Acetamide
1,2-Epoxybutane	Acetonitrile
1,2-Propylenimine	Acetophenone
1,3-Butadiene	Acrolein
1,3-Dichloropropene	Acrylamide
1,3-Propane sultone	Acrylic acid
1,4-Dichlorobenzene	Acrylonitrile
1,4-Dioxane	Allyl chloride
2,2,4-Trimethylpentane	Ammonia*
2,3,7,8-Tetrachlorodibenzo-*p*-dioxin	Aniline
2,4,5-Trichlorophenol	Anthracene*
2,4,6-Trichlorophenol	Antimony*
2,4-D, *n*-Butyl ester*	Antimony oxide*
2,4-Dichlorophenoxyacetic acid (2,4-D)*	Antimony potassium tartrate*
2,4-Dinitrophenol	Aroclor 1221*
2,4-Dinitrotoluene	Aroclor 1232*
2,4-Toluene diisocyanate	Aroclor 1242*
2,4-Toluenediamine	Aroclor 1248*
2-Acetylaminofluorene	Aroclor 1254*
2-Chloroacetophenone	Aroclor 1260*
2-Nitropropane	Aroclor 1262*
2-Picoline*	Aroclor 1268*
3,3′-Dichlorobenzidine	Arsenic*
3,3′-Dimethoxybenzidine	Arsenic chloride*
3,3′-Dimethylbenzidine	Arsenic trioxide*
3-Picoline*	Arsine*
4,4′-Methylenebis(2-chloroaniline)	Asbestos
4,4′-Methylenedianiline	Benzene
4,6-Dinitro-*o*-cresol, and salts	Benzidine

* Representatives of classes (these were selected as representative chemicals because the Clean Air Act listed a broad class of chemicals rather than specific chemicals).

TABLE C.4 *(cont.)*
Hazardous air pollutants

Compound	Compound
Benzo[*a*]pyrene*	Chromium trioxide*
Benzotrichloride	Chrysene*
Benzyl chloride	Cobalt*
Beryllium*	Cobalt sulfate heptahydrate*
Beryllium sulfate tetrahydrate*	Cobaltocene*
β-Propiolactone	Coke oven emissions
Biphenyl	Copper(I) cyanide*
bis(2-Ethylhexyl) phthalate	Cresols/cresylic acid (isomers and mixture)
bis(Chloromethyl) ether	Cumene
bis(Cyclopentadienyl) chromium*	DDE
Bromoform	Diazomethane
Cadmium*	Dibenzofuran
Cadmium chloride*	Dibutyl phthalate
Cadmium oxide*	Dichloroethyl ether
Calcium chromate dihydrate*	Dichlorvos
Calcium chromate, anhydrous*	Diethanolamine
Calcium cyanide*	Diethyl sulfate
Caprolactam	Dimethyl phthalate
Captan	Dimethyl sulfate
Carbaryl	Dimethylaminoazobenzene
Carbon disulfide	Dimethylcarbamoyl chloride
Carbon tetrachloride	Epichlorohydrin
Carbonyl sulfide	Ethyl acrylate
Catechol	Ethyl carbamate
Chloramben	Ethyl chloride
Chlordane	Ethylbenzene
Chlorine	Ethylene dibromide
Chloroacetic acid	Ethylene dichloride
Chlorobenzene	Ethylene glycol
Chlorobenzilate	Ethylene glycol diethyl ether*
Chloroform	Ethylene glycol dimethyl ether*
Chloromethyl methyl ether	Ethylene glycol monobutyl ether*
Chloroprene	Ethylene glycol monophenyl ether*
Chromium*	Ethylene oxide
Chromium carbonyl*	Ethylene thiourea

TABLE C.4 (*cont.*)
Hazardous air pollutants

Compound	Compound
Ethyleneimine	Methyl chloride
Ethylidene dichloride	Methyl chloroform
Formaldehyde	Methyl ethyl ketone
Heptachlor	Methyl iodide
Hexachlorobenzene	Methyl isobutyl ketone
Hexachlorobutadiene	Methyl isocyanate
Hexachlorocyclopentadiene	Methyl mercury(II) chloride*
Hexachloroethane	Methyl mercury(II) hydroxide*
Hexamethylene-1,6-diisocyanate	Methyl methacrylate
Hexamethylphosphoramide	Methyl *tert*-butyl ether
Hexane	Methylene chloride
Hydrazine	Methylenediphenyl diisocyanate
Hydrochloric acid	Methylhydrazine
Hydrogen cyanide*	Mineral fibers (fine)
Hydrogen fluoride	*N,N*-Diethylaniline
Hydrogen sulfide*	*N,N*-Dimethylaniline
Hydroquinone	*N,N*-Dimethylformamide
Isophorone	*N*-Nitroso-*N*-methylurea
Lead*	*N*-Nitrosodimethylamine
Lead acetate*	*N*-Nitrosomorpholine
Lead dioxide*	Naphthalene
Lead subacetate*	Nickel*
Lindane (all isomers)	Nickel cyanide*
m-Cresol	Nickel oxide*
m-Xylene	Nickel subsulfide*
Maleic anhydride	Nickel sulfate*
Manganese*	Nickel sulfate hexahydrate*
Manganese(II) sulfate monohydrate*	Nickelocene*
Mercuric chloride*	Nitrobenzene
Mercury*	*o*-Anisidine
Mercury((*O*-carboxyphenyl)thio)	*o*-Cresol
Methanol	*o*-Toluidine
Methoxychlor	*o*-Xylene
Methyl, sodium salt*	*p*-Cresol
Methyl bromide	*p*-Phenylenediamine

* Representatives of classes (these were selected as representative chemicals because the Clean Air Act listed a broad class of chemicals rather than specific chemicals).

TABLE C.4 (*cont.*)

Hazardous air pollutants

Compound	Compound
p-Xylene	Selenium*
Parathion	Selenium disulfide*
Pentachloronitrobenzene	Selenium sulfide*
Pentachlorophenol	Silver cyanide*
Phenanthrene*	Silver potassium cyanide*
Phenol	Sodium arsenite*
Phosgene	Sodium cyanide*
Phosphine	Styrene
Phosphorus	Styrene oxide
Phthalic anhydride	*t*-Butyl chromate*
Polychlorinated biphenyls	Tetrachloroethylene
Potassium cyanide*	Titanium tetrachloride*
Potassium dichromate*	Toluene
Propionaldehyde	Toxaphene
Propoxur	Trichloroethylene
Propylene dichloride	Triethylamine
Propylene oxide	Trifluralin
Pyrene*	Vinyl acetate
Pyridine*	Vinyl bromide
Quinoline	Vinyl chloride
Quinone	Vinylidene chloride
Radium*	Xylenes (isomers and mixture)
Radon	Zinc cyanide*
Selenious acid*	

APPENDIX D

TOXIC RELEASE INVENTORY CHEMICALS AND CHEMICAL CATEGORIES

The requirements of the Toxic Release Inventory for the State of Ohio in 1996 applied to the following chemicals and chemical categories. The chemicals are listed in alphabetical order with their associated Chemical Abstracts Service (CAS) registry number. The chemical categories are listed in alphabetical order and do not have CAS registry numbers.

Chemical name	CAS number	Chemical name	CAS number
Acetaldehyde	75-07-0	Acetonitrile	75-05-8
Acetamide	60-35-5	2-Acetylaminofluorene	53-96-3
Acetone	67-64-1	Acrolein	107-02-8

*C.I. = Colour Index (compiled by the Society of Dyers and Colourists, Bradford, England, and the American Association of Textile Chemists and Colorists, Research Triangle Park, North Carolina.

(*cont.*)

Chemical name	CAS number	Chemical name	CAS number
Acrylamide	79-06-1	Butyl benzyl phthalate	85-68-7
Acrylic acid	79-10-7	1,2-Butylene oxide	106-88-7
Acrylonitrile	107-13-1	Butyraldehyde	123-72-8
Aldrin	309-00-2	C.I.* Acid Green 3	4680-78-8
Allyl alcohol	107-18-6	C.I. Basic Green 4	569-64-2
Allyl chloride	107-05-1	C.I. Basic Red 1	989-38-8
Aluminum (fume or dust)	7429-90-5	C.I. Direct Black 38	1937-37-7
Aluminum oxide (fibrous)	1344-28-1	C.I. Direct Blue 6	2602-46-2
2-Aminoanthraquinone	117-79-3	C.I. Direct Brown 95	16071-86-6
4-Aminoazobenzene	60-09-3	C.I. Disperse Yellow 3	2832-40-8
4-Aminobiphenyl	92-67-1	C.I. Food Red 5	3761-53-3
1-Amino-2-methylanthraquinone	82-28-0	C.I. Food Red 15	81-88-9
Ammonia	7664-41-7	C.I. Solvent Orange 7	3118-97-6
Ammonium nitrate (solution)	6484-52-2	C.I. Solvent Yellow 3	97-56-3
Ammonium sulfate (solution)	7783-20-2	C.I. Solvent Yellow 14	842-07-9
Aniline	62-53-3	C.I. Solvent Yellow 34 (aurimine)	492-80-8
o-Anisidine	90-04-0	C.I. Vat Yellow 4	128-66-5
p-Anisidine	104-94-9	Cadmium	7440-43-9
o-Anisidine hydrochloride	134-29-2	Calcium cyanamide	156-62-7
Anthracene	120-12-7	Captan	133-06-2
Antimony	7440-36-0	Carbaryl	63-25-2
Arsenic	7440-38-2	Carbon disulfide	75-15-0
Asbestos (friable)	1332-21-4	Carbon tetrachloride	56-23-5
Barium	7440-39-3	Carbonyl sulfide	463-58-1
Benzal chloride	98-87-3	Catechol	120-80-9
Benzamide	55-21-0	Chloramben	133-90-4
Benzene	71-43-2	Chlordane	57-74-9
Benzidine	92-87-5	Chlorine	7782-50-5
Benzoic trichloride	98-07-7	Chlorine dioxide	10049-04-4
(benzotrichloride)		Chloroacetic acid	79-11-8
Benzoyl chloride	98-88-4	2-Chloroacetophenone	532-27-4
Benzoyl peroxide	94-36-0	Chlorobenzene	108-90-7
Benzyl chloride	100-44-7	Chlorobenzilate	510-15-6
Beryllium	7440-41-7	Chloroethane	75-00-3
Biphenyl	92-52-4	Chloroform	67-66-3
bis(2-Chloroethyl)ether	111-44-4	Chloromethane	74-87-3
bis(Chloromethyl)ether	542-88-1	Chloromethyl methyl ether	107-30-2
bis(2-Chloro-1-methylethyl) ether	108-60-1	Chloroprene	126-99-8
bis(2-Ethylhexyl)adipate	103-23-1	Chlorothalonil	1897-45-6
Bromochlorodifluoromethane	353-59-3	Chromium	7440-47-3
(halon 1211)		Cobalt	7440-48-4
Bromoform (tribromomethane)	75-25-2	Copper	7440-50-8
Bromomethane (methyl bromide)	74-83-9	Creosote	8001-58-9
Bromotrifluoromethane (halon	75-63-8	*p*-Cresidine	120-71-8
1301)		Cresol (mixed isomers)	1319-77-3
1,3-Butadiene	106-99-0	*m*-Cresol	108-39-4
Butyl acrylate	141-32-2	*o*-Cresol	95-48-7
n-Butyl alcohol	71-36-3	*p*-Cresol	106-44-5
sec-Butyl alcohol	78-92-2	Cumene	98-82-8
tert-Butyl alcohol	75-65-0	Cumene hydroperoxide	80-15-9

(*cont.*)

Chemical name	CAS number
Cupferron	135-20-6
Cyclohexane	110-82-7
2,4-D	94-75-7
Decabromodiphenyl oxide	1163-19-5
Diallate	2303-16-4
2,4-Diaminoanisole	615-05-4
2,4-Diaminoanisole sulfate	9156-41-7
4,4-Diaminodiphenyl ether	101-80-4
Diaminotoluene (mixed isomers)	5376-45-8
2,4-Diaminotoluene	95-80-7
Diazomethane	334-88-3
Dibenzofuran	132-64-9
1,2-Dibromo-3-chloropropane (DBCP)	96-12-8
1,2-Dibromoethane (ethylene dibromide)	106-93-4
Dibromotetrafluoroethane (halon 2402)	124-73-2
Dibutyl phthalate	84-74-2
Dichlorobenzene (mixed isomers)	25321-22-6
1,2-Dichlorobenzene	95-50-1
1,3-Dichlorobenzene	541-73-1
1,4-Dichlorobenzene	106-46-7
3,3-Dichlorobenzidine	91-94-1
Dichlorobromomethane	75-27-4
Dichlorofluoromethane (CFC-12)	75-71-8
1,2-Dichloroethane (ethylene dichloride)	107-06-2
1,2-Dichloroethylene	540-59-0
Dichloromethane (methylene chloride)	75-09-2
2,4-Dichlorophenol	120-83-2
1,2-Dichloropropane	78-87-5
2,3-Dichloropropene	78-88-6
1,3-Dichloropropylene	542-75-6
Dichlorotetraethane (CFC-144)	76-14-2
Dichlorvos	62-73-7
Dicofol	115-32-2
Diepoxybutane	1464-53-5
Diethanolamine	111-42-2
Di-(2-ethylhexyl) phthalate (DEHP)	117-81-7
Diethyl phthalate	84-66-2
Diethyl sulfate	64-67-5
3,3-Dimethoxybenzidine	119-90-4
4-Dimethylaminoazobenzene	60-11-7
3,3-Dimethylbenzidine (*o*-tolidine)	119-93-7
Dimethylcarbamyl chloride	79-44-7
1,1-Dimethyl hydrazine	57-14-7
2,4-Dimethylphenol	105-67-9
Dimethyl phthalate	131-11-3
Dimethyl sulfate	77-78-1
m-Dinitrobenzene	99-65-0
o-Dinitrobenzene	528-29-0
p-Dinitrobenzene	100-25-4
4,6-Dinitro-*o*-cresol	534-52-1
2,4-Dinitrophenol	51-28-5
2,4-Dinitrotoluene	121-14-2
2,6-Dinitrotoluene	606-20-2
Dinitrotoluene (mixed isomers)	25321-14-6
n-Dioctyl phthalate	117-84-0
1,4-Dioxane	123-91-1
1,2-Diphenylhydrazine (hydrazobenzene)	122-66-7
Epichlorohydrin	106-89-8
2-Ethoxyethanol	110-80-5
Ethyl acrylate	140-88-5
Ethylbenzene	100-41-4
Ethyl chloroformate	541-41-3
Ethylene	74-85-1
Ethylene glycol	107-21-1
Ethyleneimine (aziridine)	151-56-4
Ethylene oxide	75-21-8
Ethylene thiourea	96-45-7
Fluometuron	2164-17-2
Formaldehyde	50-00-0
Freon 113	76-13-1
Heptachlor	76-44-8
Hexachlorobenzene	118-74-1
Hexachloro-1,3-butadiene	87-68-3
Hexachlorocyclopentadiene	77-47-4
Hexachloroethane	67-72-1
Hexachloronaphthalene	1335-87-1
Hexamethylphosphoramide	680-31-9
Hydrazine	302-01-2
Hydrazine sulfate	10034-93-2
Hydrochloric acid	7647-01-0
Hydrogen cyanide	74-90-8
Hydrogen fluoride	7664-39-3
Hydroquinone	123-31-9
Isobutyraldehyde	78-84-2
Isopropyl alcohol (only persons who manufacture by the strong acid process are subject, no supplier notification)	67-63-0
4,4-Isopropylidenediphenol	80-05-7
Isosafrole	120-58-1
Lead	7439-92-1
Lindane	58-89-9
Maleic anhydride	108-31-6
Maneb	12427-38-2

(*cont.*)

Chemical name	CAS number
Manganese	7439-96-5
Mercury	7439-97-6
Methanol	67-56-1
Methoxychlor	72-43-5
2-Methoxyethanol	109-86-4
Methyl acrylate	96-33-3
Methyl *tert*-butyl ether	1634-04-4
4,4-Methylenebis(2-chloroaniline) (MBOCA)	101-14-4
4,4-Methylenebis(*N*,*N*-dimethyl) benzenamine	101-61-1
Methylenebis(phenylisocyanate) (MBI)	101-68-8
Methylene bromide	74-95-3
4,4-Methylenedianiline	101-77-9
Methyl ethyl ketone	78-93-3
Methyl hydrazine	60-34-4
Methyl iodide	74-88-4
Methyl isobutyl ketone	108-10-1
Methyl isocyanate	624-83-9
Methyl methacrylate	80-62-6
Michler's ketone	90-94-8
Molybdenum trioxide	1313-27-5
Monochloropentafluoroethane (CFC-115)	76-15-3
Mustard gas	505-60-2
Naphthalene	91-20-3
α-Naphthylamine	134-32-7
β-Naphthylamine	91-59-8
Nickel	7440-02-0
Nitric acid	7697-37-2
Nitrilotriacetic acid	139-13-9
5-Nitro-*o*-anisidine	99-59-2
Nitrobenzene	98-95-3
4-Nitrobiphenyl	92-93-3
Nitrofen	1836-75-5
Nitrogen mustard	51-75-2
Nitroglycerin	55-63-0
2-Nitrophenol	88-75-5
4-Nitrophenol	100-02-7
2-Nitropropane	79-46-9
p-Nitrosodiphenylamine	156-10-5
N,*N*-dimethylaniline	121-69-7
N-Nitrosodi-*n*-butylamine	924-16-3
N-Nitrosodiethylamine	55-18-5
N-Nitrosodimethylamine	62-75-9
N-Nitrosodiphenylamine	86-30-6
N-Nitrosodi-*n*-propylamine	621-64-7
N-Nitrosomethylvinylamine	4549-40-0
N-Nitrosomorpholine	59-89-2
N-Nitroso-*N*-ethylurea	759-73-9
N-Nitroso-*N*-methylurea	684-93-5
N-Nitrosonornicotine	6543-55-8
N-Nitrosopiperidine	100-75-4
Octachloronaphthalene	2234-13-1
Osmium tetroxide	0816-12-0
Parathion	56-38-2
Pentachlorophenol (PCP)	87-86-5
Peracetic acid	79-21-0
Phenol	108-95-2
p-Phenylenediamine	106-50-3
2-Phenylphenol	90-43-7
Phosgene	75-44-5
Phosphoric acid	7664-38-2
Phosphorus (yellow or white)	7723-14-0
Phthalic anhydride	85-44-9
Picric acid	88-89-1
Polychlorinated biphenyls (PCBs)	1336-36-3
Propane sultone	1120-71-4
β-Propiolactone	57-57-8
Propionaldehyde	123-38-6
Propoxur	114-26-1
Propylene (propene)	115-07-1
Propyleneimine	75-55-8
Propylene oxide	75-56-9
Pyridine	110-86-1
Quinoline	91-22-5
Quinone	106-51-4
Quintozene	82-68-8
Saccharin (only persons who manufacture are subject, no supplier notification)	81-07-2
Safrole	94-59-7
Selenium	7782-49-2
Silver	7440-22-4
Styrene	100-42-5
Styrene oxide	96-09-3
Sulfuric acid	7664-93-9
1,1,2,2-Tetrachloroethane (perchloroethylene)	79-34-5
Tetrachloroethylene	127-18-4
Tetrachlorvinphos	961-11-5
Thallium	7440-28-0
Thioacetamide	62-55-5
4,4-Thiodianiline	139-65-1
Thiourea	62-56-6
Thorium dioxide	1314-20-1
Titanium tetrachloride	7550-45-0
Toluene	108-88-3
Toluene-2,4-diisocyanate	584-84-9

(*cont.*)

Chemical name	CAS number	Chemical name	CAS number
Toluene-2,6-diisocyanate	91-08-7	Trifluralin	1582-09-8
Toluenediisocyanate (mixed isomers)	26471-62-5	1,2,4-Trimethylbenzene	95-63-6
		Tris(2,3-dibromopropyl) phosphate	126-72-7
o-Toluidine	95-53-4	Urethane (ethyl carbamate)	51-79-6
o-Toluidine hydrochloride	636-21-5	Vanadium (fume or dust)	7440-62-2
Toxaphene	8001-35-2	Vinyl acetate	108-05-4
Triaziquone	68-76-8	Vinyl bromide	593-60-2
Trichlorfon	52-68-6	Vinyl chloride	75-01-4
1,2,4-Trichlorobenzene	120-82-1	Vinylidene chloride	75-35-4
1,1,1-Trichloroethane (methyl chloroform)	71-55-6	Xylene (mixed isomers)	1330-20-7
		m-Xylene	108-38-3
1,1,2-Trichloroethane	79-00-5	*o*-Xylene	95-47-6
Trichloroethylene	79-01-6	*p*-Xylene	106-42-3
Trichlorofluoromethane (CFC-11)	75-69-4	2,6-Xylidine	87-62-7
2,4,5-Trichlorophenol	95-95-4	Zinc (fume or dust)	7440-66-6
2,4,6-Trichlorophenol	88-06-2	Zineb	12122-67-7

Chemical categories in alphabetical order:

Antimony compounds: includes any unique chemical substance that contains antimony as part of that chemical's infrastructure.

Arsenic compounds: includes any unique chemical substance that contains arsenic as part of that chemical's infrastructure.

Barium compounds: includes any unique chemical substance that contains barium as part of that chemical's infrastructure.

Beryllium compounds: includes any unique chemical substance that contains beryllium as part of that chemical's infrastructure.

Cadmium compounds: includes any unique chemical substance that contains cadmium as part of that chemical's infrastructure.

Chlorophenols:

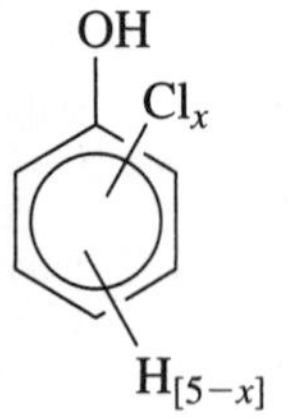

OH

Cl_x

$H_{[5-x]}$

where $x = 1$ to 5.

Chromium compounds: includes any unique chemical substance that contains chromium as part of that chemical's infrastructure.

Cobalt compounds: includes any unique chemical substance that contains cobalt as part of that chemical's infrastructure.

Copper compounds: includes any unique chemical substance that contains copper as part of that chemical's infrastructure.

Cyanide compounds: XCN where X = H or any other group where a formal dissociation can be made, for example, KCN or Ca(CN).

Glycol ethers: includes mono- and di-ethers of ethylene glycol, diethyllene glycol, and triethylene glycol:

$$R-(OCH_2CH_2)_n-OR'$$

where $n =$ 1, 2, or 3
$R =$ alkyl or aryl groups
$R' =$ R, H, or other groups which, when removed, yield glycol ethers with the structure

$$R-(OCH_2CH_2)_n-OH$$

Polymers are excluded from this category.

Lead compounds: includes any unique chemical substance that contains lead as part of that chemical's infrastructure.

Manganese compounds: includes any unique chemical substance that contains manganese as part of that chemical's infrastructure.

Mercury compounds: includes any unique chemical substance that contains mercury as part of that chemical's infrastructure.

Nickel compounds: includes any unique chemical substance that contains nickel as part of that chemical's infrastructure.

Polybrominated biphenyls (PBBs):

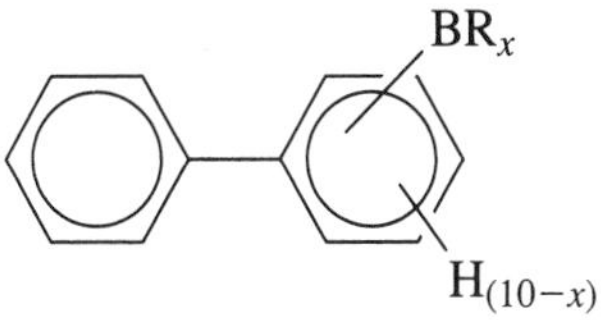

BR_x

$H_{(10-x)}$

where $x =$ 1 to 10.

Selenium compounds: includes any unique chemical substance that contains selenium as part of that chemical's infrastructure.

Silver compounds: includes any unique chemical substance that contains silver as part of that chemical's infrastructure.

Thallium compounds: includes any unique chemical substance that contains thallium as part of that chemical's infrastructure.

Zinc compounds: includes any unique chemical substance that contains zinc as part of that chemical's infrastructure.

APPENDIX E

PRESENT WORTH OF A $1.00 INVESTMENT

Year	Annual interest rate, compounded yearly			
	5%	7.5%	10%	12.5%
0	1	1	1	1
1	0.95238	0.93023	0.90909	0.88889
2	0.90703	0.86533	0.82654	0.79012
3	0.86384	0.80496	0.75131	0.70233
4	0.82270	0.74880	0.68301	0.62430
5	0.78353	0.69656	0.62092	0.55493
6	0.74622	0.64796	0.56447	0.49327
7	0.71068	0.60275	0.51316	0.43846
8	0.67684	0.56070	0.46651	0.38974
9	0.64461	0.52158	0.42410	0.34644
10	0.61391	0.48519	0.38554	0.30795
11	0.58468	0.45134	0.35049	0.27373
12	0.55684	0.41985	0.31863	0.24332
13	0.53032	0.39056	0.28966	0.21628
14	0.50507	0.36331	0.26333	0.19225
15	0.48102	0.33797	0.23939	0.17089
16	0.45811	0.31439	0.21763	0.15190
17	0.43630	0.29245	0.19784	0.13502
18	0.41552	0.27205	0.17986	0.12002
19	0.39573	0.25307	0.16351	0.10668
20	0.37689	0.23541	0.14864	0.09483
21	0.35894	0.21899	0.13513	0.08429
22	0.34185	0.20371	0.12285	0.07493
23	0.32557	0.18950	0.11168	0.06660
24	0.31007	0.17628	0.10153	0.05920
25	0.29530	0.16398	0.09230	0.05262
26	0.28124	0.15254	0.08391	0.04678
27	0.26785	0.14190	0.07628	0.04158
28	0.25509	0.13200	0.06934	0.03696
29	0.24295	0.12279	0.06304	0.03285
30	0.23138	0.11422	0.05731	0.02920

APPENDIX F

PHYSICAL CONSTANTS

Physical constant	Value
Acceleration due to gravity, g	9.80665 m/s^2 (value varies with latitude) 32,174 ft/s^2 (value varies with latitude)
Avogadro's number	6.0221×10^{-23} molecules/g·mol
Faraday constant, F	96,487 C/g equiv
Latent heat of fusion of water (°C and 1 atm)	334 J/g 144 Btu/lb
Latent heat of vaporization of water (100°C and 1 atm)	2,258 J/g 971 Btu/lb
1 angstrom, Å	10^{-10} m
1 bar	10^5 N/m^2 14,504 lb$_f$/in^2
1 gram-mole	22.4 L for ideal gas at STP of 0°C and 1 atm
1 pound-mole	359 ft^3 for ideal gas at STP of 0°C and 1 atm
Specific weight of water (20°C)	1.0 kg/L 62.4 lb/ft^3
Standard conditions	
Standard room temperature	20°C
Standard temperature and pressure, STP	20°C and 1 atm

(*cont.*)

Physical constant	**Value**
Temperature (absolute)	
Kelvin, K	273.0 + °C
Rankine, °R	459.6 + °F
Universal gas constant	8.3145 J/g mol·K
	1.987 cal/g mol·K
	8.206×10^{-2} L-atm·K/g mol
	4.968×10^{4} $lb_m \cdot ft^2$/lb mol·°R
	4.972×10^{4} ft-lb_f/slug·°R
Volume occupied by an ideal gas (0°C and 1 atm)	22.4146 L/g·mol
	359 ft^3/lb·mol

APPENDIX G

PHYSICAL PROPERTIES OF WATER AT 1 ATMOSPHERE

Temperature, °C	Density, ρ, kg/m^3	Specific weight, γ, kN/m^3	Dynamic viscosity, μ, mPa·s	Kinematic viscosity, ν, μm^2/s
0	999.842	9.805	1.787	1.787
3.98	1,000.000	9.807	1.567	1.567
5	999.967	9.807	1.519	1.519
10	999.703	9.804	1.307	1.307
12	999.500	9.802	1.235	1.236
15	999.103	9.798	1.139	1.140
17	998.778	9.795	1.081	1.082
18	998.599	9.793	1.053	1.054
19	998.408	9.791	1.027	1.029
20	998.207	9.789	1.002	1.004
21	997.996	9.787	0.998	1.000
22	997.774	9.785	0.955	0.957
23	997.542	9.783	0.932	0.934
24	997.300	9.781	0.911	0.913
25	997.048	9.778	0.890	0.893
26	996.787	9.775	0.870	0.873
27	996.516	9.773	0.851	0.854
28	996.236	9.770	0.833	0.836
29	995.948	9.767	0.815	0.818
30	995.650	9.764	0.798	0.801
35	994.035	9.749	0.719	0.723
40	992.219	9.731	0.653	0.658
45	990.216	9.711	0.596	0.602
50	988.039	9.690	0.547	0.554
60	983.202	9.642	0.466	0.474
70	977.773	9.589	0.404	0.413
80	971.801	9.530	0.355	0.365
90	965.323	9.467	0.315	0.326
100	958.366	9.399	0.282	0.294

Key: Pa·s = (mPa·s) × 10^{-3}
m^2/s = (μm^2/s) × 10^{-6}

APPENDIX H

PROPERTIES OF AIR

TABLE H.1
Properties of air at standard conditions*

Property	Value
Molecular weight, M	28.97
Gas constant, R	287 J/kg·K
Specific heat at constant pressure, c_p	1005 J/kg·K
Specific heat at constant volume, c_v	718 J/kg·K
Density, ρ	1.185 kg/m^3
Dynamic viscosity, μ	1.8515×10^{-5} Pa·s
Kinematic viscosity, ν	1.5624×10^{-5} m^2/s
Thermal conductivity, k	0.0257 W/m·K
Ratio of specific heats (c_p/c_v), k	1.3997
Prandtl number, Pr	0.720

* Measured at 101.325 kPa pressure and 298 K temperature.

Source: Davis and Cornwell, 1998.

APPENDIX I

USEFUL CONVERSION FACTORS

Multiply	Unit	By	To obtain	Unit
		Acceleration		
Feet/second squared	ft/s^2	0.3048	Meters/second squared	m/s^2
Meters/second squared	m/s^2	3.2808	Feet/second squared	ft/s^2
		Area		
Acre	acre	43,560	Square feet	ft^2
Acre	acre	0.4047	Hectare	ha
Acre	acre	4.0469×10^{-3}	Square kilometers	km^2
Hectare	ha	10,000	Square meters	m^2
Hectare	ha	2.4711	Acres	acre
Square centimeter	cm^2	0.1550	Square inches	in^2
Square feet	ft^2	9.2903×10^{-2}	Square meters	m^2
Square inch	in^2	6.4516	Square centimeters	cm^2
Square kilometer	km^2	0.3861	Square miles	mi^2
Square kilometer	km^2	247.105	Acres	acre
Square meter	m^2	10.764	Square feet	ft^2
Square meter	m^2	1.1960	Square yards	yd^2

(*cont.*)

Multiply	Unit	By	To obtain	Unit
		Concentration/density		
Grams/cubic centimeter	g/cm^3	1.9422	Slugs/square foot	$slugs/ft^2$
Grams/liter	g/L	8.345×10^{-3}	Pounds/gallon	lb/gal
Kilograms/cubic meter	kg/m^3	0.06243	Pounds/cubic foot	lb/ft^3
Pounds/cubic foot	lb/ft^3	16.0185	Kilograms/cubic meter	kg/m^3
Pounds/million gallons	lb/Mgal	0.11983	Milligrams/liter	mg/L
Slugs/cubic foot	$slugs/ft^3$	0.51541	Grams/cubic centimeter	g/cm^3
		Diffusion coefficient		
Square centimeters/second	cm^2/s	0.155	Square inches/second	in^2/s
Square inches/second	in^2/s	6.452	Square centimeters/second	cm^2/s
		Energy		
British thermal unit	Btu	1.0551	Kilojoule	kJ
Foot-pounds force	$ft\text{-}lb_f$	1.3558	Joules	J
Horsepower-hour	hp-h	2.6845	Megajoules	MJ
Joule	J	0.7376	Foot-pounds force	$ft\text{-}lb_f$
Joule	J	1.000	Watt-second	W-s
Joule	J	0.2388	Calorie	cal
Kilojoule	kJ	0.9478	British thermal units	Btu
Kilojoule	kJ	2.778×10^{-4}	Kilowatt-hour	kWh or kW-h
Megajoule	MJ	0.3725	Horsepower-hour	hp-h
Kilowatt-hour	kWh	3.600×10^{6}	Joule	J
		Force		
Kilogram-meter/second squared	$kg\text{-}m/s^2$	1.000	Newton	N
Newton	N	1.000	Kilogram-meter/second squared	$kg\text{-}m/s^2$
Newton	N	0.2248	Pound force	lb_f
Pound force	lb_f	4.4482	Newton	N
		Flow rate		
Cubic feet/second	ft^3/s	2.832×10^{-2}	Cubic meters/second	m^3/s
Cubic meters/day	m^3/day	264.172	Gallons/day	gal/day
Cubic meters/second	m^3/s	35.315	Cubic feet/second	ft^3/s
Cubic meters/second	m^3/s	22.825	Million gallons/day	Mgal/day
Cubic meters/second	m^3/s	15,850.3	Gallons/minute	gal/min
Gallons/day	gal/day	4.381×10^{-5}	Liters/second	L/s
Gallons/day	gal/day	3.785×10^{-3}	Cubic meters/day	m^3/day
Gallons/minute	gal/min	6.309×10^{-5}	Cubic meters/second	m^3/s

(*cont.*)

Multiply	Unit	By	To obtain	Unit
		Flow rate (*cont.*)		
Gallons/minute	gal/min	6.309×10^{-2}	Liters/second	L/s
Liters/second	L/s	22,824.5	Gallons/day	gal/day
Million gallons/day	Mgal/day	43.813	Liters/second	L/s
Million gallons/day	Mgal/day	3.785×10^{-3}	Cubic meters/day	m^3/day
Million gallons/day	Mgal/day	4.381×10^{-2}	Cubic meters/second	m^3/s
		Length		
Centimeter	cm	0.3937	Inch	in
Feet	ft	0.3048	Meter	m
Inch	in	2.54	Centimeter	cm
Inch	in	0.0254	Meter	m
Inch	in	25.4	Millimeter	mm
Kilometer	km	0.6214	Mile	mi
Meter	m	39.37	Inch	in
Meter	m	3.281	Feet	ft
Mile	mi	5280	Feet	ft
Mile	mi	1.6093	Kilometer	km
Millimeter	mm	0.0394	Inch	in
		Mass		
Gram	g	0.0353	Ounce	oz
Gram	g	0.0022	Pound	lb
Kilogram	kg	2.2046	Pound	lb
Kilogram	kg	0.0685	Slug	slug
Ounce	oz	28.344	Gram	g
Pound	lb	4.536×10^{2}	Gram	g
Pound	lb	0.4536	Kilogram	kg
Ton (short: 2000 lb)	ton	0.9072	Metric tonne (10^3 kg)	tonne
Ton (long: 2240 lb)	ton	1.0160	Metric tonne (10^3 kg)	tonne
		Power		
British thermal units/second	Btu/s	1.0551	Kilowatts	kW
Foot-pound force/second	ft-lb_f/s	1.3558	Watts	W
Horsepower	hp	0.7457	Kilowatts	kW
Kilowatts	kW	0.9478	British thermal units/second	Btu/s
Kilowatts	kW	1.3410	Horsepower	hp
Watts	W	1.000	Joules/second	J/s
Watts	W	0.7376	Foot-pounds/second	ft-lb_f/s
Watts	W	0.2939	Calories/second	cal/s

(*cont.*)

Multiply	Unit	By	To obtain	Unit
		Pressure		
Atmosphere (standard)	atm	101.33	Kilopascals	kPa
Inches mercury	in Hg	3.377×10^3	Kilopascals	kPa
Inches water	in H_2O	2488	Pascals	Pa
Kilopascal	kPa	1.000	Kilonewtons/square meter	kN/m^2
Kilopascal	kPa	0.1450	Pounds/square inch	lb_f/in^2
Kilopascal	kPa	0.0099	Atmosphere (standard)	atm
Pounds/square foot	lb_f/ft^2	47.880	Pascals	Pa
Pounds/square inch	lb_f/in^2	6.695	Kilopascals	kPa
Pascal	Pa	1.000	Newtons/square meter	N/m^2
Pascal	Pa	1.450×10^{-4}	Pounds/square inch	lb_f/in^2
Pascal	Pa	2.089×10^{-2}	Pounds/square foot	lb_f/ft^2
Pascal	Pa	2.961×10^{-4}	Inches mercury	in Hg
Pascal	Pa	4.019×10^{-3}	Inches water	in H_2O
		Temperature		
Degrees Fahrenheit	°F	0.555(°F + 32)	Degrees Centigrade	°C
Degrees Fahrenheit	°F	0.555(°F + 459.67)	Kelvins	K
Degrees Centigrade	°C	1.8(°C) + 32	Degrees Fahrenheit	°F
Kelvins	K	1.8(°C) − 459.67	Degrees Fahrenheit	°F
		Velocity		
Feet/second	ft/s	0.3048	Meters/second	m/s
Meters/second	m/s	2.2369	Miles/hour	mi/h
Meters/second	m/s	3.2808	Feet/second	ft/s
Miles/hour	mi/h	0.4704	Meters/second	m/s
		Viscosity, absolute		
Kilograms/meter-second	kg/m-s	0.0209	Pounds-second/square foot	$lb\text{-}s/ft^2$
Poise	poise	1.000	Grams/centimeter-second	g/cm-s
Pound-seconds/square foot	$lb\text{-}s/ft^2$	47.88	Kilograms/meter-second	kg/m-s
Pounds/foot-second	lb/ft-s	14.88	Poise	poise
		Viscosity, kinematic		
Square feet/second	ft^2/s	929	Stoke	stoke
Square feet/second	ft^2/s	0.0929	Square meters/second	m^2/s
Square meters/second	m^2/s	6.4516	Square centimeters/second	cm^2/s
Square meters/second	m^2/s	10.764	Square feet/second	ft^2/s
Stoke	stoke	1.000	Square centimeters/second	cm^2/s
Stoke	stoke	1.076×10^{-3}	Square feet/second	ft^2/s

(*cont.*)

Multiply	Unit	By	To obtain	Unit
		Volume		
Cubic centimeter	cm^3	0.0610	Cubic inches	in^3
Cubic feet	ft^3	28.3168	Liters	L
Cubic feet	ft^3	2.8317×10^{-2}	Cubic meters	m^3
Cubic inches	in^3	16.387	Cubic centimeter	cm^3
Gallons	gal	3.785×10^{-3}	Cubic meters	m^3
Gallons	gal	3.785	Liters	L
Liters	L	0.2642	Gallons	gal
Liters	L	0.0353	Cubic feet	ft^3
Liters	L	33.8150	Ounces (U.S. fluid)	oz
Cubic meters	m^3	35.3147	Cubic feet	ft^3
Cubic meters	m^3	264.172	Gallons	gal
Ounces (U.S. fluid)	oz	2.9573×10^{-2}	Liters	L

SUBJECT INDEX

Absorption
column design, 210
defined, 55
purpose of, 209
transferring in, 210
types of, 209
vapor-liquid contacting schemes, 209
Acid. *See also* Acid-base ionization
classification definition, 50
ionization constants for, 52
Acid rain
acidity of, 88
amount of, 156
aquatic systems effects, 89–90
controlling, 90
countries with, 89
defined, 87–88
ecosystem effects, 89
emission reduction requirements, 156
geographic areas with, 87
natural sources of, 88
negative effects, 89
petroleum-based fuels, 88
phase reductions and, 156
versus unpolluted rain, 89
Acid-base ionization
classification definitions, 50
diprotic acids, 50–51
forms of, 50
hydrogen ion concentration, 51
ionization constants for, 51–53
monoprotic acid, 50
ph solution, 51
Acrylonitrile manufacturing, 239–242
Activated carbon system, 490–491
Activated carbon treatments
adsorption process, 473
adsorption rate, 473
contaminants breakthrough, 474–475
examples, 474
organic contaminants removal, 473
wastewater use of, 474
Activated sludge systems
biomass population and, 490
process variations, 490–491
purpose of, 489
schematic of, 489
waste sludge quantity, 490
Adsorption
activated carbon treatments and, 473–475
defined, 55, 213
industry uses of, 213–214
types of, 473
Adsorption effects
adsorbent properties, 57
categories of, 55
defined, 55
equilibrium process of, 55
isotherm models, 58–64
between phase partitioning, 56
sorption equilibrium concentrations, 57
surface sorption and, 56
types of, 56
volatilization and, 56
Adsorption systems
gas removal, 503
uses of, 504
working of, 503–504
Advective processes
consequences of, 48
defined, 48
source concentration preventing, 48
Agenda 21. *See* Earth Summit
Agricultural wastes, locations for, 107
Air pollutants
CAA and, 83
carbon dioxide, 85
carbon monoxide, 83
classifications of, 82–83
defined, 82
HAPs, 84–85
hydrocarbons, 83
lead, 84
nitrogen oxides, 84
particulate, 83–84
photochemical oxidants, 84
sources of, 84
sulfur dioxide, 83
Air pollution
acid rain, 87–90
atmosphere, 80–85
damaging effects of, 79–80
examples of, 79
global warming, 91–99
ozone depletion, 99–101
problem complexity, 79
smog formation, 85–87
Air pollution control
gas removal, 503–504
particulate control, 499–503
pollution eliminating as, 499
Airsheds
air quality, 82
problems from, 82
purpose of, 82
Aldehydes. *See also* Ketones
characteristics of, 30

defined, 30
examples, 31
naming basis, 30
uses of, 30
Aliphatic compounds
alkames, 25–32
defined, 24
as organic compounds, 24
Alkaline chlorination, 470
Alkanes
characteristics of, 25
chlorofluorocarbons, 27
isomer compounds, 25
mercaptan compounds, 27
normal compounds, 26
paraffin compounds, 25
types of, 25
Alkenes
characteristics of, 28
formula for, 28
naming basis, 28
Alkynes, defined, 28
Allied Chemical Corporation, 120
American Fiber Manufacturers Association, 254
American National Standards Institute (ANSI), 177
American Paper Institute, 400
American Petroleum Institute, 464, 514
Amoco, 3
Ancillary equipment, pollution prevention of, 215
Aqueous reaction systems, change impact of, 375
Arkwright, Richard, 8
Aromatic compounds
benzene hydrogen substitution systems, 34–35
characteristics, 32–33
defined, 24–25
naming basis for, 33–34
simplified diagramming, 33
Arsenic
characteristics, 41
classification of, 41
drinking water and, 41
toxic nature of, 40
uses of, 40–41
Ashai Chemical Industry Company, 364
Atmosphere
air pollutants, 82–85
airsheds, 82
characteristics of, 80
composition of, 82
global air movements, 81
heat transferring, 81
layers of, 80–81
Attached growth systems, 488
Automobiles
BMW recycling, 391
CAA focus, 155
gasoline formulas, 155–156
pollutants from, 155
production mandates for, 156
recycling of, 390–392
recycling targets for, 392

Baghouse filter
efficiency of, 502–503
filter efficiency, 500
particle removal, 500
purpose of, 500
workings of, 500–501
Base. *See also* Acid-base ionization
classification definition, 50
ionization constants, 52
BASF Company, 437
Batch distillation, defined, 200
Batch reactors
characteristics of, 187
examples, 189–190
first-order reaction calculating, 188
pollution minimizing, 188–189
reaction rate calculating, 188
retention time, 187
second-order reaction calculating, 188
uses of, 187
waste neutralization, 458
zero-order reaction calculating, 188
Benefit-cost ratio, calculation, 316
Best management practices (BMP)
categories of, 536–537
POTW technical assistance programs and, 544
prevention focus, 536
success from, 536
Biocatalysis
catalytic antibodies research, 373–374
characteristics of, 371
chiral compounds, 372
chirality influences, 372–373
enzymes use in, 371
optically pure compounds, 373
science complexity, 374
sterospecific catalysts, 371–372
strict specificity in, 371
Biodiversity
environmental health indicator as, 574–575
natural habitat destruction, 574
Biofilm system
fluidized bed reactor, 494–495
functions of, 491–492
oxygen concentration, 492
rotating biological contactor, 494
tickling filter method, 492–494
trickling filters as, 491
Biological transformation
biodegradation by bacteria, 70
biodegradation preferences, 72
cometabolism, 71
complex organic compounds conversions, 71
enzyme use, 70
literature on, 72
oxidation reactions, 71
recalcitrant compounds biodegradation, 71
research on, 72
synthetic organics, 70–71
Biological waste treatment
activated sludge systems, 489–491
aerobic biodegradation processes, 486
aerobic oxidation kinetics, 486–487
as biodegradation of organics, 485
biofilm systems, 491–495
destructive technologies for, 485
energy sources, 486
growth rate, 487–488
land applications, 496–498
microorganism growth nutrients, 485–486
organic materials and, 485
stabilization ponds, 495–496
system types, 488–489
Biomass
availability of, 365
biological feedstock advantages, 365
chemicals available from, 366
conversion of, 136
costs in, 136
current uses of, 367
examples, 365
versus petroleum-based feedstocks, 365
pollution from, 136
types of, 365
uses of, 367
wood burning as, 136
Biot, Jean-Baptist, 371
BMW, 391
Boiling water reactor
steam turbines, 133
water use in, 133
Boulton, Matthew, 7
BP America, 3
Brownfields Initiative
defined, 170
EPA program for, 170
goal of, 170
hazardous waste land selling, 169
incentive types, 170
property owner responsibilities, 169–170
Brundtland Commission
purpose of, 12
Rio Declaration and, 581
sustainability definition, 12
sustainable development definition, 584
Brundtland, Gro Harlem, 580
Buffalo Sewer Authority
purpose of, 546
uses of, 546, 546–547
voluntary reduction program, 546

C&H Sugar Company, 233, 408–409

Cadmium
characteristics, 41
drinking water standards, 41
plant nutrients, 41
uses of, 41
zinc relationships, 41
California Energy Commission, environmental cost model, 322
Canada, eco-labeling in, 286
Carbon dioxide
controlling, 95–96
global warming and, 85
greenhouse gases, 93–95
turnover rates, 95
Carbon monoxide
characteristics of, 83
effects of, 83
sources of, 83
Carcinogens
arsenic as, 40
benzene and, 363
HAPs control of, 153
molecule changes, 382
U.S., releases, 349
Carson, Rachel, 115, 572
Cartwright, Edward, 8
Catalysis
analgesic ibuprofen production, 370–371
industrial classes of, 369
industry applications, 369
methanol production, 371
objectives of, 369
pollution prevention from, 368
product development states, 370
role of, 368–369
transition metal use, 370
treatment technologies, 369
types of, 369–370
Catalytic converters, 86
Center for Clean Industry and Treatment Technologies, 219
Center for Waste Reduction Technologies, 219
Characterization
critical dilution volume, 267
defined, 265
developing nature of, 267
equivalency model, 266
generic exposure, 266
impact analysis categories, 265
inherent chemical properties, 266
loading technique, 266
site models, 266
site-specific exposure, 266
Chemical design examples, 384
Chemical Manufacturers Association, 17, 277
Responsible Care program, 17–18
Chemical oxidation
agents for, 468
chlorine, 471
defined, 468
hydrogen peroxide, 470–471
ozone, 468–469
reduction reactions with, 468
Chemical reactors
batch processes, 187–190
continuous-flow reactors, 190–192
reactor uses, 186
waste material sources, 186
Chemical reductions
chromium removal, 471
sludge production, 471–473
uses of, 471
Chemical safe designs
approaches to, 377
chemical design example, 384
chemist knowledge areas, 378
defined, 377
design focus, 377
general principle, 378
hazardous industry replacement, 377
isosteric replacements, 382–383
retrometabolic design, 384
structure-activity relationships, 381–382
toxicity reducing, 378–381
Chernobyl, nuclear accident, 132
Chicago Board of Trade, 410
Chlorinated aromatic compounds
naming of, 37–38
production ban, 38
uses of, 37
Chlorine
cyanides and, 471
uses of, 471
Chlorofluorocarbons (CFCs)
black market for, 101
energy-trapping capacity, 95
freon-type replacements, 101
ozone destruction, 27
ozone effects, 95
ozone hole from, 100–101
recycling of, 156
safe chemical replacements, 381
substituted alkane class, 27
uses of, 95
Chromium
characteristics, 42
drinking standards, 42
uses of, 42
wastewater removal of, 471
City of Albuquerque
POTW limit setting, 551
reporting requirements, 545
silver reduction program components, 546
technical assistance program, 545
voluntary reduction program, 546
City of Palo Alto
mass-based limit program, 551–552
nonpermitted industrial discharges, 552
nonregulated industries, 552
objectives of, 545–546
pollution plans, 549
POTW limits setting, 551
sewer ordinance program, 548
sewer use charge incentives, 555
technical assistance program, 545
City of San Francisco
technical assistance program, components of, 545
waste minimization audits, 545
Classification
conceptual framework for, 264
defined, 263
environmental issues and, 265
impact categories, 263
stressor lists, 263–264
Clean Air Act Amendments (CAAA), fugitive emissions determinations, 510
Clean Air Act (CAA)
acid rain and, 90, 156
critical pollutants, 151–153
global climate protections, 156–157
hazardous air pollutants, 153–155
incineration, 109
laws effecting, 151
mobile sources, 155–156
permits, 155
smog reduction requirements, 87
Clean Air Act (CAA) permits
accidental release plans, 155
cleanup goals, 155
offset concept in, 155
states issuing, 155
Clean air laws
acid rain, 156
amendment dates, 150
criteria pollutants, 151–153
emission standards enforcement, 151
global climate protection, 156–157
hazardous air pollutants, 153–155
mobile sources, 155–156
objective of, 151
permits, 155
rewriting of, 150
Clean manufacturing
attitudes of, 607
benefits of, 605
characteristics, 607
democratic approach to, 606
holistic approach to, 606
integrated approach to, 606
marketing and, 607
precautionary approach to, 606
preventative approach to, 606
Clean Process Advisory System, 219
Clean Water Act (CWA)
components of, 536
focus of, 536
NPDES permits and, 536
pollution control and, 536–537
purpose of, 536
source control requirements, 539
underground tanks and, 166–167
Clean water laws
Clean Water Act, 157–158
dredged material discharges, 158
effluent limitations, 157
National Environmental Policy Act, 159
NPDES and, 157–158
objective of, 157

Oil Pollution Act, 158–159
purpose of, 157
rewriting of, 157
Safe Drinking Water Act, 159–160
stormwater control, 158
Cleaning improvements
aqueous cleaning, 227
cleaning devices, 228
cleaning processes, 226
cold cleaning, 227
manufacturing process modifying, 226
organic solvents, 226
pollutants generated during, 226
purpose of, 225
selecting criteria, 29
solvent vapors, 228
ultrasonic cleaning, 228
vapor degreasing, 227
water quantity in, 226–227
Cleaning water reduction
case study of, 430
purpose of, 422
recycling, 428–429
rinse tank configurations, 422–424
rinse tank uses, 422
rinse water requirements, 424–428
Clinton, President Bill
Council on Sustainable Development, 597–600
Executive Order on Federal Acquisition, Recycling and Waste Prevention, 284
Executive Order sustainable development, 597
Green Chemistry Challenge, 362
Coagulation
alkalinity increasing, 465
coagulant aids, 465
flocculants wastewater, 465–466
role of, 465
surface electrical charge, 464–465
upflow solids contact clarifier, 466
uses of, 465
wastewater and, 465
Coal
acid rain and, 88
energy from, 127
environmental concerns of, 129
flow of, 128
fossil fuel development, 127
as nonrenewable resource, 127
reserves of, 129
types of, 127
U.S. production, 88
Coca-Cola Company, 253
Code of Federal Regulations, 149
hazardous waste listing, 164
ComEd, life-cycle cost model, 323
Compatible materials
design issues, 401
plastic compatibility, 401
polymer mingling, 401
supply classes, 402
Completely mixed reactors, product concentrations, 190
Compliance audit, purpose of, 343
Component standardization
internal accessing, 397
recycling and, 397
Composting
aerobic process in, 110
hazardous components and, 409
heat requirement, 110
items in, 110
problems with, 110
as recycling, 409
Comprehensive Environmental Response, Compensation, and Liability Act (CERCLA). *See also* Superfund Amendments and Reauthorization Act
Brownfields Initiative, 169–170
cleanup schedules, 168
focus of, 168
hazardous material definition, 168
hazardous waste sites abandoned, 168
Love Canal and, 168
National Contingency Plan, 169
prevention promotion, 534
SARA amendments to, 168
Computer-aided design
demanufacturing designing, 407
studies on, 407
Concurrent engineering. *See also* Designing for X
objectives of, 185
origin of, 185
purpose of, 185
steps in, 185–186
Concurrent engineering process for, 185
Conservation
defined, 15
focus of, 114
philosophy of, 15
World Conservation Strategy, 612
Contact stabilization, activated sludge system, 490
Contaminant concentrations
criteria for, 44
expressing methods, 45–46
Contaminant transport
contaminant concentrations, 44–46
environmental quality evaluating, 44
partitioning process, 49–67
transformation process, 45, 67–72
transporting process, 44–45, 46–49
Contaminants
environmental impact assessments, 23
environmental impact knowledge, 23
inorganic nonmetals, 39–44
metals, 39–44
organic chemicals, 23–39
product pricing and, 22
transporting, 44–67
waste materials origins, 22
Continuous distillation
enrichment sections in, 202
flash distillation, 201
reflux and, 201
stripping section in, 202
workings of, 201
Continuous-flow reactors
characteristics of, 190
cleaning of, 192
completely mixed reactors and, 190–191
design improvements, 222
equipment cleaning, 229
plug flow reactors and, 190–191
waste neutralization, 458
Corrosive waste, as hazardous waste, 163
Cost assessments
capital budgeting issues, 321
case study, 323–325
conventional cost categories and, 321
cost assessments process, 323
economically viable, 320
environmental accounting implications, 320–321
evaluation time period, 321
external costs, 321–322
financial analysis, 320
life-cycle costing, 322–323
Council for Chemical Research, 384–385
Critical dilution volume
development of, 267
uses of, 267
Critical pollutants
enforcement results, 153
EPA and, 151–152
NAAQS for, 151–152
standards enforcement, 152–153
Crompton, Samuel, 8
Crust, 139
Crystallization
defined, 198
formation process of, 198–199
removal process, 199
types of, 199
uses of, 198
Cyanides
characteristics, 44
cyanide salts uses, 44
drinking water standards, 44
Cyclic aliphatic compounds, 31–32
Cyclones
efficiency of, 502–503
purpose of, 500
workings of, 500–501

D-glucose
adipic acid two-step synthesis process, 368
characteristics of, 367
cost of, 368
uses of, 367
Deep ecology
defined, 16
philosophy of, 16

Defense Advanced Research Projects Agency, 185
Degreasing improvements. *See also* Cleaning improvements
  pollutants generated during, 226
  purpose of, 225
Dehydrodehalogenation
  defined, 69
  purpose of, 69
Department of Agriculture
  administrative jurisdiction of, 150
  FIFRA and, 174
Department of Commerce, life-cycle data, 262
Department of Energy
  life-cycle assessments and, 253
  life-cycle data, 262
  pollution prevention software, 220
Department of the Interior, administrative jurisdiction of, 150
Department of Transportation
  hazardous materials transportation steps, 172–173
  HMTA enforcement, 172
Design for Environment, defined, 355
Designing for X
  categories of, 356
  concurrent engineering and, 185
  defined, 355
  design factors, 356
Diffusion, defined, 48
Diffusional processes
  contaminant release impact assessing, 49
  modeling of, 49
  molecular diffusion, 48
  purpose of, 48
Disassembly designing
  composting, 409
  computer-aided design, 407
  need for, 386
  recycle hierarchy, 388–389
  recycle legislation and, 390–393
  recycle versus reuse, 386–388
  requirements for, 393–402
  reuse barriers, 409–410
  strategy for, 402–407
  waste exchanges, 407–409
Disassembly ease
  assembly operations simplification, 396–397
  benefits of, 394
  cost factors of, 396
  materials separation techniques, 395
  methods of, 394
  polymer recycling facility, 396
  recycling location, 394
  welded products, 394
Disassembly strategy
  abandoned product calculations, 406
  component recovery, 402
  disassembling parts and selling calculations, 406
  disassembly analysis methodology, 403
  hierarchy in, 407
  objectives of, 402–403
  option balancing, 405–406
  pathway alternatives, 404
  sequences, 405
  shredded materials selling calculations, 406
  steps in, 403–404
Discount rate
  inflation rate versus, 311
  selecting, 312
  varying of, 312
Discrete settling
  characteristics of, 461
  settler size examples, 462–463
  tank size, 462
Dispersive processes
  causes of, 48
  computer models for, 48
  consequences of, 48
  defined, 48
Distillation
  continuous distillation column, 201–209
  industries using, 199
  model types of, 200
  working process of, 199
Distillation columns
  mass balance formula, 204
  material balance formulas, 204–208
  product recovery by, 202
  reflux ratio calculations, 203
  separation amounts, 202
  uses of, 209
Drinking water. *See also* National Primary Drinking Water Standards; Safe Drinking Water Act
  arsenic standards, 41
  cadmium standards, 41
  chromium standards, 42
  cyanide standards, 44
  industrial waste and, 119
  mercury standards, 44
Drying
  characteristics, 197
  environmental problems, 198
  heat functions, 197–198
  purpose of, 197
Duell, Charles H., 9

E-factor
  purpose of, 613
  total quality management themes, 613
Earth Day, 115
  environmental ethics, 15
Earth First, philosophy of, 16
Earth Summit
  future development principles, 581
  purpose of, 580
  Rio declaration from, 581
  Rio principles, 582
  social themes of, 580–581
  sustainable development, 581
Earth's structure
  core of, 139
  elements in, 140
  layers of, 139
  rock's classifications, 140
Ecoidustrial parks
  cooperative opportunities from, 609–610
  purpose of, 609
Ecolabeling
  countries with, 284–286
  environmental labeling, 284
  environmental product profiles, 284
  life-cycle assessments and, 284
  regulations for, 284
  third party evaluations, 284–285
  U.S. and, 284, 286–287
Ecological concepts
  characteristics of, 586
  ecosystem, 585
  environmental impact statements, 586–587
  resilience and, 585
  sustainability, 585–586
Economic concepts, sustainability and, 585
Economic growth measures
  green accounting, 615–616
  gross domestic product, 613–615
  gross national product, 613–615
Economic systems, sustainability and, 576
Ecopoint method
  ecological shortage, 268
  life-cycle impact analysis, 268
  origin of, 268
Ecosystem, defined, 585
Education
  POTW outreach programs and, 544
  sustainability framework, 604
Electric Power Research Institute, life-cycle cost model, 322–323
Electricity
  efficiency of, 136
  flow of, 137
  industrial demands, 136
  production sources, 136
Electrostatic precipitator
  efficiency of, 502–503
  purpose of, 502
  uses of, 502
  working of, 502
Elimination reactions
  chlorinated organic removal, 69
  dehydrodehalogenation, 69
  process of, 68
Emergency Planning and Community Right-to-Know Act (EPCRA). *See also* Toxic release inventory
  annual report requirement, 172, 346
  goal of, 346, 535

hazardous release notification, 171
local emergency plans, 171
pollution prevention and, 536
PPA and, 535–536
purpose of, 170
SARA as, 168, 538
storage and, 236
subtitles of, 170–171
voluntary planning, 330
Emission fees
inadequacies of, 308–309
pollution fee calculating, 308
theory of, 308
Emission reduction incentives
PPA and, 310
state incentive programs, 310–311
Emissions inventory
direct discharges, 346
purpose of, 345
waste audits and, 345
Empowerment Zone and Enterprise Community, 170
Energy conservation, process inefficiencies, 136–136
Energy Consumption
capita energy consumptions, 123
developing nations, 123
sources of, 123
U.S. usage, 125–127
Energy reserves
extraction for, 127
fossil fuels, 127
Energy usage
conservation, 136–138
consumption, 123–127
electricity, 136
fossil fuels, 127–132
historical perspective, 121–122
human consumption, 122
industrial revolution focus, 122
nuclear energy, 132–134
oil fuels, 122
renewable sources, 134–136
reserves, 127
society needs, 121–122
Engineering economies
discount rate, 311
investment alternative comparing, 314–317
optimum operating point, 311
present worth, 312–314
Environment Management Standard. *See also* ISO, 14000
environmental protection and, 178
ISO 14000 and, 177–178
purpose of, 178
Environmental accounting
"green GNP" calculating, 616
sustainability principles and, 616
Environmental audits
characteristics, 344
data needed in, 345
defined, 343
as emissions inventory, 343, 345–346
EPA policy and, 343
management information from, 344
methods of, 344
objective of, 343
types of, 343–344
Environmental design
disassembly design, 386–410
examples, 353–354
green chemistry, 357–386
green designing, 354–356
introduction, 353–357
packaging, 410–416
Environmental effects
life-cycle impact analysis, 268
origin of, 268
uses of, 268
Environmental ethics
CMA Responsibility Care Program, 17–18
company classifications, 18–19
concerns of, 15
conservationism philosophies, 15
corporate ethics, 16–17
deep ecology philosophies, 16
defined, 15
green organization involving, 17
preservationism philosophies, 15–16
regulatory pressure, 15
social ecology, 16
stakeholder management, 17
versus sustainability, 16
unlimited economic wealth concept, 16
Environmental impact statements
NEPA requirement, 159
uses of, 586–587
Environmental liability audit, purpose of, 343
Environmental management systems
company requirements, 341
components of, 341–342
defined, 341
documents supporting, 340–341
expectations of, 342
implementing, 342
ISO 14000 standards, 340
purpose of, 340
standard types, 342
Environmental performance bonds
advantages of, 612
bond expenditures, 612
uses of, 611
Environmental priority strategies
life-cycle impact analysis, 268
origin of, 268
uses of, 268
Environmental problems
poverty and, 575
sustainability and, 575
Environmental Protection Act (EPA), effluence limitations, 157
Environmental Protection Agency (EPA)
administrative jurisdiction of, 150
assumptions of, 518
Brownfield pilot program, 170
conflicting role of, 307–308
correlations for, 518
design software by, 220
environmental audits policy, 343
environmental impact statements, 586–587
fugitive emissions estimating methods, 513–514
green chemistry promoting, 362
hazardous waste categories, 163
major stationary source defined, 511
National Primary Drinking Water Regulations, 159–160
petroleum industry and, 518
pollution prevention definition, 11
pollution prevention office, 176
premanufacture notification of toxic substances, 160
project XL and, 596
purpose of, 518
Environmental regulations
clean air laws, 150–151
clean water laws, 157–160
environmental regulations, 150–178
hazardous materials laws, 160–174
introduction, 147–148
pollution laws, 175–178
product laws, 174–175
quantity of, 150
rate of, 147
regulatory process, 149–150
uses of, 150
Equipment cleaning
continuous-flow reactors and, 229
frequency of, 229
Equipment modifications
compressors, 522
goal of, 519
leaking valves, 519–522
pipe connectors, 522
pumps, 522
Esters
examples, 30
naming basis, 29
reactions forming, 29–30
uses of, 29–30
European Union, eco-labeling, 285–286
Evaporation
multiple-effect evaporators, 197
purpose of, 197
types of, 197
working of, 197
Executive Orders. *See also* President Bill Clinton
examples of, 176–177
federal agency binding, 176
Extended aeration process, activated sludge system, 490
External floating roof storage tanks
characteristics, 524
emission control methods, 529–530

Externalities, defined, 300
Externality controls
  emission fees, 308–309
  marketable permits, 309–310
  regulations, 307–308
  types of, 307
Extraction
  efficiency calculations, 211
  liquid-liquid extractors, 210
  natural product extraction, 213
  organic solvents, 211
  phases of, 210
  purpose of, 210
  solids washing, 210
  solvent extraction effectiveness, 210–211
  supercritical extractions, 212–213
Exxon Valdez, 158

Farming, sustainability framework, 604
Federal Insecticide, Fungicide, and Rodenticide Act (FIFRA)
  objective of, 174
  origin of, 174
  uses of, 174
Federal Register, 149
Federal Trade Commission, 284
Feedstock alternatives
  benefits of, 363
  benzene alternatives, 364
  biomass and, 365–367
  d-glucose, 367–368
  petroleum alternatives, 363
  phosgene alternatives, 364–365
  processes modification, 368
Fiber Box Association, 401
Fick's Law
  formula, 48
  molecular diffusion and, 48
Filtration
  graded quartz sand filter types, 480
  granular media filters, 479
  processes as, 479
  uses of, 479
Fixed-roof storage tanks
  characteristics of, 524
  emission control methods, 529
  emission loss calculations, 526–527
  fugitive emissions from, 524
Floc. *See also* Coagulation
  defined, 465
Flocculent (type 2) settling, characteristics of, 463–464
Flotation
  air flotation, 464
  emulsified oils, 464
  versus sedimentation, 464
  uses of, 464
  water flotation, 464
Fluidized bed reactors, characteristics of, 494–495
Ford, Henry, 8
Ford Motor Company, 156, 308
Fossil fuels. *See* Coal; Natural gas; Oil
Freundlich isotherm equations, 58–64
  formula for, 57
  as sorption models, 57
Frictionless Bearings, Incorporated, 296
Fugitive emission controlling
  equipment modifications, 519–522
  fixed-roof tanks, 529
  floating roof tanks, 529–530
  leak detection, 523–524
  organic liquid evaporation, 529
  techniques for, 519
Fugitive emission estimations
  examples, 527–529
  fixed-roof tank calculations, 526–527
  procedures for, 526
  software programs for, 526
Fugitive emission factor average approach
  defined, 514–515
  examples of, 515–516
  factors for, 514
  organic carbon mass emission calculating, 515
  purpose of, 515
Fugitive emission measuring
  average emission factor approach, 514–516
  CAAA requirement for, 512
  data quality maintaining, 513
  EPA correlations, 518
  estimating method alternatives, 513–514
  gas and vapor definitions, 513
  goal of, 513
  mass emissions and, 512–513
  screening ranges approach, 516–517
  techniques for, 512
  unit-specific correlations, 518–519
Fugitive emission sources
  cumulative impact of, 511–513
  equipment leaks, 511
  hazardous compounds, 511
  primary sources, 511
Fugitive emissions
  amounts of, 511–512
  CAAA and, 510
  component leaking, 209
  computerized programs for, 211
  controlling, 519–524
  defined, 47
  disposal of, 530–531
  environmental compliance, 509–510
  HON requirements, 510
  introductions, 509–511
  leak monitoring regulations, 510
  measuring, 512–519
  sources of, 511–512
  stationary sources, 510–511
  storage tanks, 524–530
  waste treatments, 530–531
Fulton, Robert, 8

Gas defined, 513
Gas removal
  adsorption systems, 503–504
  control methods, 503
  incineration, 504
  purpose of, 503
  wet scrubbers, 503
General Dynamics, 3
General Electric, 323
General principle
  chemical toxicity requirements, 378
  molecular structures modifying, 378
Georgia-Pacific Corporation, 596
Geothermal power
  locations for, 135
  as nonpolluting energy, 134
German Recycling and Waste Management Act, 390
Germany
  Blue Angel Program, 285
  eco-labeling in, 285
Global climate protection
  CFC recycling, 156
  market-based approach to, 156
  Montreal Protocol, 156
  phaseout schedule, 156–157
Global warming
  carbon dioxide and, 85, 93
  climate factor interactions, 91
  defined, 79
  effects of, 91
  greenhouse effect, 93–94
  greenhouse gases, 95
  heat trapping gases, 93
  industrial revolution impacts, 92–93
  technological options, 95–98
Goodhealth Pharmaceutical Company, 336
Great Lakes Water Quality Initiative
  goal of, 539
  purpose of, 538
  standard benchmarks in, 538–539
  states effecting, 538
Green accounting
  defined, 615
  environmental accounting, 616
  GDP and, 615
  recycling, 616
Green chemistry
  awareness of, 358
  chemical selection, 357
  concepts of, 357
  defined, 357
  industry goals, 357
  reaction condition alternatives, 374–377
  research needs, 384–386
  safe chemical design, 377, 377–384
  synthetic pathways alternatives, 363–374

waste sources, 358–362
Green chemistry research
areas for, 385
aromatic amine dyes molecular designs, 385
computer assistance in, 386
Green companies
as classification system, 18
environmental performance of, 19
Green designing
goals, 355
phase origin, 355
pollution reduction from, 355
Green Seal
category standards, 286
eco-labeling, 286
life-cycle assessments use, 284
standard development, 286–287
testing for, 286–287
Greenhouse effect
carbon dioxide concentrations, 93
effects of, 93, 93–94
name origin, 93
Greenhouse gases
chlorofluorocarbons, 95
methane emissions, 95
nitrous oxide, 95
Greenhouse technological options
atmosphere stabilizing, 97
carbon dioxide control, 95–96
global warming, 96
industries processes and, 96–97
reforestation, 98
vegetation increases, 97–98
Gross domestic product (GDP). *See also* Gross national product
depreciation and, 614
environmental considerations and, 614
environmental effects calculating, 615
green accounting and, 615
measurement methods, 613–614
net domestic product and, 615
sustainable income measuring, 615
Gross national product (GNP). *See also* Gross domestic product
losses registered to, 614
positive adjustments to, 614
purpose of, 614
quality and, 614
uses of, 614
Groundwater protection
industry impact of, 160
SDWA and, 160
state responsibility, 160

Ham, Robert K., 12
Hargraves, James, 7
Hazardous air pollutants (HAPs)
chromium electroplating rules, 154–155
defined, 84–85, 153
emission control programs, 85
health problems from, 153
MACT standards, 154
sources of, 153
Hazardous Material Transportation Act (HMTA)
enforcement of, 172
material classes, 172–173
purpose of, 172
shipper responsibility in, 172
steps in, 172–173
Hazardous Materials Transportation Uniform Safety Act, 172
Hazardous materials and waste laws
Comprehensive Environmental Response, Compensation and Liability Act, 168
Emergency Planning and Community Right-to-Know Act, 170–172
Federal Insecticide, Fungicide and, Rodenticide Act, 174
Hazardous Material Treatment Act, 172–173
Resource Conservation and Recovery Act, 162–168
Toxic Substances Control Act (TSCA), 160–162
Hazardous Organic National Emission Standards (HON)
emission standards defining, 510
requirements of, 510
Hazardous and Solid Waste Amendments (HSWA)
hazardous waste minimization, 534, 537
preventive focus, 535
purpose of, 535
RCRA reauthorization as, 534
source generation elimination, 535
Hazardous waste permits
cradle-to-grave management, 166
responsible party for, 166
shipping manifest system, 166
Hazardous waste treatment
EPA and, 165
ownership of, 165
regulation areas, 165
small business and, 165
Hazardous wastes
characteristics of, 114. 163
defined, 114, 162–163
derived-from rule, 165
industrial producing of, 114–115
liquid waste and, 164
mixture rule, 165
quantity of, 114
resource conservation, 117–118
solid waste defined, 162
solid waste exempted from, 163
solid wastes and, 114
states producing, 115
Superfund, sites, 115–117
waste "delisting," 165
Heat exchanger network (HEN)
analysis purpose, 431
design rules, 436–437
energy inputs, 431–432
exchanger network system examples, 432–433
heat transfer rates, 432
thermal pinch analysis, 431
Heat exchangers
coefficients of, 194
design of, 193
examples, 196–197
heat transfer calculating, 194
heat transfer occurring, 194
logarithmic mean temperature differences calculations, 195
purpose of, 192
temperature profiles, 195
tube exchangers and, 192
waste generation contributions, 195–196
waste reduction methods, 196
workings of, 192–199
Heath care, sustainability framework, 604
Heavy liquid, defined, 513
Heavy metals. *See* Metals, heavy
Henry's Law
defined, 64
formula for, 64
partitioning, 64
Hewlett-Packard, 398–399, 401, 414
High-rate activated sludge system, 490
Hocker Chemical Company, 115
Housekeeping
awareness lacking in, 237
benefits of, 237
common sense and, 238
employee education, 237
overlooked practices, 237–238
preventive maintenance, 238
wastewater pollutions, 237
Howe, Elias, 8
Hydrocarbons
effects of, 83
quality of, 83
sources of, 83
Hydrochlorofluorocarbons, safe chemical replacements, 381
Hydrogen peroxide oxidizing
catalyst with, 470
cyanide oxidizing with, 470
Hydrolysis
chemical reactions, 68
ph dependence, 68
process of, 68

IBM, 280
Ignitable wastes, as hazardous waste, 163
Impact analysis
characterization, 265–267
classification, 263–265
phases of, 263
quantitative examination, 263
role of, 263
valuation, 267–269
Improvement analysis
analysis difficulty, 269
objective of, 269
outcomes of, 269

Incineration
  advantages of, 110
  cost of, 110
  countries using, 109
  disadvantages of, 109
  gas removal, 504
  uses of, 109, 504
  wastes consumption, 109
  working of, 504
Industrial contaminants
  environmental fate, 47
  loading processes for, 46–47
Industrial ecology
  defined, 609
  ecoindustrial parks, 609–610
  ecology and, 608
  ecosystem diversity, 609
  elements of, 609
  evolving of, 608
  goals of, 608–609
  materials cycling, 609
  principles of, 610–613
Industrial ecology principles
  conservation strategies, 612
  e-factor, 613
  environmental performance bonds, 611–612
  industrial metabolism, 612
  intergenerational equity principle, 611
  intragenerational equity principle, 611
  polluter-pays principle, 611
  precautionary principle, 610–611
  soft material paths, 612
  subsidiarity principle, 611
  3Rs, 613
  user-pays principle, 611
Industrial metabolism
  industrial processing and, 612
  origin of, 612
Industrial Revolution
  electricity productions, 8
  patent activity, 9
  power-driven machinery, 7
  steam engines, 8
  textile industry mechanizations, 7–8
Industrial wastes
  air pollution, 79–101
  effects of, 78–79
  energy usage, 121–138
  hazardous wastes, 114–118
  introduction, 78–79
  resource depletion, 138–142
  solid wastes, 102–114
  sources of, 107
  waste quantity, 78
  wastewater sludges, 107
  water pollution, 118–121
Industrial wastewater treatment
  POTWs and, 121
  prevention versus treatments, 121
  timing of, 121
Industrialization impacts
  benefits, 9
  environmental costs, 9
  social structure, 9
Industry
  circular structure, 605
  clean manufacturing benefits, 605
  clean production goals, 605
  linear structure of, 605
  sustainability role, 604–605
Institute of Packaging Professionals, 401
Intelligen, Incorporation, 220
Intergenerational equity principle, uses of, 611
Internal floating roof storage tank
  characteristics, 524
  fugitive emissions, 524
Internal rate of return, uses of, 316
International Chamber of Commerce, 18
International Organization for Standardization (ISO). *See also* ISO, 14000
  environmental ethics, 17
  mission of, 177
  objective of, 177
International Union of Nature and Natural Resources (IUCN)
  Ottawa Conference on Environment and Development, 579
  World Conservation Strategy, 579
International Union of Pure and Applied Chemists (IUPAC), nomenclature system, 23–24
Intragenerational equity principle, uses of, 611
Inventory analysis
  assessment from, 262
  basic steps in, 259
  data accuracy, 262
  data consistency, 262
  data gathering, 261
  data sources, 261–262
  decision areas, 259–260
  focus of, 259
  inventory checklist, 260–261
  purpose of, 259
  report presentation, 263
  systematic approach to, 259
Investment alternatives comparing
  benefit-cost ratios, 316
  internal rate of return, 316
  investment dilemmas, 314
  methods for, 315
  net present value, 316–317
  payback, 316
  pollution prevention options, 314–315
Ion exchange
  cation exchange equations, 476
  cation exchange selectivity, 477
  examples, 478–479
  fixed-bed columns for, 477
  hydrogen ion replacements, 476
  hydrogen-based resins, 477
  purpose of, 475
  removal mechanisms, 477–478
  results of, 476
  workings of, 475–476
Ionic liquids
  defined, 376
  uses of, 376
ISO 1043-1 Plastics Indentification System, 398
ISO 14000. *See also* Environmental management systems
  adoption of, 340
  eco-labeling and, 286
  guidance document for, 177–178
  international law, 150
  life-cycle assessments and, 255
  manufacturing standards, 177
  objective of, 340
  pollution planning mandate, 330
  purpose of, 177
  subjects covered by, 340
Isomers, defined, 25
Isosteric replacements
  carcinogenic molecule changes, 382
  defined, 382
  organosilanes, 383
  purpose of, 382
Isotherm equations
  defined, 57
  types of, 58
  uses of, 58

Jewell Electric Instruments, 430

Kepone incident
  banning of, 120
  characteristics of, 120
  cleanup findings, 120–121
  origin of, 120
  process wastes pollution impact, 120
Ketones. *See also* Aldehydes
  examples, 31
  naming basis, 30
  uses of, 30
Kyoto Protocol
  basis for, 98
  developing country exemptions, 98
  greenhouse gas reduction requirements, 98
  industrial versus developing nations, 98
  U.S. ratification and, 98
Kyoto Summit, marketable permits, 310

Land applications
  biological waste treatment, 496
  comparative characteristics, 498
  system types of, 496–498
  uses of, 498
Land disposal
  hazardous constituents migration, 166
  hazardous waste and, 166
  uncontained liquid waste, 166

Land Disposal Program Flexibility Act, 166
Langmuir isotherm
equations, 58–64
formula, 57
as sorption models, 57
Law of demand, defined, 300
Law of supply, defined, 300
Laws. *See* Environmental regulations
Lead
air qualities, 84
characteristics, 42
environmental problems, 42
health problems from, 42
recycling infrastructure for, 42
sources of, 84
Leak detection
component sampling frequency, 523–524
cost-effective control, 523
equipment types for, 523
LDAR program and, 523
monitoring procedures, 523
Life Science Products Company, 120
Life-cycle assessment applications
benefits of, 277
corporate strategic planning, 277–278
decision-making process, 278
ecolabeling, 284
examples, 278
marketing claims, 283
pollution prevention operating strategy, 278
process modifications, 280–283
process selection, 280–283
product development, 278–279
responsible care program, 277
Life-cycle assessment computer models
categories of, 287–288
database system linking, 288
examples of, 288
limitations of, 288
purpose of, 287
review of, 288
spreadsheets format, 288
Life-cycle assessment goal definition
complexity and, 258
energy requirements, 258
forethought benefits, 258
process streamlining, 259
purpose of, 257–258, 258
scope of, 258
Life-cycle assessment history
European "green movement," 254
indirect impacts, 254
net energy analysis and, 253
product packaging focus, 254
recycling disadvantages, 254
Life-cycle assessment methodologies
"code of practice" for, 257
components of, 257
data access, 255–256
goal definition, 257–259
impact analysis, 263
improvement analysis, 269
interdependent stages, 257
inventory analysis, 259–263
scoping stage, 257–259
systems approach to, 255
Life-cycle assessment streamlining
approaches to, 270
critical environmental impact and, 269
information omitting, 270
meaning of, 269
methods for, 270
minimum standards, 270
phases for, 270–271
reasons for, 269
selection criteria, 270
Life-cycle assessments
applications of, 277–287
computer model use, 287–288
defined, 252
European use of, 254
evaluation scope, 252
executive order for, 254–255
history of, 253–254
ISO 14000 and, 255
legislative debates, 255
methodology, 255–269
motivations for, 253
overview of, 251–253
pollution impact understanding, 251
pollution prevention factors, 271–277
product impact examining, 252
purposes of, 253
regulations and, 254
regulatory process and, 254–255
states of, 252
streamlining, 269–271
systematic approach to, 252
uses of, 252
waste management operations, 288–292
Life cycle cost assessment case study
annual rate of return calculating, 324
evaluation areas, 323
option rankings, 324
P2 options evaluation, 323–324
Life cycle cost assessment process
case study, 323–325
data availability, 323
standardized form for, 323
Life cycle costing
electric utility industry use, 322
models for, 322–323
Life-style, sustainability framework and, 603, 603–604
Light liquid, defined, 513
Linear isotherm equations, 58–64
Linnhoff, Bodo, 430
Liquid, defined, 513
Liquid, heavy, defined, 513
Little, Arthur D., 253
Living standards, sustainability and, 575–576
Loading processes
defined, 46
industrial contaminants, 46–47
pollution sources, 47
process of, 46
waste load responsibilities, 47
Local discharge standards
authority for, 541–542
POTW discharges, 541
purpose of, 541
Local government sustainability role
leadership areas for, 600–601
motivation for, 601
progress of, 601
Long-term cleanup estimating
capital costs and, 317
examples, 318–320
liability cost estimating, 318–320
waste treatment costs, 317–318
Love Canal
chemical waste dumping, 116
contaminant containment approach, 116
health hazards from, 116
history of, 115–116
laws from, 168
legislation from, 116
RCRA regulations and, 149
solid waste disposal and, 504

Maleic anhydride production, 242–247
Management
commitment of, 237
housekeeping, 237–238
record keeping, 239
training, 238–239
Management audits, purpose of, 343
Mandatory program components
fines levied, 566
prevention requirement imposed, 566
purpose of, 566
reduction areas identifying, 566
required prevention implementation, 566
Mandatory program problems
government's role, 553–554
incentives focus, 554
mass-based limits, 554
Mantle, 139
Manufacturing operations
business dilemmas, 180
environmental regulation costs, 180
going green, 180–181
industrial processes, 181
introduction, 180–182
manufacturing processes, 183–215
pollution minimizing, 181–182
pollution prevention elements, 182
pollution prevention examples, 239–247

Manufacturing operations (*cont.*)
process changes, 220–239
process design, 215–220
process development, 215–220
waste minimizing, 181
wastewater treatment costs, 181
Manufacturing processes
absorption, 209–210
adsorption, 213–214
ancillary equipment, 215
chemical reactors, 186–192
concurrent engineering, 185–186
crystallization, 198–199
distillation, 199–209
drying, 197–198
enterprise structures, 183
evaporating, 197–198
examples, 186
extraction, 210–213
heat exchangers, 192–197
industrial process criteria, 186
manufacturing processes, 186
other operations, 214–215
process change benefits, 186
product design steps, 183–184
sequential engineering, 184
Marginal cost
cost of production, 303
fixes costs, 303
manufacturing purpose, 303
marginal cost curve, 303
production decision benefits, 303–305
variable costs, 303
Market cost, defined, 303
Market equilibrium point, defined, 301
Market externalities
common good and, 305
competitive economy and pollution, 306–307
cost-benefit analysis, 306
defined, 305
government policy and, 305
market system effects, 305
taxes, 305
treatment level determining, 306
Market mechanisms
externalities and, 300
government involvement in, 299, 299–300
imperfect competition, 299
market purpose, 299
product development, 299
production methods, 299
taxes and, 300
Marketable permits
greenhouse use, 310
objective of, 309–310
regulation combinations, 310
studies of, 310
working of, 309
Marketable wastewater permits
financial advantage of, 556
purpose of, 555
Market-based programs
incentive types, 555
incentives for, 554–555
marketable wastewater permits, 555–556
pollution prevention grants, 555
small business and, 555
Marketing claims
life-cycle assessments supporting, 283
single criteria advertising, 283
Mass-based limits
City of Palo Alto program, 551–552
pollutant actual loads, 551
pollution prevention and, 551
POTWs and, 551
Material Safety Data Sheets (MSDS), 171
Materials identification
confusion over, 397
identification standards, 397
plastics recycling as, 397–399
Materials selection
criteria for, 402
environmental impacts, 402
life-cycle design stage, 402
objectives of, 402
Maximum allowable control technology (MACT)
emission reduction forecast, 154
standard implementation, 154
standards for, 154
Membrane processes
examples, 484–485
industry uses of, 484
purpose of, 480
ranges of, 480
reverse osmosis, 481–484
separation processes in, 481
uses of, 481
workings of, 480–481
Mercaptans, disagreeable odor, 27
Mercury
characteristics, 43
chlorine productions, 43–44
common ion in, 43
drinking water standards, 44
fish concentrations, 43
mercury salts, 43
Minimata disease, 119–120
uses of, 43
Metals
anthropogenic metals, 39–40
arsenic, 40–41
bioaccumulation of, 40
cadmium, 41
chromium, 42
cyanides, 44
defined, 39
heavy metals, 39–40
lead, 42
mercury, 43–44
nutrients as, 40
Metals, heavy
defined, 39
environmental concerns, 39
Microeconomics
control measures, 307–311
defined, 298
engineering economies, 311–317
free-market system and, 298
marginal benefits, 303–305
marginal costs, 303–305
market externalities, 305–307
market mechanisms, 299–300
supply and demand, 300–303
Midwest Research Institutes, 253
Mineral resources
consuming rate of, 141
metal ore types, 141
reserves of, 142
Mineralization
complex organic compound conversions, 71
defined, 71
Minimata disease
containment, 119
industrial effluent predicting, 120
as mercury poisoning, 119
origins of, 119
Mining wastes, reclaiming mining areas, 107
Mobile air pollutants. *See* Automobiles
Monsanto Chemical Company, 3, 445
Montreal Protocol, global climate protection, 156
Motorola, 278
Municipal P2 program qualities
POTWs and, 558–561
program options, 558
Municipal pollution prevention programs
introduction, 533–534
P2 initiative implementing, 533
POWT programs, 533–534
pretreatment permit programs, 539–568
prevention acceptance, 533
regulatory basis for, 534–539
source control in, 539–568
Municipal solid wastes
amounts of, 104–105
composition of, 105, 105–106
contents of, 104
potential recovery from, 111
recovery of, 113
recyclable materials use, 106–107
recycling, 106

Naess, Arne, 16
National Ambient Air Quality Standards (NAAQS)
critical pollutants and, 152
versus emission standards, 151
establishing of, 151
standard types in, 152
National Center for Manufacturing Sciences, 219
National contingency plan
cleanup cost, 169
cleanup times, 169
goal of, 169

hazardous waste site ranking, 169
purpose of, 169
remediation plans, 169
National Environmental Policy Act (NEPA)
provision of, 159
record of decisions, 159
National Institute for Occupational Safety and Health (NIOSH), purpose of, 174
National Materials Exchange Network (NMEN), 408
National Pollutant Discharge Elimination System (NPDES)
versus POTW use, 157–158
purpose of, 157
National Pollution Discharge Elimination System (NPDES). *See also* Best management practices
CWA and, 536
permit from, 536
prevention practices of, 536
National Primary Drinking Water Standards
contaminant standards, 161
contaminants regulations, 159–160
National Priority List, 169
Superfund cleanup and, 117
Natural gas
flaring off of, 130
flow of, 131
locations of, 132
types of, 130
use strategies, 132
Natural product extraction, uses of, 213
Nature Conservancy, partnership of, 596
Net present value
disadvantages of, 317
prediction safety, 317
purpose of, 316–317
Netherlands, eco-labeling in, 285
New Hampshire Department of Environmental Services
sewer ordinance model, 548
strategies of, 548–549
Newcomen, Thomas, 8
Nitrogen oxide
characteristics of, 84
emissions, 90
sources of, 84
Nonaqueous Phase Liquids (NAPLs), defined, 54
Noncompliant industry programs
compliance schedules, 553
corrective action plans, 553
enforcement settlements, 553
SEP programs, 553
violation fines, 553
Nonmetals, defined, 39
Nonpermitted industry controlling
City of Palo Alto and, 552
noncompliant industries, 552–553
permit class for, 552
prevention programs for, 552
Normal compounds
names of, 26
naming basis, 26
straight-chain alkanes as, 26
North American Free Trade Agreement, 150
Nuclear energy
accidents from, 132
cost of, 132
pollution impact, 134
predictions of, 132
radioactive management, 133–134
reactor research, 133
Nuclear wastes
disposal needs, 134
reactor design and, 133
spent fuel assemblies, 134

Occidental Petroleum Company, 116
Occupational Safety and Health Act (OSHA)
origin of, 174
reactions to, 175
responsibility of, 174
results from, 175
standards developing, 174
worker safety issues, 175
Office of Technology Assessment, 355
Oil
environmental impact of, 130
extraction cost, 129
locations of, 129–130
multiple energy from, 130
origins of, 129
total amount of, 129
Oil Pollution Act (OPA)
origin of, 158
programs for, 158
purpose of, 158
spill prevention incentives, 158
trust fund for, 158–159
Oil Spill Liability Trust Fund, 158
Organic acids
common names for, 29
uses of, 28–29
Organic chemicals
characteristics of, 23
organic compounds nomenclature, 23–39
Organic compounds nomenclature
aliphatic compounds, 25–32
aromatic compounds, 32–39
carbon atoms and, 24
IUPAC system in, 24
types of, 24–25
Ottawa Conference on Environment and Development, outcomes of, 579
Oxidation ditch, activated sludge system, 490
Oxidation-reduction reactions
biological reactions, 69
characteristics of, 69
examples of, 69–70
organic compound transformation, 69
process of, 69
Ozone
chemical destruction phaseout schedule, 157
creating of, 85
defined, 84
hydrocarbons effects, 83
photochemical effects of, 84
smog and, 85–86
ultraviolet energy absorbing, 93
Ozone depletion
chemical reaction as, 99
chlorine reductions, 101
chlorofluorocarbon impact, 100–101
freon-type replacements, 101
international concerns, 101
ozone hole, 100–101
ozone role, 99–100
Ozone hole
causes of, 100
location, 100
reduction, 101
size of, 100
Ozone oxidation
advanced processes, 469
as biodegradable, 469
characteristics of, 468
examples, 468–469
gas from, 468

P2 factors, defined, 272
Packaging
degradable, 415–416
life-cycle assessment concepts, 411
minimizing of, 412–415
purpose of, 411
quantity reductions of, 411
solid waste and, 410–411
types of, 411
Packaging, degradable
degradation requirements, 415
plastics and, 415–416
Packaging, minimizing
benefits, 412
concentrates marketing, 413–414
examples, 414–415
manufacturer packaging, 414
objectives of, 412
pollution prevention, 414
project design changes, 412–413
snack bag components, 413
soft drink weight reductions, 413
strategies for, 412
water reducing, 414
weight reduction of, 412
Papin, Denis, 8
Paraffins, defined, 25
Particulate control
baghouse filter, 500
cyclone, 500
electrostatic precipitator, 502–503

Particulate control *(cont.)*
scrubber, 500–502
settling chamber, 499–500
Particulates
defined, 83
effects of, 83–84
sources of, 83
Partitioning process
absorptive effects, 55–64
acid based ionization, 50–53
causes of, 45
criteria effecting, 50
defined, 45
dispersion rate factors, 49
solubility, 53–55
properties effecting, 50
volatilization, 64–67
Pasteur, Louis, 371
Payback period, 316
uses of, 316
Permits
Clean Air Act, 155
hazardous waste, 166
marketable, 309–310
marketable wastewater, 555–556
storage, 236
Pharmaceutical companies, sustainability framework, 604
Photochemical oxidants
sources of, 84
types of, 84
Photochemical reactions
effects of, 70
pollution prevention, 70
smog as, 70
types of, 70
Photovoltaic cells
concept of, 135
cost of, 135
storage battery development, 135
Physicochemical processes
activated carbon treatments, 473–475
chemical oxidation, 468–471
chemical reduction, 471–473
coagulation, 464–466
filtration, 479–480
flotation, 464
ion exchange, 475
membrane processes, 480–485
precipitation, 466–468
reduction processes, 468–471
sedimentation, 461–464
supercritical oxidation, 471
waste neutralization, 458–461
Pinch analysis
characteristics of, 431
defined, 231
goal of, 431
process emissions, 447–449
purpose of, 430
thermal, 431–439
thermodynamic principles in, 431
water use, 439–447
Pinch analysis composite curves
examples, 437–438
grand composite curve, 437
graphing curves, 434
heat plotting, 434
heat requirement meeting, 434
heat waste use, 439
HEN design rules, 436–437
pinch temperature, 434
temperature differences, 434–436
Pinch analysis process emissions
examples, 448–449
procedures for, 447–448
Pinch analysis water use
contaminant extraction process, 440–441
contaminants removing, 440
examples, 441–442
objectives, 439
process operation improvements, 440
reuse with recycling, 442–445
reuse with regeneration, 445–447
water profiles, 441
water reusing, 440
water use reductions, 440
Pinchot, Gifford, 15
Pipe connectors, fugitive emission from, 522
Plastic recycling, identification systems, 397
Plastics
biodegradable plastic, 416
defined, 416
degradable process of, 415
quantity of, 415
recycling of, 416
Plug flow reactors
first-order reactions calculating, 190
operating cost, 191
Point sources, defined, 47, 346
Polluter-pays principle, uses of, 611
Pollution, historical perspectives
ancient periods and, 7
industrial revolution, 7–9
industrialization impacts, 9
Renaissance period and, 7
Pollution control, clean production attitudes and, 607
Pollution planning process
elements in, 330
goals of, 332
implementing, 337–338
plan development, 334
preliminary assessments, 332–334
program organization, 330–332
progress measuring, 338–339
project developing, 334–337
team knowledge of, 332
Pollution preliminary assessment
factors in, 332
objectives of, 332
P2 initiative investigations, 333
supplemental operations investigations, 333
targeted process familiarity, 332–333
waste streams prioritizing considerations, 333
Pollution prevention
corporate goals for, 3
defined, 11
elements in, 11
environmental ethics, 15–19
EPA and, 11
historical perspectives, 6–9
holistic approach to, 4
industrial waste management philosophy, 9
industry practices, 5
names for, 10, 12
premise underlying, 11
prevention, 3–6, 9–13
prevention hierarchy, 13–14
public knowledge of, 4–6
recycling versus, 14–15
source reduction methods, 11
sustainability, 12–13
3M experience, 1–3
types of, 12
waste definition, 10
waste production acceleration, 4
Pollution Prevention Act (PPA)
approaches in, 537
emission reduction incentives, 310
as enabling act, 538
end-of-pipe focus, 175
EPCRA and, 535–536, 538
features of, 176
goal of, 176
industry voluntary compliance, 176
managed hierarchy system, 175–176
multimedia focus, 538
purpose of, 537
source reductions, 537–538
state laws supplementing, 176
voluntary planning, 330
waste preventing versus, 175
Pollution prevention economics
categories of, 298
decision criteria, 298
economic overviews, 296–298
engineering, 311–317
engineering economies, 296–297
examples, 296–297
life-cycle assessments and, 297
long-term cleanup estimating, 317–320
microeconomics, 298–311
total cost assessments, 320–325
Pollution prevention examples
acrylonitrile manufacturing, 239–242
maleic anhydride production, 242–247
manufacturing operations, 239
Pollution prevention factors
defined, 272
examples of, 272–277
need for, 271–272
P2 calculating methodology, 272

P2 evaluation criteria, 275
P2 factor scoring criteria, 273
P2 lithograph printing scores, 276
purpose of, 272
uses of, 272
Pollution prevention hierarchy
national policy and, 13–14
pollution elimination, 14
Pollution prevention laws
Executive Orders, 176–177
ISO, 14000, 177–178
Pollution Prevention Act, 175–176
Pollution prevention measuring
calculating, 338
data obtaining, 339
need for, 338
normalizing data, 338–339
process and facility factors, 339
quantitative basis for, 338
toxicity reductions, 339
versus waste production, 338
Pollution Prevention Office, EPA and, 176
Pollution prevention plan implementing
areas effected by, 337
challenges in, 337
as "living document," 337–338
management approval, 337
Pollution prevention planning
company understanding for, 329–330
environmental audits, 343–350
environmental system management, 340–342
introduction, 329–330
mandates for, 330
P2 projects in, 329
structure of, 330–339
voluntary nature of, 330
Pollution prevention plans
components of, 334
goal of, 334, 549
implementation requirements, 549
implementation schedules, 334
outsiders use in, 334
POTWs requirements, 549
permit process in, 549–550
prevention assessment opportunities, 549
prevention opportunities, 550
Pollution prevention programs
implementation, 562–567
mandatory program triggers, 562
phase implementing, 562
program recommendations, 561–562
sewer use ordinances, 561–562
voluntary programs, 562
Pollution prevention programs regulations
CWA and, 536–537
EPCRA and, 535–536
hazardous waste minimizing, 534
impetus for, 534
PPA and, 537–538
RCRA and, 535
regional prevention initiatives, 538–539
statutory focus, 534
Pollution prevention projects
detailed assessments, 335
option analysis, 337
option rating weighted-sum method, 335–337
option screening, 335
prioritizing, 335
Polychlorinated dibenzodioxins
characteristics, 38–39
as unwanted byproducts, 38
Polycyclic aromatic hydrocarbons
configuration of, 36
naming of, 37
sources of, 36
Precautionary Principle
decision criteria, 610
defined, 610
future predicting with, 611
government adoption of, 610
origin of, 610
uses of, 610
Precipitation
defined, 466
heavy metals and, 468
ph variable solubility, 467
uses of, 466–467
Premanufacture notification, EPA submission to, 160
Present worth
annual cost calculations, 313
examples of, 313
future revenue calculations, 312–313
importance of, 313
need for, 312
present worth tables, 314
Preservationism
defined, 15
philosophy, 15
Presidential Green Chemistry Challenge, 362
President's Council on Sustainable Development
Executive Order for, 597
mission of, 598
recommendations of, 599–600
size of, 597
task force duties, 598–599
Pressure storage tanks
characteristics, 526
fugitive emissions, 526
Pressurized water reactors
containment costs, 133
water use in, 133
Prevention progress measuring
benefit of, 556
industry needs and, 557
loss tracking systems, 557
POTWs responsibility for, 557
procedures for, 556–557
reporting data uses, 556
Price elasticity of demand, calculating, 302
Process selection
design manufacturing process analysis, 280–283
examples, 280
functional analysis, 281
inventory analysis results, 282
life-cycle assessments and, 280
Process changes
advanced technologies, 221–233
management, 236
product changes, 233–234
storage, 234–236
Process design computer tools
CPAS system, 219
Department of Energy and, 220
industrial-scale process tools, 218
planning tools, 218
SWAMI software, 220
types of, 217
uses of, 217–219
Process development
computer tools, 217–220
constraints in, 216
decision data, 216
emission decisions, 216–217
environmental objectives identifying, 215–216
objectives in, 216
pollution prevention benefits, 216
pollution prevention processes, 215
purpose of, 215
Process reactors, design of, 192
Process technologies, advanced
cleaning improvements, 225–229
degreasing, 225–229
equipment cleanings, 229
pollution minimizing strategies, 222
powder coating, 221–222
reaction process investigating, 221
reactor control improvements, 223–224
reactor design improvements, 222–223
recovered materials, 231–233
recycling, 229–231
separation process improvements, 224–225
temperature controlling, 222
Procter and Gamble, 110, 413
Product changes, effectiveness of, 233–234
Product development
benefits of, 279
environmental performance improvements, 279
life-cycle assessments incorporating, 278
waste avoidance, 278
Product laws, Occupational Safety and Health Act, 174–175

Programme for Survival, outcomes of, 579
Publicly owned treatment works (POTW)
  as CWA element, 157
  average daily flows, 559
  effectiveness of, 558–559
  facility expansions, 533
  federal pretreatment regulations, 540
  industrial user evaluation of, 559
  industrial wastewater treatment, 121
  liability of, 547
  limit basis of, 550
  local discharge standards, 541
  local limits for, 550
  mass-based limits, 551
  noncompliance industries and, 452–453
  non-permit user targeting, 560
  nonregulated industries and, 540–541
  NPDES versus, 157–158
  P2 programs, 534
  permit programs, 560
  pollution prevention programs at, 533–534
  pretreatment programs, 540
  pretreatment regulations, 158
  prevention programs evaluation, 561
  prevention progress measuring, 556–557
  regulatory compliance, 559–560
  survey of, 560–561
  system overloading, 533
  user fees, 158
Publicly owned treatment works (POTW) implementation phase one
  mandatory program components, 565–566
  voluntary program components, 565
Publicly owned treatment works (POTW) implementation phase two
  compliance options, 567
  compliance triggers, 567
  mandatory compliance options, 566
  prevention last line as, 5、7
Publicly owned treatment works (POTW) internal pollution prevention
  internal activities for, 568
  operations examinations, 567
  pollutant transferring, 567–568
  programs for, 568
Publicly owned treatment works (POTW) prevention programs
  focus of, 542
  industrial users, 542
  market-based programs, 554–555
  prevention measuring, 555–557
  regulatory programs, 547–554
  strategy for, 542
  voluntary programs, 543–547
Publicly owned treatment works (POTW) program implementation
  flowcharts, 562–563
  industries targets, 564
  phase one, 565–566
  phase two, 566–567
  program goals, 564
  purpose of, 564
  sewer ordinance evaluation, 564
  staff tasks, 564
Publicly owned treatment works (POTW) voluntary programs
  33/50 program, 543
  benefits of, 543
  educational outreach, 544–546
  emphasis of, 543
  market stimulation by, 543
  problems with, 547
  purpose of, 543–544
  technical assistance, 544–546
  voluntary reduction programs, 546–547
Publicly owned treatment works (POTW) wastewater treating
  benefits of, 456–457
  costs of, 456
  predischarge treatments, 457
  pretreatment regulations, 457
  processing, 456
  prohibited discharge standards, 457
Pumps
  fugitive emission from, 522
  pressure relief values, 522
  seal failures, 522

Quantitative structure-activity relationships, sample equations, 381–382

Raoult's Law
  calculations, 64
  partitioning use, 64
Reaction conditions alternatives
  organic solvents environmental consequences, 374
  solvents, 374–377
  synthesizing conditions, 374
Reactive waste, as hazardous waste, 163
Reactor control improvements
  control system efficiencies, 223–224
  environmental controls, 223
  process control categories, 224
  robotics, 223
Reactor design improvements
  cleaning access, 222–223
  continuous-flow models, 222
  cost balancing of, 223
  high-efficiency heat exchanges, 222
  piping systems, 223
  stirred-tank reactors, 222
  water use impact, 223
Record of Decisions
  hazardous waste remediation plans, 169
  purpose of, 159
Record keeping
  benefits of, 239
  documentation areas, 239
  inventory control and, 239
Recoverable resources
  natural resource categories, 141
  proven resources types, 140–141
  recoverable ability of, 140
Recovered material
  environmental life-cycle assessments and, 232
  example of, 233
  market development, 233
Recycled paper labeling
  limitations of, 400
  symbols for, 400
  systems for, 401
Recycling
  accounting and, 616
  defined, 387
  economic factors effecting, 387–388
  effective use of, 229–230
  factory use of, 387
  options, 230
  perspective of, 616
  versus pollution prevention, 14, 14–15
  types of, 229
  viability of, 230
  waste steam uses, 230–231
  water minimizing, 231–232
  water reuse, 231
Recycling hierarchy
  demanufacturing successes, 389
  priorities in, 388–389
Recycling legislation
  automobiles, 390–392
  dual-system, 390
  foreign legislation, 390
  household packaging, 392
  packaging and, 390
  recycling economics, 390
  solid waste incinerators, 392–393
  "take-back" law, 390
Recycling municipal solid waste, 106
  product use, 106–107
  public education, 106
  reprocessing facilities, 106
  technologies for, 106
Recycling requirements
  compatible material use, 401–402
  disassembly designing, 393
  disassembly ease, 394–397
  factors, 393–394
  material identification, 397–401
  material selecting, 402
  product design, 393

standardizing components, 397
"Red" companies
as classification system, 18
environmental performance, 19
Regional pollution prevention initiatives, Great Lakes Water Quality Initiative, 538–539
Regulations
discharge limits, 307
emission reduction incentives, 310–311
enforcement of, 307
EPA role confusing, 307–308
as government control, 307
issue addressing, 308
political lobbying, 307
Regulator process, life-cycle assessments and, 254–255
Regulatory process
administrative jurisdictions, 150
broad language in, 149
contents of, 149
development of, 149
guidance documents, 149
international laws, 150
law types, 149
state laws, 150
Regulatory programs
discharge controlling, 552
local permits, 550–551
mass-based limits, 551–552
modified sewer use ordinances, 548–549
noncompliance industry programs, 552–553
pollution prevention plans, 549–550
pollution prevention requirements, 547
prevention opportunities, 547–548
problems with, 553–554
Reilly, William, 1
Renewable energy sources
biomass, 136
commercial energy uses, 134
geothermal power, 134–135
solar power, 135
water power, 134
wind power, 135
Residuals management
air pollution control, 499–504
introduction, 454–455
residual wastes disposition, 455
solid waste disposal, 504
sources of, 454–455
treatment options, 455
waste water treatment, 455–499
Resilience, defined, 585
Resource conservation
cost of, 118
disposal alternatives, 117–118
generation reductions, 118
industrial process waste reductions, 117
treatment alternatives, 117
Resource Conservation and Recovery Act (RCRA)
disposal of, 165
hazardous waste definition, 114, 162–165
HSWA amendments to, 162
land disposal restrictions, 166
permit system, 166
prevention focus, 537
purpose of, 162, 535
reauthorized as HSWA, 534
regulation development for, 149
regulatory scheme from, 535
storage of, 165
treatment of, 165
underground storage tank regulations, 166–168
underground storage tanks, 162
underground tanks and, 166–167
Resource depletion
mineral resource structures, 139–140
mineral resources, 141–142
rate of, 138
recoverable resources, 140–141
resource types, 138–139
types of, 138–139
Resource recovery
items recovered, 111–112
mandatory trend, 110–111
municipal waste potential, 111
recovery rates, 112–113
recycle material increasing, 112
success keys, 112
Resource scarcity, sustainability and, 575
Retrometabolic design
aerobic biodegradation resistant molecules, 384
drug manufacturing, 384
molecular structure changes, 384
purposes of, 384
Reuse
characteristics of, 386
defined, 386
demanfacturing segment as, 386–387
discarding reasons, 387
items for, 387
markets lacking for, 387
returnable bottles as, 387
Reuse barriers
Chicago Board of Trade and, 410
economic value of, 410
questions on, 409–410
scientific knowledge lacking, 409
Reverse osmosis
examples, 484–485
industry uses of, 484
membrane characteristics, 481–482
membrane types, 483
osmotic pressure, 481
surface area size, 482–484
wastewater temperature, 482
wastewater treatment versus, 481
working pressure for, 481
workings of, 481
Rinse tanks mass balances. *See also* Cleaning water reductions
countercurrent rinse tanks calculations, 425–426
dead rinse countercurrent tanks, 426
example of, 427–428
flowing rise tanks calculations, 425
live rinses, dead rinse following, 426–427
single tank calculations, 424
Rio Declaration
Brundtland Commission and, 581
Copenhagen Declaration on Social Development, 581
Earth Summit and, 581
principles of, 582
Stockholm Declaration and, 581
Rotating biological contactor
anaerobic conditions, 494
characteristics of, 494
upflow anaerobic biofilters, 494

Safe Drinking Water Act (SDWA). *See also* Groundwater protection; National Primary Drinking Water Regulations
public water system defining, 159
purpose of, 159
rewriting of, 159
Sanitary landfills
compacted layer coverings, 108
declining number of, 108
gas collection systems, 108
impervious liners, 108
leachate collection systems, 108
Scott Paper Company, 278
Screening ranges approach
assumptions of, 516
calculating, 516–517
disadvantages, 517
equipment analysis, 517
estimating factors, 516
examples, 517
Scrubber. *See also* Wet scrubbers
efficiency of, 502–503
particle removal, 502
purpose of, 500
workings of, 500–502
Sedimentation
discrete settling, 461–463
flocculent settling, 463–464
suspended solids, 461
zone settling, 464
Separation process improvements
energy consumption processes, 224
separation processes for, 225
waste generation and, 224
Sequential engineering
defined, 184
inefficiently of, 184
Settling chamber
efficiency of, 502–503
particle removal, 500

Settling chamber *(cont.)*
particle velocity calculation, 499
workings of, 499
Sewage sludges, quantity of, 107
Sewer use ordinances
integrating permitting schemes, 549
mandatory prevention program, 548
models for, 548–549
permit process benefits, 548
Shisso Chemical Plant, 119
Sierra Club, 15
Sludge management
biological waste treatments, 498
characteristics of, 498
sources of, 498
untreated results, 499
Smith, Adam, 299
Smog formation
automobile exhaust, 85–86
CAA requirements, 87
elimination actions, 86
examples, 86–87
geographic conditions, 86
name origin, 85
sources of, 85
Social ecology, philosophy of, 16
Society of Automotive Engineers, 397
Society of Environmental Toxicology and Chemistry, 257
Society of Plastics Industry, 398
plastic categories, 399
Soft material paths, concepts of, 612
Solar energy
concepts of, 135
energy storage, 135
Solid waste
conservation, 114
defined, 104, 162
hazardous waste and, 162–163
local versus national problem, 102
management of, 107–114
packaging materials, 102–103
pollution prevention opportunities, 103–104
quantity of, 102–103
recycling activities, 103–104
sources and composition, 104–107
Solid waste disposal
hazardous waste, 504
methods for, 504
Solid waste management
collection costs, 108
composting, 110
disposition of, 107–108
historical approaches, 107
incineration, 109–110
sanitary landfills, 108
source recovery, 110–113
Solid waste sources
agricultural wastes, 107
industrial wastes, 107
mining wastes, 107
municipal wastes, 104–107
sewer sludges, 107
Solubility
defined, 53
factors controlling, 54
importance of, 53–54
normal aliphatics and, 55
organic compounds factors, 55
polar compounds, 54
undissolved fraction, 54
Solventless chemistry, microwave activation, 375
Solvents
activation energy, 374–375
chemical industry use, 374
organic chemical synthesis alternations, 375–376
reusing, 377
Source control
CWA and, 539
development of, 561–567
federal pretreatment regulations, 540
industrial pretreatment programs, 540
internal pollution prevention, 567–568
local discharge standards, 541–542
mandatory programs, 539–541
nonregulated industries, 540–541
P2 program qualities, 558–561
pollution prevention implementation, 568
POTA-administrated programs, 542–557
POTWs operations and, 539
POTWs pretreatment programs, 540
requirement for, 539
Source Reduction Clearinghouse, 176
South Coast Air Quality Management District, life-cycle cost model, 322
Stability framework
clean manufacturing elements, 606–607
individual's role, 603
industry role, 604–605
influence categories, 603–604
resource conservation, 604
society characteristics, 603
Stabilization ponds
categories of, 495–496
characteristics of, 495
design data for, 496
uses of, 496
Step aeration, activated sludge system, 490
Stirred-tank reactors
design improvements, 222
uses of, 192
Stockholm Conference, outcomes of, 579
Stokes' Law
discrete settling calculations, 461
particle velocity, 499
particle velocity calculating, 499
Storage
accidental release minimizing, 236
approaches to, 234–235
common deficiency in, 235
local community committee, 236
permits for, 236
plan enforcement, 236
pollution sources, 234
problems eliminating, 235, 235–236
SPCC plan, 236
Storage tank
corrosion protection, 168
cost and, 168
exempted tanks, 167
leak detecting devices, 167
leaking magnitude, 167
owner responsibilities, 167
RCRA amendments for, 167
spill protection, 168
tank registration, 167
tank testing, 166–167
underground tank laws, 166–167
Storage tank fugitive emissions
controlling, 529–530
estimating, 526–529
industries using, 524
tank designs, 524–526
Structure-activity relationships
examples of, 381
physicochemical factor governing, 382
purpose of, 381
quantitative structure-activity relationships, 381–382
Subsidiary Principle, uses of, 611
Sulfur dioxide
characteristics, 83
effects of, 83
emissions, 90
sources of, 83
Supercritical extraction
carbon dioxide, 213
characteristics of, 212
examples of, 213
fluid constants, 213
Supercritical fluids
advantages of, 376
reaction conditions for, 376
types of, 376
Supercritical oxidation, uses of, 471
Superfund Amendments and Reauthorization Act (SARA). *See also* Comprehensive Environmental Response, Compensation, and Liability Act
as EPCRA, 168
site inventory, 169
underground storage tanks and, 167
Superfund sites
contaminants at, 117
history of, 115
Love Canal, 115–116

sites other, 117
Times Beach, 116–117
Superior Coating Incorporated, 353
Supply and demand
market domination, 302
pollution prevention activities, 302
pollution taxes, 302–303
price elasticity of demand, 302
price establishing, 300–302
Suspended growth systems, 488
Sustainability
development goal of. 12–13
initiative elements, 617
meaning of, 12
prerequisites for, 13
steps to, 616–617
Sustainability, achieving
community initiatives, 597
EPA steps, 596
partnerships developing, 596
pregnancy prevention programs, 596–597
SURE organization and, 597
Sustainability concept
activities leading to, 579
ecological concepts, 585–587
economic concepts, 585
U.N. conferences, 578
Sustainability development
achieving of, 596–597
applications of, 593–594
Brundtland Commission definition, 584
business enterprises and, 585
components of, 584–585
framework for, 588–592
goals of, 585
indicators of, 594–596
objectives, 588
precepts of, 584
Sustainability framework
Canadian national strategy, 591–592
elements of, 588–591
international development, 589–590
local development, 589–590
national development, 589–590
principles of, 591
Sustainability future
beneficial change and, 617
integrated policies for, 617
knowledgeable public, 618
objective of, 617–618
Sustainability history
concept origin, 578
environmental disconnection, 576–577
population growth, 577
social stratification, 577
standard of living, 577
technology growth, 577
U.N. conferences, 578–581
uncertainty, 577–578
Sustainability impediments
apathy, 587
consumption levels, 587
developing nations, 587
economic system inadequacy, 576
environmental problems, 575
knowledge lacking, 588
living standards falling, 575–576
nation's conflict of interest, 575
poverty, 575
public awareness levels, 588
resource scarcity, 575
technology lacking, 576
Third World nations, 575
uncertainty magnitude, 588
Sustainability indicators
characteristics, 595
choosing, 596
efficiency measurements, 595
purpose of, 594
role of, 595
source criteria for, 594–595
traditional measurements, 594
usefulness of, 596
Sustainability meanings
as acronym, 584
conceptualization of, 585–587
definitions, 583–584
development meanings, 584–585
hurdles to, 587–588
technology and, 583
Sustainability problem defining
biodiversity, 574–575
sustainability impediments, 575–576
Sustainability strategies
air, 593
Canadian national strategy, 591–592
concepts of, 593
energy, 594
housing, 593
land, 593
water, 593
Sustainability in United States
local government's role, 600–601
President's Council on, 597–600
Sustainable societies
adoption steps of, 616–617
components of, 586
concept developing, 573
depletion effects, 572–573
development achievements, 588–597
economic growth measurements, 613–616
environmental degradation causes, 574
environmental interaction, 573–574
framework for, 603–607
history of, 576–581
industrial ecology, 608–613
introduction to, 572–574
meanings of, 581–588
natural resource depletion, 572
outlook for, 617–618
people and landscape relationships, 573
problem defining, 574–576
problem magnitude, 573
society changes, 573
Third World nations, 601–603
U.S. and, 597–601
Sustainable Urban/Rural Enterprise (SURE), 597
Swiss Federal Ministry for the Environment, 267
Synthetic pathways alternatives
biocatalysis, 371–374
catalysis, 368–371
feedstocks alternatives, 363–368
undesirable intermediates elimination, 363

Technical assistance programs
BMP adoptions, 544
City of Albuquerque and, 545
City of Palo Alto, 545–546
City of San Francisco and, 545
facility inspections, 544
focus of, 544
POTW and, 544
waste minimization audits, 545
Teens Teaching Teens, 596–597
Tennessee Valley Authority, 280
Thermal pinch analysis
component curves, 434–439
HEN analysis, 431–433
HEN objectives, 431
origin of, 431
Third World nation macroeconomic management models
computable general equilibrium model, 603
minimum standard model, 603
public sector planning and management information system, 603
Third World nations
macroeconomic management of, 602–603
sustainability and, 575
sustainability barriers, 601–602
Third World nations sustainability barriers
corruption, 602
development categories, 601–602
population growth, 602
resource exploitation in, 602
tourism economy, 602
3M Corporation, 3, 320
accomplishment of, 2–3
environmental policy goals, 2
legislation effecting, 2
pollution prevention programs, 2
waste creation, 1–2
Three Mile Island, nuclear incident, 132
3Rs
concepts used with, 613
defined, 613
quality standards, 613
Times Beach
containment approaches, 116–117

Times Beach *(cont.)*
dirt road suppressant spraying, 116
Toxic Release Inventory (TRI)
annual report publishing, 172
chemicals in, 348–349
contents of, 346
data problems, 349–350
EPCRA and, 346, 535
EPCRA annual reports and, 172
intent of, 346
life-cycle data, 262
release defining, 346
voluntary pollution planning, 330
Toxic Release Inventory (TRI) chemicals
characteristics of, 348
off-site transfers, 348
OSHA carcinogens released, 349
U.S. chemicals release, 349
U.S. releases in, 348
U.S. transfers in, 348
Toxic Release Inventory (TRI) data
estimated data in, 349–350
P2 progress measuring, 350
pollution prevention with, 350
POTWs needs and, 350
problems, 349
threshold quantity exceeding, 349
Toxic Release Inventory (TRI) reporting. *See also* Emergency Planning and Community Right-to-Know Act
company classifications, 346–347
information in, 347–348
Toxic Substances Control Act (TSCA)
asbestos and, 160–162
exemptions to, 160
polychlorinated biphenyls and, 160
prescreening chemicals, 160
purpose of, 160
Toxic waste, as hazardous waste, 163
Toxicity Characteristic Leaching Procedure, 163
Toxicity reducing
absorption prevention, 378–379
electrophilic reactive intermediates, 380
electrophilic substituents, 380
molecular modifications, 379
safe chemical substitution, 381
structure modification, 379–380
Training
goal of, 238
maintenance procedures documentation, 238
pollution prevention goals, 239
topics in, 238
waste reduction goals, 238–239
worker safety, 239
Transformation process
biological transformations, 70–72
causes of, 45
elimination reactions, 68–69
hydrolysis, 68
metal compounds, 67–68
organic compounds and, 67
oxidation-reduction reactions, 69–70
photochemical reactions, 70
reactions types, 68
Transport process
advective processes, 48
disffusional processes, 48–49
dispersive processes, 48
industrial contaminants fate, 47
knowledge of, 46
loading processes, 46–48
partitioning process, 45
purpose of, 44–45
Transportation, sustainability framework, 604
Triad Energy Resources, 233, 408–409
Trickling filters
contents of, 492
design criteria for, 493
design variations, 493–494

Underwriter's Laboratory, 286
Unemployment, sustainability framework and, 603
Unilever Company, 444
Unit operations, types of, 214
Unit-specific correlation approach
cost of, 518
mass emission data, 519
workings of, 518
United Nations (U.N.)
agenda, 21, 580–581
Conference on Human Environment, 580
Earth Summit, 580–581
International Development Strategy, 578
Rio Declaration, 581
Stockholm Conference, 579
sustainability defined, 580
World Charter for Nature, 579
United Nations, Framework Convention on Climate Change, 310
United States
coal flow in, 128
eco-labeling, 286–287
electricity flow, 137
energy consumption changes, 124
energy flow, 126
energy usage, 125–127
geothermal locations, 135
GNP and energy use, 124
Green Seal, 286–287
natural gas resources, 132
nuclear generating units, 133
oil resources of, 129–130
petroleum flow in, 131
wind power, 135
User-pays Principle, uses of, 611

Valuation
decision making, 267
ecopoint methods, 268
environmental effects, 268
environmental impact weighting systems, 267
environmental priority strategies, 268–269
issue judgements, 267–268
life-cycle value, 269
purpose of, 267
Valves leaking
component monitoring, 519
fugitive emissions from, 519
initial repairs, 520
live load springs, 520–521
packing deterioration, 519
process value emissions, 521–522
Variable vapor space storage tanks
characteristics, 525–526
fugitive emissions, 525
Vasuki, N. C., 12
Volatilization
container solubility, 66
defined, 64
fugitive emissions from, 64
gas and liquid phase partitioning, 64
generalizations of, 66
Henry's law, 64–66
non-ionic containment, 67
ph solutions, 67
sorption effects, 66
vapor pressure and, 64
Voluntary program components
compliance publicly recognized, 565
compliance time, 565
nonpermit users, 565
POTW initiation of, 565
reduction mandates, 565
small contributor classifications, 565
waste minimization audit, 565
Voluntary program problems
disincentive barriers, 547
industry competition, 547
POTW liability, 547
Voluntary reduction programs
Buffalo Sewer Authority, 546–547
City of Albuquerque, 546

Waste definitions
common, 10
industrial waste, 10
legal, 10
prevention responsibilities, 10
product energy, 10
Waste exchanges
available quantities for, 408
cost effectiveness of, 408
end user defining, 408
examples, 408–409
hazardous waste laws, 408
NMEN and, 408
residual materials waste, 407–408

workings of, 408
Waste management contractor audits, purpose of, 343–344
Waste management life-cycles assessments
clean process advisory system, 289
complexity of, 291
components of, 290
environmental impacts of, 290
recycling costs, 289
recycling true costs, 290
recycling versus virgin material use, 290
solid waste recycling, 288–289
system boundary definitions, 291–292
uses of, 288
Waste neutralization
acidic reagents, 460
acidic wastewater, 459
alkalinity costs, 459
defined, 458
examples, 460
ph acceptable ranges, 458
ph reaction monitoring, 458–459
purpose of, 458
reagent amounts, 458
Waste sources
chemical production, 359
chemical synthesis process, 358
contaminants separations, 362
green chemistry promoting, 362
organic compound wastes, 361
other chemicals creations as, 361
pollution creation, 358–360
reaction sequence wastes, 361
secondary by-products, 360–361
sources of, 358
Waste treatment fugitive emissions
causes of, 531
computer modes for, 530
covering, 531
pathways for, 530
sources of, 530
Wastewater treatment
biological waste treatments, 485
options in, 456–458
physicochemical processes, 458–485
process diversity, 456
sewer line mixing, 455
sludge management, 498–499
varieties of, 455
Wastewater treatment options
disposal options, 456
municipal treatment options, 458
POTW processing, 456–457
pretreatment regulations, 457
Water
cleaning water reductions, 422–430
cost of, 421–422
immiscible separations, 421
introduction, 421–422
pinch analysis, 430–449
reagent dissolving, 421
Water pollution
cooling water control, 118–119
effects of, 119
industrial sources, 118
industrial wastewater treatment, 121
Kepone incident, 120–121
Minimata disease, 119–120
nonpoint pollution sources, 119
treatment costs, 118
Water power
environmental impact from, 134
hydroelectric facilities construction, 134
Water reuse
contaminant decreasing, 445
cost saving, 445
examples of, 444–447
recycling, 442–444
regeneration uses, 445
Watt, James, 8
Wedgwood, Josiah, 7
Wet scrubbers. *See also* Scrubbers
gas removal, 503
uses of, 503
workings of, 503
White, Peter, 13
Wind power
aesthetics objections, 135
energy demand met by, 135
locations for, 135
Woonsocket Regional Wastewater Commission, 553
World Charter for Nature, outcomes of, 579
World Conservation Strategy, outcomes of, 579

Xerox, 3, 409

"Yellow" companies
as classification system, 18
environmental performance, 19

Zone settling, characteristics of, 464